T0338603

Principles of Colloid and Surface Chemistry

Principles of Colloid and Surface Chemistry

Third Edition, Revised and Expanded

Paul C. Hiemenz

California State Polytechnic University
Pomona, California

Raj Rajagopalan

University of Florida
Gainesville, Florida

CRC Press
Taylor & Francis Group
Boca Raton London New York

CRC Press is an imprint of the
Taylor & Francis Group, an **informa** business
A TAYLOR & FRANCIS BOOK

Published 1997 by CRC Press
Taylor & Francis Group
6000 Broken Sound Parkway NW, Suite 300
Boca Raton, FL 33487-2742

© 1997 by Taylor & Francis Group, LLC
CRC Press is an imprint of Taylor & Francis Group, an Informa business

No claim to original U.S. Government works

ISBN 13: 978-0-8247-9397-5 (hbk)

Library of Congress catalog number: 97-4015

Visit the Taylor & Francis Web site at
http://www.taylorandfrancis.com

and the CRC Press Web site at
http://www.crcpress.com

Library of Congress Cataloging-in-Publication Data

Catalog record is available from the Library of Congress

Preface to the Third Edition

The face of colloid and surface chemistry has changed dramatically in the 10 years since the last edition of this book appeared in print. Advances in instrumentation now make it possible for us to "see"—and, indeed, to *manipulate*—individual atoms on a surface. Molecular engineering of polymers, surfactants, and particles is now within reach for fabricating novel advanced materials and for preparing capsules (liposomes) for drug delivery and gene therapy. Direct measurements of colloidal and surface forces are now commonplace, and colloidal dispersions are now used as models for studying equilibrium and non-equilibrium thermodynamics and flow behavior of *atomic* (as well as *geological*) systems—phenomena that are otherwise not easily accessible to the experimentalist. Colloid and surface chemistry is a truly interdisciplinary subject today, and its content and significance have no bounds.

This richness has its rewards, as has been documented in a number of advanced monographs and a few graduate-level books in recent years. However, it has its drawbacks as well, as those who use colloids and surface chemistry are drawn from increasingly diverse backgrounds. Introductory textbooks that are suitable for such a diverse group of students and professionals remain almost as rare today as they were a decade ago. This need for an introductory book accessible to everyone, regardless of background, even from the perspectives of the traditional applications of colloid and surface chemistry, cannot be overemphasized. For example, in the chemical industries alone, almost two-thirds of the operations or products involve powders and suspensions. A report from Du Pont researchers [Ennis, B., Green, J., and Davies, R., *Chemical Engineering Progress*, 90(4), 32–43 (1994)] notes that of the roughly 3000 products manufactured by Du Pont, about 60% are marketed as powders, dispersions, or suspensions and that the processing of another 20% involves particles or suspensions. Industries employ engineers and scientists of all types, and the rather tight curriculum requirements in most areas of science and engineering seldom allow even a minimal preparation of university graduates for working with materials in particulate form.

Factors such as the above have shaped our approach to the present edition. The philosophy behind the first two editions was that most of the material in the book should be a natural extension of an introductory physical chemistry course at the undergraduate level and should be accessible to anyone (students as well as practicing engineers and scientists) with such a background. This remains the rule in the present edition as well. In our opinion, this orientation is essential for giving a firm foundation in colloids and surfaces to the diverse audience, which typically includes individuals from almost all branches of engineering and the sciences.

Keeping these factors in mind, we have updated many topics and have introduced new features that draw attention to exciting developments that go beyond the scope of the present volume. Among the major revisions and additions made are the following:

1. A major portion of Chapter 1 has been rewritten. In addition to providing a broad overview of the basic issues of importance in colloids and to introducing much of

what is presented in the rest of the book, Chapter 1 now includes modern developments in instrumentation such as scanning probe microscopy and spectroscopy, and atomic and surface force measurements; a brief introduction to "model" colloids and their uses; and an introduction to fractal dimensions of aggregates and surfaces.

2. A number of vignettes are sprinkled throughout Chapter 1 to provide motivation for and to illustrate the concepts introduced in the chapter and in the rest of the book. These vignettes also serve as vehicles for conveying the excitement of the subject and as introductions to advanced concepts that are beyond the scope of the book. They cover a rather broad array of disciplines ranging from environmental remediation and separation technologies to biological and medical sciences and polymer composites, and they thus reveal the breadth of the present as well as the potential applications. In addition to the vignettes in Chapter 1, each of the other 12 chapters contains a vignette highlighting the material covered in the chapter and establishing the connections to other chapters or advanced developments.

3. Chapter 4 now goes beyond an introduction to Newtonian viscosity and flow of Newtonian dispersions. In addition to a basic introduction to viscosity and the Navier-Stokes equation, the chapter now includes the Krieger-Dougherty relation for concentrated dispersions, elements of non-Newtonian rheological behavior of dispersions, electroviscous effects, and a summary of the definitions and physical significance of dimensionless groups essential for estimating and quantifying rheological properties of dispersions.

4. Chapter 5, on scattering techniques, has been updated to include measurement of fractal dimensions of aggregates, intraparticle and interparticle structure, dynamic light scattering and relation between light scattering and complementary measurements based on scattering of X-ray and neutrons.

5. Chapter 7 now includes an introduction to structural transitions in Langmuir layers and examples of applications of Langmuir and Langmuir–Blodgett films.

6. Chapter 8 has been revised to include a discussion of the critical packing parameter of surfactants and its relation to the structure of resulting surfactant aggregates. This simple geometric basis for the formation of micelles, bilayers, and other structures is intuitively easier to understand for a beginning student.

7. Chapters 9 and 10 of the second edition (dealing with physical adsorption at gas/solid interfaces and microscopy, spectroscopy and diffractometry of metal surfaces, respectively) have been pared down and consolidated into one chapter (new Chapter 9, on adsorption on gas/solid interfaces). The outdated materials or materials extraneous to the main thrust of the book from the old Chapter 10 have been eliminated and only the discussion of low-energy electron diffraction (LEED) has been retained.

8. Chapters 11–13 of the second edition, which discussed van der Waals forces (old Chapter 11), electrical double layers (old Chapter 12), electrokinetic phenomena (old Chapter 13), and colloid stability (old Chapters 11 and 12), have been restructured and new materials on colloid stability and polymer/colloid interactions have been added. For example:

    • In the new version, Chapter 10 focuses exclusively on van der Waals forces and their implications for macroscopic phenomena and properties (e.g., structure of materials and surface tension). It also includes new tables and examples and some additional methods for estimating Hamaker constants from macroscopic properties or concepts such as surface tension, the parameters of the van der Waals equation of state, and the corresponding state principle.

    • Chapters 11 and 12 in the present edition focus exclusively on the theories of electrical double layers and forces due to double-layer interactions (Chapter 11) and electrokinetic phenomena (Chapter 12). Chapter 11 includes expressions for interacting spherical double layers, and both chapters provide additional examples of applications of the concepts covered.

- The new Chapter 13 collects the material on colloid stability previously distributed between old Chapters 11 and 12 and integrates it with new material on stability ratio and slow and fast coagulation, polymer-induced forces, and polymer-induced stabilization and destabilization.

This rearrangement makes the presentation of colloidal forces, colloid stability, and electrokinetic phenomena more logical and pedagogically more appealing.

Numerous cross-references link the materials in the various chapters; however, many of the chapters are self-contained and can stand alone, thus offering the instructor the flexibility to mix and match the topics needed in a particular course. Moreover, we have also introduced a number of other features to make the present edition more useful and convenient as a textbook *and* for self-study. For example:

1. A collection of Review Questions has been added at the end of each chapter. These should be useful not only for class assignments but also as a self-assessment tool for students.

2. The references at the end of each chapter have been grouped into two sets, one containing annotated general references (usually other textbooks, monographs or review articles) and the second with special references of narrower scope. The level of each annotated reference (undergraduate, graduate, or advanced) has also been included in the annotations, as an additional aid to students and instructors.

3. All illustrations have been redrawn, and a number of new illustrations, tables, and examples have been added. (Appendix D presents a list of all the worked-out examples and the corresponding page numbers.)

4. The presentation in each chapter has been divided into a number of sections and subsections so that the headings of the sections and subsections can serve as guideposts for students.

As it is often said, when it comes to textbooks, one never "finishes" writing a book – one just "abandons" it! There is a lot we would like to do further with this book, but these changes must wait for the next edition. We would be pleased indeed to hear comments, suggestions, and recommendations for potential revisions and additions from anyone who uses this book, so that everyone can benefit from the collective wisdom of the community at large.

Paul C. Hiemenz
Raj Rajagopalan

Acknowledgments

A number of colleagues have been kind enough to send us comments or recent research papers, reviews, or related materials for the third edition. The information we received has shaped our thinking and has influenced the contents and the organization of the present edition, and we would like to recognize the assistance and interest of these individuals here. In particular, we would like to thank Prof. Nicholas Abbott (UC, Davis), Prof. John Anderson (Carnegie Mellon), Dr. Rafat R. Ansari (NASA Lewis, Cleveland, OH), Prof. John C. Berg (Univ. of Washington), Prof. Daniel Blankschtein (MIT), Dr. Krishnan Chari (Kodak, Rochester, NY), Dr. Reg Davies (Du Pont, Wilmington, DE), Prof. Menachem Elimelech (UCLA), Prof. H. Scott Fogler (Michigan), Prof. George Hirasaki (Rice), Dr. Norio Ise (Rengo Corp., Fukui, Japan), Prof. Jacob N. Israelachvili (UC, Santa Barbara), Prof. Charles Knobler (UCLA), Prof. Josip Kratohvil (Clarkson), Prof. J. Adin Mann (Case Western), Prof. Jacob Masliyah (Alberta), Prof. Egon Matijevic (Clarkson), Dr. S. Mehta (ARCO, Plano, TX), Prof. Clarence Miller (Rice), Dr. Ian Morrison (Xerox, Rochester, NY), Prof. Brij M. Moudgil (Florida), Dr. Cherry A. Murray (AT&T, Murray Hill, NJ), Prof. R. Nagarajan (Penn State), Prof. Robert Ofoli (Michigan State), Prof. Kyriakos Papadopoulos (Tulane), Prof. Dennis C. Prieve (Carnegie Mellon), Prof. Clay Radke (UC, Berkeley), Dr. Mihail C. Roco (NSF), Prof. Robert Rowell (Univ. of Massachusetts), Prof. William B. Russel (Princeton), Prof. Robert Schechter (UT, Austin), Prof. Ken Schmitz (Univ. of Missouri, Kansas City), Prof. Dinesh Shah (Florida), Prof. P. Somasundaran (Columbia), Prof. Frank M. Tiller (Houston), Prof. Theo van de Ven (McGill), Prof. Carol Van Oss (SUNY, Buffalo), Prof. Darsh T. Wasan (IIT, Chicago), and Prof. Mark Wiesner (Rice).

Some of RR's colleagues at the University of Houston, especially Profs. Kishore K. Mohanty, Richard Pollard, Jay D. Schieber, Cindy L. Stokes, Frank M. Tiller, and Richard C. Willson, read and offered comments on some of the vignettes used in the book. Their interest and assistance have contributed significantly to the clarity of the final versions that appear in the book.

RR would also like to thank his students and coworkers, who played a major role throughout the preparation of the present edition. Mr. Roger Seow did much of the laborious task of scanning the second edition into the computer and fixing all the errors generated by the usually less-than-perfect optical character recognition programs. His remarkable attention to details and organizational skills made the initial stages of the preparation of the text appear almost effortless. Mr. Nikhil Joshi stepped in subsequently to attend to the endless printing and reprinting of the figures, tables, and text for the final stage. We are greatly indebted to both. Dr. Jean François Guérin, Dr. Yongmei Wang, Ms. Carlisa Harris, Ms. Sheryl L. Horton, Mr. Randy Diermeier, Mr. Jorge Jimenez, Mr. Dang Nhan, Mr. K. Srinivasa Rao, Mr. Antonio Rodriguez, and Mr. Sameer Talsania have been extremely patient despite the fact that their research projects frequently took second place to the preparation of the book.

The Media Center at the University of Houston deserves special recognition for its excellent work in preparing the initial versions of the drawings of most of the illustrations.

Finally, we would like to thank the folks at Marcel Dekker, Inc. — particularly Ted Allen and Eric F. Stannard, Production Editors, and Sandy Beberman, Assistant Vice President — for their infinite patience and, of course, for their characteristically friendly and courteous assistance.

Preface to the Second Edition

In the preface to the first edition, I stated that this is "primarily a textbook, written with student backgrounds, needs, and objectives in mind." This continues to be true of the second edition, and many of the revisions I have made are attempts to make the book even more useful than its predecessor to student readers. In addition, colloid and surface science continue to flourish. In preparing the second edition I have also attempted to "open up" the coverage to include some of the newer topics from an ever-broadening field.

A number of differences between the first and second editions can be cited which are readily traceable to either one or both of the foregoing considerations.

Two new chapters have been added which explore — via micelles and related structures and metal surfaces under ultra-high vacuum — both "wet" and "dry" facets of colloids and surfaces. Although neither of these areas is new, both are experiencing an upsurge of activity as new instrumentation is developed and new applications are found.

A number of chapters have been overhauled so thoroughly that they bear only minor resemblance to their counterparts in the first edition. The thermodynamics of polymer solutions is introduced in connection with osmometry and the drainage and spatial extension of polymer coils is discussed in connection with viscosity. The treatment of contact angle is expanded so that it is presented on a more equal footing with surface tension in the presentation of liquid surfaces. Steric stabilization as a protective mechanism against flocculation is discussed along with the classical DLVO theory.

Solved problems have been scattered throughout the text. I am convinced that students must work problems to gain mastery of the topics we discuss. Including examples helps bridge the gap between the textbook presentation of theory and student labors over the analysis and mechanics of problem solving.

SI units have been used fairly consistently throughout the book. Since the problems at the ends of the chapters are based on data from the literature and since cgs units were commonly used in the older literature, the problems contain a wider assortment of units. A table of conversions between cgs and SI units is contained in Appendix B.

I am very much aware of the many important topics that the book fails to cover or, worse yet, even mention. However, lines must be drawn somewhere both to keep the book manageable in size and cost and to have it useful as the basis for a course. As it is, I have added a good deal of new material without deleting anywhere near as much of the old. I have tried to select for inclusion topics of fundamental importance which could be developed with some internal coherence and with some continuity from a (prerequisite) physical chemistry course.

A number of users of the first edition communicated with me, pointing out errors and offering suggestions for improvement. I appreciate the feedback of these correspondents, and hope that the revisions I have made at least partially reflect their input.

I want to express my thanks to Carol Truett for expertly drawing the new illustrations and giving a "new look" to the old ones. I also appreciate the assistance of Lisa Scott in the preparation of the manuscript. Thanks, also, to Reuben Martinez for helping me with the proofreading and indexing. Lastly, let me again invite users to call errors and/or obscurities to my attention, and to thank them in advance for doing so.

Paul C. Hiemenz

Preface to the First Edition

Colloid and surface chemistry occupy a paradoxical position among the topics of physical chemistry. These are areas which have traditionally been considered part of physical chemistry and are currently enjoying more widespread application than ever due to their relevance to environmental and biological problems. At the same time, however, colloid and surface chemistry have virtually disappeared from physical chemistry courses. These topics are largely absent from the contemporary general chemistry course as well. It is possible, therefore, that a student could complete a degree in chemistry without even being able to identify what colloid and surface chemistry are about.

The primary objective of this book is to bridge the gap between today's typical physical chemistry course and the literature of colloid and surface chemistry. The reader is assumed to have completed a course in physical chemistry, but no prior knowledge of the topics under consideration is assumed. The book is, therefore, introductory as far as the topic subjects are concerned, although familiarity with numerous other aspects of physical chemistry is required background.

Since physical chemistry is the point of departure for this presentation, the undergraduate chemistry major is the model reader toward whom the book is addressed. This in no way implies that these are the only students who will study the material contained herein. Students majoring in engineering, biology, physics, materials science, and so on, at both the undergraduate and graduate levels will find aspects of this subject highly useful. The interdisciplinary nature of colloid and surface chemistry is another aspect of these subjects that contributes to their relevance in today's curricula.

This is primarily a textbook, written with student backgrounds, needs, and objectives in mind. There are several ways in which this fact manifests itself in the organization of this book. First, no attempt has been made to review the literature or to describe research frontiers in colloid and surface chemistry. A large literature exists which does these things admirably. Our purpose is to provide the beginner with enough background to make intelligible the journals and monographs which present these topics. References have been limited to monographs, textbooks, and reviews which are especially comprehensive and/or accessible. Second, where derivations are presented, this is done in sufficient detail so that the reader should find them self-explanatory. In areas in which undergraduate chemistry majors have minimal backgrounds or have chronic difficulties — for example, fluid mechanics, classical electromagnetic theory, and electrostatics — the presentations begin at the level of general physics, which may be the student's only prior contact with these topics. Third, an effort has been made to facilitate calculations by paying special attention to dimensional considerations. The cgs-esu system of units has been used throughout, even though this is gradually being phased out of most books. The reason for keeping these units is the stated objective of relating the student's experiences to the existing literature of colloid and surface chemistry. At present, the cgs-esu system is still the common denominator between the two. A fairly detailed list of conversions between cgs and SI units is included in Appendix C. Finally, a few problems are included in each chapter. These provide an opportunity to apply the concepts of the chapter and indicate the kinds of applications these ideas find.

Not all who use the book will have the time or interest to cover it entirely. In the author's course, about two-thirds of the material is discussed in a one-quarter course. With the same level of coverage, the entire book could be completed in a semester. To cover the amount of material involved, very little time is devoted to derivations except to answer questions. Lecture time is devoted instead to outlining highlights of the material and presenting supplementary examples.

The underlying unity which connects the various topics discussed here is seen most clearly when the book is studied in its entirety and in the order presented. Time limitations and special interests often interfere with this ideal. Those who choose to rearrange the sequence of topics should note the subthemes that unify certain blocks of chapters. Chapters 1 through 5 are primarily concerned with particle characterization, especially with respect to molecular weight; Chapters 6 through 8, with surface tension/free energy and adsorption; and Chapters 9 through 11, with flocculation and the electrical double layer. Subjects of special interest to students of the biological sciences are given in Chapters 2 to 5, 7, and 11.

Colloid chemistry and surface chemistry each span virtually the entire field of chemistry. The former may be visualized as a chemistry whose "atoms" are considerably larger than actual atoms; the latter, as a two-dimensional chemistry. The point is that each encompasses all the usual subdivisions of chemistry: reaction chemistry, analytical chemistry, physical chemistry, and so on. The various subdivisions of physical chemistry are also represented: thermodynamics, structure elucidation, rate processes, and so on. As a consequence, these traditional categories could be used as the basis for organization in a book of this sort. For example, "The Thermodynamics of Surfaces" would be a logical chapter heading according to such a plan of organization. In this book, however, no such chapter exists (although not only chapters but entire volumes on this topic exist elsewhere). The reason goes back to the premise stated earlier: These days most undergraduates know more about thermodynamics than about surfaces, and this is probably true regardless of their thermodynamic literacy/illiteracy! Accordingly, this book discusses surfaces: flat and curved, rigid and mobile, pure substances and solutions, condensed phases and gases. Thermodynamic arguments are presented — along with arguments derived from other sources — in developing an overview of surface chemistry (with the emphasis on "surface"). A more systematic, formal presentation of surface thermodynamics (with the emphasis on "thermodynamics") would be a likely sequel to the study of this book for those who desire still more insight into that aspect of two-dimensional chemistry. Similarly, other topics could be organized differently as well. Only time will tell whether the plan followed in this book succeeds in convincing students that chemistry they have learned in other courses is also applicable to the "in between" dimensions of colloids and the two dimensions of surface chemistry.

The notion that molecules at a surface are in a two-dimensional state of matter is reminiscent of E. A. Abbott's science fiction classic, *Flatland*.* Perusal of this book for quotations suitable for Chapters 6, 7, and 8 revealed other parallels also: the color revolt and light scattering, "Attend to Your Configuration" and the shape of polymer molecules, and so on. Eventually, the objective of beginning each chapter with a quote from *Flatland* replaced the requirement that the passage cited have some actual connection with the contents of the chapter. As it ends up, the quotes are merely for fun: Perhaps those who are not captivated by colloids and surfaces will at least enjoy this glimpse of *Flatland*.

Finally, it is a pleasure to acknowledge those whose contributions helped bring this book into existence. I am grateful to Maurits and Marcel Dekker for the confidence they showed and the encouragement they gave throughout the entire project. I wish to thank Phyllis Bartosh, Felecia Granderson, Jennifer Woodruff, and, especially, Mickie McConnell and Lynda Parzick for making my sloppy manuscript presentable. My appreciation also goes to Bob Marvos, George Phillips, and, especially, Dottie Holmquist for their work on the figures,

---

*E. A. Abbott, *Flatland* (6th ed.), Dover, New York, 1952. Used with permission.

which are such an important part of any textbook. I also wish to thank Michael Goett for helping with proofreading and indexing. Finally, due to the diligence of the class on whom this material was tested in manuscript form, the book has 395 fewer errors than when it started. For the errors that remain, and I hope they are few in number and minor in magnitude, I am responsible. Reports from readers of errors and/or obscurities will be very much appreciated.

Paul C. Hiemenz

Contents

1

Colloid and Surface Chemistry:

Scope and Variables

Next . . . come the Nobility, of whom there are several degrees, beginning at Six-Sided Figures, . . . and from thence rising in the number of their sides till they receive the honorable title of Polygonal. . . . Finally when the number of sides becomes so numerous, and the sides themselves so small, that the figure cannot be distinguished from a circle, he is included in . . . the highest class of all.

From Abbott's *Flatland*

1.1 INTRODUCTION

"Yesterday, I couldn't define colloid chemistry; today, I'm doing it." This variation of an old quip could apply to many recent chemistry and engineering graduates on entering employment in the "real world." Two facts underlie this situation. First, colloid and surface science, although traditional parts of physical chemistry, have largely disappeared from introductory physical chemistry courses. Second, in research, technology, and manufacturing, countless problems are encountered that fall squarely within the purview of colloid and surface science. In this section we enumerate some examples that illustrate this statement. The nine "vignettes" included in this chapter also illustrate the importance of colloid and surface science in a broad range of scientific and technological areas.

The paradoxical situation just described means that it is entirely possible for a science or an engineering student to have completed a course in physical chemistry and still not have any clear idea of what colloid and surface science are about. A book like this one is therefore in the curious position of being simultaneously "advanced" and "introductory." Our discussions are often advanced in the sense of building on topics from physical chemistry. At the same time, we have to describe the phenomena under consideration pretty much from scratch since they are largely unfamiliar. In keeping with this, this chapter is concerned primarily with a broad description of the scope of colloid and surface science and the kinds of variables with which they deal. In subsequent chapters different specific phenomena are developed in detail.

1.1a Colloid and Surface Chemistry: Some Definitions

1.1a.1 Definition of Colloids

Our first tasks are to define what we mean by colloid science and how this is related to surface science. For our purposes, any particle that has some linear dimension between 10^{-9} m (10 Å) and 10^{-6} m (1 μm or 1 μ)* is considered a colloid. For us, linear dimensions rather than

---

*In this book SI (International System of Units) units are used fairly consistently in keeping with current practice. Some quantities are traditionally expressed in hybrid units—for example, the specific area is usually measured in $m^2 g^{-1}$—and we continue this practice. The older literature uses cgs (centimeter-gram-second) units almost exclusively, so the reader must be cautious in consulting other sources. Appendix B contains a list of conversion factors between SI and cgs units.

1

particle weights or the number of atoms in a particle will define the colloidal size range; however, other definitions may be encountered elsewhere. It should be emphasized that these limits are rather arbitrary. Smaller particles are considered within other branches of chemistry, and larger ones are considered within sciences other than chemistry. The statement above may be expanded still further. Colloid science is interdisciplinary in many respects; its field of interest overlaps physics, biology, materials science, and several other disciplines. It is the particle dimension—not the chemical composition (organic or inorganic), sources of the sample (e.g., biological or mineralogical), or physical state (e.g., one or two phases)—that consigns it to our attention. With this in mind, it is evident that colloid science is the science of both large molecules and finely subdivided multiphase systems.

1.1a.2 Surfaces and Interfaces

It is in systems of more than one phase that colloid and surface science meet. The word *surface* is thus used in the chemical sense of a phase boundary rather than in a strictly geometrical sense. Geometrically, a surface has area but not thickness. Chemically, however, it is a region in which the properties vary from those of one phase to those of the adjoining phase. This transition occurs over distances of molecular dimensions at least. For us, therefore, a surface has a thickness that we may imagine as shrinking to zero when we desire a purely geometric description. The term *interface* is also used in this context. This term simply highlights the fact that the surface of interest is the dividing region between two phases.

It is self-evident that the more finely subdivided a given weight of material is, the higher the surface area will be for that weight of sample. In the following section we discuss this in considerable detail since it is the basis for combining a discussion of surface and colloid science in a single book.

1.1a.3 Other Concepts and Classification of Colloids

In subsequent sections of this chapter, we discuss further the distinction between macromolecular colloids and multiphase dispersions (Section 1.3), the use of the term *stability* in colloid science (Section 1.4), the size and shape of colloidal particles, the states of aggregation among particles, and the distribution of particle sizes that is typical of virtually all colloidal preparations (Section 1.5). The fact that particles in the colloidal size range are not all identical in size also requires a preliminary discussion of statistics, which is the subject of Section 1.5c and Appendix C.

VIGNETTE I.1 ENVIRONMENTAL SCIENCES: Colloid-Enhanced Transport in Unconsolidated Media

Contaminated bed sediments exist at numerous locations in the United States and around the world. These result mainly from past indiscriminate pollution of our aquatic environments and consist of freshwater and marine bodies including streams, lakes, wetlands, and estuaries. The bed sediments contain many hydrophobic organic compounds and metal ions that in the course of time act as sources of pollutants of the overlying aqueous phase. There are a number of transport pathways by which pollutants are transferred to the aqueous phase from contaminated sediments. One of the lesser known, but potentially important, modes of transport of pollutants from bed sediments is by diffusion and advection of contaminants associated with colloidal-size dissolved macromolecules in pore water. These colloids are measured in the aqueous phase as dissolved organic compounds (DOCs). (These are defined operationally as particles with a diameter smaller than 0.45 micrometer.)

The facilitated transport of compounds by colloids, illustrated schematically in Figure 1.1, is important in several areas and especially in the study of the fate and transport processes of hydrophobic organic compounds and metal ions in the environment. This facilitated transport also has implications in other areas in which colloid diffusion through porous media is important, such as those processes utilizing porous sorbents to clean up water and *in situ* flushing of subsurface soil with surfactant solutions for both oil recovery and cleanup of soils contaminated with oil. The understanding of facilitated transport of hydrophobic compounds from unconsolidated media such as bed sediments and soils is therefore very important and requires a knowledge of colloid and surface science of pollutant/particle interactions and transport.

In general, colloids are known to have a large capacity to bind hydrophobic compounds and have been reported to transport contaminants to very large distances in groundwater. It has been shown using simple mathematical models that orders of magnitude larger fluxes of contaminants are possible in the presence of colloids as compared to simple molecular diffusion of contaminants in the absence of any association with colloids. It is also known that contaminant binding to colloids is a fast, equilibrium process with partition constants larger than that between the bed sediment and water. In the literature there exists some information on the possible effects of colloids on the transport of pollutants from bed sediments. However, detailed studies on the flux of contaminants by colloids under controlled laboratory conditions are not available.

This vignette, in fact, captures many of the concepts and phenomena we discuss in this book and highlights them in the context of a very practical problem of considerable significance to our environment (see also the legend for Fig. 1.1).

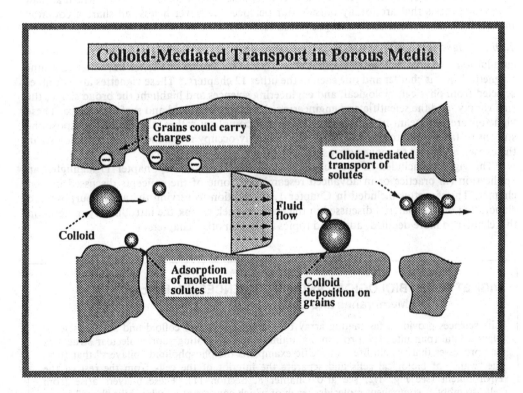

FIG. 1.1 A schematic illustration of colloid-mediated transport in porous media. The sketch illustrates the transport of molecular solutes by colloidal particles. The extent of such transport and its importance are determined by a number of factors, such as the extent of adsorption of molecular solutes on the colloids and on the grains, the deposition and retention of colloids in the pores, the influence of charges on the colloids and on the pore walls, and so on.

1.1b Impact of Colloid and Surface Chemistry in Science, Engineering, and Technology

One of the basic premises underlying the selection of topics included in this book is that areas of similarity between diverse fields should be stressed. This is not to say, of course, that differences are unimportant. Rather, it seems more valuable to point out to the beginner that useful methods and insights are frequently part of the well-established procedures of other disciplines that deal with related phenomena. Too provincial a viewpoint, especially at the beginning, is apt to isolate the student from many potentially valuable sources of information. In the long run, this seems like a greater loss than the loss of time that occurs when one concludes too hastily that a technique that works well in one system should work equally well in another system in which the particles are larger or smaller by several orders of magnitude. Errors of this last sort are generally discovered quickly enough!

1.1b.1 Examples of Applications

Any attempt to enumerate the areas in which surface and colloid chemical concepts find applicability is bound to be incomplete and quite variable with time because of changing technology. Nonetheless, we conclude this section with a partial listing of such applications. If there is any difficulty in doing this, it is because of the abundance rather than scarcity of such examples.

It should be evident from the partial list of examples summarized in Table 1.1 how many materials or phenomena of current scientific or everyday interest touch on colloid and surface science to some extent. Many of these areas, of course, have enormous technological and/or theoretical facets that are totally outside our perspective. Nevertheless, all share a common interest in small particles and/or large molecules.

1.1b.2 Vignettes

In addition to Table 1.1 and the applications discussed in this book, we have included nine "vignettes" in this chapter and one each in the other 12 chapters.* These vignettes are examples selected from physical, biological, and engineering sciences and highlight the broad scope, the rich variety, and the scientific and engineering challenges in colloid and surface science. These vignettes are not meant to be comprehensive and complete; rather, the idea is to expose the student to fascinating topics and ideas that currently occupy researchers and technologists in the chosen areas and to convey a sense of excitement about the subject.

The vignette included in each chapter (with the exception of Chapter 1) highlights an application (in practice or in advanced research) of some of the concepts discussed in that chapter. The vignettes included in Chapter 1, in addition to serving the above purpose and elaborating on the material discussed in the chapter, seek to link the introductory material in this chapter to more detailed, advanced topics treated in other chapters.

VIGNETTE I.2   BIOLOGICAL AND LIFE SCIENCES: Biological Membranes and Cells

Life sciences provide a fascinating array of examples in which colloid and surface science plays a vital (pun intended!) role in maintaining and promoting supramolecular structures and processes that sustain life. A specific example is the phospholipid "bilayers" that form the "walls" of biological cells and separate the interior of the cells from the rest of the environment (see Fig. 1.2; see also Chapter 8, Section 11). These bilayers arise from self-assembly of component molecules, each of which consists of a hydrophilic "head" group

---

*For the purpose of cross-referencing, the vignettes in this chapter are numbered sequentially as Vignette I.1, I.2, and so on, whereas those in the other chapters carry only the chapter number as a Roman numeral (e.g., Vignette X for the vignette in Chapter 10).

and hydrophobic "tails." The nature of the forces that create such layers (or membranes) and maintain their structure and functionality is a major topic of research today and is pursued by cell biologists, physiologists, biophysicists, biochemists, and engineers, among others. However, a bilayer by itself is insufficient to create and maintain a "living" cell. The cell membrane is actually a mosaic of a number of functional units that include protein molecules embedded in the membrane. The protein molecules themselves contain hydrophobic and hydrophilic components and provide the pathways for life-sustaining ions and polar molecules to move across the cell membrane. The arrangements of the hydrophilic and hydrophobic parts of a protein molecule are far more complex than those in a simpler structure such as a bilayer. The configuration of the hydrophilic and hydrophobic parts of a protein and how a protein is embedded in a bilayer are important for the molecular events that contribute to the functions of a biological cell (see Fig. 1.2 for some additional details). It therefore comes as no surprise that colloid and surface science plays a central role in such phenomena in life sciences (see Alberts et al. 1989; Tanford 1989; Bergethon and Simons 1990; Goodsell 1993; Morowitz 1992*).

In Figure 1.2, the hydrophobic parts of the protein molecules form the outer surface of the cylinders shown and hold the molecules in place, away from water, in the hydrocarbon part of the bilayer. The inner surfaces of the cylinders are hydrophilic and allow the transfer of ions and polar molecules from one side of the bilayer to the other. These pathways remain closed until an appropriate internal or external stimulus triggers them to open to allow transport across the membrane.

In addition to illustrating the importance of colloid and surface science in biological and life sciences, this vignette draws our attention to the importance of phenomena peculiar to surfactant systems. We discuss in Chapters 7 and 8 the behavior of surfactants in solutions and their tendency to self-assemble when dissolved in water or in water-oil mixtures.

Protein   **Bilayer**

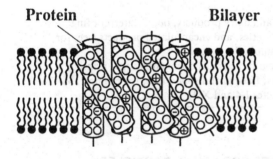

Lipid molecule

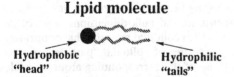

Hydrophobic   Hydrophilic
"head"   "tails"

FIG. 1.2 A simplified sketch of a hypothetical protein molecule embedded in a bilayer (a biological membrane). The bilayer shown is a two-dimensional cross section of a membrane. The bundle of cylinders shown represents the "helices" of a protein. The cylinders are part of the same protein and are joined together by other segments (not shown) of the protein protruding out of the bilayer on either side.

---

*Morowitz presents an interesting (and controversial) theory of the beginning of cellular life; the theory is based on the spontaneous condensation of amphiphilic molecules to form vesicles ("protocells").

TABLE 1.1 Some Examples of Disciplines and Topics for which Colloids and Colloidal Phenomena Are Important

Discipline	Examples
Analytical chemistry	Adsorption indicators, ion exchange, nephlometry, precipitate filterability, chromatography, and decolorization (see also Vignette I.8)
Physical chemistry	Nucleation; superheating, supercooling, and supersaturation; and liquid crystals
Biochemistry and molecular biology	Electrophoresis; osmotic and Donnan equilibria and other membrane phenomena; viruses, nucleic acids, and proteins; and hematology (see also Vignettes I.2 and I.3)
Chemical manufacturing	Catalysis, soaps and detergents, paints, adhesives, and ink; paper and paper coating; pigments; thickening agents; and lubricants (see Vignettes I.9 and IX)
Environmental science	Aerosols, fog and smog, foams, water purification and sewage treatment; cloud seeding; and clean room technology (see also Vignettes I.1 and VIII)
Materials science	Powder metallurgy, alloys, ceramics, cement, fibers, and plastics of all sorts (see Vignettes I.5 and I.6 for some modern examples)
Petroleum science, geology, and soil science	Oil recovery, emulsification, soil porosity, flotation, and ore enrichment
Household and consumer products	Milk and dairy products, beer, waterproofing, cosmetics, and encapsulated products (see, for example, Vignette I.7)
Imaging technology	Photographic emulsions; xerography; printing inks; flat-panel displays (see, for example, Vignettes I.4 and XII)

1.2 THE IMPORTANCE OF THE SURFACE FOR SMALL PARTICLES

As we see in the rest of the book, many of the interesting properties of colloids are the result of their dimension, which lies between atomic dimensions and bulk dimensions. Two of the important consequences of the size range of colloids are (a) colloidal materials have enormous surface areas and "surface energies," and (b) the properties of colloidal "particles" are not always those of the corresponding bulk matter or those of the corresponding atoms or molecules. Let us use a simple exercise or a "thought experiment" to illustrate these points.

The contemporary science student is probably aware that the concept of the atom is traceable to early Greek philosophers, notably Democritus. More than likely, however, few have bothered to follow through the hypothetical subdivision process that led to the original concept of an atom. The time has come to remedy this situation since, as mentioned above, the colloidal size range lies between microscopic chunks of material and individual atoms.

1.2a Increase in Surface Area and Energy with Decrease in Size

Consider a spherical particle of some unspecified material in which the sphere has a convenient radius of, say, 1.0 cm. Let us "reapportion" this fixed quantity of material by subdividing it repeatedly into arrays of spheres, each with a radius half that of the original sphere. The results of such an exercise are summarized in Table 1.2. These results are purely geometrical and therefore are (assumed to be) independent of any characteristic of the material. For example, it is implied in the calculations reported in Table 1.2 that the density of the material, whatever it may be, remains the same throughout the subdivision process. This is an assumption that breaks down when we reach sizes of the order of atomic dimensions; it is one of the points we consider below.

We can learn two lessons of importance in colloid and surface science from this simple exercise of subdividing a bulk particle. The first is that the properties of a material can be sensitive to the amount of sample under consideration. To see this point, assume that the material from which these spheres are made is water at a density of, say, 1.0 g cm^{-3}. Using this we can convert the particle volumes in Table 1.2 into the number of water molecules per sphere after each cut, as shown in Table 1.3 corresponding to the cuts used in Table 1.2. It is evident that by the time we reach spheres of radius 10^{-8} cm (i.e., typical atomic dimensions), we have begun chopping up the water molecules. As a matter of fact, calculating by this procedure we reach one water molecule per sphere after 25.62 halvings of all spheres, starting from one sphere of $R_s = 1.0$ cm. This corresponds to a radius of 0.193 nm for the water molecules if we use the formulas from Table 1.2 to evaluate the radius of the spheres. The radius of a water molecule estimated from the value of the van der Waals coefficient b (Atkins 1994) is about 0.145 nm, so a substantial discrepancy arises when a bulk property such as density is applied all the way down to molecular dimensions. It is not our purpose here to arrive at a precise estimate of molecular dimensions, but the above discrepancy does point out the fact that the characterization of a material may be sensitive to the size of the sample under consideration. A property such as density depends not only on the mass and volume of the molecules, but also on their packing in a bulk sample.

The second lesson we learn from this exercise is concerned with the extent of increase in surface area and the surface energy as we go to smaller and smaller particles. Let us first calculate the average number of water molecules that reside at the surface of the spheres in

TABLE 1.2 The Radius, Area, and Volume per Particle, Number of Particles, and Total Area for Any Array of Spheres After n "Cuts," Where a Cut is Defined to Be the Reapportionment of Materials into Particles With Radius that Is Half the Starting Value

Cut number	Radius (cm)	Number of spheres	Volume per sphere (cm^3)	Area per sphere (cm^2)	Total area (cm^2)
Original	1	1	4.19	1.26×10^1	1.26×10^1
Original, symbol	R_s	N_0	V_0	A_0	$A_{T,0}$
1	5×10^{-1}	8	5.24×10^{-1}	3.14	2.51×10^1
2	2.5×10^{-1}	6.4×10^1	6.55×10^{-2}	$7.86 \times^{-1}$	5.03×10^1
3	1.25×10^{-1}	5.12×10^2	8.18×10^{-3}	$1.96 \times^{-1}$	1.01×10^2
.	.	.	.	.	.
.	.	.	.	.	.
.	.	.	.	.	.
n	$\left(\dfrac{1}{2}\right)^n R_s$	$8^n N_0$	$\left(\dfrac{1}{8}\right)^n V_0$	$\left(\dfrac{1}{4}\right)^n A_0$	$2^n A_{T,0}$
.	.	.	.	.	.
.	.	.	.	.	.
19.93	10^{-6}	10^{18}	4.2×10^{-18}	1.26×10^{-11}	1.26×10^7
23.25	10^{-7}	10^{21}	4.2×10^{-21}	1.26×10^{-13}	1.26×10^8
26.58	10^{-8}	10^{24}	4.2×10^{-24}	1.26×10^{-15}	1.26×10^9

TABLE 1.3 Total Number of Water Molecules per Sphere and Number at Surface for Spheres of Water After n Cuts (Also Total Surface Energy of the Array of Spheres of Water)

Cut number	Radius (cm)	Number of water molecules per sphere	Number of water molecules at surface	Fraction of total water molecules at surface	Total surface energy (J)
0	1.0	1.38×10^{23}	1.26×10^{15}	9.13×10^{-8}	9.07×10^{-5}
1	5.0×10^{-1}	1.75×10^{22}	3.14×10^{15}	1.79×10^{-7}	1.81×10^{-4}
2	2.5×10^{-1}	2.18×10^{21}	5.03×10^{16}	3.64×10^{-7}	3.62×10^{-4}
3	1.25×10^{-1}	2.73×10^{20}	1.01×10^{17}	7.32×10^{-7}	7.27×10^{-4}
.	.	.	.	.	.
.	.	.	.	.	.
.	.	.	.	.	.
19.93	10^{-6}	1.40×10^{5}	1.26×10^{22}	9.13×10^{-2}	9.07×10^{1}
23.25	10^{-7}	1.40×10^{2}	1.26×10^{23}	9.13×10^{-1}	9.07×10^{2}
26.58	10^{-8}	1.40×10^{-1}	1.26×10^{24}	9.13	9.07×10^{3}

Tables 1.2 and 1.3 by assuming* that the area occupied by each molecule at the surface is about 0.10 nm^2. Using this estimate and the total area at various stages of subdivision from Table 1.2, we obtain an estimate of the number of molecules in the surface at each stage of the process (see Table 1.3). This quantity is also reported in the table as a fraction of the total number of molecules present. Note that this fraction approaches unity as molecular dimensions are approached. That is, as we decrease in size, a larger and larger number of atoms become surface atoms.

Now, let us consider the last column in Table 1.3. In this column the total surface energy of the array of spherical water droplets is reported at each stage of the process. We see in Chapter 6 that the surface tension of a substance is the energy required to make a unit area of new surface. For water, this quantity is about 72 mJ m^{-2} at room temperature. This value of the surface tension of water and the total areas from Table 1.2 can be used to obtain the values listed under "total surface energy" in Table 1.3. It should be recalled that a fixed amount of material (4.19 cm^3 water = 4.19 g water = 0.23 mole water) is involved throughout this entire process. It should also be noted that surface tension, like density, is a macroscopic property; its applicability is highly dubious for very small particles. Nevertheless, as the dimensions of the subdivided units decrease, the total energy associated with the formation of the surface takes on values comparable to other chemical energies. At $R_s = 10^{-4}$ cm, the surface energy is about 0.9 J/0.23 mole or about 4 J mole^{-1}. When we reach $R_s = 10^{-7}$ cm, this quantity equals 4 kJ mole^{-1}.

1.2b Specific Surface Area

The increasing importance of the surface area as the linear dimensions of particles decrease is stated concisely in a quantity known as the *specific surface area* A_{sp} of a substance. This quantity is determined as the ratio of the area divided by the mass of an array of particles. If the particles are uniform spheres, as we have assumed throughout this section, this ratio equals

$$A_{sp} = A_{tot}/m_{tot} = (n\,4\pi R_s^2)/[n(4/3)\pi R_s^3 \rho] \tag{1}$$

where n is the number of spheres having a radius R_s and made of a material of density ρ. Simplifying Equation (1)† leads to the result

---

*Our interest is in the order of magnitude of these quantities, so we need not worry about the cross-sectional shape or the surface packing efficiency of the water molecules in these calculations.
†This manner of referencing is used for equations occurring in the same chapter.

$$A_{sp} = 3/(\rho R_s) \tag{2}$$

This formula generalizes the conclusion reached in Tables 1.2 and 1.3. It shows clearly that for a fixed amount of material, the surface area is inversely proportional to the radius for uniform, spherical particles. At the same time, the formula reminds us that some lower limit for R_s must be imposed since the relationship is undefined for $R_s = 0$. If SI units were used consistently, A_{sp} would be expressed in m^2 kg^{-1}; however, m^2 g^{-1} are the most commonly used units for this quantity. In the event of nonuniform or nonspherical particles, alternate expressions for Equation (2) have to be used. The following example considers the case of cylindrical particles.

* * *

EXAMPLE 1.1   *Variation of Specific Surface Area with Geometry.*   A material of density ρ exists as uniform cylindrical particles of radius R_c and length L. Derive an expression for A_{sp} for this material and examine the limiting forms when either R_c or L is very small.

Solution: The area of each cylindrical particle equals the sum of the areas of both ends and the cylindrical surface: $A = 2(\pi R_c^2) + 2\pi R_c L$.
  The volume of each cylindrical particle equals $\pi R_c^2 L$ and its mass is given by $\rho \pi R_c^2 L$.
  For an array of n cylindrical particles, the total area per total mass equals A_{sp} and is given by

$$\begin{aligned}
A_{sp} &= n(2\pi R_c^2 + 2\pi R_c L)/(n\rho\pi R_c^2 L) \\
&= [2(R_c^2 + R_c L)]/\rho R_c^2 L \\
&= (2/\rho)(1/R_c + 1/L)
\end{aligned}$$

For a thin rod, $L \gg R_c$:

$$A_{sp} \approx \frac{2}{\rho}\frac{1}{R_c}$$

For a flat disk, $R_c \gg L$:

$$A_{sp} \approx \frac{2}{\rho}\frac{1}{L}$$

Note that in both of these limits, it is the *smaller* dimension that affects A_{sp}, with A_{sp} increasing as the smaller dimension decreases. ∎

* * *

Both Example 1.1 and Equation (2) show that the surface plays an increasingly important role as the dimensions of the particles decrease.

  The concept of specific area defined by Equation (1) is important because this is a quantity that can be measured experimentally for finely divided solids without any assumptions as to the shape or uniformity of the particles. We discuss the use of gas adsorption to measure A_{sp} in Chapter 9. If the particles are known to be uniform spheres, this measured quantity may be interpreted in terms of Equation (2) to yield a value of R_s. If the actual system consists of nonuniform spheres, an average value of the radius may be evaluated by Equation (2). Finally, even if the particles are nonspherical, a quantity known as the radius of an *equivalent sphere* may be extracted from experimental A_{sp} values. This often proves to be a valuable way of characterizing an array of irregularly shaped particles. We have a good deal more to say about average dimensions in this chapter and about equivalent spheres in Chapter 2.

  In the discussion above we emphasized two-phase* colloidal systems, in which the concept of surface plays an important role. However, as we saw in Section 1.1, our definition of the colloidal range is based on the linear dimensions of particles, and there are numerous natural

---

*We occasionally use the term *two phase* in this chapter to refer to colloids when the colloidal particles and the fluid in which they are suspended are distinct (e.g., solid particles in a liquid or liquid droplets in another, immiscible liquid or in a gas).

and synthetic polymer molecules with dimensions that, considered individually, fall within this range, as we illustrate in the following section. It is clear when we deal with single molecules that the concept of surface is greatly different from when we consider a particle made of many molecules. If, in a given situation, we tend to concentrate on the surface characteristics of a material, then the focus is on *surface* science. If, however, we look at the subdivided sample as an array of particles, then the focus is colloid science. As far as we are concerned, these two fields differ primarily in point of view. Their mutual concern with finely subdivided material is the common denominator that connects the two disciplines.

1.3 CLASSIFICATION OF COLLOIDS BASED ON AFFINITY TO CARRIER FLUID

In the preceding section, we saw that either large molecules or finely subdivided bulk matter could be considered colloids inasmuch as both may consist of particles in the range 10^{-9} to 10^{-6} m in dimension. The difference between these two situations lies in the relationship that exists between the colloidal particle and the medium in which it is embedded. Macromolecular colloids are true *solutions* in the thermodynamic sense. Subdivided bulk matter, on the other hand, forms a two-phase (at least) system with the medium. We have already noted that the word *surface* connotes the existence of a phase boundary and therefore has a specific chemical meaning in the multiphase case; this is, of course, inapplicable to macromolecular colloids. The intent of this book is to discuss as wide a variety of colloidal phenomena as possible from a unified point of view and using a single set of terms. We use the words *continuous phase* and *dispersed phase* to refer to the medium and to the particles in the colloidal size range, respectively. This distinction is somewhat vague in the case of macromolecular solutions. (See Table 1.4 for a summary of the descriptive names for two-phase colloidal systems.)

Colloids are also often classified on the basis of the affinity of the *surfaces* of the particles to the continuous phase. This classification is also ambiguous in some respects, but deserves a brief mention.

The terms used to distinguish colloidal "particles" on the basis of their affinity to the fluid in which they are dispersed are *lyophilic* and *lyophobic*. These terms mean, literally, "solvent loving" and "solvent fearing," respectively. When water is the medium or solvent, the terms *hydrophilic* or *hydrophobic* are often used. This terminology is very useful when considering surface activity such as wettability of a surface; however, when used to classify colloids, the distinction is not always clear-cut. We consider these two types of colloids separately in the following subsections.

1.3a Lyophilic Colloids

The classical use of the term *lyophilic colloids* refers to soluble macromolecular materials in which the individual particles (macromolecules such as synthetic polymer chains or proteins)

TABLE 1.4 Summary of Some of the Descriptive Names Used to Designate Two-Phase Colloidal Systems

Continuous phase	Dispersed phase	Descriptive names
Gas	Liquid	Fog, mist, aerosol
Gas	Solid	Smoke, aerosol
Liquid	Gas	Foam
Liquid	Liquid	Emulsion
Liquid	Solid	Sol, colloidal solution, gel, suspension
Solid	Gas	Solid foam
Solid	Liquid	Gel, solid emulsion
Solid	Solid	Alloy

are of colloidal dimensions. However, there are macromolecules of colloidal dimensions containing both lyophobic and lyophilic components (e.g., proteins with hydrocarbon [hydrophobic] portions and hydrophilic [peptide and carboxyl groups] portions, as mentioned in Vignette I.2). Another such example is the case of micelles (Chapter 8), which are clusters of small molecules that form spontaneously in aqueous solutions (mostly) of certain compounds (therefore these are often called *association colloids*). The onset of micellization occurs at a well-defined concentration — known as the *critical micelle concentration* — which makes micelle formation very much like a phase separation. However, the individual small molecules retain their identity in the micelles, which are not covalent entities like polymers, so their surface is problematic. We see in Chapter 8 that both chemical equilibrium and phase equilibrium can be used to discuss micelle formation, so the classification of micelles remains fuzzy on that basis also. By the time we reach Chapter 8, however, we will be more experienced with colloids and less dependent on the "black and white" categories of this section.

Above we used the words *continuous phase* and *dispersed phase* to refer to the medium and to the particles, respectively, in the colloidal size range. It should be understood that these are solvent and solute in lyophilic systems. In micellar systems, the micelles are dispersed in an aqueous continuous phase. Furthermore, the system as a whole is generally called a *dispersion* when we wish to emphasize the colloidal nature of the dispersed particles. This terminology is by no means universal. Lyophilic dispersions are true solutions and may be called such, although this term ignores the colloidal size of the solute molecules.

VIGNETTE I.3   COLLOIDAL CARGO CARRIERS—FROM COSMETICS TO MEDICINE AND GENETIC ENGINEERING: Liposomes with a Molecular Cargo and a Mission

Wouldn't it be nice if we could package appropriate doses of medicine in physiologically friendly capsules that can deliver the medicine specifically to the organs that need it while at the same time keeping it away from areas that may find the medicine toxic? Or, perhaps, our interest is much more mundane and all we need — to satisfy our vanity — is a method to trap perfumes in our skins so that the fragrance lasts longer. These were hardly the questions that engaged the attention of Alec Bangham, a British scientist who discovered surfactant capsules known as *liposomes* in the early 1960s while studying the effect of lipid molecules on the clotting of blood.

Liposomes are colloidal-size containers made of lipid bilayers (Fig. 1.3). A lipid molecule consists of a polar, hydrophilic head that is attached to (one or two) hydrophobic, hydrocarbon tails. At appropriate concentrations, the lipid molecules in water "self-assemble" to form bilayers since the hydrophobic tails like to avoid contact with the water. When such bilayers are broken up into small pieces, the fragments wrap themselves into closed structures known as liposomes and encapsulate some of the water inside. The potential applications of liposomes in cosmetics, pharmaceutical and medical technology, and genetic engineering (for studying basic properties of genes by isolating them inside a liposome or for developing schemes for gene- or protein-replacement therapies) are numerous.

Why should we be interested in using liposomes for drug delivery? In many cases, drugs administered in "free form" cause "side effects" because of their toxicity to areas of the body that are not affected by the disease or disorder. It appears possible to improve the effectiveness of drugs and minimize their toxicity by encapsulating the drugs in liposomes and delivering them efficiently and specifically to the affected organs. It is also possible to design liposomes that avoid detection* by "hunters" such as macrophages in the body so that

---

*These are known by their registered tradename, Stealth liposomes (see Lasic 1993, p. 284). As discussed by Lasic, the "stealth effect" is the result of a "steric" layer surrounding the liposome, and these liposomes are also known as sterically stabilized liposomes. We discuss steric stabilization in Chapter 13.

the drugs can be delivered to the cells that need treatment. Excellent descriptions of these and related issues in liposome technology are presented by Lasic in a lucidly written review in the *American Scientist* (Lasic 1992) and in a monograph (Lasic 1993).

The above discussion illustrates that liposome technology lies at the interface of colloid and surface chemistry, physics, biology, and medicine. There are a number of problems of interest to colloid scientists and engineers in this context, ranging from, for example, synthesis of new (short-chain as well as polymeric) surfactants for developing special membranes, artificial skins, pharmaceutical lotions, and (of course) drug delivery systems. Although many of these problems are outside the scope of our book, we consider some basic issues that are relevant to the above problems. For example, we discuss the formation of surfactant (e.g., lipid) layers in Chapter 7 and study some of their properties. Formation of surfactant micelles and the relation between the molecular architecture of the surfactants and the shapes of the self-assembled structures that result from the surfactants is covered in Chapter 8. In Chapter 8, we also touch on the special opportunities afforded by the microenvironments inside surfactant micelles and liposomes for studying catalysis and material synthesis.

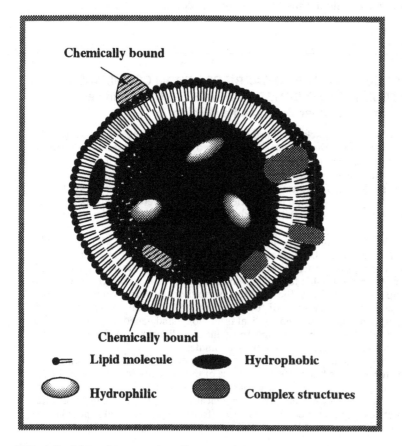

FIG. 1.3 Molecular cargo in a liposome. The cargo molecules are carried in different parts of the liposome depending on their chemical nature. Hydrophobic molecules are carried inside the hydrophobic part of the bilayer, whereas hydrophilic molecules reside in the interior. More complex molecules are wholly or partly embedded in the bilayer or chemically bound to the interior or exterior surface.

1.3b Lyophobic Colloids

Lyophobic colloids are known by a variety of terms, depending on the nature of the phases involved. Some of these are listed in Table 1.4. Some of the terms (e.g., aerosol, gel) are somewhat ambiguous, so the reader is warned to make certain that the system is fully understood, particularly when the original literature is consulted. Remember that a common feature of all systems we consider is that some characteristic linear dimension of the dispersed particles falls in the range defined in Section 1.1a. When we deal with two-phase colloids in this book, we are primarily concerned with systems in which the dispersed phase is solid and the continuous phase is liquid.

As in the case of lyophilic colloids, the use of the adjective *lyophobic* does *not* necessarily mean that the surfaces of the colloids are uniformly "liquid repelling." For example, ceramic sols such as silica and alumina powders in liquids do have surfaces with varying degrees of affinity to the liquid. Despite the ambiguities in the use of the terms *lyophobic* and *lyophilic*, such a classification is convenient.

Next, we consider another difference between lyophobic and lyophilic colloids in addition to the presence or absence of surfaces between the continuous and dispersed species. This difference deals with the "stability" of the dispersion, and we examine the meaning(s) of this term in more detail below.

VIGNETTE I.4 IMAGING SCIENCE AND TECHNOLOGY: Colloid-Based Electrophoretic Imaging Devices*

Colloids hold a considerable potential for applications that are unusual in the classical sense. Most of us are familiar with imaging devices such as the picture tube in a television set. These tubes are bulky and consume large amounts of electrical power. There is, therefore, a large incentive to develop compact imaging devices, known as *flat-panel devices*, that are easily portable and have lower power requirements. (Displays based on liquid crystal technology fall in this class.)

Another possible flat-panel device is one known as an *electrophoretic image display*, or EPID. EPIDs contain submicron-size particles of pigments dispersed in a liquid along with a dye that provides contrast. When an electrical potential is applied to the system, pigment particles are driven to the interface between the suspending liquid and a viewing plate, usually made of glass. There they can be seen under normal illumination. EPIDs have the potential of providing an image that has extremely high optical contrast under normal lighting, that is legible over a wide range of viewing angles, that is inherently retained on the display (as opposed to an image needing constant refreshing as on a TV picture tube), and that requires low voltage and power.

The EPID concept can be combined with other developments in imaging technology to produce optical devices such as light valves and x-ray imagers. An example of an electrophoretic x-ray imager is illustrated in Figure 1.4.

One of the most crucial aspects of this "chemical" display technology is the liquid dispersion of pigment particles and dye. The dispersion must be stable even when the particles are compressed (to concentrations as much as 10 times higher than in the bulk form) to form an image at the viewing plate. The pigment particles must be able to retain their charges after numerous switching operations. The fluid must allow a fast response time. Unwanted migration of particles when an electrical field is applied and other electrohydrodynamic effects must be controlled.

Many of the crucial problems for research and development in this area are the same as those encountered in other areas of colloid and surface science. There are questions that need to be addressed. How do particles interact when they are repeatedly and forcefully packed

---

*This vignette is a slightly revised version of the one prepared by one of the authors (RR) for a National Research Council report entitled, *Frontiers in Chemical Engineering: Research Needs and Opportunities*, published by the National Academy Press (Washington, DC, 1988).

and unpacked in electrical fields? How are particles, counterions, polymers, and surfactants transported across an EPID cell? How does fluid motion in an EPID device affect this overall transport? How fast do structures formed by charged particles dissipate in applied fields? These are questions central to many of the problems of interest in colloid and surface science.

1.4 CONCEPT OF STABILITY OF COLLOIDAL SYSTEMS

As mentioned in the last section, lyophilic colloids form true solutions, and true solutions are produced spontaneously when solute and solvent are brought together. In the absence of chemical changes or changes of temperature, a solution is stable indefinitely. Finely subdivided dispersions of two phases do not form spontaneously when the two phases are brought together. As a matter of fact, if such a dispersion is allowed to stand long enough, the reverse process would spontaneously occur. For example, oil and water can be vigorously mixed to form a nontransparent, heterogeneous mass; however, on standing, the mixture will separate into two clear, homogeneous layers.

We know from thermodynamics that spontaneous processes occur in the direction of decreasing Gibbs free energy. Therefore we may conclude that the separation of a two-phase dispersed system to form two distinct layers is a change in the direction of decreasing Gibbs

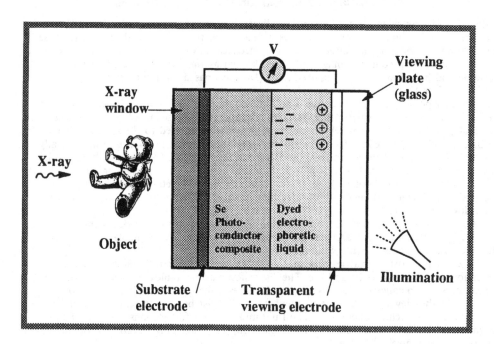

FIG. 1.4 A large-area, solid-state x-ray receptor with an electrophoretic image display. When a voltage is applied across the image cell, pigment particles and counterions in the liquid separate. Most of the voltage drop occurs across the Se layer. X-ray exposure under this condition leads to the creation of a charge-image at the photoconductor-composite/liquid interface due to the generation of x-ray-induced charges in the Se. After the x-ray exposure, the applied voltage is reduced to zero, and the pigment particles are driven to the viewing plate. The image becomes visible on illumination.

free energy. Using the logic employed in the discussion of Table 1.3, one can say that there is more surface energy in a two-phase system when the dispersed phase is in a highly subdivided state than when it is in a coarser state of subdivision. This suggests a correlation between the inherent instability of a highly dispersed lyophobic system and the thermodynamics of the surface. This is discussed in more detail in Chapter 6. For the present it is sufficient to note that lyophobic systems "dislike" their surroundings enough to want to separate out. Lyophilic systems, on the other hand, are perfectly "happy" in a solution. The two categories differ radically in their definitions of stability. We elaborate these individually in the following subsections.

1.4a   Kinetic Stability

Remember that equilibrium thermodynamics has nothing to say about the rate at which processes occur. It is a fact that many two-phase dispersions appear unchanged over very long periods of time, but the situation is analogous to the thermodynamic instability of diamond with respect to graphite. The kinetics of the diamond-graphite reaction are slow enough that the thermodynamic instability is of very little practical consequence. Likewise, many colloidal dispersions have *kinetic* stability, even though they are unstable thermodynamically. This type of colloidal "solution" (Table 1.4) resembles a true solution. This is one way in which two-phase dispersions and solutions of macromolecules are very similar and explains in part how the two can be grouped together in our study of colloidal phenomena.

Above we noted that two-phase dispersions will always spontaneously change into a smaller number of large particles given sufficient time. However, many solutions of macro-molecules do not undergo spontaneous separation into two phases. Common usage tempts us to describe the first as "unstable" and the second as "stable." Although these terms are used very frequently in colloid science, the reader should realize that *the words are meaningless unless the process to which they are applied has been clearly defined.* The situation is some-what analogous to the insistence in thermochemistry that one always keep in mind the bal-anced chemical equation to which the thermodynamic quantities (ΔH, ΔC_p, etc.) apply. The coarsening described here is only one of a variety of possible processes that a dispersion might undergo.

The coarsening process of a thermodynamically unstable dispersion is called *coalescence* or *aggregation*. It is important that we differentiate between coalescence and aggregation. By *coalescence* we mean a process by which two (or more) small particles fuse together to form a single larger particle. The central feature of coalescence is the fact that total surface area is reduced. *Aggregation* is the process by which small particles clump together like a bunch of grapes (an aggregate), but do not fuse into a new particle. In aggregation there is no reduction of surface, although certain surface sites may be blocked at the points at which the smaller particles touch. The term *coagulation* is also used to describe the process of aggregation. A colloid that is stable against coalescence or aggregation is called *kinetically stable*. This is illustrated in Figure 1.5. The classical use of the term "colloid stability" thus represents kinetic stability. In this sense, the word *stability* describes the extent to which small particles remain uniformly distributed throughout a sample.

Incidentally, when small particles coalesce all evidence of the smaller particles is erased. Only the new, larger particle remains. With aggregation, however, the small particles retain their identity; only their kinetic independence is lost. The aggregate moves as a single unit. Likewise, the clusters that form as the products of the process may be called *aggregates*. The individual particles from which the aggregates are assembled are called *primary particles*. A system may be relatively stable in the kinetic sense with respect to one of these processes, say, coalescence, and be unstable with respect to the other, aggregation.

Finally, remember that whether a stable or unstable dispersion is needed depends on the application. A dispersion that is too fine to settle out may be a source of great frustration in some parts of a manufacturing process. On the other hand, a dispersion that settles too

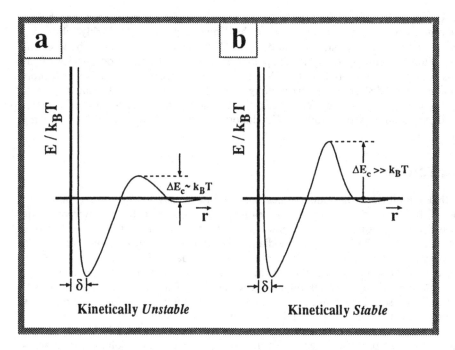

FIG. 1.5 Schematic illustration of kinetic stability of colloids. The figure shows the interaction energy (free energy) E as a function of the surface-to-surface separation r between two particles (k_B and T are the Boltzmann constant and the absolute temperature of the dispersion, respectively). (a) The free energy will reach the global minimum if the two particles can come close enough ($r = \delta$). However, the energy barrier against coagulation, ΔE_c, is of the order of the thermal energy $k_B T$, and therefore the dispersion is kinetically *unstable*. (b) The energy barrier $\Delta E_c \gg k_B T$, and the dispersion is kinetically stable since the thermodynamically favored separation distance is not reachable. (See Chapters 11 and 13 for more details.)

rapidly may be equally troublesome to those who have to pump it around. The chemist doing gravimetric analyses wants coarse precipitates; one who uses adsorption indicators wants finely subdivided particles to form. We have seen an example in Vignette I.4 in which stability of the dispersion is essential. Vignette I.5 illustrates another example from colloidal processing of ceramics in which stability of the dispersion needs to be maintained even as we "consolidate" the dispersion to form dense structures.

VIGNETTE I.5   ADVANCED MATERIALS: Colloid Stability and Ceramic Processing

Considerable recent activity in the area of ceramic processing is aimed toward the formulation of materials with high strengths, comparable to the room temperature strength of metal alloys, at high temperatures (of the order of 2000 K). The impetus comes from the significant gains made in the last 20 years with materials formed from submicron powders of silicon nitride and silicon carbide and the promise of similar improvements in the near future.

The problems with the ceramics lie as much in reproducibility as in absolute strength. With brittle materials, microscopic flaws concentrate stress and cause failure at stresses, which vary according to the classical crack theory as $\delta^{-1/2}$, where δ is the characteristic dimension of the flaw. For materials formed from powders, the flaws represent structural imperfections, introduced in the initial forming stage, that survive the subsequent processing steps. Hence variability in the initial powder or incomplete control of the fabrication steps lead to unreliable performance.

Of the many methods that are explored, one class of techniques relies on what is known as the *colloidal processing* route (Brinker and Scherer 1990). Here, one starts with a dispersion of precursor powders and densifies the dispersion to an appropriate consistency. The resulting slurry may be slip cast to produce a so-called green specimen in the required shape of the product. Slip casting employs a porous mold that imbibes the fluid but excludes the particles. The type of operation used may employ drain, pressure, or vacuum casting and is determined by the geometry desired and the time constraints. Regardless of the choice of casting method used, a pressure difference draws the fluid into the mold and leaves a dense solid casting adjacent to the wall and a fluid slip in the interior. The physical mechanism bears a strong resemblance to filtration and sedimentation. A broad outline of the above process is illustrated in Figure 1.6.

The structure of the green body depends on the processing technique and conditions as well as the underlying colloid science. For example, the dispersions are often stabilized using polymer additives (Chapter 13) so that the van der Waals attractive forces between the particles (Chapter 10) are minimized by polymer "brushes" adsorbed on the particle surfaces. This prevents "fractal" aggregates (Section 1.5b) that give rise to large voids and defects in the green body on consolidation.

The questions of interest to an engineer in this case are: How do the initial concentration, the particle size, and the nature of the interparticle potential affect the structure of the dispersion, the structure of the final specimen, and the processing time? How long does the process take? What kinds of chemical additives are suitable? The permeability and the capillary suction in the mold determine the rate of production of the specimens. How does one adjust the two to optimize production? These questions require a basic understanding of colloid and surface science and phenomena.

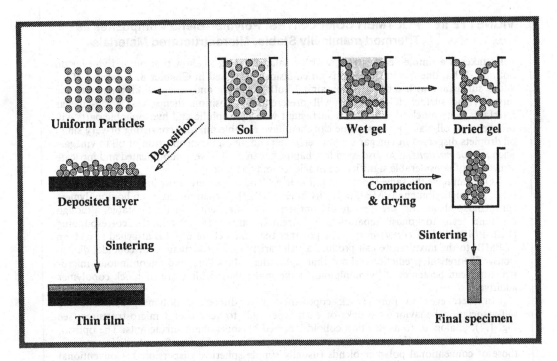

FIG. 1.6 A schematic diagram of colloidal processing of ceramic specimens. The figure illustrates some of the ways in which a dispersion is densified and transformed into porous or compact films or bulk objects. (Adapted and modified from Brinker and Scherer 1990.)

1.4b Thermodynamic Stability

In contrast to the above-described kinetic stability, colloids may also be *thermodynamically* stable. A stable macromolecular solution is an example we have already discussed. Formation of micelles beyond the critical micelle concentration is another example of the formation of a *thermodynamically stable* colloidal phase. However, when the concentration of the (say, initially spherical) micelles increases with addition of surfactants to the system, the spherical micelles may become thermodynamically *unstable* and may form other forms of (thermodynamically stable) surfactant assemblies of more complex shapes (such as cylindrical micelles, liquid-crystalline phases, bilayers, etc.).

Similarly, charged solid particles (such as latex spheres) — kinetically stable lyophobic colloids — may exist in colloidal crystalline phases (with body-centered or face-centered cubic structures) as a consequence of thermodynamically favored reduction in free energies (see Chapter 13). Even neutrally charged spherical particles ("hard spheres") undergo a phase transition from a "liquidlike" isotropic structure to face-centered cubic crystalline structures due to entropic reasons. In this sense, the stability or instability is of *thermodynamic* origin.

It is, therefore, important that we keep several things in mind when the word *stability* is used in colloid science. First, whenever we describe a two-phase dispersion in these terms, the words are being used relatively and often in a kinetic sense. Second, there is little unanimity among workers about the nomenclature of various processes. Finally, whether a stable or unstable system is desirable depends entirely on the context.

Other examples of thermodynamically stable colloidal structures are highlighted in Vignette I.6.

VIGNETTE I.6   POLYMER COMPOSITES: Polymer-Blend Composites as Thermodynamically Stable, Microstructured Materials

All it takes is a simple demonstration with a salad dressing to show that oil and vinegar do not mix. What one has in this case is an emulsion, discussed in Chapter 8. However, if we add a surfactant with molecules containing both oil-liking and vinegar-liking parts to the mixture, the surfactant molecules will preferentially position themselves at oil-vinegar interfaces, very much like the way a surfactant with hydrophilic and hydrophobic parts will behave in an oil-water mixture. One can then have a stable mixture consisting of very small oil droplets dispersed in vinegar or vice versa, depending on the proportion of oil to vinegar. Such *microemulsions*, also discussed in Chapter 8, can exist in very complicated and intricate structures of considerable significance in science and technology.

One can have the same type of situation in a blend of two mutually immiscible polymers (e.g., polymethylbutene [PMB], polyethylbutene [PEB]). When mixed, such homopolymers form coarse blends that are nonequilibrium structures (i.e., only kinetically stable, although the time scale for phase separation is extremely large). If we add the corresponding (PEB-PMB) *diblock copolymer* (i.e., a polymer that has a chain of PEB attached to a chain of PMB) to the mixture, we can produce a rich variety of microstructures of colloidal dimensions. Theoretical predictions show that cylindrical, lamellar, and bicontinuous microstructures can be achieved by manipulating the molecular architecture of block copolymer additives.

In fact, even in pure block copolymer (say, diblock copolymer) solutions the self-association behavior of blocks of each type leads to very useful microstructures (see Fig. 1.7), analogous to association colloids formed by short-chain surfactants. The optical, electrical, and mechanical properties of such composites can be significantly different from those of conventional polymer blends (usually simple spherical dispersions). Conventional blends are formed by quenching processes and result in *coarse* composites; in contrast, the above materials result from *equilibrium* structures and *reversible* phase transitions and therefore could lead to "smart materials" capable of responding to suitable external stimuli.

The colloidal structures described above are dictated by thermodynamics, and the resulting structures are thermodynamically stable. Similar thermodynamically stable structures can develop even in a copolymer *melt* (i.e., there is no other polymer or solvent). Such colloidal systems differ from kinetically stable lyophobic dispersions of the type discussed in Vignettes I.4 and I.5.

1.5 SOME PHYSICAL CHARACTERISTICS OF COLLOIDS

As has been emphasized in the sections above, one of the most important features of colloidal particles is their physical dimension, the defining characteristic of colloids. Many of the properties of colloids of scientific and industrial importance (e.g., specific surface area, viscosity, aggregation behavior, and microstructure) are strongly influenced by the dimension, as well as other characteristics such as surface charge and the chemical affinity of the particles to dissolved matter, among others. Clearly, uniform-size particles of spherical geometry are the easiest to deal with in this respect, but colloidal particles come in all sizes and shapes. Even in the case of spherical particles of uniform size, other physical and chemical properties (e.g., surface charge) may not be uniform due to surface heterogeneities at the molecular scale.

What these imply is that, at the minimum, we need to be aware of the complexities in representing quantitatively particle sizes, shapes, and the extent of their variations. It is also important to be aware of some of the standard experimental tools that are available for this purpose. In this section and the next, we consider some of these items. For convenience, we restrict ourselves primarily to some of the most common methods and techniques.

1.5a Particle Size and Shape

Figure 1.8 shows an electron micrograph of latex particles made from polystyrene cross-linked with divinylbenzene. Note that these latex particles are not the same as simple polystyrene molecules in a true solution. The particles shown in the figure display a remarkable degree of homogeneity with respect to particle size. Such a sample is said to be *monodisperse* (in *size*), in contrast to *polydisperse* systems, which contain a variety of particle sizes. We have a good

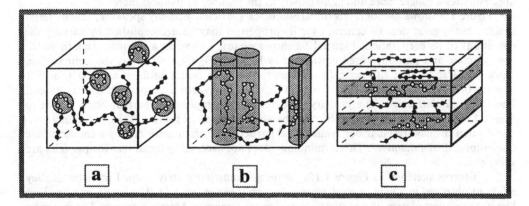

FIG. 1.7 Some of the microstructures produced by the self-association behavior of diblock copolymer solutions. The figure illustrates the (a) spherical, (b) cylindrical, and (c) lamellar structures (among others) that are possible in such solutions. Each diblock polymer chain consists of strings of white beads (representing one type of homopolymer) and strings of black beads (representing the second type of homopolymer). (Redrawn from A. Yu. Grosberg and A. Khokhlov, *Statistical Physics of Macromolecules*, AIP Press, New York, 1994.)

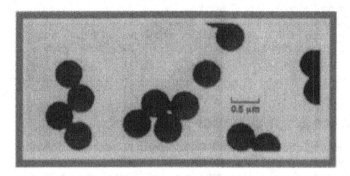

FIG. 1.8 Electron micrograph of cross-linked monodisperse polystyrene latex particles. The latex is a commercial product ($d = 0.500$ μm) sold as a calibration standard. (Photograph courtesy of R. S. Daniel and L. X. Oakford, California State Polytechnic University, Pomona, CA.)

deal more to say about polydisperse systems in other sections of this chapter. The particles of Figure 1.8 are also perfectly spherical. Except for the nature of the material involved, this figure could be a photograph of the hypothetical array of spheres discussed in Section 1.2.

Monodisperse spheres are not only uniquely easy to characterize, but also very rarely encountered. Polymerization under carefully controlled conditions allows the preparation of the polystyrene latex shown in Figure 1.8. Latexes of this sort are used as standards for the size calibration of optical and electron micrographs (also see Section 1.5a.3). However, in the majority of colloidal systems, the particles are neither spherical nor monodisperse, but it is often useful to define convenient "effective" linear dimensions that are "representative" of the sizes and shapes of the particles. There are many ways of doing this, and whether they are appropriate or not depends on the use of such dimensions in practice. There are excellent books devoted to this topic (see, for example, Allen 1990) and, therefore, we consider only a few examples here for the purpose of illustration.

1.5a.1 Characterizing Variations in Shape: Particles with a High Degree of Symmetry

Many solid particles are not actually spherical, but are characterized by a high degree of symmetry like a sphere and are often approximated as spheres. For example, a polyhedron approximates a sphere more and more closely as the number of its faces increases.

Figure 1.9 shows micrographs of carbon black particles. Broadly speaking, this material is soot, but a great deal of control over its properties may be accomplished by varying the conditions of its preparation. Figure 1.9a shows what is known as a thermal black in which both discrete and partially fused particles may be seen. Figure 1.9b shows the same carbon black preparation (not the same field of particles) after heat treatment at 2700°C in the absence of oxygen. The particles take on a distinctly polyhedral shape with this treatment, known as *graphitization*. The primary particles of the graphitized thermal black shown in Figure 1.9b are sufficiently symmetrical to be approximated as spheres. Likewise, many substances that display irregular but symmetrical particles are often described by a characteristic dimension called a *diameter*. This terminology does not necessarily mean that the particles are spherical.

The forms sketched in Figure 1.10a represent some irregularly shaped particles as they might be observed in a light or electron micrograph (see Section 1.6). The length of a line that bisects the projected area of a particle is a parameter known as *Martin diameter*. The direction along which the Martin diameter is measured is arbitrary, but it should be used consistently to avoid subjective bias. The lines sketched in the figure are intended to represent this quantity. The Martin diameter is most easily measured on micrographs, although movable crosshairs in the eyepiece of a microscope also permit such distances to be measured by direct observation.

Another method of characterizing irregular particles consists of reporting the diameter of a circle that projects the same cross section as the particle in question. This is done by

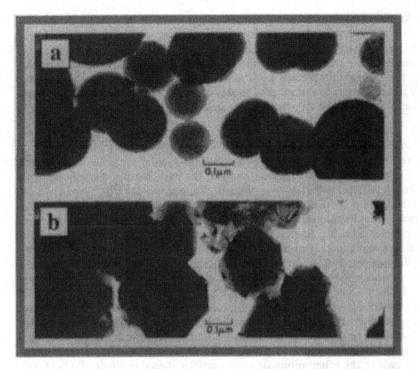

FIG. 1.9 Electron micrograph (150,000×) of carbon black particles: (a) before heat treatment; and (b) after heating to 2700°C in the absence of oxygen. (Adapted from F. A. Heckman, *Rubber Chem. Technol.* **37**, 1243 (1964.)

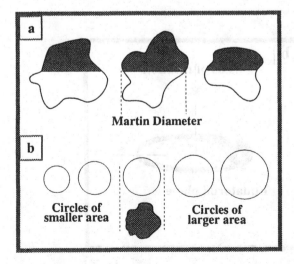

FIG. 1.10 Characterization of the "size" of irregular particles: (a) a schematic illustration of Martin diameters. (b) the use of a graticule to estimate the characteristic dimension.

inserting an object known as a *graticule* (an assortment of circles of different sizes etched on a transparent slide) into the eyepiece of a microscope and deciding which circle most closely approximates the projected area of the particle (see Fig. 1.10b). Both the graticule dimension and the Martin diameter are extremely tedious to evaluate since a large number of particles must be examined for the values to have any statistical significance, but instruments are available to classify images by size automatically. However, the fact that the labor can be done electronically does not decrease the importance of recognizing what is involved in particle sizing.

1.5a.2 Characterizing Variations in Shape: Particles with a Low Degree of Symmetry

The sphere is favored above all other shapes as a model for actual particles because it is characterized by a single parameter. Sometimes, however, the particles of a dispersion are so asymmetrical that no *single* parameter, however defined, can begin to describe the particle. In this case, the next best thing is to describe the particle as an ellipsoid of revolution. An ellipsoid of revolution is the three-dimensional body that results from the complete rotation of an ellipse around one of its axes. We define a as the radius of the ellipsoid measured along the axis of rotation and b as the radius measured in the equatorial plane. Obviously, if these two measurements of radius are equal for a particle, that particle is spherical. If $a > b$, the particle is called a *prolate ellipsoid*; if $a < b$, it is an *oblate ellipsoid*. These two geometries are illustrated in Figure 1.11.

The ratio (a/b), called the axial ratio of the ellipsoid, is frequently used as a measure of the deviation from sphericity of a particle. It plays an important role, for example, in our discussions of sedimentation and viscosity in Chapters 2 and 4, respectively. In the event that $a \gg b$, the prolate ellipsoid approximates a cylinder and, as such, is often used to describe rod-shaped particles such as the tobacco mosaic virus particles shown in Figure 1.12a. Likewise, if $a \ll b$, the oblate ellipsoid approaches the shape of a disk. Thus, even the irregular clay platelets of Figure 1.12b may be approximated as oblate ellipsoids.

One other particle geometry deserves mention. Suppose we were to take a length of string or other flexible material and allow it to tumble freely for a while in a large container. The string would certainly be expected to emerge from this treatment as a tangled jumble. Many long-chain molecules have sufficient flexibility to take on a random configuration like this under the influence of thermal jostling. This "random coil" is likely to be symmetrical rather than stretched out. We accordingly refer to the "radius" of such a coil. The random coil is discussed in detail in Chapter 2, Section 7.

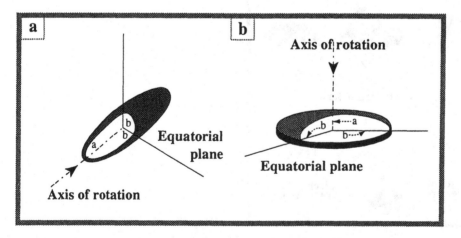

FIG. 1.11 Ellipsoids of revolution: (a) a prolate ($a > b$) ellipsoid; and (b) an oblate ($a < b$) ellipsoid. The figure shows the relationship between the semiaxes and the axis of revolution.

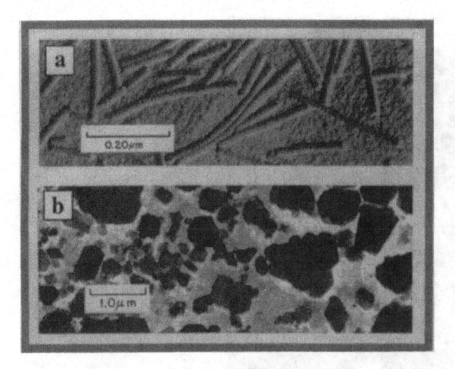

FIG. 1.12 Electron micrograph of two different types of particles that represent extreme variations from spherical particles: (a) tobacco mosaic virus particles (Photograph courtesy of Carl Zeiss, Inc., New York); and (b) clay particles (sodium kaolinite) of mean diameter 0.2 μm (by matching circular fields). In both (a) and (b), contrast has been enhanced by shadow casting (see Section 1.6a.2a and Figure 1.21). (Adapted from M. D. Luh and R. A. Bader, *J. Colloid Interface Sci.* **33**, 539 (1970).

1.5a.3 *"Model Particles" of Various Shapes and Sizes*

So far, we have focused on how to deal with the variations in the sizes and shapes of colloidal particles encountered in practice. However, in recent years another equally important and highly useful perspective in the use of colloids has emerged from viewing the variations in size and shape (and surface chemistry) as controllable parameters. The question is, Can we control the shapes and sizes of the particles to produce "model" particles that can be used to study properties of colloids and to develop new uses for colloids? The answer is "yes" (within reason).

We have already mentioned the use of the type of polystyrene latex particles shown in Figure 1.8 for the calibration of magnification in electron microscopy. Such latex particles are also used as "supports" in microbiological assay techniques. However, depending on the crystal (or amorphous) structure of the material used, we can also produce particles of many different shapes and sizes by controlling the nucleation and growth conditions. Two such examples are shown in Figures 1.13 and 1.14. More details on the preparation of such colloids are available in numerous review articles (see, e.g., Matijevic 1993 and references therein). Such colloids are often called *model colloids* since they are useful as model systems (i.e., because of their controlled size, shape, and surface chemistry) for studying a variety of phenomena both within and outside colloid science, as illustrated by the following partial list:

1. Spherical latex particles with a reasonably well-defined number of charges per particle can be synthesized and used to study the non-Newtonian behavior of charged dispersions and related electroviscous phenomena (described in Chapter 4). The surface

FIG. 1.13 Spherical and cubic "model" particles with crystalline or amorphous microstructure: (a) spherical zinc sulfide particles (transmission electron microscopy, TEM, see Section 1.6a.2a); x-ray diffraction studies show that the microstructure of these particles is crystalline; (b) "cubic" lead sulfide particles (scanning electron microscopy, SEM, see Section 1.6a.2a); (c) amorphous spherical particles of manganese (II) phosphate (TEM); and (d) crystalline cubic cadmium carbonate particles (SEM). (Reprinted with permission of Matijević 1993.)

    chemistry of the particles and the chemical constituents in the solution can be controlled to produce neutral "hard-spherelike" particles or charged particles.

2. Such particles can also be used as models of atomic fluids or atomic solids of interest in physics, chemistry, and materials science (see, e.g., Section 13.2, Figs. 13.3 and 13.4; see also Murray and Grier 1995).

3. Since needlelike particles of controlled shape and size can be prepared (see Fig. 1.14, for example), the effects of asymmetry of the particles on flow and optical properties of dispersions can be studied in a systematic manner.

4. Model particles can also be used to develop processing methods for the fabrication of advanced "structural" ceramics and composites of high strength (see, e.g., Vignette I.5).

5. The magnetic, optical, and electrical properties of materials often depend on the microstructural details and the morphology of materials. Even if the final state is not a colloid, many products pass through colloidal processing routes prior to the final stage. The availability of methods to produce model particles allows us to study and control the desired properties of the final product.

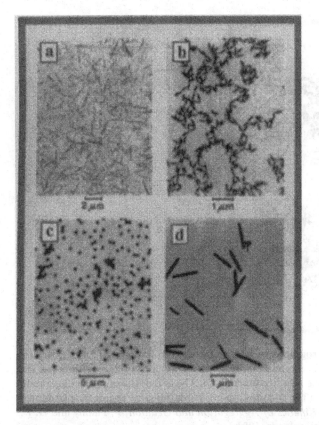

FIG. 1.14 "Model" particles of different shapes with the same or different chemical compositions: (a) rodlike particles of akageneite (β-FeOOH); (b) ellipsoidal particles of hematite (α-Fe$_2$O$_3$); (c) cubic particles of hematite; and (d) rodlike particles of mixed chemical composition (α-Fe$_2$O$_3$ and β-FeOOH). All are TEM pictures. (Reprinted with permission of Matijević 1993.)

Many additional examples of the above types and others are given in the references cited in Matijevic (1993).

1.5b Particle Aggregates

We have already introduced the idea that the primary particles of a dispersed system tend to associate into larger structures known as aggregates. The nature of the interparticle forces responsible for this aggregation is one of the most examined areas of colloid science. We defer our discussion of the aggregation (or coagulation) process until Chapter 13, but a few remarks about aggregates — the kinetic units that result from that process — and how their dimensions are represented quantitatively are in order at this time.

In many situations the dispersed phase is present as aggregates, not as primary particles. In such cases, it is the size, shape, and concentration of the aggregates that determine the properties of the dispersion itself. As a matter of fact, some substances — for example, those carbon blacks known as channel and furnace blacks — possess rigidly fused, aggregatelike structures as their primary particles.

Figure 1.15 shows some micrographs of aggregates of gold primary particles (with an average diameter of about 14 nm). The pictures are obtained using transmission electron microscopy (see Section 1.6a.2a), and the figure shows four aggregates at four different resolutions. Notice that these pictures show large amounts of open space within the aggregates

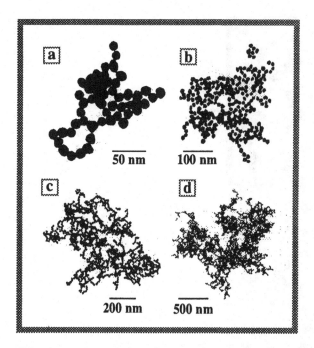

FIG. 1.15 Transmission electron micrographs of aggregates of gold particles. These aggregates were made from a gold colloid to study the relation between the kinetics of aggregation and the resulting structures of the aggregates (see also Section 1.5b.2). The a–d portions of the illustration show aggregates at various resolutions. (The pictures are not from the same aggregate.) (Adapted from D. A. Weitz and J. S. Huang, in *Kinetics of Aggregation and Gelation*, F. Family and D. P. Landau, Eds., Elsevier, Amsterdam, Netherlands, 1984.)

despite the fact that the pictures are two-dimensional projections of three-dimensional objects.[9] This open structure is related to what is known as the *fractal* nature of such aggregates. We consider this in the next subsection and discuss the relation between aggregation kinetics and the structure of the aggregates subsequently.

1.5b.1   Fractal Dimension of an Aggregate

Suppose we wish to measure the characteristic dimensions of aggregates or aggregatelike particles using micrographs. If the aggregation process results in irregular but, on the average, symmetrical particles, we might characterize the aggregates in terms of the dimension of an inscribing boundary, such as the Martin diameter or the diameter of an equivalent circle determined from a graticule. Then suppose we wish to evaluate the mass of the aggregate

---

*When an electron micrograph shows evidence of aggregation, we must remember that this may be an artifact arising from the preparation of the sample for microscopy. In other words, the amount of aggregation that a colloid displays in its dispersed state and the amount that appears in an electron micrograph made from the same preparation may be quite different. Optical microscopy is safer in this regard since the actual dispersion may be examined without first evaporating the continuous phase to dryness. Also, in both optical and transmission electron microscopy, it is the *projected* image of the particle that is observed, and microscopic observation alone is often inadequate to distinguish a particle such as that shown in Figure 1.15 from a true aggregate in which the structure is fairly readily disrupted. However, one can use other, "nondestructive" techniques to quantify the structure of aggregates; the use of light, neutron, and x-ray scattering is one such method and is illustrated in Section 5.6a.

enclosed within this boundary. To convert a linear dimension into a particle mass, the particle shape and density must be known.

We have already commented on the approximations involved in treating irregular particles as spheres. Here, however, we encounter an additional problem besides the geometrical approximation already discussed. The question is, What do we use for the density of an aggregate to convert the particle volume into its mass? If the aggregate dimensions have been measured, it is clearly the aggregate density that must be used. The latter is intermediate between the densities of the dispersed and continuous phases, the exact value depending on the structure of the aggregate. Of course, one might attempt to estimate the number of primary particles in an aggregate and then use their size and density as an alternate means of evaluating the mass of an aggregate. The main point of this, however, is the following. Whenever the dispersed particles are in an aggregated form, the properties of the dispersed units are intermediate between those of the two different phases involved. We encounter this difficulty again in Chapter 2, in which the density of the settling unit, whatever it may be, is involved in sedimentation.

Assume for the moment that the aggregate is a solid particle (i.e., without any empty interstitial space) with the same density ρ as that of the primary particles. Then, the mass $m(r)$ of the aggregate enclosed within a distance r from a suitably chosen center will be given by

$$m(r) = \rho(4/3)\pi r^3 \tag{3}$$

If we account for the fact that the region enclosed within the sphere of radius r is not completely filled by particles but contains empty spaces, then the actual mass will be less than that given by the above expression. In fact, it turns out that in many cases, one can write

$$m(r) \propto \rho(4/3) \pi r^{d_f} \tag{4}$$

where $d_f < 3$, the spatial dimension. The quantity d_f is known as the *fractal dimension* of the aggregate. In fact, it is known more precisely as the *mass fractal dimension* since it is the mass of the aggregate that we have used to specify the fractal dimension. This qualification is also meant to draw one's attention to the fact that other properties can be used to specify the fractal dimension (e.g., surface area), and one would generally expect the fractal dimension to depend on the property used to determine it. In Section 1.5c, we see that this situation frequently arises in characterizing particle distributions. Fractal objects are often called *self-similar* objects. The adjective self-similar draws attention to the fact that such objects look similar when viewed over a range of length scales (see Example 5.4 in Chapter 5). This can be seen from the pictures of aggregates presented in Figure 1.15, which show that the aggregates look similar for a range of resolutions. The fractal dimension, which is a measure of self-similarity, can be specified exactly for "mathematically" produced fractal objects, but for real objects such as the ones shown in Figure 1.15 one obtains the fractal dimension by fitting the mass (or an appropriate property) of the object with the size using an equation like Equation (4).

One way of measuring the fractal dimension of aggregates is discussed in Chapter 5 (See Section 5.6a and Example 5.4). In the example below, we illustrate the relation between the fractal structure of aggregates and the surface area of the aggregates.

* * *

EXAMPLE 1.2 *Surface Area of Fractal Aggregates.* An aerosol "reactor" is used to grow relatively large catalyst particles from monodispersed, spherical primary particles of diameter $d_p = 100$ nm. The specific gravity of the primary particles is 1.5. An examination of the aggregates using an optical microscope shows that the aggregates are essentially spherical. The aggregates produced by the reactor are sieved, and aggregates in five classes of diameters d_i (see Section 1.5c) are selected for measurement of surface areas. A cylindrical column of volume $V_{cyl} = 10$ cm^3 is packed with fractionated aggregates having a narrow distribution with the average diameters shown below. Nitrogen adsorption experiments (see Chapter 9) are then conducted to measure the total area of the aggregates in each class.

$d_i \times 10^6$, m	10	50	100	500	1000
$A_{Total}(d_i)$, m^2	91	56	46	28	23

Assume that the actual volume of the aggregates in the nitrogen adsorption experiments is 60% of the total volume (with the rest accounting for pore space between aggregates in the packing). Calculate the fractal dimension of the aggregates.

An independent x-ray and light scattering analysis (see Section 5.6a and Example 5.4) of a dispersion of the aggregates suggests that the aggregates have a fractal structure with a fractal dimension of 2.65. Is this confirmed by your result?

Solution: Assume that the aggregates are fractals with the mass $m_a(d)$ of an aggregate of diameter d being given by

$$m_a(d) = Cd^{d_f}$$

where C is a constant and d is specified in meters. (Notice that the numerical value of the "constant" of proportionality [i.e., C] depends on the dimensions used for the diameter d.]

The number $N_a(d)$ of primary particles in an aggregate of diameter d is simply the mass of the aggregate divided by the mass of the primary particle:

$$N_a(d) = \frac{6C}{\pi \, d_p^3 \, \rho_p} \, d^{d_f}$$

where ρ_p is the density of the primary particle.

If we assume that the surface area πd_p^2 lost by the contacts between the primary particles during the process of aggregation is negligible, the surface area of the aggregate $A_a(d)$ will simply be the number of primary particles in the aggregate times the surface area of the primary particle:

$$A_a(d) = \frac{6C}{d_p \rho_p} \, d^{d_f} = CA_{sp}d^{d_f}$$

where $A_{sp} = (6/d_p\rho_p)$ is the specific surface area of the primary particles, see Equation (2). The above equation shows that the surface area of a fractal aggregate is also a fractal of dimension d_f.

The total surface area of the aggregates in the volume V_{cyl} in the adsorption experiments is given by the number of aggregates in the cylinder times the area of each aggregate:

$$A_{Total}(d) = \frac{V_{cyl}(1-\varepsilon)}{(\pi d^3/6)} A_a(d) = \frac{6V_{cyl}(1-\varepsilon)}{\pi} CA_{sp}d^{d_f-3}$$

For the given data, V_{cyl} is 10^{-5} m^3; ε, the porosity of the packing, is 0.4; and A_{sp} is $6/d_p\rho_p = 4 \cdot 10^4$ m^2/kg. The measured area can, therefore, be plotted against the diameter on a *log-log* graph to check if the aggregates are fractals and to determine the constant C. (Remember, the slope of a log-log plot gives the exponent.)

Figure 1.16 shows a plot of the given data. It is clear that ln A_{Total} is linear with respect to ln d, with a slope equal to -0.3, giving a d_f value of 2.7. Notice that if the "aggregates" had been smooth, solid spheres, the surface area of the aggregate would have been proportional to d^2, i.e., $d_f = 2$. The fact that $d_f > 2$ is an indication of the porous nature of the aggregates.

This result shows that the conclusion reached from scattering results is consistent with surface area measurements.

The intercept can be obtained by extrapolating the line to $d = 1$ m (i.e., at ln $d = 0$) and is 2.85 m$^{2.3}$. Therefore,

$$A_{Total}(d) = \frac{6V_{cyl}(1-\varepsilon)}{\pi} CA_{sp}d^{d_f-3} = 2.85 \, d^{-0.3}, \qquad \text{with } d \text{ in meters.}$$

from which one gets $C = 6.3$ kg/m$^{2.7}$.   ∎

* * *

The fractal dimension 2.7 of the surface area of the aggregates in the above example is larger than 2, expected for a smooth, non-fractal surface. In contrast, the fractal dimension as defined in Equation (4) is less than the dimension of the space. Its value relative to the spatial dimension is a measure of the structure of the aggregate. For instance, the closer d_f is to 3 in Equation (4), the denser (i.e., the more tightly packed) is the aggregate. There are ways in which one can measure the fractal dimension experimentally (e.g., using light, x-ray, or neutron scattering techniques; see Chapter 5, Section 6), but we are not concerned with that in

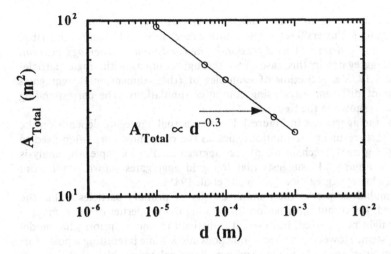

FIG. 1.16 The total area measured versus the diameter of the aggregate on a log-log scale for the data given in Example 1.2.

this book. It is, however, important to note that fairly successful attempts have been made to link the fractal dimension of aggregates to the processing techniques used in forming the aggregated particles (Brinker and Scherer 1990, Chapter 3).

Computer simulations combined with experiments have also shown that one can deduce from the fractal dimension the nature of nucleation and growth of particles and what chemical and physical mechanisms control the formation of particle aggregates. We consider this briefly before proceeding to other topics.

1.5b.2 Relating Aggregate Structure to Growth Mechanisms

A very interesting and fast-growing approach to research concerning aggregates is the technique of computer simulation. By this method, aggregates are "assembled" by a computer, which uses random numbers to determine the coordinates from which each primary particle approaches the growing aggregate. The model that has been most studied consists of spherical primary particles, although linear sets of spheres have been used to simulate asymmetrical primary particles.

The probability of adhesion on contact (say, p_a) can be made a variable quantity in such simulations. The probability of adhesion on contact can be thought of as the effectiveness of the "reaction" of forming an N-particle aggregate when a primary particle collides with an ($N - 1$)-particle aggregate.

One can also introduce in the simulation other physical mechanisms such as the rate of diffusion of the particles. If adhesion probability ("reaction") controls the rate of formation of the aggregate, the aggregation process is known as a *reaction-limited aggregation* (RLA); if diffusion controls the rate of aggregation (i.e., the rate of diffusional transport of a particle to another particle or an aggregate is slow relative to adhesion on contact), the process is known as *diffusion-limited aggregation* (DLA). In DLA, one assumes that $p_a = 1$ (the particles adhere on contact), whereas in RLA p_a is less than unity (i.e., the particles do not always stick to each other on contact). As might be expected intuitively, more open aggregates result when the probability of adhesion at initial contact is high. If, on the other hand, the added particle is permitted to roll along the surface of the growing aggregate before adhering, a more compact structure results.

Variations of the above concepts are also possible to simulate other situations. For example, one variation is to permit the joining together of small aggregates to form larger ones rather than restricting the addition to primary particles only. This leads to structures that are even more expanded than those resulting from the addition of primary particles alone (see Fig.

1.17). Such a procedure is known as *diffusion-limited cluster-cluster aggregation*, or DLCCA, if diffusion of the aggregates ("clusters") is the rate-limiting mechanism. (The above-described DLA is then more accurately described as *diffusion-limited monomer-cluster aggregation*, or DLMCA, since the aggregates in this case grow through contacts with single particles ["monomers"]). Figure 1.17 is a collection of examples of (three-dimensional) aggregates assembled in a number of different ways using computer simulations. The corresponding fractal dimensions are also shown in the figure.

The resemblance of the aggregates in Figure 1.17 to an actual aggregate depends on the growth mechanism. In fact, results of simulations such as the ones shown are often used to model or to interpret the growth mechanisms of real aggregates. For example, the analysis of the clusters shown in Figure 1.15 suggests that the gold aggregates shown result from diffusion-limited cluster-cluster aggregation (see Weitz et al. 1985).

These computer simulations permit the number density of primary particles within the aggregate to be evaluated, important information for relating the properties of the aggregate to its composition. As might be expected, however, it is difficult to know a priori what model to use for a particular system. However, this technique does allow some interesting a posteriori interpretations of known structures to be made. Another closely related problem that has been

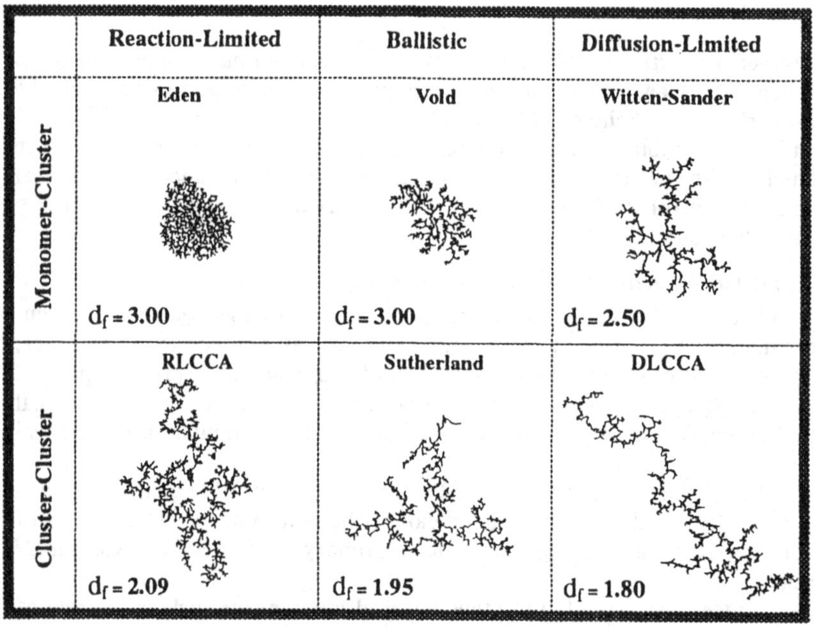

	Reaction-Limited	Ballistic	Diffusion-Limited
Monomer-Cluster	Eden $d_f = 3.00$	Vold $d_f = 3.00$	Witten-Sander $d_f = 2.50$
Cluster-Cluster	RLCCA $d_f = 2.09$	Sutherland $d_f = 1.95$	DLCCA $d_f = 1.80$

FIG. 1.17 Aggregates obtained through computer simulations using various growth models. The figure shows typical aggregates produced in the simulations under a number of conditions. The results show two-dimensional renditions of three-dimensional simulations. The column headings identify the "controlling step" in the aggregation process (i.e., the type of particle motion and probability of adhesion p_a). (*Reaction limited* implies $p_a < 1$; *ballistic* implies that the "particle" motion is rectilinear [with the added assumption that $p_a = 1$]; *diffusion limited* implies that the "particle" motion is a random walk [with the assumption that $p_a = 1$].) The row labels specify which type of collision is considered (i.e., monomer-cluster or cluster-cluster). The names associated with the models are also shown. For example, diffusion-limited monomer-cluster aggregation (DLMCA) is known as the Witten-Sander model. (RLCCA signifies reaction-limited cluster-cluster aggregation.) (Redrawn from D. W. Schaefer, *MRS Bulletin* **8**, 22 (1988). Simulations are from P. Meakin, in *On Growth and Form*, H. E. Stanley and N. Ostrowsky, Eds., Martinus-Nijhoff, Boston, 1986.)

studied by computer simulation is the volume occupied by a sediment. As with aggregates, it is found that sediments become more voluminous as the probability of adhesion on contact increases.

This discussion of aggregates leads us to another important characteristic of dispersions we have not yet considered in sufficient detail: *polydispersity*. Monodisperse systems are the exception rather than the rule. Even in those rare cases in which a monodisperse system exists, any aggregation that occurs will result in a distribution of particle sizes because of the random nature of the aggregation process.

1.5c Polydispersity

It is difficult to imagine an array of particles that would be easier to describe than the latex particles of Figure 1.8. A single parameter such as the radius of the spheres is sufficient to characterize the dispersed phase in terms of a linear dimension. However, the only realistic attitude to take toward the dispersed systems we are interested in is to assume that they are polydisperse (see, e.g., Vignette I.7, which presents an extreme case). Even the particles in Figure 1.8, which appear remarkably uniform, have a narrow distribution of particle sizes. There are rare cases in which the distribution of dimensions is of negligible width, but, generally speaking, a statistical approach is required to describe a colloidal dispersion. A brief introduction to statistical analysis of polydispersed colloids, definitions of moments, and theoretical distribution functions is given in Appendix C. Here, we are mainly concerned with illustrating some of the essential ideas.

VIGNETTE I.7 FOOD SCIENCES: Structure and Processing of Food Products

Most food products and food preparations are colloids. They are typically multicomponent and multiphase systems consisting of colloidal species of different kinds, shapes, and sizes and different phases. Ice cream, for example, is a combination of emulsions, foams, particles, and gels since it consists of a frozen aqueous phase containing fat droplets, ice crystals, and very small air pockets (microvoids). Salad dressing, special sauce, and the like are complicated emulsions and may contain small surfactant clusters known as micelles (Chapter 8). The dimensions of the "particles" in these entities usually cover a rather broad spectrum, ranging from nanometers (typical micellar units) to micrometers (emulsion droplets) or millimeters (foams). Food products may also contain macromolecules (such as proteins) and gels formed from other food particles aggregated by adsorbed protein molecules. The texture (how a food feels to touch or in the mouth) depends on the structure of the food.

One of the major concerns of a food chemist or technologist is the stability of food products with time (Dickinson 1992). Requirements change with the type of food and may depend on nutritional requirements, governmental regulations, or economy of production. For example, products such as margarine, cream, and spreads require specific textural and flow characteristics. Beers may require additives in order to produce froth on dispensing. Sauces and gravies need to maintain their structure at fairly high temperatures. In practical terms, these imply that one is concerned with the microstructure of the constituents; how it varies with chemical additives, temperature, etc.; and how to produce products with appropriate "shelf life" and structural integrity.

In general terms these problems are not unlike those encountered in the manufacture of other colloidal products such as paints, face creams, and printing inks and toners, for example. The difference, of course, is that the inks and toners are hardly appetizing and one seldom cares about how they taste! (Of course, physical stability alone does not determine the taste or the worthiness of a food product. Chemical reactions play a role in taste and in determining the structure, and other considerations such as nutritional value and esthetic appeal may also apply.)

TABLE 1.5 A Hypothetical Distribution of 400 Spherical Particles[a]

Class boundaries $<d<$ (μm)	Class mark d_i (μm)	Number of particles n_i	Fraction of total number in class $f_{n,i}$	Total number with $d<d_i$, $n_{T,i}$
0–0.1	0.05	7	0.018	7
0.1–0.2	0.15	15	0.038	22
0.2–0.3	0.25	18	0.045	40
0.3–0.4	0.35	28	0.070	68
0.4–0.5	0.45	32	0.080	100
0.5–0.6	0.55	70	0.175	170
0.6–0.7	0.65	65	0.163	235
0.7–0.8	0.75	59	0.148	294
0.8–0.9	0.85	45	0.113	339
0.9–1.0	0.95	38	0.095	377
1.0–1.1	1.05	19	0.048	396
1.1–1.2	1.15	4	0.010	400

[a]The data are classified into twelve classes: the class marks, numbers, and fractions of particles per class and the total number of particles up to and including each class are listed. See text for clarification.

As described in Appendix C, if we have a discrete set of measurements of a quantity such as the diameters of particles, one can represent the results in terms of a few useful quantities like the average diameter and standard deviation. Table 1.5 represents the frequency distribution for a hypothetical array of spheres; all the numerical examples of this section are based on this sample of 400 particles. The particles have been sorted into categories called *classes* with a narrower range of dimensions. Each class is represented by the midpoint of the interval, a quantity called the class mark, symbolized by d_i for class i. Similarly, we define the number of particles in each class as n_i.

A common graphical representation of a frequency distribution is the *histogram*, a bar graph in which the class marks are plotted as the abscissa and the height of the bar is proportional to the number of particles in the class. Sometimes the ordinate is defined as the fraction of particles in the class $f_{n,i}$, with the first subscript n representing the fact that the fraction is based on the number of particles. Figure 1.18a is a plot of the histogram of the data in Table 1.5. Obviously, as the number of classes approaches infinity, the width of each interval approaches zero and the histogram approaches a smooth curve. Analytical distribution functions give the equation for such smooth curves. However, in practice, a bar graph is a convenient approximation to the smooth function.

Another way in which these kinds of data are sometimes represented is as a *cumulative* curve in which the total number (or fraction) of particles $n_{T,i}$ having diameters less (sometimes more) than and including a particular d_i are plotted versus d_i. Figure 1.18b shows the cumulative plot for the same data shown in Figure 1.18a as a histogram. The cumulative curve is equivalent to the integral of the frequency distribution up to the specified class mark. Cumulative distribution curves are used in Chapter 2 in connection with sedimentation.

1.5c.1 Average Diameters and Standard Deviation

Although the histogram is a convenient pictorial way to present data, a more concise representation is often required. Most students will quickly identify the *average* as such a representation, and we calculate this quantity here for the data in Figure 1.18. First, however, let us write a very general definition of "average" since this word will take on a considerably wider meaning than usual before we are finished with it. For classified data of the type presented in Table 1.5, average may be defined as follows:

$$\text{"average"} = \sum_i (\text{weighting factor})_i \, (\text{quantity being averaged})_i \qquad (5)$$

This formulation may look somewhat unfamiliar, but we shall see that both the weighting factor and the quantity being averaged can mean many different things in colloid science. Appendix C presents a more detailed presentation of basic statistical concepts for both discrete and continuous distributions for those who desire more information on these topics.

For the data in Table 1.5, the first kind of average we examine is what is known as the *number average* $\bar{d}_n$ because the weighting factors $f_{n,i}$ are the *number fraction* of particles in each class:

$$f_{n,i} = n_i / \sum_i n_i \qquad (6)$$

In evaluating the number average diameter $\bar{d}_n$, the d_i values are the quantities being averaged. With these substitutions, Equation (5) becomes

$$\bar{d}_n = \sum_i \frac{n_i}{\sum_i n_i} d_i \qquad (7)$$

This quantity is also known as the *mean* of the distribution. For the data in the table, $\bar{d}_n = 0.64 \ \mu m$; this value has been marked in Figure 1.18 (and is denoted by $\bar{d}$).

In experiments, often a single property is measured for a large number of particles—such as the surface area for a fine powder—rather than for individual particles as in microscopy, and an effective size may be "backed out" from the quantity measured. This is the situation, for example, when Equation (2) is used to evaluate the radius of an equivalent sphere from experimental A_{sp} values.

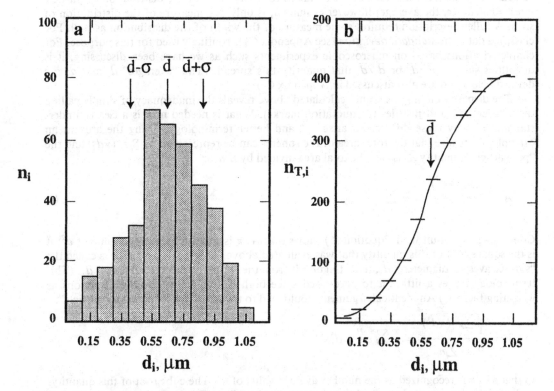

FIG. 1.18 Graphical representation of the data in Table 1.5. Data are presented as (a) a histogram and (b) a cumulative distribution curve.

When polydispersity is present, we have no problem labeling the calculated equivalent radius an "average". We have used quotation marks around this label to alert us to the possibility that this average may be something other than the mean.

To explore this further, we present some additional data about the 400 spheres in Table 1.5, namely, that the sample possesses a total surface area of $5.85 \cdot 10^2 \; \mu\text{m}^2$ or $5.85 \cdot 10^2 / 400 = 1.46 \; \mu\text{m}^2$ per particle. Likewise, the total volume of the 400 spheres is $76 \; \mu\text{m}^3$ or $0.19 \; \mu\text{m}^3$ per particle. We see in Chapter 9 how the surface areas of actual powdered samples are measured and the volume is readily available from mass when the density of the bulk material is known. Now let us calculate the "average" diameter of an equivalent sphere from these data.

1. Since $A = \pi d^2$, the "average" diameter based on average area of the particle $= (A/\pi)^{1/2} = 0.68 \; \mu\text{m}$.

2. Since $V = (\pi/6)d^3$, the "average" diameter based on average volume of the particle $= (6V/\pi)^{1/3} = 0.72 \; \mu\text{m}$.

3. For this same population of spheres, the "average" diameter based on the number fraction $f_{n,i}$ of the particles in each class $i = \bar{d}_n = 0.64 \; \mu\text{m}$, as we have already seen.

These calculations illustrate a very general feature of polydisperse systems: Different experimental approaches (e.g., measuring d's versus measuring areas) give different averages for what nominally (hence the quotation marks) might be called the same quantity. This divergence between "average" values determined by different methods can be a cause of consternation for the uninitiated, who might expect corroboration of a previous result determined by a different method. In fact, the only time these different procedures would ever result in the same value would be in the event that all spheres in the sample had the same diameter.

Herein lies the value of these different "averages": the divergence between the "averages" calculated by different methods offers a clue as to the breadth of the distribution of particle sizes. Remember, the average, however evaluated, is only one measure of the distribution of sizes. A fuller description requires some measure of the width of the distribution as well. For classified data, the *standard deviation* (see Appendix C) is routinely used for this purpose. For characterizations based on macroscopic experiments such as we have been discussing, it is quantities such as $\bar{d}_s/\bar{d}_n$ or $\bar{d}_v/\bar{d}_s$ that quantify this spread. (The "averages" $\bar{d}_s$ and $\bar{d}_v$ are defined below and are also discussed in Appendix C.)

The disparity among the values calculated above reveals the inadequacy of simply calling them "average" (and justifies the quotation marks). What is needed now is a clearer understanding of why these differences arise . . . and better terminology. Using the area as an example, we realize that the total area of the spheres can be represented by $\Sigma_i \, n_i \, (\pi d_i^2)$ and the "per particle" quantity $\bar{A}$ is just the total area divided by $\Sigma_i \, n_i$ or

$$\bar{A} = \frac{\sum_i n_i (\pi d_i^2)}{\sum_i n_i} = \pi \sum_i \frac{n_i}{\sum_i n_i} d_i^2 \qquad (8)$$

Comparing this result with Equation (5) shows that $\bar{A}/\pi$ is simply the average value of d^2; it is the square root of this quantity that was evaluated above. This kind of average is called the "surface average diameter," $\bar{d}_s$, to distinguish it from the "number average" diameter $\bar{d}_n$. (This terminology leaves a little bit to be desired since both $\bar{d}_s$ and $\bar{d}_n$ use number fractions as weighting factors.) An identical argument would lead to the result for volume "per particle":

$$\bar{V} = \frac{1}{6} \frac{\sum_i n_i (\pi d_i^3)}{\sum_i n_i} = \frac{\pi}{6} \sum_i \frac{n_i}{\sum_i n_i} d_i^3 \qquad (9)$$

so that $6\bar{V}/\pi$ is recognized as the number average value of d^3. The cube root of this quantity, evaluated above, is called the "volume average diameter," $\bar{d}_v$. These parameters are summa-

TABLE 1.6 Some of the More Widely Encountered Size "Averages" in Surface and Colloid Science

Name	Symbol	Definition	Quantity averaged	Weighting factor
Number average, or mean	$\bar{d}_n$ or $\bar{d}$	$\dfrac{\Sigma_i n_i d_i}{\Sigma_i n_i}$	Diameter	Number in class
Second moment about origin	$\overline{d^2}$	$\dfrac{\Sigma_i n_i d_i^2}{\Sigma_i n_i}$	Square of diameter	Number in class
Surface average	$\bar{d}_s$	$(\overline{d^2})^{1/2}$	Square of diameter	Number in class
Third moment about origin	$\overline{d^3}$	$\dfrac{\Sigma_i n_i d_i^3}{\Sigma_i n_i}$	Cube of diameter	Number in class
Volume average	$\bar{d}_v$	$(\overline{d^3})^{1/3}$	Cube of diameter	Number of class
Radius of gyration	$(\overline{R_g^2})^{1/2}$	$\left(\dfrac{\Sigma_i m_i r_i^2}{\Sigma_i m_i}\right)^{1/2}$	Square of radius	Mass in class

rized in Table 1.6. Note that the relative magnitudes of the number, surface, and volume averages are given by the sequence

$$\bar{d}_n < \bar{d}_s < \bar{d}_v \qquad (10)$$

for a polydisperse system. Only for a monodisperse system would all three parameters have identical values. Therefore, the divergence from unity of the ratio of any two of these, measured independently, is often taken as an indication of the polydispersity of the dispersion. We see in the next section that this comparison of averages evaluated by different techniques finds particular application in the characterization of molecular weight distributions.

1.5c.2 Radius of Gyration

Table 1.6 also lists the radius of gyration. This is an average dimension often used in colloid science to characterize the spatial extension of a particle. We shall see that this quantity can be measured for polydisperse systems by viscosity (Chapter 4) and light scattering (Chapter 5). It is therefore an experimental quantity that quantifies the dimensions of a disperse system and deserves to be included in Table 1.6. Since the typical student of chemistry has probably not heard much about the radius of gyration since general physics, a short review seems in order.

We assume that the particle with the radius of gyration under discussion may be subdivided into a number of volume elements of mass m_i. Then, the moment of inertia I about the axis of rotation of the body is given by

$$I = \sum_i m_i r_i^2 \qquad (11)$$

where r_i is the distance of the ith volume element from the axis of rotation.

Regardless of the shape of the particle, there is a radial distance at which the entire mass of the particle could be located such that the moment of inertia would be the same as that of the actual distribution of the mass. This distance is the *radius of gyration* R_g. According to this definition, it is clear that

$$R_g^2 \sum_i m_i = I = \sum_i m_i r_i^2 \tag{12}$$

Therefore, we see that the value of R_g for an array of volume elements is the average value of r^2, with mass fraction (rather than number fraction) as the weighting factor:

$$R_g^2 = \frac{\Sigma_i m_i r_i^2}{\Sigma_i m_i} \tag{13}$$

We see in the following subsection that this type of weighting factor gives one of the common molecular weight averages for polydisperse systems.

The radius of gyration is a parameter that characterizes particle size without the need to specify particle shape. However, the relationship between R_g and the actual dimensions of a particle depends on the shape of the particle. Such relationships are derived in most elementary physics texts for rigid bodies of various geometries. To translate a radius of gyration into an actual geometrical dimension, some shape must be assumed. For example, for a sphere of radius R_s, $R_g = (3/5) R_s^2$. It should be emphasized, however, that the radius of gyration in itself is a perfectly legitimate way of describing the dimensions of a particle. The specification of particle geometry is really optional.

1.5c.3   Molecular Weight Averages

The averages discussed above can also be extended to molecular weights. In Section 1.6b.2, we will discuss a technique called *size-exclusion chromatography* (SEC) that can be used to obtain number fractions and weight fractions of particles in various molecular weight classes. From this sort of information, the two most common molecular weight averages, number-average molecular weight and weight-average molecular weight, can be readily determined.

1.5c.3a   Number-Average Molecular Weight.   Suppose a dispersion is classified into a set of categories in which there are n_i particles of molecular weight M_i in the ith class. Then the number-average molecular weight equals

$$\overline{M}_n = \frac{\Sigma_i n_i M_i}{\Sigma_i n_i} = \sum_i f_{n,1} M_i \tag{14}$$

where, as before, $f_{n,i}$ is the number fraction of particles in class i.

1.5c.3b   Weight-Average Molecular Weight.   Alternatively, we could define the average in such a way that the weight of particles in each class w_i rather than their number is used as the weighting factor. This results in an average known as the *weight-average molecular weight* $\overline{M}_w$

$$\overline{M}_w = \frac{\Sigma_i w_i M_i}{\Sigma_i w_i} = \sum_i f_{w,i} M_i \tag{15}$$

Note that $f_{w,i} = w_i/\Sigma_i w_i$ is the *weight fraction* of particles in class i. Since the weight of material in a particular size class is given by the product of the number of particles in the class and their molecular weight, Equation (15) may be written as

$$\overline{M}_w = \frac{\Sigma_i (n_i M_i) M_i}{\Sigma_i n_i M_i} = \frac{\Sigma_i n_i M_i^2}{\Sigma_i n_i M_i} = \frac{\Sigma_i f_{n,i} M_i^2}{\Sigma_i f_{n,i} M_i} \tag{16}$$

We shall see that measurement of the osmotic pressure of a polydisperse system permits the experimental evaluation of $\overline{M}_n$ (Chapter 3), and light scattering experiments enable us to measure $\overline{M}_w$ (Chapter 5). It follows from the definition of these various averages that

$$\frac{\overline{M}_w}{\overline{M}_n} \geq 1 \tag{17}$$

by analogy with the inequalities in Equation (10). The ratio given in Equation (17) is known as the *polydispersity index* of the material. Only when the system is monodisperse does the equality apply in Equation (17).

Here, too, the deviation of this ratio from unity may be taken as a measure of polydispersity. The relationship between the ratio $\overline{M}_w/\overline{M}_n$ and the standard deviation of the molecular weight distribution is easily seen as follows. From Equation (16), it is clear that

$$\sum_i f_{n,i} M_i^2 = \overline{M}_w \sum_i f_{n,i} M_i = \overline{M}_w \overline{M}_n \tag{18}$$

From the general procedure for defining the mean, the left-hand side of Equation (18) may also be written as $\overline{M^2}$. Substituting this result into Equation (C.5) of Appendix C (with M in place of ϕ), we can write the standard deviation σ of the molecular weight distribution as

$$\sigma = (\overline{M}_n \overline{M}_w - \overline{M}_n^2)^{1/2} = \overline{M}_n \left(\frac{\overline{M}_w}{\overline{M}_n} - 1 \right)^{1/2} \tag{19}$$

Therefore the square root of the amount by which the molecular weight ratio exceeds unity measures the standard deviation of the distribution relative to the number average molecular weight.

Example 1.3 illustrates these relationships for a hypothetical polymer.

* * *

EXAMPLE 1.3 *Polydispersity of a Synthetic Polymer.* Columns (1) and (2) of Table 1.7 list the number of moles and the molecular weight, respectively, for eight fractions of a synthetic polymer. Calculate $\overline{M}_n$ and $\overline{M}_w$ from these data and evaluate σ using

$$\sigma = \left(\frac{\Sigma_i n_i (M_i - \overline{M}_n)^2}{\Sigma_i n_i} \right)^{1/2} \tag{A}$$

(see Appendix C) and Equation (19).

Solution. For each class of molecules calculate the quantities listed in columns (3)–(6) in Table 1.7.

The sum of the values in column (1) equals $\Sigma_i n_i = 0.136$ mole.

The product of columns (1) and (2) is w_i and values for this quantity are listed in column (3). The sum of the entries in column (3) equals $n_i M_i = 6700$ g.

Dividing 6700 g by 0.136 mole gives $\overline{M}_n = 49{,}300$ g mole^{-1}.

The product of columns (2) and (3) equals $w_i M_i$, and values for this quantity are listed in column (4); $\Sigma_i w_i M_i = 337.8 \cdot 10^6$ g^2 mole^{-1}.

Dividing $3.378 \cdot 10^8$ by 6700 gives $\overline{M}_w = 50{,}400$ g mole^{-1}.

The square of the difference between M_i and $\overline{M}_n$ is given in column (5), and column (6) lists $n_i (M_i - \overline{M}_n)^2$;

$$\Sigma_i n_i (M_i - \overline{M}_n)^2 = 7.68 \cdot 10^6$$

TABLE 1.7 Number of Moles and Molecular Weights for Eight Classes of a Hypothetical Fractionated Polymer (Remaining Quantities Calculated in Example 1.3)

(1) n_i (mole)	(2) M_i (g mole^{-1})	(3) w_i (g)	(4) $w_i M_i \times 10^{-6}$	(5) $(M_i - \overline{M})^2 \times 10^{-8}$	(6) $n_i (M_i - \overline{M})^2 \times 10^{-6}$
0.003	30,000	90	2.70	3.72	1.12
0.007	35,000	245	8.58	2.04	1.43
0.015	40,000	600	24.00	0.86	1.29
0.024	45,000	1080	48.60	0.18	0.43
0.040	50,000	2000	100.00	0.00	0.00
0.032	55,000	1760	96.80	0.32	1.02
0.010	60,000	600	36.00	1.14	1.14
0.005	65,000	325	21.10	2.50	1.25
$\Sigma = 0.136$		$\Sigma = 6700$	$\Sigma = 337.78$		$\Sigma = 7.68$

TABLE 1.8 The Most Common Molecular Weight Averages, Their Definitions, and Their Methods of Determination

Average	Definition	Methods
$\overline{M}_n$	$\dfrac{\Sigma_i n_i M_i}{\Sigma_i n_i}$	Osmotic pressure; other colligative properties
$\overline{M}_w$	$\dfrac{\Sigma_i f_{n,i} M_i^2}{\Sigma_i f_{n,i} M_i}$	Light scattering; sedimentation velocity
$\overline{M}_{vis}$	$\left(\dfrac{\Sigma_i n_i M_i^{1+a}}{\Sigma_i n_i M_i}\right)^{1/a}$	Intrinsic viscosity

Dividing $7.68 \cdot 10^6$ by 0.136 gives σ^2 by Equation (A); therefore $\sigma^2 = 5.65 \cdot 10^7$ and $\sigma = 7500$.

The ratio $\overline{M}_w/\overline{M}_n = 1.022$ and, by Equation (19), $\sigma/\overline{M}_n = (0.022)^{1/2}$, or $\sigma = 0.148(49,300) = 7300$.

The discrepancy between the two values of σ is a matter of significant figures and is not meaningful. ∎

* * *

As in the case of particle sizes, averages of molecular weights can also be defined in terms of properties of the dispersions. Here we define one such average based on viscosities of the dispersions.

1.5c.3c   Viscosity–Average Molecular Weight.   The *viscosity average molecular weight* $\overline{M}_{vis}$ is written as

$$\overline{M}_{vis} = \left(\frac{\Sigma_i n_i M_i^{1+a}}{\Sigma_i n_i M_i}\right)^{1/a} \tag{20}$$

The logic of this definition follows the logic used in defining average diameters based on surface area or volume discussed above and, as its name implies, $\overline{M}_{vis}$ is a molecular weight average determined from viscosity measurements on polydisperse polymer samples. The exponent a in this definition is called the *Mark-Houwink coefficient*, generally $0.5 < a < 1.0$. The exponent is characteristic of the polymer-solvent-temperature conditions of the experiment. We see how these experiments are conducted and the significance of the Mark-Houwink coefficient in Chapter 4. For now, we may take Equation (20) as merely defining another kind of molecular weight average. Note that $\overline{M}_{vis} = \overline{M}_w$ when $a = 1$.

The various molecular weight averages we have discussed, their definitions, and the experimental methods that measure them are listed in Table 1.8. All these different "averages" are admittedly confusing. Without this information, however, it would be far more confusing to try to rationalize the discrepancy between two different molecular weight determinations on the same sample by methods that yield different averages. The divergence between such values is a direct consequence of polydispersity in the sample. As we have seen, it is not only unavoidable but also informative as to the extent of polydispersity.

1.6   SOME CLASSICAL AND EMERGING EXPERIMENTAL TOOLS

In this section, we present a few examples of instruments available for visual observation and imaging of colloids and surfaces, for measurement of "sizes" and for surface force measurements. Such a presentation can hardly be comprehensive; in fact, that is not our purpose here. Throughout the book, we discuss numerous other techniques such as osmotic pressure measurements, light and other radiation scattering techniques, surface tension measurements, and electrokinetic measurements, among others. There are many more tools for which there is no space here. Some of the techniques described in other chapters can provide the type of information described in this section or serve as adjuncts to the techniques described here.

The main objective of the present section, however, is to begin with a very standard technique such as optical microscopy and to use it to illustrate why colloids are difficult to see and what modern developments have emerged in recent years to allow us to "see" and do things that were considered impossible until a decade ago. We also use this opportunity to review briefly some new techniques that are currently available to measure interaction forces between particles directly. We appeal to some of these techniques in other chapters when we discuss colloidal forces.

1.6a "Visual" Observation and Imaging

1.6a.1 Optical Microscopy

Because of the particle sizes involved, classically the optical microscope has been the instrument of choice especially for lyophobic colloids. Excellent books and manuals are available (Bradbury 1991; Cherry 1991; Schaeffer 1953) on the numerous variations of optical microscopy, and we do not go into all the details. Our purpose here is merely to point out some very elementary principles that make this method ideally suited for direct examination of colloids. We also use this introduction as a first step in pointing out modern techniques that fall under the class of "microscopy" but use principles (e.g., electron tunneling; see Vignette I.8) and radiation (e.g., electron or x-ray) other than those used in optical microscopy.

1.6a.1a Contrast. For a particle to be visible optically, there must be an acceptable difference between its refractive index and that of its surroundings; this is known as *contrast*. This requirement has nothing to do with particle size: A glass rod can be made to "disappear" by immersing it in a liquid of matching refractive index. The fact that a technique is not applicable to all possible systems does not invalidate it; this only means that one must have access to more than one technique in order to deal with a variety of problems. Contrast is a generic term, and how the contrast between a particle and its surrounding arises depends on

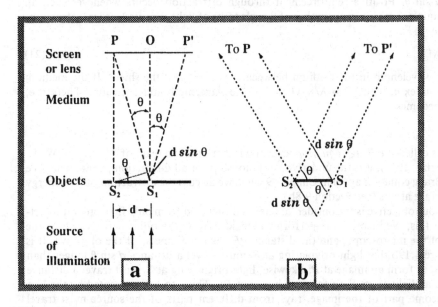

FIG. 1.19 Basic optical principle governing the operation of an optical microscope: (a) the geometry on which the resolving power d of a microscope is based; (b) detail showing how light from both sources must be intercepted by the lens to become part of the image.

the radiation and probing technique used. For example, in the case of scattering techniques, discussed in Chapter 5, contrast arises from the refractive index difference between the particle and surroundings when light scattering is used, whereas in neutron scattering it is the difference in the interaction between the neutrons and the atomic nuclei that give rise to the contrast.

1.6a.1b Resolution and Magnification.   The first thing we tend to think of in connection with microscopes is the magnification they achieve. More important, however, is a quantity known as the *resolving power*, or *limit of resolution*, of the microscope. Magnification determines the size of an image, but the resolving power determines the amount of distinguishable detail. Enlargement without detail is of little value. For example, a row of small spherical particles will appear simply as a line if the row is enlarged with an instrument of poor resolving power. Further magnification would increase the thickness of the line, but would not reveal its particulate nature. If the resolving power is increased, however, the individual spheres would be discernible. One may then choose a magnification that is convenient. Both the depth and area of the in-focus field decrease as the magnification is increased, so one pays a price for enlargement even though the amount of perceptible detail is not affected much by the magnification.

A beam of light is always diffracted at the edges of an object to produce a set of images of the edge known as a diffraction pattern. The diffracted light is what our eye receives from an object and from which our image of the object is constructed. The diffracted light contains sufficient information to assemble an image; any light that is not incorporated into the image will result in a loss of detail in the image. To estimate the efficiency with which an image reproduces the object, let us consider the situation in which all first-order diffracted light is intercepted by the lens, but light from all higher orders of diffraction is assumed to be lost.

Suppose we have a set of pinholes in an opaque shield and that this shield is illuminated by a source far enough away that the incident light may be regarded as a set of parallel rays, as sketched in Figure 1.19a. To an observer on the opposite side of the shield, each pinhole will function as a light source from which a hemispherical wave front seems to emerge. If adjoining pinholes are separated by a distance d, the wave from one source, say S_1, must travel a longer distance than light from S_2 to reach a distant screen at point P. If the distance between the shield and the screen is large compared to the wavelength, the extra distance may be equated with $d \sin \theta$. Positive reinforcement through diffraction occurs whenever such an extra distance equals some integral number of wavelengths of light. For first-order diffraction, then, we require

$$d \sin \theta = \lambda' \tag{21}$$

where λ' is the wavelength in the medium between the screen and the shield. If the medium has a refractive index n, then $\lambda' = \lambda/n$ where λ is the wavelength under vacuum. Therefore, Equation (21) becomes

$$d \sin \theta = \lambda/n \tag{22}$$

Equation (21) is called the *Bragg equation* after the father-and-son team of W. H. and W. L. Bragg (Nobel Prize, 1915); it is the underlying relationship for all diffraction phenomena. We encounter the Bragg equation again in Chapter 9 when we discuss the diffraction of low-energy electrons by surface atoms (see Section 9.8b).

Recall that our objective is to consider an image constructed from only first-order diffracted light. To do this, we identify the perforated shield S_1S_2 as the object, the screen as the objective lens of the microscope, and the distance PP' as the diameter of the objective. It is clear from Figure 1.19b that light originating at S_1 must travel a distance $d \sin \theta$ longer than the light from S_2 to form an image at P; likewise, light originating at S_2 must travel a distance $d \sin \theta$ longer than light from S_1 to form an image at P'. Thus, to be intercepted by the lens and thereby become part of the image, rays from different parts of the source must travel paths that differ by $2d \sin \theta$. In order for this to happen and still consist of only first-order diffracted light, the difference in path lengths must equal λ'. The resolving power is defined to be the magnitude of the separation between objects that is required to produce discernibly different images when the angle subtended by the microscope is 2θ. Therefore, the resolving power is identical to d in Figure 1.19 and is estimated to be

$$d = \lambda/(2n \sin \theta) \tag{23}$$

which is sometimes written

$$d = \lambda/2(NA) \tag{24}$$

where NA (i.e., $n \sin \theta$) is called the *numerical aperture* of the lens.

The significance of Equation (23) is that any points closer together than the distance d will produce a first-order diffraction image with light having a wavelength equal to λ/n at some angle greater than θ. This light would not be intercepted by the lens PP', so a significant amount of detail about distances of this magnitude is lost. It is the wave nature of light that imposes a limit to the amount of detail an image may possess. Equation (23) shows that the resolving power is decreased by increasing θ or by decreasing λ/n. The subtended angle 2θ is increased by increasing the diameter of the lens and by decreasing the distance between the object and the lens; the design of the lens limits the range of these parameters. The ratio λ/n may be decreased by decreasing λ or by increasing n. Although shorter wavelengths improve resolving power, visible light is almost always used in microscopy, primarily because of the absorption of shorter wavelengths by glass. Since the refractive index of some oils is 50% higher than that of air, a significant improvement in the resolving power is achieved by filling the gap between object and lens with so-called immersion oil.

We may use Equation (23) to estimate the particle size that will be clearly resolved microscopically. As a numerical example, we may take λ to be 500 nm, $n = 1.5$, and $2\theta = 140°$, all of which are attainable and are optimal values for these quantities. This leads to a resolving power of 142 nm, which represents the lower limit of resolved particle size under completely ideal circumstances. Under closer to average circumstances, a figure twice this value may be more typical. These figures reveal that direct microscopic observation is feasible only for particles at the upper end of the colloidal size range and then only if the particle contrasts sufficiently in refractive index with its surroundings. This is the reason for exploring the use of other radiations such as x-ray and neutron and other techniques to "visualize" submicroscopic particles and features of interest in colloid science. We briefly consider a rather straightforward extension of optical microscopy, i.e., electron microscopy, in Section 1.6a.2.

1.6a.1c Dark-Field Microscopy. There are many variations of optical microscopic techniques in use today. One variation of direct microscopic examination extends the range of microscopy considerably but at the expense of much detail. By this technique, known as *dark-field microscopy*, particles as small as 5 nm may be detected under optimum conditions, with about 20 nm as the lower limit under average conditions. In dark-field microscopy, the sample is illuminated from the side rather than from below as in an ordinary microscope. If no particles were present, no light would be deviated from the horizontal into the microscope, and the field would appear totally dark. The presence of small dispersed particles, however, leads to the scattering of some light from the horizontal into the microscope objective, a phenomenon sometimes called the *Tyndall effect*. The presence of colloidal particles is indicated by minute specks of light in an otherwise dark field.

In dark-field microscopy, the particles are only a blur; no details are distinguishable at all. Some rough indication of the symmetry of the particles is afforded by the twinkling that accompanies the rotation of asymmetrical particles, but this is a highly subjective observation. However, the technique does permit the rate of particle diffusion to be observed. We see in Chapter 2 how to relate this information to particle size and shape. The number of particles per unit volume may also be determined by direct count once the area and depth of the illuminated field have been calibrated. This is an important technique for the study of coagulation kinetics, a topic we discuss in Chapter 13.

In dark-field, as in direct observation microscopy, the refractive index difference between the particles and the medium is one of the crucial factors that influences the feasibility of the method. Very good resolving power is obtained with metallic particles, which offer maximum contrast in refractive index. The technique of dark-field microscopy played an important historic role in colloid science, particularly in the study of metallic colloids. R. Zsigmondy (Nobel Prize, 1925), for example, made extensive use of this technique in his study of colloidal gold. Note that it is the deviation of a beam of light or scattered light that is utilized in dark-field microscopy. We have a good deal more to say about light scattering in Chapter 5.

1.6a.2 Electron Microscopy and Scanning Probe Microscopies

There are numerous modern developments that have made atomic-scale resolution possible in recent years. In fact, some of these developments in instruments can also be used to measure forces between particles and surfaces. These developments for force measurements are discussed briefly in Section 1.6c and in Vignette I.8. In this section, we review electron and scanning probe microscopies (SPMs), which allow atomic-scale visualization of surfaces and particles.

1.6a.2a Transmission and Scanning Electron Microscopy. It is clear from Equation (23) that the best prospect for extending the range of microscopy lies in the extension by orders of magnitude of the wavelength used to produce the image. The wave-particle duality principle of modern physics shows us how to make this extension. According to the *de Broglie equation* (see Atkins 1994), the wavelength of a particle is inversely proportional to its momentum:

$$\lambda = h/mv \tag{25}$$

where h is Planck's constant and m and v are the particle mass and velocity, respectively. In an electron microscope, a beam of electrons replaces light in producing an image. An electron beam is produced by a hot filament, accelerated by an electron gun, and focused by electric or magnetic fields that function as lenses. In this section, we consider primarily a technique based on the transmitted electron beam. Accordingly, the method is called *transmission electron microscopy* (TEM). Some brief remarks are made near the end of this section about a technique known as *scanning electron microscopy*.

A schematic comparison of light and electron microscopes is shown in Figure 1.20. Although wavelengths on the order of 10^{-12} m (1 picometer or 1 pm) are easily achieved in electron microscopes, the numerical apertures of such microscopes are low. Accordingly, the resolving power of a conventional electron microscope is generally about 1 nm. However, because of this remarkable resolution, small features may be enormously magnified without loss of detail.

The intensity of the electron beam that is transmitted through the specimen under observation in an electron microscope depends on the thickness of the sample and the concentration of atoms in the sample. Thus, unless there are large differences in these characteristics between the particles of the sample and the support on which they rest, a very low contrast image will be produced. A poor-quality image results from these low-contrast situations; the image is analogous to a landscape viewed from an airplane with the sun directly overhead. The most common way of overcoming this difficulty is by means of a technique known as *shadow casting*. The sample to be examined in the electron microscope is placed in a chamber, which is then evacuated. Next, some gold or other metal is vaporized in the same chamber. The vapor condenses on all cooler surfaces, including the surface of the sample. If the vapor source is positioned to the side of the sample, the condensed vapor will deposit unevenly, effectively casting a shadow over the sample. The situation is shown schematically in Figure 1.21; Figure 1.12 shows actual photomicrographs enhanced by this technique. With shadow casting, the field shows much more detail, just as the view from an airplane does when the sun is lower in the sky.

When the angular position of the vapor source relative to the sample is known, the thickness of the particle may be evaluated from the length of its shadow by a simple trigono-

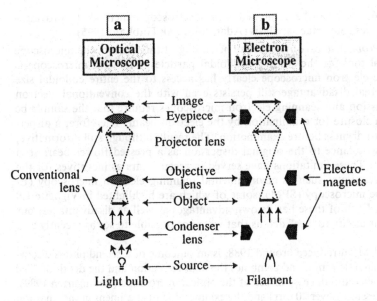

FIG. 1.20 Schematic comparison of (a) light and (b) electron microscopes showing components that perform parallel functions in each.

metric calculation. Alternatively, the length of the shadow cast by a spherical reference particle, introduced for calibration, may be used as a standard to calculate the thickness of particles.

In contrast to the TEM, the scanning electron microscope (SEM) works somewhat like a TV cathode-ray tube. An electron gun is used to scan a surface with the help of deflection coils and focusing "lenses" in a raster pattern, and the "reflected" electron signal is used synchronously to generate an image of very high magnification on a screen. The SEM has been in

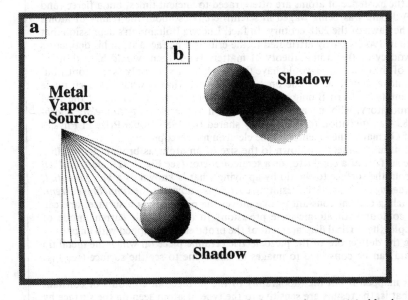

FIG. 1.21 Shadowing of a spherical particle using metal vapor: (a) side view; and (b) top view.

routine use since the 1960s and can be combined with spectroscopic methods to provide a powerful array of techniques for surface analysis (Adamson 1990; Hubbard 1995).

1.6a.2b Scanning Probe Microscopy.   At the beginning of this section, the microscope was proposed as an ideal tool for the study of colloidal particles. The light microscope is limited in range, but the electron microscope clearly has access to the entire colloidal size range. However, a very real disadvantage still persists even with the conventional electron microscope. The transmission and scanning electron microscopes require that the sample be placed in an evacuated enclosure for examination by the electron beam. Therefore, a dispersion must be evaporated to dryness before examination. The result, although still informative, bears about as much resemblance to the original dispersion as a pressed flower bears to a blossom on a living plant. These limitations have served in recent years as incentives for the development of newer techniques such as environmental scanning electron microscopy (E-SEM) and scanning probe microscopy (SPM), some of which are highlighted in Vignette I.8 (see also Hubbard 1995). Each of these has its own advantages as well as disadvantages, but they all have increased our ability to "see" details that were not possible to see as recently as a decade ago.

For example, the E-SEM, introduced around 1988, is an outcome of a serendipitous discovery of a new method of signal detection and combines the high resolution and the depth of field of the SEM with the flexibility and the ease of use of the optical microscope (Baumgarten 1989). It allows the use of high pressures (over 20 torr) and the examination of specimens in their natural state so that biological specimens and dynamic processes such as wetting and drying can be studied in real time. Some of the SPMs, on the other hand, require very low pressure environments, but others allow natural environments (see Vignette I.8 and Section 1.6c).

In subsequent chapters, we discuss some *in situ* techniques for the characterization of colloidal particles, especially with respect to particle size, structure, and molecular weight.

VIGNETTE I.8   NEW EXPERIMENTAL TOOLS: Scanning Probe Microscopes for Surface Analysis at the Atomic Scale

Speculations about the existence of atoms are often traced to ancient times, but a fierce (and sometimes vitriolic) debate was being waged in scientific circles about the reality of atoms even as recently as the dawn of the 20th century. In fact, Ludwig Boltzmann's depression that drove him to suicide in 1906 is partly attributed to the criticism he faced from his detractors for his staunch advocacy of the atomic theory of matter. Who, then, would have thought that before the end of the century we would have the capability to not only "see" atoms, but to manipulate and move them one by one on a surface? Yet, that is what has been made possible by the ingenuity of Gerd Binnig and Heinrich Rohrer, two physicists at the IBM Zürich Research Laboratory, who invented what is known as the *scanning tunneling microscope* (STM) in 1982—an invention for which they shared the 1986 Nobel Prize in Physics with Ernst Ruska of Germany, the inventor of the electron microscope.

When a probe, usually a metal tip down to the size of an atom, is brought to within a nanometer from the surface of a conductor or a semiconductor (see Fig. 1.22), electrons can be made to jump from the surface to the tip by applying what is known as a *bias potential*; this is called the *tunneling effect*, and the resulting current is known as the *tunneling current*. By keeping the tunneling current constant as one moves the probe along a surface, one can follow the surface contour with atomic-scale precision! In the constant current mode of operation, for example, the vertical displacement of the probe can be mapped with angstrom-level precision using the deflections of the piezoelectric ceramic piece on which the metal tip is mounted. The data can be converted to images to allow one to see the surface (see Fig. 1.22, inset).

In fact, what is mapped is not the topography of the surface but contours of constant electron densities; that is, the results are sensitive to the type of atom seen on the surface by the probe. This implies that the STM can serve as a powerful and highly specific and sensitive local probe to study a surface (and materials adsorbed on a surface) with high precision.

Vacuum tunneling of electrons is not the only option, however. The invention of STM has triggered the development of a whole family of techniques known collectively as *scanning probe microscopy* (SPM) that are not restricted to tunneling as the mechanism or to only conducting or semiconducting surfaces. One type of SPM, known as the *atomic force microscopy* (AFM; see Section 1.6c.2), can be used to measure forces such as van der Waals forces, ion-ion interactions, and hydration forces on polymer-coated surfaces, biologically relevant bilayers, and so on. In the force-measuring mode, an SPM is often referred to as a surface *force* microscope (SFM), but the label SPM signifies the more general capability of these devices. For example, atoms on surfaces can be manipulated and rearranged individually using STM and SPM, and conformation of polymers and biological macromolecules on surfaces can be probed.

In addition, the STM can be combined with suitable spectroscopic techniques to design powerful local probes of a surface to a precision not possible through other techniques; these are known as *scanning probe spectroscopy*. Because of its precision, the STM/SPM family can provide details that are often missed by other techniques. For example, we discuss a diffraction technique known as *low-energy electron diffraction* (LEED) in Chapter 9 (Section 9.8) for mapping the arrangement of atoms on a surface. Sharp diffraction spots obtained through LEED are taken to be indicative of ordered arrangements of the atoms on a surface. However, we know now from STM studies that even a surface showing clear and sharp diffraction spots in LEED experiments can in fact be highly defective. (This illustrates why more than one experimental technique [especially independent techniques] is often necessary in research.)

The development of SPMs has significantly changed the "landscape" of the instrumentation available to colloid and surface scientists, and the possibilities are endless!

1.6b Size and Molecular Weight Measurements

1.6b.1 Particle Size Measurement

Particle size measurement is one of the essential requirements in almost all uses of colloids. However, our discussion in Section 1.5 makes it clear that this is no easy task, especially since even the definition of *particle size* is difficult in many cases. A number of techniques have been developed for measuring particle size and are well documented in specialized monographs (e.g., Allen 1990). Optical and electron microscopy described in the previous section can be used when *ex situ* measurements are possible or can be acceptable, but we also touch on a few nonintrusive methods such as static and dynamic light scattering (Chapter 5) and field-flow fractionation (see Vignette II; Chapter 2) in other chapters.

In this chapter, we restrict our discussion to a chromatographic technique normally used for molecular weight measurements. The chromatographic concept can also be used for direct size (instead of molecular weights) measurement in the case of rigid particles, as we illustrate in our description of field-flow fractionation methods in Chapter 2.

1.6b.2 Molecular Weights

Lyophilic colloids, especially those of synthetic origin, also possess a distribution of particle sizes, but these are generally characterized by molecular weight rather than by linear dimension (except when R_g is measured for such systems). We discuss the experimental methods used for molecular weight determination of particles in the colloidal size range in Chapters 2–5. These methods yield the kinds of molecular weight averages discussed in Section 1.5c.3. Here, we consider a colloidal solution, say polystyrene in toluene, and discuss a chromatographic method for its fractionation.

1.6b.2a Chromatographic Techniques. To begin, a polymer like polystyrene is formed by the addition reaction in which styrene monomers $CH_2 = CHC_6H_5$ combine into long chains that may be thousands of repeat units long. This reaction is one of random addition, so not all

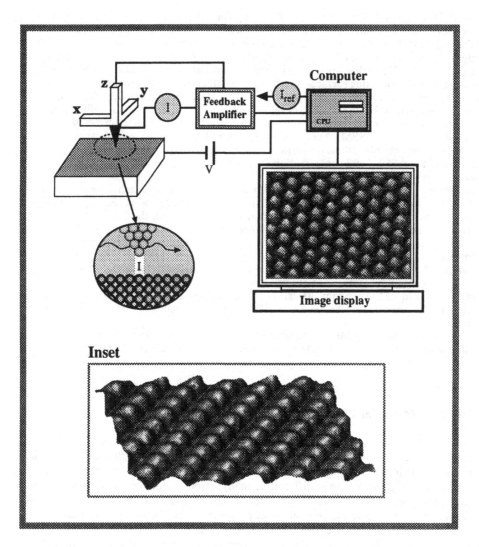

FIG. 1.22   One type of operation of a scanning tunneling microscope (STM). A tunneling current *I* flows between the sharp tip of the probe and the surface when a bias voltage *V* is applied to the sample. A computer monitors the tunneling current and adjusts the distance between the probe and the surface such that a constant tunneling current I_{ref} is maintained. The resulting changes in the position of the tip are then recorded and converted to an image such as the one shown on the monitor or the one shown in the inset. The image shown in the inset is that of an atomically smooth nickel surface. The periodic arrangement of the atoms on the surface can be seen clearly in the STM image. (Adapted with permission of C. M. Lieber, *Chem. & Eng. News* **72**(16), 28 (1994). The inset is adapted from T. A. Hoppenheimer, in *A Positron Named Priscilla: Scientific Discovery at the Frontier*, M. Bartusiak, B. Burke, A. Chaikin, A. Greenwood, T. A. Heppenheimer, M. Hoffman, D. Holzaman, E. J. Maggio, and A. S. Moffat, Eds., National Acad. Press, Washington, DC, 1994.)

çhains are identical in length. Furthermore, chains containing slightly different numbers of repeat units will not differ enough in their properties to allow complete separation. Nevertheless, procedures exist in which an initially polydisperse system is divided into fractions of narrower molecular weight distribution. One such method is called *size-exclusion chromatography* (SEC) because it segregates molecules in the distribution on the basis of their spatial extensions (see Section 1.6b.2b).

In chromatography, a cylindrical column is packed with porous particles and then filled with solvent. Next, a portion of the colloidal solution is layered on the solvent, and the liquid is allowed to pass through the column. The eluted liquid is monitored for the colloid by an appropriate detection method; spectrophotometry and refractive index are probably the most widely used methods of detection. What results is a *chromatogram* showing the detector output as a function of the volume of liquid eluted through the column. For a synthetic colloid like polystyrene, a broad peak results. Samples of biological origin often contain several components of quite different molecular weight. Under optimum conditions, these may emerge as distinct peaks in the chromatogram.

The volume at which a particular colloidal fraction emerges from the column is called the *retention volume* V_R of that fraction. Normally, a particular experimental system is calibrated with a colloidal solute of known molecular weight M. A plot of log M versus V_R is generally linear over several orders of magnitude in molecular weight. Not only the solute but also the solvent, the column packing, and the operating conditions affect the calibration. Therefore, the calibration is set up on a system-by-system basis using molecular weights that have been determined by one of the absolute methods we discuss elsewhere in this book. Thus, the calibration process can be time consuming, but, once set up, chromatography provides a rapid method for fractionating polydisperse systems.

Figure 1.23 shows how log M and the detector output vary with V_R. By reading from the chromatogram to the calibration curve, the molecular weight of the fraction emerging at a

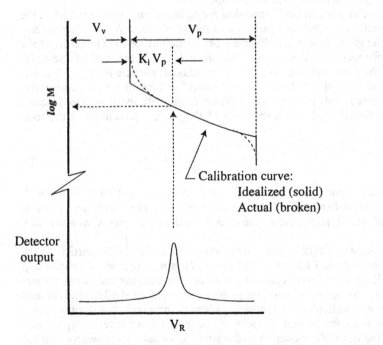

FIG. 1.23 Plot of log M and detector output versus retention volume for size-exclusion chromatography. Also shown is the relation among V_R, V_V, V_P, and $K_i V_p$ as discussed in the text. (Redrawn with permission of P. C. Hiemenz, *Polymer Chemistry: The Basic Concepts*, Marcel Dekker, New York, NY, 1984.)

particular value of V_R can be determined. Furthermore, for a molecular weight fraction M_i, the height h of the detector output signal is directly proportional to the mass m_i of the solute eluted in that particular fraction. Since $m_i \propto h_i$, $\Sigma_i h_i \propto \Sigma_i m_i$ and $m_i/\Sigma_i m_i = h_i/\Sigma_i h_i$. Thus, the weight fractions of different molecular weight classes in the distribution are readily obtained by this method. Likewise, if n_i is the number of moles of component i, $h_i/M_i \propto n_i$ and $(h_i/M_i)/\Sigma_i h_i/M_i = n_i/\Sigma_i n_i$. If we assume that suitable calibration exists, the chromatogram produced by SEC can be interpreted in terms of either number fractions or weight fractions of the various molecular weight classes. This kind of information can be used to calculate molecular weight averages (see Section 1.5c.3).

Incidentally, this method is known by various names among workers in different fields. In biochemistry, the technique is called *gel filtration chromatography* (GFC), and in polymer chemistry, it is *gel permeation chromatography* (GPC).

1.6b.2b   Size-Exclusion Chromatography.   Now, however, let us briefly consider the mechanism by which molecules are fractionated by SEC. The basic idea is very simple. Large molecules cannot penetrate into the pores of the packing medium (the gel) and hence are eluted first. The smaller particles are distributed in the pores as well as in the voids between the gel particles and emerge later.

The total volume of solvent in the column can be divided into two categories: V_V is the volume of solvent in the voids between the gel particles, and V_p is the volume of solvent in the pores. Not all of the pore volume is accessible to particles in a given molecular weight class. We define $K_i V_p$ as the pore volume into which molecules from the ith class can permeate. Thus, K_i describes what fraction of a pore is accessible to molecules in class i. The fraction K_i is zero for very large particles and unity for very small particles and varies over this range for particles of intermediate size. The retention volume for molecules in class i is given by

$$V_R = V_V + K_i V_p \tag{26}$$

Figure 1.23 shows the relationship among V_R, V_V, V_P, and K_i.

It is fairly easy to arrive at a theoretical expression for K_i based on a simple model of the permeation process. We picture the colloidal particle as a sphere of radius R_s and the pore as a cylinder of radius R_c and length ℓ, as shown in Figure 1.24a (Giddings 1991). The center of the solute particle cannot approach any closer than a distance R_s from the wall of the pore. Therefore, the radius of the pore that is accessible to the colloidal particle is $(R_c - R_s)$. The layer of solution adjacent to the walls of the pore is "off limits" to the solute, so the concentration of the colloid in the pore is only a fraction of its value in the bulk solution. This fraction is K_i for the sphere of radius R_s and equals the ratio of the accessible volume to the total volume of the pore:

$$K_i = \frac{\pi (R_c - R_s)^2 (\ell - R_s)}{\pi R_c^2 \ell} \approx \left(\frac{R_c - R_s}{R_c}\right)^2 = \left(1 - \frac{R_s}{R_c}\right)^2 ; \ell >> R_s \tag{27}$$

Equation (27) is an approximation valid when the pore is long enough so that $(\ell - R_s) \approx \ell$, so that the fraction on the right-hand side of the equation ranges between zero and unity as required for R_s equal to R_c and 0, respectively, and thus gives the parameter K_i as a function of particle size.

The next stage of developing a theory would be to relate the radius of the particle to the molecular weight. Thus expanded, Equations (26) and (27) would provide a relationship between M and V_R that—if the theory were successful—would eliminate the need for empirical calibration of SEC. For some systems, such as rigid globular proteins, establishing the connection between R_s and M is not difficult. For synthetic polymers, however, the flexible chains exist as random coils with an extension that depends on the interaction between polymer and solvent. In this case, relating R_s to M is a complicated matter, so we are content with empirical calibration to interpret V_R in terms of M. This, of course, is not the only modification that needs to be examined in a fully developed theory for K_i. Models based on more complicated geometries for both the colloidal particles and the pores have been considered, but we shall

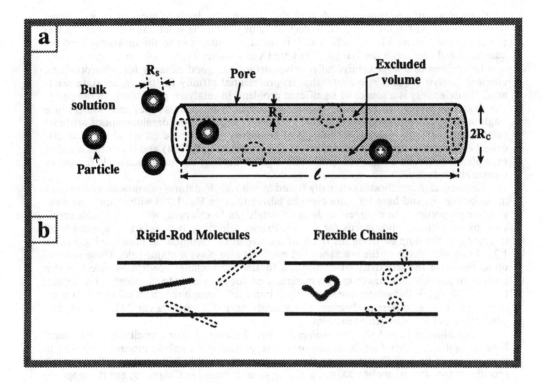

FIG. 1.24 Size exclusion of a particle in a pore: (a) exclusion of a spherical particle in a cylindrical pore; and (b) exclusion of macromolecules. Particles shown in dashed lines indicate positions or orientations that are excluded. (Redrawn with modifications from Giddings 1991.)

not pursue these. Yau et al. (1979) discuss additional theoretical models and experimental procedures in detail for those desiring more information on this topic.

It is important to note that the implication of "size exclusion" is more general than its use in SEC might imply. The exclusion of particles based on size and shape (as illustrated for a simple case in Fig. 1.24b) finds applications in other important contexts. For instance, specially designed zeolite catalysts are used in industry for accomplishing highly selective reactions, and the accessibility of the reactants to the sites on the surfaces of such catalysts is controlled using size-exclusion (and shape-exclusion) principles. This is illustrated in Vignette I.9. The size-exclusion and shape-exclusion properties of the zeolites of the type described in Vignette I.9 are also taken advantage of in industrial separation processes (the zeolites are often called *molecular sieves* for this reason).

VIGNETTE I.9   SIZE EXCLUSION IN HETEROGENEOUS CATALYSIS AND SEPARATION PROCESSES: Zeolites—Molecular Cages for Selectivity in Separation and Catalysis

The adjective *heterogeneous* in heterogeneous catalysis draws attention to the fact that an adsorbent in the form of one phase (e.g., solid) assists in reactions between molecules in another phase (e.g., gas). When a molecule adsorbs on a surface, whether by chemisorption or physisorption (see Chapter 9), one or more of its atoms form strong (chemisorption) or relatively weak (physisorption) bonds with the surface. This leads to breaking or weakening of other bonds within the adsorbed molecules, thus promoting reactions with other species present in the system. The adsorbent in this instance contributes to the disassembly of the adsorbate and becomes a catalyst (see Vignette IX in Chapter 9).

In addition to a large catalytically active surface and good capacity for adsorption, an efficient catalyst requires *selectivity*, that is, preferential affinity for the appropriate reactants. Nonselectivity is a source of significant problems in catalysis, particularly in the petrochemical industry, which has to deal with hydrocarbons of various types and isomers in a single stream. This situation has provided a strong incentive for the development of artificial (man-made) catalysts that offer the type of selectivity unthinkable on metal surfaces discussed in the last section of Chapter 9 (see, for example, Ball 1994) and illustrates another example of molecular design (or "molecular engineering") of advanced materials for use in science and industry.

Zeolites, aluminosilicates originally found as minerals in nature, can now be synthesized in various forms and have intricate cagelike labyrinths (see Fig. 1.25) with shape- and size-selective properties. The channels in these materials can be enlarged, with remarkable precision, by substituting phosphorus for silicon. Precise selectivity is achieved by a number of mechanisms, the simplest being the result of the size of the channels, as illustrated in Figure 1.25. Molecular shape of the reactants and products also plays a major role. These mechanisms influence the selectivity of absorption (notice the spelling; *absorption* refers to the amount of gas molecules taken in by a material and not necessarily to the ability of a surface to adsorb the gas). But *adsorptive* selectivity is imparted through the chemical elements used in the material, as different chemical constituents have different "talents" (Ball 1994) for adsorption and for promoting reactions.

The zeolites are also known as *molecular sieves* because of their capacity to discriminate between molecules; they find numerous uses in separation and catalytic processes. Although they appear to be "solid" particles to the naked eye, they are highly porous, with a typical specific surface area of about 1000 m^2/g. Catalysis is discussed in Chapter 9, but the scope of that chapter does not permit detailed discussions of the various types of catalysts and the role of physisorption and chemisorption in catalysis; this vignette provides a glimpse of the rationale used in the molecular design of new materials of interest in surface chemistry and how the concepts introduced in Chapter 1 and Chapter 9 fit into the larger scheme.

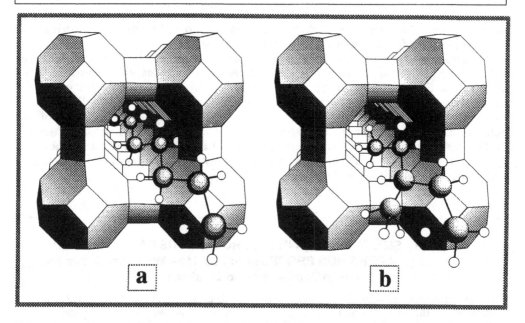

FIG. 1.25 Shape selectivity of a zeolite cage. The cage allows (a) straight-chain hydrocarbons to snake their way into the pores while (b) preventing branched-chain hydrocarbons from entry. (Adapted from Ball 1994.)

1.6c Force Measurements

We have already seen in the sections above and in the vignettes that interaction forces between surfaces and among particles are important in colloidal systems. Stability of colloids, adhesion and wetting phenomena, xerography and printing, and the like depend on the types of forces that exist between surfaces and how they vary with physical and chemical conditions. Such forces may be approached or understood on at least two different levels. The first is a coarser level, e.g., the magnitude and strength of the surface forces as they manifest themselves in macroscopic phenomena. Examples of these include the magnitude of the repulsive barrier needed to stabilize a dispersion from coagulation and the strength of adhesive force needed for an adhesive film to meet the performance requirement. The second is on a finer scale, as in obtaining a detailed "force law" for interaction between two surfaces as a function of distance of separation. We discuss examples of these in Chapters 10 and 11.

What is needed or sufficient depends on the application. For engineering applications, the first level might often be sufficient, whereas for developing new materials or understanding chemical and biological phenomena involving surfaces (see the vignette in Chapter 12), a more detailed understanding is necessary.

A number of techniques have been developed over the years to determine colloidal and surface forces as well as interatomic and intermolecular forces of interest in colloid and surface chemistry. These methods can be divided into two groups: (a) indirect methods and (b) direct methods. We discuss examples of both in other chapters. It is therefore useful to consider these methods here briefly.

1.6c.1 Indirect Methods

Indirect methods are those in which we use some macroscopic property or phenomenon to deduce information about (colloidal or intermolecular) interaction forces. These methods are known as *indirect* since additional assumptions or approximations about the relation between the forces and the property or phenomenon measured are needed. Some examples of these methods are illustrated in Figure 1.26 and are described below (Israelachvili 1991).

1. Figures 1.26a and 126b illustrate particle detachment and peeling experiments for estimating information on attractive short-range forces that are important in particle adhesion and adhesion energies between solid surfaces in contact. Such experiments provide information of importance in powder technology, xerography, ceramic processing, and the making of adhesive films.

2. As we see in Chapter 6, surface tension and contact angle measurements provide information on liquid-liquid and solid-liquid adhesion energies (Fig. 1.26c). Contact angles measured under different atmospheric environments or as a function of time provide valuable insights into the states of surfaces and adsorbed films and of molecular reorientation times at interfaces.

3. The thicknesses of free soap films and liquid films adsorbed on surfaces (Figs. 1.26d and 1.26e), which can be measured using optical techniques such as reflected intensity, total internal reflection spectroscopy, or ellipsometry as functions of salt concentration or vapor pressure, can provide information on the long-range repulsive forces stabilizing thick wetting films. We see an example of this in Chapter 11.

4. More sophisticated methods such as nuclear magnetic resonance (NMR), light scattering, x-ray scattering, and neutron scattering can be used to measure dynamic interparticle separations and motions in liquids (Figs. 1.26f and 1.26g). In such experiments, the "particles" can be globular or spherical (e.g., micelles, vesicles, colloidal particles, latex particles, viruses), sheetlike (e.g., clays, lipid bilayers), or rodlike (e.g., DNA). The interparticle forces can be studied as functions of the solution conditions and the mean separation between particles by changing the hydrostatic or osmotic pressure via a semipermeable membrane. Notable among these techniques is the compression cell or osmotic pressure technique from which the interaction force between particles can be obtained from the deviations from ideality in the PVT data (see Chapter 10

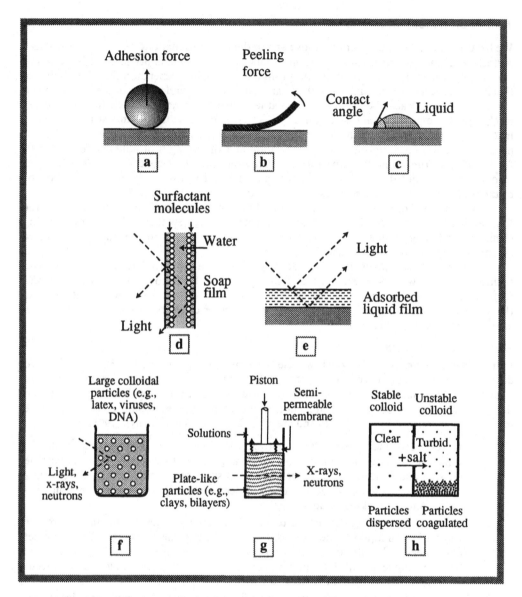

FIG. 1.26 Examples of various types of measurements that provide information on the forces between particles and surfaces: (a) adhesion measurements; (b) peeling measurements; (c) contact angle measurements (see Chapter 6); (d) equilibrium thickness of thin free films; (e) equilibrium thickness of thin adsorbed films (examples of practical applications include wetting of hydrophilic surfaces by water, adsorption of molecules from vapor, protective surface coatings and lubricant layers, photographic films; see Chapters 6 and 9); (f) interparticle spacing in liquids (examples of applications include colloidal suspensions, paints, pharmaceutical dispersions; see Chapter 13); (g) sheetlike particle spacings in liquids (examples of practical applications include clay and soil-swelling behavior, microstructure of soaps and biological membranes); (h) coagulation studies (e.g., basic experimental technique for testing the stability of colloidal preparations; see Chapter 13). (Redrawn with permission of Israelachvili 1991.)

for an example using van der Waals equation of state). These techniques are usually limited to measuring only the repulsive parts of a force law.

5. Coagulation of colloidal dispersions (Fig. 1.26h) as a function of salt concentration, pH, or temperature of the suspending liquid medium can also be used to obtain information on the interplay of repulsive and attractive forces between particles in pure liquids as well as in surfactant and polymer solutions.

The above methods are not usually sufficient when detailed information on the strength of interaction forces as a function of separation distance between surfaces is needed. Moreover, methods such as osmotic pressure measurements involve collective interactions of many molecules or particles so that the data obtained are of a thermodynamic nature. Translating this information to interaction force as a function of distance is not straightforward and is often ambiguous. Direct force measurements have been made possible in recent years by developments such as the scanning force microscopy (SFM) mentioned in Vignette I.8 and discussed further in Section 1.6c.2b and have contributed to a significant change in the field of colloid and surface science. We consider two important examples of direct force measurement techniques below.

1.6c.2 Direct Methods

As pointed out by Israelachvili (1991), the principle of direct force measurements is usually very straightforward, but the challenge is in measuring very weak forces at very small intermolecular or surface separations that must be controlled and measured to within 0.1 nm. Following Israelachvili (1991), we divide our description into two parts, namely, surface force measurements and interatomic force measurements.

1.6c.2a Surface Forces: Surface Force Apparatus. The surface force apparatus (SFA) was originally developed in the late 1960s by Tabor, Winterton, and Israelachvili (see Israelachvili 1991) for measuring the van der Waals forces between molecularly smooth mica surfaces in air or vacuum. The apparatus, which has since been modified for making measurements in liquids, consists of two curved surfaces of mica, the interaction forces between which are measured using highly sensitive force-measuring springs (see Fig. 1.27a). The two surfaces are in a crossed cylinder configuration and geometrically correspond locally to a sphere near a flat surface or to two spheres close together. The separation between the two surfaces can be measured by use of an optical technique (which employs multiple beam interference fringes), from microns down to molecular contact, usually to better than 0.1 nm. In addition, the exact shapes of the two surfaces and the refractive index of the liquid (or material) between them can also be measured. The ability to measure refractive index allows one to determine the quantity of material (e.g., lipid or polymer) deposited or adsorbed on the surfaces.

The force is measured through the expansion or contraction of the piezoelectric crystal by known amounts and by determining optically how much the two surfaces have actually moved. The difference in the two values is then multiplied by the stiffness of the force-measuring spring to obtain the force difference between the initial and final positions. Thus, both repulsive and attractive forces can be measured as functions of the distance between interacting surfaces, with a sensitivity of the order of 10^{-8} N. Figure 1.27b illustrates an example of force measurement for the case of attraction.

In the past, mica has been the material of choice for the interacting surfaces because of the ease of handling and since molecularly smooth surfaces can be fabricated; mica surface coated with a thin film of other materials (e.g., lipid monolayers or bilayers, metal films, polymer films, or other macromolecules such as proteins) can also be used. The use of alternative materials such as molecularly smooth sapphire and silica sheets and carbon and metal oxide surfaces is also being explored.

The invention and refinement of the SFA have been among the most significant advances in experimental colloid science and have allowed researchers to identify and quantify most of the fundamental interactions occurring between surfaces in aqueous solutions as well as nonaqueous liquids. Attractive van der Waals and repulsive electrostatic double-layer forces, oscillatory (solvation or structural) forces, repulsive hydration forces, attractive hydrophobic

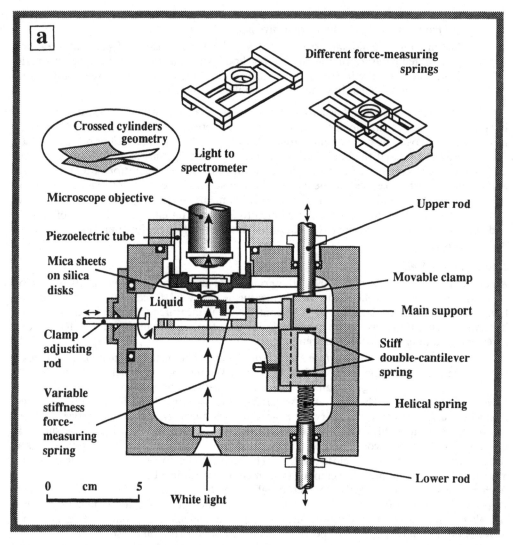

FIG. 1.27 The surface force apparatus (SFA) and an example of direct force measurement. (a) The surface force apparatus for direct measurement of the forces between surfaces in liquids or vapors at the angstrom resolution level. With the SFA, two atomically smooth surfaces immersed in a liquid are brought toward each other, with the surface separation being controlled to 0.1 nm, and the forces between the two surfaces can be measured. Moreover, the surfaces can be moved laterally past each other and the shear forces also measured during sliding. (Redrawn with permission of Israelachvili 1991.) (b) A schematic illustration of how the force between two surfaces is measured. As the two surfaces (inset) are brought together from separations x larger than x_2, the spring deflects toward contact to reach mechanical equilibrium (i.e., the spring force matches the attractive force between the surfaces). However, beyond the point x_2 at which the gradient of the interaction force equals the spring constant K, mechanical equilibrium is no longer possible, and the separation distance jumps to x_4. When the movement is reversed, the same thing happens again and x jumps from x_3 to x_1. By using springs of different spring constant K, one can also measure the force between the surfaces for $x_2 > x > x_3$. (Adapted from W. B. Russel, *The Dynamics of Colloidal Systems*, University of Wisconsin Press, Madison, WI, 1987.)

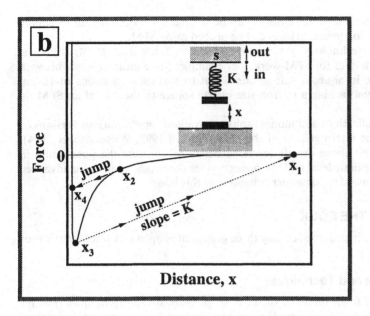

FIG. 1.27 Continued

forces, steric interactions involving polymeric systems, and capillary and adhesion forces have all been measured using this apparatus or variations of it (we see an example in Chapter 10). Moreover, the scope of phenomena that can be studied using the SFA has also been extended to measurements of dynamic interactions and time-dependent effects such as the viscosity of liquids in very thin films, shear and frictional forces, and the fusion of lipid bilayers.

The impact of the SFA and similar direct measurement forces goes beyond simple testing of theories of intermolecular forces. For instance, such measurements are also useful for explaining more complex phenomena such as the unexpected stability of certain colloidal dispersions in high salt, the crucial role of hydration and ion correlation forces in clay swelling and ceramic processing, and the deformed shapes of adhering particles and vesicles (see Israelachvili 1991). Further, both static (i.e., equilibrium) and dynamic (e.g., viscous) forces can now be studied with remarkable precision and accuracy for obtaining information on the structure of liquids adjacent to surfaces and related interfacial phenomena. Such studies demonstrate that properties of ultrathin films are profoundly different from those of bulk liquids.

1.6c.2b Atomic Forces: Scanning Force Microscope. As we have already seen in Vignette I.8, the SFM is an offshoot of the scanning tunneling microscope (STM) and is a hybrid between a surface profilometer and the SFA (Wiesendanger 1994). The difference between an SFM and the SFA is that forces at the atomic level can be measured using the former with the help of an atomically sharp tip that scans the surface (for this reason, the SFM with atomic resolution is often called the *atomic force microscope* [AFM]). The tip radii can be as small as one atom or larger than 1 mm, but the smaller tips are more popular since they allow one to measure directly the force between an individual atom and a surface or even between two individual atoms.

In SFM, the probe tip is mounted on a highly sensitive, cantilever-type spring. The force of interaction between the sample and the tip can be calculated from the spring constant and the measured deflection of the cantilever. The deflection is sensed using the STM principle (Vignette I.8) or capacitance or optical methods. The SFM can be operated in the "contact regime" or like the SFA. In the latter mode, one can measure van der Waals forces (see Chapter 10), ion-ion repulsion forces (see Chapter 11), and capillary forces and frictional forces, among others. In contrast to STM, the SFM can be used for both conductors and insulators. The fact that nonconducting samples can be studied is very significant to colloid science since biological and polymeric surfaces can be probed using SFM.

Further, a whole new technology has arisen devoted to fabricating highly sensitive, micron-size, force-sensing devices for SFM work. For example, force measurements between a flat surface and a sphere in aqueous salt solutions out to surface separations of 60 nm have been made recently by attaching a micron-size quartz sphere to the end of an SFM tip (Israelachvili 1991).

Numerous research publications and monographs are available on the various versions of SFM (and STM) and on the interpretation of the data (Bonnell 1993; Wiesendanger 1994), but for our purpose here the above description is sufficient to illustrate the possibilities in force measurements at the atomic level and the opportunities they open up in more advanced studies of the materials discussed in subsequent chapters of this book.

1.7   AN OVERVIEW OF THE BOOK

The rest of the book can be divided into roughly three groups of chapters. A brief overview of these is given below.

1.7a   Basic Phenomena and Techniques

Chapters 2–6 and Chapter 9 focus on many of the basic phenomena and techniques relevant to colloid and surface chemistry. These provide additional theoretical concepts and experimental tools routinely employed in the area. These include

1.   The competition between diffusion and sedimentation and how this is used to measure particle size, solvation of particles, etc. (Chapter 2)
2.   Osmometry and Donnan equilibrium (Chapter 3)
3.   Flow and rheology of dispersions and viscosity of polymer solutions and how viscosity is used to characterize dispersions (Chapter 4)
4.   Turbidity measurements, static and dynamic light scattering, and a brief introduction to x-ray and neutron scattering (Chapter 5)
5.   Surface tension and contact angle, wetting phenomena, effects of the curvature of the surface on capillarity and phase equilibria, and porosimetry (Chapter 6)
6.   Adsorption at solid-gas surfaces, derivation of adsorption isotherms, surface area measurement, and structural analysis of surfaces and adsorbed layers using low-energy electron diffraction (Chapter 9)

1.7b   Association Colloids

Chapters 7 and 8 discuss surfactant solutions:

1.   In the Chapter 7, formation of monolayers in air-liquid interfaces and the resulting film pressure and phase transitions are discussed. This chapter also includes a brief discussion of adsorption on solid surfaces from solutions.
2.   Chapter 8 discusses self-assembly of surfactants to form micelles, models of micellization, use of micelles in catalysis and solubilization, and oil-in-water and water-in-oil microemulsions.

1.7c   Colloidal Forces, Electrokinetic Phenomena, and Colloidal Stability

The final group deals with colloidal forces and their applications to colloid stability and deposition phenomena. These include

1.   The van der Waals forces at the atomic level as well as those between macroscopic bodies (Chapter 10)
2.   Theories of electrical double layers and forces due to double-layer overlap (Chapter 11)

3. Electrokinetic phenomena such as electroosmosis, streaming potential, and viscoelectric effects (Chapter 12)
4. The Derjaguin-Landau-Verwey-Overbeek (DLVO) theory of coagulation, polymer-induced forces, and steric stabilization (Chapter 13)

As mentioned above, each chapter also includes a vignette that provides a glimpse of advanced concepts or techniques that are beyond the scope of this book.

REVIEW QUESTIONS

1. What is the definition of a *colloid*?
2. Name at least five industrial products in colloidal form and discuss what colloidal properties are essential for each of them.
3. Name at least five examples in which a large surface area is desirable either from the processing standpoint or from the functionality of the end product.
4. What are *lyophilic* colloids? What are *lyophobic* colloids? Give some examples.
5. What are the difficulties with the characterization of colloids as *lyophilic* or *lyophobic*?
6. Why is a *foam* considered a colloid?
7. What is the meaning of *stability* as used in the context of colloids? What is the difference between *kinetic stability* and *thermodynamic stability* of a colloid?
8. Give at least five examples of colloids that are *thermodynamically* stable.
9. Define *coalescence* and *aggregation*. Give examples of processes or products in which aggregation (i.e., coagulation) is desirable. Give some examples for which coagulation is not desirable.
10. What is meant by the *fractal dimension* of an aggregate? What is the upper bound of the fractal dimension? Name at least one use of the fractal dimension of an aggregate.
11. What are some of the measures of *polydispersity*?
12. Why is the average diameter of a collection of polydisperse spheres different from the average diameter determined from the average of a property such as area or volume?
13. What is meant by the term *polydispersity index* in the case of molecular weights of polymers?
14. What is the *limit of resolution* of an optical microscope? Why?
15. What is *dark-field microscopy*? How does it compare in terms of resolution and contrast with conventional optical microscopy?
16. What is the limit of resolution of an *electron microscope*? Why?
17. What is the difference between *atomic* forces and *surface* forces? Can you deduce the details of atomic forces from surface force measurements? Why or why not?
18. Describe some of the recent developments for imaging surfaces and particles? What principles do they use? Name some advantages and limitations of each of the techniques.
19. What is meant by *scanning probe microscopy*?
20. Why is the measurement of atomic forces and surface forces important? Name some techniques for measuring atomic and surface forces.
21. Explain how scanning probe techniques can be modified to provide spectroscopic information.

REFERENCES

General References (with Annotations)

Atkins, P., *Physical Chemistry*, 5th ed., W. H. Freeman, New York, 1994. (Undergraduate level. This is an excellent, introductory-level book on physical chemistry and should be valuable for reviewing basic concepts of physical chemistry not described in detail in this chapter and elsewhere in this book.)

Ball, P., *Designing the Molecular World: Chemistry at the Frontier*, Princeton University Press, Princeton, NJ, 1994. (Undergraduate level. A highly readable, popular introduction in the *Scientific American* style to the many facets of molecular design in chemistry. Chapter 2 discusses chemical reactions, heterogeneous catalysis, and separation. Chapter 7 is an excellent introduction to self-organization of surfactant molecules into micelles, vesicles, Langmuir layers, and biological cells. This chapter also discusses the formation of polymer gels.)

Dickinson, E., *An Introduction to Food Colloids*, Oxford University Press, Oxford, England, 1992. (Undergraduate level. A readable introduction to the application of colloids in food science. Many of the topics should be accessible to undergraduate students.)

Giddings, J. C., *Unified Separation Science*, Wiley, New York, 1991. (Undergraduate level. A textbook on chromatographic and other analytical and preparative separation techniques. Some of the techniques discussed are relevant to the concepts we introduce in Chapters 2 and 3 of this book.)

Israelachvili, J. N., *Intermolecular and Surface Forces*, 2d ed., Academic Press, New York, 1991. (Undergraduate and graduate levels. The best reference available currently on the topic. Many examples of the application of surface forces in biological systems. The links between molecular forces and surface forces and the relation between molecular forces and bulk properties of materials are discussed in a manner accessible to advanced undergraduate students.)

Kruyt, H. R. (Ed.), *Colloid Science*, Vols. 1 and 2, Elsevier, Amsterdam, Netherlands, 1949, 1952. (Undergraduate and graduate levels. One of the classic references on colloids.)

Murray, C. A., and Grier, D. G., Colloidal Crystals. *American Scientist*, **83**, 238 (1995). (Graduate-level concepts, but some of the general ideas presented are accessible to undergraduates. This article is a moderately technical, but qualitative, overview of the use of charged latex particles for studying phase transitions in atomic solids and alloys. The color illustrations presented provide a glimpse of the numerous new avenues made possible by "model colloids" for studying structure of materials.)

Myers, D., *Surfaces, Interfaces, and Colloids: Principles and Applications*, VCH Publ., New York, 1991. (Undergraduate level. A qualitative overview of topics in colloid science. Numerous applications are discussed.)

Shaw, D. J., *Introduction to Colloid and Surface Chemistry*, 4th ed., Butterworth-Heinemann, Oxford, England, 1992. (Undergraduate level. A somewhat condensed treatment of many of the topics discussed in this book.)

Tanford, C., *Ben Franklin Stilled the Waves*, Duke University Press, Durham, NC, 1989. (A nontechnical, popular introduction to surfactant films and their applications in biology by a pioneer in the so-called hydrophobic effect.)

Vold, R. D., and Vold, M. J., *Colloid and Interface Chemistry*, Addison-Wesley, Reading, MA, 1983. (Graduate and undergraduate levels. A textbook at a more advanced level than this book.)

Other References

Adamson, A. W., *Physical Chemistry of Surfaces*, 5th ed., Wiley, New York, 1990.

Alberts, B., Bray, D., Lewis, J., Raff, M., Roberts, K., and Watson, J. D., *Molecular Biology of the Cell*, 2d ed., Garland Publishing, New York, 1989.

Allen, T., *Particle Size Measurement*, 4th ed., Chapman and Hall, New York, 1990.

Baumgarten, N., *Nature*, **341**, 81 (1989).

Bergethon, P. R., and Simons, E. R., *Biophysical Chemistry: Molecules to Membranes*, Springer-Verlag, New York, 1990.

Bonnell, D. A., *Scanning Tunneling Microscopy and Spectroscopy: Theory, Techniques, and Applications*, VCH Publ., New York, 1993.

Bradbury, S., *Basic Measurement Techniques for Light Microscopy*, Oxford University Press, Oxford, England, 1991.

Briggs, A., *Acoustic Microscopy*, Oxford University Press, Oxford, England, 1992.

Brinker, C. J., and Scherer, G. W., *Sol-Gel Science: The Physics and Chemistry of Sol-Gel Processing*, Wiley, New York, 1990.

Cherry, R., *New Techniques of Optical Microscopy and Microspectroscopy*, CRC Press, Boca Raton, FL, 1991.

Goodsell, S., *The Machinery of Life*, Springer-Verlag, New York, 1993.

Hubbard, A. T. (Ed.), *The Handbook of Surface Imaging and Visualization*, CRC Press, Boca Raton, FL, 1995.

Lasic, D., *American Scientist*, **80**, 20 (1992).

Lasic, D., *Liposomes: From Physics to Applications*, Elsevier Scientific, Amsterdam, Netherlands, 1993.

Matijevic, E., *Chem. Mater.*, **5**, 412 (1993).

Morowitz, H. J., *Beginnings of Cellular Life: Metabolism Recapitulates Biogenesis*, Yale University Press, New Haven, CT, 1992.

Schaeffer, H. F., *Microscopy for Chemists*, Dover, New York, 1953.

Weitz, D. A., Huang, J. S., Lin, M. Y., and Sung, J., *Phys. Rev. Lett.*, **54**, 1416 (1985).

Wiesendanger, R., *Scanning Probe Microscopy and Spectroscopy: Methods and Applications*, Cambridge University Press, Cambridge, England, 1994.
Yau, W. W., Kirkland, J. J., and Bly, D. D., *Modern Size Exclusion Liquid Chromatography*, Wiley, New York, 1979.

PROBLEMS[*]

1. The specific area of dust particles from the air over Pittsburgh, Pennsylvania, has been determined by gas adsorption[†]

Treatment	$A_{sp}(m^2g^{-1})$
4 h under vacuum at 200°C	5.61
8 h under vacuum at 25°C	2.81

   a. Calulate the radius of these particles if they are assumed to be uniform spheres of density $2.2 \ g \ cm^{-3}$.
   b. Propose an explanation for the effect of degassing on particle size.
   c. What kind of average for the radius is obtained by this procedure?

2. Colloidal palladium particles in an alumina matrix catalyze the hydrogenation of ethene to ethane. The following data describe various catalyst preparations:[‡]

Diameter of Pd particles (Å)	55	75	75	115	145
Parts per million Pd in catalyst	170	250	200	250	250
Percentage conversion per 25 mg catalyst	50	45	40.5	38.5	29

   a. Calculate the activity of these catalysts on the basis of the weight of Pd and the area of Pd in the preparations.
   b. Does the catalytic role of Pd seem to be a bulk or surface phenomenon?

3. The accompanying table shows how the trace metal content of coal-ash aerosols depends on particle size:[¶]

Range of diameters (μm)	Trace element (μg/g ash)					
	Pb	Tl	Sb	Cd	Se	As
30–40	300	5	9	<10	<15	160
5–10	820	20	25	<10	<50	800
1.1–2.1	1600	76	53	35	59	1700

   Discuss the implications of these results on human health in view of the following considerations:
   a. The intrinsic toxicity of these trace elements
   b. small particles travel to the lung, larger particles are stopped in the nose, pharynx, and so on
   c. Trace elements are absorbed from the alveoli 7–10 times more efficiently than from the upper respiratory spaces
   d. A possible mechanism for the effect of particle size on trace element content

4. Select a field containing about 10 particles from Figure 1.12b and measure the Martin diameters of the population parallel to the bottom of the micrograph (a photocopy can be made and

---

[*]The data for many of the problems in this book are taken from graphs appearing in the original literature. As a result, the values given do not necessarily reflect the accuracy of the original experiments. Likewise, the number of significant figures cited may not be justified in terms of the approximations involved in graph reading.
[†]Corn, M., Montgomery, T. L., and Reitz, R. J. *Science* **159**, 1350 (1967).
[‡]Turelevich, J., and Kim, G. *Science*, **169**, 873 (1970).
[¶]Natusch, D.F.S., Wallace, J. R., and Evans, C. A., Jr. *Science* **183**, 202 (1974).

the particles checked off to avoid duplication and omissions). Classify the data and calculate the mean and standard deviation. Repeat, measuring the Martin diameter parallel to the side of the micrograph. Discuss the agreement or discrepancy between the two means in terms of:
   a. Bias in the choice of the field
   b. Systematic orientation effects
   c. The size of the population

5. Suppose that the particles in Figure 1.8 were actually oblate ellipsoids (all in their preferred orientation) rather than spheres. Would their volume be over- or underestimated if the particles were assumed to be spheres? In terms of their axial ratio, calculate the factor by which the mass is under- or overestimated when the particles are assumed to be spheres. (Consult a handbook for the volume of an ellipsoid.)

6. Mixtures of 50 g of ZnO-TiO_2 are each shaken with 250 ml of water and allowed to settle. After 14 days of equilibration, it is found that the sediment volume is 1.65 times larger when the weight ratio of ZnO to TiO_2 is 1.0 than when the ratio is 100.[*] Some particle characteristics are the following:

	Diameter (μm)	Density (g ml^{-1})	Charge in water
ZnO	1.0	5.6	Positive
TiO_2	2.2	4.2	Negative

   a. Assuming the particles are uniform spheres, calculate the ratio of the number of particles of ZnO to TiO_2 for each of the weight ratios given.
   b. Propose an explanation for the more voluminous sediment that results when the weight ratio is 1.0 than when it is 100.

7. The following data describe the particle size distribution in a dispersion of copper hydrous oxide particles in water:[†]

d_i (μm)	0.426	0.401	0.376	0.351	0.326	0.301	0.276	0.251	0.226	0.201
n_i	1	0	6	6	17	14	11	12	6	6

Calculate $\bar{d}_n$, σ, $\bar{d}_s$, and $\bar{d}_v$ for this dispersion.

8. A graticule was used to size sand particles and glass spheres.[‡] The percentage by weight of particles less than the stated size was found to be as follows:

d (μm)	0.4	0.8	1.6	2.4	3.0	4.0	8.0	12.0
Sand	0.01	0.07	0.23	0.56	1.23	2.35	11.77	18.06
Glass	0.01	0.11	0.26	0.43	0.72	1.43	17.84	28.07
d (μm)	16.0	20.0	30.0	40.0	60.0	75.0	90.0	120.0
Sand	24.62	32.11	52.33	64.14	83.60	–	98.7	100.0
Glass	36.89	47.68	59.47	61.78	91.49	100.0	–	–

Plot the results on probability and log-probability coordinates. Use the best representation to evaluate (the appropriate) average and standard deviation for the samples (see Appendix C).

9. Particle size distributions were measured on aerosols collected in a New York highway tunnel, and the following results were obtained:[¶]

Cumulative % mass $<d_i$ (μm)	30	40	50	60	70	80
d_i weekend	0.5	1.0	2.50	5.0	–	–
d_i, weekday	–	–	0.07	0.2	0.9	4.0

---

[*]Princen, L. H., and DeVena, M. J. *J. Am. Oil Chem Soc.* **39**, 269 (1962).
[†]McFadgen, P., and Malijević, E. *J. Colloid Interface Sci.* **44**, 97 (1973).
[‡]Fairs, G. L. *Chem. Ind.* **62**, 374 (1943).
[¶]Lee, R. E., Jr., *Science* **178**, 567 (1972).

Plot these results on log-probability coordinates and estimate the mean and standard deviation for each distribution (see Appendix C). What kind of "averages" are these quantities? How do the weekend and weekday particle size distributions compare with respect to the location and width of the maximum? The weekend results are attributed to automobile exhaust, whereas the weekday results are assumed to be "diluted" by aerosols from outside the tunnel.

10. The mass of bull sperm heads in a sample was determined by interference microscopy, and the following results were obtained:[*]

n_i	4	2	27	37	32	26	20	8	3	3	1
$w_i \times 10^{12}$ (g)	5	6	7	8	9	10	11	12	13	14	15

Calculate the number average and weight average weights of these particles.

11. Yau et al.[†] used polystyrene samples of known molecular weight in tetrahydrofuran to calibrate the retention volume of a size-exclusion chromatograph. They obtained the following results:

$M \times 10^{-3}$ (g mole^{-1})	4.0	50.0	110	179	390	1800
V_R (ml)	16.3	14.5	13.9	13.6	13.1	12.1

Do these results display the expected correlation between M and V_R? The flow rate in the instrument used in this research was 1.0 ml min^{-1}; how long would it take for samples of molecular weight 10^4, 10^5, and 10^6 to emerge from the column? Is this order of elution times qualitatively consistent with Equations (26) and (27)?

*Bahr, G. F., and Zeitler, E. *J. Cell Biol*. **21** , 175 (1964).
†Yau, W. W., Jones, M. E., Ginnard, C. R., and Bly, D. D. In *Size Exclusion Chromatography* (T. Provder, Ed.), American Chemical Society, Washington, DC, 1980.

2

Sedimentation and Diffusion
and Their Equilibrium

*Even if you had completed your third year . . . in the University, and were perfect in the
theory of the subject, you would still find that there was need of many years of experience,
before you could move in a fashionable crowd without jostling against your betters.*

From Abbott's *Flatland*

2.1 INTRODUCTION

2.1a Sedimentation and Diffusion: Why Are They Important?

Sedimentation and diffusion affect many colloidal phenomena. Sedimentation occurs under
the influence of gravity and, considerably faster, in a centrifuge. Many analytical and preparative techniques in colloid science (particularly in biophysics and biochemistry) take advantage
of centrifugal force for physical characterization or purification of materials in the colloidal
size range. In fact, because of the small sizes of the particles or macromolecules involved, the
use of centrifugation is indispensable. In view of these, basic principles of (gravitational and
centrifugal) sedimentation are an essential part of colloid science.

Diffusion, of molecular species as well as colloidal particles, plays perhaps a more dominant role in many topics of interest to us. For example, without diffusion of ions we will not
have the diffuse electrical double layers next to charged surfaces (discussed in Chapter 11). At
the colloidal level, diffusion plays a central role in the transport and collision of particles in
colloidal stability (discussed in Chapter 13). There are many more such examples.

The equilibrium between gravitational or centrifugal force and diffusion is routinely taken
advantage of in colloid science, as illustrated in Vignette II. Our objective in this chapter is to
examine the effects of sedimentation and diffusion, first taken separately and then combined,
on particles in the colloidal size range.

VIGNETTE II   ANALYTICAL SEPARATIONS AND CHROMATOGRAPHY:
Sedimentation Field Flow Fractionation

Were it not for the never-ending, gentle tussle between gravity and diffusion, our planet
would not have an atmosphere, nor would we be here to reflect on it! The *barometric
equation*, which describes this "balance of power" between the above two well-known phenomena, is derived in most introductory physical chemistry books and is mentioned in the
closing paragraph of this chapter. There are many more life-sustaining processes that are

---

affected by or rely on sedimentation and diffusion, but frequently it is the more mundane "practical" consequences of these phenomena that attract our attention.

One such consequence is their use in the physical characterization of colloidal dispersions and macromolecular solutions. Let us highlight one such application through one element of a class of analytical separation techniques known as *field flow fractionation* (FFF).

The name field flow fractionation stands for a family of techniques, invented in the 1960s, that take advantage of the response of colloids and macromolecules to electrical, thermal, flow, or centrifugal fields to produce a chromatography-type separation of the particles (Giddings 1966, 1991, 1993). In a typical set up, a suitable force field is applied in a direction normal to the axis of a thin chamber that contains the dispersion. The field forces the particles against one of the walls of the chamber, and, at steady state, a concentration profile is set up in the direction of the applied field as a consequence of the differences in the responses of the various species in the dispersion to the applied field (see Fig. 2.1a). The particles are then eluted by flowing an elutant fluid through the chamber. The fluid velocity decreases progressively from the axis toward the accumulation wall because of friction at the wall (see Chapter 4, Section 4a). As a consequence, the particles are carried along the axis at different velocities depending on their distance from the accumulation wall. For example, the component closest to the accumulation wall lags behind the one near the center of the chamber (see Fig. 2.1b). Samples can now be eluted through a detector or collection device. The detection is usually based on changes in standard properties such as refractive index or light absorption.

One of the more advanced of the FFF techniques is sedimentation FFF (SdFFF), in which the applied field is a centrifugal force (see Fig. 2.1b). A typical separation achieved through SdFFF is also illustrated in Figure 2.1b. The SdFFF is suitable for species with molecular weights larger than about 10^6 and has proved useful for a large number of biocolloids (e.g., subcellular particles), polymers, emulsions, and natural and industrial colloids (Giddings 1991).

It is our objective in this chapter to outline the basic concepts that are behind sedimentation and diffusion. As we see in this chapter, gravitational and centrifugal sedimentation are frequently used for particle-size analysis as well as for obtaining measures of solvation and shapes of particles. Diffusion plays a much more prevalent role in numerous aspects of colloid science and is also used in particle-size analysis, as we see in Chapter 5 when we discuss dynamic light scattering. The equilibrium between centrifugation and diffusion is particularly important in analytical and preparative ultracentrifuges.

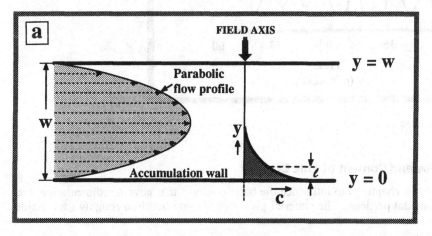

FIG. 2.1 Sedimentation field flow fractionation (SdFFF): (a) an illustration of the concentration profile and elutant velocity profile in an FFF chamber; and (b) a schematic representation of an SdFFF apparatus and of the separation of particles in the flow channel. A typical fractionation obtained through SdFFF using a polydispersed suspension of polystyrene latex spheres is also shown. (Adapted from Giddings 1991.)

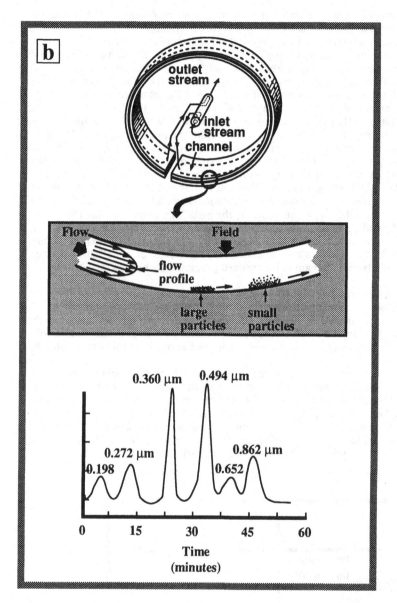

FIG. 2.1 Continued

2.1b Purpose and Content of This Chapter

The purpose of this chapter is to introduce the basic concepts that govern sedimentation and diffusion of colloidal particles. The material presented is organized into roughly four major groups.

1. The first group, consisting of Sections 2.2–2.4, covers sedimentation. After some preliminaries, we discuss Stokes's law, a hydrodynamic equation that will appear again when we discuss electrokinetic phenomena in Chapter 12 and the kinetics of coagulation in Chapter 13. Stokes's law is a key relationship in understanding the rate of sedimentation and is used in the derivation of the sedimentation equation for spherical particles. Following this, the equation for the sedimentation coefficient, a

standard measure of the steady-state velocity of a particle in a centrifugal field, is given and its use is illustrated through an example.

2. The second major topic, diffusion, is covered in Section 2.5 from a classical perspective, through Fick's laws of diffusion and thermodynamic arguments. In discussing diffusion, we focus our attention on the diffusion coefficient, which is defined experimentally by Fick's laws and theoretically by two equations derived by Einstein.

3. The statistical basis of diffusion requires arguments that may be familiar from kinetic molecular theory. Elementary concepts from the theory of random walks and its relation to diffusion form the third topic, which is covered in Section 2.6. As is well known, the random walk statistics can also be used for describing configurational statistics of macromolecules under some simplifying assumptions; this is outlined in Section 2.7.

4. Finally, we turn our attention to equilibrium between sedimentation and diffusion (Section 2.8). As our educated intuition might lead us to expect, larger particles sediment more rapidly and diffuse more slowly than smaller particles. The effects of sedimentation and diffusion, therefore, are not of comparable magnitude for all sizes of particles. There is a range of particle sizes, however, for which the two are comparable and equilibrium between them is established. This equilibrium between sedimentation and diffusion has been studied extensively by means of the ultracentrifuge. Since many of the particles thus investigated are of biological importance, we frequently use protein molecules as our examples in this chapter.

Much of our discussion of sedimentation alone and of sedimentation combined with diffusion focuses on the determination of the mass or molecular weight of the dispersed particles.

2.2 SEDIMENTATION: SOME BASIC CONSIDERATIONS

To see how sedimentation occurs, consider the gravitational forces that operate on a particle of volume V and density ρ_2 that is submerged in a fluid of density ρ_1. The situation is shown in Figure 2.2a for a spherical particle, but the discussion is independent of the actual shape of

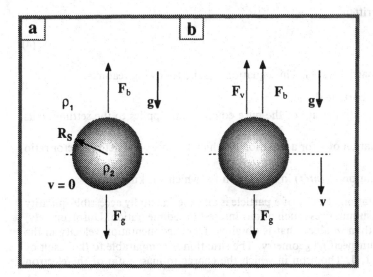

FIG. 2.2 The forces acting on a spherical particle due to gravity and viscosity: (a) due to gravity alone; and (b) due to gravity and the viscosity of the medium ($\rho_2 > \rho_1$).

the particle. The particle experiences a force $\mathbf{F_g}$ due to gravity, taken to be positive in the downward direction. At the same time a buoyant force $\mathbf{F_b}$ acts in the opposite direction. A net force equal to the difference between these forces results in the acceleration of the particle:

$$\mathbf{F_{net}} = \mathbf{F_g} - \mathbf{F_b} = V(\rho_2 - \rho_1)\mathbf{g} \tag{1}$$

This force will pull the particle downward—that is, $\mathbf{F_{net}}$ will have the same sign as $\mathbf{g}$—if $\rho_2 > \rho_1$, and the particle is said to sediment. If, on the other hand, $\rho_1 > \rho_2$, then the particles will move upward, which is called *creaming*.

2.2a   Friction Factor and Stationary Settling Velocity

As the net velocity of the particle is increased, the viscous force $\mathbf{F_v}$ opposing its motion also increases. Soon this force, shown in Figure 2.2b, equals the net driving force responsible for the motion. Once the forces acting on the particle balance, the particle experiences no further acceleration and a stationary state velocity is reached. It may be shown that, under stationary state conditions and for small velocities, the force of resistance is proportional to the stationary state velocity $\mathbf{v}$:

$$\mathbf{F_v} = f\mathbf{v} \tag{2}$$

where the proportionality constant f is called the *friction factor*. The friction factor has the dimensions mass time^{-1}, kg s^{-1} in SI units. We consider some aspects of the proof of Equation (2) for spherical particles in the following section. However, Equation (2) is independent of any particular geometry. The stationary state velocity is positive for sedimentation and negative for creaming.

Since the net force of gravity and the viscous force are equal under stationary state conditions, Equations (1) and (2) may be equated to give

$$V(\rho_2 - \rho_1)\mathbf{g} = f\mathbf{v} \tag{3}$$

The stationary state is quite rapidly achieved, so Equation (3) describes the velocity of a settling particle during much of its fall. This velocity is known as the *stationary settling velocity*.

2.2b   Mass-to-Friction-Factor Ratio

Equation (3) may also be written

$$m \left(1 - \frac{\rho_1}{\rho_2}\right) \mathbf{g} = f\mathbf{v} \tag{4}$$

where m is the mass of the particle ($\rho_2 V$). This equation has the following features:

1. It is independent of particle shape.
2. It assumes that the bulk density of the pure components applies to the settling units (i.e., no solvation).
3. It permits the evaluation of $\mathbf{v}$ for a situation in which the mass-to-friction-factor ratio (m/f) is known.
4. It permits the evaluation of (m/f) in a situation for which v is known.

The stationary sedimentation velocity of a particle is an experimentally accessible quantity for some systems, so item 4 summarizes much of our interest in sedimentation. Unfortunately, it is the ratio (m/f) rather than m alone that is obtained from sedimentation velocity in the general case of particles of unspecified geometry. The situation is comparable to the result of the classical experiment of J. J. Thomson in which the charge-to-mass ratio of the electron was determined.

What is needed is a method for arriving at one of the quantities in the ratio independently. This information, in addition to the ratio, permits the evaluation of both the mass and the friction factor of the particle. In general, it takes two experiments to evaluate numerically these two parameters that characterize the dispersed particles. There are two ways of proceeding to overcome this impasse:

1. The particle may be assumed to be a sphere, in which case its friction factor may be calculated theoretically and thus eliminated from Equation (4). We discuss the friction factor of spherical particles in Section 2.3b.

2. An experiment may be conducted that permits the evaluation of f from measured quantities. Diffusion studies are ideally suited for this purpose, as we see further in this chapter.

2.2c Sensitivity of Sedimentation to Density Differences

Before turning to the two procedures above to eliminate f from the sedimentation equation, one other consideration inherent in the use of Equation (4) should be discussed. As noted above, Equation (4) uses the bulk density of the pure components. If the continuous and bulk phases are totally noninteracting, this may be justified. However, for aggregates or solvated lyophilic particles, the density of the settling unit is intermediate between the densities of the two pure components. In these cases, choosing an appropriate density for the settling particle can be a real problem.

Equation (3) shows that the sedimentation velocity increases with the density difference between the particle and the medium. Any situation that brings the density of the settling unit closer to that of the solvent will decrease the sedimentation velocity. To an observer who is unaware of its derivation, however, the smaller velocity would be interpreted by Equation (4) as indicating a smaller value of (m/f). Since the actual mass of colloidal material is unaffected by the solvation, it is more correct to attribute the reduced sedimentation velocity to an increase in the value of the friction factor.

In the next section, we see how to deal quantitatively with solvation. Until now, the friction factor has been merely a proportionality factor of rather ill-defined origin. We shall not undertake a derivation of Equation (2) in any general sense. In the next section, however, we outline the derivation of an important result due to G. G. Stokes—the friction factor for an unsolvated sphere.

2.3 GRAVITATIONAL SEDIMENTATION

2.3a Stokes's Law for Spheres

We begin our discussion of the Stokes equation by considering a single spherical particle of radius R_s in a state of relative motion with respect to the surrounding fluid. For the purposes of this derivation, it does not matter whether the fluid is stationary and the particle is moving through it or whether the particle is stationary with the fluid moving around it. Although the former describes sedimentation, it is somewhat more convenient to discuss the phenomenon in terms of the fluid flowing in the $+z$ direction with a velocity v_z; this is the same as the particle moving in the $-z$ direction with a velocity of the same magnitude. This situation is represented schematically in Figure 2.3.

If the spherical particle were not present in Figure 2.3, the volume elements of the flowing fluid would move upward in straight lines. In the presence of the particle, however, the flow profile is distorted around the sphere in the manner suggested by Figure 2.3. It is apparent that the velocity of any volume element passing the sphere is a function of both time and location and must be described as such in any quantitative treatment. The trajectory of such a volume element is called the flow *streamline* function. For spherical particles, this was analyzed by G. G. Stokes in 1850.

The streamline function is the solution of a differential equation, the details of which we will not pursue. We will have more to say about flow streamlines in our discussion of viscosity in Chapter 4. For the present, however, it is sufficient to note that an important part of

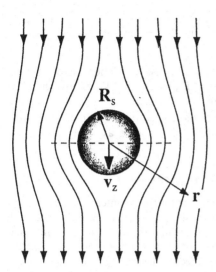

FIG. 2.3   Distortion of flow streamlines around a spherical particle.

solving any differential equation is the adequate incorporation of the boundary conditions that describe the problem. Since the disturbance of the flow streamlines in Figure 2.3 is centered at the particle, it is convenient to discuss the boundary conditions in terms of the spatial variable r originating at the center of the sphere. Several aspects of the problem can be described in terms of this variable.

1.   The thickness of each successive layer in the fluid is the infinitesimal mathematical increment dr. Since the streamlines roughly follow the outline of the particle and are thin compared to the radius of the sphere R_s, we can think of each flow layer as moving tangentially to the surface of the sphere.

2.   The velocity varies from layer to layer in the fluid near the surface of the sphere; this variation can be described in terms of a velocity gradient dv/dr. Any integration over this gradient must be able to assign two different values of velocity to two different r values.

3.   When $r = \infty$, the disturbing influence of the particle has been damped away and the streamlines behave as if the particle was not present.

4.   When $r = R_s$, the velocity is zero. This is described as a *nonslip* (or *no-slip*) condition between the stationary surface of the particle and the layer of fluid adjacent to it. There is nothing particularly self-evident about the nonslip condition between the solid and the fluid, but it is an experimental fact. Some commonplace evidence that suggests this is the layer of dust that accumulates on the blades of a fan. However stiff the breeze may be some distance in front of the blades, the air is still and travels with the surface of the blades.

The fact that the velocity of a fluid changes from layer to layer is evidence of a kind of friction between these layers. The layers are mathematical constructs, but the velocity gradient is real and a characteristic of the fluid. The property of a fluid that describes the internal friction or resistance to flow is the viscosity of the material. Chapter 4 is devoted to a discussion of the measurement and interpretation of viscosity. For now, it is enough for us to recall that this property is quantified by the coefficient of viscosity η of a material. The coefficient of viscosity has dimensions of mass length^{-1} time^{-1}, kg m^{-1}s^{-1} in SI units. In actual practice, the cgs unit of viscosity, the poise (P), is widely used. Note that pure water at 20°C has a viscosity of about 0.01 P $= 10^{-3}$ kg m^{-1} s^{-1}

Because of internal friction, some of the translational kinetic energy of flow dissipates over a period of time. For the physical situation pictured in Figure 2.3, Stokes was able to show that the rate of energy dissipation dE/dt is given by

$$dE/dt = 6\pi\eta v_z^2 R_s \tag{5}$$

The product of the viscous force F_v exerted by the particle on the fluid and the velocity v_z of the fluid also equals the rate at which work is done on the fluid by the particle. Remember, a force times a distance equals energy, so a force times a velocity equals the rate of energy dissipation. Thus, we have two different ways of expressing the rate of energy dissipation associated with the presence of a spherical particle in the flowing liquid. Equating these two expressions gives

$$F_v v_z = 6\pi\eta v_z^2 R_s \tag{6}$$

It is clear that the force acting on the fluid acts in opposition to the flow. Therefore, if the fluid is flowing in the positive direction, the sphere resists the flow by a force in the negative direction. Conversely, if the particle is moving in one direction, the fluid resists the movement by a force in the opposite direction. This force is given by *Stokes's law*,

$$\mathbf{F} = 6\pi\eta R_s \mathbf{v} \tag{7}$$

where the subscripts for $\mathbf{F}$ and $\mathbf{v}$ are no longer needed, provided we remember that $\mathbf{F}$ and $\mathbf{v}$ are in opposing directions. (Note that $\mathbf{F}$ and $\mathbf{v}$ are vectors and, therefore, a direction as well as a magnitude must be specified for each.)*

2.3b Sedimentation Equation

Equation (7) expresses, for a spherical particle, the general result presented as Equation (2). Two aspects of the result are important:

1. For small, stationary-state velocities, the viscous force on a particle is directly proportional to the velocity.

2. For a spherical particle, the friction factor is given by

$$f = 6\pi\eta R_s \tag{8}$$

It was just noted that the case of spherical particles is one in which the friction factor can be eliminated from Equation (4), yielding a result that permits the mass of a spherical particle to be evaluated from sedimentation data alone. To see how this works, we return to Equation (3), using Equation (8) to evaluate f and $(4/3)\pi R_s^3$ as a substitution for V. Thus, for a spherical particle, Equation (3) becomes

$$\frac{4}{3} \pi R_s^3 (\rho_2 - \rho_1) \mathbf{g} = 6\pi\eta R_s \mathbf{v} \tag{9}$$

Equation (9) may be solved for the stationary-state sedimentation velocity

$$\mathbf{v} = \frac{2}{9} \frac{R_s^2(\rho_2 - \rho_1)\mathbf{g}}{\eta} \tag{10}$$

and for the radius of the spherical particle

$$R_s = \left(\frac{9\eta v}{2(\rho_2 - \rho_1)g} \right)^{1/2} \tag{11}$$

Since the particle to which it applies is a sphere, Equation (11) may also be used to calculate the mass and friction factor of the particle:

---

*Vectors are denoted by bold letters and their magnitudes by corresponding roman letters. Such a differentiation is not used when only the magnitudes are important.

$$m = \rho_2 \frac{4}{3} \pi R_s^3 = \frac{4}{3} \pi \rho_2 \left(\frac{9\eta v}{2(\rho_2 - \rho_1)g} \right)^{3/2} \tag{12}$$

and

$$f = 6\pi\eta \left(\frac{9\eta v}{2(\rho_2 - \rho_1)g} \right)^{1/2} \tag{13}$$

Equation (9) is an important result since it describes the relationship among R_s, v, η, and $\Delta\rho$, the density difference. Any one of these quantities may be evaluated by Equation (9) when the other three are known. Thus, Equation (9) can be used to determine the density difference between two phases or to determine the viscosity of a liquid. In this chapter, however, our interest is in the characterization of colloidal particles by means of observations of their sedimentation behavior. Therefore, we are primarily concerned with Equations (11) and (12), which are specifically directed toward this objective.

2.3c   Effects of Nonsphericity and Solvation

Stokes's law and the equations developed from it apply to spherical particles only, but the dispersed units in systems of actual interest often fail to meet this shape requirement. Equation (12) is sometimes used in these cases anyway. The lack of compliance of the system to the model is acknowledged by labeling the mass, calculated by Equation (12), as the mass of an "equivalent sphere." As the name implies, this is a fictitious particle with the same density as the unsolvated particle that settles with the same velocity as the experimental system. If the actual settling particle is an unsolvated polyhedron, the equivalent sphere may be a fairly good model for it, and the mass of the equivalent sphere may be a reasonable approximation to the actual mass of the particle. The approximation clearly becomes poorer if the particle is asymmetrical, solvated, or both. Characterization of dispersed particles by their mass as equivalent spheres at least has the advantage of requiring only one experimental observation, the sedimentation rate, of the system. We see in sections below that the equivalent sphere calculations still play a useful role, even in systems for which supplementary diffusion studies have also been conducted.

Actual particles can deviate from the Stokes model by being either solvated, asymmetrical, or both, in which case f increases so that a particle of mass m will display a smaller sedimentation velocity than it would if it were an unsolvated sphere. The justification for this statement is based on Figure 2.4, which shows an unsolvated sphere and represents its radius as R_o. If this particle were solvated, it would swell to a larger radius R_s associated with a larger friction factor. Small asymmetrical particles rotate through every possible orientation during settling because of Brownian motion. Figure 2.4 shows that an encompassing sphere drawn around the tumbling particle will also have a radius larger than R_o.

An increase in f is associated with this situation as well. To deal with these ideas quantitatively, it is convenient to consider the following ratios:

$$f/f_0 = (f/f^*)(f^*/f_0) \tag{14}$$

1. In Equation (14), f is the friction factor of the actual particle.
2. The f_0 is the friction factor for an unsolvated sphere given by Stokes's law to equal $6\pi\eta R_o$ in the present notation. This is the lowest friction factor possible for a particle with the required mass.
3. The ratio f/f_0 measures the amount by which the actual friction factor exceeds the minimum value.
4. The f^* is the friction factor for a spherical particle having the same volume as the *solvated* particle of mass m.
5. The ratio f^*/f_0 measures the increase in f due to solvation.
6. The ratio f/f^* measures the increase in f due to asymmetry.

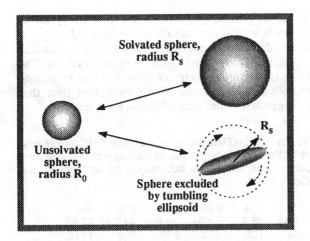

FIG. 2.4 The effects of solvation and asymmetry on the effective radius of a particle. The figure illustrates how solvation and asymmetry have qualitatively equivalent effects in increasing the effective radius of an unsolvated sphere of radius R_o. (Redrawn with permission of P. C. Hiemenz *Polymer Chemistry: The Basic Concepts*, Marcel Dekker, New York, 1984.)

We see presently that both of the ratios on the right-hand side of Equation (14) can be expanded in terms of appropriate models, so the experimental f value—compared to f_0 for a particle of the same mass—can be analyzed in terms of solvation and asymmetry.

Prior to this, however, let us consider how the mass of an equivalent sphere can be extracted from experimental data since this quantity is an important concept in its own right.

2.3d Sedimentation Measurements

We begin our discussion of the experimental aspects of sedimentation by considering settling caused by gravity; in Section 2.4, we examine centrifugation. Substitution of numerical values into Equation (10) reveals that the range of particle sizes that can be studied by sedimentation under gravity is quite limited. For example, taking $\Delta\rho/\eta = 100 \text{ g cm}^{-3} \text{ P}^{-1}$ yields values of v between $2.2 \cdot 10^{-12}$ and $2.2 \cdot 10^{-6} \text{ m s}^{-1}$ for spherical particles of R_s between 1.0 nm and 1.0 μm. This value of $\Delta\rho/\eta$ corresponds physically to sulfur particles in water at 20°C, in which $\Delta\rho$ and v are both positive, and to gas bubbles in water, in which $\Delta\rho$ and v are negative. Although sedimentation under gravity is feasible only at the upper size limit of the usual colloidal range, there are still a number of important applications of the technique. We consider only two of the procedures that have been developed for such studies.

If a particle is large enough to be visible to the unaided eye or in a traveling microscope, its velocity can be measured directly; however, smaller particles present more of a problem. If a dispersion consists of particles of uniform size, all will settle at the same rate, so their positions relative to each other do not change until they reach the bottom of the container. This means that a sharp boundary will exist between the domain occupied by the settling particles and that part of the system already swept free of particles. Even though the individual particles may not be visible, this boundary may be visible if the particles differ sufficiently in color or refractive index from the continuous phase. The velocity of the boundary and the velocity of the particles are the same. If the particles differ in size, no such boundary will develop. Each size fraction in a polydisperse system will settle at its own characteristic velocity, so that at any given time one end of the column may be free of some size fractions but not of others. The "boundary zone" between the domain that clearly contains settling particles and the domain that is totally free of particles will appear to be a diffuse region across which the concentration of the disperse phase gradually diminishes. Since no sharp boundary develops, some other method must be devised to follow sedimentation velocity. Clearly, we would like as much information as possible about the particle size distribution since the system is polydisperse. The following procedure shows how this information may be obtained.

Suppose the pan of a balance is positioned at some convenient location below the surface of a dispersion. As sedimentation occurs, the settled material collects on the balance pan. The total weight W of the material on the pan is measured at various times t, either by adding counterweights to a second pan or by noting the displacement of a calibrated fiber that supports the pan. The following example suggests how this kind of data is analyzed.

* * *

EXAMPLE 2.1 *Analyzing Cumulative Sedimentation Data for Most Probable Settling Velocity.* The following data show—as a function of time—the weight (as percentage of total) of suspended clay particles W, which has accumulated on a plate submerged 20 cm beneath the surface in a sedimentation experiment (Oden 1915).

Time (s)	34	57	115	175	254	294	337	360	406
W (%)	7.42	11.13	16.7	21.44	28.7	33.04	38.85	41.75	47.55
Time (s)	458	503	558	648	705	773	862	982	1112
W (%)	53.35	57.7	62.05	67.85	70.75	73.65	76.55	79.45	82.35
Time (s)	1245	1318							
W(%)	83.8	84.52							

Criticize or defend the following proposition: The data give the time required for particles to fall 20 cm, making it easy to convert time to sedimentation velocity for each point. Equation (11) may then be used to convert the velocity into the radius of an equivalent sphere. The resulting graph of W versus radius is a cumulative distribution function similar to that shown in Figure 1.18b.

Solution: The proposition is on the right track, but needs refinement to be quantitatively correct. Remember, at any given time the weight of material that has accumulated equals the weight of all particles large enough to have settled onto the pan in the time of the experiment. Such particles come from two categories; they include all those particles large enough to have fallen through the full length of the container and a fraction of smaller particles that have only fallen from a fraction of the column but still land on the pan. Both of these contribute to the total weight observed at any time.

The two contributions can be calculated as follows: The quantity dW/dt at any instant t is the slope of the cumulative curve at a particular point. It represents the rate of deposition of particles in a size range that has not already settled out completely in the bottom during the time t. Since this size range has been settling at the same rate dW/dt from the beginning of the experiment (i.e., $t = 0$), the weight of the particles in this size range that has collected on the pan until time t is given by $t(dW/dt)$. Therefore, the remainder of the particles on the pan, that is, $W - [t(dW/dt)]$, represents the weight w corresponding to larger particles that have already settled out by time t. That is, the total weight W at any time t can be written as

$$W = w + t\frac{dW}{dt}$$

The weight w represents the *cutoff size* corresponding to time t, and dW/dt represents the rate of settling of particles smaller than the above cutoff size.

Figure 2.5 shows this graphically; note that the intercepts of the tangent lines, drawn at different times, give w, the weight of particles in the size class associated with time under consideration. It is a plot of this last quantity against time that gives a representation similar to Figure 1.18b.

In this example, about one-third of the solids are coarse enough to have settled out in 5 minutes. The remaining particles have velocities of $6.7 \cdot 10^{-4}$ m s^{-1} ($= 20 \cdot 10^{-2}$ m/300 s) or less, corresponding to equivalent spheres with radii of 12 μm or less. Incidentally, the equivalent sphere is a poor model for these clay particles, as can be seen from the electron micrographs in Figure 1.12b.   ■

* * *

A slight variation of this method for analyzing particle size distributions is particularly convenient if the system contains particles of several discretely different particle sizes rather than a

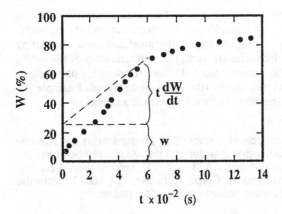

FIG. 2.5 Sedimentation of clay particles. Cumulative weight (as percentage) versus time for the data given in Example 2.1. The figure also shows the graphical construction explained in the example for obtaining the cutoff weight w associated with any time t. (Data from Oden, 1915.)

continuous distribution of sizes. A thin band of such a dispersion is layered on top of a column of pure solvent. Particles in the different size categories will settle through the solvent at different rates so that, if the column is long enough, the first fraction collected at the bottom of the column will contain only the largest particles, with smaller sizes showing up in progressively later fractions. The clear solvent is literally the column within which the various particle sizes are resolved. The "resolving power" of such an arrangement increases with the length of the column, at least up to some optimum length. It might be noted that there is a certain formal similarity between this situation and column chromatography.

Provided the column is long enough to separate the different sizes of particles, this method gives w versus time directly. Any additional interpretations of the results are made in a manner entirely analogous to the one just described. Since only a narrow band of the dispersion is used in this method, the weight of the dispersed phase in each fraction will be relatively low. Fairly sensitive analytical techniques are required for the fractions collected.

Sedimentation runs should be conducted at a constant temperature, not only so that $\Delta\rho$ and η are known, but also to minimize disturbances due to convection. Any sort of disturbance will obviously disrupt the segregation of the particles by size that has occurred as a result of sedimentation. An intrinsic difficulty with the balance method lies in the fact that the liquid below the balance pan is less dense than the liquid with dispersed particles above the pan. Thus, there is a tendency for a counterflow of pure solvent to arise, which would introduce an error in the particle size analysis.

The methods just discussed are only two of a wide variety of techniques that provide essentially the same kinds of information. In general, any measurement that gives (a) the amount of suspended material a fixed distance below the surface at various times or (b) the amount of material at various depths at any one time can be interpreted in terms of particle size distribution. Pressure, density, and absorbance are additional measurements that have been analyzed this way.

Instruments are commercially available that automatically carry out the analysis presented above. Although much of the tedium of the experiment is thereby relieved, automation does not alter the basic assumptions and/or approximations of the method. In one commercial instrument, a collimated beam of light or low-energy x-rays scans the settling compartment, measuring the absorbance of the suspended particles at various depths. Because the detector does not have to wait for settling particles to arrive at a fixed position, this method is more rapid than direct observation of sedimentation. A built-in computer converts the absorbance at a particular distance below the top of the sample to concentration after some time has elapsed from the start of settling. The distance-time combination is reduced to a velocity, which is converted into the radius of an equivalent sphere by Equation (11). All of this is calculated automatically, and a recorder plots the cumulative weight percent of settling material versus the equivalent radius. The sample compartment is thermostated, and suspending media with a range of known densities and viscosities are used; the characteristics of the media are part of the input data of the operation. By utilizing the full range of operating parameters, this instrument allows equivalent diameters from 0.1 to 100 μm to be analyzed. Example 2.2 considers the application of Stokes's law to the output of such a particle analyzer.

* * *

EXAMPLE 2.2 *Use of the Sedimentation Equation for Particle Size Determination.* A titanium dioxide pigment of density 4.12 g cm^{-3} is suspended in water at 33°C. At this temperature, the density and viscosity of water are 0.9947 g cm^{-3} and 7.523 · 10^{-3} P, respectively. A particle size analyzer (SediGraph, Micromeritics Instruments Corp., Norcross, GA 30093) plots the following data for cumulative weight percent versus equivalent spherical radius:

Cumulative wt %

| 95 | 90 | 80 | 70 | 60 | 50 | 40 | 30 | 20 | 10 | 5 |

Radius of an equivalent sphere (μm)

| 0.60 | 0.45 | 0.33 | 0.37 | 0.28 | 0.27 | 0.24 | 0.22 | 0.21 | 0.18 | 0.15 |

To what sedimentation velocity does the most abundant component in this distribution correspond?

Solution: These cumulative percentages are of the same form as in Figure 2.5; therefore, the particle size distribution peaks where the cumulative curve increases most steeply. This occurs at about 0.29 μm for this distribution. Equation (10) permits the sedimentation velocity for particles of this size to be calculated:

$$v = \frac{2}{9} \frac{R_s^2(\rho_2 - \rho_1)g}{\eta} = \frac{2(2.9 \cdot 10^{-5})^2 (4.12 - 0.99) \, 980}{9(7.52 \cdot 10^{-3})}$$
$$= 7.62 \cdot 10^{-5} \, cm \, s^{-1} = 7.62 \cdot 10^{-7} \, m \, s^{-1}$$

The distance-time combination corresponding to this weight percent is determined by the particle size analyzer and converted to R_s. ∎

* * *

A basic limitation of all these methods is the narrow range of particle sizes that can be investigated by sedimentation under gravity. Therefore, we turn next to a consideration of centrifugation, particularly the ultracentrifuge, as a means of extending the applicability of sedimentation measurements.

2.4 CENTRIFUGAL SEDIMENTATION

A well-known fact from elementary physics is that a particle traveling in a circular path of radius r at an angular velocity ω (in radians per second) is subject to an acceleration in the radial direction equal to $\omega^2 r$. Since there are 2π radians per revolution, ω equals the number of revolutions per second (rps) times 2π, or the number of revolutions per minute (rpm) times the ratio $2\pi/60$. It is not particularly difficult to produce accelerations by centrifugation that are many times larger than the acceleration due to gravity. It is conventional, in fact, to describe the radial acceleration achieved by a centrifuge as some multiple of the standard gravitational acceleration or as being so many g's, where g is about 9.8 m s^{-2}.

2.4a Ultracentrifuge

The ultracentrifuge is an instrument in which a cell is rotated at very high speeds in a horizontal position. As we see below, the gravitational acceleration is easily increased by a factor of 10^5 in such an apparatus. Accordingly, the particle size that may be studied by sedimentation is decreased by the same factor. Two types of ultracentrifuges, analytical and preparative, are common, particularly in biological analyses. The analytical ultracentrifuges require very small amounts of samples (about 0.1 to 1 ml) and use optical systems designed to measure concentration or concentration gradients in the samples directly. Preparative ultracentrifuges typically use larger volumes of samples (5 to 100 ml) and are run in "batch mode," that is, the contents are removed from the centrifuge after a fixed amount of time, the purpose being fractionation for preparation or purification of samples. The ultracentrifuge has been used extensively for the characterization of colloidal materials, particularly those of biological origin, such as proteins, nucleic acids, and viruses (Cantor and Schimmel 1980).

Because of the extreme importance of the ultracentrifuge, it seems appropriate to describe it in some detail. Although the particulars differ from instrument to instrument, the essential features are present in all ultracentrifuges. The actual sedimentation takes place in a cell mounted within an aluminum or titanium rotor. The cell is sector shaped; its side walls converge toward the center along radial lines. Since the radial acceleration is proportional to the distance from the axis of rotation, we see that this quantity varies from top to bottom in the cell. Although this variation is considered explicitly in a section below, it is sufficient for the present to consider the average acceleration at the midpoint of the cell, which is typically located $6.5 \cdot 10^{-2}$ m from the center of the rotor. For speeds of 10,000, 20,000, and 40,000 rpm, accelerations of

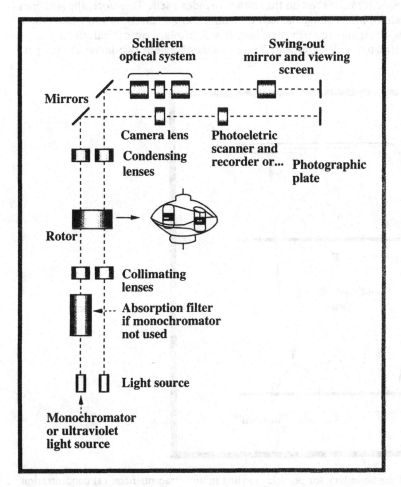

FIG. 2.6 Schematic of optical systems in the Spinco Model E Ultracentrifuge (Beckman Instruments, Inc., Spinco Division, Palo Alto, CA).

$7.13 \cdot 10^4$, $2.85 \cdot 10^5$, and $1.14 \cdot 10^6$ m s^{-2} are produced at this radius. These correspond to $7.27 \cdot 10^3$, $2.91 \cdot 10^4$, and $1.16 \cdot 10^5$ times the acceleration of gravity, respectively.

An important part of the ultracentrifuge is the optical system that makes observation during operation a possibility; it is shown schematically in Figure 2.6. The sample compartment fits into a hole in the rotor and is positioned to intersect the light path of the optics. Those faces of the cell perpendicular to the light path are transparent to visible and ultraviolet light so that various optical methods of chemical analysis may be employed to measure the distribution of material along the sedimentation path. Schlieren refractometry, interferometry, and spectrophotometry are all employed for this purpose, the last being particularly useful when very low concentrations are involved. The location of the settling particles may be followed by one of these methods, and the results recorded either photographically or on a chart recorder. Figure 2.7a shows how the concentration profile varies with radial location in the ultracentrifuge as the particles settle. Although the boundary between the solution and the solvent that has been swept free of solute seems clear in this representation, the use of schlieren refractometry makes the sedimentation even more easy to visualize.

The schlieren system of optics is an analytical method that is particularly well suited to following the location of a chemical boundary with time. It is routinely employed in ultracentrifuges and also in electrophoresis experiments, as we see in Chapter 12. Schlieren optics produces an effect that depends on the way the refractive index varies with position, that is, the refractive index *gradient* rather than on the refractive index itself. Therefore, the schlieren effect is the same at all locations along the axis of sedimentation, except at any place where the refractive index is changing. In such a region, it will produce an optical effect that is proportional to the refractive index gradient. The boundary between two layers is thus per-

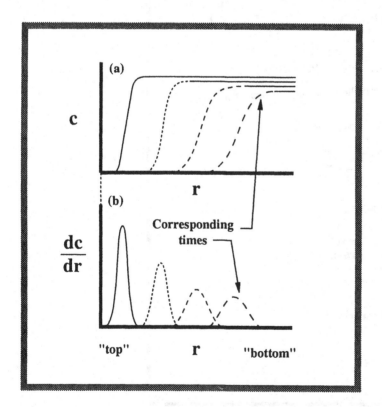

FIG. 2.7 Location of the boundary for particles settling in an ultracentrifuge: (a) concentration profile; and (b) derivative profile as revealed by schlieren optics. (Redrawn with permission of P. C. Hiemenz, *Polymer Chemistry: The Basic Concepts*, Marcel Dekker, New York, 1984.)

ceived by schlieren optics as a sharp peak on a flat baseline. The displacement of such a peak with time measures the velocity of the boundary. Likewise, a band of settling particles is seen as a broad schlieren peak with a width and velocity that measure the width and velocity of the band. The schlieren optical system works by using a diaphragm to cut off from a photographic plate the light that is deviated from the optical path by the refractive index gradient. The same effect may be produced on a ground-glass screen for instantaneous viewing by an ingenious system that combines a diagonal slit and a cylindrical lens to produce an image of the refractive index gradient. Figure 2.7b shows how the same data presented in Figure 2.7a would appear with schlieren optics. It is thus an easy matter to monitor the peak location as the sedimentation proceeds.

The usual precautions regarding temperature and vibration control also apply to the ultracentrifuge. To overcome air resistance and the attendant frictional heating, the compartment in which the rotor spins is evacuated and may be thermostated over a wide range of temperatures. The rotor is mounted on a flexible drive shaft that minimizes the need for precise balancing as a requirement for vibration-free operation. Finally, the rotor assembly is enclosed in an armored steel chamber for safety. At these speeds, a runaway rotor is deadly!

2.4b Sedimentation Coefficient

The results of a sedimentation experiment in a centrifugal field are conventionally reported in terms of what is known as a *sedimentation coefficient*. This quantity is defined as

$$s = \frac{dr/dt}{\omega^2 r} \tag{15}$$

that is, it equals the sedimentation velocity per unit of centrifugal acceleration. In the SI system, this ratio has the units $m\ s^{-1}/m\ s^{-2}$ or seconds. In practice, the quantity 10^{-13} s is defined to be 1.0 *svedberg* (1.0 S) after T. Svedberg, the originator of the ultracentrifuge and pioneer in its use (Nobel Prize, 1926). Sedimentation coefficients are usually reported in this unit. If the location of a particle along its settling path is measured as a function of time, the sedimentation coefficient is readily evaluated by integrating Equation (15) using the following limits of integration. The component is at radial position r_1 at time t_1 and at r_2 at t_2. Therefore,

$$\omega^2 s \int dt = \int \frac{dr}{r} \tag{16}$$

or

$$s = \frac{\ln (r_2/r_1)}{\omega^2 (t_2 - t_1)} \tag{17}$$

The sedimentation coefficient depends on concentration; consequently, it is usually measured at several different concentrations, and the results are extrapolated to zero concentration. It is customary to designate this limiting value by a superscript zero. Experimental values are also generally labeled with respect to temperature, so a value listed as s_{20}^0 corresponds to a sedimentation coefficient measured at (or corrected to) 20°C and extrapolated to zero concentration. Under stationary state conditions, the force due to the centrifugal field and the viscous force of resistance will be equal. Therefore, $\omega^2 r$ replaces g in Equation (4) to give

$$m\left(1 - \frac{\rho_1}{\rho_2}\right) \omega^2 r = f \frac{dr}{dt} \tag{18}$$

where f is the friction factor for the settling particle. Equations (18) and (15) may be combined to yield

$$\frac{m}{f}\left(1 - \frac{\rho_1}{\rho_2}\right) = s \tag{19}$$

which shows that the sedimentation coefficient is directly proportional to the ratio of the mass to the friction factor. As with sedimentation under the acceleration of gravity, any further interpretation of m/f depends either on independent determination of the friction factor from diffusion or on the assumption of spherical particles, with Equation (8) used to evaluate f. Thus, experimental sedimentation coefficients may be analyzed to yield the mass, radius, and friction factor of an equivalent sphere. In the absence of supplementary data, this is as far as sedimentation alone can be interpreted. Example 2.3 illustrates the use of these relationships.

* * *

EXAMPLE 2.3 *Sedimentation in an Ultracentrifuge.*   What should be the speed of an ultracentrifuge so that the boundary associated with the sedimentation of a particle of molecular weight 60,000 g mole^{-1} moves from $r_1 = 6.314$ cm to $r_2 = 6.367$ cm in 10 min? The densities of the particle and the medium are 0.728 and 0.998 g cm^{-3}, respectively, and the friction factor of the molecule is $5.3 \cdot 10^{-11}$ kg s^{-1}.

Solution: First we calculate the particle mass by dividing the molecular weight by Avogadro's number:

$$m = 60,000/6.02 \cdot 10^{23} = 9.97 \cdot 10^{-20} \text{ g molecule}^{-1}$$
$$= 9.97 \cdot 10^{-23} \text{kg molecule}^{-1}$$

Equation (19) allows us to calculate the sedimentation coefficient:

$$s = (m/f) [1 - (\rho_1/\rho_2)] = (9.97 \cdot 10^{-23}/5.3 \cdot 10^{-11}) [1 - (0.998/0.728)]$$
$$= 6.98 \cdot 10^{-13} s = 6.98 \text{ S}$$

Equation (17) can now be solved for ω^2

$$\omega^2 = \frac{\ln(r_2/r_1)}{s(t_2 - t_1)} = 1.997 \cdot 10^7 s^{-2}$$

or $\omega = 4.47 \cdot 10^3$ rad s^{-1}. Dividing by 2π converts this to revolutions per second: $\omega = 711$ revolutions per second, or 42,700 rpm.   ∎

* * *

In the preceding sections of this chapter, we have considered sedimentation as if it were the only process that influenced the spatial distribution of particles. If this were the case, all systems of dispersed particles, even gases, would eventually settle out. In practice, convection currents arising from temperature differences keep many systems well stirred. Even in carefully thermostated laboratory samples, however, there is another factor operating that prevents the complete sedimentation of small particles, namely, *diffusion*.

Diffusion and sedimentation are opposing processes inasmuch as the former tends to keep things dispersed, whereas the latter tends to collect them in one place. Diffusion is more important for smaller particles: remember that sedimentation is negligible for gases. For larger particles, diffusion is negligible. Of course, there is a range of particle sizes for which both effects are comparable; we examine the combined effects of sedimentation and diffusion in Section 2.8. Even for very small particles, the balance between diffusion and sedimentation becomes important if "sedimentation" is induced through centrifugal forces, as in the case of ultracentrifuges or separation techniques such as the sedimentation field flow fractionation described in Vignette II. For the present, however, let us examine the phenomenon of diffusion.

2.5   DIFFUSION

If external forces such as gravity can be neglected, the composition of a single equilibrium phase will be macroscopically uniform throughout. This means that the concentration or density is constant throughout the phase. It should be noted that we are talking about the macroscopic description of the phase. At the molecular level, there will be local fluctuations from the mean value, a fact important in, for example, light scattering, as we see in Chapter 5 when we discuss Rayleigh scattering from solutions (Section 5.3b). However, for the present,

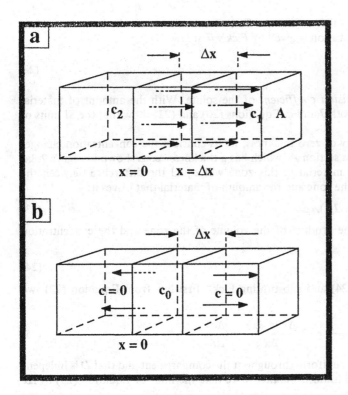

FIG. 2.8 Schematic of a diffusion process through a volume element of thickness Δx: (a) the volume element separates two solutions of concentrations c_1 and c_2; and (b) the element contains a solution at concentration c_0 and separates two regions of pure solvents.

we restrict ourselves to the macroscopic description. Fundamentally, it is the second law of thermodynamics that is responsible for the uniform distribution of matter at equilibrium since entropy is maximum when the molecules are distributed randomly throughout the space available to them. If for some reason a nonuniform distribution of matter should exist, the particles of the system will experience a force that tends to distribute them uniformly. Consider, for example, the situations shown in Figure 2.8a, in which two solutions of different concentrations are separated by a hypothetical porous plug of thickness dx. We can set $dx = 0$; in this case, there will be a migration of solute from the high-concentration side to the low-concentration side until a uniform distribution is obtained, that is, equilibrium is reached. The resulting transport of matter is called *diffusion*. In the following sections, we set up the equations that describe the diffusional flux and the consequent readjustment of concentration and will develop an equation that relates the diffusion coefficient to the friction coefficient.

2.5a Fick's Laws of Diffusion

Now let us define the following quantities. Suppose Q represents the amount of material that flows through a cross section of area A in Figure 2.8. The rate of change in the quantity Q/A is called the flux of solute across the boundary; that is, the flux J is defined as

$$J = \frac{d(Q/A)}{dt} \tag{20}$$

Therefore, the amount of material that crosses A in a time interval Δt is

$$\Delta Q = AJ\Delta t \tag{21}$$

The phenomenological equation that relates the flux of material across the boundary to the concentration gradient at that location is given by *Fick's first law*:

$$J = -D\frac{\partial c}{\partial x} \tag{22}$$

where D is defined as the *diffusion coefficient* of the solute. With the amount of material measured in the same units in both J and c, Equations (20) and (22) show that the SI units of D are square meters per second.

Now, instead of a boundary of zero thickness, let us consider the concentration changes that occur within a zone of cross section A, which has a thickness Δx as shown in Figure 2.8a. Any change in the amount of material in this zone will equal the difference between the amount of material that enters the zone and the amount of material that leaves it:

$$\Delta Q = Q_{in} - Q_{out} = (J_{in} - J_{out})A\Delta t \tag{23}$$

The quantity ΔQ also equals the product of the volume of the zone and the concentration change that occurs in it:

$$\Delta Q = A\Delta x\Delta c \tag{24}$$

Equating Equations (23) and (24) and substituting Fick's first law from Equation (22), we obtain

$$-D\frac{[(\partial c/\partial x)_{x=0} - (\partial c/\partial x)_{x=\Delta x}]}{\Delta x} = D\frac{\Delta(\partial c/\partial x)}{\Delta x} = \frac{\Delta c}{\Delta t} \tag{25}$$

assuming that the cross section is uniform throughout the compartment and that D is independent of small changes of concentration. Finally, if we take the limit of Equation (25) as Δx and Δt approach zero, we obtain

$$\frac{\partial c}{\partial t} = D\frac{\partial^2 c}{\partial x^2} \tag{26}$$

a result known as *Fick's second law*.

Equation (26) is a differential equation with a solution that describes the concentration of a system as a function of time and position. The solution depends on the boundary conditions of the problem as well as on the parameter D. This is the basis for the experimental determination of the diffusion coefficient. Equation (26) is solved for the boundary conditions that apply to a particular experimental arrangement. Then, the concentration of the diffusing substance is measured as a function of time and location in the apparatus. Fitting the experimental data to the theoretical concentration function permits the evaluation of the diffusion coefficient for the system under consideration.

Rather than getting deeply involved in the mathematics of differential equations, we use a statistical model to find a solution to Equation (26) for a system with simple boundary conditions. This will be sufficient to illustrate the experimental technique by which diffusion coefficients are determined and will also lead to a better understanding of the random processes underlying diffusion. This statistical discussion and the experimental procedure it suggests are addressed in Sections 2.6c and 2.6d.

Next, it will be helpful to anticipate a description of experimental procedures and consider the magnitude of measured diffusion coefficients. The self-diffusion coefficients for ordinary liquids with small molecules are of the order of magnitude 10^{-9} m^2 s^{-1}; for colloidal substances, they are typically of the order 10^{-11} m^2 s^{-1}. In the next section, we see that for spherical particles the diffusion coefficient is inversely proportional to the radius of the sphere. Therefore, every increase by a factor of 10 in size decreases the diffusion coefficient by the same factor. Qualitatively, this same inverse relationship applies to nonspherical particles as well. Once again, we see that diffusion decreases in importance with increasing particle size, precisely those conditions for which sedimentation increases in importance. For larger particles, for which D is very small, the diffusion coefficient also becomes harder to measure. For spherical particles, the time required for the particle to diffuse a unit distance is directly proportional to its radius. For small molecules, this time is experimentally accessible; it becomes experimentally inconvenient for particles at the upper end of the colloidal range.

2.5b Diffusion Coefficient and Friction Factor: Thermodynamic Description of Diffusion

As already noted, the driving force underlying diffusion can be thought of as thermodynamic in origin. A very general way to describe a force is to write it as the negative gradient of a potential. In the context of diffusion, the potential to be used is the chemical potential μ_i, the partial molal Gibbs free energy of the component of interest. Thus, the magnitude of the driving force (per particle) diffusion may be written as

$$F_{diff} = -\frac{1}{N_A}\frac{\partial \mu_i}{\partial x} \tag{27}$$

This is often called the thermodynamic force for diffusion. It is necessary to divide by Avogadro's number N_A since μ_i is a molar quantity. Thermodynamics show that

$$\mu_i = \mu_i^0 + RT \ln a_i = \mu_i^0 + RT \ln (\gamma_i c_i) \tag{28}$$

where a_i, c_i, and γ_i are the activity, concentration, and activity coefficient, respectively, of the ith component. Since we are interested in infinitely dilute systems, the activity coefficient may be assumed to equal unity. Substituting Equation (28) into Equation (27) gives

$$F_{diff} = k_B T \frac{\partial \ln c_i}{\partial x} = -\frac{k_B T}{c_i}\frac{\partial c_i}{\partial x} \tag{29}$$

where k_B is *Boltzmann's constant* R/N_A. Under stationary state conditions, this force will be equal to the force of viscous resistance, given by fv according to Equation (2). Therefore, the magnitude of the velocity of diffusion equals

$$v = -\frac{k_B T}{fc}\frac{\partial c}{\partial x} \tag{30}$$

where the subscript has been omitted from the concentration of the solute c because this is now the only quantity involved in the relationship. Finally, we make the following observation. The flux of material through a cross section equals the product of its concentration and its diffusion velocity:

$$J = cv_{diff} \tag{31}$$

Combining Equations (30) and (31) and comparing with Equation (22) leads to the important result

$$D = \frac{k_B T}{f} \tag{32}$$

It should be noted that this derivation contains no assumptions about the shape of the particles. However, when the particles are assumed to be spherical, we can substitute Equation (8) for f, and the resulting equation for the diffusion coefficient is the well-known *Stokes-Einstein relation*.

Many of the relationships of this chapter have involved the friction factor f, which, until now, has been an unknown quantity except for spherical particles. Equation (32) breaks this impasse and points out the complementarity between sedimentation and diffusion measurements. For example, substitution of Equation (32) into (4) gives

$$m = \frac{k_B T v}{D[1 - (\rho_1/\rho_2)]g} \tag{33}$$

and substituting Equation (32) into Equation (19) yields

$$m = \frac{k_B T s}{D[1 - (\rho_1/\rho_2)]} \tag{34}$$

Equations (33) and (34) show that diffusion studies combined with sedimentation studies, either under the force of gravity or in a centrifuge, yield information about particle masses with no assumptions about the shape of the particle.

Human hemoglobin, for example, has a sedimentation coefficient of 4.48 S and a diffusion coefficient of $6.9 \cdot 10^{-11} \, m^2 \, s^{-1}$ in aqueous solution at 20°C. The density of this material is $1.34 \, g \, cm^{-3}$. Substituting these values into Equation (34) shows the particle mass to be

$$m = (1.38 \cdot 10^{-23})(293)(4.48 \cdot 10^{-13})/(6.9 \cdot 10^{-11})(1 - 1.0/1.34)$$
$$= 1.03 \cdot 10^{-22} \, kg \, particle^{-1} \tag{35}$$

or, in terms of molecular weight,

$$M = (1.03 \cdot 10^{-22})(6.02 \cdot 10^{23}) = 62,300 \, g \, mol^{-1} \tag{36}$$

Since the friction factor was measured experimentally, this value is correct regardless of the state of solvation or ellipticity of the hemoglobin molecules in solution. We see below how the combination of these two experimental approaches may be interpreted further to yield some information about solvation and ellipticity.

2.5c Effect of Solvation

We have already seen that the ratio f/f_0 describes the effect on the friction factor of either solvation, ellipticity, or both. This ratio equals unity for an unhydrated sphere and increases with both the amount of bound solvent and the axial ratio of the particles. We are now in a position to see how this ratio may be evaluated experimentally. The steps of the procedure are the following:

1. The diffusion coefficient and sedimentation velocity (or sedimentation coefficient) are used to evaluate m by Equation (33) or (34).
2. The friction factor is evaluated from the diffusion coefficient by Equation (32).
3. If we assume the particle to be unsolvated, we can determine its volume by dividing its mass by the density of the dry material.
4. The radius of the particle is calculated from its volume if we assume the particle to be a sphere.
5. The friction factor f_0 of the equivalent unsolvated sphere is evaluated by Equation (8).
6. The ratio of the experimental friction factor to f_0 is determined.

As noted, the larger this ratio is than unity, the more the particle deviates from the unsolvated spherical shape assumed to calculate f_0. Although this statement is qualitatively accurate, we realize that the f/f_0 value reflects some actual quantitative condition of the particles, and we search for additional ways to interpret this ratio.

The factoring of f/f_0 into two contributions in Equation (14) was introduced with this interpretation in mind. It turns out that the effect of solvation is easily handled by a physically reasonable model through f^*/f_0. Dealing with particle asymmetry through f^*/f is more complicated, but this has also been analyzed according to a model that is plausible for many situations. Let us now consider each of these developments in turn.

When particles are solvated, a certain volume of the solvent must be counted as part of the dispersed phase rather than the continuous phase. In dilute solutions, the effect of this reclassification of some solvent is negligible for the remaining solvent (component 1), but the effect on the solute (component 2) may be considerable. The effect of the attached solvent on the volume of the solute particles may be calculated if some model is assumed for the mode of attachment. To assume the solvation occurs uniformly throughout the particle is a plausible model for the solvation of lyophilic colloids. For example, protein molecules have ionic and polar substituent groups distributed at random along the polymer chain. Such groups are expected to be extensively hydrated in an aqueous solution.

If we assume that solvation occurs at numerous positions along the molecule, the volume of the solvated particle may be written as the sum of the volume of the unsolvated particle and the volume of the bound (subscript b) solvent:

$$\text{volume of solvated particle} = V_2 + V_{1,b} \tag{37}$$

This can be rearranged as

$$V_2 + V_{1,b} = V_2 \left(1 + \frac{V_{1,b}}{V_2}\right) = V_2 \left[1 + \frac{m_{1,b}}{m_2}\left(\frac{\rho_2}{\rho_1}\right)\right] \tag{38}$$

where $m_{1,b}$ is the mass of bound solvent and the last equality requires the density of the bound solvent to be identical to that of the free solvent. In the event that the solvated specie differs in density from its bulk counterpart, the ratio ρ_2/ρ_1 may be replaced by $(\bar{v}_1/\bar{v}_2)$, where the $\bar{v}_i$'s are the partial specific volumes of the components, and measure the volume per unit mass of the indicated components in the solution at the indicated concentration and hence reflect any volume changes that accompany the interaction. We continue to discuss solvation in terms of Equation (38), using the more familiar densities.

We are explicitly looking for the effect of solvation on the friction factor of a sphere; particle asymmetry is handled by the second ratio f/f^*. By Stokes's law, the ratio of the friction factors of solvated to unsolvated spheres f^*/f_0 is equal to the ratio of their radii a_{solv}/a_o. This ratio equals the cube root of the volume ratio of the solvated to unsolvated spheres, or

$$\frac{f^*}{f_0} = \left[1 + \frac{m_{1,b}}{m_2}\left(\frac{\rho_2}{\rho_1}\right)\right]^{1/3} \tag{39}$$

The ratio f^*/f_0 equals unity when $m_{1,b} = 0$ and increases with increasing solvation, as required. We examine an application of this relationship below.

Asymmetry as well as solvation can cause a friction factor to have a value other than f_0. Next let us consider the ratio f/f^*, which, according to Equation (14), accounts for the effect of particle asymmetry on the friction factor. We saw in Section 1.5 that ellipsoids of revolution are reasonable models for many asymmetric particles.

Jean Perrin derived expressions for the ratio f/f^* for ellipsoids of revolution in terms of the ratio of the equatorial semiaxis to the semiaxis of revolution b/a. The following expressions were obtained:

First, for prolate ellipsoids ($b/a < 1$),

$$\frac{f}{f^*} = \frac{\left[1 - \left(\frac{b}{a}\right)^2\right]^{1/2}}{\left(\frac{b}{a}\right)^{2/3} \ln\left(\frac{1 + [1 - (b/a)^2]^{1/2}}{b/a}\right)} \tag{40}$$

Second, for oblate ellipsoids ($b/a > 1$),

$$\frac{f}{f^*} = \frac{\left[\left(\frac{b}{a}\right)^2 - 1\right]^{1/2}}{\left(\frac{b}{a}\right)^{2/3} \tan^{-1}\left[\left(\frac{b}{a}\right)^2 - 1\right]^{1/2}} \tag{41}$$

Equations (40) and (41) allow the ratio f/f^* to be evaluated for any axial ratio. The reverse problem, finding an axial ratio that is consistent with an experimental f/f^* ratio, is best accomplished by interpolating from plots of Equations (40) and (41). These two equations, along with Equations (14) and (39), allow the states of solvation and ellipticity compatible with an experimental friction factor to be established. Example 2.4 considers this for the human hemoglobin molecule.

* * *

EXAMPLE 2.4 *Solvation and Ellipticity of Human Hemoglobin from Sedimentation Data.* The diffusion coefficient of the human hemoglobin molecule at 20°C is $6.9 \cdot 10^{-11}$ m^2 s^{-1}. Use this value to determine f for this molecule. Evaluate f_0 for hemoglobin using the particle mass calculated in Equation (35). Indicate the possible states of solvation and ellipticity that are compatible with the experimental f/f_0 ratio.

Solution: According to Equation (32), $f = k_B T/D = (1.38 \cdot 10^{-23})(293)/6.9 \cdot 10^{-11} = 5.86 \cdot 10^{-11}$ $kg\ s^{-1}$.

Equation (35) shows that the mass of one of these molecules is equal to $1.03 \cdot 10^{-22}$ kg.

Dividing this by the density (1.34 g cm^{-3}) gives $7.69 \cdot 10^{-26}$ m^3 as the particle volume. The radius of an unsolvated sphere of this volume is given by $(3V/4\pi)^{1/3} = 2.64 \cdot 10^{-9}$ m.

Using the viscosity of water, 0.01 P, the friction factor f_0 is calculated by Equation (8):

$$f_0 = (6\pi)(10^{-3})(2.64 \cdot 10^{-9}) = 4.98 \cdot 10^{-11}\ kg\ s^{-1}$$

Since these calculations involve only empirical data, we can say the experimental value of f/f_0 $= 5.86 \cdot 10^{-11}/4.98 \cdot 10^{-11} = 1.18$.

Next we recall Equation (14), which interprets f/f_0 as the product

$$(f/f^*)(f^*/f_0)$$

If $f/f^* = 1.00$, the particle is a sphere and $f^*/f_0 = 1.18$. Using the bulk density of water and the protein and Equation (39), we can show that $m_{1,b}/m_2 = 0.48$ g water per g hemoglobin.

If $f^*/f_0 = 1.00$, the particle is unsolvated and $f/f^* = 1.18$. This corresponds to an axial ratio $b/a = 0.24$, according to Equation (40), if the particle is a prolate ellipsoid, or to $b/a = 4.0$, according to Equation (41), if the particle is oblate. These values are summarized in Table 2.1.

■

* * *

It is apparent that there are many combinations of f/f^* and f^*/f_0 that form the product 1.18, which we attempted to explain in Example 2.4. Setting either one of these contributions equal to unity merely establishes an upper limit for the other. Intermediate cases between solvated spheres and unsolvated ellipsoids can also be calculated that are consistent with a given axial ratio.

TABLE 2.1 A Summary of the Characterization of Human Hemoglobin at 20°C Based on Sedimentation and Diffusion Measurements

Quantity	Value	Determination
Sedimentation coefficient, s	4.48×10^{-13}s	Experimental
Diffusion coefficient, D	6.9×10^{-11} $m^2 s^{-1}$	Experimental
Density, ρ	1.34 g cm^{-3}	Experimental
Mass of particle, m	1.03×10^{-22} kg molecule^{-1}	Eq. (35)
Molecular weight	6.23×10^4 g mol^{-1}	Eq. (36)
f	5.86×10^{-11} kg s^{-1}	Eq. (32)
Volume of particle	7.69×10^{-26} m^3	$V_{unsolv} = m/\rho$
Radius of equivalent sphere	2.64×10^{-9} m	$R_{sph} = (3V/4\pi)^{1/3}$
f_0	4.98×10^{-11} kg s^{-1}	Eq. (8)
f/f_0	1.18	—
$m_{1,b}/m_2$, if spherical	0.48 g H_2O (g protein)$^{-1}$	Eq. (39)
a/b if unsolvated and prolate	4.0	Eq. (40)
a/b if unsolvated and oblate	0.24	Eq. (41)

Source: Data from E. J. Cohn, and J. T. Edsall, *Proteins, Amino Acids and Peptides*, American Chemical Society Monograph, reprinted by Hafner, New York, 1965.

Figure 2.9 is a plot of possible combinations of hydration and asymmetry for protein particles in water. Similar curves could be drawn for other materials as well. For the human hemoglobin molecule discussed in Table 2.1, the combination of sedimentation and diffusion measurements gives an f/f_0 value that lies within the domain defined by the 1.15 and 1.20 contours of Figure 2.9. The current picture of the structure of human hemoglobin, deduced from x-ray diffraction studies, suggests that the molecule may be regarded as an ellipsoid with height, width, and depth equal to 6.4, 5.5, and 5.0 nm, respectively. Applying these dimensions to the dispersed unit leads us to describe the particle as being hydrated to the extent of about 0.4–0.5 g water (g protein)$^{-1}$.

Sedimentation and diffusion data allow for the unambiguous determination of particle mass and also allow the suspended particles to be placed on a contour in a plot such as that of Figure 2.9. This is as far as these experiments can take us toward the characterization of the particles. Of course, additional data from other sources, such as the x-ray diffraction results just cited, may lead to still further specification of the system. One such source of information is intrinsic viscosity data for the same dispersion. In Chapter 4 we discuss the complementarity between viscosity data and sedimentation-diffusion results (see Section 4.7b).

2.6 BROWNIAN MOTION AND DIFFUSION

A liquid that is totally homogeneous on a macroscopic scale undergoes continuous fluctuations at the molecular level. As a result of these fluctuations, the density of molecules at any location in the liquid varies with time and at any time varies with location in such a way that the mean density of the sample as a whole has its bulk value. This pattern of "flickering" molecular densities will produce continually varying pressures on the surface of any particle

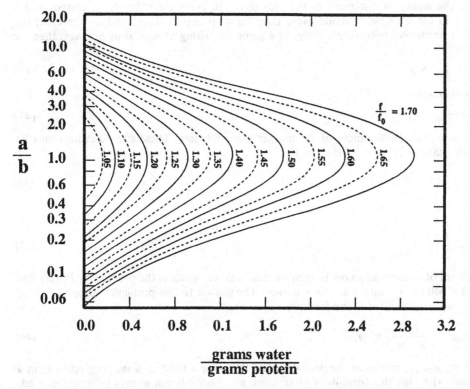

FIG. 2.9 Variations of the ratio f/f_0 with asymmetry and hydration for aqueous protein dispersions. (Redrawn from L. Oncley, *Ann. NY Acad. Sci.* **41**, 121 (1941).

submerged in the liquid. Since the fluctuations are confined to domains of the order of molecular dimensions, this randomly variable pressure is quite small. A small particle will be displaced, however, by the resulting force imbalance at its surface. The pattern of its displacements will also be a totally random thing, a reflection of the fluctuations that cause the motion. As the size of the submerged particles increases, the effect of fluctuations on them decreases. In this section we consider the trajectory of a particle in the colloidal size range that is engaged in this pattern of motion. Such movements have been studied microscopically and are called *Brownian motion* after Robert Brown, a British biologist who described them in 1828.

In this section we review random walk statistics and their relation to diffusion. More elaborate discussions of these and related topics are available in Berg (1993); the collection of the original papers of Einstein (reprinted in Einstein 1956) is another excellent source of this material.

2.6a   Random Walk and Random Walk Statistics

In general, a dispersed particle is free to move in all three dimensions. For the present, however, we restrict our consideration to the motion of a particle undergoing random displacements in one dimension only. The model used to describe this motion is called a one-dimensional *random walk*. Its generalization to three dimensions is straightforward.

To begin, the statistical nature of this phenomenon should be apparent. We might watch the pattern of displacements of thousands of otherwise identical particles and find no uniformity in the zigzag steps they follow. Only statistical quantities such as the "average" displacement after a certain number of steps or after a certain elapsed time make any sense. Let us consider how to calculate such a quantity.

Suppose we consider a game in which a marker is moved back and forth along a line, say, the x axis, in a direction determined by the toss of a coin. The rules of the game provide that we move the marker a distance ℓ in the plus direction every time "heads" is tossed, and a distance ℓ in the negative direction every time "tails" is tossed. If n_+ and n_- represent the number of heads and tails, respectively, in a game consisting of a total of n tosses, then we may write

$$n_+ + n_- = n \tag{42}$$

We may also write

$$x = (n_+ - n_-)\ell \tag{43}$$

where x is the net displacement of the marker after n tosses. These two equations may be solved simultaneously for n_+ and n_- to give

$$n_+ = \frac{1}{2}\left(n + \frac{x}{\ell}\right) \tag{44}$$

and

$$n_- = \frac{1}{2}\left(n - \frac{x}{\ell}\right) \tag{45}$$

The problem of interest may now be expressed as follows. What is the probability $P(n,x)$ that the marker will be at position x after n moves? The answer to this problem is supplied by the well-known binomial distribution formula

$$P(n,x) = \frac{n!}{n_+! \, n_-!} p_+^n p_-^n \tag{46}$$

in which p_+ and p_- represent the probability of throwing a head or a tail, respectively, in a single toss. Although this formula is often used, its validity is not always fully appreciated. Accordingly, let us briefly examine the origin of Equation (46).

Each time we toss an unbiased coin, the outcome is independent of all other tosses. The probability of tossing n_+ heads is therefore $p_+^{n_+}$ since the probabilities compound by multiplication when we require a specified set of outcomes. Therefore n_+ events, each of which occurs with a probability p_+, will occur with a probability $p_+^{n_+}$. If we specify the outcome further by requiring n_- tails in addition to the heads already specified, then the probability is given by $p_+^{n_+} p_-^{n_-}$. What we have calculated is the probability of a particular, fully specified sequence of outcomes such as the following for $n = 6$: HHHTTT. However, the same net number of heads and tails could come about in other ways. For example, HTTTHH, HHTHTT, and HTHTHT are also consistent with $n = 6$ and $n_+ = n_- = 3$, along with the distribution previously given and other possibilities. Therefore we must multiply the probability of one specified sequence by the number of other sequences that have the same net composition of heads and tails.

At first glance, this factor may appear to be $n!$, the number of different ways the n items can be rearranged, giving a probability of $n!\, p_+^{n_+} p_-^{n_-}$. This quantity overcounts, however, since not all rearrangements are recognizably different. For example, we can interchange the first two members of the sequence TTHHHT to produce the sequence TTHHHT; the interchange obviously leaves the sequence unaltered. Therefore we must divide $n!\, p_+^{n_+} p_-^{n_-}$ by the number of ways that identical members of the series can be interchanged without altering the sequence. Any of the n_+ heads (or the n_- tails) may be interchanged among themselves with this result. There are $n_+!$ permutations of the heads and $n_-!$ permutations of the tails. Therefore these are the factors by which the previous result overcounted. Dividing by these factors leaves us with the binomial equation. The student should be so familiar with this argument as to be able to justify each factor in Equation (46).

Since the probability of tossing either a head or a tail is equal to 1/2 for a fair coin, Equation (46) may be rewritten

$$P(n,x) = \frac{n!}{\left[\frac{1}{2}\left(n + x/\ell\right)\right]!\left[\frac{1}{2}\left(n - x/\ell\right)\right]!}\left(\frac{1}{2}\right)^n \tag{47}$$

by incorporating Equations (44) and (45). The length ℓ by which the marker is moved in our hypothetical game is equivalent to the displacement of a particle in a single fluctuation. Because of the nature of the fluctuations underlying the whole process, these individual steps are very small. Observable diffusion is always the result of a very large number of such steps. For the case in which n is a large number, the factorials of Equation (47) may be expanded according to what is known as *Sterling's approximation*:

$$\ln y! = y \ln y - y \tag{48}$$

which is valid for large values of y. Taking the logarithm of Equation (47) and applying Approximation (48) leads to the result

$$-\ln P = \frac{n\ell + x}{2\ell} \ln\left(1 + \frac{x}{n\ell}\right) + \frac{n\ell - x}{2\ell} \ln\left(1 - \frac{x}{n\ell}\right) \tag{49}$$

The net displacement x is always small compared to the total distance traveled $n\ell$ since a good deal of the back-and-forth motion cancels out. Therefore the logarithmic terms in Equation (49) may be expanded as a power series (see Appendix A) in $(x/n\ell)$, with all terms higher than second order in $x/n\ell$ regarded as negligible. This leads to the result

$$\ln P \approx -\frac{x^2}{2\, n\, \ell^2} + \cdots \tag{50}$$

or

$$P(n, x) = C \exp\left(-\frac{x^2}{2\, n\, \ell^2}\right) \tag{51}$$

where the factor C represents a normalization constant.

A well-behaved probability function adds up to unity when the probabilities for all possi-

TABLE 2.2 Integrals Related to Gamma Functions
Encountered in This Chapter

$$\int_0^\infty x^n e^{-ax^2} dx =$$

$\frac{1}{2}\sqrt{\pi/a}$ if $n = 0$	$\frac{1}{2a}$ if $n = 1$
$\frac{1}{4a}\sqrt{\pi/a}$ if $n = 2$	$\frac{1}{2a^2}$ if $n = 3$
$\frac{3}{8a^2}\sqrt{\pi/a}$ if $n = 4$	$\frac{1}{a^3}$ if $n = 5$

ble outcomes are totaled. Therefore, in order to evaluate the constant C in Equation (51), we should integrate Equation (51) over all possible values of x—which is equivalent to summing the probabilities—and set the result equal to unity. This leads to the expression

$$C = \left[\int_{-\infty}^\infty \exp\left(-\frac{x^2}{2\,n\,\ell^2}\right) dx \right]^{-1} \tag{52}$$

The integral is a gamma function, the value of which can be determined from the list of related integrals in Table 2.2. Evaluating Equation (52) gives

$$C = (2\pi n\ell^2)^{-1/2} \tag{53}$$

Substitution of this result into Equation (51) gives a continuous analytical expression for $P(n,x)$, which equals the binomial result for large values of n:

$$P(n, x) = (2\,\pi\,n\,\ell^2)^{-1/2} \exp\left(-\frac{x^2}{2\,n\,\ell^2}\right) \tag{54}$$

Strictly speaking, we should multiply both sides of Equation (54) by dx. The equation now expresses the probability of a displacement between x and $(x + dx)$ after n random steps of length ℓ.

2.6b Diffusion Coefficient from Random Walk Statistics

The application of Equation (54) to at least one set of boundary conditions for a diffusion problem is easy. We let a modification of Figure 2.8a describe the system. This time, instead of having a solution on one side of the barrier and pure solvent on the other side, suppose we imagine that both sides of the barrier are filled with solvent. Furthermore, suppose that the solute under investigation is introduced into the system in the pores of the plug, assuming the latter to be infinitesimally thin. Thus at the beginning of the experiment all the material is present at $x = 0$, in the notation of the derivation above, at a concentration c_0. With the passage of time, the material will gradually diffuse outward; the number of diffusion steps taken will be directly proportional to the elapsed time:

$$n = Kt \tag{55}$$

Equation (54) may be transformed into an expression that gives the probability as a function of x and t by substituting Equation (55) into Equation (54):

$$P(x, t)\, dx = (2\,\pi\,K\,t\,\ell^2)^{-1/2} \exp\left(-\frac{x^2}{2\,K\,t\,\ell^2}\right) dx \tag{56}$$

This suggests that the concentration as a function of x and t in the diffusion cell just described is given by multiplying c_0 by $P(x,t)$:

$$c(x, t) \ dx = c_0 \ (2 \ \pi \ K \ t \ \ell^2)^{-1/2} \ \exp\left(-\frac{x^2}{2 \ K \ t \ \ell^2}\right) dx \tag{57}$$

Since $P(x,t)$ is normalized, its integral over all values of x equals unity. Likewise, integrating Equation (57) over all values of x gives c_0: The same quantity of solute is present at all times whether it is concentrated at the origin or is spread out from $-x$ to $+x$. The function given by Equation (57) is a solution to Fick's second law, shown in Equation (26), for a one-dimensional problem in which all the material is present initially at $x = 0$ and at a concentration c_0.

The only ambiguity remaining at this point is the value of K in Equations (55)–(57). This constant is easily evaluated as follows. If Equation (57) is indeed a solution to Equation (26), then the right- and left-hand sides of Equation (26) must be equal when the indicated differentiations are carried out. Differentiating Equation (57), substituting into Equation (26), and simplifying lead to the result

$$K = \frac{2D}{\ell^2} \tag{58}$$

Substituting this expression into Equation (57) yields

$$\frac{c(x, t)}{c_0} \ dx = (4 \ \pi \ D \ t)^{-1/2} \ \exp\left(-\frac{x^2}{4 \ D \ t}\right) dx \tag{59}$$

This relationship describes the diffusion of a solute in the x direction when it is concentrated initially in an infinitesimally thin layer at the origin.

Equation (59) may be developed somewhat further as follows. We define a parameter z to be

$$z = \frac{x}{(2 \ D \ t)^{1/2}} \tag{60}$$

and then rewrite Equation (59) in terms of this quantity:

$$\frac{c(x, t)}{c_0} \ dx = (2 \ \pi)^{-1/2} \ \exp\left(-\frac{z^2}{2}\right) dz = P(z) \ dz \tag{61}$$

Thus the concentration ratio c/c_0 is seen to be described at all times as a function of the single parameter z. The function $P(z)$ defined by Equation (61) is the normal or Gaussian distribution function, Equation (C.10).* Example 2.5 considers how the concentration profile of the diffusing species changes with time according to the normal distribution function.

*   *   *

EXAMPLE 2.5 *Unsteady State Variation of Concentration Profiles Due to Diffusion: Gaussian Distribution.* By consulting tables of the normal distribution function, draw curves that show the broadening of a band of material with time if the substance is initially at concentration c_0 and in a plug of infinitesimal thickness at $x = 0$. Assume that the diffusion coefficient has the value $5 \cdot 10^{-11} \ m^2 \ s^{-1}$ for this material. Use $t = 10^6$ and $t = 3 \cdot 10^6$ s to see how the concentration profile changes with time.

Solution: Tables of the normal distribution function list values of $P(z)$ for different values of z. As examples, we cite the following entries from the tables (Beyer 1987):

z	0.3	0.7	1.0	1.3	1.7	2.0	2.3	2.7
$P(z)$	0.3814	0.3123	0.2420	0.1714	0.0941	0.0540	0.0283	0.0104

---

*This manner of referencing is used to describe equations occurring in other chapters or in the appendices. The number (or the letter) to the left of the point is the chapter number (or appendix).

Actual displacements x are obtained by multiplying z by $(2Dt)^{1/2}$; this means that the ordinate must be divided by this factor so that the area under the curve remains normalized:

For $t = 10^6$ s, $(2Dt)^{1/2} = [2(5 \cdot 10^{-11})10^6]^{1/2} = 10^{-2}$ m

Therefore $x = 10^{-2} z$.

$x(m)$	0.003	0.007	0.010	0.013	0.017	0.020	0.023	0.027
$P(x) \cdot 10^{-2}$	0.381	0.312	0.242	0.171	0.094	0.054	0.028	0.010

For $t = 3 \cdot 10^6$ s, $(2Dt)^{1/2} = 1.73 \cdot 10^{-2}$ m. Again, z is multiplied by this quantity to give x, and $P(z)$ is divided by the same factor to give the relative value of the ordinate, $P(x)$:

$x(m)$	0.0052	0.0121	0.0173	0.0225	0.0294	0.0346	...
$P(x) \cdot 10^{-2}$	0.221	0.181	0.140	0.099	0.054	0.031	

These values are plotted in Figure 2.10, in which the factor 10^{-2} for $P(x)$ has been used as a scale factor for the ordinate to show the relative shapes of the curves.   ∎

* * *

Figure 2.10 indicates a gradual approach toward equilibrium by the fact that the concentration tends to be more nearly uniform as time increases. In Appendix C the function z is defined to be δ/σ, where δ is the deviation of a particular value from the mean of a distribution and σ is the standard deviation of the distribution. Since the net displacement of a diffusing particle is analogous to δ, we may infer that the quantity $\sqrt{2Dt}$ is also analogous to the standard deviation. The point of this is the following.

It is well known that the width of the normal error curve at its inflection point (where the

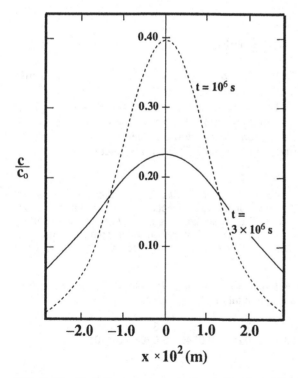

FIG. 2.10 Variation of concentration with time. The figure shows the variations of c/c_0 with distance from the origin at $t = 10^6$ and $t = 3 \cdot 10^6$ s if $D = 5 \times 10^{-11}$ m^2 s^{-1} for the data developed in Example 2.5.

second derivative changes sign or the curve changes from concave to convex) is equal to σ. Therefore we can conclude that the width of the curves shown in Figure 2.10, measured at their inflection points, increases in proportion to $\sqrt{t}$. This is one method whereby D could be measured in an experiment that corresponds to the boundary conditions specified above.

2.6c Random Walk Statistics and Experimental Measurement of the Diffusion Coefficient

With the mathematics of the one-dimensional random walk as background, we may visualize the following experimental arrangement by which D could be measured. Suppose that a narrow band of dispersion is layered between two portions of solvent in a long tube. We shall not worry about the practical difficulties in doing this since, in practice, other initial conditions that are easily obtained are actually used. The narrow band we have pictured, however, approximates the initial state of the system described by Equation (59). We might further imagine observing this band by means of a schlieren optical system (Section 2.4a). The two edges of the band, where the refractive index gradient is large, would define the positively and negatively sloped branches of the schlieren peak. With the passage of time, the material in the band diffuses outward; the schlieren peak would also broaden. In other words, the schlieren pattern observed at successive times would generate a family of curves—for example, Figure 2.7b—that very much resemble the theoretical curves of Figure 2.10.

The curves in Figure 2.10 are drawn for an arbitrary value of the diffusion coefficient. The experimental profiles produced by the schlieren optics are characterized by the diffusion coefficient of the experimental system. The remaining question is how to extract the appropriate D value from the experimental observations.

Remember that the normal distribution function has an inflection point (where the second derivative changes sign) at $z = 1.0$. Therefore the x value at which the inflection point occurs at any time equals $\sqrt{2Dt}$ according to Equation (60). By locating the inflection point at different times during a diffusion experiment, the appropriate D value may be evaluated for the diffusing species. For example, on the $t = 10^6$ s curve in Figure 2.10, the inflection point lies at $x = 0.010$ m. Substituting $x = 0.010$ m and $t = 10^6$ s when $z = 1$ into Equation (60) enables us to calculate the value of the diffusion coefficient that was used ($D = 5 \cdot 10^{-11}$ m^2 s^{-1}) to draw the curves in Figure 2.10. A similar analysis may be conducted on experimental curves obtained by the schlieren method.

As an experimental procedure, this method is less precise than others developed for the evaluation of D. It does point out, however, the intimate connection between diffusion and the random events at the molecular level that cause it.

A more practical experimental method for the determination of D is based on Figure 2.8. The theoretical arrangement represented by Figure 2.8 is implemented experimentally in an apparatus like that sketched in Figure 2.11. One side of the sintered glass barrier contains solution; the other side contains pure solvent. The entire apparatus is thermostated and both compartments are magnetically stirred. Samples are withdrawn at various times, and the quantity of material that has diffused into the solvent compartment is measured. The primary difficulty with this procedure is the tendency of the pores to plug owing to the entrapment of air bubbles, clogging by solid particles, or adsorption of the diffusing molecules themselves. Nevertheless, what is obtained experimentally is a record of the approach toward a uniform distribution of material on both sides of the barrier.

The situation represented by the apparatus in Figure 2.11 has also been analyzed theoretically. The function $c(x,t)$, which satisfies Fick's second law when there is a solution of concentration c_0 on one side of a boundary at $t = 0$ and pure solvent on the other side is given by

$$c = \frac{c_0}{2}\left[1 - \int_0^z P(z)\,dz\right] = \frac{c_0}{2}[1 - 2\,Erf\,(z)] \tag{62}$$

We can verify the plausibility of this expression as follows. Recall that z and t are inversely related and that the integral gives the area under one-half of the error curve between its

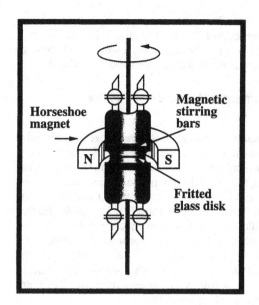

FIG. 2.11 Laboratory apparatus equivalent to that in Figure 2.8. The entire apparatus is rotated between the poles of a magnet for stirring.

midpoint ($z = 0$) and some specified value of z. Therefore, at $t = 0$, z is infinite, the integral equals 1/2, and $c = 0$. At $t = \infty$, z is zero, the integral equals zero, and $c = 1/2\ c_0$. Thus Equation (62) makes sense at either extreme. The detailed profile of c/c_0 versus z is obtained by reading values of the integral in Equation (62) from tables that give the area under the normal error curve. The results of this procedure are plotted in Figure 2.12, in which c/c_0 is shown versus x at several different times. The approach toward a uniform distribution of material is evident.

As in the case of diffusion from an initially thin layer, experimental concentration data and theoretical concentration profiles may be compared. From this comparison, the value of

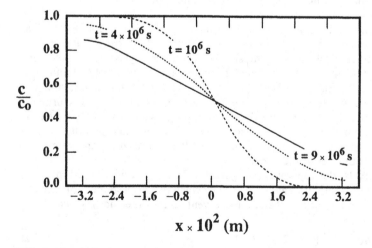

FIG. 2.12 Diffusion of solute from solution into pure solvent. Here c/c_0 is plotted versus distance at $t = 10^6$, $4 \cdot 10^6$, and $9 \cdot 10^6$ s for $D = 5 \cdot 10^{-11}$ m^2 s^{-1}. The plot has been drawn according to Equation (62).

D that is consistent with the observed diffusion behavior may be evaluated. The diffusion coefficient is a function of concentration; therefore, it is measured at a series of different concentrations and extrapolated to zero concentration; the limiting value is given the symbol D^0. Diffusion coefficients are temperature dependent, and it is sometimes necessary to adjust a value measured at one temperature to some other temperature. Example 2.6 considers how this is done.

* * *

EXAMPLE 2.6 *Temperature Dependence of Diffusion Coefficients.* Suppose the diffusion coefficient of a material is measured in an experiment (subscript *ex*) at some temperature T_{ex} at which the viscosity of the solvent is η_{ex}. Show how to correct the value of D to some standard (subscript *s*) conditions at which the viscosity is η_s. Take 20°C as the standard condition and evaluate D^0_{20} for a solute that displays a D^0 value of $4.76 \cdot 10^{-11}$ m^2 s^{-1} in water at 40°C. The viscosity of water at 20 and 40°C is $1.0050 \cdot 10^{-2}$ and $0.6560 \cdot 10^{-2}$ P, respectively.

Solution : By Equation (32), D is proportional to T/f and, by Equation (8), f is proportional to η. Therefore $D \propto T/\eta$, or

$$\frac{D^0_{20}}{D^0_{40}} = \frac{T_s/\eta_s}{T_{ex}/\eta_{ex}} = \frac{T_s}{T_{ex}} \frac{\eta_{ex}}{\eta_s}$$

Applying this relationship to the data provided gives

$$D^0_{20} = (293/313)(0.6560/1.0050)(4.76 \cdot 10^{-11}) = 2.91 \cdot 10^{-11}\, m^2\, s^{-1}.$$

The temperature dependence of the viscosity is seen to have the largest effect on the temperature dependence of D. ∎

* * *

2.6d "Average" Displacements from Random Walk Statistics

Having examined the connection between the phenomenological equations of diffusion and the statistics of the random walk, let us now return to the random walk as a model for *Brownian motion*. The problem we wish to consider is the "average" displacement of a marker after an *n*-step, one-dimensional random walk. The foregoing discussion already supplies the answer to this problem. We have seen that the probability function for the one-dimensional walk is symmetrical about the origin, implying a mean displacement of zero. This simply reflects the fact that on the average the number of heads and tails will be equal. While we accept this result, we feel that there is something unsatisfactory about it. The average displacement is zero because positive and negative displacements are equally probable and effectively cancel one another. It is not because the marker scarcely moves from its initial location, a conclusion suggested by the value of the mean displacement.

This shows that the mean displacement is simply not a useful parameter to characterize the trajectory of the particle and suggests we should seek an alternate quantity. Instead of averaging the displacements directly, suppose we first square them to eliminate the differences in sign, then average them and take the square root. This quantity, called the *root mean square* (rms) *displacement*, will give a better measure of the meanderings of the marker since the sign differences have been eliminated.

The calculation of the rms displacement is quite simple. Any average is calculated by multiplying the quantity to be averaged by an appropriate probability function and then integrating the result over all possible values of the variable. This is a direct extension of Equation (C.3) using a continuous function for the weighting factor. Applying this procedure to the problem at hand gives the result

$$\overline{x^2} = \int_{-\infty}^{\infty} x^2\, P(n, x)\, dx \tag{63}$$

in which $P(n,x)$ is given by Equation (54). Making this substitution gives

$$\overline{x^2} = (4 \pi Dt)^{-1/2} \int_{-\infty}^{\infty} x^2 \exp\left(-\frac{x^2}{4Dt}\right) dx \tag{64}$$

the value of which is found in Table 2.2. Evaluating the integral leads to the conclusion that

$$\overline{x^2} = 2Dt \tag{65}$$

Note that we have taken the initial position to be $x = 0$. Therefore, the left-hand side of Equation (65) is the average of the square of the *displacement* from the initial position. This important equation, also derived by Einstein, provides us with a means for measuring the diffusion coefficient for particles that are visible in a microscope. This is a particle size range for which measurement of D by following concentration changes with time is very difficult. Instead, the actual displacement of a particle in time t is measured microscopically. The rms displacement, evaluated from a statistically meaningful number of observations, permits D to be calculated. Jean Perrin (Nobel Prize, 1926) interpreted observations of a Brownian motion in terms of Equations (8), (32), and (65) as a means of determining the first precise value of Avogadro's number.*

Equation (65) also permits us to assign a physical interpretation to the diffusion coefficient in addition to the macroscopic meaning it has from Fick's laws. Rearranging and factoring in a way that admittedly ignores the averaging procedure, we write Equation (65) as

$$D = \frac{1}{2}\left(\frac{x}{t}\right) x \tag{66}$$

Since D is a constant for a particular substance, the ratio x/t must vary inversely with x. The distance traveled by a diffusing particle divided by the time required for the displacement x/t gives the apparent diffusion velocity of the particle. Equation (66) shows that the diffusion velocity is inversely proportional to the length of the path over which it is measured. This apparently paradoxical conclusion becomes less mysterious when we concentrate on the distinction between the diffusion velocity and the actual velocity of the particle as it travels along its zigzag path. The actual velocity reflects the average translational kinetic energy of the particle; its average value depends on the absolute temperature but is independent of the distance traveled. The diffusion velocity, on the other hand, decreases as the distance traveled increases. This simply means that large displacements are so much less probable than small displacements that they require disproportionately longer times to occur. Note that if $x = 1$ m, D has the significance of being equal to half the diffusion velocity measured over that distance. This is a result that was anticipated at the end of Section 2.5a.

2.7   THE RANDOM COIL AND RANDOM WALK STATISTICS

The dimensions of a randomly coiled polymer molecule are a topic that appears to bear no relationship to diffusion; however, both the coil dimensions and diffusion can be analyzed in terms of random walk statistics. Therefore we may take advantage of the statistical argument we have developed to consider this problem.

2.7a   End-to-End Distance in a Polymer Coil

Suppose we visualize a polymer molecule as consisting of n segments of length ℓ connected by a completely flexible linkage at each joint. Imagine, furthermore, that the placement of each successive segment is determined by some sort of purely random criterion. We could anchor

---

*Jean Perrin's 1913 monograph, *Les Atomes*, on Brownian motion and some of the related topics on the molecular nature of matter has ⁀n reprinted recently (Perrin 1990) and is an interesting source of information on the evolutioɪ ɟf ideas on diffusion and determination of Avogadro's number. This monograph also contains some of Perrin's sketches of random walks executed by colloidal spheres in his experiments.

one end of the chain to the origin of a hypothetical coordinate system, for example, and position successive segments as follows. Suppose we toss a single die and agree to orient the next segment in the $+x$ direction if the die shows a 1 and in the $-x$ direction if a 6 shows. Likewise, 2, 5, 3, and 4 correspond to $+y$, $-y$, $+z$, and $-z$, respectively. Our problem is this: On the *average*, what will be the end-to-end distance L for a chain of n segments? To calculate this, we must assume that n is very large, and we must ignore the sites excluded by previously positioned segments.

We recognize first that the end-to-end distance can be resolved into x, y, and z components that obey the relationship

$$L^2 = x^2 + y^2 + z^2 \tag{67}$$

The probability that the loose end of the chain is in a volume element located at x, y, and z is given by

$$P(x,y,z) \, dx \, dy \, dz = P(x)P(y)P(z) \, dx \, dy \, dz \tag{68}$$

where $P(x)$ is the probability that the x coordinate has a value between x and $x + dx$, with similar definitions for $P(y)$ and $P(z)$. Equation (54) gives the expression for $P(x)$, except we must remember that now only $n/3$ segments will be aligned with the x axis since each of the directions is independent and equally probable. Since the x, y, and z directions are equivalent, the same expression holds for $P(y)$ and $P(z)$ with the appropriate change of variables. Therefore, on incorporating Equation (54), Equation (68) becomes

$$P(x, y, z) \, dx \, dy \, dz = \left(2\pi \, \frac{n}{3}\ell^2\right)^{-3/2} \exp\left(-\frac{3 \, L^2}{2 \, n \, \ell^2}\right) dx \, dy \, dz \tag{69}$$

This expression gives the probability that the loose end of the molecule is in a volume element located at some particular values of x, y, and z, as shown in Figure 2.13a. Our interest is not in any specific x, y, z coordinates, but in all combinations of x, y, z coordinates that result in the end of the chain being a distance L from the origin. This can be evaluated by changing the volume element in Equation (69) to spherical coordinates and then integrating over all angles.

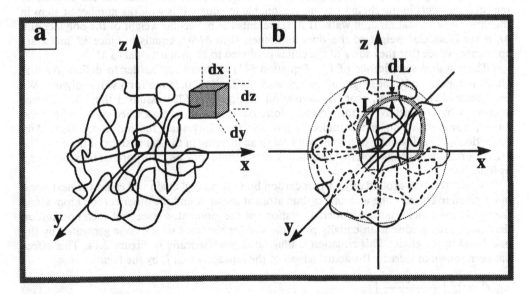

FIG. 2.13 Coordinate systems for a flexible random coil. One end of a flexible random coil lies at the origin and the other end is (a) in a volume element $dx \, dy \, dz$ and (b) in a spherical volume element $4\pi L^2 dL$.

This amounts to replacing the volume element in Equation (69) with the volume of a spherical shell of radius L and thickness dL:

$$dx\, dy\, dz \rightarrow 4\pi L^2 dL \tag{70}$$

A geometrical representation of this situation is shown in Figure 2.13b. Incorporating this expression into Equation (69) gives the probability that the loose end of the molecule lies between L and $(L + dL)$, irrespective of the direction:

$$P(L)\, dL = 4\pi \left(2\pi \frac{n}{3}\ell^2\right)^{-3/2} L^2 \exp\left(-\frac{3 L^2}{2 n \ell^2}\right) dL \tag{71}$$

With this distribution function, it is an easy matter to calculate average quantities. Again, since positive and negative displacements are equally probable, we evaluate the average of L^2. Following the same procedure as used in Equation (63), we write

$$\overline{L^2} = \left(\frac{2\pi n \ell^2}{3}\right)^{-3/2} 4\pi \int_0^\infty L^4 \exp\left(-\frac{3 L^2}{2 n \ell^2}\right) dL \tag{72}$$

Evaluation of this integral, using Table 2.2, leads to the result

$$\overline{L^2} = n\ell^2 \tag{73}$$

which is the desired quantity.

Equation (73) shows that the rms end-to-end distance in a polymer coil equals the square root of the number of steps times the length of each step. We might ask, therefore, what is the physical significance of n and ℓ for a polymer molecule? A general formula for a typical vinyl polymer is

$$\left(\begin{array}{c} \text{H} \quad \text{H} \\ | \quad\;\; | \\ -\text{C}-\text{C}- \\ | \quad\;\; | \\ \text{H} \quad \text{X} \end{array}\right)_n$$

where n is called the degree of polymerization. Since this quantity measures the number of repeated segments in the chain, it seems reasonable to equate this with the number of steps in the three-dimensional random walk. If M represents the molecular weight of the polymer and M_0 is the molecular weight of the vinyl monomer, then M/M_0 equals n. Since M_0 and ℓ are constants, we see that the radius of the coil is predicted to be proportional to $M^{1/2}$.

The physical significance of ℓ in Equation (73) is somewhat harder to define. At first glance it appears to be the length of the repeating unit, about 0.25 nm for a vinyl polymer. We must remember, however, that the derivation of Equation (73) assumed that the coil was connected by completely flexible joints. Molecular segments are attached at definite bond angles, however, so an actual molecule has less flexibility than the model assumes. Any restriction on the flexibility of a joint will lead to an increase in the dimensions of the coil. The effect of fixed bond angles on the dimensions of the chain may be incorporated into the model as follows.

A 360° rotation around any carbon-carbon bond in a vinyl chain will cause the next bond in the chain to trace a cone with one carbon atom at the apex and the other carbon atom along the rim of the cone. Ignoring hindered rotation for the moment, we see that each position on the rim of such a cone is an equally probable site for the apex of the cone generated by the next bond in the chain. This situation is illustrated schematically in Figure 2.14. This effect has been shown to increase the actual length of the repeating unit ℓ_0 by the factor

$$\ell^2 = \ell_0^2 \left(\frac{1 - \cos\theta}{1 + \cos\theta}\right) \tag{74}$$

θ is the bond angle. For a tetrahedral bond angle, $\cos\theta = -(1/3)$, so the additional factor in Equation (74) equals 2.

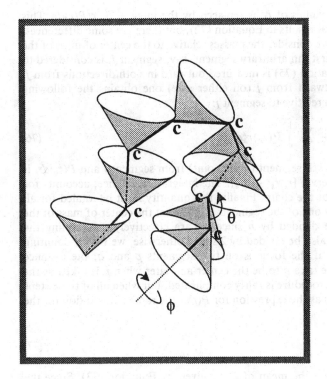

FIG. 2.14 The effect of fixed bond angle in restricting the flexibility of a polymer coil. For tetrahedral bonds, $\theta = 109°$.

In Figure 2.14, it is implied that the terminal carbon atom can occupy any position on the rim of the cone; that is, there is assumed to be perfectly free rotation around the penultimate carbon-carbon bond. This is equivalent to saying that all values of ϕ, the angle that describes the rotation (see Fig. 2.14), are equally probable. Any hindrance to free rotation will block certain configurations, expanding the coil dimensions still further.

Finally, interactions between the polymer segments and the solvent molecules may introduce a bias that tends to position segments as close or as far as possible from other segments, depending on the nature of the interaction. As a matter of fact, those properties of polymer solutions that are sensitive to the dimensions of the molecules, such as viscosity, vary widely with solvent "goodness." A good solvent may be defined as one in which polymer-solvent contacts are favored; a poor solvent is one in which polymer-polymer contacts are favored. Therefore the value of ℓ in Equation (74) will be increased in a good solvent and decreased in a poor solvent. In discussing size-exclusion chromatography in Section 1.6b, we anticipated this conclusion by noting that polymer-solvent interactions complicate the relationship between M and L for a random coil.

2.7b   Radius of Gyration of a Polymer Coil

The statistical relationships of this section can also be applied to Equation (1.13), which defines the *radius of gyration*. If the masses in that equation are identified with the mass m_0 of the repeat units of the polymer—of which there are n in the molecule—then Equation (1.13) can be written as

$$\overline{R_g^2} = \frac{m_0 \Sigma_i r_i^2}{n\, m_0} = \frac{1}{n} \Sigma_i r_i^2 \tag{75}$$

Next, the summation over discrete values of r_i^2 is replaced by the integral of $r^2P(r)dr$. This sounds exactly like the procedure that results in Equation (72), but there are some differences. To evaluate the radius of gyration, we consider the masses relative to the center of mass of the swarm. We proceed in two steps. First, an arbitrary segment, say, segment j, is considered to be the center of mass, and r in Equation (75) is measured outward in both directions from j. Using k to index the segments outward from j (on either side) one obtains the following expression for the radius of gyration relative to segment j:

$$(R_g^2)_j = \frac{1}{n}\left(\sum_{k=1}^{j}\int P(r_k)\, r_k^2 dr_k + \sum_{k=1}^{n-j}\int P(r_k)r_k^2 dr_k\right) \tag{76}$$

where r_k denotes the distance of the kth segment (mass point) from segment j and $P(r_k)dr_k$ is the probability that r_k lies in the interval $[r_k, r_k + dr_k]$. The integral, in essence, accounts for the various possible configurations of the chain. Finally, this quantity must be totaled for all values of j between 1 and n since any one of the segments might lie at the center of mass of the coil. This summation must also be divided by n since we are effectively determining the average value for j. The result must also be divided by 2 since, otherwise, we will be counting each distance twice. (For example, if the focus is on two segments p and q, the distance between p and q is counted when we take q to be the center and later when p is taken as the center.) Mathematically, the above procedure is fairly complicated, but when all of these steps are carried out—using Equation (69) as the expression for $P(r)$—the radius of gyration for the random coil may be shown to equal

$$\overline{R_g^2} = \frac{1}{6}\, n\, \ell^2 \tag{77}$$

which is simply one-sixth the value of the mean of L^2 as given by Equation (73). Since the radius of gyration can be measured by viscosity and light scattering for random coils, it is apparent that Equation (77) may be used to measure ℓ. In this way, the effects of hindered rotation around bonds and of imbibed solvent may be evaluated quantitatively.

2.7c   Relation to the Kinetic Theory of Gases

Finally, we note that all the statistical equations of this chapter could have been borrowed directly from the kinetic theory of gases by simply changing the variables. We illustrate this now by going in the opposite direction. For example, if we replace the quantity $3/n\ell^2$ by m/k_BT and replace L by v in Equation (69), we obtain the *Boltzmann distribution* of molecular velocities in three dimensions. If we make the same substitutions in Equation (73), we obtain an important result from kinetic molecular theory:

$$\overline{v^2} = \frac{3\, k_B T}{m} \tag{78}$$

This shows that Equation (73) occupies a position for the random coil that is analogous to the position of average kinetic energy in the kinetic molecular theory. This is not just a fortuitous similarity, but a reflection of the statistical basis of both. As a little self-test: Did you recognize the similarity between the formalisms of the last few sections and kinetic theory as we went along?

2.8   EQUILIBRIUM BETWEEN SEDIMENTATION AND DIFFUSION

We have already noted that sedimentation and diffusion are opposing processes, the first tending to collect and the second to scatter. Let us now consider the circumstances under which these two tendencies equal each other. Once this condition is reached, of course, there will be no further macroscopic changes; the system is at equilibrium. In order to formulate this problem, consider the unit cross section shown in Figure 2.15, in which the x direction is in the direction of either a gravitational or a centrifugal field. Suppose this field tends to pull

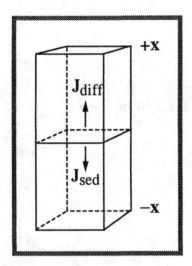

FIG. 2.15 The relationship between the flux due to sedimentation and that due to diffusion. At equilibrium, the two are equal.

the particles in the $-x$ direction. Gradually the concentration of the particles will increase in the region below the cross section of interest.

Back-diffusion occurs at a rate that increases with the buildup of a concentration gradient. When equilibrium is finally reached, we may write

$$J_{sed} = -J_{diff} \tag{79}$$

where J_{sed} is the flux across the area due to sedimentation and J_{diff} is the flux due to diffusion. The J_{diff} is given by Equation (22), and J_{sed} by

$$J_{sed} = vc \tag{80}$$

in which v is the sedimentation velocity and c is the concentration at the plane. Substituting Equations (22) and (80) into Equation (79) gives

$$v\,c = D\frac{dc}{dx} \tag{81}$$

If we substitute the value for the rate of sedimentation under gravity, Equation (4), we obtain

$$\frac{m}{f}\left(1 - \frac{\rho_1}{\rho_2}\right)g\,c = D\frac{dc}{dx} \tag{82}$$

If the sedimentation occurs in a centrifugal field, on the other hand, g must be replaced by $\omega^2 x$ in Equation (4):

$$\frac{m}{f}\left(1 - \frac{\rho_1}{\rho_2}\right)\omega^2 x\,c = D\frac{dc}{dx} \tag{83}$$

Equations (82) and (83) are easily integrated to produce expressions that give c as a function of x at equilibrium. Defining c_1 and c_2 to be the equilibrium concentrations at x_1 and x_2, respectively, and then integrating, we obtain for Equation (82)

$$\frac{m}{fD}\left(1 - \frac{\rho_1}{\rho_2}\right)g\,(x_2 - x_1) = \ln\left(\frac{c_2}{c_1}\right) \tag{84}$$

and for Equation (83)

$$\frac{m}{2fD} \left(1 - \frac{\rho_1}{\rho_2} \right) \omega^2 (x_2^2 - x_1^2) = \ln \left(\frac{c_2}{c_1} \right) \tag{85}$$

Finally, we recall Equation (32), which permits us to substitute $k_B T$ for Df, giving

$$\ln \left(\frac{c_2}{c_1} \right) = \frac{m}{k_B T} \left(1 - \frac{\rho_1}{\rho_2} \right) g \ (x_2 - x_1) \tag{86}$$

for sedimentation equilibrium under gravity and

$$\ln \left(\frac{c_2}{c_1} \right) = \frac{m}{2 \ k_B T} \left(1 - \frac{\rho_1}{\rho_2} \right) \omega^2 (x_2^2 - x_1^2) \tag{87}$$

for sedimentation equilibrium in a centrifuge. Note that sedimentation equilibrium studies permit the evaluation of particle mass with no assumptions about particle shape. More elaborate mathematical formulations of sedimentation/diffusion equilibria may be found in advanced textbooks (see, for example, Probstein 1994).

It should now be apparent why the ultracentrifuge is such an important tool in molecular biology. Although the method is in no way restricted to particles of biological significance, these particles are of a size and density that are ideally suited to the ultracentrifuge. An ultracentrifuge permits the evaluation of particle mass through equilibrium studies—see Equation (87)—and the evaluation of the ratio m/f through studies of the rate of sedimentation— see Equation (18). Combining these data permits the separate evaluation of f. From the mass and density of the material, the volume and radius of an equivalent sphere, and hence f_0, can be calculated. Then, Figure 2.9 may be consulted to determine particle characterizations that are consistent with the ratio f/f_0. Although both utilize the same instrument, sedimentation rate and sedimentation equilibrium are two different experiments that complement one another very nicely.

It might be noted that sedimentation equilibrium is approached very slowly; however, techniques that permit equilibrium conditions to be estimated from preequilibrium measurements have been developed by W. J. Archibald. Equations (86) and (87) predict a linear semilogarithmic plot of c versus x or x^2 for gravitational and centrifugal studies, respectively. The slope of such a plot is proportional to the mass of the particles involved. Remember that monodispersity was assumed in the derivation of these equations. If this condition is not met for an experimental system, the plot just described will not be linear. If each particle size present is at equilibrium, however, each component will follow the equations and the experimental plot will be the summation of several straight lines. Under certain conditions these may be resolved to give information about the polydispersity of the system. In any event, nonlinearity implies polydispersity once true equilibrium is reached.

We conclude this chapter with a final observation about Equation (86). If the particles in question are gas molecules instead of suspended particles, then the concentration ratio equals the ratio of pressures measured at two locations, and the particle mass requires no correction for buoyancy. Under these conditions, Equation (86) becomes

$$\ln \left(\frac{p_2}{p_1} \right) = \frac{m \ g \ (x_2 - x_1)}{k_B T} = \frac{M \ g \ (x_2 - x_1)}{RT} \tag{88}$$

This familiar equation gives the variation of barometric pressure with elevation and is known as the *barometric equation*. Once again we are reminded of the connection between the material of this chapter and kinetic molecular theory.

REVIEW QUESTIONS

1. List a few examples from your experience of phenomena in which sedimentation and diffusion play major roles, individually or together.
2. Describe the reasons for hydrodynamic friction. What is the physical significance of the friction factor? What are its units?
3. What is *Stokes's law*? Under what conditions is it valid?
4. Describe what types of information can be obtained from the sedimentation equation? What are the restrictions on the applicability of the equation derived in the text?
5. How is the expression for the sedimentation coefficient derived? What are the units of the sedimentation coefficient?
6. State *Fick's laws of diffusion*. Describe at least two different techniques to use Fick's laws to measure the diffusion coefficient.
7. Sketch qualitatively the variation of concentration profiles with time that you would expect from the solution of Fick's second law. Assume suitable boundary and initial conditions.
8. How would you combine sedimentation and diffusion measurements to determine the mass of a particle without any information on the shape of the particle.
9. Describe how the random walk statistics are used to relate the random walk to the diffusion coefficient.
10. What assumptions are made in the development of the random walk statistics discussed in the text, and what do they imply for the correspondence between random walk and diffusion?
11. What is the difference between the mean of the displacement and rms displacement? Can they be equal under any circumstance?
12. What is the relation between the diffusion coefficient and the rms displacement?
13. Describe an experiment to determine Avogadro's number from the average root mean square displacement of a particle due to random walk.
14. What is the *Stokes-Einstein equation*? Suggest at least two uses for that equation.
15. Why can one use random walk statistics to derive expressions for the end-to-end distance of a polymer chain? Under what conditions can this be done?
16. What is the difference between the radius of gyration of a polymer chain and the end-to-end distance?
17. What is the physical significance of the segment length used in the derivation of polymer chain statistics?
18. Describe the relation between sedimentation/diffusion equilibrium and the barometric equation.

REFERENCES

General References (with Annotations)

Berg, H. C., *Random Walks in Biology*, expanded ed. Princeton University Press, Princeton, NJ, 1993. (Undergraduate level. A lucid introduction to random walk statistics. Discusses translational and rotational diffusion, self-propelled motion, random walk, and sedimentation. Relevant to biological species of colloidal dimensions, but no background in biology is needed.)

Bird, R. B., Stewart, W. E., and Lightfoot, E. N., *Transport Phenomena*, Wiley, New York, 1960. (Undergraduate and graduate levels. A classic textbook on the formulation and solutions of equations of transport phenomena in engineering. Contains both undergraduate-level and graduate-level topics and problems.)

Einstein, A. *Investigations in the Theory of the Brownian Movement.*, Dover, New York, 1956. (Undergraduate level [now!]. A collection of the original papers of Einstein on Brownian movement. Accessible to undergraduates. A window to an era in which atomic theory of matter was still under hot debate.)

Giddings, J. C. *Unified Separation Science*, Wiley, New York, 1991. (Undergraduate level. A good introduction to many examples of separation processes in which diffusion plays an important role.)

Perrin, J. *Atoms*, Ox Bow Press, Woodbridge, CT, 1990. (Undergraduate level today; a research monograph when it originally appeared! The translation of the 1913 French original, *Les Atomes*, by the French physicist recognized by the 1926 Nobel Prize in physics for experimental observation of Brownian motion. The first five chapters deal with atomic theory, Brownian motion, fluctuations, and experimental observation of Brownian motion. Another window to the past for those interested in history of science.)

Probstein, R. F. *Physicochemical Hydrodynamics: An Introduction*, 2d ed., Wiley-Interscience, New York, 1994. (Primarily graduate level. Despite the confusing title, this is a textbook on topics in colloidal phenomena.)

Other References

Beyer, W. H. (Ed.). *CRC Handbook of Mathematical Sciences*, 6th ed., CRC Press, Boca Raton, FL, 1987.

Cantor, C. R., and Schimmel, P. R. *Biophysical Chemistry. Part II: Techniques for the Study of Biological Structure and Function*, W. H. Freeman, San Francisco, CA, 1980.

Giddings, J. C. *Separation Science*, **1**, 123 (1966).

Giddings, J. C. *Science*, **260**, 1456 (1993).

Oden, S. *Proc. R. Soc. Edinburgh*, **36**, 219 (1915).

Svedberg, T., and Pederson, K. O. *The Ultracentrifuge*, Oxford University Press, Oxford, England, 1940.

PROBLEMS

1. The following results describe the rate of accumulation of rutile (TiO_2; $\rho = 4.2$ g cm^{-3}) particles on a submerged balance plan.*

Time t (min)	W (g)	Time t (min)	W (g)
0.0	0.000	35	0.310
1.5	0.045	60	0.403
2.0	0.060	84	0.480
2.5	0.075	127	0.610
3.0	0.090	159	0.702
4.0	0.110	226	0.880
5.5	0.125	274	0.998
8.0	0.160	327	1.110
14.0	0.210	384	1.240
24.0	0.260	420	1.320

(a) Plot W versus t and use this to evaluate w as a function of time. Prepare a plot of w versus t.

(b) At what time (t_{max}) does the greatest increment in w appear to occur?

(c) Calculate the radius of an equivalent sphere corresponding to t_{max} ($\rho_{soln} = 0.997$ g cm^{-3}, $\eta = 0.00894$ P, $h = 12$ cm).

2. A preparation of reduced and carboxymethylated Mouse-Elberfeld virus protein particles in water reached sedimentation equilibrium at 25°C after 40 hr at 12,590 rpm. The following data show the recorder displacement (proportional to concentration) versus r for this protein:†

c (arbitrary units)   2.29 2.51 2.79 3.09 3.51 3.89 4.47 5.01 5.89 6.61 7.41 8.51

r (cm)   6.55 6.58 6.60 6.65 6.67 6.69 6.71 6.74 6.76 6.79 6.81 6.84

---

*Sullivan, W. F., and Jacobson, A. E. *Symposium on Particle Size Measurements*, ASTM Publication No. 234, 1959.

†Rueckert, R. R. *Virology* **26**, 345 (1965).

‡Vinograd, J., Bruner, R., Kent, R., and Weigle, J. *Proc. Nat. Acad. Sci., USA* **49**, 902 (1963).

(a) Use these results to evaluate the mass of the particles present ($\rho_{protein}$ = 1.370 g cm^{-3}) and estimate R_0 and f_0 for the molecules. Does the sample appear to be monodisperse?

(b) The sedimentation coefficient is known to be 2.7 S for this preparation. Evaluate f and f/f_0.

(c) What can be said about the possible axial ratio–hydration combinations of this protein in terms of Figure 2.9?

3. Southern bean mosaic virus (SBMV) particles are centrifuged at 12,590 rpm and the absorbance at 260 nm is measured along the settling direction as a function of time. The center of the absorption band varies with distance from the center of the rotor as follows:[‡]

t (min)	R (cm)
16	6.22
32	6.32
48	6.42
64	6.52
80	6.62
96	6.72
112	6.82
128	6.92
144	7.02

4. Phosphatidylcholine micelles are spherical particles having a molecular weight of 97,000 g mole^{-1}. Assuming that the density of the dry lipid (ρ = 1.018 g cm^{-3}) applies to the micelles, calculate the radius R_s and the diffusion coefficient D for these particles in water at 20°C. The experimental value of the diffusion coefficient is 6.547 × 10^{-7} cm^2 s^{-1} under these conditions.[#] Evaluate f/f^* and estimate the hydration of the lipid.

5. Calculate the diameter of a spherical particle (ρ = 4 g cm^{-3}) for which the rms displacement due to diffusion at 25°C is 1% the distance of sedimentation in a 24-hr period through a medium for which ρ = 1 g cm^{-3} and η = 9 × 10^{-3} P. For what diameter is the diffusion distance 10% of the settling distance?

6. The diffusion of alkyl ammonium ions into clay pellets has been studied by bringing the pellet into superficial contact with an isotopically labeled salt and then, after a suitable time, using a microtome to slice the pellet. The radioactivity is then measured in successive thin slices of the pellet. Assuming that Equation (62) describes the diffusion process, estimate how long it would take for 1% of the initial activity of each of the ions to appear in the 15th slice inward from the exposed surface of the dry clay if each slice is 40-μm thick. The diffusion coefficients for the methyl and trimethyl ammonium cations under these conditions are 7.03 × 10^{-12} and 2.65 × 10^{-11} cm^2 s^{-1}, respectively.[†]

7. Suppose two reservoirs 4-cm apart are cut into an agar gel in a petri dish. Solutions of Pb(NO$_3$)$_2$ and Na$_2$CrO$_4$ are introduced simultaneously into the two reservoirs. At what distance into the gel does PbCrO$_4$ precipitate if the diffusion coefficients of Pb^{2+} and CrO$_4^{2-}$ in agar are 0.657 × 10^{-5} and 0.752 × 10^{-5} cm^2 s^{-1}, respectively?[‡] Where would Prussian blue precipitate if the reservoirs contained Fe^{3+} (D = 0.434 × 10^{-5} cm^2 s^{-1}) and Fe(CN)$_6^{4-}$ (D = 0.557 × 10^{-5} cm^2 s^{-1})?

[#]Hasser, H. In *Water, a Comprehensive Treatise*, Vol. 4 (F. Franks, Ed.), Plenum, New York, NY, 1975.

[†]Gast, R. G., and Morfland, M. M. *J. Colloid Interface Sci.* **37**, 80 (1971).

[‡]Lee, R. E., and Meeks, F. R. *J. Colloid Interface Sci.* **35**, 584 (1971).

8. Use a compass to inscribe a circle around the aggregate shown on the cover of the book. Such a "graticule" might be used to define an average diameter for the aggregate. Compare the radius of the inscribed circle to that of the primary particles in the aggregate. How does the ratio of these radii compare with the value that would be obtained for the case $n = 76$ if the aggregate were built up according to a random walk model? Compare the features of the two models with the objective of accounting for the relative magnitude of the aggregate radius relative to the primary particle radius in the two cases.

9. The molecular weights and sedimentation coefficients of human plasminogen and plasmin ($\rho = 1.40$ g cm^{-3}) are as follows:*

	Plasminogen	Plasmin
M (g mole^{-1})	81,000	75,400
s at 20°C (S)	4.2	3.9

    (a) Calculate the diffusion coefficient for each.
    (b) Prepare a graph such as that of Figure 2.10 showing quantitatively how an initially thin band of these proteins widens with time. Show at least three different times.

10. Colloids (casein micelles) of two different particle sizes are isolated from skim milk by centrifugation under different conditions. The sedimentation and diffusion coefficients of the two preparations are as follows:†

Preparation conditions	$D \times 10^8$ (cm^2 s^{-1})	$s \times 10^{13}$ (s)
5 min at 5,000 rpm	0.97	2200
40 min at 20,000 rpm	2.82	800

    Calculate the mass per particle and the gram molecular weight of the two micelle fractions, assuming $\rho = 1.43$ g cm^{-3} for the dispersed phase.

11. The following data give the number of gold particles (as $\log_{10}$) versus depth beneath the surface for an aqueous dispersion allowed to reach sedimentation equilibrium under the influence of gravity:‡

Depth (mm)	4.44	5.06	5.67	6.30	6.90	7.53	8.15	8.65
$\log n$	10.36	10.51	10.63	10.75	10.89	11.05	11.22	11.39

    Calculate the radius of the gold particles ($\rho_{Au} = 19.3$ g cm^{-3}), treating the dispersed units as equivalent spheres.

12. At equilibrium at 20°C the concentration of tobacco mosaic virus (TMV) shows a linear semilogarithmic graph when plotted against the square of the distance to the axis of rotation in a centrifuge. Evaluate the molecular weight of the TMV particles if $\ln c = -1.7$ at $r^2 = 44.0$ cm^2 and $\ln c = -2.8$ at $r^2 = 41.2$ cm^2 when the dispersion is centrifuged at 6.185 rps. The density of the TMV is 1.36 g cm^{-3}.§

13. Verify that the expansion (see Appendix A) of Equation (49) leads to Equation (50). Retain no terms higher than the second order in expansions. Verify that the integration (see Table 2.2) of Equations (64) and (72) leads to Equations (65) and (73), respectively.

*Barlow, G. H., Summaria, L., and Robbins, K. C. *J. Biol. Chem.* **244**, 1138 (1969).
†Morr, C. V., Lin, S. H. C., Dewan, R. K., and Bloomfield, V. A. *J. Dairy Sci.* **56**, 415 (1973).
‡McDowell, C. M., and Usher, F. L. *Proc. R. Soc. London* **138A**, 133 (1932).
§Weber, F. N., Jr., Elton, R. M., Kim, H. G., Rose, R. D., Steere, R. L., and Kupke, D. W. *Science* **140**, 1090 (1963).

3

Solution Thermodynamics
Osmotic and Donnan Equilibria

All faults or defects, . . . Pantocyclus attributed to some deviation from perfect Regularity in the bodily figure, caused perhaps . . . by some collision in a crowd; by neglect to take exercise, or by taking too much of it; or even by a sudden change of temperature.

From Abbott's *Flatland*

3.1 INTRODUCTION

3.1a What Are Osmotic Pressure and Donnan Equilibria?

Osmotic pressure is a property of solutions of organic or inorganic solutes and appears in a number of contexts in the study of colloidal systems. As we see below, it is usually defined using a real or thought experiment involving a solution separated from the pure solvent by a membrane permeable only to the solvent. In such a setup, the pure solvent would flow through the membrane in an attempt to equalize the concentration difference. The pressure necessary to stop this flow is known as the *osmotic pressure* (from the Greek word *osmos*, meaning "to push or thrust"). In thermodynamic terms, the concentration difference created by the membrane represents a difference in the chemical potentials in the two compartments separated by the membrane, and the osmotic pressure counteracts the difference in the chemical potential.

It is important to note that the concept of osmotic pressure is more general than suggested by the above experiment. In particular, one does not have to invoke the presence of a membrane (or even a concentration difference) to define osmotic pressure. The osmotic pressure, being a property of a solution, always exists and serves to counteract the tendency of the chemical potentials to equalize. It is not important how the differences in the chemical potential come about. The differences may arise due to other factors such as an electric field or gravity. For example, we see in Chapter 11 (Section 11.7a) how osmotic pressure plays a major role in giving rise to repulsion between electrical double layers; here, the variation of the concentration in the electrical double layers arises from the electrostatic interaction between a charged surface and the ions in the solution. In Chapter 13 (Section 13.6b.3), we provide another example of the role of differences in osmotic pressures of a polymer solution in giving rise to an effective attractive force between colloidal particles suspended in the solution.

A related phenomenon occurs when the membrane in the above-mentioned experiment is permeable to the solvent and small ions but not to a macroion such as a polyelectrolyte or charged colloidal particles that may be present in a solution. The polyelectrolyte, prevented from moving to the other side, perturbs the concentration distributions of the small ions and gives rise to an ionic equilibrium (with attendant potential differences) that is different from what we would expect in the absence of the polyelectrolyte. The resulting equilibrium is known as the *Donnan equilibrium* (or, the *Gibbs-Donnan equilibrium*) and plays an important role in a number of problems of interest in colloid and surface chemistry, electrochemistry, and biology, as illustrated through an example in Vignette III.

3.1b   Importance of Osmotic Pressure and Donnan Equilibrium

The osmotic pressure can be measured accurately for colloidal solutes, and one molecular parameter of interest that is readily determined by osmometry is the number average molecular weight of the solute. Molecular weights determined by osmometry are absolute values; no calibration with known standards or any assumed theoretical models is required. Even the assumption of solution ideality is not a problem, since results are extrapolated to zero solute concentration before calculations are made.

Equilibrium thermodynamics of solutions of colloidal solutes determines the quantitative details of osmotic and Donnan equilibria and occupies our concern in this chapter. In particular, our discussion here centers on the osmotic pressure.

VIGNETTE III   BIOPHYSICS AND PHYSIOLOGY: Donnan Equilibrium and the "Resting States" of Nerve Cells

A major function of biological cell membranes is to regulate the transport of ionic and nonionic nutrients and reject metabolic wastes while maintaining the intracellular ionic atmosphere necessary for life-sustaining functions. The transfer of materials often has to be accomplished *against* concentration gradients. Understanding how the cells regulate such transport is essential in health care.

A cell is enclosed by a lipid bilayer known as the plasma membrane. In Vignette I.2 in Chapter 1 we discussed an example of a membrane, a complex structure with a mosaic of embedded or adsorbed moieties such as proteins. It is these membranes that protect the intracellular contents from the exterior environment of the cells and regulate the transport of materials into and out of the cells. They can also act as signal transducers and control the electrical excitation in the nervous system by altering the (membrane) permeability to particular ions in response to stimuli. Such electrical activities can propagate over long distances and represent one of the most spectacular of the membrane functions.

For our purpose here, we are interested in the following question: How do the membranes maintain the ionic environments in the cell over long periods of time while simultaneously allowing transmembrane passage of the needed ions? The mechanism that makes this possible is the so-called Donnan equilibrium; we discuss the details of this equilibrium in this chapter, but a qualitative picture of the Donnan effect can be obtained with the help of a simplified model of a cell shown in Figure 3.1 (See, also, Section 3.5a.).

Figure 3.1a shows a membrane that is permeable to water and K^+ and Cl^- ions but impermeable to colloidal electrolytes (polyelectrolytes such as charged proteins). Let α denote the interior of the cell and β the extracellular region. In the absence of the polyelectrolyte, water, K^+ and Cl^- partition themselves into the two sides such that the chemical potentials of each species are the same inside as well as outside, as thermodynamics would demand. Moreover, the requirement of electroneutrality in both α and β demands that the concentrations of each species K^+ and Cl^- be the same on either side of the partition.

Now, what can we expect if polyelectrolyte ions P^{z-}, each with z negative charges, are present in α? Since the electroneutrality in α is disturbed by the presence of the polyelectrolyte, some of the Cl^- ions will be pushed out to the extracellular region; similarly, there will be an accompanying increase in the concentration of the K^+ ions inside the cell. Eventually the system will reach a state illustrated in Figure 3.1b, with a higher concentration (hence, higher chemical potential) of Cl^- outside and a higher concentration of K^+ inside. This effect is the *Donnan effect* and is behind the maintenance of what is known as the "resting state" of the cell, that is, the polyelectrolyte maintains the concentrations in its environment against the chemical potential gradients. This is one of the reasons why cell and molecular biology books begin with a description of the Donnan equilibrium and the attendant osmotic pressure effects in their discussions of cellular functions.

Elementary and advanced treatments of such cellular functions are available in specialized monographs and textbooks (Bergethon and Simons 1990; Levitan and Kaczmarek 1991; Nossal and Lecar 1991). One of our objectives in this chapter is to develop the concepts necessary for understanding the Donnan equilibrium and osmotic pressure effects. We define osmotic pressures of charged and uncharged solutes, develop the classical and statistical thermodynamic principles needed to quantify them, discuss some quantitative details of the Donnan equilibrium, and outline some applications.

3.1c Focus of This Chapter

As mentioned above, the primary focus of this chapter is on osmotic pressure and its basis in solution thermodynamics. We consider both classical and statistical thermodynamic interpretations of osmotic pressure. The next three sections are devoted to this. The last two sections describe osmotic effects in charged systems and a few applications of osmotic phenomena.

1. The thermodynamic preliminaries and concepts needed for defining osmotic pressure are discussed in Sections 3.2a–c. The nonideality of colloidal solutions can be appreciable since the solvent and solute particles are so different in size. Classical thermo-

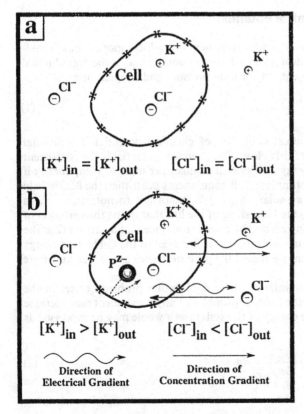

FIG. 3.1 Donnan equilibrium and regulation of ionic concentrations resulting from the presence of a semipermeable membrane: (a) conditions in the absence of polyelectrolyte molecules; and (b) ionic concentrations and electrical gradients in the presence of the polyelectrolyte. The membrane is permeable to water and K^+ and Cl^-, but not to the polyelectrolyte P^{z-}.

dynamics allows this nonideality to be quantified in terms of the so-called virial coefficients, as outlined in Section 3.2d.

2. Some of the experimental details of osmometry and the use of osmometry for determining molecular weights and second virial coefficients are discussed in Section 3.3.

3. If we turn from phenomenological thermodynamics to statistical thermodynamics, then we can interpret the second virial coefficient in terms of molecular parameters via a model. We pursue this approach for two different models, namely, the excluded-volume model for solute molecules with rigid structures and the Flory-Huggins model for polymer chains, in Section 3.4.

4. We conclude the chapter with a discussion of the Donnan equilibrium and the thermodynamic behavior of charged colloids, particularly with respect to osmotic pressure and molecular weight determination (Section 3.5), and some applications of osmotic phenomena (Section 3.6).

Although this is the first place in this book that we have devoted any attention to charged particles, it is not the last. Chapters 11-13, in particular, devote a good deal of attention to such systems. A part of Chapter 4 is devoted to the effects of charges on particles on viscous behavior of dispersions.

3.2 OSMOTIC PRESSURE: THERMODYNAMIC FOUNDATIONS

3.2a Gibbs Free Energy and Chemical Potential

An extremely useful quantity in the thermodynamic treatment of multicomponent phase equilibria is the chemical potential. The chemical potential for component i, μ_i, is the partial molal Gibbs free energy with respect to component i at constant pressure and temperature:

$$\mu_i = \left(\frac{\partial G}{\partial n_i}\right)_{p,T,n_{j\neq i}} \tag{1}$$

Although the notation of this mathematical definition of chemical potential is somewhat cumbersome, the physical significance is fairly clear. The chemical potential is the coefficient that describes the way the Gibbs free energy of a system changes per mole of component i if the temperature, pressure, and number of moles of all components other than the ith are held constant. Although it is expressed on a molar basis, it is important to note that μ_i is a differential quantity; that is, it represents the local slope of the line that shows the variation of G with n_i. The line itself arises from slicing across the complex surface that describes G at the specified values of p, T, and n_j. For a pure substance, μ_i is identical to the Gibbs free energy per mole of that substance. For the present we regard the pure substance as the standard state for μ_i and represent it by the symbol μ_i^o.

The great utility of the chemical potential in phase equilibrium problems arises in the following way. In open systems in which the number of moles of any component may increase or decrease, any change in the Gibbs free energy of the system as a whole may be expressed as the sum of the following contributions:

$$dG = \left(\frac{\partial G}{\partial T}\right)_{p,n} dT + \left(\frac{\partial G}{\partial p}\right)_{T,n} dp + \sum_i \left(\frac{\partial G}{\partial n_i}\right)_{p,T,n_j} dn_i \tag{2}$$

This equation may be written

$$dG = -S\,dT + V\,dp + \sum_i \mu_i dn_i \tag{3}$$

by substituting into Equation (2) the definition of chemical potential and the familiar relationships

$$\left(\frac{\partial G}{\partial T}\right)_{p,n} = -S \tag{4}$$

and

$$\left(\frac{\partial G}{\partial p}\right)_{T,n} = V \tag{5}$$

For any equilibrium that occurs at constant temperature and pressure, the first two terms from the right-hand side of Equation (3) equal zero, reducing the equation to the form

$$dG = \sum_i \mu_i dn_i \tag{6}$$

If the total system consists of several different phases, which we designate by Greek subscripts $\alpha, \beta, \dots$, then we may also write

$$dG = dG_\alpha + dG_\beta + \dots \tag{7}$$

Finally, the equilibrium condition requires that

$$dG = 0 \tag{8}$$

Therefore, the equilibrium between two phases means that

$$dG = 0 = dG_\alpha + dG_\beta = \sum_i \mu_{i,\alpha} dn_{i,\alpha} + \sum_i \mu_{i,\beta} dn_{i,\beta} \tag{9}$$

That is, for each of the components, the following holds:

$$\mu_{i,\alpha} dn_{i,\alpha} + \mu_{i,\beta} dn_{i,\beta} = 0 \tag{10}$$

If the entire system consists of only two phases, the hypothesis here, the conservation of matter requires that any substance lost from one phase must appear in the other:

$$dn_{i,\alpha} = -dn_{i,\beta} \tag{11}$$

Combining these last two results leads to the conclusion that the condition for phase equilibrium at constant temperature and pressure is

$$\mu_{i,\alpha} = \mu_{i,\beta} \tag{12}$$

It is important to realize that the chemical potential of each component must be the same in all equilibrium phases, although the value for this quantity will, in general, be different for each component.

3.2b Thermodynamic Activities of Components

Perhaps the most basic equation involving the chemical potential is the one that relates this quantity to the activity of component i in the solution, a_i:

$$\mu_i = \mu_i^0 + RT \ln a_i \tag{13}$$

For the present a_i is expressed in mole fraction units. We see, therefore, that μ_i approaches μ_i^0 as a_i approaches unity. Furthermore, since μ_i is the partial molal Gibbs free energy, Equation (5) also applies to μ_i provided we replace V with $\overline{V}_i$, the partial molal volume of component i:

$$\left(\frac{\partial \mu_i}{\partial p}\right)_{T,n} = \overline{V}_i \tag{14}$$

Suppose we apply these relationships to the equilibrium of a liquid mixture and its vapor. At equilibrium, μ_i must have the same value for each component in both the liquid and vapor phases. Therefore

$$\mu_{i,L} = \mu_{i,V} \tag{15}$$

or

$$\left(\frac{\partial \mu_{i,L}}{\partial p}\right)_{T,n} = \left(\frac{\partial \mu_{i,V}}{\partial p}\right)_{T,n} = \overline{V}_i = \frac{RT}{p_i} \tag{16}$$

where the last relationship assumes the vapor to behave ideally. If Equation (13) is used to evaluate the left-hand side of this equation, we obtain

$$RT\,\partial\ln a_i = RT\frac{\partial p_i}{p_i} \tag{17}$$

Recalling that $p_i = p_i^0$, the *normal vapor pressure* of the pure liquid, when $a_i = 1$, we may integrate Equation (17) to give

$$a_i = \frac{p_i}{p_i^0} \tag{18}$$

This relationship constitutes the basic definition of the activity. If the solution behaves ideally, $a_i = x_i$ and Equation (18) define *Raoult's law*. Those four solution properties that we know as the *colligative properties* are all based on Equation (12); in each, solvent in solution is in equilibrium with pure solvent in another phase and has the same chemical potential in both phases. This can be solvent vapor in equilibrium with solvent in solution (as in vapor pressure lowering and boiling point elevation) or solvent in solution in equilibrium with pure, solid solvent (as in freezing point depression). Equation (12) also applies to osmotic equilibrium as shown in Figure 3.2.

3.2c Osmotic Equilibrium: Relating Chemical Potentials and Compositions to Osmotic Pressure

Figure 3.2 shows schematically two liquid phases—one solution, the other pure solvent—separated by a partition known as a *semipermeable* membrane. In many ways this membrane is the central feature of osmometry; we have more to say about it, but for now it is sufficient to define a semipermeable membrane as one that is permeable to the solvent and impermeable

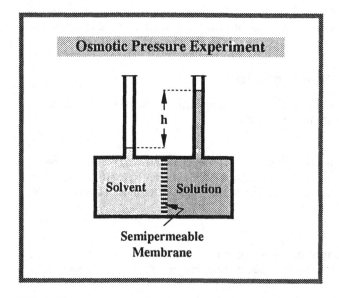

FIG. 3.2 Schematic representation of an osmotic pressure experiment.

to the solute. Thus the semipermeable membrane in Figure 3.2 prevents the two liquids from mixing and at the same time allows both sides of the membrane to come to equilibrium.

In order for solvent and solution to be in equilibrium in an apparatus such as that shown in Figure 3.2, the solution side must be at a higher pressure than the solvent side. This excess pressure is what is known as the *osmotic pressure* of the solution. If no external pressure difference is imposed, solvent will diffuse across the membrane until an equilibrium hydrostatic pressure head has developed on the solution side. In practice, to prevent too much dilution of the solution as a result of the solvent flow into it, the column in which the pressure head develops is generally of a very narrow diameter. We return to the details of osmotic pressure experiments in the next section. First, however, the theoretical connection between this pressure and the concentration of the solution must be established.

Since the two sides of the membrane are in true isothermal equilibrium in an osmotic pressure experiment, the chemical potential of the solvent must be the same on both sides of the membrane. On the side containing pure solvent, μ_i equals μ_i^0. On the solution side of the membrane, the chemical potential of the solvent must also equal the same value according to the equilibrium criterion of Equation (12).

Equation (13) reminds us that the chemical potential has its greatest value, μ_i^0, for a pure substance. Any value of a_i less than unity will cause μ_i to be altered from μ_i^0 by an amount $RT \ln a_i$, which will be negative for $a_i < 1$. Second, any pressure on a liquid that exceeds p_i^0 increases μ_i, above μ_i^0. This is seen from the combination of Equations (13) and (18). Thus consideration of the chemical potential of the solvent makes it clear how osmotic equilibrium comes about. The presence of a solute lowers the chemical potential of the solvent. This is offset by a positive pressure on the solution, the osmotic pressure π, so that the net chemical potential on the solution side of the membrane equals that of the pure solvent on the other side of the membrane. This is summarized by the expression

$$\mu_1^0 = \mu_1^0 + RT \ln a_i + \int_{p_1^0}^{p_1^0 + \pi} \overline{V}_1 dp \qquad (19)$$

In which subscript 1 indicates the solvent. We use the subscripts 1 and 2 to indicate solvent and solute, respectively, throughout this chapter. In order to relate π to the concentration of the solution, then, we must find a way to integrate Equation (19). The easiest way of doing this is to assume that $\overline{V}_1$ is constant. This approximation is justified because the solution is a condensed phase and shows negligible compressibility. Making this assumption and integrating Equation (19) gives

$$\ln a_i = -\frac{\pi \overline{V}_1}{RT} \qquad (20)$$

Combining Equations (13) and (20) permits us to express the chemical potential in terms of osmotic pressure instead of activity:

$$\mu_1 = \mu_1^0 - \pi \overline{V}_1 \qquad (21)$$

This relationship will be useful in Chapter 5.

Equation (20) provides the relationship we have sought between osmotic pressure and concentration. If the solution is ideal, we may replace activity by mole fraction. Then Equation (20) becomes

$$\ln x_1 = -\frac{\pi \overline{V}_1}{RT} = \ln (1 - x_2) \approx -x_2 - \frac{x_2^2}{2} - \cdots \qquad (22)$$

where the approximation arises from the expansion of the logarithm (see Appendix A). Therefore for ideal, two-component solutions, we write

$$x_2 = \frac{\pi \overline{V}_1}{RT} \qquad (23)$$

Since real solutions tend toward ideality as the solute concentration decreases,

$$x_2 = \frac{n_2}{n_1 + n_2} \approx \frac{n_2}{n_1} \tag{24}$$

in which the n's are the number of moles of the indicated component. The approximate form of Equation (24) applies in the case of dilute solutions for which $n_2 \ll n_1$. Introducing the dilute solution approximation of Equation (24) into Equation (23) yields

$$n_2 = \frac{n_1 \pi \overline{V}_1}{RT} = \frac{\pi V_1}{RT} \tag{25}$$

where V_1 is simply the volume of the solvent in the solution, $n_1 \overline{V}_1$. This relationship, known as the *van't Hoff equation* (after J. H. van't Hoff, recipient of the first Nobel Prize in chemistry, 1901), is analogous in form to the ideal gas law and, like the ideal gas law, is a limiting law that applies perfectly only in the limit of $n_2 \rightarrow 0$ or $n_2/V_1 \rightarrow 0$. Equation (25) shows that osmotic pressure experiments provide a means of measuring the number of solute particles in a solution. If the weight of solute in the solution is also known, this information may be used to evaluate molecular weights. We consider the application of osmometry to problems of molecular weight determination in Example 3.1. For now it is sufficient to note that even high molecular weight solutes—for which the number of molecules per weight of sample is orders of magnitude less than for low molecular weight compounds—produce appreciable osmotic pressures. For example, if a 1% aqueous solution is assumed to be ideal at 25°C, then Equation (25) shows that solutes of molecular weight 10^3, 10^4, 10^5, and 10^6 have osmotic pressures of 2530, 253, 25.3, and 2.53 mm of solution, respectively. Thus even very high molecular weight solutes generate pressures that result in easily measured liquid columns. We see below how to get around the assumption of ideality in Equation (25) so this result can be applied with confidence to real solutions.

3.2d Nonideality: Virial Expansion for the Osmotic Pressure

After reaching Equation (20) we abandoned a completely general discussion of osmotic pressure in favor of the simpler assumption of ideality. The ideal result, Equation (25), applies to real solutions in the limit of infinite dilution. The objective of this section is to examine the extension of Equation (20) to nonideal solutions or, more practically, to solutions with concentrations that are greater than infinitely dilute.

Since both the osmotic pressure of a solution and the pressure-volume-temperature behavior of a gas are described by the same formal relationship of Equation (25), it seems plausible to approach nonideal solutions along the same lines that are used in dealing with nonideal gases. The behavior of real gases may be written as a power series in one of the following forms for n moles of gas:

$$\frac{pV}{nRT} = 1 + B\rho + C\rho^2 + \ldots \tag{26}$$

or

$$\frac{pV}{nRT} = 1 + B\left(\frac{n}{V}\right) + C\left(\frac{n}{V}\right)^2 + \ldots \tag{27}$$

Equations of this type are known as virial equations, and the constants they contain are called the virial coefficients. It is the second virial coefficient B that describes the earliest deviations from ideality. It should be noted that B would have different but related values in Equations (26) and (27), even though the same symbol is used in both cases. One must be especially attentive to the form of the equation involved, particularly with respect to units, when using literature values of quantities such as B. The virial coefficients are temperature dependent and vary from gas to gas. Clearly, Equations (26) and (27) reduce to the ideal gas law as $p \rightarrow 0$ or as $n/V \rightarrow 0$. Finally, it might be recalled that the second virial coefficient in Equation (27) is related to the van der Waals constants a and b as follows:

$$B = b - \frac{a}{RT} \tag{28}$$

This last relationship points out that for gases the second virial coefficient arises both from the finite size and from the interactions of the molecules of the gas since these are the origins, respectively, of b and a. We have more to say about this in Chapter 10 (see, for example, Sections 10.4 and 10.5).

As we extend these ideas to nonideal solutions, a similar set of statements will apply to the resulting power series:

1. The second virial coefficient is our primary concern since we focus attention on the first deviations from ideality.
2. The value of B will depend in part on the units chosen for concentration, as well as on the temperature and the nature of the system.
3. The virial equation for osmotic pressure must reduce to the van't Hoff equation in the limit of infinite dilution.
4. The second virial coefficient may be expected to reflect both the finite size and the interactions of the molecules.

Each of these points is taken up in the following discussion.

The easiest way to extend these considerations to the osmotic pressure of nonideal solutions is to return to Equation (22), which relates π to a power series in mole fraction. This equation applies to ideal solutions, however, since ideality is assumed in replacing activity by mole fraction in the first place. To retain the form and yet extend its applicability to nonideal solutions, we formally include in each of the concentration terms a correction factor defined to permit the series to be applied to nonideal solutions as well:

$$\frac{\pi \overline{V}_1}{RT} = A'x_2 + \frac{1}{2} B'x_2^2 + \ldots \tag{29}$$

The coefficients A', B', . . . must all equal unity in ideal solutions in order to recover Equation (22). Since the van't Hoff equation is a limiting law, the coefficient A' must equal unity in all solutions. Therefore Equation (29) becomes

$$\frac{\pi \overline{V}_1}{RT} = x_2 + \frac{1}{2} B'x_2^2 + \ldots \tag{30}$$

Next, let us consider the transformation of mole fraction concentration units into other units more appropriate for use with solutes with molecular weights that might be unknown. In working with unknowns, mass volume^{-1} units offer the greatest flexibility since only a balance and a volumetric flask are needed to quantitatively characterize a solution in these units. We represent this unit of concentration by c in this chapter. For the reasons presented here, this same concentration unit is used in viscosity and light scattering work with solutes of unknown molecular weight. If V is the volume of a solution of components 1 and 2, it may be written as $n_1\overline{V}_1 + n_2\overline{V}_2$, where the $\overline{V}_i$'s are partial molar volumes. Remember, the partial molar volumes give the volume occupied by a mole of the indicated component in a mixture; their precise value depends on the concentration of the solution. Therefore, if m_2 is the mass of solute in the solution,

$$c = \frac{m_2}{n_1\overline{V}_1 + n_2\overline{V}_2} \approx \frac{m_2}{n_1\overline{V}_1} \tag{31}$$

where the approximation applies to dilute solutions in which $n_2 \ll n_1$. Since m_2 is equal to the product of n_2 and M_2,

$$c = \frac{n_2 M_2}{n_1\overline{V}_1} \approx x_2 \frac{M_2}{\overline{V}_1} \tag{32}$$

in which the dilute solution form of Equation (24) has been used to introduce x_2. Substituting this result into Equation (30) gives

$$\frac{\pi \overline{V}_1}{RT} = \frac{\overline{V}_1}{M_2} c + \frac{1}{2} B' \left(\frac{\overline{V}_1}{M_2}\right)^2 c^2 + \ldots \tag{33}$$

Ignoring higher order terms, we can rearrange Equation (33) to yield

$$\frac{\pi}{RTc} = \frac{1}{M_2} + \frac{1}{2} \frac{B' \overline{V}_1}{M_2^2} c = \frac{1}{M_2} + B c \tag{34}$$

The quantity π/c is called the *reduced osmotic pressure*, and $(\pi/c)_0$ the limiting reduced osmotic pressure. Osmotic pressure results are most commonly encountered as plots of reduced osmotic pressure — with or without the RT — versus c. This form suggests that a plot of π/RTc versus c should be a straight line, the intercept and slope of which have the following significance:

$$\text{intercept} = \left(\frac{\pi}{RTc}\right)_0 = \frac{1}{M_2} \tag{35}$$

$$\text{slope} = B = \frac{1}{2} \frac{B' \overline{V}_1}{M_2^2} \tag{36}$$

Note that the value of the intercept, the value of π/RTc at infinite dilution, obeys the van't Hoff equation, Equation (25). At infinite dilution even nonideal solutions reduce to this limit. The value of the slope is called the second virial coefficient by analogy with Equation (27). Note that the second virial coefficient is the composite of two factors, B' and $(1/2) \overline{V}_1/M_2^2$. The factor B' describes the first deviation from ideality in a solution; it equals unity in an ideal solution. The second cluster of constants in B arises from the conversion of practical concentration units to mole fractions. Although it is the nonideality correction in which we are primarily interested, we discuss it in terms of B rather than B' since the former is the quantity that is measured directly. We return to an interpretation of the second virial coefficient in Section 3.4.

3.3   OSMOMETRY: SOME APPLICATIONS

3.3a   Experimental Considerations

To carry out an osmotic pressure experiment, we need to prepare a solution, find a suitable semipermeable membrane, achieve isothermal equilibrium, and measure the equilibrium pressure. Aside from noting that the pressures produced by colloidal solutes are large enough to be measurable, we have not yet considered any of the experimental aspects of osmometry. This is our present task.

First, a suitable solvent and membrane must be found. The solvent must dissolve enough solute to produce an adequate pressure. The results of measurements made at relatively high concentrations may be extrapolated, as discussed in Section 3.2d and illustrated below, to zero concentration, so we need not worry about the effects of nonideality. As low a solvent viscosity as possible is desirable to minimize the time required for equilibration.

It is important that the materials be free of contaminants in osmotic pressure experiments. Suppose, for example, that the solvent contains a small amount of impurity that, like the colloidal solute, is retained by the membrane. Then, as far as the osmotic pressure is concerned, that impurity will contribute to the osmotic pressure in the same way that the colloid does. Since the osmotic pressure responds to the number of solute particles present, a low molecular weight impurity in extremely small amounts may contain as many or more particles as a dilute solution of a colloidal solute of very high molecular weight. Quite large errors in molecular weight may arise in this way. The confusion may be compounded if the same system is investigated using a different membrane material. It is conceivable that another membrane would be permeable to the impurity. This would result in a different apparent molecular weight for the colloid. We have approached the issue of impurities from the viewpoint of the solvent. Actually, the colloid is more likely to be the source of the impurity since these materials are often difficult to purify. We discuss some additional aspects of this problem in the sections on average molecular weight (Section 3.3c), charged colloids (Section 3.5b), and dialysis (Section 3.6a).

The membrane is the source of most of the difficulties in osmometry. There is no general way to select a membrane material that will be permeable to one component and impermeable to another for any conceivable combination of chemicals. Very high molecular weight components are generally more easily retained, however, so we have a slight advantage in this regard. The membrane must be sufficiently thin to permit equilibration at a reasonable rate. At the same time, it must be strong enough to withstand the considerable pressure differences that may exist across it. This problem may generally be overcome, at least in part, by suitable mechanical support of the membrane. A more serious problem is the preparation of thin membranes that are free from minute imperfections that would constitute a "leak" between the two compartments. Such a leak would totally invalidate the experimental results. One peculiarity of membranes is their tendency to display what is known as an asymmetry pressure; that is, an equilibrium pressure difference may exist across a membrane even when there is solvent on both sides. This must be measured and subtracted as a "blank correction."

Equilibrium osmometry is a thermodynamic phenomenon. As such, it makes no difference what mechanism the membrane uses to retain the solute, nor do we learn anything about the mechanism from equilibrium studies. It is easy to visualize that some solutes, especially those in the colloidal size range, are retained by a sieve effect; that is, the molecules are simply too large to pass through the pores in the membrane material. Another possibility is a mechanism by which the membrane displays selective solubility. This means that the membrane dissolves the solvent but not the solute. In this way the solvent can pass through the membrane, while the solute is retained. An analogous mechanism for charged particles may arise by the membrane repelling (and thus retaining) particles of one particular charge.

A great many different materials have been used in osmotic pressure experiments. Various forms of cellophane and animal membranes are probably the most common membrane materials. Various other polymers, including polyvinyl alcohol, polyurethane, and polytrifluorochloroethylene, have also been used along with such inorganic substances as $CuFe(CN)_6$ precipitated in a porous support.

Once a suitable membrane and solvent are selected, an experimental arrangement must be devised that measures the equilibrium pressure under isothermal conditions. Many variations in apparatus design have been studied. Two particularly instructive pieces of apparatus are shown in Figure 3.3. The assembly shown in Figure 3.3a consists of an inner solution compartment with a relatively large opening at the membrane end and a capillary at the small end. The entire solution chamber is then immersed in a tube containing the solvent. Once assembled, the entire apparatus is placed in a constant temperature bath for equilibration.

Another osmometer design is shown assembled in Figure 3.3b and in detail in Figure 3.3c. The membrane is pressed between two grooved faces that contain solvent on one side and solution on the other. The grooves are attached to filling tubes and to capillaries in which the pressure head develops. This apparatus permits a large contact area between liquids with a minimum volume of liquids involved.

Several times in this discussion we have noted the importance of experimental conditions that permit as rapid an equilibration as possible. The implication of these remarks is that osmotic equilibrium is reached slowly. In some cases as much as one week may be required for equilibrium to be achieved. To shorten this time, procedures based on measuring the rate of approach to equilibrium have been developed. The osmometer of Figure 3.3b is especially suited for this procedure.

In successive runs, the capillary on the solution side of the membrane is filled with solution to some initial setting that will be above or below the equilibrium location of the meniscus. At various times after the initial settings, the height of the liquid column is mea-

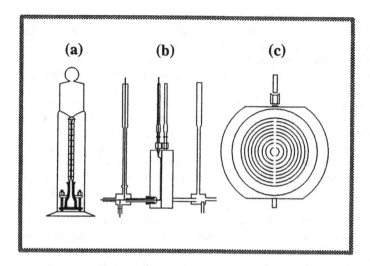

FIG. 3.3 Two types of osmometers: (a) the solution compartment is submerged in solvent. (Adapted from D. M. French and R. H. Ewert, *Anal. Chem.*, **19**, 165 (1947)); and (b) the solution and solvent occupy grooves on opposite faces of the central unit, as shown in detail in (c) (Adapted from R. M. Fuoss and D. J. Mead, *J. Phys. Chem.*, **47**, 59 (1943)).

sured. It is found that the ascending and descending branches of the curve converge to the same point – a value that equals the equilibrium osmotic pressure. This is shown in Figure 3.4. The rate at which equilibrium is achieved decreases as equilibrium is approached. Therefore it is desirable to bracket the true value as narrowly as possible to take full advantage of this approach-to-equilibrium extrapolation procedure. With some judicious planning, the time for an osmotic experiment may be shortened considerably by this dynamic method.

Once equilibrium has been reached, the height difference between the two liquid surfaces is all that remains to be measured. The primary factor to note here is that capillaries are used to minimize the dilution effects. This means that corrections for capillary rise must be taken into account unless the apparatus allows the difference between two carefully matched capillaries to be measured. We discuss capillary rise in Chapter 6, Sections 6.2 and 6.4. Finally, there is an extremely important practical reason, in addition to the theoretical requirement of isothermal conditions, for good thermostating in osmometry experiments. The apparatus consists of a large liquid volume attached to a capillary and therefore has the characteristics of a liquid thermometer: The location of the meniscus is quite sensitive to temperature fluctuations.

3.3b Measurement of Molecular Weight Using Osmometry

We now illustrate the use of the equations developed in Section 3.2 for interpreting osmotic pressure data. Figure 3.5 shows examples of two plots of π/RTc versus concentration. In Figure 3.5a the data all describe different molecular weight fractions of the same solute, cellulose acetate, in acetone solutions. Since the lines in this plot all have essentially the same slope, B must be the same for each.

Figure 3.5b shows data for a sample of nitrocellulose in three different solvents. All show the same intercept corresponding to a single molecular weight as required. Note, however, that the slopes are different, including even a negative slope, indicating that wide variations in B are possible. We examine the factors that determine B in Section 3.4.

Example 3.1 considers the molecular weights of the polymers in Figure 3.5 in terms of Equation (35).

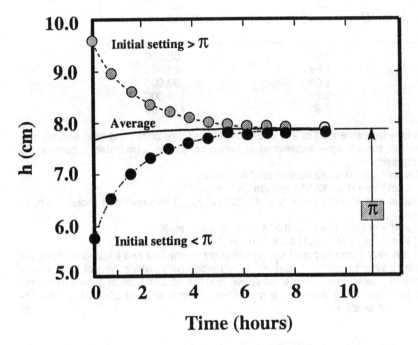

FIG. 3.4 Data showing the approach to osmotic equilibrium from initial settings above and below the equilibrium column height. (Adapted from R. U. Bonnar, M. Dimbat, and F. H. Stross, *Number Average Molecular Weights*, Wiley, New York, 1958.)

* * *

EXAMPLE 3.1 *Molecular Weights from Osmotic Pressure Measurements.*   To what molecular weights do the limiting reduced osmotic pressures obtained from Figure 3.5 correspond? The data in Figure 3.5 are presented in such a way that unit problems do not enter the picture. We are not usually so lucky! Consider some possible combinations of units for π, V, and R that are compatible with the units of the ordinate in Figure 3.5.

Solution: According to Equation (35), the molecular weights of the various polymers are simply the reciprocals of the limiting reduced osmotic pressures in the units that have been used for the ordinate in Figure 3.5, namely, mole g^{-1}, Therefore

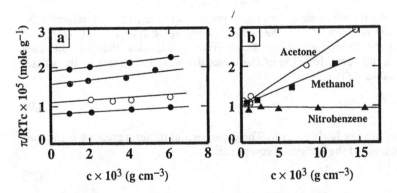

FIG. 3.5. Plots of π/RTc versus concentration: (a) various cellulose acetate fractions in acetone (data from A. Bartovics and H. Mark, *J. Am. Chem. Soc.*, **65**, 1901 (1943)); and (b) nitrocellulose in three different solvents (data from A. Dobry, *J. Chem. Phys.*, **32**, 50 (1935)).

Part in Figure 3.5	$(\pi/RTc)_0 \cdot 10^{-5}$ (mole g^{-1})	M (g mole^{-1})
(a)	1.92	52,000
(a)	1.59	63,000
(a)	1.09	92,000
(a)	0.79	126,000
(b)	0.90	111,000

Osmotic pressures may be expressed in any of the usual pressure units (and then some!) and volumes are often—but not always—expressed in cubic centimeters (milliliters); compatible units of R must be chosen:

If π is in atm and V in cm^3, use $R = 82.05$ atm cm^3 K^{-1} mole^{-1}.

If π is in torr and V in cm^3, use $R = 62,360$ torr cm^3 K^{-1} mole^{-1}.

If π is in mm of solution and V in cm^3, use $R = 62,360 \, (\rho_{Hg}/\rho_{Soln})$ mm cm^3 K^{-1} mole^{-1}, where the ρ's are densities.

If π is in dyne cm^{-2} and V in cm^3, use $R = 8.314 \cdot 10^7$ erg K^{-1} mole^{-1}.

If π is in N m^{-2} and V in m^3, use $R = 8.314$ J K^{-1} mole^{-1}

If the masses (in c) are expressed in g and kg, respectively, in the last two situations, then the units of the ordinate reduce to cm^2 s^{-2} and m^2 s^{-2}, respectively, which do not bear much resemblance to molecular weight units. Note, however, that the acceleration of gravity in appropriate units can be factored out of these latter quantities to leave units of cm and m, respectively, for reduced osmotic pressure.

* * *

3.3c   Osmometry and Polydispersity of Molecular Weights

As we saw in Section 1.5c, the condition of polydispersity is quite normal with colloidal solutes. Our discussion of osmometry has shown that it is possible to evaluate the molecular weight of a colloidal solute by osmotic pressure measurements. Next we consider the fact that the sample on which such a measurement is made is more than likely polydisperse and that the molecular weight obtained from such an experiment is some average quantity. The objective of this section is to show that it is the *number average molecular weight* that is determined in an osmotic pressure experiment on a polydisperse system. For the purposes of this demonstration, the solution is assumed to be ideal. (Notice, however, that, in the case of surfactant aggregates, i.e., micelles, discussed in Chapter 8, osmometry can lead to weight average molecular weight $\overline{M}_w$, as shown by Puvvada and Blankschtein in 1989.)

Experimental results from a polydisperse system may be related as follows:

$$\pi_{exp} = \frac{c_{exp}RT}{\overline{M}} \tag{37}$$

where π_{exp} and c_{exp} represent the experimental osmotic pressure and concentration, respectively, and $\overline{M}$ is the average molecular weight. It is the method of averaging in Equation (37) that we seek to determine. This relationship applies to the observable quantities. Precisely the same equation may be written, however, for each of the molecular weight fractions of the solute, designated here by the subscript i:

$$\pi_i = \frac{c_iRT}{M_i} \tag{38}$$

Two additional relationships are fairly evident. The experimental osmotic pressure is the sum of the pressure contributions of the individual components:

$$\pi_{exp} = \sum_i \pi_i \tag{39}$$

and the experimental concentration is the sum of the concentrations of the components:

$$c_{exp} = \sum_i c_i \tag{40}$$

Next, Equations (37)–(40) can be combined as follows:

$$\pi_{exp} = \sum_i \pi_i = RT \sum_i \frac{c_i}{M_i} = RT\frac{c_{exp}}{\overline{M}} = RT\frac{\Sigma_i c_i}{\overline{M}} \tag{41}$$

Equation (41) may be simplified to yield

$$\overline{M} = \frac{\Sigma_i c_i}{\Sigma_i (c_i/M_i)} \tag{42}$$

This result still fails to resemble any of the standard averages listed in Table 1.8. However, if we introduce the expression

$$c_i = \frac{n_i M_i}{V} \tag{43}$$

we obtain

$$\overline{M} = \frac{(1/V)\,\Sigma_i n_i M_i}{(1/V)\,\Sigma_i\,(n_i M_i/M_i)} = \frac{\Sigma_i n_i M_i}{\Sigma_i n_i} \tag{44}$$

Equation (44) shows the average molecular weight determined from osmometry to be the number average molecular weight as defined by Equation (1.14).

This same conclusion may also be reached by the following argument. The product $n_i M_i$ in Equation (43) equals the weight of component i in the solution; the total weight of solute in the solution equals $\Sigma_i n_i M_i$. The experimental osmotic pressure depends on and therefore measures the total number of moles of solute $\Sigma_i n_i$. The ratio of the total weight to the total number of moles of solute defines the number average molecular weight.

Any experiment that "counts" the number of molecules in a given weight of polydisperse material can be interpreted in terms of a number average molecular weight. All of the colligative properties have this as a feature of their shared thermodynamic origin. Another technique, known as end-group analysis, may be used to determine the molecular weight of certain polymers. Example 3.2 makes it clear that this too yields a number average molecular weight for polydisperse samples.

* * *

EXAMPLE 3.2 *Degree of Polymerization and Molecular Weight Distribution.* Polycaprolactam—otherwise known as nylon 6—with a degree of polymerization n has the following molecular structure: $H_2N(CH_2)_5CO[NH(CH_2)_5CO]_{n-2}NH(CH_2)_5COOH$. A 1.06-g sample of this material is dissolved in an appropriate solvent and titrated with an alcoholic KOH solution of normality 0.0250 N. Exactly 5.00 ml of base are required to neutralize the carboxyl groups of the sample. What is the molecular weight of the sample? What is n? Criticize or defend the following proposition: This method can only be used for molecules that contain just one of the analyzed functional groups per molecule.

Solution: This problem is really no different from the molecular weight determinations of unknown acids that are often conducted in general chemistry lab courses. What is important to recognize is that there is one carboxyl group per molecule or one equivalent per mole. Therefore the molecular weight of the polymer is given by

$$\overline{M} = \frac{1.06\,g}{(5.0)(0.025)\,mEq} \cdot \frac{10^3 mEq}{Eq} \cdot \frac{1\,Eq}{1\,mole} = 8480\,g\,mole^{-1}$$

This is the same average molecular weight that would be determined by an osmotic pressure experiment on the same sample. Since the molecular weight of the repeat unit is 113 g mole^{-1}, the degree of polymerization $n = 8480/113 \approx 75$.

The proposition recognizes the importance of knowing the number of analyzed groups per molecule, but incorrectly requires that there be only one such group. Some polymers might contain the same functional group at both ends of a linear chain. In this case there are 2 Eq/mole. A Y-shaped molecule could have three such groups, an X-shaped molecule four, and so forth. These last cases would have 3 and 4 Eq/mole, respectively. The number of groups does not matter as long as that number is known. ∎

* * *

That end-group analysis and osmometry give number averages for polydisperse samples is simply because the number of solute molecules is counted in each case. We see in Chapter 5 that light scattering effectively weighs the molecules rather than counts them (see, for example, Section 5.4d). Hence light scattering gives a weight average value for M. The fact that osmotic pressure "counts" solute particles has some interesting consequences for charged colloids, as we see in Section 3.5b.

3.4 STATISTICAL FOUNDATIONS OF SOLUTION THERMODYNAMICS

By analogy with Equation (28), we might expect the second virial coefficient to depend on the size and/or the interactions of the molecules in solution. Although this expectation is basically correct, we must not take the form of Equation (28), a *gas* equation, too literally in discussing *solutions*. Furthermore, it must be recalled that, in gases, only interactions between the molecules of the gas are possible. In solution, we may consider solvent-solvent, solute-solute, and solvent-solute interactions. This is a good reminder that the analogy with gases cannot be pushed too far.

By itself, *classical* thermodynamics provides us with no information on the molecular origin of the second virial coefficient, which is merely a phenomenological coefficient from an exclusively thermodynamic viewpoint. *Statistical* thermodynamics undertakes the task of providing molecular interpretations to such quantities. The added complication is that the statistical thermodynamic approach requires the use of models, and models inevitably over-simplify things. Nevertheless, we have arrived at that point at which we add insight only to the extent that we do some modeling. Thus, for example, we can learn something about solute-solute interactions and solute-solvent interactions from experimental B values, but only if we are willing to accept the models on which these interpretations are based. Toward this end, it is important to know what goes into these models. A model that is very plausible in one system may not make any sense for another.

In Section 3.4a we examine a model for the second virial coefficient that is based on the concept of the excluded volume of the solute particles. A solute-solute interaction arising from the spatial extension of particles is the premise of this model. Therefore the potential exists for learning something about this extension (i.e., particle dimension) for systems for which the model is applicable. In Section 3.4b we consider a model that considers the second virial coefficient in terms of solute-solvent interaction. This approach offers a quantitative measure of such interactions through B. In both instances we only outline the pertinent statistical thermodynamics; a somewhat fuller development of these ideas is given in Flory (1953). Finally, we should note that some of the ideas of this section are going to reappear in Chapter 13 in our discussions of polymer-induced forces in colloidal dispersions and of coagulation or "steric" stabilization (Sections 13.6 and 13.7).

3.4a Excluded-Volume Effect

Before considering how the excluded volume affects the second virial coefficient, let us first review what we mean by excluded volume. We alluded to this concept in our model for size-exclusion chromatography in Section 1.6b.2b. The development of Equation (1.27) is based on the idea that the center of a spherical particle cannot approach the walls of a pore any closer than a distance equal to its radius. A zone of this thickness adjacent to the pore walls is a volume from which the particles — described in terms of their centers — are denied entry because of their own spatial extension. The volume of this zone is what we call the *excluded volume* for such a model. The van der Waals constant b in Equation (28) measures the excluded volume of gas molecules; for spherical molecules it equals four times the actual volume of the sphere, as discussed in Section 10.4b, Equation (10.38).

3.4a.1 Statistical Entropy and Entropy of Mixing: Definitions
A point of entry for statistical considerations into thermodynamics is the Boltzmann entropy relationship

$$S = k_B \ln \Omega \tag{45}$$

in which Ω is called the thermodynamic weight or the partition function and k_B is the Boltzmann constant, which is replaced by R when we are working on a "per mole" rather than a "per molecule" basis. The thermodynamic weight is the tricky part of Equation (45); it counts the number of ways a particular state can come about. On a qualitative basis, chemists learn to use this relationship almost intuitively. One associates higher entropy with states that are more disordered, and disordered states can come about in a larger number of ways than states of higher order. The familiar analogy that is often used here is to note that there are more ways for a deck of cards to exist as "shuffled" than as "arranged by suits." We can readily see that Equation (45) encompasses the third law of thermodynamics: There is only one way to organize a perfect crystal at absolute zero, hence $\Omega = 1$ and $S = 0$.

Entropy changes can also be developed in terms of Equation (45):

$$S_2 - S_1 = k_B \ln \left(\frac{\Omega_2}{\Omega_1} \right) \tag{46}$$

Our interest is in the entropy of mixing—that is, ΔS for $1 + 2 \rightarrow$ mixture—which may be written

$$\Delta S_m = k_B \ln \left(\frac{\Omega_{mix}}{\Omega_1 \Omega_2} \right) \tag{47}$$

by virtue of Equation (46). Note that we use the subscripts m for the mixing process and *mix* for the mixture itself. What remains to be done is to count the ways in which N_1 molecules of component 1 and N_2 molecules of component 2 exist when mixed together or, more specifically, the factor by which this number exceeds the number of ways the separate molecules can exist.

The derivation we follow proceeds through four stages:

1. We must devise an expression for Ω_{mix}.
2. From Ω_{mix} a straightforward application of Equation (45) is required to obtain an equation for S_{mix}. It takes a bit of additional argument to convert this to an expression for ΔS_m.
3. A single approximation will enable us to go from ΔS_m to ΔG_m, but the implications of this assumption deserve some comment.
4. Some mathematical manipulations convert ΔG_m into an expression for $\mu_i - \mu_1^0$, which is directly related to π through Equation (21).

3.4a.2 Development of Partition Function

We begin by considering the number of ways a solute molecule of excluded volume u can be placed in an otherwise empty volume V. Suppose we imagine V to be sectioned off a number of sites (say, N_s), each of which can accommodate the volume u. Since V is intended to represent the volume of the solution—a macroscopic quantity—and u is a molecular parameter, the number of such sites is large. The first solute molecule to enter V can occupy any one of these sites, and we represent by ω_1 the number of possible placements for the first solute molecule. Because the total volume is partitioned into these sites, it follows that

$$\omega_1 = KV = N_s \tag{48}$$

where K is an appropriate proportionality constant, equal to $(1/u)$ in our present model. (In general, K may differ from $1/u$ since small amounts of unfilled [unfillable] space may exist even when the N_s units of volume $1/u$ each are closely packed, i.e., $N_s u$ can be less than the total volume V. Therefore, for the sake of generality, we develop the equations here in terms of K also, instead of restricting ourselves to $K = 1/u$ only.)

The second molecule to be placed can go into any one of the remaining sites, hence the number of ways to place the second molecule is given by

$$\omega_2 = K(V - u) = (N_s - 1) \tag{49}$$

For the third molecule, this quantity is given by $\omega_3 = K(V - 2u)$, and for the jth it is

$$\omega_j = K[V - (j - 1)u] = [N_s - (j - 1)] \tag{50}$$

The number of ways of placing the first, the second, the third, and so on to the ith is

$$\Omega \propto \omega_1\omega_2\omega_3 \ldots \omega_i = N_s (N_s - 1)(N_s - 2) \ldots [N_s - (i - 1)] = \frac{N_s!}{(N_s - i)!} \tag{51}$$

We have written Equation (51) as a proportion rather than as an equality since the given expression actually overcounts the number of possible placements. If there is a total of N_2 solute molecules, then there are $N_2!$ different permutations of these molecules, and Equation (51) should be divided by $N_2!$ because of the interchangeability of the molecules. Incorporating this idea and Equation (50) into Equation (51) gives

$$\Omega = \frac{1}{N_2!} \prod_{j=1}^{N_2} K[V - (j - 1)u] = \frac{N_s!}{N_2!(N_s - N_2)!} \tag{52}$$

where $\prod$ represents the product of terms. Letting $i = j - 1$ allows Equation (52) to be written more concisely as

$$\Omega = \frac{1}{N_2!} \prod_{i=0}^{N_2-1} K(V - iu) = \frac{1}{N_2!} \prod_{i=0}^{N_2-1} KV\left(1 - i\frac{u}{V}\right) \tag{53}$$

After the N_2 solute molecules are placed, all remaining sites are filled with solvent molecules. Since our model is specifically interested in the solute-solute excluded-volume effect, we may say that there is only one way for the solvent molecules to be placed. Although this overlooks details about the solvent, we see presently that such details would eventually be subtracted, so we lose nothing by this simplification. Equation (53) therefore gives the expression for Ω_{mix} we sought as the first step of our derivation.

3.4a.3 Expression for Entropy of Mixing

Substituting Equation (53) into Equation (45) gives a statistical expression for the solute contribution to the configurational entropy of the mixture:

$$\frac{S_{mix}}{k_B} = -\ln N_2! + \sum_{i=0}^{N_2-1} \ln\left[KV\left(1 - i\frac{u}{V}\right)\right] \tag{54}$$

The summation replaces the product in going from Equation (53) to Equation (54) since we are dealing with logarithms in the latter. Note that the configurational entropy refers explicitly to the entropy associated with the mixture itself; the internal entropy of the molecules themselves is clearly not included. Next a series of mathematical manipulations will transform Equation (54) into a more useful form:

1. Divide the summation in Equation (54) into two terms:

$$\sum \ln\left[KV\left(1 - i\frac{u}{V}\right)\right] = \sum \ln(KV) + \sum \ln\left(1 - i\frac{u}{V}\right)$$

2. Note that $\sum \ln(KV) = N_2 \ln(KV)$ since there are N_2 identical terms in the summation.

3. Since $u/V << 1$ as already noted, $\ln(1 - iu/V)$ can be expanded as a series (see Appendix A) with only the leading term retained to give

$$\sum \ln\left(1 - i\frac{u}{V}\right) \approx \sum\left(-i\frac{u}{V}\right) \approx -\frac{u}{V}\sum i$$

4. The sum of integers from 0 to y is $y(y + 1)/2$; hence

$$\sum_{i=0}^{N_2-1} i = (N_2 - 1)\frac{N_2}{2} \approx \frac{N_2^2}{2}$$

since N_2 is large.

5. We can change the equation from dealing with numbers of molecules (N) to numbers of moles (n) by dividing both sides by Avogadro's number N_A and writing $k_B N_A = R$.

Applying these manipulations to Equation (54) gives

$$\frac{S_{mix}}{R} = -\frac{1}{N_A} \ln N_2! + n_2 \ln(KV) - \frac{1}{2} N_A \, n_2^2 \frac{u}{V} \tag{55}$$

Replacing V by the mole-weighted sum of the partial molar volumes, as was done in developing Equation (31), we write

$$\frac{S_{mix}}{R} = -\frac{1}{N_A} \ln N_2! + n_2 \ln[K(n_1 \overline{V}_1 + n_2 \overline{V}_2)] - \frac{1}{2} \frac{N_A n_2^2 \, u}{n_1 \overline{V}_1 + n_2 \overline{V}_2} \tag{56}$$

If we set n_2 equal to zero in Equation (56), the physical system described corresponds to pure solvent, and the equation reduces to $S_1 = 0$. If we set n_1 equal to zero, we are describing pure solute, and Equation (56) becomes

$$\frac{S_2}{R} = n_2 \ln K + \ln(n_2 \overline{V}_2) - \frac{1}{N_A} \ln N_2! - \frac{N_A n_2 \, u}{2 \overline{V}_2} \tag{57}$$

Subtracting these expressions for S_1 and S_2 from S_{mix} gives ΔS_m, which was the goal of the second stage of our derivation:

$$\frac{\Delta S_m}{R} = n_2 \ln\left(\frac{n_1 \overline{V}_1 + n_2 \overline{V}_2}{n_2 \overline{V}_2}\right) - \frac{1}{2} N_A u \frac{n_2^2}{n_1 \overline{V}_1 + n_2 \overline{V}_2} + \frac{1}{2} N_A \frac{n_2 u}{\overline{V}_2} \tag{58}$$

Note that whatever configurational entropy we had attributed to the solvent would have disappeared at this point.

As outlined above, our next objective is to write an expression for ΔG_m. Since $\Delta G = [\Delta H - T \Delta S]$ for a constant temperature process such as this, we can immediately write

$$-\frac{\Delta G_m}{RT} = n_2 \ln\left(\frac{n_1 \overline{V}_1 + n_2 \overline{V}_2}{n_2 \overline{V}_2}\right) - \frac{1}{2} N_A u \frac{n_2^2}{n_1 \overline{V}_1 + n_2 \overline{V}_2} + \frac{1}{2} N_A \frac{n_2 u}{\overline{V}_2} \tag{59}$$

by assuming $\Delta H_m = 0$. In making this assumption we imply that the energetic interactions between solvent-solvent pairs, solute-solute pairs, and solvent-solute pairs are all the same. Thus, removing a solvent molecule from a bulk sample of the liquid and replacing it with a molecule of the solute does not involve an enthalpy change. It is the introduction of this assumption that leaves the excluded volume as the sole contributor to the second virial coefficient. In the next section we assume a nonzero value for ΔH_m and examine the effect this has on B. For now, it is enough to note that the present model attributes solution nonideality to molecular size and not to the energetics of molecular interactions. The model is clearly inappropriate for any system in which the molecular interactions are important.

3.4a.4 Chemical Potential, Osmotic Pressure, and Second Virial Coefficient

The final stage of our derivation converts Equation (59) to an expression for $(\mu_1 - \mu_2)$ by differentiating Equation (59) with respect to n_1 according to Equation (1):

$$\mu_1 - \mu_1^0 = -RT\left[\frac{n_2 \overline{V}_1}{V} + \frac{1}{2} N_A \, \overline{V}_1 \left(\frac{n_2}{V}\right)^2 u\right] \tag{60}$$

Equation (21) shows how to convert Equation (60) into an expression for osmotic pressure:

$$\pi = \frac{RT}{\overline{V}_1}\left[\frac{n_2 \overline{V}_1}{V} + \frac{1}{2} N_A u \, \overline{V}_1 \left(\frac{n_2}{V}\right)^2\right] \tag{61}$$

Since $n_2/V = c/M_2$, Equation (61) may be written in practical concentration units:

$$\pi = RT \left(\frac{c}{M} + \frac{1}{2} \frac{N_A u}{M^2} c^2 \right) \qquad (62)$$

Comparing this result with Equation (34) shows that

$$B = \frac{1}{2} \frac{N_A u}{M^2} \qquad (63)$$

according to this model.

The essence of this model for the second virial coefficient is that an excluded volume is defined by surface contact between solute molecules. As such, the model is more appropriate for molecules with a rigid structure than for those with more diffuse structures. For example, protein molecules are held in compact forms by disulfide bridges and intramolecular hydrogen bonds; by contrast, a randomly coiled molecule has a constantly changing outline and imbibes solvent into the domain of the coil to give it a very "soft" surface. The present model, therefore, is much more appropriate for the globular protein than for the latter. Example 3.3 applies the excluded-volume interpretation of B to an aqueous protein solution.

*   *   *

EXAMPLE 3.3 *Excluded Volume of Bovine Serum Albumin from Osmotic Pressure Measurements.* A plot of π/c versus c for an aqueous solution of the bovine serum albumin molecule at 25°C and pH = 5.37 is shown in Figure 3.6. The molecule is known to be nearly spherical and uncharged at this pH. Evaluate the molecular weight and the excluded volume of this protein from the intercept and slope of this line, 0.268 torr $(g\ kg^{-1})^{-1}$ and $1.37 \cdot 10^{-3}$ torr $kg^2\ g^{-2}$, respectively. From the particle mass and volume, estimate the partial specific volume of the solute in solution. The specific volume of the unsolvated protein is about 0.75 $cm^3\ g^{-1}$; does the solute appear to be solvated?

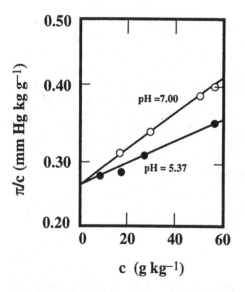

FIG. 3.6   Plot of π/c versus concentration for bovine serum albumin in 0.15 M NaCl at pH 7.00 and 5.37. (Adapted from G. Scatchard, A. C. Batchelder, and A. Brown, *J. Am. Chem. Soc.*, **68**, 2320 (1946).)

Solution: First, convert to SI units, assuming the density of the solution to be 1.0 g cm^{-3}.
   For the intercept

$$0.268 \text{ torr (g kg}^{-1})^{-1} \cdot \frac{1.01 \cdot 10^5 \text{ N m}^{-2}}{760 \text{ torr}} \cdot \left(\frac{10^{-3} \text{ m}^3}{\text{kg}}\right) \cdot \left(\frac{10^3 \text{ g}}{\text{kg}}\right) = 35.6 \text{ N m}^{-2} (\text{kg m}^{-3})^{-1}$$

For the slope

$$1.37 \cdot 10^{-3} \text{ torr kg}^2 \text{g}^{-2} \cdot \frac{1.01 \cdot 10^5 \text{ N m}^{-2}}{760 \text{ torr}} \cdot \left(\frac{10^{-3} \text{ m}^3}{\text{kg}}\right)^2 \cdot \left(\frac{10^3 \text{ g}}{\text{kg}}\right)^2 = 0.812 \text{ N m}^4 \text{ kg}^{-2}$$

Calculate the molecular weight by Equation (35):

$$M = RT(\pi/c)_0^{-1} = (8.314)(298)/35.6 = 69.6 \text{ kg mole}^{-1} = 69,600 \text{ g mole}^{-1}$$

The mass per particle is obtained by dividing by N_A:

$$69.6/6.02 \cdot 10^{23} = 1.16 \cdot 10^{-22} \text{ kg molecule}^{-1}$$

Divide the slope by RT to get B:

$$0.182/(8.314)(298) = 7.35 \cdot 10^{-5} \text{ m}^3 \text{ kg}^{-2} \text{ mole}$$

Use Equation (63) to calculate the excluded volume:

$$u = \frac{2 B M^2}{N_A} = \frac{2(7.35 \cdot 10^{-5})(69.6)^2}{6.02 \cdot 10^{-23}} = 1.18 \cdot 10^{-24} \text{ m}^3 \text{ molecule}^{-1}$$

Since the actual volume of a spherical particle is one-quarter the excluded volume, the volume of the molecule is $2.95 \cdot 10^{-25} \text{ m}^3$ molecule^{-1}.
   The partial specific volume is given by the ratio of the molecular volume to the molecular mass:

$$2.95 \cdot 10^{-25}/1.16 \cdot 10^{-22} = 2.54 \cdot 10^{-3} \text{ m}^3 \text{ kg}^{-1} = 2.54 \text{ cm}^3 \text{ g}^{-1}$$

Comparing this with the nonsolvated value of 0.75 cm^3 g^{-1} definitely suggests that the particle is hydrated.
   Note that the volume of the spherical molecule may also be converted to a molecular radius, which equals 4.11 nm for this molecule.   ∎

* * *

The second set of data points in Figure 3.6—measured for the same solute but at a pH = 7.00—shows a steeper slope and therefore a larger B value. It would be incorrect to analyze this slope by the procedure used in the example above, however. The reason for this is that the protein acquires a charge in going from pH 5.37 to 7.00. The charge must be explicitly taken into account to interpret B in this case. We discuss this in Section 3.5b. It is pertinent to note, however, that charged particles require counterions to give them electrical neutrality. These counterions occupy a region in the solution that surrounds the charged colloid; we describe them as setting up an ion atmosphere around the central particle. It makes sense that the particle plus its ion atmosphere should have a larger excluded volume than the uncharged particle, which entrains no ion atmosphere.

3.4b   Inclusion of Energetic Interactions: Flory-Huggins Theory

In many colloidal solutions the solute is best described as a random coil in which the domain of the molecule contains both polymer segments and solvent molecules. Vinyl-type synthetic polymers in organic solvents are especially well suited for this representation. Depending on the energetics of the interaction between solvent molecules and polymer chain segments, the solvent may be imbibed into the coil domain to a greater or lesser extent, and the spatial extension of the chain depends on these interactions, as well as on the size of the molecule itself. A solvent that swells the coil dimensions is called a *good solvent*, and one that causes the coil to shrink is called a *poor solvent*. The former situation arises when solvent-solute contacts are favored, and the latter when solute-solute contacts are favored. The "goodness" of a solvent for a particular polymer depends not only on the nature of the species involved, but also on the temperature. Generally speaking, lowering the temperature causes a decrease in solvent goodness.

In this section we look at a statistical model for a solution that allows for both the random coil geometry of the solute molecule and the variability of interactions between the solute and different solvents. We merely sketch the general outline of the theory; in the next section we examine the implications of this model for interpreting the second virial coefficient as a measure of solvent goodness. The approach we adopt is generally known as the Flory-Huggins theory, after P. J. Flory and M. L. Huggins, whose independent efforts merged in the final result. Flory was awarded the Nobel Prize in 1974 for his numerous contributions to polymer chemistry.

3.4b.1 Entropy of Mixing

The Flory-Huggins theory begins with a model for the polymer solution that visualizes the solution as a three-dimensional lattice of N sites of equal volume. Each lattice site is able to accommodate either one solvent molecule or one polymer segment since both of these are assumed to be of equal volume. The polymer chains are assumed to be monodisperse and to consist of n segments each. Thus, if the solution contains N_1 solvent molecules and N_2 solute (polymer) molecules, the total number of lattice sites is given by

$$N = N_1 + nN_2 \tag{64}$$

Without spelling out the geometry of the lattice explicitly, we further assume that each site in the lattice is surrounded by z neighboring sites, making z the coordination number of the lattice.

On the basis of this model, an expression for ΔS_m can be derived. We will not go through the details of the derivation, but merely note the following similarities and differences between this derivation and the one that leads to Equation (58) for the excluded-volume model:

1. Instead of counting the ways to place solute molecules, this derivation counts the ways to place solute segments.
2. The placement of successive segments is subject to a restriction that did not arise in the previous derivation, namely, segments from the same chain must occupy adjacent lattice sites because they are covalently bonded together.
3. The derivation considers the number of ways ω_i of placing each of the n segments in, say, the ith solute molecule, then goes on to count the number of ways of scaling this up for N_2 solute molecules. The thermodynamic probability of the mixture Ω_{mix} results from these steps.
4. The entropy of the mixture is calculated from this by the Boltzmann entropy equation, Equation (45). By separately letting N_2 and N_1 equal zero, the configurational entropies of the solvent and the solute, respectively, are obtained from S_{mix}. Finally, by subtracting S_1 and S_2 from S_{mix}, an expression is obtained for ΔS_m.

Except for the complication of positioning connected segments in adjacent sites, the general outline of this derivation clearly parallels the previous derivation.

For a mole of solution—i.e., $N_1 + N_2 = N_A$—the theory outlined above results in the following expression for ΔS_m:

$$\frac{\Delta S_m}{R} = -\left[\frac{N_1}{N_A} \ln\left(\frac{N_1}{N}\right) + \frac{N_2}{N_A} \ln\left(\frac{nN_2}{N}\right)\right] \tag{65}$$

The ratios N_1/N_A and N_2/N_A are mole fractions of the two components since N_A is the total number of molecules in a mole of solution. On the other hand, N is the total number of lattice sites and N_1 and nN_2 are the numbers of sites occupied by solvent and solute, respectively. Since all of the lattice sites have equal volumes, the ratios in the logarithms are the volume fractions occupied by the two components. Letting x_i be the mole fraction of component i and ϕ_i the volume fraction of that component, Equation (65) can be written

$$\Delta S_m = R[x_1 \ln \phi_1 + x_2 \ln \phi_2] \tag{66}$$

Note that $\phi_2 \rightarrow x_2$ as $n \rightarrow 1$; that is, Equation (66) reduces to the expression for the entropy of mixing for an ideal solution in the event that the solute molecule is no larger than the solvent molecule.

3.4b.2 Energetic Interactions and Enthalpy Change Due to Mixing

At this point we could proceed as in the previous derivation of an expression for B: Evaluate ΔG_m, μ_m and π by setting $\Delta Hm = 0$. Doing this, however, would sacrifice this model's ability to deal with different interaction energies between solvent and solute segments. Hence our next objective is to find an expression for ΔH_m in terms of the lattice model we have developed. Once this is done, we can combine ΔS_m and ΔH_m into an expression for ΔG_m and then proceed as above.

To do this we assign an energy of interaction w_{11} to a pair of solvent molecules and w_{22} to a pair of polymer segments. The latter arises from the intermolecular forces between segments and not from the covalent bonds between them. In the same fashion, we define w_{12} to be the energy of the solvent-segment interaction.

As noted above, a lattice site in the polymer solution has z nearest neighbors that may be occupied by either solvent molecules or solute segments. If ϕ_i is the volume fraction of the solution occupied by species i, then we assume that this fraction also applies to the z sites that adjoin the specific site under consideration. Hence any site is surrounded, on the average, by $z\phi_1$ solvent molecules and $z\phi_2$ chain segments. If the central site is occupied by a solute segment (component 2), then the intermolecular forces between this segment and its neighbors contribute $(z\phi_1 w_{12} + z\phi_2 w_{22})$ to the energy of the system. Since the lattice consists of N sites of which $\phi_2 N$ are occupied by solute segments interacting in the manner just described, we write $[(1/2)N\phi_2(z\phi_1 w_{12} + z\phi_2 w_{22})]$ for the total interaction energy of all such segments. The factor 1/2 enters the expression because each pair of interacting species is counted twice by this procedure. If the lattice is completely filled with polymer segments, then the interaction energy is simply given by $[(1/2)z\phi_2 N w_{22}]$.

Precisely the same series of steps gives the energy contribution by sites occupied by solvent molecules:

1. The contribution of interactions with its neighbors made by each solvent molecule is $(z\phi_1 w_{11} + z\phi_2 w_{12})$.

2. The number of sites occupied by solvent molecules is $\phi_1 N$.

3. The total energy contribution by all solvent molecules is $(1/2)N\phi_1(z\phi_1 w_{11} + z\phi_2 w_{12})$.

4. If the lattice is completely filled with solvent, then the interaction energy is given by $(1/2)z\phi_1 N w_{11}$.

Now consider the change in interaction energies that accompanies the mixing process $1 + 2 \rightarrow$ mixture. We must total the interactions of all chain segments and all solvent molecules to obtain the interaction energy of the mixture and then subtract from this the interactions corresponding to pure solute and pure solvent. Assembling the required values from above, we write

$$\Delta H_m = \frac{1}{2} z N [\phi_2(\phi_2 w_{22} + \phi_1 w_{12}) + \phi_1(\phi_1 w_{11} + \phi_2 w_{12}) - \phi_1 w_{11} - \phi_2 w_{22}] \tag{67}$$

Since $\phi_1 + \phi_2 = 1$, Equation (67) can be rearranged as

$$\Delta H_m = -\frac{1}{2} z N \phi_1 \phi_2 (w_{11} + w_{22} - 2w_{12}) \tag{68}$$

In terms of pairwise interactions, the solution process can be represented

$$(1,1) + (2,2) \rightarrow 2(1,2) \tag{69}$$

and, following the usual thermodynamic notation, we can write Δw for the process as

$$\Delta w = 2w_{12} - w_{11} - w_{22} \tag{70}$$

Combining Equations (68) and (70) gives

$$\Delta H_m = \frac{1}{2} z N \phi_1 \phi_2 \Delta w = \frac{1}{2} z N_1 \phi_2 \Delta w \tag{71}$$

The forces between molecules that we measure by these w's are (usually) forces of attraction and, by convention, are represented by negative numbers. Thus when 1,2 attractions are stronger than 1,1 and 2,2 attractions, Δw and ΔH_m are both negative. Since a negative enthalpy of mixing makes a favorable contribution to a negative value for ΔG_m, this sign convention makes sense. Conversely, if the 1,1 and 2,2 attractions are stronger (more negative), then Δw and ΔH_m are positive. The case in which ΔH_m is zero is called *athermal mixing*; Equation (70) shows that this corresponds to a situation in which solute-solute, solvent-solvent, and solute-solvent interactions are all equivalent in energy. This was assumed to be the case in the excluded-volume model for solution nonideality discussed in the last section.

An additional development of Equation (70) can be made by assuming that w_{12} is the geometric mean of w_{11} and w_{22}. It makes sense that the energy of interaction between unlike molecules is somehow related to the homogeneous interactions; this manner of averaging the latter has advantages that will be evident below. By assuming $w_{12} = (w_{11}w_{22})^{1/2}$, Equation (70) becomes

$$\Delta w \propto w_{11} + w_{22} - 2(w_{11}w_{22})^{1/2} \propto (\sqrt{w_{11}} - \sqrt{w_{22}})^2 \tag{72}$$

This development cannot result in a negative value for Δw or ΔH_m and is therefore definitely inapplicable for systems in which the solute and solvent display some specific type of interaction, such as hydrogen bonding. However, when purely physical interactions are involved, Equation (72) has proved to be quite useful.

The utility of this approach lies in the fact that Equation (72) describes the mixing process in terms of homogeneous interactions, which are readily measured for pure liquids. The heat of vaporization, for example, is a liquid property that increases as the strength of intermolecular attractions increases. Rather than working with molar heats of vaporization, it is more convenient to divide the molar heat by the molar volume to define what is known as the *cohesive energy density* (CED) of a material. As the name implies, the CED measures the energy per unit volume that holds the molecules of a liquid together. As such, it is directly proportional to the w's of our discussion and can be evaluated from readily available data. Introducing the concept of cohesive energy density into Equation (72) enables us to write

$$\Delta H_m \propto \Delta w \propto \left[\sqrt{(CED)_1} - \sqrt{(CED)_2} \right]^2 \tag{73}$$

Polymers decompose before they evaporate, so it appears that the concept of CED is not applicable to these materials. However, by finding a solvent with which a particular polymer mixes athermally, we can assign to the polymer by Equation (73) the same CED as that solvent. Thus cohesive energy densities for a number of polymers, as well as low molecular weight solvents, have been determined. Table 3.1 lists some representative examples of such data.

Cohesive energy densities can be used on a limited basis to give quantitative meaning to the chemist's rule of thumb, "like dissolves like." Specifically, the more alike a solvent and a polymer are in CED, the more nearly athermal their mixing will be. The more different the two are in this property, the more endothermic the mixing process will be. Remember that Equation (72) makes no provision for exothermic mixing. In the next section we see how such information might be used. Remember that Equation (71) is not limited to endothermic

TABLE 3.1 Values of the Square Root of the Cohesive Energy Density for Some Polymers and Low Molecular Weight Solvents

Solvent	$(CED)^{1/2}$, $(cal\ cm^{-3})^{1/2}$	Polymer	$(CED)^{1/2}$, $(cal\ cm^{-3})^{1/2}$
n-Decane	6.6	Poly(tetrafluoro	
Cyclohexane	8.2	ethylene) (Teflon)	6.2
Toluene	8.9	Polyethylene	7.7–8.2
Acetone	9.9	Polystyrene	8.5–9.1
Cyclohexanol	11.4	Poly(methyl	
Ethanol	12.7	methacrylate)	9.1–9.5
Methanol	14.5	Polypropylene	9.2–9.4
Water	23.4	Poly(vinyl chloride)	9.7–9.9
		Poly(ethylene	
		terephthalate)	10.7
		Poly(acrylonitrile)	12.3–12.8

Source: H. Burrell in *Polymer Handbook*, 2d ed., (J. Brandrup and E. H. Immergut, Eds.), Wiley, New York, 1975.

situations. In the next section we use the more general form of ΔH_m to consider the Flory-Huggins theory as it applies to the second virial coefficient.

3.4b.3 Chemical Potential, Osmotic Pressure, and Second Virial Coefficient

Equations (65) and (71), respectively, give ΔS_m and ΔH_m according to the Flory-Huggins theory. From these components, ΔG_m can be assembled directly, and by differentiation with respect to N_1 the Flory-Huggins expression for $(\mu_1 - \mu_1^0)$ may be obtained:

$$\mu_1 - \mu_1^0 = RT \ln \phi_1 + RT \left(1 - \frac{1}{n}\right) \phi_2 + \frac{1}{2} z \Delta w \phi_2^2 \tag{74}$$

It is conventional to let $1/2\ z\Delta w = \chi RT$, where χ is called the *Flory-Huggins interaction parameter*. Note that $1/2\ \Delta w$ is the energy change per 1,2 pair according to Equation (69); therefore, with the coordination number z absorbed, the parameter χ measures this in units of RT. Finally, Equation (21) establishes the connection between chemical potential and osmotic pressure; according to this equation,

$$-\pi \overline{V}_1 = RT \left[\ln \phi_1 + \left(1 - \frac{1}{n}\right) \phi_2 + \chi \phi_2^2\right] \tag{75}$$

By expressing all volume fractions in terms of the solute, the first term on the right-hand side of Equation (75) becomes $\ln (1 - \phi_2)$, which may be expanded (see Appendix A) as $(-\phi_2 - \phi_2^2/2)$. With this modification, Equation (75) becomes

$$\frac{\pi \overline{V}_1}{RT} = \phi_2 + \frac{1}{2} \phi_2^2 - \left(1 - \frac{1}{n}\right) \phi_2 - \chi \phi_2^2 \tag{76}$$

All that remains to be done to complete our derivation of the second virial coefficient in terms of the Flory-Huggins theory is convert volume fractions into practical concentration units. First, we can express the volume fraction of the solute in terms of partial molar volumes:

$$\phi_2 = \frac{n_2 \overline{V}_2}{n_1 \overline{V}_1 + n_2 \overline{V}_2} \approx \frac{n_2}{n_1} \frac{\overline{V}_2}{\overline{V}_1} \tag{77}$$

where the approximate form applies to dilute solutions. If we recall Equations (24) and (32), Equation (77) may be written

$$\phi_2 \approx x_2 \frac{\overline{V}_2}{\overline{V}_1} = c_2 \frac{\overline{V}_2}{M_2} \tag{78}$$

Introducing practical concentration units to Equation (78), the Flory-Huggins theory yields

$$\frac{\pi}{RTc} = \frac{\overline{V}_2}{M_2 n \overline{V}_1} - \frac{\frac{1}{2} - \chi}{\overline{V}_1} \left(\frac{\overline{V}_2}{M_2}\right)^2 c = \frac{1}{M_2} + \frac{\frac{1}{2} - \chi}{\overline{V}_1} \left(\frac{\overline{V}_2}{M_2}\right)^2 c; \, c = c_2 \tag{79}$$

The last simplification is possible because the solute molecule is n times larger than a solvent molecule, and the same size relationship applies to the partial molar volumes. According to Equation (79), the second virial coefficient is given by

$$B = \frac{\frac{1}{2} - \chi}{\overline{V}_1} \left(\frac{\overline{V}_2}{M}\right)^2 \tag{80}$$

Since χ measures Δw for the process described by Equation (69), we see that Equation (80) accomplishes what we set out to do, namely, relate the second virial coefficient to differences in the interaction energies between various pairs of molecules.

3.4b.4   Solvent Goodness and Theta Temperature of Polymer Solutions

One of the first things to observe about Equation (80) is the fact that it allows the second virial coefficient to be positive, negative, or zero, depending on whether χ is less than, greater than, or equal to 1/2, respectively. Figure 3.5b reveals that positive and negative slopes are both observed in plots of reduced osmotic pressure versus concentration. We have more to say about this. For now, it is enough to note that under conditions with $\chi = 1/2$, $B = 0$ and solutions that are not too concentrated behave ideally. This is an advantage from the point of view of molecular weight determination since a polymer-solvent system that meets this requirement satisfies the van't Hoff equation. This means that the molecular weight can be determined from a single solution without the need to do a series of experiments and extrapolate to infinite dilution.

The condition of $B = 0$ also marks the demarcation between good and poor solvent conditions. Positive values of the second virial coefficient characterize good solvents, and negative values characterize poor solvents. This state of affairs — which is usually called the Θ condition — is very important in polymer chemistry. We encounter it again in our discussions of viscosity (Chapter 4) and light scattering (Chapter 5). Before examining the significance of the Θ condition any further, let us first consider the two states on either side of it.

A positive B value indicates a good solvent. According to Equation (80), this is guaranteed for negative (or small positive) values of χ. Since χ is proportional to Δw and ΔH_m, a positive value for the second virial coefficient corresponds to an exothermic (or small endothermic) enthalpy of mixing. Conversely, a negative value for the second virial coefficient corresponds to a positive (or small negative) value of χ and an endothermic (or small exothermic) enthalpy of mixing. Classifying the solvent as good or poor on the basis of the slope of a plot of π/c versus c is therefore consistent with the contribution of ΔH_m to favorable mixing.

The Flory-Huggins expression for ΔG_m can be related to other quantities besides osmotic pressure. One of the things that can be done is to calculate solubility limits, and therefore miscibility diagrams, for various polymer-solvent systems. Although we shall not pursue this in detail, it is of interest to note that this approach leads to the conclusion that $\chi = 1/2$ is a critical value for this parameter; that is, the critical point on a miscibility diagram corresponds to $\chi = 1/2$ for a polymer of infinite molecular weight ($n = \infty$). What is significant about this is the following: Suppose that we could somehow adjust the "goodness" of a particular solvent-polymer system, decreasing this quality from an initially good state. At $\chi = 1/2$ the solvent would go from good to poor for a polymer of infinite molecular weight, and that particular molecular weight fraction would undergo phase separation. A polymer of somewhat lower molecular weight does not undergo phase separation until the value of χ is somewhat larger than 1/2. The shorter the polymer chain, the more χ must exceed 1/2 to reach the threshold of miscibility. Remember that $\chi > 1/2$ corresponds to "poor" solvent conditions in terms of the second virial coefficient. The extension of these ideas to miscibility limits shows that systems of this sort are on their way to phase separation. It is only because the molecules have finite molecular weights that they remain in solution at all.

In the remarks above we considered adjusting solvent goodness as if it were an imaginary process. In fact, it can be carried out physically in two different ways for a particular polymer solution. One way is to lower the temperature of the system; another way is to dilute the initial system with a poor solvent. An application such as this is one in which the cohesive energy densities listed in Table 3.1 can be put to use. The utility of changes in solvent goodness lies in the possibility of fractionating a specimen with respect to molecular weight by such a variation. Since synthetic polymers are almost always highly polydisperse, the addition of a poor solvent to a solution of the polymer (or lowering its temperature) causes the highest molecular weight fraction to separate out of solution. The separated phase can be physically removed and the process repeated until a series of fractions is obtained. The same thing can be accomplished by temperature variations for a fixed polymer-solvent system.

Since the goodness of a polymer-solvent system can be adjusted by changing the temperature, it is desirable to recast Equation (80) in a way that shows this effect explicitly. Toward this end, we define the following identity to introduce a temperature variable into Equation (80):

$$\left(\frac{1}{2} - \chi\right) = \Psi(1 - \Theta/T) \tag{81}$$

We discuss this more fully below, but one thing to note immediately is that $T = \Theta$ describes the same state as described by $\chi = 1/2$, namely, the condition of $B = 0$. It is apparent that Θ is a temperature—variously known as the *theta temperature* or the *Flory temperature*. Introducing this parameter indicates why the $B = 0$ situation is called the Θ condition. In order to justify the equivalence of the two sides of Equation (81), consider the following steps:

1. The term 1/2 on the left-hand side of Equation (81) enters the Flory-Huggins theory as part of the series expansion of $\ln(1 - \phi_2)$ in the transition between Equations (75) and (76).

2. Thus $(1/2) R$ is an entropy contribution to the second virial coefficient.

3. Our expression for ΔS_m—on which this term is based—was derived by assuming purely random placement of polymer segments and solvent molecules. This may not be fully justified because of intermolecular forces that we did not consider in arriving at Equation (66). We take advantage of this opportunity to allow for some bias in the placement of particles on the lattice and replace $1/2 R$ with ΔS_Ψ as the contribution of entropy to the second virial coefficient.

4. We recall from the transition between Equations (74) and (75) that χRT is the enthalpy contribution to the second virial coefficient. In the present context, we designate this ΔH_Ψ.

5. By multiplying the numerator and denominator of the terms by the same factors, the left-hand side of Equation (81) can be transformed as follows:

$$\frac{1}{2} - \chi = \frac{\frac{1}{2} R}{R} - \frac{\chi RT}{RT} = \frac{\Delta S_\Psi}{R} - \frac{\Delta H_\Psi}{RT}$$

6. If $\Delta S_\Psi/R$ is factored out of the last version, we obtain

$$\frac{\Delta S_\Psi}{R}\left(1 - \frac{1}{T}\frac{\Delta H_\Psi/R}{\Delta S_\Psi/R}\right) = \Psi\left(1 - \frac{1}{T}\frac{\Delta H_\Psi/R}{\Delta S_\Psi/R}\right)$$

where $\Delta S_\Psi/R$ has been written Ψ.

7. The ratio of ΔH_Ψ to ΔS_Ψ has kelvin units; this quantity is called the theta temperature. Substituting Θ for this ratio gives the right-hand side of Equation (81).

These manipulations may appear to add little except for needless complication to an interpretation of the second virial coefficient for random coils. Recall, however, that Equation (81) allows the variation of solvent goodness caused by temperature changes to be described quantitatively. Thus the interaction parameter χ is used to describe how B changes when a polymer is dissolved in different solvents. By contrast, Θ is used to describe the variation in B when a given polymer-solvent system is examined at different temperatures. This has been done for the polystyrene-cyclohexane system at three different temperatures; the results are discussed in Example 3.4.

* * *

EXAMPLE 3.4 *Theta Temperature of A Polymer Solution from Second Virial Coefficient Data.* Values of the second virial coefficient along with some pertinent volumes are tabulated below for the polystyrene-cyclohexane system at three temperatures.

T (K)	$B \cdot 10^5$ (cm^3 g^{-2} mole)	$\overline{V}_1$ (cm^3 mole^{-1})	$\overline{V}_2/M_2$ (cm^3 g^{-1})
303	-4.55	109.5	0.930
313	4.45	110.9	0.935
323	9.01	112.3	0.940

Use these data to estimate Θ and Ψ for this system. Does ΔS_Ψ agree with the expected entropy contribution to the second virial coefficient?

Solution: The theta temperature is that value of T at which $B = 0$. It is apparent that B changes sign (i.e., passes through zero) about midway between 303 and 313K. Equations (80) and (81) can be combined to give

$$B = \frac{(\overline{V}_2/M_2)^2}{\overline{V}_1} \Psi \left(1 - \frac{\Theta}{T}\right)$$

which can be solved for Ψ once Θ is known. Using $\Theta = 308K = 35°C$, the following values of Ψ can be calculated using the volumes provided:

T (K)	$1 - \Theta/T$	$\overline{V}_1 B(\overline{V}_2/M_2)^{-2}$	Ψ
303	-0.017	$-5.76 \cdot 10^{-3}$	0.339
313	0.016	$5.65 \cdot 10^{-3}$	0.353
323	0.046	$1.15 \cdot 10^{-2}$	0.249

For reasons not discussed, it is correct to use the value of Ψ interpolated to Θ conditions rather than, say, average the divergent values. Thus we estimate $\Psi = 0.34$, or $\Delta S_\Psi = 0.34\,R$. Since the Flory-Huggins theory predicts a contribution of $1/2\,R$ to the second virial coefficient, it seems that an additional entropy effect—given by $-0.16\,R$—must be included in order to account for the experimental B value. The fact that this "correction" to ΔS_m is negative implies that the entropy of mixing has been slightly overestimated by assuming random placement of the constituents on the lattice. ■

* * *

We have now looked at two models for the second virial coefficient of uncharged colloidal solutes. In Section 3.5b we see that B depends on the magnitude of the particle charge for polyelectrolyte solutes.

3.5 OSMOTIC EQUILIBRIUM IN CHARGED SYSTEMS

We have had no occasion as yet in this book to note that colloidal solutes may possess an electrical charge just like their low molecular weight counterparts. Portions of Chapter 4 and Chapters 11–13 are concerned with those properties of colloids that are direct consequences of the charge of the particles. For the present we introduce the idea of charged particles by examining the effect of the charge on the osmotic pressure of the system.

The charge on a colloidal particle may originate either from the dissociation of functional groups that are covalently bonded to the colloid or from the preferential adsorption of ions to the surface. The charge of a colloid cannot be regarded as a fixed quantity like molecular weight, but must be treated as a variable with a value that depends on the nature and concentration of other components of the system. For example, proteins are positively charged at very low pH levels and negatively charged at very high pH levels; the point of electroneutrality varies from one protein to another. In the case of proteins, it is clearly the ionization of acidic and basic functional groups attached to the polypeptide chain that is primarily responsible for the charge characteristics of the molecules.

At this point we shall not concern ourselves any further with the origin of the charge of a colloidal system; rather, our attitude is that charge is one more characteristic that must be measured and understood in order to characterize certain systems fully.

Although it is not particularly difficult to formulate the thermodynamics of charged systems in perfectly general terms, the resulting notation is cumbersome. Instead of the completely general form, therefore, we consider a very specific case. The principal features will emerge clearly from this example; other situations may be readily derived by parallel arguments. The system we are concerned with consists of three components: Component 1 is the solvent, usually water; component 2 is the colloidal electrolyte; and component 3 is a low molecular weight uni-univalent electrolyte MX. We (arbitrarily) designate the colloidal electrolyte PX_z, consisting of a positively charged macroion having a valence number $+z$, paired with $z\,X^-$ ions. It could be the negative ion of the colloidal electrolyte that is the macroion (as in Vignette III) and the low molecular weight solute could have a different stoichiometry, but the essential features would remain the same.

In physical chemistry it is convenient to express concentrations as molalities and to use molality units to express the activity of the components. This is the convention we follow in this section. Accordingly, the standard state for a component consists of a solution in which that component has an activity of 1.0 mole (kg solvent)$^{-1}$.

3.5a Donnan Equilibrium

The specific situation we wish to consider is the osmotic equilibrium that develops in an apparatus that has a semipermeable membrane impermeable to the macroion only. That is, the membrane is assumed to be permeable not only to the solvent but also to both of the ions of the low molecular weight electrolyte, but not to the colloidal ion P^{z+}. At equilibrium the low molecular weight ions will be found on both sides of the membrane, but not in equal concentrations, because of the presence of the macroions on one side of the membrane. We have already come across an example of such a situation in the vignette at the beginning of this chapter on the role of Donnan equilibrium on the so-called resting states of nerve cells.

For the purpose of our discussion, we designate the side of the membrane that contains the macroions as the α phase and the solution from which the macroions are withheld as the β phase. Equation (12) continues to describe the equilibrium condition; applying it to component 3 leads to the following:

$$\mu_{3,\alpha} = \mu_{3,\beta} \tag{82}$$

Substituting Equation (13) for the β phase and Equation (19) for the α phase, which is under an osmotic pressure π, yields

$$\mu_3^0 + RT \ln a_{3,\beta} = \mu_3^0 + RT \ln a_{3,\alpha} + \int_0^\pi \overline{V}_3 dp \tag{83}$$

At sufficiently low concentrations of the macroion, the osmotic pressure term will be negligible compared with $RT \ln a_{3,\alpha}$, so Equation (83) becomes

$$a_{3,\alpha} = a_{3,\beta} \tag{84}$$

It may be recalled from physical chemistry that the activity of a 1 : 1 electrolyte is given by the product of the activities of the positive and negative ions of the compounds; therefore

$$(a_{M,\alpha})(a_{X,\alpha}) = (a_{M,\beta})(a_{X,\beta}) \tag{85}$$

This expression describes what is known as the *Donnan equilibrium*. It does *not* say that the activity of M^+ and/or X^- is the same on both sides of the membrane, but that the ion activity product is constant on both sides of the membrane. In the sense that an ion product is involved, the Donnan equilibrium clearly resembles all other ionic equilibria.

Remembering that $a_+ = \gamma_+ m_+$ and $\gamma_\pm^2 = \gamma_+ \gamma_-$, where $\gamma_\pm$ is the mean ionic activity coefficient (appropriate to molality units), enables us to rewrite Equation (85) as

$$(m_{M,\alpha})(m_{X,\alpha})\gamma_{\pm,\alpha}^2 = (m_{M,\beta})(m_{X,\beta})\gamma_{\pm,\beta}^2 \tag{86}$$

Of course, in the limit of infinite dilution $\gamma_\pm \to 1$. For the present we restrict our attention to sufficiently dilute solutions so that activity coefficients may be neglected and molalities may be used instead of activities. It might also be noted that in dilute solutions, for which this simplification is apt to be valid, molality and molarity are almost equal.

Another factor that we have not yet taken into account is the requirement that both sides of the membrane be electrically neutral. For the α phase, which contains the macroion, this condition is expressed by

$$zm_{P,\alpha} + m_{M,\alpha} = m_{X,\alpha} \tag{87}$$

In the β phase, which contains only low molecular weight ions, electroneutrality requires

$$m_{M,\beta} = m_{X,\beta} \tag{88}$$

The significance of the Donnan equilibrium is probably best seen as follows. Combining Equations (86) and (88) yields

$$m_{M,\beta}^2 = m_{X,\beta}^2 = (m_{M,\alpha})(m_{X,\alpha}) \tag{89}$$

Next, we use Equation (87) to substitute for either $m_{M,\alpha}$ or $m_{X,\alpha}$ in Equation (89), obtaining the following quadratic equations for $m_{M,\alpha}$ and $m_{X,\alpha}$:

$$m_{M,\alpha}^2 + zm_P m_{M,\alpha} - m_{M,\beta}^2 = 0 \tag{90}$$

and

$$m_{X,\alpha}^2 - zm_P m_{X,\alpha} - m_{X,\beta}^2 = 0 \tag{91}$$

where we have dropped the subscript α from $m_{P,\alpha}$, for convenience. These expressions permit us to evaluate the concentration of the low molecular weight ions in the compartment with the macroions, the α phase, in terms of z and the concentration of ions in the other compartment.

The situation is most easily understood by considering a numerical example. Table 3.2 lists values of $m_{M,\alpha}$ and $m_{X,\alpha}$ calculated using Equations (90) and (91). These values have been determined for two different values of $m_{M,\beta} = m_{X,\beta}$: 10^{-3} and 10^{-2}. The parameter zm_P has been selected at six evenly spaced intervals between 10^{-3} and 10^{-2}. A solution containing 1 g of colloidal electrolyte of molecular weight 10^5 per 100 g of water, for example, would have a value of $m_P = 10^{-4}$; if the macroion carries a charge of $+10$, the parameter zm_P equals 10^{-3}. It is evident from Table 3.2 that the concentration of low molecular weight positive ions is larger in the β phase than in the α phase (which contains the macroions), and that the situation is reversed for the negative ions. The requirement of electroneutrality brings this about. To show better the uneven distribution of low molecular weight ions on the two sides of the membrane, Table 3.2 also lists the ratio of the concentrations on both sides of the membrane for both the positive and the negative ions.

Two conclusions may be drawn readily from an inspection of the results of Table 3.2. First, the uneven distribution of the simple ions becomes more pronounced as the quantity zm_P increases. The low molecular weight ions are free, after all, to pass through the membrane; it is only electroneutrality that holds them back. The more macroions present or the higher

TABLE 3.2 Values of $m_{M,\alpha}$ and $m_{X,\alpha}$ and the Ratios $(m_\alpha/m_\beta)_M$ and $(m_\alpha/m_\beta)_X$ for Two Values of m_β and a Range of Values of zm_P*

zm_P	$m_{M,\beta} = m_{X,\beta} = 10^{-3}$				$m_{M,\beta} = m_{X,\beta} = 10^{-2}$			
	$m_{M,\alpha}$	$m_{X,\alpha}$	$(m_\alpha/m_\beta)_M$	$(m_\alpha/m_\beta)_X$	$m_{M,\alpha}$	$m_{X,\alpha}$	$(m_\alpha/m_\beta)_M$	$(m_\alpha/m_\beta)_X$
10^{-3}	6.18×10^{-4}	1.62×10^{-3}	0.62	1.62	9.51×10^{-3}	1.05×10^{-2}	0.95	1.05
2×10^{-3}	4.14×10^{-4}	2.41×10^{-3}	0.41	2.41	9.05×10^{-3}	1.11×10^{-2}	0.91	1.11
4×10^{-3}	2.36×10^{-4}	4.24×10^{-3}	0.24	4.24	8.20×10^{-3}	1.22×10^{-2}	0.82	1.22
6×10^{-3}	1.62×10^{-4}	6.16×10^{-3}	0.16	6.16	7.44×10^{-3}	1.34×10^{-2}	0.74	1.34
8×10^{-3}	1.23×10^{-4}	8.12×10^{-3}	0.12	8.12	6.77×10^{-3}	1.48×10^{-2}	0.67	1.48
10^{-2}	9.9×10^{-5}	1.01×10^{-2}	0.10	10.09	6.18×10^{-3}	1.62×10^{-2}	0.62	1.62

*All concentrations are in moles per kilogram of solvent. The α phase contains positive macroions.

charge they carry, the more asymmetrically the simple electrolyte will be distributed. Table 3.2 also shows that the uneven distribution of low molecular weight electrolyte becomes less pronounced as the concentration of this electrolyte is increased. We return to this point in the next section when we discuss the osmotic pressure of charged systems.

The combined effects of electroneutrality and the Donnan equilibrium permits us to evaluate the distribution of simple ions across a semipermeable membrane. If electrodes reversible to either the M^+ or the X^- ions were introduced to both sides of the membrane, there would be no potential difference between them; the system is at equilibrium and the ion activity is the same in both compartments. However, if calomel reference electrodes are also introduced into each compartment in addition to the reversible electrodes, then a potential difference will be observed between the two reference electrodes. This potential, called the *membrane potential*, reflects the fact that the membrane must be polarized because of the macroions on one side. It might be noted that polarized membranes abound in living systems, but the polarization there is thought to be primarily due to differences in ionic mobilities for different solutes rather than the sort of mechanism that we have been discussing. We return to a more detailed discussion of the electrochemistry of colloidal systems in Chapter 11.

3.5b   Osmotic Pressure of Charged Colloids

Now let us turn our attention to the osmotic pressure generated by the macroion in this system. Since we have already restricted ourselves to dilute solutions, it is adequate for our purposes to use Equation (35), making allowance for the fact that we have been expressing concentrations as molality in this section. The volume of 1 kg of solvent equals $1000\,V_1^0/M_1$, so Equation (35) becomes

$$\pi \frac{1000\,V_1^0}{M_1} = m\,R\,T \tag{92}$$

where m is the molality of the solute responsible for the osmotic pressure and V_1^0 is the molar volume of the solvent. Since there are solute molecules on both sides of the membrane, the osmotic pressure will be due to the excess solute on the side of the membrane that carries the macroion (the α phase). Therefore we may replace m in Equation (92) by

$$m = m_{P,\alpha} + m_{M,\alpha} + m_{X,\alpha} - m_{M,\beta} - m_{X,\beta} \tag{93}$$

Substituting from the electroneutrality Equations (87) and (88) transforms Equation (93) into

$$m = m_{P,\alpha} + m_{M,\alpha} + z m_{P,\alpha} + m_{M,\alpha} - 2m_{M,\beta} \tag{94}$$

This is further modified by the Donnan relationship, Equation (86), also rewritten to include electroneutrality, to give

$$m = m_{P,\alpha}(1 + z) + 2\,m_{M,\alpha} - 2\,[m_{M,\alpha}\,(z\,m_{P,\alpha} + m_{M,\alpha})]^{1/2} \tag{95}$$

Combining Equations (92) and (95) yields the following for the osmotic pressure of the system:

$$\pi = \frac{M_1 RT}{1000\,V_1^0}\left[(1 + z)m_P + 2\,m_M - 2\,m_M\left(1 + \frac{z\,m_P}{m_M}\right)^{1/2}\right] \tag{96}$$

It should be noted that all concentrations have been expressed in terms of the molality of the solute in the compartment that contains the macroion, so the subscript α is no longer necessary.

Although Equation (96) is rather awkward as written, several highly informative variations of it are obtained by considering different limiting situations. These are summarized in Table 3.3 for the cases in which $m_M = 0$, $m_M \gg m_P$, and $m_M > m_P$. In the table, the expressions for π that follow from Equation (96) in each of these cases are written both in terms of the molality of the macroion m_P and with the concentration of the macroion expressed in weight per unit volume c_2, specifically grams of colloid per liter of solution.

TABLE 3.3 Special Cases of Equation (96) for the Osmotic Pressure of a Charged Colloid in the Presence of a Low Molecular Weight Salt

	m_P (mole kg^{-1})	c_2 (g liter^{-1})
$m_M = 0$	$\pi = \dfrac{M_1 RT}{1000 V_1^0}(1 + z)m_P$	$\pi = \dfrac{RT(1 + z)c_2}{M_2}$
$m_M \gg m_P$	$\pi = \dfrac{M_1 RT}{1000\, V_1} m_P$	$\pi = \dfrac{RTc_2}{M_2}$
$m_M > m_P$	$\pi = \dfrac{M_1 RT}{1000\, V_1^0}\left(m_P + \dfrac{z^2}{4}\dfrac{m_P^2}{m_M}\right)$	$\pi = \dfrac{RTc_2}{M_2} + \dfrac{z^2}{4}\dfrac{1000\, V_1^0\, RT\, c_2^2}{M_1 M_2^2 m_M}$
		$\dfrac{\pi}{RTc_2} = \dfrac{1}{M_2} + \dfrac{z^2\, 1000\, c_2}{4\rho_1 M_2^2 m_{MX}}$

We do not present the algebraic manipulations that lead to the various forms presented in Table 3.3; however, two relationships involved in generating these forms might be noted. First, the quantity $(1 + zm_P/m_M)^{1/2}$ in Equation (96) may be approximated by the binomial series expansion (see Appendix A) to give

$$\left(1 + \frac{z\, m_P}{m_M}\right)^{1/2} = 1 + \frac{1}{2}\frac{z\, m_P}{m_M} - \frac{1}{8}\left(\frac{z\, m_P}{m_M}\right)^2 + \ldots \tag{97}$$

When $m_M \gg m_P$, only the first two terms are retained; when $m_M > m_P$, the first three are used. Second, the relationship between molality and grams per liter units is given by the following in dilute solutions:

$$m_P = \frac{1000\, V_1^0}{M_1 M_2} c_2 \tag{98}$$

All the forms presented in Table 3.3 are readily obtained from Equation (96) by incorporating Equation (97) or (98) or both into Equation (96). Now, let us consider the physical significance of the resulting special cases.

The two extreme values of m_M, $m_M = 0$ and $m_M \gg m_P$, are especially interesting. The former corresponds to the case of no added salt (since the macroion is positive) and the latter to a large excess or "swamping" amount of added salt. Now suppose an osmotic pressure experiment were conducted on two solutions of the same colloid, assumed to have a fixed charge z, with the objective of determining the molecular weight of the colloid. Further suppose that the two determinations differ from one another in the sense that one corresponds to zero added salt and the second to *swamping electrolyte* conditions. Finally, suppose the results are simply interpreted in terms of Equation (35) to yield the molecular weight of the colloid. Comparing Equation (35) with the results listed in Table 3.3 reveals that the correct molecular weight would be obtained for the charged colloid under swamping electrolyte conditions, but an apparent molecular weight less than the true weight by a factor $z + 1$ is obtained in the absence of salt.

How are we to understand this odd result? The answer is easy when we remember that osmotic pressure counts solute particles. The macroion cannot pass through the semipermeable membrane. In the absence of added salt, its counterions will not pass through the membrane either since the electroneutrality of the solution must be maintained. Therefore the equilibrium pressure is that associated with $(z + 1)$ particles. Failure to consider the presence of the counterions will lead to the interpretation of a low molecular weight for the colloid. As we already saw, the presence of increasing amounts of salt leads to a leveling off of the ion concentrations on the two sides of the membrane. The effect of the charge on the macroion is essentially "swamped out" with increasing electrolyte.

One interesting aspect of the limiting case of swamping electrolyte is the fact that the conclusion is totally independent of the specific nature of the ions. This is a partial justification for an assumption that was implicitly made at an earlier point. In writing Equations (87) and (88), we assume that the only ions present are M^+, X^-, and P^{z+}. In aqueous solutions, however, H^+ and OH^- are always present also, but have clearly been assumed to be negligible in writing Equations (87) and (88). The swamping electrolyte concentration may always be chosen to justify neglecting these contributions.

Table 3.3 also includes an approximation for the case in which the concentration of the salt exceeds that of the colloid, but not to the swamping extent, $m_M > m_P$. Comparison of that case with the result given in Equation (34) suggests that the contribution of charge to the second virial coefficient of the solution is given by

$$B = \frac{1000 \, z^2}{4 \, M_2^2 \, \rho_1 \, m_{MX}} \tag{99}$$

Strictly speaking, this contribution should be added to the excluded volume of the particles. The excluded volume of the particles becomes more important as the concentration of the salt increases, a conclusion that may be seen by considering two facts:

1. Charged colloids behave as if they were uncharged under swamping electrolyte conditions.

2. The electrostatic contribution to B, Equation (99), is inversely proportional to salt concentration.

This effect may be qualitatively understood as follows. In a charged system the colloid consists of the macroion and its low molecular weight counterions, which are, of course, distributed through a portion of the solution in the neighborhood of the macroion. Thus we visualize the colloidal ion as being surrounded by an ion atmosphere, the same sort of model that is invoked in the Debye-Hückel theory of electrolyte nonideality. The "excluded volume" that is required, therefore, includes both the volume of the colloidal particle and the volume of that part of the solution that contains the counterions. The precise distribution of counterions around a charged particle is the subject matter of Chapter 11 and, to a lesser extent, Chapter 12. In those chapters, we see that the extent of the domain over which the ion atmosphere extends increases as the electrolyte concentration decreases. Therefore, it is reasonable to expect the volume of the ion atmosphere to be sufficiently larger than the macroion in dilute solutions so that the volume actually excluded by the colloidal particle can be neglected.

We have already seen that the second virial coefficient may be determined experimentally from a plot of the reduced osmotic pressure versus concentration. Since all other quantities in Equation (99) are measurable, the charge of a macroion may be determined from the second virial coefficient of a solution with a known amount of salt. As an illustration of the use of Equation (99), we consider the data of Figure 3.6 in Example 3.5.

*   *   *

EXAMPLE 3.5 *Evaluation of Charges of Macroions from Osmotic Pressures.* In Example 3.3, we evaluated M and B for bovine serum albumin at pH = 5.37, at which the molecule is known to be uncharged. Use the data in Figure 3.6 to evaluate B and, from it, the charge of the molecule at pH = 7.00. The data in Figure 3.6 were measured in 0.15 M NaCl.

Solution: The slope of the line at pH = 7.00 is $2.28 \cdot 10^{-3}$ torr kg^2 g^{-2}; we convert to SI units as in Example 3.3:

$$2.28 \cdot 10^{-3} \, \text{torr kg}^2 \, \text{g}^{-2} \cdot \frac{1.01 \cdot 10^5 \text{N m}^{-2}}{760 \, \text{torr}} \cdot \left(\frac{10^{-3} \, \text{m}^3}{\text{kg}}\right)^2 \cdot \left(\frac{10^3 \, \text{g}}{\text{kg}}\right)^2 = 0.303 \, \text{N m}^4 \, \text{kg}^{-2}$$

Division by RT gives B:

$$0.303/(8.314)(298) = 12.23 \cdot 10^{-5} \, \text{m}^3 \, \text{kg}^{-2} \, \text{mole}$$

As noted above, it is the difference between this value and the B value for the uncharged molecule (Example 3.3) that is interpreted by Equation (99). Also recall that M was determined to be 69,600 g mole $^{-1}$ in the example cited. Therefore

$$(12.23 - 7.35)10^{-5} = \frac{1000z^2}{4M_2^2\rho_1 m_{MX}}$$

$z^2 = 4(4.88 \cdot 10^{-5} \text{ m}^3 \text{ kg}^{-2} \text{ mole})(69.6 \text{ kg mole}^{-1})^2 \cdot (1.0 \text{ g cm}^{-3})(0.15 \text{ mole kg}^{-1})(100 \text{ cm m}^{-1})^3 \div 1000 \text{ g kg}^{-1} = 142$

and $z \approx \pm 12$.

Since the pH is higher than that at which the molecule is uncharged, the albumin must be negatively charged. Hence we identify z as -12 in this case. ∎

* * *

3.6 SOME APPLICATIONS OF OSMOTIC PHENOMENA

3.6a Dialysis

Substances with particles in the colloidal size range are often obtained in a form that contains low molecular weight impurities. For example, enzymes are separated from homogenized tissue samples by extraction in a buffer solution. The enzyme preparation, therefore, is "contaminated" by the components of the buffer. Likewise, synthetic high polymers generally contain unreacted monomer, initiator, and catalyst. There are many experiments in which traces of low molecular weight impurities would have no effect, as, for example, in sedimentation. This chapter has shown clearly, however, that the presence of low molecular weight solutes may have large effects in an osmotic pressure experiment if these substances are retained by the membrane, either directly or through the electroneutrality condition.

Experiments with semipermeable membranes not only point out the need for purification, but also suggest a means by which this may be accomplished. The procedure of dialysis is one technique for removing low molecular weight salts or nonelectrolytes from a colloid. The method consists merely of enclosing the colloid to be purified in a bag made of some semipermeable material. The sealed bag is then placed in a quantity of the solvent. The membrane must be permeable to the solvent and to any low molecular weight impurities present, but impermeable to the colloid. As a result of the semipermeability of the bag, the impurities will distribute themselves through both compartments. The outer portion is replaced often or even continuously, so the low molecular weight impurities are gradually flushed away.

Since the membrane is also permeable to the solvent, the solvent simultaneously diffuses into the bag, diluting the colloid. Ample air space must be present in the bag at the beginning, otherwise it will rupture owing to the pressure developed by the solvent imbibed. There is also a danger that the porosity of the membrane will increase if the bag is stretched as a result of internal pressure buildup. Cellophane tubing is most commonly used as the membrane material. It is sold in rolls for this purpose and may be cut to length and tied at the ends to make the required bags.

Purification by dialysis is a slow process. Its rate is increased, however, by increasing the surface area of the membrane since that is where the exchange of solute between the two phases takes place. Stirring and frequent replacement of the solvent accelerate the process by maintaining the maximum gradient of concentration across the membrane. With ionic contaminants, the rate of dialysis may be enhanced by placing electrodes in the compartment surrounding the enclosed colloid and taking advantage of the migration of the ions in an electric field. This modification is known as *electrodialysis*.

Finally, it might be noted that colloids may be concentrated by a slight modification of the dialysis procedure. The liquid against which the colloid of interest is being dialyzed may itself be a concentrated colloid. With aqueous dispersions, for example, polyethylene oxide solutions may be used as the second colloid.

The second colloid is prepared at higher activity; therefore the solvent is drawn toward the more concentrated phase. This increases the concentration of the colloid of interest. Alternatively, the concentration increase may be accomplished by allowing the solvent to evaporate from the outer surface of the bag.

Just as with osmotic pressure, the membranes in dialysis must be carefully selected to be compatible with the system under study. Specifically, this amounts to impermeability with respect to the colloid(s) involved and permeability with respect to low molecular weight components.

3.6b Reverse Osmosis

As we discussed in Section 3.2, samples of solution and solvent separated by a semipermeable membrane will be at equilibrium only when the solution is at a greater pressure than the solvent. This is the osmotic pressure. If the solution is under less pressure than the equilibrium osmotic pressure, solvent will flow from the pure phase into the solution. If, on the other hand, the solution is under a pressure greater than the equilibrium osmotic pressure, the pure solvent will flow in the reverse direction, from the solution to the solvent phase. In the last case, the semipermeable membrane functions like a filter that separates solvent from solute molecules. In fact, the process is referred to in the literature by the terms *hyperfiltration* and *ultrafiltration*, as well as *reverse osmosis* (Sourirajan 1970); however, the last term is enjoying common use these days.

As we have discussed already, the property of membrane semipermeability applies to all sorts of systems. Likewise, reverse osmosis may be applied to a wide variety of systems. An application that has attracted a great deal of interest in recent years is the production of potable water from saline water. Since no phase transitions are involved as, for example, in distillation, the method offers some prospect of economic feasibility in coastal regions.

Cellulose acetate seems to be the most thoroughly investigated of many possible membrane materials. Cellulose acetate membranes are capable of yielding 96–98% retention of NaCl, for example, and of delivering about $0.2 \text{ cm}^3 \text{ s}^{-1} \text{ atm}^{-1} \text{ m}^{-2}$. This amounts to about 50 gal day^{-1} ft^{-2} at 100 atm. Note that in this application of osmometry, it is not the solute but the membrane that makes up the colloidal system. In this case, the solid portion of the membrane would be the continuous phase and the pores, necessarily small if the membrane is to be effective, the "dispersed" phase. According to this point of view, numerous aspects of membrane technology become part of the interests of colloid and surface chemists. Practical desalination by reverse osmosis depends on a membrane that (a) has enough contact area to process large volumes of solution and (b) is thin enough to do so rapidly, yet (c) is sturdy enough to withstand these pressures. Since some of these points work in opposition to each other, the technical problem is one of optimization.

It is the rate of separation rather than the efficiency of salt retention that is the primary practical issue in the development of reverse osmosis desalination. In addition to a variety of other factors, the rate of reverse osmotic flow depends on the excess pressure across the membrane. Therefore the problem of rapid flow is tied into the technology of developing membranes capable of withstanding high pressures. The osmotic pressure of sea water at 25°C is about 25 atm. This means that no reverse osmosis will occur until the applied pressure exceeds this value. This corresponds to a water column about 840-ft high at this temperature.

Note that a solution more concentrated than the original one also results from the reverse osmosis process. This means that the method of reverse osmosis may also be used as a method for concentrating solutions. Fruit juices and radioactive wastes, for example, have been concentrated by this method.

REVIEW QUESTIONS

1. Describe in simple terms what *osmotic pressure* is. Does a solute have to have charges for it to have an osmotic pressure?
2. What is the relation between the chemical potential of a solute in a solution and its osmotic pressure?
3. Is the presence of a semipermeable membrane separating solutions at two different concentrations needed for osmotic pressure to exist?
4. Describe some examples of phenomena or processes that are influenced by the osmotic pressure.

5. Describe an experiment that can be used to measure osmotic pressures of colloids or macromolecular solutions. Can the osmotic pressure of a colloid or a polymer solution exceed atmospheric pressures? If yes, under what conditions? If not, why?

6. Write the equation of state for the osmotic pressure of a solution that behaves ideally.

7. How does the validity of the *van't Hoff equation* depend on the structure or the nature (e.g., solid particle, flexible coil, etc.) of the solute? Why or why not?

8. How would you correct for deviations from ideality?

9. What is *reduced osmotic pressure*?

10. What is a *virial expansion*?

11. What is the physical significance of the *second virial coefficient*? Of what is it a function?

12. Is the second virial coefficient a function of density? Explain.

13. Sketch the variation of the second virial coefficient with temperature.

14. Is the second virial coefficient positive or negative, or can it be either? What is the physical significance of its sign?

15. How does one use osmotic pressure measurements to determine the molecular weight of a solute? How does polydispersity in molecular weight affect such a measurment?

16. What is the meaning of the term *excluded-volume interaction*?

17. Outline the logic used in deriving expressions for the osmotic pressure and second virial coefficient due to excluded-volume interactions?

18. Outline how the *Flory-Huggins theory* accounts for enthalpy of mixing.

19. What is the physical significance of the χ parameter in the Flory-Huggins theory? How is it related to solute/solute and solute-solvent interactions?

20. What is *theta temperature* (or, the *Flory temperature*)? What are the relative magnitudes of the excluded-volume interactions and the energetic interactions in a dilute polymer solution at its theta temperature?

21. What is *Donnan equilibrium*? Give at least three examples of applications for which the Donnan equilibrium is important.

22. How does added salt affect the osmotic pressure of a charged colloid?

23. Give a few examples of the use of semipermeable membranes in practice.

REFERENCES

General References (with Annotations)

Crow, D. R., *Principles and Applications of Electrochemistry*, 4th ed., Blackie A & P, London, 1994. (Undergraduate level. A treatment of electrochemistry; requires only basic physical chemistry as a prerequisite.)

Flory, P. J., *Principles of Polymer Chemistry*, Cornell University Press, Ithaca, NY, 1953. (Undergraduate level. A classic textbook by a pioneer in the field.)

Hiemenz, P. C., *Polymer Chemistry: The Basic Concepts*, Marcel Dekker, New York, 1984. (Undergraduate level. An introduction to polymer chemistry. Discusses many of the topics relevant to the present chapter.)

Other References

Bergethon, P. R., and Simons, E. R., *Biophysical Chemistry: Molecules to Membranes*, Springer-Verlag, New York, 1990.

Levitan, I. B., and Kaczmarek, L. K., *The Neuron: Cell and Molecular Biology*, Oxford University Press, New York, 1991.

Morawetz, H., *Macromolecules in Solution*, Wiley, New York, 1965.

Nossal, R. J., and Lecar, H., *Molecular and Cell Biophysics*, Addison-Wesley, Redwood City, CA, 1991.

Puvvada, S., and Blankschtein, D., *J. Phys. Chem.*, **93**, 7753 (1989).

Richards, E. G., *An Introduction to the Physical Properties of Large Molecules in Solution*, Cambridge University Press, Cambridge, England, 1980.

Sourirajan, S., *Reverse Osmosis*, Academic Press, New York, 1970.

Tanford, C., *Physical Chemistry of Macromolecules*, Wiley, New York, 1961.

PROBLEMS

1. Criticize or defend the following proposition: As proof that no low molecular weight fractions of polymer have passed through the membrane in an osmotic pressure experiment, the following test may be performed. A quantity of "poor" solvent is added to an aliquot taken from the solvent side of the membrane. The absence of precipitate proves that no low molecular weight polymer passed through the membrane.

2. The osmotic pressure of solutions of a fractionated, atactic poly(isopropylacrylate) solution was measured at 25°C with the following results:*

π (g cm^{-2})	1.39	2.46	4.20	6.52
$c \times 10^2$ (g cm^{-3})	0.47	0.69	1.05	1.36

   Prepare a plot of π/c versus c for these results and evaluate $(\pi/c)_0$. Calculate M and B for this system. Define what is meant by an "atactic" polymer and compare with syndiotactic and isotactic polymers. List reference(s) consulted for these definitions.

3. An important assumption made in truncating Equation (34) is that the third virial coefficient is small. It is known† that the third virial coefficient depends strongly on the second so that it approaches zero in poor solvents even faster than B does. In fact, if Γ_2 is defined as the product BM_2, it is known that Γ_3 is approximately $0.25\Gamma_2^2$ in a good solvent; that is, Equation (34) may be written with one additional term as

$$\frac{\pi}{RTc} = \frac{1}{M}(1 + \Gamma_2 c + 0.25\Gamma_2^2 c^2)$$

   Describe how this result may be used to facilitate the evaluation of M and B in the event that a plot of reduced osmotic pressure versus c still contains too much curvature to permit a meaningful straight line to be drawn.

4. Osmotic pressures for aqueous solutions of n-dodecylhexaoxyethylene monoether, $C_{12}H_{25}$ $(OC_2H_4)_4OC_2H_5$, were measured at 25°C. At concentrations below 0.038 g liter^{-1} no osmotic pressure develops, indicating complete membrane permeability. Above this concentration a pressure develops, indicating the presence of impermeable species. In the following data these pressures are reported for various $c - c_0$ values, where $c_0 = 0.038$ g liter^{-1}, the threshold for an osmotic effect:‡

π (cm)	4.90	6.53	7.62	10.58
$c - c_0$ (g liter^{-1})	29.72	38.12	43.90	58.46

   Plot $\pi/(c - c_0)$ versus $c - c_0$ to evaluate the molecular weight and B for the species responsible for the osmotic pressure.

5. The data of the preceding problem may be interpreted by assuming the following model. Above 0.038 g liter^{-1}, solute molecules associate into aggregates. By comparing the M value obtained in Problem 4 with the molecular weight of the original ether, calculate the number of molecules in the aggregate according to this model. Assuming the colloidal particles are spherical, use the second virial coefficient evaluated in Problem 4 to estimate the molar volume and radius of the aggregates. Do the quantities calculated in this problem seem reasonably self-consistent?

6. The cohesive energy density of a low molecular weight liquid is given by the heat of vaporization of that liquid, expressed per unit volume. Verify the value given for the CED of acetone in Table 3.1 from the facts that $\Delta H_v = 30.2$ kJ mole^{-1} and $\rho = 0.792$ g cm^{-3}. Would it be better to use the change in internal energy on vaporization ΔU_v rather than ΔH_v as a measure of intermolecular attraction? Does it make much difference quantitatively whether ΔU_v or ΔH_v is used?

---

*Mark, J. E., Wessling, R. A., and Hughes, R. E., *J. Phys. Chem.*, **70**, 1895 (1966).
†Flory, P. J., *Principles of Polymer Chemistry*, Cornell University Press, Ithaca, NY, 1973.
‡Attwood, D., Elworthy, P. H., and Kayne, S. B., *J. Phys. Chem.*, **74**, 3529 (1970).

7. The osmotic pressure of polystyrene (PS) solutions in toluene and methylethyl ketone (MEK) was measured* at 25°C, and the results were analyzed to give B values of 4.59×10^{-4} and 1.39×10^{-4} cm^3 g^{-2} mole, respectively, for the two solutions. Use these results to criticize or defend the following proposition: According to Table 3.1, $\sqrt{CED} = 8.5–9.1$ for polystyrene. For toluene and MEK, $\sqrt{CED}$ equals 8.9 and 9.04, respectively. Since the MEK solution has a smaller B value than the toluene solution, it appears that the best value to use for the $\sqrt{CED}$ for polystyrene is at the upper end of the range of values given and close to the value of MEK. In this way the quantity $[(\sqrt{CED})_{MEK} - (\sqrt{CED})_{PS}]^2$ will be smaller than the same quantity for polystyrene in toluene. This is consistent with the order of the B values.

8. The solvent activity in a solution of polybutadiene in benzene was determined by measuring the vapor pressure p_1 of benzene over solutions containing various concentrations of polymer.† A plot of $\ln(p_1/p_1^0) - \ln \Phi_1 - (1 - 1/n)\Phi_2$ versus Φ_2^2—in which p_1^0 is the vapor pressure of the pure benzene—yields a straight line having an intercept of zero and a slope equal to 0.33. Evaluate the interaction parameter χ from this result. Is the Θ temperature above or below the experimental temperature? Explain.

9. Krigbaum and Geymer (1959)‡ measured the osmotic pressure of polystyrene in cyclohexane at several different temperatures. The following is a sample of their results for a single fraction of polymer:

c (g cm^{-3})	$T = 24°C$ $(\pi/RTc) \times 10^6$ (mole g^{-1})	c (g cm^{-3})	$T = 34°C$ $(\pi/RTc) \times 10^6$ (mole g^{-1})	c (g cm^{-3})	$T = 44°C$ $(\pi/RTc) \times 10^6$ (mole g^{-1})
0.0976	8.0	0.0081	13.3	0.0959	18.6
0.182	6.0	0.0201	14.2	0.1780	28.1
0.259	8.7	0.0964	14.2	0.2550	40.0
		0.1800	18.7		
		0.2570	26.2		

Plot all of these points on the same graph as π/RTc versus c. Although considerable nonideality exists, estimate the limiting slope at each temperature. Interpretation is assisted by realizing that each set of data approaches the same intercept as $c \to 0$. What is the approximate molecular weight of the polymer? How do the values of B estimated from the limiting slopes compare with the values given for the same system over an overlapping range of temperatures in Example 3.4?

10. Solutions of bovine serum albumin in 0.15 M NaCl were studied at other pH levels in addition to those shown in Figure 3.6. The following data are examples of additional measurements:§

pH	m_P (g protein kg^{-1})	π (mm Hg)
6.19	57.71	21.48
6.64	56.17	21.40

Since the limiting value of π/m_P shown in Figure 3.6 applies to these data also, it is possible to evaluate the second virial coefficient at these pH levels. Evaluate B and z, the effective protein charge, at the pH values shown. (Note that a slight variation in NaCl concentrations at these different pH levels should be taken into account for a more accurate determination of z.)

11. The osmotic pressure of salt-free (electrodialyzed) bovine serum albumin solutions was measured at pH $= 5.37$ ($\pm$?).§ At this pH the net charge of the protein molecules is zero. The following data were obtained in different runs:

*Bawn, C., Freeman, R., and Kamaliddin, A., *Trans. Faraday Soc.*, **46**, 862 (1950).

†Jessup, R. S., *J. Res. Nat. Bur. Stand.*, **60**, 47 (1958).

‡Krigbaum, W. R., and Geymer, D. O., *J. Am. Chem. Soc.*, **81**, 1859 (1959).

§Scatchard, G., Batchelder, A. C., and Brown, A., *J. Am. Chem. Soc.*, **68**, 2320 (1946).

π (mm Hg)	4.26	7.44	8.14	12.97	11.31	19.62
m_P (g protein kg^{-1})	19.56	19.71	40.88	46.05	60.81	63.25

Discuss the reasons why it is so difficult to obtain meaningful osmotic pressure data in salt-free solutions. Consider specifically the reconciliation of the electrolyte-free aspect of the experiment with the accurate control of pH.

12. In reverse osmosis both solvent and solute diffuse because of gradients in their chemical potentials. For the solvent there is no gradient of chemical potential at an osmotic pressure of π; at applied pressures p greater than π, there is such a gradient that is proportional to the difference $p - \pi$. To a first approximation, the gradient of the solute chemical potential is independent of p and depends on the difference between concentrations on opposite sides of the membrane. This leads to the result that the fraction of solute retained varies as $[1 + \text{const.}/(p - \pi)]^{-1}$. Verify that the following data* for a reverse osmosis experiment with 0.1 M NaCl and a cellulose acetate membrane follow this relationship:

Applied p (atm)	10	13	20	38	51	75
Percent salt retained	63	79	88	94	95	97

(π is about 2.6 atm for 0.1 M NaCl.)

---

*Data of J. E. Breton, Jr., cited by H. K. Lonsdale, in *Desalination by Reverse Osmosis* (U. Merton, Ed.), MIT Press, Cambridge, MA, 1966.

4

The Rheology of Dispersions

Imagine that your Tradesman drags behind his regular and respectable vertex, a parallelogram of twelve or thirteen inches in diagonal: What are you to do with such a monster sticking fast in your house door?

From Abbott's *Flatland*

4.1 INTRODUCTION

4.1a What Is Rheology?

The way liquids flow is one of their most obvious properties. We use a variety of terms in everyday language to describe this aspect of fluidlike substances. Thus we speak of the "thickness" of cream, the "weight" of oil, and the "leveling" of paint to describe the flow behavior and properties of such materials. The science student will probably recognize that all these terms allude in one way or another to a property known as the *viscosity* of the liquid.

However, substituting a technical term does little to make this somewhat elusive subject more intelligible. Moreover, viscosity is just one indicator of the flow characteristics of materials. In order to understand fully the flow behavior of materials, it is important that we understand some of the basic physical laws or rules that materials follow when subjected to external forces so that we can formulate theories to predict the deformation and flow of matter under a given condition. The study of the *flow and deformation of materials* is known as *rheology*. Rheology, a term coined by Professor E. C. Bingham of Lafayette College in Indiana in 1929 in consultation with a colleague from the Department of Classics, comes from the Greek root *rheos* meaning *stream* (Barnes et al. 1989) and is the subject of our focus in this chapter.

4.1b Why Is Rheology Important?

Materials in a colloidal state are frequently preferred in industrial processing operations because their large surface areas per unit volume enhance chemical reactivity, adsorptive capacity, heat transfer rates, and so on. Therefore, one cannot overlook the importance of the flow behavior and properties of colloids since they exert a significant influence on the performance, efficiency, and economy of the process. Note that some examples of this (e.g., ceramic processing, electrophoretic display devices, and food colloids) were mentioned in the vignettes presented in Chapter 1. In addition, one often uses the flow properties and behavior of the products as measures of the microstructure (or, "morphology") of the products* and as a means of quality control (e.g., printing inks, toners, paints, skin creams, blood substitutes,

---

*Although it is often said that "Morphology from rheology is theology!"

gels used as drug-delivery systems, etc.) and even to ensure consumer appeal and marketability of the products. Vignette IV provides an example of rheology that most people are likely to appreciate.

From a technical standpoint, it is also important to note that colloids display a *wide* range of rheological behavior. Charged dispersions (even at very low volume fractions) and sterically stabilized colloids show *elastic* behavior like solids. When the interparticle interactions are not important, they behave like ordinary liquids (i.e., they flow easily when subjected to even small shear forces); this is known as *viscous behavior*. Very often, the behavior falls somewhere between these two extremes; the dispersion is then said to be *viscoelastic*. Therefore, it becomes important to understand how the interaction forces and fluid mechanics of the dispersions affect the flow behavior of dispersions.

**VIGNETTE IV   RHEOLOGY IN MANUFACTURING AND QUALITY
CONTROL: Rheology of Chocolate**

Rheology is of major importance in processing operations in industry. The rheological behavior of process streams and dispersions determines the pumping and transportation costs, the ease of mixing operations in reactors, and the "quality" of the final product in many cases. What we do not often realize is that it is also one of the factors that determine the esthetic and sensual appeal of certain products, especially in the case of food products and cosmetics! Differences in the "feel" of a face lotion or a skin cream or the "consistency" of a ketchup can make or break the market for the product!

Take chocolate, for example. The way chocolate feels in the mouth (in addition to its taste, of course) makes a difference in its appeal to chocolate enthusiasts, and much of it has to do with the rheological changes due to the melting of the fat in the mouth. From a processing standpoint, the intricacies of the molds and coating patterns require that the chocolate be free-flowing at high shear stresses but have negligible flow at low stresses (Dickinson 1992, Chapter 3); that is, it should be *shear-thinning*—a concept we discuss in this chapter.

Why are colloids and the rheology of colloids relevant to chocolate? Melted chocolate is a complex, multiphase fluid consisting of solid nonfat particles (about 70% by volume, mostly sugar granules and the rest from crushed cocoa bean) and cocoa butter. (Milk chocolate includes, in addition, some milk solids and milk fat. In both cases, the particle size range is broad, and the shapes are irregular.) Melted chocolate is a non-Newtonian fluid that is shear-thinning and has a yield stress (see Chapter 4, Section 4.1c). The viscosity could be adjusted by adding more cocoa butter, but that is an expensive proposition. A much less expensive and quite effective way is to add lecithin or a suitable emulsifier made of small molecules. The rheology of melted chocolate is very sensitive to lecithin, which first reduces the viscosity and yield stress but increases them at larger concentrations. The optimum amount needed is less than one-tenth of the amount of cocoa butter that would be needed for similar results. The current theory is that lecithin adsorbs onto the sugar particles and prevents them from forming agglomerates, thereby influencing the rheological behavior. This is the domain of the colloid scientist, who is more likely to excel in the needed task if also a chocolate enthusiast! (The monograph of Dickinson cited in the last paragraph is a very good source of colloidal issues of importance in food science.)

Of course, the processing engineer or the colloid chemist seldom worries about the fundamental aspects of the microstructure of the product and its relation to rheology, but many of the topics we discuss in this chapter are useful for gaining the conceptual basis necessary to deal with the practical aspects of dispersion rheology.

4.1c   Focus of This Chapter

Our objective in this chapter is to discuss the flow behavior of fluids and dispersions and introduce some of the basic ideas and theories that are essential for understanding the structure and flow properties of colloids. We particularly focus on the property "viscosity" in terms of several different models and assumptions for noninteracting as well as interacting dispersions at low and high volume fractions.

1. We begin with a brief discussion of *Newton's law of viscosity* and follow this with a discussion of Newtonian flow (i.e., the flow of liquids that follow Newton's law) in a few standard configurations (e.g., cone-and-plate geometry, concentric cylinders, and capillaries) under certain specific boundary conditions. These configurations are commonly used in viscometers designed to measure viscosity of fluids.

2. Next, we take a closer look at the equation of motion for fluids, known as the *Navier-Stokes equation*. Our intention here is to discuss the general concepts that go into setting up the equation of motion rather than to present a comprehensive treatment of the Navier-Stokes equation.

3. The next topic is Einstein's theory of viscosity of dispersions of rigid, spherical particles. This theory is the starting point for most of the current approaches to flow properties of colloids and plays a practical, pedagogical, as well as historical role.

4. We examine some major conditions under which Einstein's theory breaks down. For our purpose, the two important reasons are (a) the effect of the concentration of the dispersion and (b) the effects of interparticle forces, particularly the electrostatic repulsive forces or polymer additives. This leads us next to the *non-Newtonian* behavior of dispersions.

5. Whether a substance behaves like a liquid or a solid is often a matter of the time during which the behavior is observed and the magnitudes of the applied forces. Even mountains and continents deform and move in geological time scales. Metals and rocks are rigid below a critical stress (force per unit area) known as the *yield stress*, but show significant deformation when stresses exceed this limit. What is interesting about colloids (and what makes them complicated) is that they can behave like solids or liquids in time spans and at stress levels common in daily life. It is therefore important to define some useful time scales and energy scales that relate the physicochemical properties of colloids (e.g., charges, surface potentials, thickness of adsorbed polymer layers, sizes of the particles, etc.) to the flow properties. The section on non-Newtonian behavior reviews, in addition to the basic types of non-Newtonian behavior, the time scales and dimensionless groups of importance in this context. Some sample results for viscosity of charged dispersions and coagulated dispersions are also presented to illustrate the departure of non-Newtonian systems from the simpler systems discussed. It is, however, important to remember that the contents of this chapter barely touch on the vast area of dispersion rheology, which is an area of active research currently.

6. We then conclude the chapter with a brief discussion of the viscosity of polymer solutions.

From the outset, it is helpful to be aware of the differences among the models we consider here. As an example, fluids are treated as continuous matter in discussing experimental viscometry; in contrast, we use a statistical, particulate model in discussing polymer solutions. Another example is the development of different models for lyophobic and lyophilic systems. The *Einstein theory of viscosity* is based on a model of rigid spherical particles. As such, it is most appropriate for two-phase systems or dispersions containing particles with rigid structures (although, as we also point out, they describe the viscosity of dispersions of not-so-rigid particles under certain circumstances). The *Kirkwood-Riseman theory* (discussed in Chapter 4, Section 4.9b), in contrast, applies to flexible chains and is suitable for synthetic polymers in solution. A final example is the assumption — made through the bulk of the chapter — that flow has no effect on the size, shape, or structure of the dispersed units. In Chapter 4, Section 4.8, we relax this constraint and consider the viscous behavior of coagulated or flocculated systems in which the flow may disrupt the flocs.

Both the fluid mechanics and the statistical mechanics on which some of the key theoretical results of the chapter are based are sufficiently complicated that we only sketch the highlights of these topics. We attempt to impart some physical plausibility to these theories, however, by using both force and energy perspectives in discussing the viscous resistance to flow.

4.2   NEWTON'S LAW OF VISCOSITY

The coefficient of viscosity was introduced in Chapter 2, Section 2.3a, but this parameter is elusive enough to warrant further comment. In this section we examine the definition of the coefficient of viscosity — the viscosity, for short — of a fluid. This definition leads directly to a discussion of some experimental techniques for measuring viscosity; these are discussed in the following sections.

Imagine two parallel plates of area A between which is sandwiched a liquid of viscosity η. If a force F parallel to the x direction is applied to one of these plates, it will move in the x direction as shown in Figure 4.1. Our concern is the description of the velocity of the fluid enclosed between the two plates. In order to do this, it is convenient to visualize the fluid as consisting of a set of layers stacked parallel to the boundary plates. At the boundaries, those layers in contact with the plates are assumed to possess the same velocities as the plates themselves; that is, $v = 0$ at the lower plate and equals the velocity of the moving plate at that surface. This is the *nonslip condition* that we described in Chapter 2, Section 2.3. Intervening layers have intermediate velocities. This condition is known as *laminar flow* and is limited to low velocities. At higher velocities, turbulence sets in, but we do not worry about this complication.

We can imagine within the fluid two layers separated by dy, over which distance the velocity changes by an amount dv. Therefore dv/dy defines a velocity gradient; *Newton's law of viscosity* states that the shear stress, $\tau = F/A$, is proportional to dv/dy. The viscosity η of the sandwiched fluid is the factor of proportionality:

$$\tau = \frac{F}{A} = \eta \frac{dv}{dy} = \eta \dot{\gamma} \tag{1}$$

Since the velocity v may be written as dx/dt and dx/dy defines the *shear strain* (i.e., flow deformation per unit length) acting on the sample, the velocity gradient may be written as $d(shear)/dt$ and is called the *rate of shear*, often denoted by $\dot{\gamma}$. Note that (dv/dy), or γ', has as units time^{-1}. Thus a dimensional statement of Equation (1) gives (mass length time^{-2} length^{-2}) = η time^{-1}, which shows that η has dimensions (mass length^{-1} time^{-1}) or (kg m^{-1} s^{-1}) in SI units. This is called a pascal-second (abbreviated as Pa.s). We observed in Section 2.3a that the cgs unit of viscosity, the poise (abbreviated as P), is widely used and that 10 P = 1 kg m^{-1} s^{-1} (i.e., 10 P = 1 Pa.s).

Equation (1) merely hypothesizes that the shear stress is linearly proportional to the rate of strain. Fluids that obey the form predicted by Equation (1) are said to be *Newtonian*. Figure 4.2 is a sketch of F/A versus the velocity gradient for several different modes of behavior. For a Newtonian fluid, this representation gives a straight line of zero intercept and

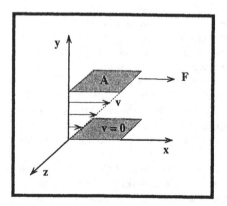

FIG. 4.1   The relationship between applied force per unit area and fluid velocity.

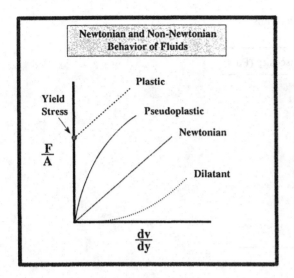

FIG. 4.2 Comparison of Newtonian liquids with several forms of non-Newtonian behavior.

slope equal to η. *Non-Newtonian* fluids generally show nonlinear plots; their "viscosity," the slope of the tangent to the curve at various points, is a function of the rate of shear. Most actual representations of experimental results display the data with the coordinates interchanged from the way they are shown in Figure 4.2. In that case, it is the cotangent of the angle that describes the slope of the line at any point that determines the true (if Newtonian) or apparent (if non-Newtonian) viscosity of the system. For the present, Newtonian behavior is our concern. We discuss some examples of non-Newtonian behavior in Section 4.8.

A second interpretation of η is as valid as Equation (1) and perhaps more illuminating. To arrive at this alternative, we multiply both sides of Equation (1) by dv/dy:

$$\frac{F}{A}\frac{dv}{dy} = \eta \left(\frac{dv}{dy}\right)^2 \tag{2}$$

Now consider the following points to interpret $(F/A)(dv/dy)$:

1. Writing the velocity v as dx/dt, one can regroup the variables on the left-hand side of Equation (2) as $d[Fdx/Ady]/dt$.
2. A force times an increment of distance Fdx equals an increment of energy dE.
3. An area times an increment of distance Ady equals an increment of volume dV.
4. Since the force under consideration measures viscous resistance to flow, the quantity dE/dV measures the energy dissipated per unit volume.
5. Dividing the change in dE/dV, i.e., $d(dE/dV)$, by dt gives the *rate of energy dissipation per unit volume*, which we denote by E_V for convenience.

Based on these ideas, Equation (2) can be rewritten as

$$\frac{dE_V}{dt} = \eta \left(\frac{dv}{dy}\right)^2 \tag{3}$$

which shows that the volume rate of energy dissipation is proportional to the square of the velocity gradient, with the viscosity of the fluid as the factor of proportionality.

Equations (1) and (3) are equivalent as definitions of the viscosity of a fluid. To convince ourselves that η as defined by these expressions does indeed measure resistance to flow, consider two liquids of widely different viscosity, say, water and molasses:

TABLE 4.1   Viscosities of Some Familiar Materials
at Room Temperature

Liquid	Approximate viscosity (Pa.s)
Glass	10^{40}
Molten glass (500°C)	10^{12}
Bitumen	10^{8}
Molten polymers	10^{3}
Golden syrup	10^{2}
Liquid honey	10^{1}
Glycerol	10^{0}
Olive oil	10^{-1}
Bicycle oil	10^{-2}
Water	10^{-3}
Air	10^{-5}

Source: Barnes et al., 1989.

1. Imagine an arrangement like that shown in Figure 4.1 and consider the force that must be applied to sustain a velocity gradient of, say, 1 s^{-1}. The higher the viscosity of the fluid is, the greater will be the force. Molasses would require more force than water.

2. Imagine the rate at which energy is dissipated in maintaining a unit velocity gradient. The higher the viscosity is, the faster will be the energy dissipation per unit volume. Molasses would dissipate energy more rapidly than water.

Table 4.1 presents typical values of viscosities of some common materials. One can see from this table that viscosity varies over several orders of magnitude. One may also note that, although the table lists a viscosity for glasses, the magnitude of the viscosity clearly suggests that the deformation of glasses at room temperatures will be extremely small so glasses are best treated as solids under normal conditions. (Recall the discussion in Section 4.1c of time scales and their relation to whether a substance is defined as a liquid or a solid.) Typical rates of shear for some familiar processes are shown in Table 4.2.

In the next section we consider an experimental approach to viscosity. We generate the apparatus of interest by wrapping—in our imagination—the fluid in Figure 4.1 into a closed ring around the z axis. The two rigid surfaces then describe concentric cylinders, and the instrument is called a *concentric-cylinder viscometer*.

4.3   CONCENTRIC-CYLINDER AND CONE-AND-PLATE VISCOMETERS

4.3a   Concentric-Cylinder Viscometers

Imagine a viscous fluid enclosed in a gap between two concentric cylinders as shown in Figure 4.3a. If one of the cylinders is caused to rotate, a viscous resistance to the rotation will be transmitted through the fluid to the nonrotating cylinder and produce a torque on the nonrotating cylinder. An apparatus that is easily visualized—although not widely used—consists of a cup centered on a turntable with a bob concentrically suspended in it (Fig. 4.3a). In addition to the characteristics of the apparatus, the torque on the suspending wire depends on the viscosity of the fluid—it would be greater for molasses than for water, all other things being equal. Such a torque can be measured, and our objective is to see how the viscosity of the fluid can be evaluated from such data.

We define r to be the distance variable in the radial direction and ω to be the velocity of rotation, with ω measured in radians/second. This means the velocity of a cylindrical layer of

TABLE 4.2 Typical Shear Rates for Some Familiar Processes

Situation	Typical range of shear rates (s^{-1})	Application
Sedimentation of fine powders in a suspending liquid	10^{-6}–10^{-4}	Medicines, paints
Leveling due to surface tension	10^{-2}–10^{-1}	Paints, printing inks
Draining under gravity	10^{-1}–10^{1}	Painting and coating, toilet bleaches
Extruders	10^{0}–10^{2}	Polymers
Chewing and swallowing	10^{1}–10^{2}	Foods
Dip coating	10^{1}–10^{2}	Paints, confectionery
Mixing and stirring	10^{1}–10^{3}	Manufacturing liquids
Pipe flow	10^{0}–10^{3}	Pumping, blood flow
Spraying and brushing	10^{3}–10^{4}	Spray drying, painting, fuel atomization
Rubbing	10^{4}–10^{5}	Application of creams and lotions to the skin
Milling pigments in fluid bases	10^{3}–10^{5}	Paints, printing inks
High-speed coating	10^{5}–10^{6}	Paper
Lubrication	10^{3}–10^{7}	Gasoline engines

Source: Barnes et al., 1989.

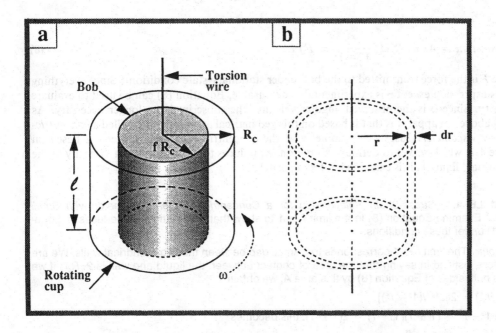

FIG. 4.3 Schematic representation of a concentric-cylinder viscometer: (a) geometry of a cup and bob; and (b) volume element within a liquid gap.

fluid a distance r from the axis of rotation is $r\omega$. There is a velocity gradient across the gap in this apparatus; this can be written $r(d\omega/dr)$. Next we represent a layer of fluid as a cylindrical shell of length ℓ and thickness dr as shown in Figure 4.3b and consider the viscous force acting on such a shell. By Equation (1), this force is given by

$$F = \eta A \frac{dv}{dr} = \eta (2\pi r \ell)\left(r\frac{d\omega}{dr}\right) \tag{4}$$

where $2\pi r\ell$ is the area of the shell. Torque is given by the product of a force and the distance through which it operates, so this element of viscous force must be multiplied by r to give a torque T:

$$T = Fr = 2\pi\eta\,\ell\,r^3\frac{d\omega}{dr} \tag{5}$$

When the apparatus begins to rotate, the fluid experiences an initial acceleration, but a stationary state is rapidly attained in which forces balance and acceleration is zero. Equation (5) gives a generalized expression for torque; under stationary-state conditions it must be independent of r. If this were not the case, forces would be different in different parts of the fluid, and acceleration would occur. Accordingly, we set the torque on this volume element equal to a constant:

$$(2\pi\,\eta\,\ell\,r^3)\left(\frac{d\omega}{dr}\right) = constant \tag{6}$$

Now suppose the outer cylinder has a radius R_c and rotates with an experimental velocity ω_{ex}. Furthermore, assume the inner cylinder has a radius fR_c — where f is a fraction close to unity for small gaps — and is stationary. We can integrate Equation (6) between these limits to obtain a value for the constant torque that characterizes the experiment:

$$2\pi\,\eta\,\ell\int_0^{\omega_{ex}} d\omega = constant\int_{fR_c}^{R_c} r^{-3}\,dr = -\frac{const.}{2}\left(\frac{1}{R_c^2} - \frac{1}{(fR_c)^2}\right) \tag{7}$$

or

$$constant = 4\pi\,\eta\,\ell\,R_c^2\,\omega_{ex}\frac{f^2}{1-f^2} = FfR_c \tag{8}$$

where F is the force transmitted to the bob under stationary-state conditions. Since everything in this expression except η is experimentally measurable, Equation (8) can be used to evaluate viscosity. Since ω is the observed angular velocity, the subscript is no longer necessary. As noted above, an apparatus that is based on this geometrical arrangement is called a *concentric-cylinder viscometer*. Just as the geometry of the concentric-cylinder viscometer is based on Figure 4.1, which serves to define η, Equation (8) reduces to Newton's law of viscosity in the appropriate limit. This is examined in Example 4.1.

* * *

EXAMPLE 4.1 *Stress-Strain Relationship for a Concentric-Cylinder Viscometer with Small Gaps.* Examine Equation (8) in the limit $f \rightarrow 1$ to show that the relationship reduces to Equation (1) under these conditions.

Solution: The limit $f \rightarrow 1$ corresponds to a small gap between the two cylindrical walls. We are therefore justified in saying that the area of contact between the liquid and wall is $2\pi R_c\ell$; if we divide both sides of Equation (8) by this area A, we obtain

$(F/A) = 2\eta\omega[f^2/(1 - f^2)]$

Since $1 - f^2 = (1 + f)(1 - f) \approx 2(1 - f)$, this becomes

$F/A = \eta\omega f^2/(1 - f)$

Multiplying the numerator and denominator of this expression by R_c and letting $v_{max} = R_c\omega$ gives

$$(F/A) = \eta \omega_{max} f^2/[R_c(1 - f)] = \eta(v_{max} - 0)f^2/[R_c(1 - f)] = \eta f^2 \Delta v/\Delta R_c$$

since the velocity at the stationary inner wall is zero. In the limit we obtain

$$\lim_{f \to 1} (F/A) = \eta \, dv/dr$$

which is equivalent to Equation (1), the defining equation for η. When the separation between the cylinders is negligible compared to their radius of curvature, the concentric cylindrical surfaces approximate the infinite parallel plates of the model used to define the coefficient of viscosity. ∎

* * *

One of the appealing features of the concentric-cylinder viscometer is the fact that the rate of shear can be varied by adjusting either the width of the gap or the angular velocity. This is a valuable capability in light of the kinds of non-Newtonian behavior that colloidal systems may display (as shown in Fig. 4.2). Equation (6), however, reminds us that this capability must be viewed with caution. That relationship shows that it is $r^3(d\omega/dr)$ that is constant across the gap, not dv/dr itself. Thus, even though we may change either the speed or the dimensions of the viscometer and thereby achieve different *average* rates of shear, the latter remains an average. The local velocity gradient varies from place to place within the fluid, although this variation is small for narrow gaps.

A variety of commercial viscometers is available that embodies the essential features of the device we have analyzed here. In most commercial instruments, it is the inner cylinder that rotates, even though there is a theoretical advantage to having the outer cylinder rotate rather than the inner one. The reason is that laminar flow is stable when the outer cylinder rotates, whereas centrifugal forces tend to induce turbulent flow in the reverse case. By rotating the inner member, however, interchangeable spindles can be used and the sturdiness and versatility of such a design offsets the theoretical advantage of the rotating cup arrangement. In actual practice, instruments may deviate from the concentric-cylinder arrangement, and their use to evaluate η depends on calibration with known standards rather than mathematical analysis of the instrument geometry.

4.3b Cone-and-Plate Viscometers

We conclude this section with a few remarks about the cone-and-plate type of viscometer, sketched schematically in Figure 4.4. In this viscometer, the fluid is placed between a stationary plate and a cone that touches the plate at its apex. This apparatus also possesses cylindrical symmetry, but this time in order to indicate a location within the fluid we must specify not only r, the distance from the axis of rotation, but also the location within the gap between the cone and the plate, as measured by θ, the angle from the vertical (see Fig. 4.4). Mathematical analysis of this apparatus leads to the result

$$\frac{dv}{dr} \approx \omega \frac{\cos \theta}{\cos \theta_0} \tag{9}$$

where θ_0 is the angle between the vertical and the wetted surface of the cone. This approximation holds for the case in which both θ and θ_0 are close to 90°. This condition is always met in actual practice, for which the angle $(90 - \theta_0)$ is generally less than 5°. Some commercial concentric-cylinder viscometers have interchangeable parts; this means that a conical rotor may be substituted for a cylindrical one. Equation (9) shows that the primary velocity gradient within the fluid is independent of radial position to a very good approximation. This feature distinguishes the cone-and-plate viscometer from the concentric-cylinder viscometer and is a significant advantage in any study that requires accurate knowledge of the rate of shear to which a sample is subjected.

To determine the viscosity of the fluid, the torque T necessary to turn the cone at an angular velocity ω is measured. This torque depends on the viscosity of the fluid according to the equation

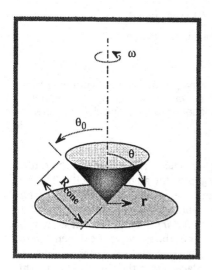

FIG. 4.4 Schematic representation of a cone-and-plate viscometer (the angle is greatly exaggerated).

$$T = \frac{\frac{4}{3}\pi R_{cone}^3 \eta \omega \sin \theta_0}{\cot \theta_0 + \frac{1}{2} \ln \left[(1 + \cos \theta_0)/(1 - \cos \theta_0) \right] \sin \theta_0} \tag{10}$$

where R_{cone} is the radius of the cone measured along the wetted surface. Like the concentric-cylinder viscometer, this apparatus can be operated, in principle, at a variety of different angular velocities, and thus η can be studied at different rates of shear. Furthermore, this last quantity is reasonably constant throughout the fluid in the apparatus.

As noted above, dilute colloidal systems display Newtonian behavior; that is, their apparent viscosity is independent of the rate of shear. Accordingly, the capability to measure η under conditions of variable shear is relatively superfluous in these systems. However, non-Newtonian behavior is commonplace in charged colloids and coagulated colloids (see Section 4.8).

Both the concentric-cylinder and the cone-and-plate viscometers are widely used, especially in cases of non-Newtonian behavior. Liquids of low molecular weight and dilute solutions in such solvents generally display Newtonian behavior. In the last case there is no advantage in being able to vary the shear rate, nor is it a disadvantage that this varies locally within the apparatus. Accordingly, a simpler viscometer is often used for such liquids. A capillary viscometer takes advantage of the fact that a velocity gradient exists whenever a fluid flows past a stationary wall. In the next section we examine the capillary viscometer and Poiseuille's law, by which it is analyzed.

4.4 THE POISEUILLE EQUATION AND CAPILLARY VISCOMETERS

4.4a Flow Through Cylinders: The Poiseuille Flow

Figure 4.5 shows a portion of a cylindrical capillary of radius R_c and length ℓ. Our interest is in the flow of a viscous liquid through such a capillary. This arrangement has the same cylindrical symmetry as the viscometers we discussed in the last section and, once again, it is convenient to focus attention on a cylindrical shell of fluid of radius r and thickness dr, as shown in Figure 4.5. Because of the nonslip condition at the wall of the capillary, the liquid shell adjacent to that wall has a velocity equal to zero. The velocity increases for shells of

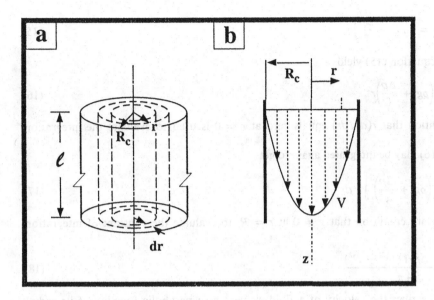

FIG. 4.5 Flow in a cylindrical capillary: (a) a volume element in the flowing liquid; and (b) the parabolic flow profile.

decreasing r so that a velocity gradient exists. Let us attempt to write an expression for v as a function of r, the radius of the cylindrical shell. Since stationary-state conditions hold within the volume element, then any nonviscous forces must exactly balance the viscous force.

The increment in viscous force acting on this element is the difference between the viscous forces on the outer and inner surfaces of the element, with each of these given by Equation (1), with $A = 2\pi r \ell$:

$$\Delta F_{vis} = (F_{vis})_{out} - (F_{vis})_{in} = 2\pi (r + dr)\eta \ell \left(\frac{dv}{dr}\right)_{r+dr} - 2\pi r \eta \ell \left(\frac{dv}{dr}\right)_r \tag{11}$$

Next, we must relate $(dv/dr)_{r+dr}$ to $(dv/dr)_r$. The following expression accomplishes this, provided dr is small:

$$\left(\frac{dv}{dv}\right)_{r+dr} = \left(\frac{dv}{dr}\right)_r + \left(\frac{d^2v}{dr^2}\right)_r dr \tag{12}$$

If this result is substituted into Equation (11), expanded, and only terms linear in dr retained, the expression becomes

$$\Delta F_{vis} = 2\pi \eta \ell \left(r\frac{d^2v}{dr^2} dr + \frac{dv}{dr} dr\right) = 2\pi \eta \ell \frac{d}{dr}\left(r\frac{dv}{dr}\right) \tag{13}$$

This force is counterbalanced by the increments in gravitational and pressure forces:

$$\Delta(F_g + F_{press}) = 2\pi\ell\rho grdr + 2\pi\Delta prdr \tag{14}$$

where the first term equals the weight of the shell and the second is the force on the shell if a pressure difference of Δp exists across the ends of the tube. Note that Δp as defined here is the difference between the pressure at the "exit" of the capillary and the one at the entrance; therefore, Δp is negative since the pressure at the entrance has to be higher for the flow to occur toward the exit. We have assumed (as evident from the figure) that the acceleration due to gravity g acts in the z direction. Setting Equations (13) and (14) equal to each other gives

$$\eta \frac{d}{dr}\left(r \frac{dv}{dr}\right) = \left(\rho g + \frac{\Delta p}{\ell}\right)r \tag{15}$$

Integration of Equation (15) yields

$$\eta r \frac{dv}{dr} = \frac{1}{2}\left(\rho g + \frac{\Delta p}{\ell}\right)r^2 \tag{16}$$

where the condition that $r(dv/dr)$ equals zero at $r = 0$ is used to evaluate the integration constant.

Equation (16) may be integrated again to get

$$\eta \int dv = \frac{1}{2}\left(\rho g + \frac{\Delta p}{\ell}\right) \int r\, dr \tag{17}$$

Using the boundary condition that $v = 0$ at $r = R_c$ to evaluate the constant of integration yields

$$v = \frac{(\rho g + \Delta p/\ell)(r^2 - R_c^2)}{4\eta} \tag{18}$$

This equation describes the velocity of a fluid element as a parabolic function of its radial distance from the center of the tube (see Fig. 4.5b).

The rate of volume flow through the tube V/t equals the summation of the flow rate through each shell, that is, the cross-sectional area of each shell multiplied by the velocity of that shell, where the latter is given by Equation (18):

$$\frac{V}{t} = \int_0^{R_c} \frac{(\rho g \ell + \Delta p)}{4\eta\ell}(r^2 - R_c^2)2\pi r\, dr \tag{19}$$

or

$$\frac{V}{t} = \frac{(\rho g \ell + \Delta p)\pi R_c^4}{8\eta\ell} \tag{20}$$

Equation (20), known as the *Poiseuille equation*, provides the basis for the most common technique for measuring the viscosity of a liquid or a dilute colloidal system, namely, the *capillary viscometer*.

4.4b Capillary Viscometers

The capillary viscometers take advantage of the Poiseuille equation for determining viscosity of fluids. Most capillary viscometers are designed with a relatively large bulb at both ends of the capillary, as shown in Figure 4.6. A constant volume in the upper bulb is designated by two lines etched at either end of the bulb. The viscometer is used by measuring the time required for the liquid level to drop from one line to the other as the fluid drains through the capillary. In such an apparatus, the difference in height of the two liquid columns is relatively constant during the time required for flow. Generally, the only pressure difference across the liquid is due to the weight of the liquid. Under these conditions, Equation (20) can be written

$$\eta = A\rho t \tag{21}$$

in which the constant A incorporates all the parameters that characterize the apparatus. Comparison of the flow times of two substances, one known (subscript 1) and one unknown (subscript 2), through the same apparatus provides an easy way to evaluate η for the unknown. In this case Equation (21) becomes

$$\eta_2 = \frac{\rho_2\, t_2}{\rho_1\, t_1}\eta_1 \tag{22}$$

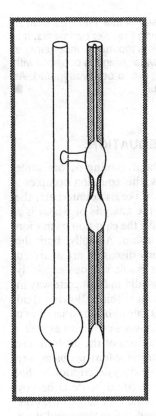

FIG. 4.6 Schematic representation of a capillary viscometer.

For greater accuracy, an additional term may be added to Equation (21) that corrects η for the fact that the Poiseuille equation does not apply exactly at the two ends of the capillary. With this correction for end effects, Equation (21) becomes

$$\eta = A \rho t - B\frac{\rho}{t} \tag{23}$$

where $B = V/8\pi$. With this correction included, both A and B can be regarded as instrument constants and evaluated by calibrating the viscometer with two liquids of known density and viscosity. Commercial capillary viscometers are generally designed to make this end correction small, often negligible.

* * *

EXAMPLE 4.2 *Comparison Between Capillary Viscometers and Concentric-Cylinder Viscometers.* Criticize or defend the following proposition: A set of capillary viscometers with different radii can be used in much the same way as a concentric-cylinder viscometer with variable speed or gap width to conduct studies in which the rate of shear is an independent variable.

Solution: Equation (16) shows that the velocity gradient is not uniform in a capillary viscometer any more than it is in a concentric-cylinder instrument. The rate of shear *dv/dr* is directly proportional to the radial distance from the axis of the cylinder. At the wall it has its maximum value, which is proportional to R_c; at the center of the tube it equals zero. Some intermediate value, say, the average, might be used to characterize the gradient in a given instrument. This quantity will be different for capillaries of different radii. All of this is similar to the situation in concentric-cylinder viscometers.

The two types of viscometers differ in the following way, however. A much wider range of shear rates is attainable in concentric-cylinder instruments with adjustable features than in capillaries. Because an average velocity gradient is used in describing these experiments, it is essential that a wide range of averages be spanned; otherwise, there is too much uncertainty in the individual values for meaningful results to be obtained. Since a wider range is possible with the concentric instrument, it is preferable when variable shear rates are to be investigated. As we saw in the last section, cone-and-plate viscometers are better yet.   ■

* * *

4.5   THE EQUATION OF MOTION: THE NAVIER-STOKES EQUATION

It is impossible to read much of the literature on viscosity without coming across some reference to the equation of motion. In the area of fluid mechanics, this equation occupies a place like that of the Schrodinger equation in quantum mechanics. Like its counterpart, the equation of motion is a complicated partial differential equation, the analysis of which is a matter for fluid dynamicists. Our purpose in this section is not to solve the equation of motion for any problem, but merely to introduce the physics of the relationship. Actually, both the concentric-cylinder and the capillary viscometers that we have already discussed are analyzed by the equation of motion, so we have already worked with this result without explicitly recognizing it. The equation of motion does in a general way what we did in a concrete way in the discussions above, namely, describe the velocity of a fluid element within a flowing fluid as a function of location in the fluid. The equation of motion allows this to be considered as a function of both location and time and is thus useful in nonstationary-state problems as well.

As is the case with all differential equations, the boundary conditions of the problem are an important consideration since they determine the "fit" of the solution. Many problems are set up to have a high level of symmetry and thereby simplify their boundary descriptions. This is the situation in the viscometers that we discussed above and that could be described by cylindrical symmetry. Note that the cone-and-plate viscometer—in which the angle from the axis of rotation had to be considered—is a case for which we skipped the analysis and went straight for the final result, a complicated result at that. Because it is often solved for problems with symmetrical geometry, the equation of motion is frequently encountered in cylindrical and spherical coordinates, which complicates its appearance but simplifies its solution. We base the following discussion on rectangular coordinates, which may not be particularly convenient for problems of interest but are easily visualized.

4.5a   The Navier-Stokes Equation: General Considerations

Figure 4.7 shows a small rectangular volume element $\Delta x \, \Delta y \, \Delta z$ located within a flowing fluid. The fluid passing through this element has a velocity $\mathbf{v}$, which may be resolved into x, y, and z components—represented by v_x, v_y, and v_z, respectively—since, in general, $\mathbf{v}$ bears no special relationship to the coordinates. Note that $\mathbf{v}$ was in the axial direction in the cylindrical problems we discussed above (i.e., $\mathbf{v} = \mathbf{i}v_x$, where $\mathbf{i}$ is the unit vector in the x direction, and $v_y = v_z = 0$); no such restriction operates here. Now let us examine the rate at which momentum accumulates in this volume element. To begin, we recognize that the rate of change of momentum equals a force $\mathbf{F}$:

$$\frac{d(m\mathbf{v})}{dt} = m\left(\frac{d\mathbf{v}}{dt}\right) = m\mathbf{a} = \mathbf{F} \tag{24}$$

where $\mathbf{a}$ is the acceleration of the fluid element. Furthermore, the net rate of change of momentum per unit volume is given by $d(\rho\mathbf{v})/dt$. Therefore the net rate of change in momentum per unit volume equals the sum of all forces acting per unit volume. This last quantity is made up of three contributions, which include external forces (such as gravity) and pressure forces in addition to viscous forces; therefore we write

$$\frac{d(\rho\mathbf{v})}{dt} = \mathbf{F}_{ext} + \mathbf{F}_{press} + \mathbf{F}_{vis} \tag{25}$$

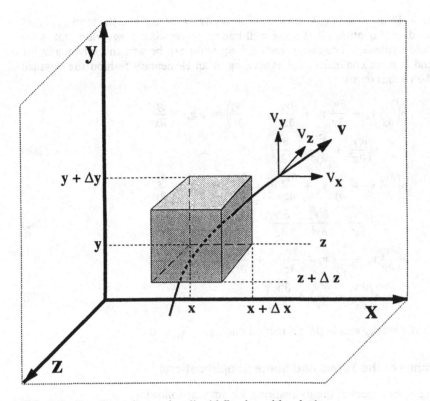

FIG. 4.7 A volume element in a liquid flowing with velocity **v**.

Treating density as a constant, the case for an incompressible fluid, enables us to write

$$\frac{d(\rho \mathbf{v})}{dt} = \rho \left(\frac{\partial \mathbf{v}}{\partial x} \frac{dx}{dt} + \frac{\partial \mathbf{v}}{\partial y} \frac{dy}{dt} + \frac{\partial \mathbf{v}}{\partial z} \frac{dz}{dt} + \frac{\partial \mathbf{v}}{\partial t} \right) \tag{26}$$

where we have used the chain rule for differentiation to expand $d\mathbf{v}/dt$ into the form shown.

The external, pressure, and viscous forces acting on the volume element can be developed more fully also. We do not go through the details of this development, but instead cite the final result followed by some explanatory remarks:

$$\rho \left(\frac{\partial \mathbf{v}}{\partial x} v_x + \frac{\partial \mathbf{v}}{\partial y} v_y + \frac{\partial \mathbf{v}}{\partial z} v_z + \frac{\partial \mathbf{v}}{\partial t} \right) = \rho\, \mathbf{g} - \mathbf{i}\, \frac{\partial p}{\partial x} - \mathbf{j}\, \frac{\partial p}{\partial y} - \mathbf{k}\, \frac{\partial p}{\partial z}$$
$$+ \eta \left(\frac{\partial^2 \mathbf{v}}{\partial x^2} + \frac{\partial^2 \mathbf{v}}{\partial y^2} + \frac{\partial^2 \mathbf{v}}{\partial z^2} \right) \tag{27}$$

where the bold **g** now indicates the *vectorial* acceleration due to gravity and **i**, **j**, and **k** are unit vectors in the x, y, and z directions, respectively. Equation (27) is the equation of motion for an *incompressible* fluid and is a vectorial equation (that is, quantities on both the left-hand side and the right-hand side are vectors and have *directions* as well as magnitudes). Remember, the solution to this differential equation for a set of specified boundary conditions gives a general expression for **v** as a function of x, y, z, and t.

Now let us briefly consider the various terms in Equation (27) with the idea of establishing their plausibility. Rewriting Equation (27) in vector notation condenses it by combining terms with similar significance; in this notation Equation (27) becomes

$$\rho (\mathbf{v} \cdot \nabla \mathbf{v}) + \rho\, \frac{\partial \mathbf{v}}{\partial t} = \rho\, \mathbf{g} - \nabla p + \eta\, \nabla^2 \mathbf{v} \tag{28a}$$

where $\nabla = \mathbf{i}\partial/\partial x + \mathbf{j}\partial/\partial y + \mathbf{k}\partial/\partial z$, $\nabla^2 = \partial^2/\partial x^2 + \partial^2/\partial y^2 + \partial^2/\partial z^2$ and the term in parentheses is a dot product. Equation (28a) is the well-known *Navier-Stokes equation* that is the centerpiece of fluid dynamics. The above vectorial equation can be written individually for each direction and is more convenient for explaining in an elementary fashion the physical significance of the viscous terms:

x direction:
$$\rho\left(\frac{\partial v_x}{\partial x}v_x + \frac{\partial v_x}{\partial y}v_y + \frac{\partial v_x}{\partial z}v_z + \frac{\partial v_x}{\partial t}\right) = \rho g_x - \frac{\partial p}{\partial x}$$
$$+ \eta\left(\frac{\partial^2 v_x}{\partial x^2} + \frac{\partial^2 v_x}{\partial y^2} + \frac{\partial^2 v_x}{\partial z^2}\right) \tag{28b}$$

y direction:
$$\rho\left(\frac{\partial v_y}{\partial x}v_x + \frac{\partial v_y}{\partial y}v_y + \frac{\partial v_y}{\partial z}v_z + \frac{\partial v_y}{\partial t}\right) = \rho g_y - \frac{\partial p}{\partial y}$$
$$+ \eta\left(\frac{\partial^2 v_y}{\partial x^2} + \frac{\partial^2 v_y}{\partial y^2} + \frac{\partial^2 v_y}{\partial z^2}\right) \tag{28c}$$

z direction:
$$\rho\left(\frac{\partial v_z}{\partial x}v_x + \frac{\partial v_z}{\partial y}v_y + \frac{\partial v_z}{\partial z}v_z + \frac{\partial v_z}{\partial t}\right) = \rho g_z - \frac{\partial p}{\partial z}$$
$$+ \eta\left(\frac{\partial^2 v_z}{\partial x^2} + \frac{\partial^2 v_z}{\partial y^2} + \frac{\partial^2 v_z}{\partial z^2}\right) \tag{28d}$$

Notice that if the gravity acts in the z direction only, $g_x = g_y = 0$.

4.5b   Significance of the Terms and Some Simplifications

Starting on the right-hand sides of these equations, we note the following:

1.   The first term on the right in each case describes the force of gravity acting on the volume element. In this development, gravity is the only external force we consider. The gravitational force per unit volume is given by $\rho\mathbf{g}$. (Notice that when the acceleration due to gravity acts in the z direction, $\mathbf{g} = \mathbf{i}0 + \mathbf{j}0 + \mathbf{k}g_z = \mathbf{k}g$, where $g_z = g$ is the magnitude of the vector $\mathbf{g}$.)

2.   The second term on the right in each case describes the force on the volume element due to pressure. The net force in the x direction due to pressure is given by $(p_{x+\Delta x} - p_x)\Delta y\Delta z$. Dividing through by $\Delta x\Delta y\Delta z$ gives the x component of force per unit volume as $[(p_{x+\Delta x} - p_x)/\Delta x]$ or $\partial p/\partial x$ in the limit $\Delta x \rightarrow 0$. Similar expressions apply to the y and z components of F_{press}. (Note that here p_x stands for the pressure p at x, i.e., the subscript x denotes the position at which the pressure p is evaluated. The pressure is a scalar quantity. In contrast, $\mathbf{v}$ and $\mathbf{g}$ are vectors, and the subscripts in those cases denote the components in the appropriate directions.)

3.   The third term on the right describes *viscous forces*. The viscous force acting on, say, the face of area $\Delta x\Delta y$ of the volume element in the y direction is given by $\eta\Delta x\Delta y(\partial v_y/\partial z)_{z+\Delta z}$ on one side of the volume element and by $\eta\Delta x\Delta y(\partial v_y/\partial z)_z$ on the other side. Dividing the difference between these two terms by the volume of the element $\Delta x\Delta y\Delta z$ and taking the limit $\Delta z \rightarrow 0$ gives the contribution to F_{vis} from the velocity gradient of v_y in the z direction, namely, $\eta(\partial^2 v_y/\partial z^2)$, in Equation (28c). Similar considerations apply to the other viscous terms in Equations (28b)–(28d).

Next, we turn our attention to the terms on the left-hand side of Equation (28a):

1.   The first term on the left is called the *inertial* term, and the second arises from the temporal variation in the velocity at any given position. For low velocities, the former may be neglected.

2.   Under conditions for which the inertial term can be neglected compared to the other terms in the equation, Equation (28a) becomes

$$\rho\frac{\partial \mathbf{v}}{\partial t} = \rho\mathbf{g} - \nabla p + \eta\nabla^2\mathbf{v}. \tag{29a}$$

This is called the *Stokes approximation* to the equation of motion. At steady-state conditions, this may be written as

$$-\rho\mathbf{g} + \nabla p = \eta\nabla^2\mathbf{v} \qquad\qquad (29b)$$

This equation, with or without the body-force term $\rho\mathbf{g}$, is called the *Stokes equation.**

3. Under stationary-state conditions, the velocity is independent of time, and Equation (28a) becomes

$$\rho\mathbf{g} - \nabla p + \eta\nabla^2\mathbf{v} = \rho(\mathbf{v}\cdot\nabla\mathbf{v}) \qquad\qquad (30)$$

This is called the steady-state *Navier-Stokes equation*.

We remarked at the beginning of this section that the equation of motion is the cornerstone of any discussion of fluid dynamics. When one considers the various coordinate systems in which it may be expressed, the vector identities that may transform it, or the approximations that may be used to simplify it, Equation (28a) takes on many forms, some of which are scarcely recognizable as the same relationship. The purpose of this section is to illustrate that—despite its complexity and variations—the equation of motion is really nothing more than a statement of *Newton's second law of motion!*

Until now, we have been primarily concerned with the definition and measurement of viscosity without regard to the nature of the system under consideration. Next we turn our attention to systems containing dispersed particles with dimensions in the colloidal size range. Viscosity measurements can be used to characterize both lyophobic and lyophilic systems; we discuss both in the order cited.

4.6 EINSTEIN'S THEORY OF VISCOSITY OF DISPERSIONS

In 1906 Albert Einstein (Nobel Prize, 1921) published his first derivation of an expression for the viscosity of a dilute dispersion of solid spheres. The initial theory contained errors that were corrected in a subsequent paper that appeared in 1911. It would be no mistake to infer from the historical existence of this error that the theory is complex. Therefore we restrict our discussion to an abbreviated description of the assumptions of the theory and some highlights of the derivation. Before examining the Einstein theory, let us qualitatively consider what effect the presence of dispersed particles is expected to have on the viscosity of a fluid.

4.6a The Effect of Particles on the Viscosity of Dispersions

The presence of a colloidal size particle in the liquid increases the viscosity because of the effect it has on the flow pattern. Two effects that readily come to mind are illustrated in Figure 4.8. In Figure 4.8a, the velocity profile near a wall is shown for a pure liquid. The variation of velocity among layers is indicated by the arrows of different lengths. In Figure 4.8b, a nonrotating particle is pictured cutting across several layers in the flowing liquid. Since the particle does not rotate (by hypothesis), it must slow down the fluid so that the layers on opposite sides of the particle have the same velocity, that of the particle itself. The overall velocity gradient is thus reduced. Since the applied force is presumably the same in both Figures 4.8a and 4.8b, the reduced velocity gradient must be offset by an increase in η.

Alternatively, we might consider a particle that is induced to rotate by its position in the velocity gradient. Such a situation is shown in Figure 4.8c. In this case some of the energy that would otherwise keep the liquid flowing is deposited in the particle, causing it to rotate. In both cases the presence of the particle increases the viscosity of the fluid.

The increase in viscosity due to dispersed particles is expected to increase with the concen-

---

*This is different from "Stokes's law" we discussed in Equation (2.7), which results from the solution of Equation (29b) for flow over spheres (Bird et al. 1960). Equation (2.7) is for the *frictional force* on a sphere and is also known as the Stokes equation.

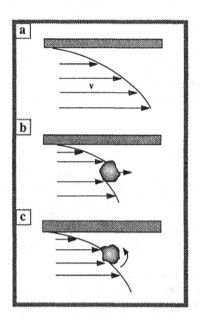

FIG. 4.8 Schematic illustration of the flow pattern near a stationary wall for a pure liquid and for a dispersion: (a) a pure liquid near a stationary wall; (b) a dispersion containing nonrotating particles; and (c) a dispersion containing rotating particles.

tration of the particles, a dependence we may tentatively describe in terms of a power series in concentration c:

$$\eta = A + Bc + Cc^2 + \ldots \tag{31}$$

In this equation A, B, C, . . . are constants with values to be determined. This much is evident: As the concentration of a dispersion goes to zero, its viscosity must go to that of the continuous phase. Therefore, $A = \eta_0$, the viscosity of the medium. Furthermore, the constants B, C, . . . might reasonably be expected to depend on the size, shape, orientation, and so on of the dispersed units.

Einstein's derivation is based on the hydrodynamic equations of the preceding section. As such, it is limited to those cases in which ρ and η for the fluid are constant and in which the flow velocity is low. Furthermore, the theory postulates an extremely dilute dispersion of rigid spheres with *no slippage of the liquid at the surface of the spheres*. Finally, the spheres are assumed to be large enough compared to the solvent molecules to permit us to regard the solvent as a continuum, but small enough compared to the dimensions of the viscometer to permit us to ignore wall effects. These size restrictions make this result applicable to particles in the colloidal size range.

Einstein considered a fluid in laminar flow through a dilute, random array of spherical particles. An obvious difficulty in applying these equations to the case in question is the enormous number of surfaces at which boundary conditions must be specified. This leads to the first reason why an infinitely dilute dispersion must be considered. If the particles are sufficiently far apart, each will modify the flow pattern in its environment as if it alone were present. This introduces two boundary conditions: no slippage at the surface and unperturbed flow at larger distances from the surface. Note the resemblance between this model and that of the Stokes law, Equation (2.7), for the drag force on a spherical particle. As in that derivation, two different expressions for the rate of energy dissipation in the dispersion are equated to give the final result. It is important to remember Equation (3) in this context since it establishes the coefficient of viscosity as the pertinent parameter in any discussion of the rate of energy dissipation per unit volume dE_V/dt.

Einstein was able to solve the equation of motion, first for the case of a single sphere present, to give dE_v/dt for a spherical volume ("cell") of dispersion of radius r centered at the spherical particle of radius R_s. The distance r is so much larger than the radius of the spherical particle that the disturbance of flow due to the particle has vanished at the surface of the hypothetical enclosing sphere. Einstein has shown the result of this integration to be

$$\left(\frac{dE_v}{dt}\right)_{Solv.+Sph.} = K\eta_0\left(\frac{8}{3}\pi r^3 + \frac{4}{3}\pi R_s^3\right) \tag{32}$$

where η_0 is the viscosity of the solvent and K is a constant with a significance that does not concern us since it cancels out. The subscripts *Solv.* and *Sph.* refer to the solvent in the cell and the sphere of radius R_s, respectively. Since $(4/3)\pi r^3$ represents the volume V of the system and $(4/3)\pi R_s^3$ represents the volume occupied by the spherical particle, we can rewrite this result as

$$\left(\frac{dE_v}{dt}\right)_{Solv.+Sph.} = K\eta_0 V(2 + \phi) \tag{33}$$

where ϕ is the volume fraction occupied by the sphere. Examining the limit of Equation (32) as $R_s \to 0$ permits us to evaluate the rate of dissipation of energy in the same volume element in the absence of the sphere; that is, if solvent alone is present,

$$\left(\frac{dE_v}{dt}\right)_{Solv.} = 2K\eta_0 V \tag{34}$$

Next, we can use Equations (33) and (34) to consider the increment per sphere in the rate of energy dissipation. Subtracting Equation (34) from Equation (33) and rearranging gives

$$\left(\frac{dE_v}{dt}\right)_{Solv.+Sph.} = \left(1 + \frac{\phi}{2}\right)\left(\frac{dE_v}{dt}\right)_{Solv.} \tag{35}$$

If the dispersion contains N spheres sufficiently far part for the effects of each to be independent of the others, we may write for the whole dispersion (subscript *Disp.*)

$$\left(\frac{dE_v}{dt}\right)_{Disp.} = N\left(\frac{dE_v}{dt}\right)_{Solv.+Sph.}$$

$$= \left(1 + \frac{\phi}{2}\right)N\left(\frac{dE_v}{dt}\right)_{Solv.} \tag{36}$$

$$= \left(1 + \frac{\phi}{2}\right)\left(\frac{dE_v}{dt}\right)_{Total\ Solv.}$$

Here, the subscript *Total Solv.* stands for all the solvent (of volume $V_{Total} = NV$, i.e., N times the solvent "cells" of volume V each). Of course, the volume fraction ϕ remains unchanged in Equation (36) since both the volume of spheres and the volume of solvent cells are multiplied by the amount N. Thus, one has from Equation (36)

$$\left(\frac{dE_v}{dt}\right)_{Disp.} = 2K\eta_0 V_{Total}\left(1 + \frac{\phi}{2}\right) \tag{37}$$

Einstein was able to derive another equation describing the rate of energy dissipation per unit volume for a dispersion of spheres:

$$\left(\frac{dE_v}{dt}\right)_{Disp.} = 2K\eta(1 - \phi)^2 V_{Total} \tag{38}$$

in which η (without a subscript) is the measured viscosity *of the dispersion* and K has the same significance as in Equation (37). Again in the derivation of this result, the particles are assumed to be far apart. Therefore, Equations (37) and (38), two expressions for the same quantity derived with comparable assumptions, may be equated:

$$2K\eta(1 - \phi)^2 V_{Total} = 2K\eta_0 V_{Total}\left(1 + \frac{\phi}{2}\right) \tag{39}$$

Rearranging this equation yields

$$\frac{\eta}{\eta_0} = \frac{(1 + \phi/2)}{(1 - \phi)^2} = \left(1 + \frac{\phi}{2}\right)(1 + \phi + \phi^2 + \cdots)^2 \tag{40}$$

where only the leading terms in ϕ have been retained since ϕ must be small to satisfy the requirement of large distances between spheres. If no term higher than first order in ϕ is retained, Equation (40) becomes

$$\eta/\eta_0 = 1 + 2.5\phi + \cdots \tag{41}$$

a result known as *Einstein's equation of viscosity of dispersions*. This derivation was not only an important accomplishment in its own right, but also served as a model for many subsequent derivations. Before considering these, however, let us examine Einstein's law from an experimental point of view.

4.6b Einstein's Theory: Experimental Tests

The Einstein equation is one of those pleasant surprises that occasionally emerges from complex theories: a remarkably simple relationship between variables, in this case the viscosity of a dispersion and the volume fraction of the dispersed spheres. A great many restrictive assumptions are made in the course of the derivation of this result, but the major ones are (a) that the particles are solid spheres and (b) that their concentration is small.

These conditions are relatively easy to meet experimentally, so Equation (41) has been tested in numerous studies. Figure 4.9 is an example of the sort of verification that has been

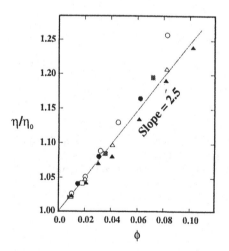

FIG. 4.9 Experimental verification of Einstein's law of viscosity for spherical particles of several different sizes (Squares are yeast particles, $R_s = 2.5$ μm; circles are fungus spores, $R_s = 4.0$ μm; triangles are glass spheres, $R_s = 80$ μm). Open symbols represent measurements in concentric-cylinder viscometers, and closed symbols represent measurements in capillary viscometers. (Data from F. Eirich, M. Bunzl, and H. Margaretha, *Kolloid Z.*, **74**, 276 (1936).)

obtained. In the work summarized in this figure, a variety of model systems were investigated using both capillary and concentric-cylinder viscometers. The solid line in the figure is the Einstein prediction; the agreement between theory and experiment is seen to be very good.

Note that the applicability of the Einstein equation seems equally good regardless of the size of the spheres used. Although the range of particle radii in Figure 4.9 is relatively narrow, this conclusion has been verified for particles as small as individual molecules and as large as grains of sand. It might also be noted that experiments of this sort are carried out in mixed solvents or electrolyte solutions with densities equal to that of the dispersed particles (i.e., in density-matched liquids) so that the spheres do not settle under the influence of gravity.

The simplicity of the Einstein equation makes it relatively easy to test, but also limits its usefulness rather sharply. With so few variables involved, the quantities we may evaluate by Equation (41) are few. Viscosity is a measurable quantity from which we try to extract information about the dispersion. All that Equation (41) offers directly in this area is the evaluation of ϕ from viscosity measurements, again provided ϕ is small and the particles are spheres.

The data in Figure 4.9 are limited to concentrations below $\phi = 0.10$. We might ask, What are the consequences of increasing the concentration to higher volume fractions of spherical particles? Equation (40) may appear to suggest that the range of applicability of Einstein's equation might be extended by retaining terms higher than first order in the power series; that is,

$$\eta/\eta_0 = 1 + 2.5\phi + k_1\phi^2 + \cdots \tag{42}$$

might give a better fit to the data from dispersions with concentrations more than infinitely dilute. There are several points about this result that should be noted:

1. Equations (31) and (42) are identical in form. Equation (42) shows that the volume fraction is the theoretically preferred unit of concentration as far as viscosity is concerned.

2. The theoretical evaluation of k_1 and the higher order coefficients in Equation (42) requires more than merely retaining additional terms in Equation (40). Einstein's derivation of Equations (32) and (38) is based on the restriction that the dispersion is very dilute, and therefore simply retaining higher order coefficients in Equation (40) will not eliminate the above restriction.

3. A number of theoretical attempts have been made to evaluate k_1 by going back through the Einstein derivation and superimposing the effects of neighboring particles rather than treating them as independent. A few of the theoretical values that have been obtained for k_1 from different models are 14.1, 12.6, and 7.35.

Rather than attempting to choose among these theoretical approaches, let us examine an empirical approach to the problem of deviations from the Einstein equation at high concentrations. Toward this end, it is convenient to rearrange Equation (42) as follows:

$$\frac{(\eta/\eta_0 - 1)}{\phi} = 2.5 + k_1\phi + \cdots \tag{43}$$

This has the effect of reducing the order of each term on the right-hand side of the equation. The term on the left-hand side of the equation is defined as the *reduced viscosity* (see Table 4.3). Now if we plot the reduced viscosity against ϕ, the result should be a straight line of slope k_1 and intercept 2.5, at least at low concentrations before still higher order terms become important.

Figure 4.10 shows the viscosity of dispersions of glass spheres of radius 65 μm plotted in the manner suggested by Equation (43). Several conclusions are evident from the data replotted in this way:

TABLE 4.3 Symbol, Common and International Union for Pure and Applied Chemistry (IUPAC) Names, and Limiting Values for a Variety of Forms Commonly Used to Present Viscosity Data

Functional form	Symbol	Common name	IUPAC name	$\lim\limits_{c \to 0}$
—	η	Viscosity	—	η_0
η/η_0	η_r	Relative viscosity	Viscosity ratio	1
$\eta/\eta_0 - 1$	η_{sp}	Specific viscosity	—	0
$(\eta/\eta_0 - 1)/c$	η_{red}	Reduced viscosity	Viscosity number	$[\eta]$
$c^{-1}\ln(\eta/\eta_0)$	η_{inh}	Inherent viscosity	Logarithmic viscosity number	$[\eta]$
$\lim\limits_{c \to 0} \eta_{red}$ or $\lim\limits_{c \to 0} \eta_{inh}$	$[\eta]$	Intrinsic viscosity	Limiting viscosity number	—

1. The intercept of the curve is clearly the Einstein coefficient, 2.5, as suggested by Equation (43).

2. The initial slope of Figure 4.10 suggests k_1 is about 10.0 for these data, a reasonable value in the light of theoretical predictions.

3. At still higher volume fractions, η increases even more steeply than predicted by the two-term version of Equation (43).

The results of experiments in viscometry are routinely reported in a variety of forms; the ones we have used in Figures 4.9 and 4.10 are only two of the possibilities. Table 4.3 lists some of the functional forms in which data are often presented. Also listed are the symbols, common and IUPAC names, and the limiting values for these quantities as the concentration of the colloid goes to zero. In Table 4.3 the symbol c is used to represent the concentration of the dispersed phase. For the present, we continue to let ϕ be the unit of concentration.

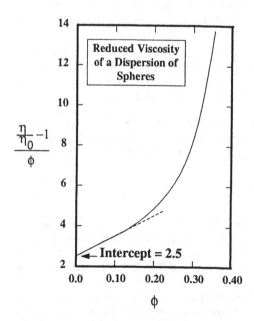

FIG. 4.10 Reduced viscosity versus volume fraction for a dispersion of glass spheres up to $\phi \approx$ 0.40. ($R_s = 6.5 \cdot 10^{-3}$ cm. (Data from V. Vand, *J. Phys. Colloid Chem.*, **52**, 300 (1948).)

The definition of the *reduced viscosity* (see Table 4.3) is somewhat analogous to the reduced osmotic pressure given in Equation (3.34), and the *intrinsic viscosity* $[\eta]$ is the limiting value of this quantity. Note that the reduced viscosity gives the relative increase in the viscosity of the solution over the solvent per unit of concentration. Since $[\eta]$ equals the limiting value of the reduced viscosity, $[\eta]$ has the significance of measuring the first increment of viscosity due to the dispersed particles. The intrinsic viscosity, therefore, is a characteristic of the dispersed particles. The Einstein equation predicts that it should have a value of 2.5 for spherical particles. Note that $[\eta]$ may also be evaluated by extrapolating inherent viscosities to $\phi = 0$. Example 4.3 demonstrates the equivalence of these two procedures for evaluating $[\eta]$.

* * *

EXAMPLE 4.3 *The Limiting Behavior of the Inherent Velocity at Low Volume Fractions.* Show that the inherent viscosity $(1/\phi) \ln(\eta/\eta_0)$ reduces to $[\eta]$—defined as the *limiting reduced viscosity*—as $\phi \to 0$.

Solution: By using volume fraction for c in Equation (31), one has $\eta = A + B\phi + C\phi^2 + \ldots$. Since $A = \eta_0$, one can rewrite this as $\eta/\eta_0 = 1 + B'\phi + C'\phi^2 + \ldots$, where $B' = (B/\eta_0)$, $C' = (C/\eta_0)$, etc. The intrinsic viscosity is the coefficient of the first-order concentration term; therefore this may be written as $\eta/\eta_0 = 1 + [\eta]\phi + C'\phi^2 + \ldots$. Replacing the argument of the logarithm in the definition of inherent viscosity by this expansion gives

$$\eta_{inh} = (1/\phi) \ln(1 + [\eta]\phi + C'\phi^2 + \ldots) = (1/\phi)([\eta]\phi + C'\phi^2 + \ldots)$$

where the second form is obtained by using the series expansion (see Appendix A) for the logarithm. Further simplification leads to

$$\eta_{inh} = [\eta] + C'\phi + \ldots$$

In the limit of $\phi \to 0$, this reduces to $[\eta]$, the limiting reduced viscosity. ∎

* * *

The Einstein theory shows that volume fraction is the theoretically favored concentration unit in the expansion for viscosity, even though it is not a practical unit for unknown solutes. As was the case in the Flory-Huggins theory in Chapter 3, Section 3.4b, it is convenient to convert volume fractions into "mass/volume" concentration units for the colloidal solute. According to Equation (3.78), $\phi = c(\overline{V}_2/M_2)$, where c has units "mass/volume" and $\overline{V}_2$ and M_2 are the partial molar volume and molecular weight, respectively, of the solute. In viscosity work, volumes are often expressed in deciliters—a testimonial to the convenience of the 100-ml volumetric flask! In this case, $\overline{V}_2$ must be expressed in these units also. The reader is advised to be particularly attentive to the units of concentration in an actual problem since the units of intrinsic viscosity are concentration^{-1} when the reduced viscosity is written as an expansion of powers of concentration c. (The intrinsic viscosity is *dimensionless* when the reduced viscosity is written as an expansion of powers of volume fraction ϕ.) With the substitution of Equation (3.78), Equation (42) becomes

$$\frac{\eta}{\eta_0} = 1 + 2.5 \frac{\overline{V}_2}{M_2}c + k_1\left(\frac{\overline{V}_2}{M_2}\right)^2 c^2 + \cdots \tag{44}$$

or

$$\frac{1}{c}\left(\frac{\eta}{\eta_0} - 1\right) = 2.5 \frac{\overline{V}_2}{M_2} + k_1\left(\frac{\overline{V}_2}{M_2}\right)^2 c + \cdots \tag{45}$$

Equation (45) shows that as long as balances, volumetric flasks, and viscometers are available, $[\eta]$ can be determined. All that is required is to measure viscosity at a series of concentrations, work up the data as $(1/c)[(\eta/\eta_0) - 1]$, and extrapolate to $c = 0$. If the experimental value of $[\eta]$ turns out to be 2.5 $(\overline{V}_2/M_2)$, then the particles are shown to be unsolvated spheres. If $[\eta]$ differs from this value, the dispersed units deviate from the requirements of the Einstein model. In the next section we examine how such deviations can be interpreted for lyophobic colloids.

4.7 DEVIATIONS FROM THE EINSTEIN MODEL

The Einstein theory is based on a model of dilute, unsolvated spheres. In this section we examine the consequences on intrinsic viscosity of deviations from the Einstein model in each of the following areas:

1. Concentrations that are not very low
2. Particle swelling due to solvation
3. Nonspherical particles approximated as ellipsoids of revolution

In addition, we introduce one other major source of deviation discussed in more detail in Section 4.8, namely, effects of interparticle interactions due to charges on the particles and polymer additives.

We consider each of these items as potential causes of deviation in the following subsections.

4.7a Effect of High Volume Fractions

We have already indicated that the coefficient k_1 in Equation (42) has been calculated for spheres by various theoretical models. While this coefficient is a measure of concentration effects, we do not pursue its derivation. Instead, we qualitatively examine the effect of particle crowding as the origin of the positive deviations from the Einstein theory that inevitably set in at higher concentrations, as seen in Figures 4.10 and 4.11.

To do this, we consider a dispersion of volume fraction ϕ and examine the increment in viscosity $d\eta$ as a small amount of particles is added to the dispersion. If we take ϕ to be small enough that the Einstein equation, Equation (41), holds, the increment $d\eta$ that accompanies the addition of particles is then given by

$$d\eta = 2.5\eta d\phi^* \tag{46}$$

where η is the viscosity of the dispersion prior to the addition of the new particles. The change in volume fraction denoted by $d\phi^*$ represents the volume of added particles divided by the volume of space available for those particles. It can be written in terms of the *total* volume fraction ϕ of all spheres in the system as $d\phi/(1 - \phi)$. Therefore, Equation (46) becomes

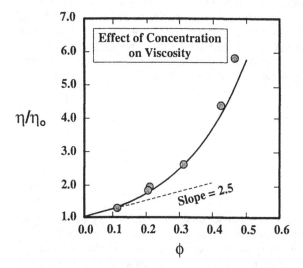

FIG. 4.11 The effect of particle crowding on viscosity. The solid line is drawn according to Equation (48); the points are experimental results. (Data from R. Roscoe, *Br. J. Appl. Phys.*, **3**, 267 (1952).)

$$d\eta = 2.5\,\eta\,\frac{d\phi}{1 - \phi} \tag{47}$$

One can now integrate this result, recalling that $\eta = \eta_0$ at $\phi = 0$, to give

$$\eta/\eta_0 = (1 - \phi)^{-2.5} \tag{48}$$

This result reduces to the Einstein equation as $\phi \to 0$. The solid line in Figure 4.11 was drawn according to Equation (48). This result shows that the upward curvature displayed in Figures 4.10 and 4.11 can be explained on the basis of particle crowding; unfortunately, this approach produces no new parameters to quantify the effect. It might be noted that the original derivation of this result was undertaken to explore the effects of polydispersity. Specifically, the different categories of spheres were assumed to be of different sizes. Since the Einstein result is independent of particle size, it is really irrelevant to the effect shown in Figure 4.11.

4.7a.1   The Krieger-Dougherty Equation

Equation (48) has been derived under the assumption that the volume fraction can reach unity as more and more particles are added to the dispersion. This is clearly physically impossible, and in practice one has an upper limit for ϕ, which we denote by ϕ_{max}. This limit is approximately 0.64 for random close packing and roughly 0.71 for the closest possible arrangement of spheres (face-centered cubic packing or hexagonal close packing). In this case, $d\phi^*$ in Equation (46) is replaced by $d\phi/[1 - (\phi/\phi_{max})]$. Equation (47) then becomes, when written in terms of the intrinsic viscosity $[\eta]$,

$$d\eta = [\eta]\eta\,\frac{d\phi}{[1 - (\phi/\phi_{max})]} \tag{49}$$

which, on integration, leads to

$$\frac{\eta}{\eta_0} = [1 - (\phi/\phi_{max})]^{-[\eta]\phi_{max}} \tag{50}$$

As noted above, we have replaced the Einstein coefficient of 2.5 by the more general intrinsic viscosity $[\eta]$ in the above equations. Equation (50) is known as the *Krieger-Dougherty equation* and has been found to be highly useful for relating viscosity data for both low shear rates ($\dot\gamma \to 0$) and high shear rates ($\dot\gamma \to \infty$). For instance, the experimental results of Krieger (1972) and de Kruif et al. (1985) lead to

$$\phi_{max} = 0.632; \qquad [\eta] = 3.13 \text{ for } \dot\gamma \to 0 \tag{51}$$

and

$$\phi_{max} = 0.708; \qquad [\eta] = 2.71 \text{ for } \dot\gamma \to \infty \tag{52}$$

Equations (51) and (52) imply an interesting feature. The respective values of ϕ_{max} suggest that at low shear the particles pack themselves randomly as the volume fraction increases (i.e., $\phi_{max} \approx 0.64$), whereas they approach ordered packing at high shear rates.

It turns out that the Krieger-Dougherty equation can also be used for intermediate shear rates with suitable modifications. Experimental data also suggest that $[\eta]$ and $\dot\gamma$ are independent of particle size, although they are stress dependent. Therefore viscosities of monodispersed colloids of different particle sizes can be represented by a single equation by suitably defining the variables. A discussion of these and other extensions may be found in Barnes et al. (1989).

4.7b   Effects of Solvation and Shapes

Figure 2.4 in Chapter 2 suggested how either solvation or ellipticity could increase the effective size of a particle as far as its friction factor is concerned. A moment's reflection will convince us that this same conclusion is qualitatively true with respect to intrinsic viscosity also.

The solvation of a sphere swells its volume above the "dry" volume, which is presumably what is used to evaluate ϕ. Therefore whatever factor describes the increase in particle volume due to solvation is absorbed into $[\eta]$. This is easily seen as follows. Suppose the mass of colloidal solute in a solution is converted to the volume of unsolvated material using the "dry" density. In this way, mass/volume units are converted to "volume/volume" units, which we label ϕ_{dry} since the unsolvated density was used in evaluating this quantity. If the solvation is uniform throughout the particle—as would be the case for, say, aqueous proteins—then the solvated particle exceeds the unsolvated particle in volume by the factor $[1 + (m_{1,b}/m_2)(\rho_2/\rho_1)]$ as shown by Equation (2.38). Recall that in this expression, $m_{1,b}$ is the mass of bound solvent, m_2 is the mass of the solute particle, and ρ_i is the density of solvent or solute, as appropriate. Still assuming ϕ_{dry} has been used to evaluate $[\eta]$, we see

$$\frac{1}{\phi_{dry}}\left(\frac{\eta}{\eta_0} - 1\right) = 2.5\left(1 + \frac{m_{1,b}}{m_2}\left[\frac{\rho_2}{\rho_1}\right]\right) + \cdots \tag{53}$$

$$\frac{1}{c}\left(\frac{\eta}{\eta_0} - 1\right) = 2.5\left(1 + \frac{m_{1,b}}{m_2}\left[\frac{\rho_2}{\rho_1}\right]\right)\left(\frac{1}{\rho_2}\right) + \cdots \tag{54}$$

in practical units. Example 4.4 considers how intrinsic viscosity measurements can be used to reach conclusions about the state of solvation of colloidal particles.

* * *

EXAMPLE 4.4 *Extent of Hydration of a Protein Molecule from Intrinsic Viscosity Measurements.* Suppose an aqueous solution of a spherical protein molecule shows an intrinsic viscosity of $3.36 \text{ cm}^3 \text{ g}^{-1}$. Taking $\rho_2 = 1.34 \text{ g cm}^{-3}$ for the dry protein, estimate the extent of hydration of the protein.

Solution: It is apparent from the units of $[\eta]$ that solute concentration has been expressed in g/cm^3. Dividing this concentration by the density of the unsolvated protein converts the concentration to "dry" volume fraction units. Since the concentration appears as a reciprocal in the definition of $[\eta]$, we must multiply $[\eta]$ by ρ_2 to obtain $(1/\phi_{dry})[(\eta/\eta_0) - 1]$. For this protein the latter is given by $(3.36)(1.34) = 4.50$. If the particles were unsolvated, this quantity would equal 2.5 since the molecules are stated to be spherical. Hence the ratio $4.50/2.50 = 1.80$ gives the volume expansion factor, which equals $[1 + (m_{1,b}/m_2)(\rho_2/\rho_1)]$. Therefore $(m_{1,b}/m_2) = 0.80 (1.00/1.34) = 0.60$. The intrinsic viscosity reveals the solvation of these particles to be $0.60 \text{ g } H_2O$ per gram of protein. ∎

* * *

We noted above that either solvation or ellipticity could cause the intrinsic viscosity to exceed the Einstein value. Simha and others have derived extensions of the Einstein equation for the case of ellipsoids of revolution. As we saw in Section 1.5a, such particles are characterized by their axial ratio. If the particles are too large, they will adopt a preferred orientation in the flowing liquid. However, if they are small enough to be swept through all orientations by Brownian motion, then they will increase $[\eta]$ more than a spherical particle of the same mass would. Again, this is very reminiscent of the situation shown in Figure 2.4.

Figure 4.12a shows plots of the intrinsic viscosity—in volume fraction units—as a function of axial ratio according to the *Simha equation*. Figure 4.12b shows some experimental results obtained for tobacco mosaic virus particles. These particles—an electron micrograph of which is shown in Figure 1.12a—can be approximated as prolate ellipsoids. Intrinsic viscosities are given by the slopes of Figure 4.12b, and the parameters on the curves are axial ratios determined by the Simha equation. Thus we see that particle asymmetry can also be quantified from intrinsic viscosity measurements for unsolvated particles.

Finally, we note that both solvation and ellipticity can occur together. The contours shown in Figure 4.13a illustrate how various combinations of solvation and ellipticity are compatible with an experimental intrinsic viscosity. The particle considered in Example 4.4 has an intrinsic viscosity of 4.50 and was calculated to be hydrated to the extent of $0.60 \text{ g } H_2O$ per gram of protein. The same value of $[\eta]$ is also compatible with nonsolvated ellipsoids of

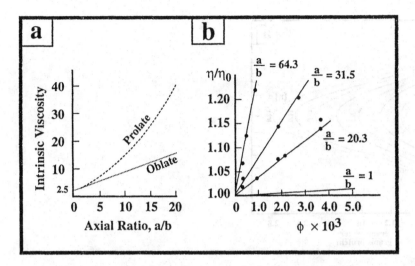

FIG. 4.12 Viscosity of dispersions of some nonspherical particles: (a) intrinsic viscosity as a function of the axial ratio a/b for oblate and prolate ellipsoids of revolution according to the Simha theory (redrawn with permission of Hiemenz 1984); (b) experimental values of relative viscosity versus volume fraction for tobacco mosaic virus particles of different a/b ratios (data from M. A. Lauffer, *J. Am. Chem. Soc.*, **66**, 1188 (1944)).

axial ratios 4.0 and 0.22. Intermediate combinations of solvation and ellipticity can also be determined from Figure 4.13a.

Contours like this are qualitatively the same sort of thing we obtain from sedimentation-diffusion experiments as shown in Figure 2.9. Therefore let us consider the relationship between the two types of data. In general, exactly the same factors affect both the intrinsic viscosity and the friction factor ratio, but the functional dependencies are somewhat different. Figure 4.13b shows how a contour of f/f_0 selected from a sedimentation-diffusion study and an intrinsic viscosity contour selected on the basis of viscosity experiments might overlap. In this case the solvation-ellipticity combination is characterized unambiguously: $a/b = 2.5$ and $(m_{1,b}/m_2) = 1.0$. Figure 4.13b shows the complementarity of viscosity and sedimentation-diffusion data.

The attentive reader will realize that we have strayed rather far from the hard spheres of the Einstein theory to find applications for it. It should also be appreciated, however, that the molecules we are discussing are proteins that—through disulfide bridges and hydrogen bonding—have fairly rigid structures. Therefore the application of the theory—amended to allow for solvation and ellipticity—is justified. This would not be the case for synthetic polymers, which are best described as random coils and for which a different formalism is employed. This is the topic of Section 4.9.

4.7c Electroviscous Effects and Viscoelectric Effects

In Chapter 3 we saw that electrical effects complicate the osmotic equilibrium for charged particles. Not surprisingly, particle charge complicates analysis of viscosity as well. Like osmotic phenomena, the viscosity of dispersions of charged particles is highly sensitive to the concentration of the electrolytes. Chapters 11–13 examine the behavior of simple ions near charged surfaces. For now, we review some of the concepts, variables, and definitions of those chapters as background for our discussion of charge effects on viscosity (and, more generally, rheology).

For simplicity, consider a lyophobic colloidal particle with a surface that carries a charge. Electroneutrality requires that an equal amount of opposite charges (counterions) be present

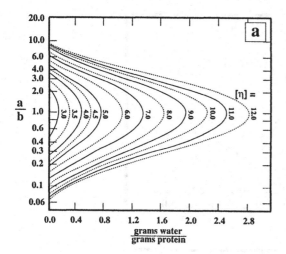

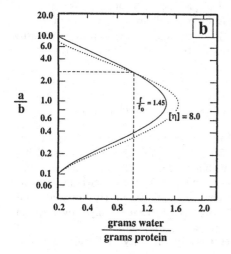

FIG. 4.13 Intrinsic viscosity of a protein solution: (a) variation of the intrinsic viscosity of aqueous protein solutions with axial ratio a/b and extent of hydration $m_{1,b}/m_2$ (redrawn from L. Oncley, *Ann. NY Acad. Sci.*, **41**, 121 (1941)); (b) superposition of the $[\eta] = 8.0$ contour from Fig. 4.13a and the $f/f_0 = 1.45$ contour from Figure 2.9. The crossover unambiguously characterizes particles with respect to hydration and axial ratio.

not too far from the charged surface. The charged surface of the particle and the counterions are referred to as an *electrical double layer*. The counterions are distributed over a region that extends some distance from the surface of the particles because of the thermal (diffusive) motion of the counterions. This region is known as the *diffuse* part of the double layer; its range—represented by the reciprocal of the so-called Debye-Hückel parameter κ—depends on the concentration and charge of the counterions and the electrical potential at the surface of the particles.

It is convenient to think of the diffuse part of the double layer as an ionic atmosphere surrounding the particle. Any movement of the particle affects the particle's ionic atmosphere, which can be thought of as being "dragged along" through bulk motion and diffusional motion of the ions. The resulting electrical contribution to the resistance to particle motion manifests itself as an additional viscous effect, known as the *electroviscous* effect. Further, when the particle moves, the surface where slippage occurs lies within the diffuse part of the double layer at ionic dimensions from the surface. The potential at that location can be measured using electrophoresis experiments, among others (Chapter 12), and is called the *zeta potential* ζ.

Two charged particles approaching each other sense the presence of each other through the overlap of their electrical double layers. This double-layer overlap results in a repulsive force between similarly charged particles.

The dispersed particles also attract each other through *van der Waals* forces (Chapter 10), which decrease with increasing interparticle separation. A material property known as the *Hamaker constant*, A, can be used as a measure of the strength of the van der Waals attraction.

The net outcome of the encounter between two particles—that is, whether attraction or repulsion prevails—depends on magnitudes of attractive and repulsive forces as the particles approach each other. Table 4.4 summarizes some pertinent relationships from Chapters 10-12 that will be useful in the following discussion of the effects of electrostatic and van der Waals forces on the rheology of dispersions.

The electroviscous effects are usually classified in three categories depending on the origin of the underlying mechanism.

1. If the fluid in capillary flows considered in Section 4.4a is an electrolyte solution, the flow will give rise to a *streaming potential*, as discussed in Chapter 12. The streaming potential causes a backflow because of electroosmotic effects (also discussed in Chapter 12), thereby causing a reduction in the net flow in the forward direction. Since the flow rate for an identical pressure drop along the length of the capillary will be larger in the absence of the above double-layer effect, the presence of the electrical double layer makes the liquid appear to have an enhanced viscosity. In the case of dilute, charged dispersions, a similar effect occurs as a consequence of the additional energy dissipation caused by the distortion of the double layer under shear. The net effect is an increase in the viscosity of the dispersions. This effect is called the *primary electroviscous effect*.

2. The *secondary electroviscous effect* refers to the change in the rheological behavior of a charged dispersion arising from interparticle interactions, i.e., the interactions between the electrical double layers around the particles.

3. The term *tertiary electroviscous effect* is applied to the changes in the conformation of polyelectrolytes that are caused by *intra*molecular double-layer interactions. It is customary to extend this definition to include all effects in which the geometry of the system is altered as a result of double-layer interactions.

TABLE 4.4 Summary of Some Colloidal Interaction Energies and Parameters from Chapters 10-12

Quantity described	Equation
Debye-Hückel Parameter κ (unit, 1/length)	$\kappa = [(1000\, e^2 N_A / \varepsilon k_B T) \sum_i z_i^2 M_i]^{1/2}$ Equation (11.41)
Van der Waals attraction between two flat surfaces (unit, energy)	$\Phi_A = -(A/12\pi)d^{-2}$ Equation (10.63)
Electrical double-layer repulsion between two flat surfaces (unit, energy)	$\Phi_R = 64\, k_B T\, n_\infty\, \kappa^{-1} \Upsilon_0^2 \exp(-d/\kappa^{-1})$ Equation (11.86)
Zeta potential ζ (unit, volts)	$\zeta = 3\, \eta\, u/2\varepsilon$ From Equation (12.27)

Note: e = electron charge; N_A = Avogadro's number; z_i = charge of ion of type i; M_i = molar concentration of ions in the bulk; ε = dielectric constant of the medium; Φ_A = energy of attraction; A = Hamaker constant; d = distance between the surfaces; Φ_R = energy of repulsion; n_∞ = ionic concentration (in number/volume); $\Upsilon_0 \approx 1$; η = viscosity of the liquid; u = electrophoretic mobility

In studying and interpreting the effects of electrical double layers on the viscosities of dispersions, one usually assumes that the viscosity of the suspending medium itself remains unchanged. However, situations do exist in which the electrical fields at a fluid/particle interface exert a significant influence on the structure of the fluid itself in the vicinity of the interface. This can modify the viscosity of the fluid. This class of electrically induced effects is known as *viscoelectric effects*. We consider this briefly in Chapter 12 in the context of its influence on the migration of charged particles in an electric field. In this chapter, however, we are concerned only with electroviscous effects and not viscoelectric effects. A good introduction to the viscoelectric effects (and a broader treatment of electroviscous effects) may be found in Hunter (1981).

Our objective in this chapter is modest, namely, to provide a general discussion of the electroviscous effects and to present a few equations that serve as guidelines for understanding the effects of colloidal forces on the viscosity of dispersions. The underlying theories are rather complicated and fall outside our scope.

The electroviscous effects and the other effects discussed in Sections 4.7a–c lead to what is called non-Newtonian behavior in the flow of dispersions. In the next section, we begin with a brief review of the basic concepts concerning deviations from Newtonian flow behavior and then move on to consider how high particle concentrations and electroviscous effects affect the flow and viscosity.

4.8   NON-NEWTONIAN BEHAVIOR

We have already devoted a considerable amount of space to our discussion of viscosity without ever venturing beyond Newtonian systems. At least as much—probably more—could be said about non-Newtonian systems.

Colloids (even those consisting of only spherical particles) do not necessarily behave like Newtonian fluids. Deviations from Newtonian behavior occur for a number of reasons, but we are concerned mainly with two effects: (a) those due to interparticle hydrodynamic interactions and (b) those due to colloidal forces such as electrostatic effects, effects due to polymer layers adsorbed on the particles, etc. The hydrodynamic effects exist even for neutral particles although they become important only for moderate to large concentrations of the particles. The colloidal forces can exert a significant influence even at low volume fractions if the ranges of colloidal attractions and repulsions are significant. The formal theoretical developments that account for the effects of colloidal and hydrodynamic forces are rather complex and are not sufficiently well established. We shall not go into any of the details of these since they require a more advanced background in fluid mechanics than introduced in the previous sections of this chapter; some introduction to the relevant ideas are available in advanced books or reviews such as Russel (1980), Tadros (1984, Chapter 6), Hunter (1990, Chapter 18), and Russel et al. (1989, Chapter 14). However, a discussion of the rheology of dispersions without at least some indication of what happens in the case of interacting particles is not only incomplete, but is also misleading. Therefore, our goal in this section is to provide a *general idea* of what happens to the flow behavior of dispersions when the above forces become important, and what physicochemical factors or features of the phenomena one should pay attention to in attempting to understand the rheology of dispersions.

4.8a   Examples of Non-Newtonian Features

Non-Newtonian fluids are generally those for which the viscosity is not constant even at constant temperature and pressure. The viscosity depends on the shear rate or, more accurately, on the previous kinematic history of the fluid. The linear relationship between the shear stress and the shear rate, noted in Equation (1), is no longer sufficient. Strictly speaking, the coefficient of viscosity is meaningful only for Newtonian fluids, in which case it is the slope of a plot of stress versus rate of shear, as shown in Figure 4.2. For non-Newtonian fluids, such a plot is generally nonlinear, so the slope varies from point to point. In actual practice, the data are traditionally represented with the rate of shear (dv/dy) as the ordinate and the stress (F/A) as the abscissa, as shown in Figure 4.14. In this case, the apparent viscosity at some particular point is given by the cotangent of the angle that defines the slope at that particular point.

The non-Newtonian fluids in shear flows may be classified broadly into three types:

1. Fluids with shear stresses that at any point depend on the shear rates only and are *independent* of time. These include (a) what are known as *Bingham plastics*, materials that require a minimum amount of stress known as yield stress before deformation, (b) pseudoplastic (or shear-thinning) fluids, namely, those in which the shear stress decreases with the shear rate (these are usually described by "power-law" expressions for the shear stress; i.e., the rate of strain on the right-hand-side of Equation (1) is raised to a suitable power), and (c) dilatant (or shear-thickening) fluids, in which the stress increases with the shear rate (see Fig. 4.2).

2. Fluids in which the shear stress/shear rate behavior depends on time, i.e., the kinematic history of the fluid. These fall in one of two categories: (a) *thixotropic* or (b) *rheopectic*. Many polymer gels display time-dependent stress/shear rate behavior. In thixotropic fluids, the microstructure of the fluid progressively breaks down and the viscosity decreases. In rheopectic fluids, the applied shear promotes gradual formation of local structure and the apparent viscosity increases. As we discuss below, thixotropic behavior is associated with flocs of asymmetrical particles. In certain instances, quicksand apparently operates through a thixotropic mechanism. The struggles of the victim merely decrease the viscosity of the trap and worsen the victim's plight. By contrast, wet beach sand is an example of a dilatant system in which the apparent viscosity increases with shear. Anyone who has wiggled their toes in the wet sand knows that under these conditions (low shear) the wet sand is very fluid. However, the same sand is hard to a firm footstep (high shear). In contrast to thixotropy, dilatancy is favored by symmetrical, nonflocculated particles. It is almost always observed at concentrations in the neighborhood of 40% dispersed particles. At this concentration and in the absence

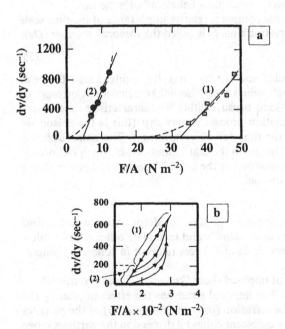

FIG. 4.14 Rate of shear versus shearing stress: (a) for 7% aqueous carbon black dispersions (data from A. I. Medalia and E. Hagopian, *Ind. Eng. Chem. Prod. Res. Dev.*, **3**, 120 (1964)); and (b) for an 11% aqueous bentonite dispersion (pH = 8.7) for which the time of the cycle is 70 sec (based on data of W. F. Gabrysh, H. Eyring, P. Lin-Sen, and A. F. Gabrysh, *J. Am. Ceram. Soc.*, **46**, 523 (1963)).

of disruptable flocs, the only way such a system can flow is by the gradual rolling of particles past one another. If the rate of shear is too great, this deformation is impossible.

3. Viscoelastic materials, as mentioned in Section 4.1, display behavior somewhere between a solid and a liquid.

Colloidal dispersions, polymer solutions, and both colloidal and polymer gels can in principle behave like any of the above-mentioned materials.

4.8b   Time Scales and Dimensionless Parameters of Importance

While in the case of noninteracting dispersions one needed to consider only the effect of the particle concentration, in interacting dispersions one needs to consider the time over which the flow behavior is observed and its magnitude relative to the time scales over which either shear or colloidal forces alter the local structure of the dispersions. What the flow behavior is, which interaction effects dominate the behavior, and how they do depend on the competing influences of the applied shear and interaction effects. In this section, we outline some of the important parameters one can formulate to judge the relative effects of various colloidal interactions and the physical significance of those parameters.

4.8b.1   The Deborah Number

We mentioned in Section 4.1 that whether a material deforms under applied stress is a matter of the magnitudes of the shear force exerted and the time of observation. It is common to use silicone putty (known as *Silly Putty*) to illustrate the above statement. If you have enough patience, you will notice that silly putty is a highly viscous material (although you may not think of it that way) that will find its own level when placed in a container. In this sense, it behaves like a "liquid." On the other hand, as its name is meant to suggest, a ball of Silly Putty will also bounce when dropped to the floor. That is, under severe and sudden deformation, it behaves like a solid. The behavior of the Silly Putty thus brings to our attention the importance of time scales and deformation rates in classifying the "flow behavior" of materials.

The dimensionless number in rheology that compares relative importance of the time scale of the deformation process t_D over the observation time t_O is called the *Deborah number* (*De*):

$$De = t_D/t_O \tag{55}$$

The time t_D is typically very small for liquids and is very large for solids. Large Deborah numbers signify solidlike behavior, and small values typify liquidlike behavior. Correspondingly, a colloid or a gel may behave like a solid because either its characteristic time is very large (we discuss this below) or the deformation process is very fast (this is the reason we compare the time for deformation t_D with the time for observation t_O). The name *Deborah* comes from the fifth chapter of the book of Judges in the Old Testament, in which Deborah is reported to have said that "the mountains flowed before the Lord." The idea is that everything flows if one has a (very) long life to observe the motion!

4.8b.2   The Peclet Number

While the Deborah number is often used to compare the time for deformation with the time of observation in experiments, it also inspires us to identify and formulate other dimensionless groups that compare the various characteristic times and forces relevant in colloidal phenomena. We discuss some of the important ones.

The Peclet number compares the effect of imposed shear (known as the *convective* effect) with the effect of diffusion of the particles. The imposed shear has the effect of altering the local distribution of the particles, whereas the diffusion (or Brownian motion) of the particles tries to restore the equilibrium structure. In a quiescent colloidal dispersion the particles move continuously in a random manner due to Brownian motion. The thermal motion establishes an equilibrium statistical distribution that depends on the volume fraction and interparticle potentials. Using the Einstein-Smoluchowski relation for the time scale of the motion, with the Stokes-Einstein equation for the diffusion coefficient, one can write the time taken for a particle to diffuse a distance equal to its radius R_s as

$$t_{Diff} \sim R_s^2/D = (6\pi\eta_0 R_s^3/k_B T) \tag{56}$$

where we have used the Stokes-Einstein equation for the diffusion coefficient D (see Equations (2.32) and (2.8)),

$$D = k_B T/6\pi\eta_0 R_S \tag{57}$$

This characterizes the time taken for the restoration of the equilibrium microstructure after a disturbance caused, for example, by convective motion, i.e., this is the relaxation time of the microstructure. The time scale of shear flow is given by the reciprocal of the shear rate, $\dot{\gamma}$. The dimensionless group formed by the ratio (t_{Diff}/t_{Shear}) is the Peclet number

$$Pe = (6\pi\eta_0 R_s^3\dot{\gamma}/k_B T) = R_s^2\dot{\gamma}/D \tag{58}$$

which specifies the relative importance of convection over diffusion. For example, if $Pe \ll 1$, the distribution of particles is only slightly altered by the flow, and the behavior is dominated by diffusional relaxation of the particles. When $Pe \gg 1$, convective (hydrodynamic) effects dominate the behavior, and experiments suggest that a dispersion of spherical particles behaves like a shear-thinning fluid.

Before leaving this discussion, it is important to note that other forms of Peclet numbers are also possible and may be more appropriate depending on the type of convective influence studied. For example, in the case of oscillatory flows (as in oscillatory viscometers), it is more useful to define the Peclet number as $(R_s^2\omega/D)$, where ω is the frequency of oscillation. Regardless of the particular definition, the general significance of the Peclet number remains the same, i.e., it compares the effect of convection relative to diffusion.

4.8b.3 Relative Energies or Lengths

In addition to the Peclet number, one can also define other dimensionless groups that compare either relevant time scales or energies of interaction. Using some of the concepts previewed in Section 4.7c and Table 4.4, one can define an electrostatic group (in terms of the zeta potential ζ and relative permittivity ε_r of the liquid) as

$$N_{El} = \varepsilon_r\varepsilon_0\zeta^2 R_s/k_B T \tag{59}$$

which is a measure of the electrostatic energy relative to the thermal energy, represented by $k_B T$. Similarly, the range of electrostatic repulsion κ^{-1} (the Debye length) relative to the particle radius R_s is represented by

$$N_{DL} = \kappa R_s \tag{60}$$

The strength of the van der Waals attraction relative to thermal energy is represented by

$$N_{vdW} = A/k_B T \tag{61}$$

where A is the effective Hamaker constant between the particles through the intervening fluid.

Other dimensionless groups that compare the thickness of the adsorbed polymer layer to the radius of the particle or the radius of gyration of the polymer to the particle radius in polymer/colloid mixtures can also be easily defined. We are mostly concerned with the volume fraction ϕ and the Peclet number Pe in our discussions in this chapter. However, the other dimensionless groups may appear in the equations for intrinsic viscosity of dispersions when the dominant effects are electroviscous or sterically induced.

4.8c Charged Particles

In Section 4.7c we outlined the types of effects one can expect in the response of charged dispersions to deformation. In this section, we present some results for the viscosity of charged colloids for which electroviscous effects could be important. As mentioned above, we shall not go into the theoretical details behind the equations since they require a fairly advanced knowledge of fluid dynamics and, in some cases, statistical mechanics. Moreover, some of the concepts, such as electrical double layers, double-layer thickness (i.e., the Debye or the Debye-Hückel length) κ^{-1} and zeta potential ζ, used in the following discussion require material from Chapters 11 and 12, and it will help to return to this section after reviewing those chapters.

4.8c.1   The Primary Electroviscous Effect

The first analysis of the primary electroviscous effect dates to 1916 when Smoluchowski presented the following equation for the intrinsic viscosity:

$$[\eta] = 2.5\left\{1 + \frac{4(\varepsilon_r\varepsilon_0\zeta)^2}{k\,\eta_0\,R_s^2}\right\} \tag{62}$$

where ε_r is the relative permittivity, ε_0 is the permittivity of free space, k is the specific conductivity of the continuous phase, and ζ is the zeta potential, which is used in place of the potential at the surface of the particle (a sphere of radius R_s). The physical significance and measurement of the zeta potential are discussed in Chapter 12. Smoluchowski's result is supposed to apply for thin electrical double layers (i.e., large κR_s), a range for which the primary electroviscous effects are smallest. A subsequent analysis by Krasny-Ergen in 1936 for the same conditions led to a correction for the second term in the brackets:

$$[\eta] = 2.5\left\{1 + \frac{6(\varepsilon_r\varepsilon_0\zeta)^2}{k\,\eta_0\,R_s^2}\right\} \tag{63}$$

These equations (and others that follow) are based on the solutions of equations of motion for the particles, as well as the electrolyte, that we use in Chapter 12 in the context of electrophoresis. It turns out that the analysis of Krasny-Ergen fails to satisfy one of the boundary conditions and does not take into account energy dissipations caused by the electric currents arising from the motion of the electrolyte. It is more appropriate for $\kappa R_s \rightarrow \infty$.

A corrected and more general analysis of the primary electroviscous effect for weak flows, i.e., for low Pe numbers (for small distortions of the diffuse double layer), and for small zeta potentials, i.e., $\zeta < 25$ mV, was carried out by Booth in 1950. The result of the analysis leads to the following result for the intrinsic viscosity $[\eta]$ for charged particles in a 1 : 1 electrolyte:

$$[\eta] = 2.5\left\{1 + 4\pi(\varepsilon_r\varepsilon_0\zeta)^2\frac{(\kappa R_s)^2(1 + \kappa R_s)^2 Z(\kappa R_s)}{k\,\eta_0\,R_s^2}\right\} \tag{64}$$

The function $Z(\kappa R_s)$ is a power series with the two limiting forms, for extended diffuse layers:

1.   Thick double layers, i.e., small κR_s:

$$Z(\kappa R_s) \approx (200\pi\kappa R_s)^{-1} + (11\kappa R_s/3200\pi) \tag{65a}$$

2.   Thin double layers, i.e., large κR_s:

$$Z(\kappa R_s) = (3/2\pi)(\kappa R_s)^{-4} \tag{65b}$$

   The substitution of Equation (65b) in Equation (64) leads to the Krasny-Ergen equation for $\kappa R_s \rightarrow \infty$, as one would expect.

Although Booth's result already indicates the influence of the charges on the particles and the electrolyte in the dispersion on the viscosity of the dispersion, one sees more complex behavior only when the effect of larger distortions of the double layer is included in the analysis. An extension for larger distortions (represented by the Peclet number Pe) of the double layer is available (Russel 1978a) and can be written as

$$[\eta] = 2.5\left\{1 + \frac{6(\varepsilon_r\varepsilon_0\zeta)^2}{k\,\eta_0\,R_s^2}\frac{1}{1 + Pe^2}\right\} \tag{66}$$

for small zeta potentials (less than about 25 mV) and for $Pe \ll \kappa R_s$ and $\kappa R_s \gg 1$. Russel's result indicates that the primary electroviscous effect leads to a shear-thinning (pseudoplastic) behavior even for moderate Pe and thin double layers with low potentials. Moreover, variations in normal stresses also occur, although they cannot be deduced from the above equation alone.

The results presented here provide a glimpse of the complications in the flow behavior that may arise as a result of the distortions of the electrical double layer even at low volume fractions of the dispersions. Extensions of Booth's result and others are available in the literature (see Hunter 1981), but we shall not go into those here.

The variations in the intrinsic viscosity predicted by the primary electroviscous effect are often small, and it is difficult to attribute variations in the experimentally observed $[\eta]$ from the Einstein value of 2.5 to the above effect since such variations can be caused easily by small amounts of aggregation. Nevertheless, Booth's equation has been found to be adequate in most cases. Further discussions of this and related issues are available in advanced books (Hunter 1981).

4.8c.2   The Secondary Electroviscous Effect

The secondary electroviscous effect is often interpreted in terms of an increase in the effective "collision diameter" of the particles due to electrostatic repulsive forces (i.e., the particles begin to "feel" the presence of other particles even at larger interparticle separations because of electrical double layer). A consequence of this is that the excluded volume is greater than that for uncharged particles, and the electrostatic particle-particle interactions in a flowing dispersion give an additional source of energy dissipation.

Let us consider situations for which the double-layer thickness is large enough (i.e., κR_s is small) and interparticle distances are large (Russel 1978b). Then, at typical interparticle distances in a quiescent dispersion, the van der Waals attraction (see Chapter 10) is insignificant. We can then use the electrostatic energy expression presented in Chapter 11 to get a dimensionless parameter α that represents the ratio of electrostatic energy to thermal energy (i.e., $k_B T$):

$$\alpha = (4\pi\varepsilon_r\varepsilon_0\zeta^2 R_s^2\kappa/k_B T)\exp(2\kappa R_s) = 4\pi N_{El}N_{DL}\exp(2N_{DL}) \tag{67}$$

where N_{El} and N_{DL} are defined in Equations (59) and (60), respectively. The parameter α is another way of expressing the strength of electrostatic energy relative to the thermal energy. For values of $\alpha \gg 1$ (i.e., the electrostatic energy is dominant), the effective collision diameter L for the particles in the absence of convective forces can be shown to be

$$L = \kappa^{-1}\ln[\alpha/\ln(\alpha/\ln\alpha)] \tag{68}$$

Under shear flow, the minimum center-to-center separation, r, between the particles will be in the interval $\{2R_s \leq r \leq L\}$; i.e., at the high-shear limit $r = 2R_s$ and at quiescent conditions $r = L$. An analysis of the flow behavior in this case leads to the following expression for the zero shear rate (i.e., $Pe \ll 1$) limit of the relative viscosity:

$$\frac{\eta(0)}{\eta_0} = 1 + [\eta]\phi + \frac{2}{5}([\eta]\phi)^2 + \frac{3}{40}\ln\left(\frac{\alpha}{\ln\alpha}\right)\left[\ln\left(\frac{\alpha}{\ln(\alpha/\ln\alpha)}\right)\right]^4\frac{\phi^2}{(\kappa R_s)^5}\cdots \tag{69}$$

The intrinsic viscosity $[\eta]$ in the above expression *includes* the primary electroviscous effect. The experimental data of Stone-Masui and Watillon (1968) for polymer latices seem to be consistent with the above equation (Hunter 1981). Corrections for α for large values of κR_s are possible, and the above equation can be extended to larger Peclet numbers. However, because of the sensitivity of the coefficients to κR_s and the complications introduced by multiparticle and cooperative effects, the theoretical formulations are difficult and the experimental measurements are uncertain. For our purpose here, the above outline is sufficient to illustrate how secondary electroviscous effects affect the viscosity of charged dispersions.

4.8c.3   The Tertiary Electroviscous Effect

As mentioned in Section 4.7c, the tertiary electroviscous effect is at least partly due to the expansion and contraction of particles arising from the conformational changes of the polyelectrolytes (adsorbed or chemically bound to the surface of the particles) with changes in the electrolyte concentration or pH of the medium. For particles stabilized by a layer of polyelectrolyte, one can calculate an effective volume fraction ϕ' (if the thickness δ of the polyelectrolyte layer is known) using

$$\phi' = \phi[1 + (\delta/R_s)]^3 \tag{70}$$

where ϕ is the actual volume fraction of the particles (of radius R_s). One can then use an equation such as Equation (50), the Krieger-Dougherty equation, to estimate the viscosity. The analysis of the tertiary electroviscous effect is complicated by the fact that factors such as the effect of the particle surface on the conformation of the polyelectrolyte chains, the structure of the chains in the layer, and the extent of the penetration of the liquid into the layer make it difficult to estimate the thickness and the hydrodynamic effects of the stabilizer layer. This area is still in its infancy. Some additional information may be obtained from Hunter (1981).

4.8d Dispersions of Aggregates

So far we have focused our attention on dispersions in which interparticle repulsions dominate. Attractive forces (see Chapter 10) have been assumed to be negligible, and as a consequence the dispersions are stable. When the van der Waals attraction begins to play a significant role (i.e., the net effect of repulsion and attraction is such that we have coagulation of the dispersion), more complicated flow behavior occurs. Such dispersions are more common in practice and deserve our attention.

Aggregated colloids are common examples of systems in which the apparent viscosity depends on the rate of shear. The existence of velocity gradients means that differences in flow velocity may extend over the dimensions of colloidal particles. If the dispersed "particle" is an aggregate of primary particles, a *floc*, then the shearing forces associated with viscosity that operate across the floc may disrupt or rearrange the aggregate. This corresponds to a dissipation of translational energy and hence contributes to the viscosity. It is clear that the extent to which this occurs may vary with the velocity gradient; this is one way in which the apparent viscosity may depend on the rate of shear.

Let us see how thixotropic behavior relates to the phenomena of coagulation. The data shown in Figure 4.14a were obtained for a 7% slurry of carbon black in water. Curve 1 shows the results obtained immediately after the dispersion was prepared in a high-speed blender. After additional mild agitation, the results shown in curve 2 are obtained; these are independent of further agitation. With shorter periods of mild agitation, a family of curves lying between 1 and 2 would be obtained.

The data presented in Figure 4.14a are consistent with the following mechanism. The dispersion that emerges from the blender is fundamentally unstable with respect to coagulation and coagulates rapidly to form a volume-filling network throughout the continuous phase. Except for the size and structure of the "chains," the situation is comparable to a cross-linked polymer swollen by solvent. In both, the liquid is essentially immobilized by the network of chains, and the system behaves as an elastic solid under low stress. The term *gel* is used to describe such systems whether the dispersed particles are lyophilic or lyophobic.

As the force applied to the surface of the gel is increased, however, a point is ultimately reached—the yield value—at which the network begins to break apart and the system begins to flow (curve 1 in Fig. 4.4a). Increasing the rate of shear may result in further deflocculation, in which case the apparent viscosity would decrease further with increased shear. Highly asymmetrical particles can form volume-filling networks at low concentrations and are thus especially well suited to display these phenomena.

As the system is subjected to ongoing, low-level mechanical agitation, the network structure is rearranged to a dispersion of more compact flocs that display both a lower yield value and a lower apparent viscosity than the initial dispersion (curve 2). A certain amount of time is required for the dispersed units to acquire a size and structure compatible with the prevailing low level of agitation. This is why intermediate cases (not shown in Fig. 4.14a) are observed before the actual stationary-state condition is obtained.

If the time for measurement of a stress/shear curve is short compared to the time required for rearrangement of the structure of the dispersed particles, then different results are obtained depending on whether the rate of shear is increasing or decreasing. Figure 4.14b is an example of such hysteresis for 11% dispersions of bentonite (a montmorillonite clay with plate-shaped primary particles) in which the entire cycle is measured in 70 sec. These data show the sensitivity of such experiments to the level of shear and to the time of observation. If the direction of the cycle is reversed along the descending branch of the curve, different results are obtained depending on whether the reversal is done at a rate of shear above (region (1) in Fig. 4.14b) or below (region (2)) about 200 s^{-1}. This shows that the structure within the colloid builds up rapidly (compared to the cycle time of 70 sec) at rates of shear below 200 s^{-1}, with negligible buildup at greater rates of shear.

These complicated observations are difficult to interpret in terms of fundamental interactions between particles; nevertheless, they have tremendous practical significance. For example, carbon black strengthens rubber against deformation, and for this reason 3–4 kg of carbon black is introduced into every tire made. This accounts for an annual worldwide consumption of carbon black of over 4 billion kg for automobile tires alone. Likewise, dispersions of clay in oils are used as lubricating greases. Obviously, the viscosity of these substances under various conditions of shear is an important consideration. Printing inks, drilling muds (circulated around the shaft in well-drilling operations to cool the bit and flush away cuttings), paper coatings, paints, and innumerable industrial slurries may all be considered examples of areas for which these considerations are vitally important.

In a paint, for example, a controlled level of aggregation is important in both the actual application of the paint and its storage in the container during application. In the former, thixotropy (the word means "changing with touch") permits the paint to "thin" under the shearing influence of the paintbrush or spraygun. Once applied, it thickens, preventing the drip or sag of the paint on the surface. In addition, the time required for this yield value to develop should be sufficient to allow for the leveling of brush marks. Thixotropy is an important property of paint in the bucket as well as on the wall. The buildup of a yield value interferes with the sedimentation of the pigment and eliminates the need to stir the paint continuously to assure uniformity. The fact that these requirements are well met by commercial paints indicates the success of paint chemists in regulating thixotropy.

One of the major difficulties in developing theories of the rheology of coagulated or flocculated dispersions is that the microstructures of the aggregates are nonequilibrium structures under shear. Understandably, the rheology of such dispersions is history dependent, as we have seen above, and requires computer simulations and nonequilibrium statistical mechanics for proper study.

4.9 VISCOSITY OF POLYMER SOLUTIONS

The viscosity of a polymer solution is one of its most distinctive properties. The spatial extension of the molecules is great enough so that the solute particles cut across velocity gradients and increase the viscosity in the manner suggested by Figure 4.8. In this regard they are no different from the rigid spheres of the Einstein model. What is different for these molecules is the internal structure of the dispersed units, which are flexible and swollen with solvent. The viscosity of a polymer solution depends, therefore, on the polymer-solvent interactions, as well as on the properties of the polymer itself.

4.9a The Staudinger-Mark-Houwink Equation

An important empirical generalization about the intrinsic viscosity of polymer solutions is given by

$$[\eta] = kM^a \tag{71}$$

in which k and a are experimentally determined constants. We shall call Equation (71) the *Staudinger-Mark-Houwink equation*. Staudinger (Nobel Prize, 1953), one of the founders of modern polymer chemistry, originally proposed this relationship with $a = 1$. It was subsequently recognized that more commonly $0.5 < a < 0.8$. The constants k and a are called the Mark-Houwink coefficients; they depend on the temperature as well as the polymer-solvent system. Values of k and a are determined for a particular system by measuring the intrinsic viscosity of polymer fractions of known molecular weight. Example 4.5 illustrates how this is done.

* * *

EXAMPLE 4.5 *Empirical Determination of the Mark-Houwink Coefficients for a Polymer Solution*. The molecular weights of various polycaprolactam preparations were determined by end-group analysis (see Example 3.2); intrinsic viscosities of the various fractions in *m*-cresol were measured at 25°C. The following values are representative of the results obtained (Reimschussel and Dege 1971):

$M \cdot 10^{-3}$	3.50	4.46	7.69	13.0	17.6	21.6	30.8
$[\eta]$(dl/g)	0.36	0.43	0.61	0.87	1.10	1.25	1.59

Determine the values of k and a that fit these data.

Solution: By taking the logarithms of Equation (71), a linear form is obtained:

$\ln [\eta] = \ln k + a \ln M$.

This could be plotted with a and $\ln k$ evaluated from the slope and intercept, respectively. Alternatively, a least squares analysis of the data can be performed. When this is carried out using the logarithms of the above results, it is found that $a = 0.683$ and $\ln k = -6.593$, or $k = 1.37 \cdot 10^{-3}$. The units of k are consistent with the concentration units "dl/g." This result can be inverted to give M directly: $M = 1.51 \cdot 10^{4}[\eta]^{1.46}$. ∎

* * *

The practical significance of the result of this example lies in the great ease with which viscosity measurements can be made. Once the k and a values for an experimental system have been established by an appropriate calibration, molecular weights may readily be determined for unknowns measured under the same conditions. Extensive tabulations of Mark-Houwink coefficients are available, so the calibration is often unnecessary for well-characterized polymers (see Table 4.5).

If intrinsic viscosity is used to evaluate the molecular weight of a polydisperse sample, the molecular weight so obtained is an average value. Equation (1.20) defined the viscosity average, which is the kind of average obtained. We are now in a better position to see how this comes about.

TABLE 4.5 Mark-Houwink Coefficients for Some Typical Polymer-Solvent Systems at the Indicated Temperatures

Polymer	Solvent	Temperature (°C)	$k \times 10^3$ (cm^{3}g^{-1})	a
Polypropylene	Cyclohexanone	$92 = \Theta$	172	0.50
Poly(vinyl alcohol)	Water	80	94	0.56
Poly(oxyethylene)	CCl$_4$	25	62	0.64
Poly(methyl methacrylate)	Acetone	30	7.7	0.70
Polystyrene	Toluene	34	9.7	0.73
Natural rubber	Benzene	30	18.5	0.74
Poly(acrylonitrile)	Dimethyl formamide	20	17.7	0.78
Poly(vinyl chloride)	Tetrahydrofuran	20	3.63	0.92
Poly(ethylene terephthalate)	*m*-Cresol	25	0.77	0.95

Source: M. Kurata, M. Iwama, and K. Kamada, in *Polymer Handbook*, 2d ed. (J. Srandrup and E. H. Immergut, Eds.), Wiley, New York, 1975.

It is apparent that the experimental (subscript *ex*) intrinsic viscosity depends on an average molecular weight $\overline{M}$; it is the *nature of this average* that we want to establish. According to Equation (71), we write

$$[\eta]_{ex} = k\overline{M}^a \tag{72}$$

The intrinsic viscosity contains a contribution of concentration that we note by writing $[\eta]_{ex} = (\eta_{sp})_{ex}/c_{ex}$ for which it is understood that this is a limiting value. If the sample is polydisperse, both $(\eta_{sp})_{ex}$ and c_{ex} are made up of contributions from molecules in different molecular weight classes, which we designate by the subscript *i*; that is, we can write

$$(\eta_{sp})_{ex} = \sum_i (\eta_{sp})_i \tag{73}$$

and

$$c_{ex} = \sum_i c_i = \frac{1}{V}\sum_i m_i \tag{74}$$

where the rightmost expression for c_{ex} recognizes that practical units are used in these experiments.

Each of the molecular weight fractions is expected to obey Equation (71) independently; therefore

$$|\eta_{sp}|_i = c_i[\eta]_i = c_i k M_i^a \tag{75}$$

Substituting Equations (73)–(75) into Equation (72) yields

$$\frac{k \sum_i c_i M_i^a}{\sum_i c_i} = \frac{(k/V)\sum_i m_i M_i^a}{(1/V)\sum_i m_i} = k\overline{M}^a \tag{76}$$

which shows that

$$\overline{M}^a = \frac{\sum_i m_i M_i^a}{\sum_i m_i} \tag{77}$$

Since $m_i = n_i M_i$, Equation (77) may be written as

$$\overline{M} = \overline{M}_v = \left(\frac{\sum_i n_i M_i^{(a+1)}}{\sum_i n_i M_i}\right)^{1/a} \tag{78}$$

which is the form given in Chapter 1 in Equation (1.20) and Table 1.8.

In this section we have looked at the Staudinger-Mark-Houwink equation as a purely empirical relationship useful for determining the molecular weight of unknown polymeric solutes. A considerable amount of work has been directed toward understanding the theoretical basis for this result. Although a detailed discussion would take us too far afield, we examine certain special cases of Mark-Houwink *a* values in the next section.

4.9b   Polymer Chain Extension and Viscosity

Two special cases of the Staudinger-Mark-Houwink equation can be justified without much difficulty: $a = 1.0$ and $a = 0.5$. In our discussion we ignore all numerical coefficients and concentrate on the variables, particularly with respect to molecular weight dependence. This is sufficient to arrive at an understanding of the significance of the Mark-Houwink exponent. We examine two models for polymer chains in solution. These models picture the polymer chain as being either unwound so that each segment experiences the flow streamlines or tightly coiled so that the flow leaves interior segments unaffected. The two models we discuss are known, respectively, as the free-draining and nondraining models. We see presently that these models—which clearly represent extremes of behavior—correspond to exponents of 1.0 and 0.5, respectively, in the Staudinger-Mark-Houwink equation. It is apparent that many polymer-solvent-temperature situations will result in chain coils of intermediate conformation; for these a is expected to take on values between the extremes we consider.

4.9b.1   Free-Draining Model

We begin by examining the free-draining model. Figure 4.15a shows a portion of a polymer chain situated in a velocity gradient. This will cause the molecule to rotate, thus converting some translational kinetic energy to rotational kinetic energy. This amounts to a dissipation of energy and thus connects with viscosity. Although not self-evident, it can be shown that the angular velocity induced in the molecule in this situation is directly proportional to the velocity gradient. Furthermore, a chain segment at a distance $r = r_i$ from the center of mass of the molecule, say, segment i, acquires a velocity $v_i = r_i\omega$ from the rotation. Figure 4.15b defines the location of the ith segment relative to the center of mass.

Next, we consider the force of viscous resistance experienced by this segment as a result of moving through its surroundings with a velocity v. Equation (2.2) relates the force of viscous resistance to the velocity through the expression $F_v = fv$, where f is the friction factor. The friction factor of a chain segment plays the same role but is given the symbol ξ to emphasize that it applies to a segment rather than the entire molecule. Therefore, for the ith segment, we write

$$F_{v,i} = \xi v_i \tag{79}$$

Both $F_{v,i}$ and v_i can be resolved into x and y components. Using the geometry defined by Figure 4.15b, we obtain

$$F_{v,i}^x = \xi r_i \omega \sin\theta \tag{80}$$

and

$$F_{v,i}^y = \xi r_i \omega \cos\theta \tag{81}$$

Since force times distance equals energy, the rate at which energy is dissipated by viscous forces on segment i is given by $F_{vi}^x v_x + F_{vi}^y v_y$, which we identify as $(dE/dt)_i$. Substituting Equations (80) and (81) gives

$$\left(\frac{dE}{dt}\right)_i = \xi r_i^2 \omega^2 (\sin^2\theta + \cos^2\theta) \propto \xi r_i^2 \left(\frac{dv}{dy}\right)^2 \tag{82}$$

For the entire polymer chain (subscript p), this result must be totaled for the n segments of the chain to produce $(dE/dt)_p$:

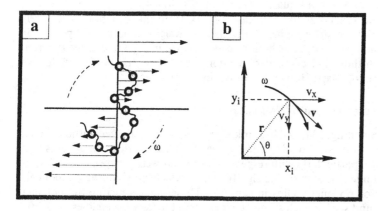

FIG. 4.15 Flow of polymer solutions: (a) velocity gradient through the center of mass of a polymer chain; (b) definition of coordinates for ith segment of rotating chain. The distance $r = r_i$ is the distance of the ith segment from the center of mass. (Redrawn with permission from Hiemenz 1984.)

$$\left(\frac{dE}{dt}\right)_p \propto \xi \sum_{i=1}^{n} r_i^2 \left(\frac{dv}{dy}\right)^2 \tag{83}$$

Since r_i is the distance from the center of mass, we use Equation (2.75) to replace $\Sigma_i r_i^2$ with $n\overline{R_g^2}$, where R_g is the radius of gyration of the molecule. Thus, Equation (83) becomes

$$\left(\frac{dE}{dt}\right)_p \propto \xi n\overline{R_g^2} \left(\frac{dv}{dy}\right)^2 \tag{84}$$

To scale up this last result from a single molecule to all of the molecules in a volume of solution V, we multiply Equation (84) by the particle concentration, given by c/M:

$$\frac{dE_V}{dt} \propto \xi n \overline{R_g^2} \frac{c}{M} \left(\frac{dv}{dy}\right)^2 \tag{85}$$

Notice that the left-hand side is now energy dissipation per unit volume. Comparing this result with Equation (3) shows that the increment in viscosity caused by the free-draining chain is given by the coefficient of $(dv/dy)^2$ and may be written

$$\eta - \eta_0 \propto \xi n \overline{R_g^2} \frac{c}{M} \propto \xi n \overline{R_g^2} \frac{c}{n} \tag{86}$$

where the second form recognizes that M is proportional to the number of segments in the chain. Finally, Equation (86) can be cast in the form of the intrinsic viscosity to yield

$$[\eta] = \frac{1}{c} \frac{(\eta - \eta_0)}{\eta_0} \propto \frac{\xi \overline{R_g^2}}{\eta_0} \tag{87}$$

which shows that the intrinsic viscosity of a solution of freely draining chains is proportional to the square of the radius of gyration of those chains. The segmental friction factor has dimensions "mass/time," and η_0 has dimensions "mass/(length time)"; therefore, the right-hand side of Equation (87) has dimensions of (length^3/mass), which are the reciprocal concentration units appropriate for intrinsic viscosity.

Independent studies of the segmental friction factor reveal it to be essentially independent of the molecular weight of the polymer. Therefore for this model the molecular weight dependence of the intrinsic viscosity is the same as that of $\overline{R_g^2}$. According to Equation (2.77), $\overline{R_g^2} \propto n \propto M$; therefore the only molecular weight dependence that survives out of all of this is a first-power dependence. We are thus led to conclude that the Mark-Houwink coefficient a equals unity for the case of free-draining chains. Since polymer chains are generally jumbled into coils, for which this may not be a good model, it is not surprising that experimental a values are usually less than this.

We discussed solvent goodness in Chapter 3, Section 3.4b.4. There are several aspects of that discussion that are pertinent here:

1. Solvent is squeezed out of the coil domain more and more as the solvent becomes poorer.
2. A change in solvent goodness can come about either from addition of a poorer solvent or from decreases in temperature.
3. Theta conditions correspond to a solvent so poor that precipitation would occur for a polymer of infinite molecular weight.
4. Theta conditions are identified experimentally as the situation in which the second virial coefficient of the osmotic pressure is zero.

In view of these considerations, it is not surprising that experimental a values vary systematically with decreasing solvent goodness. As the solvent goodness decreases, the chain becomes more tightly coiled so that flow streamlines penetrate the coil to a lesser extent. In an extreme situation we can imagine the coil so impermeable to the solvent flow that it behaves as a rigid sphere. Such a coil is said to be *nondraining* since any solvent imbibed by the coil is essentially immobilized. There are two very important things to realize about this state of affairs. First, a coil in which polymer-polymer contacts are this highly favored sounds like an alternate description of theta conditions. Second, if the coil approaches a rigid sphere in behavior, Einstein's equation for viscosity becomes appropriate again.

4.9b.1 The Kirkwood-Riseman Theory

Kirkwood and Riseman have developed a theory that allows for variable degrees of solvent drainage through the coil domain. We shall not go into this theory in any detail, except to note that it should reduce to Equation (87) in the free-draining limit and to the Einstein equation in the nondraining limit. The Kirkwood-Riseman theory can be written in the form

$$[\eta] \propto \frac{(\overline{R_{g,0}^2})^{3/2}}{n} f(X) \tag{88}$$

where the subscript 0 indicates Θ conditions and $X = [n\xi/\eta_0(\overline{R_{g,0}^2})^{1/2}]$. The function $f(X)$ in Equation (88) is our concern. For our purposes, it is enough to note that $f(X)$ approaches a value that is proportional to $[n\xi/\eta_0 (\overline{R_{g,0}^2})^{1/2}]$ in the free-draining limit (we continue to ignore numerical constants) and approaches a constant value in the nondraining limit. If we substitute the free-draining limit into Equation (88), we obtain

$$[\eta] \propto \frac{(\overline{R_{g,0}^2})^{3/2}}{n} \frac{n\xi}{\eta_0(\overline{R_{g,0}^2})^{1/2}} \propto \frac{\xi}{\eta_0} \overline{R_{g,0}^2} \tag{89}$$

The radius of gyration is expected to be different under theta and nontheta conditions since the extent of coil swelling due to imbibed solvent changes with solvent goodness. We define a coil expansion factor α* as follows:

$$\alpha = \frac{R_g}{R_{g,0}} \tag{90}$$

Although α is ordinarily greater than unity, fractional values are also possible. The range of fractional values is more limited, however, since the polymer tends to precipitate rather than squeeze out much more solvent under conditions poorer than theta conditions. Incorporating α into Equation (89) gives

$$[\eta] \propto \frac{\xi}{\eta_0} \frac{\overline{R_g^2}}{\alpha^2} \tag{91}$$

which is equivalent to Equation (87) as required.

Using the nondraining limit of the Kirkwood-Riseman theory gives

$$[\eta]_\Theta \propto \frac{(\overline{R_{g,0}^2})^{3/2}}{n} \tag{92}$$

Since Equation (2.77) shows that $\overline{R_{g,0}^2} \propto n$, Equation (92) becomes $[\eta] \propto n^{1/2} \propto M^{1/2}$. This important result shows that $a = 0.5$ is expected under Θ conditions. This expectation has been repeatedly verified by viscosity experiments under independently determined conditions. The subscript Θ has been attached to $[\eta]$ in Equation (92) in recognition of this.

For high molecular weight polymers in good solvents, $[\eta]$ exceeds $[\eta]_\Theta$ because of coil expansion under nondraining conditions; that is, as more solvent enters the coil domain than would be present under Θ conditions, Equation (92) continues to apply, with $\overline{R_g^2}$ replacing $\overline{R_{g,0}^2}$. Using Equation (90) to quantify this expansion effect, we obtain

$$[\eta] \propto \frac{\alpha^3 (\overline{R_{g,0}^2})^{3/2}}{n} \tag{93}$$

---

*Note that this α is different from the one defined in Eq. (67).

Taking the ratio of Equation (93) to Equation (92) gives

$$\frac{[\eta]}{[\eta]_\Theta} = \alpha^3 \tag{94}$$

which shows how the effect of solvent uptake on the spatial extension of polymer coils can be quantitatively determined from intrinsic viscosity experiments under theta and nontheta conditions.

We remarked above that the Einstein equation might be pertinent in the case of a nondraining coil; Equation (93) does not seem to bear this out. The Einstein theory in the form of Equation (41) predicts that $(\eta/\eta_0 - 1)$ is proportional to the volume fraction of the dispersed particles. For the nondraining coils, the volume fraction is proportional to the volume of each coil domain times the number concentration of the particles. The first of these factors is proportional to R_g^3 and the second to c/M. This leads to the prediction

$$\left(\frac{\eta}{\eta_0} - 1\right) \propto \phi \propto R_g^3 \frac{c}{M} \propto \alpha^3 (\overline{R_{g,0}^2})^{3/2} \frac{c}{M} \tag{95}$$

which is equivalent to Equation (92).

We have omitted a great deal of detail in this discussion of polymer viscosity. The interested reader will find some of the missing information supplied in Flory (1953). In particular, we have omitted all numerical coefficients, which limits us to ratios as far as computational capability is concerned. Numerical coefficients are available for Equation (92), for example, and this allows coil dimensions to be evaluated from viscosity measurements. A general conclusion that unifies all of this section is that any factor that causes a polymer chain to be more extended in space—whether by coil unfolding or swelling by solvent—tends to increase $[\eta]$. This is exactly what we expect in terms of the purely qualitative picture provided by Figure 4.8. Example 4.6 illustrates this for some actual polymers.

* * *

EXAMPLE 4.6 *Variation of Viscosity with Polymer Configuration.* For cellulose triacetate in acetone at 25°C, $k = 8.97 \cdot 10^{-5}$ dl/g and $a = 0.9$. For polyisobutene in benzene at 24°C, $k = 1.07 \cdot 10^{-3}$ dl/g and $a = 0.5$. Calculate $[\eta]$ for these systems if each of the respective polymers has $M = 10^5$. Comment on the correlation of the a values with the nature of the system in each case.

Solution: Intrinsic viscosities are calculated by direct substitution into the Staudinger-Mark-Houwink equation. For cellulose triacetate,

$$[\eta] = 8.97 \cdot 10^{-5}(10^5)^{0.9} = 2.84 \text{ dl/g}$$

For polyisobutene,

$$[\eta] = 1.07 \cdot 10^{-3}(10^5)^{0.5} = 0.34 \text{ dl/g}$$

Even though the two polymers have the same molecular weight, the cellulose triacetate has an intrinsic viscosity more than eight times greater than the polyisobutene. Note that the Mark-Houwink coefficient a is primarily responsible for this; the intrinsic viscosities would be ranked oppositely if k were responsible.

These results are fully consistent with the nature of the systems involved. Cellulose triacetate repeat units are six-member rings that carry three bulky acetate groups each. Internally, the rings are inflexible with respect to rotation and, because of the bulk of the acetate groups, are expected to be severely hindered in their rotation with respect to each other. As a consequence, such a molecule exists in solution in a highly extended form that approaches the predictions of the free-draining model in intrinsic viscosity. Polyisobutene, by contrast, is an aliphatic hydrocarbon for which low molecular weight aliphatic molecules are expected to be better solvents. Although benzene dissolves this polymer, it is a poorer solvent because of the polarizable pi electrons. It is therefore plausible that this system corresponds to theta conditions with an a value of 0.5. ∎

* * *

REVIEW QUESTIONS

1. What is meant by *rheology*? What role does viscosity of a fluid play in rheology?
2. Describe the physical significance of *Newton's law of viscosity*. Is Newton's law always applicable?
3. What is meant by a *fluid*, and how does it differ from (a) a solid and (b) a gel in terms of the flow behavior?
4. Why is rheology important in colloid science? Give as many examples as you can.
5. Explain what a *non-Newtonian* fluid is, and list the different types of non-Newtonian flow behavior.
6. Why is the shear history often important in the case of dispersions and polymer solutions?
7. Why are time scales and ratios of forces or energies important in studying flow behavior of dispersions?
8. What is the *Deborah number*? What is its physical significance? What is the *Peclet number*? Describe at least two ways of defining the Peclet number for flow of dispersions.
9. What is the *Navier-Stokes equation*? What is the physical significance of each of the terms appearing in the Navier-Stokes equation? How does the Navier-Stokes equation differ from the Stokes equation? Can you use the Navier-Stokes equation for a non-Newtonian fluid?
10. Why does the viscosity of a dispersion change, and how does it change, with the addition of colloidal particles?
11. List some of the conditions under which the Einstein equation for viscosity of dispersions fails and how one can correct the situation.
12. What is *inherent viscosity*, and what are its units?
13. What is *reduced viscosity*?
14. How do charges on particles change the viscosity of a dispersion? What are electroviscous effects? How do they differ from the *viscoelectric* effect?
15. How does the viscosity of a polymer solution differ from the viscosity of dispersions? What factors are important in the case of the former?
16. What is meant by "free-draining model" and "nondraining model" in the case of viscosities of polymer solutions?
17. Describe a way to determine the molecular weight of a polymer through viscosity measurements.

REFERENCES

General References (with Annotations)

Barnes, H. A., Hutton, J. F., and Walters, K., *An Introduction to Rheology*, Elsevier, Amsterdam, Netherlands, 1989. (Undergraduate and graduate levels. A practical introduction to the rheology of suspensions and polymer solutions. Chapter 2 presents introductory concepts and measurement techniques at the undergraduate level. Effects of colloidal interactions are not described.)

Bird, R. B., Stewart, W. E., and Lightfoot, E. N., *Transport Phenomena*, Wiley, New York, 1960. (Undergraduate and graduate levels. A classic on transport phenomena. Roughly one-third of the book is on fluid mechanics.)

Dickinson, E., *An Introduction to Food Colloids*, Oxford University Press, Oxford, England, 1992. (Undergraduate level. Although devoted to food colloids, the discussion of rheology in Chapter 3 of this volume is generally applicable to dispersions.)

Hiemenz, P. C., *Polymer Chemistry: The Basic Concepts*, Marcel Dekker, New York, 1984. (Undergraduate level. An introduction to fundamental concepts in polymer chemistry.)

Hunter, R. J., *Zeta Potentials in Colloid Science: Principles and Applications*, Academic Press, London, 1981. (Advanced. A research-level monograph on electrokinetic phenomena and electroviscous and viscoelectric effects.)

Hunter, R., *Foundations of Colloid Science*, Vol. 2, Oxford University Press, Oxford, England, 1990. (Mostly graduate level. Chapter 18 presents a graduate-level introduction to rheology of dispersions.)

Russel, W. B., Saville, D., and Schowalter, W., *Colloidal Dispersions*, Cambridge University Press, Cambridge, England, 1989. (This graduate-level treatment of colloidal dispersions devotes Chapter 14 to a brief overview of rheology and discusses the physical significance of the dimensionless groups relevant in rheology. Table 14.2 summarizes the relevant dimensionless groups.)

Other References

Booth, F., *Proc. Roy. Soc.*, **A205**, 533 (1950).

Cohn, E. J., and Edsall, J. T., *Proteins, Amino Acids and Peptides*, ACS Monograph, Hafner, New York, 1965.

de Kruif, C. G., van Iersel, E. M. F., Vrij, A., and Russel, W. B., *J. Chem. Phys.*, **83**, 4717 (1985).

Einstein, A., *Investigations on the Theory of the Brownian Movement*, Dover, New York, 1956.

Flory, P. J., *Principles of Polymer Chemistry*, Cornell University Press, Ithaca, NY, 1953.

Frisch, H. L., and Simha, R., The Viscosity of Colloidal Suspensions and Macromolecular Solutions. In *Rheology*, Vol. 1 (F. R Eirich, Ed.), Academic Press, New York, 1956.

Hermans, J. J., *Flow Properties of Disperse Systems*, North-Holland, Amsterdam, Netherlands, 1953.

Krasny-Ergen, B., *Kolloid Z.*, **74**, 172 (1936).

Krieger, I. M., *Adv. Colloid Interface Sci.*, **3**, 111 (1972).

Reimschussle, H. K., and Dege, G. J., *J. Polym. Sci.*, **9**, 2343 (1971).

Richards, E. G., *An Introduction to the Physical Properties of Large Molecules in Solution*, Cambridge University Press, Cambridge, England, 1980.

Russel, W. B., *J. Fluid Mech.*, **85**, 673 (1978a).

Russel, W. B., *J. Fluid Mech.*, **85**, 209 (1978b).

Russel, W. B., *Adv. Colloid Interface Sci.*, **24**, 287 (1980).

Stone-Masui, J., and Watillon, A., *J. Colloid and Interface Sci.*, **28**, 187 (1968).

Tadros, Th. F. (Ed.), *Surfactants*, Academic Press, London, 1984.

Tanford, C., *Physical Chemistry of Macromolecules*, Wiley, New York, 1961.

PROBLEMS

1. Gillespie and Wiley* used a cone-and-plate viscometer to measure F/A versus dv/dx for dispersions of silica and cross-linked polystyrene in dioctyl phthalate. At a volume fraction of 0.35 for both solids, the following results were obtained:

Silica	$F/A \times 10^{-3}$ (dyne cm^{-2})	2.2	1.4	1.0	0.50	0.25
	dv/dx (s^{-1})	500	325	235	125	60
Polystyrene	$F/A \times 10^{-3}$ (dyne cm^{-2})	1.6	0.80	0.55	0.25	
	dv/dx (s^{-1})	500	235	160	100	

Use these data to determine either η or the yield value for these dispersions depending on whether or not the system is Newtonian. Are these results consistent with the fact that the axial ratio was nearer unity and the particle size distribution narrower for the polystyrene than the silica? Explain.

2. An aqueous polybutyl methacrylate latex (average radius = 200 Å) has a viscosity of 50,500 cP at $\phi = 0.25$ and a viscosity of 36.7 cP at the same rate of shear when 1.71×10^{-5} g NaCl is added per gram of polymer.† Assuming that these charged particles must be surrounded by a layer of dissolved ions in solution, what conclusions can you draw about the dependence of the thickness of this layer of ions on the electrolyte content of the continuous phase?

3. A copolymer of vinylpyridine and methacrylic acid (62 and 38 mole%, respectively, in polymer) was studied in 90% methanol–10% water solution. The specific viscosity was measured‡ as a function of added NaOH or HCl, and the following results were obtained:

η_{sp}	0.1	0.25	0.28	0.24	0.21	0.21	0.20
mEq added per gram	0	2	4	6	2	4	6
			HCl			NaOH	

---

*Gillespie, T., and Wiley, R., *J. Phys. Chem.*, **66**, 1077 (1962).

†Brodnyan, J. G., and Kelley, E. L., *J. Colloid Sci.*, **19**, 488 (1964).

‡Alfrey, T., Jr., and Morawitz, H., *J. Am. Chem. Soc.*, **74**, 436 (1952).

Discuss these results in terms of the apparent effect of acid and base on the configuration of the polymer chain. Be sure your explanation is consistent with the chemical nature of the copolymer.

4. A dispersion of polydisperse spheres shows a relative viscosity of about 2.6 at a volume fraction of about 0.31.* Compare this result with the predictions of Equation (48).

5. The viscosity of cross-linked polymethyl methacrylate spheres in benzene was measured† and found to be

ϕ	0.050	0.035	0.028	0.019	0.014	0.010
η/η_0	2.15	1.61	1.41	1.19	1.18	1.12

Calculate $[\eta]$ for these spheres. Is there evidence in these results that the polymer particles may be swollen by the solvent? Explain.

6. The viscosity of uniform, cross-linked polystyrene spheres of two different diameters was measured in benzyl alcohol at 30°C:‡

$d = 0.382 \, \mu m$	ϕ	0.013	0.030	0.059	0.075
	η_{sp}	0.036	0.086	0.178	0.233
$d = 0.433 \, \mu m$	ϕ	0.02	0.04	0.08	
	η_{sp}	0.056	0.116	0.251	

Evaluate the intrinsic viscosity for each size of spherical particle and comment on the results in terms of the Einstein prediction that $[\eta]$ should be independent of particle size. Is the fact that benzyl alcohol is only a moderately good solvent for linear polystyrene consistent with the observed deviation between the experimental and theoretical values for $[\eta]$? Explain.

7. Criticize or defend the following proposition using the data of Problem 6 and the fact that the intrinsic viscosities of cross-linked polystyrene spheres in solvents such as benzene and CCl_4 (good solvents for linear polystyrene) lie in the range 5.8 to 7.5: Cross-linked polystyrene spheres swell by imbibing solvent, the effect being more extensive the better the solvent properties of the continuous phase for the non-cross-linked polymer.

8. Use the following data to evaluate the Mark-Houwink a and k constants for cellulose acetate in acetone at 25°C:§

M_n	$c(\text{g dl}^{-1})$	η_{sp}
130,000	0.094	0.289
	0.273	0.990
	0.546	2.770
86,000	0.114	0.286
	0.351	1.100
	0.703	3.120
76,000	0.118	0.247
	0.353	0.890
	0.775	2.700
61,000	0.138	0.239
	0.275	0.520
	0.428	0.880
48,000	0.152	0.209
	0.303	0.450
	0.684	1.230

*Eilers, H., *Kolloid Z.*, **97**, 313 (1941).
†Kose, A., and Hachisu, S., *J. Colloid Interface Sci.*, **46**, 460 (1974).
‡Papir, Y. S., and Krieger, I. M., *J. Colloid Interface Sci.*, **34**, 126 (1970).
§Sookne, A. M., and Harris, M., *Ind. Eng. Chem.*, **37**, 475 (1945).

9. Various molecular weight fractions of cellulose nitrate were dissolved in acetone, and the intrinsic viscosity was measured at 25°C:*

$M \times 10^{-3}$ (g mole^{-1})	77	89	273	360	400	640	846	1550	2510	2640
$[\eta]$(dl g^{-1})			1.23	1.45	3.54	5.50	6.50	10.6	14.9	30.3	31.0	36.3

Use these data to evaluate the constants k and a in the Staudinger-Mark-Houwink equation for this system.

10. The relative viscosity of solutions of cellulose nitrate in acetone was measured and extrapolated to zero rate of shear:*

η/η_0	1.45	1.53	1.67	1.89	2.31	3.41
c(g dl^{-1})	0.0151	0.0176	0.0212	0.0264	0.0352	0.0528

Use these data to evaluate the intrinsic viscosity for the polymer and use the a and k values from Problem 9 to calculate the molecular weight from the value of $[\eta]$.

11. The following intrinsic viscosity values of some high molecular weight polystyrene fractions have been reported†:

$\overline{M} \times 10^{-6}$ (g mole^{-1})	$[\eta]$ dl g^{-1})	
43.8	5.5	
27.4	4.4	Cyclohexane at 35.4°C
43.5	67.7	
26.8	36.5	Benzene at 40°C

Use these data to evaluate the constants in the Staudinger-Mark-Houwink equation. Are the values obtained consistent with the known facts that 35.4°C is the Flory (Θ) temperature for polystyrene in cyclohexane while benzene is a good solvent for polystyrene at 40°C.

12. Solutions of nylon-6,6 were studied in 90% formic acid solutions at 25°C, and the following data were obtained for two different molecular weight fractions:‡

c(g dl^{-1})	0.744	0.527	0.368	0.164
η_{sp}/c (dl g^{-1})	0.485	0.477	0.478	0.450

and

c (g dl^{-1})	0.742	0.640	0.537	0.436	0.332	0.225	0.132	0.058
η_{sp}/c (dl g^{-1})	0.897	0.892	0.886	0.876	0.864	0.847	0.805	0.778

Calculate the intrinsic viscosity for these two polymers. The Mark-Houwink constants for this system are known to be $a = 0.72$ and $k = 11 \times 10^{-4}$ dl g^{-1}; calculate the molecular weights of the two nylon fractions.

13. The radius of gyration of polymer coils can be determined independently from light scattering. Fox and Flory§ measured both R_g and $[\eta]$ for various molecular weight fractions of polystyrene in various solvents at several temperatures. The following results were obtained:

---

*Holtzer, A. M., Benoit, H., and Doty, P., *J. Phys. Chem.,* **58**, 624 (1954).
†McIntyre, D., Fetlers, L. J., and Slagowski, E., *Science,* **176**, 1041 (1972).
‡Taylor, G. B., *J. Am. Chem. Soc.,* **69**, 635 (1947).
§Fox, T. G., Jr., and Flory, P. J., *J. Am. Chem. Soc.,* **73**, 1915 (1951).

Solvent	T (°C)	$M \times 10^{-3}$ (g mole^{-1})	$[\eta]$ (dl g^{-1})	R_g(Å)
Methylethyl ketone	22	1760	1.65	437
	22	1620	1.61	414
	67	1620	1.50	400
	22	1320	1.40	367
	25	980	1.21	343
	22	940	1.17	306
	22	520	0.77	222
	25	318	0.60	194
	22	230	0.53	163
Dichloroethane	22	1780	2.60	576
	22	1620	2.78	545
	67	1620	2.83	529
	22	562	1.42	310
	22	520	1.38	278
Toluene	22	1620	3.45	527
	67	1620	3.42	523

Use these data to evaluate the factor of proportionality in Equation (93). Does this factor seem reasonably constant?

14. The intrinsic viscosity of a polystyrene solution in cyclohexane was measured* under theta conditions and found to be 0.078 dl g^{-1} for a polymer of $M = 8370$ g mole^{-1}. Use the average value of the factor of proportionality from Problem 13 to evaluate the radius of gyration of the polystyrene molecules in this solution. Calculate what this quantity is expected to be in terms of Equation (2.77), using twice the degree of polymerization as the number of steps in the random walk (since there are two C–C bonds per segment) and taking $0.154(2)^{1/2} = 0.218$ nm as the value of ℓ_0 (the C–C bond length corrected for tetrahedral bond angles). Briefly explain why the experimental value for R_g is larger than that calculated by Equation (2.77).

*Krigbaum, W. R., Mandelkern, L., and Flory, P. J., *J. Polym. Sci.,* **9,** 381 (1952).

5

Static and Dynamic Light Scattering
and Other Radiation Scattering

An interesting and oft-investigated question, "What is the origin of light?" and the solution of it, has been repeatedly attempted, with no other result than to crowd our lunatic asylums with the would-be solvers. Hence, after fruitless attempts to suppress such investigations by making them liable to a heavy tax, the Legislature . . . absolutely prohibited them.

From Abbott's *Flatland*

5.1 INTRODUCTION

5.1a What Is Radiation Scattering?

In Chapter 1 we described dark-field microscopy in which particles too small for direct microscopic observation could be detected against a dark field by horizontal illumination. Airborne dust or smoke particles show in a beam of light in an otherwise dark room in the same way. In both cases, the particles interact with the light that strikes them and deflect some of that light from its original direction. We speak of this light as being "scattered." A whole assortment of optical phenomena related to this are generally known as *light scattering* effects.

It turns out that the intensity of scattered light at any angle depends on the wavelength of the incident light, the size and shape of the scattering particles, and the optical properties of the scatterers, as well as the angle of observation. Furthermore, the functional relationship among these variables is known, at least for spherical particles and other geometries, under certain circumstances. By applying these relationships to light scattering experiments, information about the particles responsible for the scattering can be deduced. A similar statement can be made about experiments based on other forms of radiation (e.g., x-ray or neutron scattering), although the mechanism of scattering depends on the particular type of radiation.

5.1b What Are Static and Dynamic Scattering
    and Why Are They Important?

Let us focus on light scattering. Light scattering can be classified as *static* or *dynamic* depending on how the intensity is measured. In *static light scattering* the time-averaged total intensity is measured as a function of scattering angle. We see in this chapter how the weight and a characteristic linear dimension of the particle may be determined for some systems from static light scattering. Moreover, information on the internal structure and shapes of the particles as well as *inter*particle structure can be deduced by measuring the angle dependence of the intensity.

In contrast, in *dynamic light scattering* (DLS) the temporal variation of the intensity is measured and is represented usually through what is known as the intensity autocorrelation function. The diffusion coefficients of the particles, particle size, and size distribution can be deduced from such measurements. There are many variations of dynamic light scattering, and different names (e.g., *intensity fluctuation spectroscopy, quasi-elastic light scattering*, etc.) are used in the literature depending on what is measured and how it is measured (the *diffusing wave spectroscopy* [DWS] mentioned below is an example of such a variation). However, we focus on the most essential (and standard) in this chapter and leave the rest to advanced books cited below in this chapter.

The techniques discussed in this chapter generally assume that multiple scattering is negligible (i.e., each photon is scattered only once as it passes through the sample). Multiple scattering is a problem in this sense. However, one recent advance in light scattering turns this problem into an opportunity and models the random path of a multiply scattered photon as a random walk (see Chapter 2). This technique, known as *diffusing wave spectroscopy*, is highlighted in Vignette V as an example of recent advances that take the basic idea of light scattering one step (or many steps!) further from what we discuss in this chapter.

Multiple scattering is also avoided to a large extent when x-rays or neutrons are used in place of light. In fact, many of the concepts we discuss in this chapter on the use of light scattering carry over to the scattering of x-rays and neutrons. Moreover, x-rays and neutrons can probe shorter length scales because of their smaller wavelengths. As a result, x-ray and neutron scattering have become valuable adjuncts to light scattering in colloid science. Although a full discussion of x-ray and neutron scattering is beyond our scope, where appropriate we have pointed out their similarities and complementary nature.

The theory of light scattering has changed rapidly in recent years. A general relationship for the intensity of light scattered by a spherical particle was derived by Lorenz in the latter part of the 19th century and applied to colloids by Mie in 1908. Some quotations from two of the references at the end of this chapter give an indication of the historical development of light scattering since Mie's complicated theory was presented. In a book published in 1956, Stacey remarks that scattering patterns "have been tabulated in only a few cases because the computation is so laborious" (p. 56). In his 1969 book, Kerker writes of the same patterns that so many have been published "that these can hardly be coped with in the usual tabular form, much less published in the normal way" (p. 77).

During the period of slightly more than a decade between these books, the computer arrived on the scene. Things have not been the same since! A corresponding level of rapid advances has occurred in instrumental capabilities and design with the advent of lasers, fiber-optic probes, and digital electronics (see Zare et al.'s 1995 work for an introduction to the use of lasers in applications of interest here). The availability of lasers makes measurement of very small frequency shifts possible, and fiber-optic probes provide ways to limit sample volumes and to minimize multiple scattering in concentrated systems so that the theories discussed in this chapter can be used without modification. Similarly, advances in digital electronics and software have led to fast and powerful correlators for dynamic analysis (see Sections 5.4 and 5.8).

Our concern in this chapter is not primarily with complicated systems that require elaborate calculations, although we briefly discuss several specific examples of such systems. Instead, we focus attention on systems for which simplifying approximations to the general theory can be applied.

**VIGNETTE V   STRUCTURE AND STRUCTURAL TRANSITIONS IN DENSE
SYSTEMS: Single and Multiple Scattering—
The Art and Science of Seeing the Invisible**

Standard static and dynamic light scattering methods assume that there is very little multiple scattering by the particles, that is, the dispersion has to be sufficiently dilute so that the photons are scattered only once as they pass through the sample. Is there a way to "look" inside a dispersion that is cloudy or milky, such as a foam, and to extract information on the local structure and its kinetics and relaxation? Or, is it possible to "tailor" a dispersion so that some particles may be selectively made "visible," while others are invisible (and therefore do not scatter light)?

New innovations in theoretical analysis and surface chemistry provide affirmative answers to the above questions and have opened new ways to study interactions in colloidal systems.

For example, an important variant of the techniques discussed in this chapter, known as *diffusing wave spectroscopy* (DWS), takes a conceptually different view of multiple scattering in concentrated systems such as foams and turns it into an asset. The traditional methods we discuss in this chapter require single scattering so that shifts in frequency and phase in the scattered light can be related to information about the particles. In contrast, in DWS the tortuous path of a multiply scattered photon (see Figure 5.1) is approximated by a random walk, very much the same way we have modeled the random walk of a diffusing particle in Chapter 2! For instance, in the case of foams, the static transmission of light, modeled through the solution of the diffusion equation (with the diffusion coefficient written in terms of the mean free path using classical results), can be used to obtain the average bubble size and its variation as the foam coarsens (Weitz and Pine 1993). The fluctuations in the intensity of the diffusing wave provides, with a corresponding diffusion-based analysis, information on the dynamics.

In contrast to traditional dynamic light scattering (which, as we discuss in Section 5.8, probes the continuous random motion of the particles), DWS here probes spatially localized, temporally intermittent events that are characterized by relatively larger length scales. Despite this, the basic quantity measured ($g_1(t_d)$ in the notation of Section 5.8; there is no angular dependence in DWS because of multiple scattering) is essentially the same, only the method of analysis is different.

Even the traditional methods discussed in this chapter can be used for concentrated dispersions through contrast matching. For example, silica particles coated with silane coupling agents in a refractive index-matched mixture of ethanol and toluene can be used in combination with visible "probe" particles to study the dynamics of particles in dense systems. In the case of microemulsions (Chapter 8), selective deuteration of a component (oil, water, or surfactant) can be used in neutron scattering experiments even to measure the curvature of the oil-water interface.

Scattering techniques have thus become a powerful and indispensable tool for providing both routine information such as molecular weight and size and detailed structural information previously considered inaccessible. This chapter introduces the basic concepts (although restricted largely to light scattering) necessary for understanding the standard, as well as the emerging, uses of scattering in colloid science.

5.1c Focus of This Chapter

The primary focus of this chapter is on classical, static light scattering since a good understanding of this is essential for many of routine uses, such as determination of molecular weight and second virial coefficient. Static light scattering also sets the stage for the introduction of dynamic light scattering, as well as other forms of radiation scattering.

1. First we focus on static scattering. We begin with a review of some basic concepts from the theory of electromagnetic radiation (Section 5.2) since a background in this area and in the interaction of radiation with matter forms the first step in understanding scattering techniques. This review, however, is restricted to interaction of light with matter, although the chapter does discuss, briefly, the use of x-ray and neutron scattering in colloid science.

2. Next, we introduce the theory of Rayleigh scattering (Section 5.3), the first of many models covered in the chapter. The Rayleigh theory for dilute systems and solutions is developed here, with illustrative examples of the determination of molecular weight and the second virial coefficient. This is followed by a brief description of some of the basic experimental considerations and an introduction to absorbance and turbidity (Section 5.4).

3. Section 5.5 moves on to an extension of the Rayleigh theory essential for colloid science, namely, the Debye theory for particles of the order of the wavelength of the radiation source. The important concept of interference effects, the form factor, the Zimm plot, and

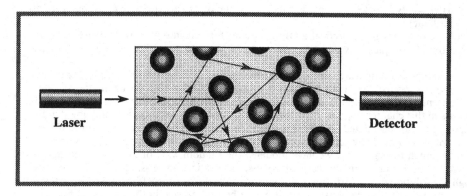

FIG. 5.1 Multiple scattering is viewed as a random walk of the photon in diffusing wave spectroscopy (DWS).

the dissymmetry ratio are introduced. Examples of the use of the form factor (often known as the *intra*particle structure factor) and its limiting behavior and the extension of the interference concept to the *inter*particle structure factor and its uses are introduced in Section 5.6. Since a combination of light scattering and x-ray or neutron scattering is needed for the latter, Section 5.6 also serves as a natural place to compare these complementary scattering techniques.

4. A brief treatment of scattering by large, absorbing particles and the concept of absorption and scattering cross sections are presented in Section 5.7 along with two examples of applications of the Mie theory (to absorbing, but small, particles) and a discussion of Tyndall spectra.

5. The final section (Section 5.8) introduces dynamic light scattering with a particular focus on determination of diffusion coefficients (self-diffusion as well as mutual diffusion), particle size (using the Stokes-Einstein equation for the diffusion coefficient), and size distribution.

This chapter is designed to provide the basic know-how for using light scattering and to set up the foundation needed for learning more advanced concepts and recent developments. For more advanced material, a number of advanced monographs (some containing state-of-the-art reviews), textbooks, and research publications containing details and new applications that are beyond the undergraduate level have been cited throughout the chapter.

5.2 INTERACTION OF RADIATION WITH MATTER

The phenomena with which we are concerned in this chapter are displayed by the entire spectrum of electromagnetic radiation. For those applications done in the visible part of the spectrum, the common designation *light* scattering is used. Visible light shares a variety of parameters and descriptive relationships with other regions of the electromagnetic spectrum. The purpose of this section is to examine briefly some of the characteristics of electromagnetic radiation, particularly those that are needed for an understanding of light scattering.

5.2a Elements of the Theory of Electromagnetic Radiation

Electromagnetic radiation consists of oscillating electrical ($\mathbf{E}$) and magnetic ($\mathbf{H}$) fields that are perpendicular to each other and perpendicular to the direction of propagation of the wave, as shown in Figure 5.2. Under vacuum, the velocity of propagation of an electromagnetic wave c is about $3 \cdot 10^8$ m s^{-1} and is independent of the wavelength of the radiation. The frequency ν, wavelength λ_0, and velocity of the radiation are related through the familiar equation

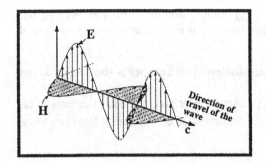

FIG. 5.2 The relationship between the electric and magnetic fields and the direction of propagation of electromagnetic radiation.

$$c = \lambda_0 \nu \tag{1}$$

If the radiation is passing through a medium other than a vacuum, its velocity and wavelength are both diminished by a factor n, the refractive index of the medium. Then Equation (1) becomes

$$\frac{c}{n} = \frac{\lambda_0}{n} \nu \tag{2}$$

or

$$v = \lambda \nu \tag{3}$$

We use the symbols c and λ_0 to refer, respectively, to the velocity and wavelength under vacuum only. In this chapter the symbol λ, without a subscript, always refers to the wavelength of the radiation in the medium.

5.2a.1 Coulomb's Law

Our discussion of light scattering centers on the oscillations of the electric field. Before turning to the oscillating aspect of the field, let us first review a few points about electric fields per se. We begin by retreating to Coulomb's law, which states that the force $\mathbf{F}$ between two charges q_1 and q_2 that are separated by a distance r is proportional to q_1q_2/r^2. In SI units, the q's are measured in coulombs, r in meters, and $\mathbf{F}$ in newtons; the proportionality factor in Coulomb's law must be dimensionally consistent. For charges under vacuum, it is traditional to write the proportionality factor as $1/(4\pi\varepsilon_0)$, where ε_0 is called the *permittivity of vacuum*. Thus Coulomb's law for $\mathbf{F}$ along the r-direction ($\mathbf{i}_r$ = unit vector) is written as

$$\mathbf{F} = \frac{1}{4\pi\varepsilon_0} \frac{q_1q_2}{r^2} \mathbf{i}_r \tag{4}$$

where the permittivity of vacuum ε_0 is equal to $8.854 \cdot 10^{-12}$ C^2 N^{-1} m^{-2}. (The SI units C^2 N^{-1} m^{-2} are equivalent to C^2 J^{-1} m^{-1} or kg^{-1} m^{-3} s^2.) The proportionality constant $1/(4\pi\varepsilon_0)$ is equal to $8.988 \cdot 10^9$ N m^2 C^{-2}. The older literature uses cgs units, for which the proportionality constant between $\mathbf{F}$ and q_1q_2/r^2 equals 1.00 dyne cm^2 $(esu)^{-2}$, where the esu is the electrostatic unit of charge defined to make this proportionality factor equal to unity.

If the charges are embedded in a medium, the electrical properties of the intervening molecules decrease the force from the value calculated by Equation (4). The relative dielectric constant of the medium ε_r measures this effect quantitatively. In surroundings other than a vacuum, the force between two charges is given by

$$\mathbf{F} = \frac{1}{4\pi\varepsilon_0\varepsilon_r} \frac{q_1q_2}{r^2} \mathbf{i}_r \tag{5}$$

Note that the force is a vector. Its direction is along the line connecting the two charges, but Equation (5) specifies only the magnitude of the force. It is important to remember several aspects of Equation (5):

1. The relative dielectric constant ε_r is dimensionless. This quantity is also known as the *relative permittivity*.
2. The product $\varepsilon_0\varepsilon_r$ is sometimes written ε (without subscripts), and ε has the units of ε_0.
3. The relative permittivity ε_r appears in the denominator of Coulomb's law even when cgs units are used.

5.2a.2 Electric Field

Next, let us apply these ideas to the electric field. By definition, an electric field **E** describes the force experienced by a unit test charge $q_t = 1$. Thus, if we let one of the charges in Equation (5) equal q_t, the following expression is obtained for the field associated with the remaining charge q:

$$E = \frac{1}{4\pi\varepsilon}\frac{q}{r^2} \tag{6}$$

The electric field, being a vector, has a direction as well as a magnitude; the direction of the field is perpendicular to the direction of propagation of the field.

The dominant characteristic of the electrical and magnetic fields that comprise electromagnetic radiation is their periodically oscillating nature, a fact that enables us to describe them by the mathematics of waves. For light scattering, it is the electric field that is of interest. The oscillating nature of an electric field propagating in the positive x direction is described by the equation

$$E = E_0 \cos\left[2\pi\left(\nu t - \frac{x}{\lambda}\right)\right] \tag{7}$$

in which E_0 is a vector with a magnitude equal to the maximum amplitude of the field. Since we have postulated the x direction as the direction of propagation, the electric field lies in the yz plane and may, in general, be resolved into y and z components since **E** is a vector. Both the y and z components of the field are described by Equation (7) when the equation is modified by the inclusion of phase angles. This is because the two components need not be in phase with each other. Accordingly, we write

$$E_y = E_{0y} \cos\left[2\pi\left(\nu t - \frac{x}{\lambda}\right) + \delta_y\right] \tag{8}$$

and

$$E_z = E_{0z} \cos\left[2\pi\left(\nu t - \frac{x}{\lambda}\right) + \delta_z\right] \tag{9}$$

in which the δ terms are the phase angles of the two components.

5.2a.3 Polarization of Light

In the most general case, the above two equations mean that the electric field vector traces an ellipse in the yz plane. There are two special cases of note in this general situation. If the phase difference between the two components of the field $(\delta_y - \delta_z)$ is zero or some integral multiple of π, the ellipse flattens to a line. If the phase difference is $\pi/2$ or any odd integral multiple of $\pi/2$ and the amplitudes of the two components are equal, the ellipse is rounded to a circle. In the former case we speak of the radiation as being *plane polarized*, and in the latter case as being *circularly polarized*.

Ordinary light is said to be unpolarized. This last term is somewhat unfortunate because all light displays some form of polarization. In ordinary light, however, all forms of polarization are present, so the individual effects cancel out. The use of various filters makes it possible to conduct experiments with radiation that show a unique state of polarization. Polaroid filters, for example, transmit plane-polarized light only. In discussing light scattering, we are concerned primarily with unpolarized light and occasionally with plane-polarized light.

An interesting example of polarization arises in the study of light reflected from a surface. Suppose we consider a beam of light incident on the planar surface of some material having a higher refractive index than the medium from which the beam approaches. At the surface some of the light will be refracted, and some will be reflected. Figure 5.3 illustrates this for the case in which the reflected and the refracted beams are separated by an angle of 90°.

In this situation, two very different results are obtained, depending on whether the incident light is linearly polarized in the plane of the figure or perpendicular to it. If the light is polarized in the plane of the figure, no light will be reflected at all. On the other hand, if it is initially polarized perpendicular to the plane of the figure, it will be reflected with the same polarization. If ordinary light—a mixture of the two types of light just mentioned—is used for the incident radiation, only one of the components contributes to the reflection. Furthermore, the reflected light is polarized perpendicular to the plane of the figure. Polaroid filters are used in photography and in sunglasses to reduce the glare of reflected light since this light is polarized.

This behavior is not observed uniquely when the angle between the two beams is 90°; rather, the intensity of the reflected beam varies continuously with the angle. At 90°, however, the polarization effect is most pronounced. We shall see that some scattering phenomena also show an angular dependence, as well as the fact that scattered light displays maximum polarization at 90°.

5.2b   Interaction Between an Electric Field and a Charge

In this section we discuss the interaction between an electric field and a charge that is free to move with the field. Such a charge experiences a force that accelerates it with the field. If the field is oscillating, the acceleration of the charge will also oscillate. One of the basic results of classical electromagnetics is that the acceleration of a charge leads to the emission of radiation.

It was the apparent violation of this requirement that led to the postulate of quantization in the Bohr theory of the hydrogen atom. However, we are concerned here with the classical result in which the charge does radiate. Our objective is to describe the emitted radiation some distance r from the emitter.

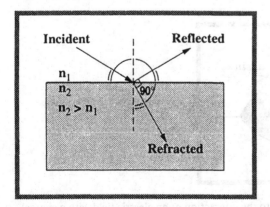

FIG. 5.3   The relationship between the incident, reflected, and refracted beams of radiation at a plane surface.

5.2b.1   Emission of Radiation by a Charge

The radiation emitted by an oscillating charge may be described by its electric field vector, which is given by

$$\mathbf{E} = \frac{q\,\mathbf{a}_p \sin \phi_z}{4\pi\varepsilon_0\,c^2\,r} \tag{10}$$

according to electromagnetic theory. In this equation, q equals the magnitude of the charge, $\mathbf{a}_p$ is its periodic acceleration, and c is the velocity of light. The coordinate system defined by Figure 5.4a will help describe this field. The origin of the coordinates is located at the emitting charge. The angles between the line of sight — along which r is measured — and the x, y, and z axes are designated ϕ_x, ϕ_y, and ϕ_z, respectively.

The oscillating charge behaves like an antenna, and Equation (10) describes the field of such an antenna as long as r is large compared to the wavelength of the radiation that induces the oscillation. It should also be noted that the antenna to which Equation (10) applies is aligned vertically (z axis) and is therefore "driven" by vertically polarized radiation.

We are using Equation (10), presented without proof, as the point of departure for our discussion of light scattering. Therefore, it is important that we find its predictions reasonable.

1.   First, let us consider the plausibility of the $\sin \phi_z$ factor. This factor ranges between 0 and 1 as ϕ_z varies from 0 to $\pi/2$. This means that the maximum field will be observed at right angles to the oscillating charge, and no field will be observed along the axis of the oscillation. It is the projection of the acceleration in the plane perpendicular to the line of sight that induces the field. The strength of the field is proportional to this projection at any location, as shown in Figure 5.4b. This factor describes the entire angular dependence of the induced field produced by a vertical driving field. Since it depends on the angle ϕ_z alone, the induced field is seen to be symmetrical with respect to the z axis.

2.   Next we note that the induced field varies inversely with r. It makes sense that the field should decrease as we get farther from the antenna, but the inverse first-power dependence may be unexpected since we are more familiar with inverse-square laws. However, it is the energy or intensity of the light that varies according to an inverse-square law. In the next section we convert this expression for $\mathbf{E}$ to an expression for energy; the more familiar r^{-2} functionality appears then.

3.   Finally, it is sufficient for our purposes to think of the remaining factors in Equation (10), $(4\pi\varepsilon_0 c^2)^{-1}$, as providing dimensional consistency to the expression. Taking a look at the SI units of the right-hand side of Equation (10), we obtain

$$\frac{(C)\,(\mathrm{m\ s^{-2}})}{(C^2\,N^{-1}\,\mathrm{m^{-2}})\,(\mathrm{m\ s^{-1}})^2\,(\mathrm{m})} = \frac{N}{C}$$

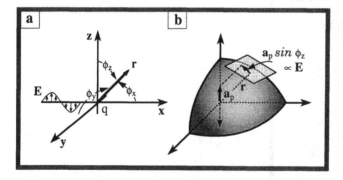

FIG. 5.4   Coordinates and acceleration relevant to the interaction of an electric field with a charge: (a) the coordinates of an electric field $\mathbf{E}$ relative to an oscillating charge located at the origin; (b) projection of the acceleration in the plane perpendicular to the line of sight.

Newtons coulomb $^{-1}$ are units of force per charge as required. Note that multiplication of both the numerator and denominator of this dimensional expression converts N C^{-1} to V m^{-1} since Nm = J and J/C = V. This last set of units for **E** is particularly useful for describing the field between electrodes and, as such, will be encountered in Chapter 12, in which we discuss electrokinetic phenomena.

Equation (10) describes the field emitted by an antenna that, in turn, is driven by another field. The oscillation of one field promotes the oscillation of a charge in the antenna, and this induces another electric field. The frequency is the same for all three.

5.2b.2   Induced Dipole Due to the Field

This description of antennas may seem more appropriate to a discussion of radio or television waves. We must realize, however, that at the molecular level dipoles behave exactly like antennas. Since molecules are made up of charged parts, a *dipole moment* μ is induced by the electric field of the radiation in any material through which radiation passes. In this discussion, the dipole moment equals the product of the effective charge displaced by the field and its distance of separation from the opposite charge. In SI, μ has units C m. We consider isotropic materials characterized by a *polarizability* α. As the name implies, this property measures the ease with which charge separation—polarity—is induced in a molecule by an electric field. For isotropic substances, the dipole moment and the field are related by the expression

$$\mu = \alpha \mathbf{E}_0 \tag{11}$$

Using Equation (11) as the basis of a dimensional analysis of α shows α has SI units given by

$$\frac{C\,m}{N\,C^{-1}} = \frac{C^2\,m}{N}$$

The quantity $\alpha/4\pi\varepsilon_0$—which is the polarizability value used in cgs units—is informative. It is examined in Example 5.1.

* * *

EXAMPLE 5.1   *Polarizability of Particles.*   Criticize or defend the following proposition on the basis of the units of $\alpha/4\pi\varepsilon_0$: The larger the volume of a particle is, the easier it is to induce polarity in that particle.

Solution: The ratio $\alpha/4\pi\varepsilon_0$ has the units $(C^2\ m\ N^{-1})/(C^2\ m^{-2}\ N^{-1}) = m^3$, which are units of volume. In fact, polarizabilities of actual molecules are on the order of $10^{-29}\ m^3$ molecule^{-1} = 0.01 nm^3 molecule^{-1}, which is the magnitude of molecular volumes. For individual atoms, we expect the polarizability to increase with atomic volume since the outermost electrons are less tightly restrained by the nucleus in such cases. Extension of this principle to covalently bonded species must be done cautiously, however, since the bonding affects the overall picture. The accuracy of the proposition, then, depends on the nature of the "particle" under consideration. Even when the principle stated in the proposition is not literally true, it offers a convenient mnemonic for the definition of polarizability. ■

* * *

5.2b.3   Electric Field Emitted by a Charge

If we imagine the molecule to lie at the origin of a coordinate system so that $x = 0$, we may substitute Equation (7) into (11) to obtain

$$\mu = \alpha \mathbf{E}_0 \cos(2\pi\nu t) \tag{12}$$

A dipole moment may be regarded as the product of the distance ξ between two charges and the magnitude of the charge q. A useful way of looking at Equation (12) is to identify the charge as

$$q = \alpha E_0 \tag{13}$$

where E_0 is the magnitude of the vector $\mathbf{E}_0$, and the separation of the charges as

$$\xi = \cos (2\pi\nu t) \tag{14}$$

Then the magnitude of the periodic acceleration of the charge is given by

$$a_p = \frac{d^2\xi}{dt^2} = -4\pi^2\nu^2\cos(2\pi\nu t) \tag{15}$$

Equations (13) and (15) may now be used in Equation (10) to give

$$E = -\frac{\alpha E_0 4\pi^2 \nu^2 \cos (2\pi\nu t) \sin \phi_z}{4\pi\varepsilon_0 c^2 r} \tag{16}$$

For maximum generality, we must remember that the field is periodic in space as well as in time; hence the cosine factor in Equation (16) is corrected by analogy with Equation (7):

$$E = -\frac{\pi\nu^2 \alpha E_0 \cos[2\pi(\nu t - r/\lambda)] \sin \phi_z}{\varepsilon_0 c^2 r} \tag{17}$$

Equation (17) describes the induced field a distance r from the dipole.

5.3 SCATTERING BY SMALL PARTICLES: THEORY OF RAYLEIGH SCATTERING

In this section we discuss the first of several light scattering theories to be considered in this chapter, Rayleigh scattering. We shall see presently that Rayleigh scattering applies only when the scattering centers are small in dimension (i.e., when the "characteristic dimension" of the scatterers is small) compared to the wavelength of the radiation, the "yardstick" used in the measurements. As such, it is severely limited in its applicability to colloidal particles, at least when visible light is the radiation involved. Rayleigh scattering is the easiest of the scattering theories to understand, however, so it is a logical place to begin. Furthermore, we shall extend its applicability to larger particles in other sections by introducing suitable correction factors.

5.3a Scattering by Single Molecules and Gases

When a beam of radiation is incident on a molecule, a certain fraction of that radiation will undergo the process described in the preceding section and be emitted by the dipole. Any light that does not interact this way will continue past the molecule along the original path. This undeviated or transmitted light will be attenuated compared to the incident light since some of the original beam of light is scattered from its initial path. Note that this attenuation has nothing to do with absorption; the effect we are considering is a classical result and does not involve transitions between quantum states.

Since Rayleigh scattering does not apply to particles in the colloidal size range, we do not present the derivation in detail; instead, Table 5.1 summarizes some key steps in the development of the *Rayleigh equation*:

$$\frac{i_s}{I_{0,u}} = \frac{2\pi^2 M}{r^2 \lambda_0^4 N_A \rho} (n - 1)^2 (1 + \cos^2 \phi_x) \tag{18}$$

in which i_s is the intensity, as measured at r and ϕ_x (see Figure 5.4), of the light scattered per unit volume by a gas of molecular weight, density, and refractive index given by M, ρ, and n, respectively. The incident light is unpolarized (subscript "u") and has an intensity $I_{0,u}$. Equation (18) was derived by Lord Rayleigh in 1871.

An interesting application of the *Rayleigh equation* is the explanation it offers for why the sky appears blue. This arises from the inverse fourth-power dependence on λ_0 for i_s. Suppose, for example, that two radiations are compared that have wavelengths that differ by a factor of 2. Then the scattered intensity of the shorter wavelength will be 16 times as great as that of the longer wavelength. Although red and blue light do not differ by quite this much in wavelength, the blue component of white light, say, sunlight, is scattered very much more

TABLE 5.1 Steps Involved in the Derivation of the Rayleigh Equation

Step	Justification	Cumulative effect
1. Evaluate the intensity of light scattered at ϕ_z	Intensity of light is proportional to the square of the electrical field	$i \propto \dfrac{\pi^2 v^4 \alpha^2 |E_0|^2 \cos^2[2\pi(vt - r/\lambda)] \sin^2\phi_z}{\varepsilon_0^2 c^4 r^2}$
2. Evaluate the intensity of incident light	No scattering factors needed in previous result	$I_0 \propto |E_0|^2 \cos^2\left[2\pi\left(vt - \dfrac{r}{\lambda}\right)\right]$
3. Evaluate the intensity ratio for light polarized in vertical plane	Ratio of preceding factors	$\dfrac{i_v}{I_{0,v}} = \dfrac{\pi^2 v^4 \alpha^2 \sin^2\phi_z}{\varepsilon_0^2 c^4 r^2}$
4. Evaluate the intensity ratio for light polarized in horizontal plane	Replace $\sin\phi_z$ by $\sin\phi_y$ in the previous equation	$\dfrac{i_h}{I_{0,h}} = \dfrac{\pi^2 v^4 \alpha^2 \sin^2\phi_y}{\varepsilon_0^2 c^4 r^2}$
5. Evaluate the intensity ratio for unpolarized light	Equal contributions from vertical and horizontal components	$\dfrac{i}{I_{0,u}} = \dfrac{\tfrac{1}{2}(i_v + i_h)}{I_{0,u}}$ $= \dfrac{1}{2}\dfrac{\pi^2 v^4 \alpha^2}{\varepsilon_0^2 c^4 r^2}(\sin^2\phi_z + \sin^2\phi_y)$
6. Replace $\sin^2\phi_y + \sin^2\phi_z$ with $(1 + \cos^2\phi_x)$ and introduce λ_0 through Equation (1)	$r\cos\phi_i$ is the projection of r on the i axis; therefore, $r^2(\cos^2\phi_x + \cos^2\phi_y + \cos^2\phi_z)$ is equal to r^2	$\dfrac{i}{I_{0,u}} = \dfrac{\pi^2 \alpha^2}{2\varepsilon_0^2 \lambda_0^4 r^2}(1 + \cos^2\phi_x)$
7. Scale up for independent scatterers: Multiply by number of sites per unit volume (subscript "s" on i)	For ideal gases, this factor equals ($N_A\rho/M$)	$\dfrac{i_s}{I_{0,u}} = \dfrac{\pi^2 N_A \rho \alpha^2}{2\varepsilon_0^2 r^2 \lambda_0^4 M}(1 + \cos^2\phi_x)$
8. Use the Clausius-Mosotti equation from physical chemistry (Atkins 1994, Chapter 22) to introduce refractive index n	For n close to unity (gases), the Clausius-Mosotti equation becomes $\alpha = \dfrac{3M\varepsilon_0}{N_A\rho}\dfrac{n^2 - 1}{n^2 + 2} \approx \dfrac{2M\varepsilon_0}{\rho N_A}(n - 1)$	$\dfrac{i_s}{I_{0,u}} = \dfrac{2\pi^2 M}{r^2 \lambda_0^4 N_A \rho}(n - 1)^2(1 + \cos^2\phi_x)$

than the red. Accordingly, the sky overhead appears blue. At sunset, we see mostly transmitted light. Since the blue has been most extensively removed from this by scattering, the sky appears red at sunset. One of the early uses of the Rayleigh equation was in the determination of the value of the Avogadro's number, as illustrated through Problem 2 at the end of this chapter.

The Rayleigh theory does not apply when the scattering molecules are absorbing or when the atmosphere contains dust particles, water drops, or other particles with dimensions that are larger than ordinary gas molecules.

5.3b   Rayleigh Scattering Applied to Solutions: Fluctuations

A crucial aspect of the transition from Step 6 to Step 7 in Table 5.1 is the requirement that the individual molecules are far enough apart to be treated as independent sources. This assumption is justified for gases, but in liquids the molecules are close enough together so that interference occurs between the waves emitted from different centers. As a matter of fact, there would be complete destructive interference of all scattered light in liquids if the molecules were randomly arranged and stationary. Nevertheless, pure liquids do scatter light. It is not the individual molecules that are the scattering centers in this case, but rather the small domains of compression or rarefaction that arise from fluctuations.

We saw in Chapter 2 that molecular motion results in small fluctuations in density at the molecular level. Although the average density of a liquid is a constant equaling the experimental density, there will be small transient domains within it that have densities larger or smaller than the mean value.

Liquid solutions also scatter light by a similar mechanism. In the case of a solution, the scattering may be traced to two sources: fluctuations in solvent density and fluctuations in solute concentration. The former are most easily handled empirically by subtracting a solvent "blank" correction from measurements of the intensity of light scattered from solutions. What we are concerned with in this section, then, is the remaining scattering, which is due to fluctuations in the solute concentration in the solution.

A fluctuation in the concentration of a small volume element δV of solution will result in a change in the properties of that volume element. We begin the analysis of this situation by defining δc and $\delta \alpha$ as the fluctuations in solution concentration and polarizability, respectively, in this element. The first thing that we recognize about these quantities is that their average values are zero since both positive and negative fluctuations occur.

Although their averages may be zero, the averages of their squares are not zero. Remember that similar situations were encountered in Chapter 2, Sections 2.6 and 2.7, in which we discussed particle displacements due to diffusion and segment displacements in a random coil. In both of these cases it was by considering the average values of the square of the displacements that meaningful quantities could be obtained. Similarly, the average values of δc and $\delta \alpha$ are zero, but their mean square values will be different from zero.

The equation in Step 6 in Table 5.1 shows that the intensity of scattered light depends on the square of polarizability. We conclude, therefore, that the way to adapt this equation to the scattering by solutions is to replace α^2 in the above equation by $\overline{(\delta \alpha)^2}$

$$\frac{i}{I_0} = \frac{\pi^2 \overline{(\delta \alpha)^2}}{2 \, \epsilon_0^2 \, r^2 \lambda^4} (1 + \cos^2 \phi_x) \tag{19}$$

Note that the wavelength λ of light in the medium rather than the value under vacuum is used in this expression since this is the light that reaches the scattering center.

Next we must consider how to extend this result to a unit volume of solution and how to relate $\delta \alpha$ to δc. The steps involved in these extensions are not difficult, but they are lengthy. Accordingly, we shall not develop the entire argument in detail. Instead, some of the key steps in the development along with a brief justification for each are summarized in Table 5.2. In this table, each major substitution is presented along with variations of Equation (19) that reflect the cumulative effects of all the substitutions. The first entry in the table, for example,

TABLE 5.2 Some Key Substitutions and Their Justification for the Transformation of Equation (19) to Equation (20)

Substitution	Justification	Cumulative effect on Equation (19)
$\dfrac{\overline{\delta\alpha^2}}{\text{volume}} = \dfrac{1}{\delta V}\ \overline{\delta\alpha^2}\ \delta V$ fluctuation $$\overline{\delta\alpha^2} = \varepsilon_0^2\,\delta V^2\left(2n\frac{dn}{dc}\right)^2\overline{\delta c^2}$$	1. $\dfrac{1}{\delta V}$ domains of volume δV can fit into 1 cm^3 of solution 1. δV replaces $\dfrac{M}{\rho N_A}$ in Clausius-Mosotti equation (Table 5.1) 2. For $n = 1$, $\alpha = \varepsilon_0\,\delta V(n^2 - 1)$ 3. $\delta\alpha = \varepsilon_0\,\delta V\,2n\,dn$ and $dn = \dfrac{dn}{dc}\,\delta c$	$$\frac{i_s}{I_0} = \frac{\pi^2\overline{\delta\alpha^2}}{r^2 2\varepsilon_0^2\lambda^4}\,\delta V\,(1 + \cos^2\phi_x)$$ $$\frac{i_s}{I_0} = \frac{2\pi^2\delta V[n(dn/dc)]^2\overline{\delta c^2}}{r^2\lambda^4}(1 + \cos^2\phi_x)$$
$$\overline{\delta c^2} = \frac{k_BT}{(\partial^2 G/\partial c^2)_0}$$ Subscript 0: evaluated at equilibrium	1. Taylor series (see Appendix A) expansion of G around equilibrium value: $$G = G_0 + \left(\frac{\partial G}{\partial c}\right)_0\delta c + \frac{1}{2}\left(\frac{\partial^2 G}{\partial c^2}\right)_0\delta c^2$$ 2. $\left(\dfrac{\partial G}{\partial c}\right)_0 = 0$ at equilibrium 3. $G - G_0 \simeq \dfrac{1}{2}k_BT$, since fluctuation is due to thermal energy $\left(\dfrac{1}{2}k_BT\text{ per degree of freedom}\right)$	$$\frac{i_s}{I_0} = \frac{2\pi^2\delta V[n(dn/dc)]^2 k_BT}{r^2\lambda^4(\partial^2 G/\partial c^2)_0}(1 + \cos^2\phi_x)$$
$$\frac{\partial^2 G}{\partial c^2} = -\frac{\delta V}{\overline{V}_1 c}\left(-\frac{\partial\mu_1}{\partial c}\right)_0$$	1. $dG = \mu_1 dn_1 + \mu_2 dn_2$ 2. Since $\overline{V}_1 dn_1 = -\overline{V}_2 dn_2$, $dG = \left(\mu_2 - \dfrac{\overline{V}_2}{\overline{V}_1}\mu_1\right)dn_2$ 3. Since $M\,dn_2 = dc\,\delta V$, $\dfrac{dG}{dc} = \left(\mu_2 - \dfrac{\overline{V}_2}{\overline{V}_1}\mu_1\right)\dfrac{\partial V}{M}$ 4. By the Gibbs-Duhem equation, $$\frac{\partial^2 G}{\partial c^2} = -\frac{\delta V}{M}\left(\frac{\overline{V}_2}{\overline{V}_1} + \frac{n_1}{n_2}\right)\frac{d\mu_1}{dc}$$ 5. $c = \dfrac{n_2 M}{n_1\overline{V}_1} + n_2\overline{V}_2$	$$\frac{i_s}{I_0} = \frac{2\pi^2[n(dn/dc)]^2 k_BT\overline{V}_1 c}{r^2\lambda^4(-\partial\mu_1/\partial c)_0}(1 + \cos^2\phi_x)$$
$$\left(\frac{\partial\mu_1}{\partial c}\right)_0 = -\overline{V}_1\left(\frac{\partial\pi_{osm}}{\partial c}\right)_0$$	1. By Equation (3.21)	$$\frac{i_s}{I_0} = \frac{2\pi^2[n(dn/dc)]^2 k_BTc}{r^2\lambda^4(\partial\pi_{osm}/\partial c)_0}(1 + \cos^2\phi_x)$$

converts Equation (19) to an expression for the relative light scattered by a unit of volume i_s/I_0 by multiplying Equation (19) by the number of volume elements in 1 cm^3: $1/\delta V$. Although the justifications in Table 5.2 are sketchy, they provide hints that will show the interested reader how to proceed in order to develop the required relationship in detail. Only the third entry in the table, the connection between δc^2 and $\partial^2 G/\partial c^2$ requires a more elaborate proof than the qualitative justification supplied.

The cumulative result of these substitutions is to replace $\overline{(\delta\alpha)^2}$ in Equation (19) by a number of other factors, all of which are experimentally measurable:

1. The refractive index gradient (dn/dc). This is simply the local slope of a plot of the refractive index of a solution versus its concentration.
2. The concentration c of the solution. This is expressed in units as grams per volume.
3. The quantity $(\partial\pi_{osm}/\partial c)_{0,T}$ is evaluated for an equilibrium solution. This is the significance of the subscript 0. The subscript T denotes that the derivative is taken at isothermal conditions.

In view of these substitutions, Equation (19) becomes

$$\frac{i_s}{I_0} = \frac{2\,\pi^2\,[n\,(dn/dc)]^2\,k_B T\,c}{r^2\lambda^4\,(\partial\pi_{osm}/\partial c)_{0,T}}\,(1\,+\,\cos^2\phi_x) \tag{20}$$

In Chapter 3 we developed expressions for the equilibrium osmotic pressure of a solution as a function of its concentration. Equation (3.34) may be written

$$\pi_{osm} = RT\left(\frac{c}{M}\,+\,B\,c^2\right) \tag{21}$$

Since Equation (21) applies at equilibrium, we may evaluate $(\partial\pi_{osm}/\partial c)_{0,T}$ from Equation (21):

$$\left(\frac{\partial\pi_{osm}}{\partial c}\right)_{0,T} = RT\left(\frac{1}{M}\,+\,2\,B\,c\right) \tag{22}$$

Combining Equations (20) and (22) yields

$$\frac{i_s}{I_0} = \frac{2\,\pi^2\,[n\,(dn/dc)]^2\,c}{N_A\,r^2\lambda^4\,(1/M\,+\,2\,B\,c)}\,(1\,+\,\cos^2\phi_x) \tag{23}$$

Before looking at the experimental aspects of light scattering, it is convenient to define several more quantities. First, a quantity known as the Rayleigh ratio R_θ is defined as

$$R_\theta = \frac{i_s\,r^2}{I_0\,(1\,+\,\cos^2\theta)} \tag{24}$$

where θ is the value of ϕ_x measured in the horizontal plane. According to Equation (23), the Rayleigh ratio should be independent of both θ and r. An experimental verification of this is one way of testing the applicability of the Rayleigh theory to the experimental data. Next it is convenient to identify the numerical and optical constants in Equation (23) as follows:

$$K = \frac{2\,\pi^2\,n^2\,(dn/dc)^2}{N_A\,\lambda^4} \tag{25}$$

With these changes in notation, Equation (23) becomes

$$R_\theta = \frac{Kc}{1/M\,+\,2\,B\,c} \tag{26}$$

or

$$\frac{Kc}{R_\theta} = \frac{1}{M}\,+\,2\,B\,c \tag{27}$$

This suggests that a plot of (Kc/R_θ) versus c should be a straight line for which the intercept and slope have the following significance:

$$\text{Intercept} = 1/M \tag{28}$$

$$\text{Slope} = 2B \tag{29}$$

Comparing Equations (28) and (29) with Equations (3.35) and (3.36) reveals that plots of (π_{osm}/RTc) versus c and (Kc/R_θ) versus c have identical intercepts, at least for monodisperse colloids (see Section 5.4d for a discussion of the average obtained for polydisperse systems), and that the slopes differ by a factor of 2, with the light-scattering results having the larger slope.

The Rayleigh ratio as defined by Equation (24) has a precise meaning, yet it is a quantity somewhat difficult to visualize physically. After we have discussed the experimental aspects of light scattering, we shall see that R_θ is directly proportional to the turbidity of the solution when turbidity is the same as the absorbance determined spectrophotometrically.

5.4 EXPERIMENTAL ASPECTS OF LIGHT SCATTERING

With the advances in highly monochromatic laser light sources and powerful desktop computers, light scattering instruments have attained a corresponding level of sophistication (see Zare et al. 1995). In general terms, all light scattering instruments contain the same components: a light source (usually a laser), a spectrometer (containing the optical components for defining the scattering angle volume), a detector (usually a photomultiplier), a signal analyzer such as a spectrum analyzer or a correlator (for dynamic measurements; see Section 5.8), and a computer with software for analysis are standard for dynamic measurements. A schematic representation of a typical arrangement is shown in Figure 5.5.

Commercial instruments vary in features and capabilities; some, for example, are restricted to a single angle (very low θ or $\theta = 45°$ or $90°$), and some have capabilities for simultaneous measurements at a fixed number of different angles. Fiber-optic cables for source and detection paths can also be used for minimizing alignment problems and multiple scattering effects in concentrated dispersions, but these are usually found in specially built research instruments rather than commercial ones. Excellent introductions to most of the hardware and software features are available in specialized books (Chu 1991; Pecora 1985), and we focus only on some general background here. Zare et al. (1995), cited above, is an excellent source of relatively simple experiments meant for the beginner.

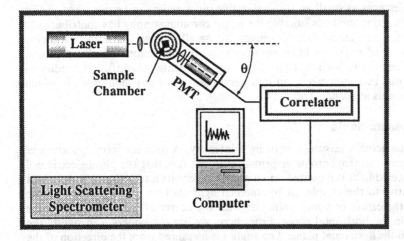

FIG. 5.5 Schematic top view of a typical light scattering instrument showing the different components and the definition of θ.

5.4a   Some Preliminary Considerations

In order to determine M and B by means of Equation (27), it is clear that all the other quantities in the equation must be measured. It is convenient to group these factors into two categories, concentration and optical terms, for the purposes of our discussion.

5.4a.1   Concentration Terms

To begin, concentration enters the light scattering expressions primarily through the equations for osmotic pressure. Hence the same conditions apply in this application as in osmometry (see Chapter 3). Specifically, light scattering should be measured under isothermal conditions. Although concentration units other than weight per unit volume (the units of c) may be used, few of the alternatives are as useful as these. In consulting the literature, however, one should be attentive to the possibility that various workers may use slightly different units for c.

5.4a.2   Optical Terms

All the remaining variables in Equation (27) are optical in origin. The factors to be provided with numerical values are the refractive index of the solution and the refractive index gradient, dn/dc in K from Equation (25), and the Rayleigh ratio from Equation (24). All these optical parameters are wavelength dependent; therefore, each should be measured at the same wavelength. It is the value of this working wavelength that is used as the numerical substitution for λ in K.

The actual measurement of the refractive index of the solution poses no difficulty, but the evaluation of the refractive index gradient is more troublesome. The assumptions of the derivation of Equation (23) restrict its applicability to dilute solutions. The refractive index of a dilute solution changes very gradually with concentration; hence a plot of n versus c, the slope of which equals dn/dc, will be nearly horizontal. Since the intensity ratio depends on the square of dn/dc, it is clear that successful interpretation of Equation (23) depends on the accuracy with which this small quantity is evaluated. Measuring the absolute refractive indices of various solutions and determining dn/dc by difference or graphically would introduce an unacceptable error. A more precise method must be used to measure this quantity.

A differential refractometer is a device that specifically measures differences in refractive indices. By means of a differential refractometer, the difference between the refractive index of a solution and that of a solvent may be measured directly with the necessary precision. A variety of instrument designs is available for doing this. Most involve directing a light beam toward a two-compartment chamber, one portion of which contains the solvent, while the other contains the solution. The light is deviated differently by each, and position-sensitive photodetectors measure the small differences in deflection and translate them into a refractive index difference. Differences as small as 10^{-7} refractive index units can be measured in this way. The search for detection methods suitable for liquid chromatography has contributed to the development of accurate techniques for measuring small changes in refractive index. Note that we have also cited refractive index measurements in connection with size-exclusion chromatography in Chapter 1, Section 1.6b, and with sedimentation/diffusion studies in Chapter 2. A differential refractometer is an indispensable part of any laboratory in which light-scattering experiments are conducted.

5.4b   Intensity Measurements

Now let us consider the actual measurement of light intensity. A light scattering photometer differs from an ordinary spectrophotometer primarily in the fact that the photoelectric cell that measures the scattered light is mounted on an arm that permits it to be located at various angular positions relative to the sample. In commercial light scattering devices, the detector rests on a turntable, the center of which coincides with the center of the sample. Thus the angle ϕ_x is measured in the horizontal plane. From here, we use the symbol θ to signify the angle of observation in the horizontal plane. The angle θ is measured from the direction of the transmitted beam where $\theta = 0°$. The incident beam therefore strikes the sample at $\theta = 180°$.

The cells in which the scattering solutions are measured should have flat windows at the angle at which the scattering is measured. Cells with octagonal cross sections (actually, only half an octagon is used) are especially convenient since they present flat faces at 0, 45, 90, 135, and 180°. Cylindrical cells have been used, but they must be corrected for reflections from the walls. Regardless of the cell geometry, it is imperative that the cells be clean; otherwise the scattering from a fingerprint may exceed that from the solute! The solvent must also be purified of all extraneous matter. Filtration through sintered glass or centrifugation is usually employed to remove any dust particles, which would also invalidate the measurement.

The easiest way to calibrate a light scattering photometer is to use a suitable standard as a reference. Although polymer solutions and dispersions of colloidal silica have been used for this purpose, commercial photometers are equipped with opal glass reference standards.

5.4c Relating Intensities to Absorbance and Turbidity

Except for the movable photomultiplier tube, a light scattering photometer is very nearly identical to an ordinary spectrophotometer, which measures the ratio of the intensity of transmitted light to the intensity of incident light I_t/I_0. The *absorbance* per unit optical path ε_{abs} is defined in terms of this quantity as

$$\varepsilon_{abs} = -\ln (I_t/I_0) \tag{30}$$

(Note that ε here does *not* stand for permittivity.) Now let us examine the relationship between absorbance and the intensity of scattered light. In a light scattering experiment with nonabsorbing materials, the intensity of the transmitted light equals the initial intensity minus the intensity of the light scattered in all directions I_s:

$$I_t = I_0 - I_s \tag{31}$$

Combining Equations (30) and (31) leads to the result

$$\varepsilon_{abs} = -\ln \left(\frac{I_0 - I_s}{I_0}\right) = -\ln\left(1 - \frac{I_s}{I_0}\right) \approx \frac{I_s}{I_0} \tag{32}$$

where the approximation arises from retaining the first term of the series expansion of the logarithm (see Appendix A). The entire development of Section 5.3 is limited to dilute solutions and small n values. Therefore the approximation in Equation (32) is applicable to the systems we have been discussing. When the light attenuation is due to scattering, the ratio (I_s/I_0) is called the *turbidity*, τ, instead of the absorbance.

The quantity I_s in Equation (32) is not the same as the light scattered to a particular point (r,ϕ_x), but equals the summation of these contributions, totaled over all angles:

$$\frac{I_s}{I_0} = \sum_{\substack{all \\ angles}} \frac{i_s}{I_0} \tag{33}$$

This summation may be replaced by an integral as follows. An element of area dA on the surface of a sphere of radius r and making an angle ϕ_x with the horizontal is

$$dA = 2\pi r \sin \phi_x (r d\phi_x) \tag{34}$$

as shown by Figure 5.6. Therefore, the total scattered intensity ratio is given by

$$\frac{I_s}{I_0} = \int_0^\pi \left(\frac{i_s}{I_0}\right) 2\pi r^2 \sin \phi_x \, d\phi_x \tag{35}$$

Substituting Equations (23) and (25) into this expression yields

$$\frac{I_s}{I_0} = \int_0^\pi \frac{K c(1 + \cos^2 \phi_x) 2\pi r^2 \sin \phi_x d\phi_x}{r^2(1/M + 2 B c)} \tag{36}$$

The factor r^2 cancels out of Equation (36) and the value of the integral over ϕ_x is (8/3). Therefore

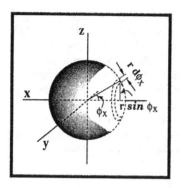

FIG. 5.6 Definition of an element of area required for the summation over all angles of the intensity of scattered light.

$$\frac{I_s}{I_0} = \tau = \frac{16\,\pi\,K\,c}{3\,(1/M + 2\,B\,c)} \qquad (37)$$

The parameter H is defined to equal the cluster of constants:

$$H = \frac{16\,\pi\,K}{3} = \frac{32\,\pi^3\,n^2\,(dn/dc)^2}{3\,N_A\,\lambda^4} \qquad (38)$$

and in terms of this quantity, Equation (37) becomes

$$\frac{H\,c}{\tau} = \frac{1}{M} + 2\,B\,c \qquad (39)$$

The formal similarity between Equations (27) and (39) helps us understand somewhat better the physical significance of the Rayleigh ratio R_θ. It is directly proportional to the attenuation of the light per unit optical path, measured as absorbance, when the attenuation is due to scattering alone. In this case absorbance is more properly called *turbidity*.

Now let us consider some actual results from light scattering experiments on systems that satisfy the assumptions of the theory.

5.4d   Results from Light Scattering Experiments: Weight-Average Molecular Weights

In the preceding section we saw how turbidity measurements are made and how they may be analyzed to yield numerical values for some of the parameters of interest in colloid chemistry. Figure 5.7 shows a plot of Hc/τ versus c for three different fractions of polystyrene in methylethyl ketone. The measurements shown in the figure were made at 25°C and at a wavelength of 436 nm. The molecular weights for the three fractions may be determined from the intercepts of the lines as illustrated in Example 5.2.

* * *

EXAMPLE 5.2   *Determination of Molecular Weights Using Rayleigh Ratios:*   Assuming that the same system of units is used throughout, what are the units for R_θ and τ and also for K and H? Verify that these lead to reciprocal molecular weight units for the combinations Hc/τ and Kc/R_θ. To what molecular weights do the three intercepts in Figure 5.7 correspond? Do the three fractions behave as expected with respect to their second virial coefficients?

Solution: Since refractive index is dimensionless, dn/dc has reciprocal concentration units. In Equation (20) the factors $\{[n(dn/dc)]^2 c/(d\pi/dc)\}$ have the units pressure^{-1} since the units of c's cancel. The product $r^2\lambda^4$ in Equation (20) has the units length^6 or volume^2; therefore, the denominator of Equation (20) has the units pressure volume^2. The product $k_B T$ in the numerator

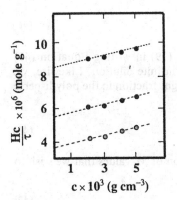

FIG. 5.7 Plots of Hc/τ versus c for three different fractions of polystyrene in methylethyl ketone. (Redrawn from B. A. Brice, M. Halwer, and R. Speiser, *J. Opt. Soc. Am.*, 40, 768 (1950).)

can also be expressed in pressure volume units. Therefore i_s/I_0 has the units volume $^{-1}$ and describes the intensity ratio per unit volume as described.

Equation (24) shows that the units of R_θ are those of $r^2(i_s/I_0)$. The units of the first factor are length^2 and those of the second length $^{-3}$. Therefore R_θ has the units length $^{-1}$. By Equations (35) and (37), the units of τ are the same as R_θ. Each of these parameters measures the light attenuation per unit path length.

By Equation (25), the units of K are concentration $^{-2}$ length $^{-4}$ = length^6 mass $^{-2}$ length $^{-4}$ = length^2 mass $^{-2}$. The fact that this quantity is divided by N_A places (mole $^{-1}$) $^{-1}$ in the units. Equation (38) shows that H has the same units.

In terms of units, $Hc/\tau = Kc/R_\theta$ = (length^2 mass $^{-2}$ mole)(mass length $^{-3}$)/(length $^{-1}$) = mole mass $^{-1}$, reciprocal molecular weight units if mass is expressed in grams, as is the case in practical concentration units.

The three intercepts in Figure 5.7 are $3.70 \cdot 10^{-6}$, $5.56 \cdot 10^{-6}$, and $8.62 \cdot 10^{-6}$ mole g $^{-1}$. The reciprocals of these numbers give the molecular weights directly: 116,000, 180,000, and 270,000 g mole $^{-1}$.

The lines in Figure 5.7 appear to be parallel and hence characterized by a single B value. This would be expected for a single polymer-solvent-temperature system. ■

* * *

At first glance it appears that light scattering experiments as described so far provide no information that is not already available from osmometry. Indeed, for monodisperse colloids this is true, at least for the experiments we discussed above. The apparent redundancy between osmotic pressure and light scattering results should not be interpreted to mean that the two procedures duplicate one another entirely. For one thing, light scattering is free from the limitations imposed on osmometry by the availability of a suitable semipermeable membrane. Furthermore, turbidity measurements do not require time for equilibration, and hence they may be used for systems that change with time in a manner that is not possible with osmometry.

In addition to these practical considerations, there are other ways in which light scattering and osmometry differ. Some of these will become apparent only in subsequent sections in which additional characteristics of light scattering are developed. Another important difference, however, arises in the type of average that is measured for polydisperse systems.

In Chapter 3 we saw that osmometry enables us to measure the number-average molecular weight for a polydisperse colloid. In view of the way the osmotic pressure enters the development of Equations (27) and (39), it appears that the same type of average is obtained from turbidity experiments also. This is not the case. The following argument shows that light scattering measures the weight-average molecular weight.

For a polydisperse system, Equation (39) relates the experimental concentration, the experimental turbidity, and the average molecular weight:

$$\frac{H c_{exp}}{\tau_{exp}} = \frac{1}{\overline{M}} \tag{40}$$

It is sufficient to consider only the leading term of Equation (39) in writing Equation (40) since the molecular weight is evaluated from the intercept at infinite dilution. Likewise, we expect that Equation (39) will also apply to each molecular weight fraction in the polydisperse system

$$\frac{H c_i}{\tau_i} = \frac{1}{M_i} \tag{41}$$

It is the relationship between the value of $\overline{M}$ and the distribution of M_i values that we wish to determine. To accomplish this we note that

$$c_{exp} = \sum_i c_i \tag{42}$$

and

$$\tau_{exp} = \sum_i \tau_i \tag{43}$$

Combining the last four equations gives

$$\overline{M} = \frac{\tau_{exp}}{H c_{exp}} = \frac{\Sigma_i \tau_i}{H \Sigma_i c_i} = \frac{H \Sigma_i c_i M_i}{H \Sigma_i c_i} = \frac{\Sigma_i c_i M_i}{\Sigma_i c_i} \tag{44}$$

Now, recalling that $c_i = n_i M_i / V$ enables us to write

$$\overline{M} = \frac{\Sigma_i n_i M_i^2}{\Sigma_i n_i M_i} = \overline{M}_w \tag{45}$$

Equation (45) corresponds to the weight-average molecular weight $\overline{M}_w$ as defined by Equation (1.16).

It will be recalled from Chapter 1 that the number of particles in a molecular weight class provides the weighting factor used to compute the number-average molecular weight. The weight of particles in a class gives the weighting factor for the weight-average molecular weight. For this reason the weight average is especially influenced by the larger particles in a distribution. Therefore the weight-average molecular weight is always larger than the number average for a polydisperse system. As we saw in Chapter 1, the ratio of the two different molecular weights is a useful measure of the polydispersity of a sample. The authors of the research shown in Figure 5.7 also measured $\overline{M}_n$ for the same samples; the average value of $\overline{M}_w/\overline{M}_n$ for the three samples was 1.16. From polymerization theory this ratio is expected to be closer to 2 for polystyrene as synthesized, suggesting that these samples had been fractionated prior to molecular weight determination.

Thus we see that the redundancy between osmometry and light scattering is only an apparent effect for polydisperse systems. In fact, the combination of the two analyses provides additional information about the characteristics of the system.

Equation (39) shows that the slope of a light scattering plot is twice the value of the slope of a comparable plot from osmometry. In addition to the factor of 2, there is a more subtle difference between the slopes arising from a difference in the two values of the second virial coefficient. Examination of Equation (3.36) reveals that the second virial coefficient is inversely proportional to the square of the molecular weight of the solute. For polydisperse systems it is the average molecular weight that appears in this expression. Since the weight-average molecular weight is larger than the number average, the second virial coefficient B will be somewhat smaller as determined by light scattering than by osmometry after the factor of 2 has been taken into account.

Aside from the difference just noted, the interpretation of the second virial coefficient in light scattering is exactly the same as that developed in Section 3.4. It should be noted, however, that Equation (39) does not apply to charged systems. The reason for this lies in the fact that the charge of macroions is also a fluctuating quantity, and this must also be considered in developing a scattering theory for charged particles. The resulting analysis shows that it is a plot of Hc/τ versus $c^{1/2}$ which is linear in this case, with the limiting slope proportional to z^2, the average value of the square of the charge. The slope is also predicted to be negative in this situation.

In concluding this section, it should be emphasized that the turbidity values plotted to interpret light scattering experiments are the solution turbidities, corrected for scattering by the solvent. Also, the entire theoretical development leading to Equation (39) is based on the assumption that the scatterers are isotropic. In this case, unpolarized incident light will produce a scattered beam that is totally polarized at $\theta = 90°$. When anisotropic particles are present, there is a depolarization of the light scattered at 90°. The ratio of the horizontally to vertically polarized scattered light may be determined by inserting a Polaroid filter between the sample and the photomultiplier. From the measured value of this depolarization ratio, a correction factor (called the *Cabannes factor*) may be introduced to allow for anisotropy. In the sample with $M = 116,000$ in Figure 5.7, for example, the ratio of the two different polarizations at 90° has a value of 0.013, for which the Cabannes factor equals 0.98. The turbidity should be multiplied by this factor to correct for the fact that the anisotropy enhances the amount of scattered light.

5.5 EXTENSION TO LARGER PARTICLES AND TO INTRAPARTICLE INTERFERENCE EFFECTS

In the remainder of this chapter we see that a good deal more information about scattering particles may be deduced, at least under some circumstances, from the study of the light scattered by a sample. In developing the Rayleigh theory and applying it to solutions, a definite model was postulated, and certain variables emerged as factors that affect the intensity of the scattered light. Before we extend light scattering theory to more complex systems, it is convenient to review the assumptions of the Rayleigh model:

1. The scattering centers are isotropic, dielectric, and nonabsorbing.
2. The scatterers have a refractive index that is not too large (see Step 8 in Table 5.1).
3. The particles are small in dimension compared to the wavelength of the light.

This last assumption is fundamental to the theory and originates as early as Equation (17), in which it is assumed that the field that drives the oscillating dipole is the same throughout the scatterer. It is generally held that particles must have no dimension larger than about $\lambda/20$ for this assumption to apply.

In general, in all radiation scattering techniques, the magnitude of the wavelength of the radiation relative to a suitable characteristic dimension, say, L_{ch}, appropriate for the particles plays an important role (as do other relevant properties of the system such as the absorptivity, refractive index, etc.). The wavelength of the radiation λ (or a combination of λ, the refractive index n, and the angle θ at which the scattered radiation is measured) may be thought of as the "yardstick" (say, L_{yd}) used in the scattering measurements. What properties of the dispersions or of the particles one can access through the scattering measurements are determined by the ratio (L_{ch}/L_{yd}).

The different theories (or "models") we discuss in this chapter and the different types of properties (e.g., osmotic pressure, particle size and shape, fractal structure of aggregates, etc.) we seek to measure using scattering correspond to appropriately different definitions of L_{ch} and L_{yd} and different ranges of magnitudes of the ratio (L_{ch}/L_{yd}). It is good to keep this in mind as we develop more complicated theories of scattering. As an aid, Table 5.3 presents a preliminary overview of some of the different definitions of L_{ch} and L_{yd} we encounter in subsequent sections of this chapter and their relation to the types of information that can be obtained using scattering techniques.

For example, Equation (39) suggests that light scattering is a technique ideally suited to the study of particles in the colloidal size range since the turbidity increases with the molecular

TABLE 5.3 Examples of the "Yardsticks" L_{yd} and the "Characteristic Lengths" L_{ch} Used in Different Theories of Scattering and the Different Properties Accessed through the Theories

Theory	L_{yd}	L_{ch}	Remarks
Rayleigh	λ	R_s, particle "size"	Applicable for $(R_s/\lambda) < 1/20$; extension of the Rayleigh equation to solutions allows the measurement of osmotic pressure, molecular weight, and turbidity of colloidal or polymer solutions; see Section 5.3
Debye	$\lambda/[4\pi \sin(\theta/2)] = s^{-1}$	Particle size or aggregate size; see Equation (57)	The information obtained depends on the magnitude of the ratio (L_{ch}/L_{yd}). Depending on the value of this ratio, details on particle shape or "fractal" dimension can be obtained; see Section 5.6
Mie	λ	R_s, particle "size"	Efficiencies of scattering and absorption are obtained as functions of $\beta = 2\pi R_s/\lambda$; see Section 5.7

Note: Consult the appropriate sections of the chapter for more comprehensive information on the assumptions, restrictions and uses of the theories.

weight of the particle. However, the assumptions underlying the derivation of Equation (39) impose a limitation. The Rayleigh theory shows that turbidity increases with molecular weight, at least until the particle is large enough to have some dimension exceeding about $\lambda/20$. In terms of the theory presented so far, we have no way of interpreting the light scattered by larger particles. The *Debye theory*, which we examine in the following section, will show us how to overcome this limitation.

The Rayleigh approximation shows that the intensity of scattered light depends on the wavelength of the light, the refractive index of the system (subject to the limitation already cited), the angle of observation, and the concentration of the solution (which is also restricted to dilute solutions). In the Rayleigh theory, the size and shape of the scatterers (M and B) enter the picture through *thermodynamic* rather than optical considerations.

A fully developed theory of light scattering that allows all the variables, including particle size and shape, to take on a full range of values is extremely complex. Because of this complexity, many treatments, such as the Rayleigh theory, are approximations that apply only to a narrow range of values of the parameters. In the Debye approximation, most of the preceding restrictions will continue to apply, except that the limitation on particle size will be relaxed considerably. At the same time, however, the stipulation of low values of the refractive index becomes even more stringent. As we see below, the Debye approximation introduces some additional complexity to the theory of light scattering and trades off some range in refractive index for extra range in particle size. From this, a positive dividend emerges: We shall be able to determine a characteristic linear dimension of the scattering particles without any assumptions about the shape of the particles. In the cases to which it applies, this information is definitely worth the price of a little additional complexity.

We omit most of the mathematical details in developing the additional theory; instead, we emphasize the major concepts of the theory, its range of applicability, and its plausibility in limiting cases.

5.5a The Debye Scattering Theory

5.5a.1 Intraparticle Interference and the Form Factor

As emphasized above, the Rayleigh approximation is restricted to particles with dimensions that are small compared to the wavelength of light. Suppose we now relax this restriction in our model of a scattering particle, allowing the particles to take on dimensions comparable to λ. Under these circumstances different regions of the same particle will behave as scattering centers. Because the distances between these various scattering centers are of the same magnitude as the wavelength of light, there will be interference between the waves of light scattered from different parts of the same particle.

Figure 5.8 shows how this comes about. The electric field of the incident light beam is out of phase when it strikes two different portions of the scatterer, designated i and j. Although the figure shows i and j as polymer segments, they could be a pair of volume elements in any material. In fact, applications of the Debye theory to polymers are the most widely encountered. The light scattered from i and j is characterized by the same field that induces the oscillation at these locations. Therefore the light scattered from the two sites will be out of phase and display interference when observed at a large distance compared to Δx, say, along BB' in Figure 5.8. As shown in the figure, this interference effect may be constructive or destructive, depending on the value of θ. We propose, therefore, that the Rayleigh ratio defined by Equation (24) must be multiplied by a correction factor $P(\theta)$, known as the *form factor* (or the *intraparticle structure factor*), to correct for the interference effects not considered previously. Several things may be anticipated about this factor:

1. As the dimensions of the particle become negligible compared to the wavelength of light ($\Delta x \rightarrow 0$), $P(\theta) \rightarrow 1$—under these conditions, no correction is necessary.

2. The correction factor is a function of the angle of observation —as implied by the notation $P(\theta)$—since it is essentially an interference effect.

3. Multiplying the denominator of both sides of Equation (27) by $P(\theta)$ corrects i_s (a theoretical quantity calculated without interference) for interference:

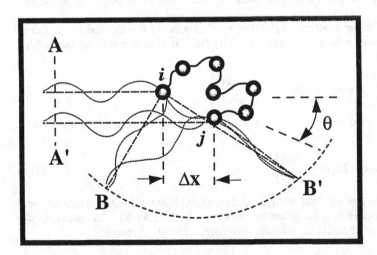

FIG. 5.8 Interference of light rays scattered by segments i and j in a polymer chain. (Redrawn with permission of P. C. Hiemenz, *Polymer Chemistry: The Basic Concepts*, Marcel Dekker, New York, 1984.)

$$\frac{Kc}{[i_s P(\theta)r^2]/[I_0(1 + \cos^2\theta)]} = \frac{1}{P(\theta)}\left(\frac{1}{M} + 2Bc\right) \tag{46}$$

4. The actual intensity of scattered light measured at a point C, denoted by i_C, equals the product i_s times $P(\theta)$. Therefore the experimental Rayleigh ratio obeys Equation (27) modified as follows:

$$\frac{Kc}{R_\theta} = \frac{1}{P(\theta)}\left(\frac{1}{M} + 2Bc\right) \tag{47}$$

In view of these considerations, it is apparent that the way to extend the previous theories to larger particles is to evaluate $P(\theta)$. To do this in a general way, two complications must be introduced into Figure 5.8:

1. Two scattering centers in a large particle are not simply displaced from one another in the x direction by an amount Δx—rather, the coordinates of one relative to the other must be described by a radial distance and two angles, for example, θ and ϕ.
2. A large particle does not consist of merely two scattering centers, but may be subdivided into several centers, the number of which increases with the size of the particle.

These two considerations must be incorporated into any general expression for $P(\theta)$.

5.5a.2 Expressions for the Form Factor

We begin our summary of the derivation of $P(\theta)$, however, by considering only the pair of scattering centers shown in Figure 5.8. To do this quantitatively, imagine that region i is at the origin of a coordinate system so the light that reaches j has to travel an additional distance Δx. The field at i and j is now represented by the following formulations of Equation (7):

$$\mathbf{E}_i = \mathbf{E}_0 \cos(2\pi\nu t) \tag{48}$$

and

$$\mathbf{E}_j = \mathbf{E}_0 \cos\left[2\pi\left(\nu t - \frac{\Delta x}{\lambda}\right)\right] \tag{49}$$

The light scattered from each of these sites will be characterized by the same field as the one that induces the oscillation. Therefore the light scattered from i and j will be out of phase by an amount $2\pi \Delta x/\lambda$.

Let us now consider the net scattered light that reaches a point C—located somewhere along BB'—a (large) distance r from the scatterer. The field at C is the sum of the fields emerging from i and j:

$$\mathbf{E}_C = \mathbf{E}_i + \mathbf{E}_j = \mathbf{E}_0\left\{\cos(2\pi\nu t) + \cos\left[2\pi\left(\nu t - \frac{\Delta x}{\lambda}\right)\right]\right\} \tag{50}$$

If we use the appropriate trigonometric formula for the sum of two cosines, this may be written as

$$\mathbf{E}_C = \left[2\cos\left(\frac{\pi \Delta x}{\lambda}\right)\right]\mathbf{E}_0 \cos\left[2\pi\left(\nu t - \frac{\Delta x}{2\lambda}\right)\right] \tag{51}$$

This equation shows that the electric field of light scattered to C is not altered in frequency or wavelength, but that the amplitude is modified by the factor $2\cos(\pi\Delta x/\lambda)$. The intensity of light depends on the square of the field amplitude; therefore, with interference,

$$i_C \propto \left[2\cos\left(\frac{\pi \Delta x}{\lambda}\right)\right]^2 |\mathbf{E}_0|^2 \tag{52}$$

and without interference ($\Delta x \to 0$),

$$i_s \propto 2^2 \, |\mathbf{E}_0|^2 \tag{53}$$

Combining Equations (52) and (53) leads to

$$\frac{i_c}{i_s} = \cos^2\left(\frac{\pi \, \Delta x}{\lambda}\right) \tag{54}$$

Since $i_c = i_s P(\theta)$, the factor $\cos^2(\pi\Delta x/\lambda)$ must equal $P(\theta)$ for this simple case.

A fair amount of straightforward but tedious trigonometry is necessary to establish the relationship between Δx and r, ϕ_x, and θ for the general case of any orientation between i and j. The result of this analysis is the expression (given without proof)

$$\Delta x = 2r \cos\phi_x \sin(\theta/2) \tag{55}$$

where ϕ_x is defined the same as in Figure 5.6 and θ continues to be the scattering angle measured in the horizontal plane.

It is not any specific value of ϕ_x in which we are interested, but in all possible values. This means that Equation (55) is to be integrated over all values of ϕ_x, a procedure that leads to the result (given without proof)

$$\frac{i_c}{i_s} = 1 + \frac{\sin[(4\pi r/\lambda)\,\sin(\theta/2)]}{(4\pi r/\lambda)\,\sin(\theta/2)} \tag{56}$$

Equation (56) benefits considerably from some simplification in notation. Accordingly, we define

$$s = \frac{4\pi}{\lambda} \sin(\theta/2) \tag{57}$$

This quantity has units of "1/length" and is the magnitude of a vector known as the *scattering vector*. For our purpose here, it is sufficient to note that s^{-1} is the "yardstick" used for measuring the distance between the scattering centers when the intensity of the scattered radiation of wavelength λ is recorded at an angle θ from the direction of the incident radiation.

Using the above notation permits us to rewrite Equation (56) as

$$\frac{i_c}{i_s} = 1 + \frac{\sin(sr)}{sr} \tag{58}$$

The right-hand side of Equation (58) correctly defines $P(\theta)$ for interference between two regions of a large particle.

The next question we must consider is how this result may be applied to a particle that consists of N, not just two, scattering centers. If the composition of the particle is such that the scattering from one part of the particle has no effect (other than interference) on the scattering from another part, then the particle may be subdivided into N scattering elements, and Equation (58) can be applied to all possible pairs. If we define r_{ij} as the distance between the ith and jth members of the set of N scattering elements, then this argument leads to the result (given without proof)

$$\frac{i_c}{i_s} \propto \sum_i \sum_j \left(\frac{\sin(sr_{ij})}{sr_{ij}}\right) \tag{59}$$

Expression (59) is written as a proportionality rather than an equation because the summation requires that a normalization factor be introduced. The ratio i_c/i_s must equal unity when r_{ij} is small since it explicitly corrects for interference effects that vanish under these circumstances. For small values of sr_{ij}, $\sin(sr_{ij})/sr_{ij}$ equals unity and $\Sigma_i\Sigma_j \sin(sr_{ij})/sr_{ij}$ equals N^2. Therefore normalization requires that we write

$$\frac{i_c}{i_s} = \frac{1}{N^2} \sum_i \sum_j \frac{\sin(sr_{ij})}{sr_{ij}} = P(\theta) \tag{60}$$

Note that the first term of Equation (58) is implicitly present in Equation (60) when r_{ij} equals zero. Equation (60) provides the general expression for $P(\theta)$ that we have sought. It also shows that $P(\theta)$ is more generally a function of s since one can vary s without changing θ by using a radiation source of a different wavelength λ. We return to this point in Section 5.6.

It is possible that the reader will recognize Equation (60) from another context. It is exactly the same expression that describes the diffraction of x-rays by polyatomic molecules, the situation for which it was derived by Debye (Nobel Prize, 1936). An important insight emerges from this realization. All interference phenomena between electromagnetic radiation and matter follow the same mathematical laws. Interference becomes important when there is some characteristic distance L_{ch} in the material under consideration that is of the same magnitude as the wavelength of the available radiation. The interference phenomena will be identical for identical values of L_{ch}/λ regardless of the separate values of L_{ch} and λ. Thus x-ray diffraction is an experimental technique that uses x-rays (for which $\lambda \approx 0.1$ nm) to measure interatomic distances that are on the order of nanometers. Likewise, we may use visible light (for which $\lambda \approx 500$ nm) to measure particles with dimensions in the colloidal size range. The scattering of microwaves by atmospheric rain and snow particles is an example of still another situation in which interference phenomena may be observed by scaling the wavelength of the radiation to suit the particle sizes under investigation. A brief comparison of x-ray, neutron, and light scattering is made in Section 5.6c.

Before turning to the applications of the Debye approximation, we should elaborate more fully on a point that was glossed over. This is the assumption—made at the outset, but explicated in going from Equation (58) to Equation (59)—that the scattering behavior of each scattering element is independent of what happens elsewhere in the particle. The approximation that the phase difference between scattered waves depends only on their location in the particle and is independent of any material property of the particle is valid as long as

$$\frac{2\pi L_{ch}}{\lambda}(n - 1) \ll 1 \tag{61}$$

It is the condition expressed by the inequality (61) that requires $(n - 1)$ to become smaller and smaller as the theory is applied to progressively larger particles. For example, when $\theta = 10°$, the Debye approximation is good to within 10% for spheres with a radius R_s that is about 62, 37, and 25 times λ at $n = 1.1$, 1.2, and 1.3, respectively. As we shall see presently, the approximation applies to a somewhat wider range of R_s and n values at smaller angles of observation and to a narrower range at larger angles.

5.5b Zimm Plots

Although Equation (60) may correct Equation (47) for the turbidity of systems in which the scattering centers are not negligible in size compared to the wavelength of light, it is not a very promising looking result. A little more mathematical manipulation will correct this impression. Specifically, suppose we examine Equation (57) for the case in which $\theta/2$ is small. This will make s small regardless of the value of r_{ij} so that $\sin (sr_{ij})$ in Equation (60) may be expressed as a power series (see Appendix A):

$$P(\theta) = \frac{1}{N^2} \sum_i \sum_j \frac{\sin (sr_{ij})}{sr_{ij}} = \frac{1}{N^2} \sum_i \sum_j \frac{sr_{ij} - (sr_{ij})^3/3! + \cdots}{sr_{ij}} \tag{62}$$

For small values of s (more accurately, for small values of sr_{ij}; we come back to this later), the expansion may be limited to the first two terms:

$$P(\theta) = \frac{1}{N^2} \sum_i \sum_j 1 - \frac{(sr_{ij})^2}{6} = 1 - \frac{s^2}{6N^2} \sum_i \sum_j r_{ij}^2 \tag{63}$$

It is actually $1/P(\theta)$ in which we are interested, so we may again take advantage of the fact that s is small to write Equation (63) as

$$\frac{1}{P(\theta)} \approx 1 + \frac{s^2}{6N^2} \sum_i \sum_j r_{ij}^2 \tag{64}$$

when the second term on the right-hand side is much smaller than unity. In spite of various simplifying approximations, Equation (64) still does not appear particularly useful; the problem is the double summation of r_{ij} terms. We have encountered summations like this before in discussing the radius of gyration in Chapter 2, Section 2.7. A little manipulation of the results of that section will show us how to replace the summations in Equation (64):

1. Equation (2.76) gives an expression for $(R_g^2)_j$, the square of the radius of gyration of a swarm of masses around the jth, assumed to be the center of mass. (Note, also, the difference in notations; in particular, we have used n instead of N in Equation (2.76) for the number of segments.)
2. The two summations in Equation (2.76) can be consolidated into one—spanning all n of the mass elements—for the purposes of this discussion. Also, the product $P(r)r^2$ in Equation (2.76) is equivalent to a specific value of r_{ij}^2.
3. In going from Equation (2.76) to Equation (2.77), we added together (i.e., integrated) n terms like those in Equation (2.76) to allow for the fact that any of the n mass elements may play the role of the jth term in the derivation. Doing this in the case of the double sum in Equation (64) counts all mass elements twice, so a factor of 1/2 must be inserted.

We summarize these observations by writing

$$\overline{R_g^2} = \frac{1}{2N^2} \sum_i \sum_j r_{ij}^2 \tag{65}$$

which may be substituted into Equation (64) to give

$$\frac{1}{P(\theta)} = 1 + \frac{1}{3} \overline{R_g^2} s^2 \tag{66}$$

Equation (66) is valid in the limit of small values of $\theta/2$. Of course, if the scattering particle is not too large, the expansions of Equations (63) and (64) will be valid at larger θ, assisted by the fact that the values of r_{ij} will be small. This explains why the range of parameters to which the Debye equation applies depends on the angle of observation, becoming narrower for larger angles and broader at small angles.

Substitution of Equation (66) into Equation (47) yields

$$\frac{Kc}{R_\theta} = \left(\frac{1}{M} + 2Bc\right)\left[1 + \frac{16\pi^2 \overline{R_g^2}}{3\lambda^2} \sin^2\left(\frac{\theta}{2}\right)\right] \tag{67}$$

Let us consider Equation (67) in three important limiting cases with the objective of developing a graphical technique for using Equation (67):

1. In the limit of $\theta = 0$, Equation (67) reduces to Equation (27); that is, there is no interference effect in the scattered light.
2. In the limit of $c \to 0$, Kc/R_θ is proportional to $\sin^2(\theta/2)$.
3. If both c and $\theta \to 0$, Kc/R_θ equals $1/M$.

These limits suggest how experimental data might be collected, plotted, extrapolated, and interpreted. The resulting graph is known as the *Zimm plot* after its originator.

Equation (67) shows clearly that i should be measured as a function of both concentration *and* angle of observation in order to take full advantage of the Debye theory. The light scattering photometer described in Section 5.4 is designed with this capability, so this requirement introduces no new experimental difficulties. The data collected then consist of an array of i/I_0 values (i needs no subscript since it now applies to small and large particles) measured over a range of c and θ values. By means of Equation (24), the i/I_0 ratios are converted to R_θ values. The results are plotted with Kc/R_θ as the ordinate and $\sin^2(\theta/2) + c$ as the abscissa.

Figure 5.9 shows this sort of plot for light scattering data collected from solutions of cellulose nitrate in acetone at 25°C. The measurements were made using 436-nm light of mercury. In the figure, [$\sin^2(\theta/2) + kc$] has been used for the abscissa. The numerical constant k—equal to 2000 in this case—spreads out the points and results in a more intelligible display. As with any scale factor in graphing, it is found by trial and error. Each of the points in Figure 5.9 corresponds to a particular pair of c,θ values. When the points measured at the same values of c and those measured at the same values of θ are connected, a grid of lines is obtained like that sketched in the figure. If all the experimental c and θ values are small enough for the theories to hold exactly, then the grid consists of two sets of parallel straight lines with different slopes. In general, however, the range of experimental c and θ values exceeds the range of validity of the theory, and the lines show some curvature.

The next step in the treatment of the data is the extrapolation of the curves drawn at constant θ to $c = 0$ and the extrapolation of those drawn at constant c to $\theta = 0$. This is done by placing a mark (the triangles in Figure 5.9) on the smooth lines drawn through the experimental points at the value of the abscissa that corresponds to the value of that coordinate at the desired limit. For example, when the limit for $c = 0$ of the $\theta = \theta_1$ line is located, it will lie on the θ_1 line, and the value of the abscissa will be $\sin^2(\theta_1/2)$. Likewise, when the limit for $\theta = 0$ of the $c = c_1$ line is located, it will lie on the c_1 line, and the value of the abscissa will be kc_1. The triangles in Figure 5.9 were positioned on this basis. Note that the triangles describe two straight lines that have a common intercept.

Let us now consider the interpretation of this graph. First, it should be remembered that interference effects vanish at $\theta = 0$. The line so labeled in Figure 5.9 corresponds to values of Kc/R_θ at different values of c, all expressed at $\theta = 0$, where Equation (27) is valid. According to that equation, the slope of this line equals $2B$, and the intercept equals $1/M$. This is the same result we obtained previously for small particles. By extrapolating to $\theta = 0$, the procedure now applies to larger particles as well. In formulas, then, we write for the $\theta = 0$ line

$$(\text{Slope})_{\theta=0} = 2B \tag{68}$$

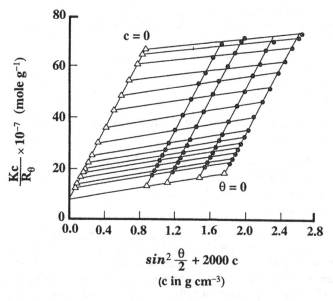

FIG. 5.9 Experimental Zimm plot for cellulose nitrate in acetone. (Redrawn from H. Benoit, A. M. Holtzer, and P. Doty, *J. Phys. Chem.*, **58**, 635 (1954).)

and

$$(\text{Intercept})_{\theta=0} = 1/M \tag{69}$$

The new feature of the Zimm plot is the second extrapolated line, corresponding to $c = 0$. This line connects values of Kc/R_θ measured at different values of θ and extrapolated to $c = 0$. Accordingly, it is described by Equation (67) with $c = 0$; that is, theory predicts the $c = 0$ line to have an intercept of $1/M$ and a slope equal to $(16\pi^2\overline{R_g^2}/3\lambda^2 M)$. It is expected, then, that the two lines will extrapolate to a common intercept $1/M$, and that the slope of the $c = 0$ line will be proportional to $\overline{R_g^2}$. Summarizing in formulas, we write for the $c = 0$ line

$$(\text{Slope})_{c=0} = \frac{16\,\pi^2\,\overline{R_g^2}}{3\,\lambda^2\,M} \tag{70}$$

and

$$(\text{Intercept})_{c=0} = 1/M \tag{71}$$

It follows, therefore, that the square of the radius of gyration equals

$$\overline{R_g^2} = \frac{3\,\lambda^2}{16\,\pi^2}\left(\frac{\text{Slope}}{\text{Intercept}}\right)_{c=0} \tag{72}$$

The Zimm plot shown in Figure 5.9 is analyzed according to these relationships in Example 5.3.

* * *

EXAMPLE 5.3 *The Use of Zimm Plots: Determination of Molecular Weight, Second Virial Coefficient, and Radius of Gyration of Cellulose Nitrate in Acetone.* The following are the values of the intercept and slope from Figure 5.9:

Intercept = $7.87 \cdot 10^{-7}$ mole g^{-1}
Slope of the $\theta = 0$ line = $5.70 \cdot 10^{-7}$ mole g^{-1}
Slope of the $c = 0$ line = $6.78 \cdot 10^{-6}$ mole g^{-1}

Evaluate M, B, and R_g for cellulose nitrate in acetone at 25°C from these data. Use 1.359 for the refractive index of acetone at this wavelength.

Solution: The molecular weight is the reciprocal of the common intercept and equals $1.27 \cdot 10^6$ g mole^{-1}.

The slope of the $\theta = 0$ line includes the factor k, which must have the units concentration^{-1} for dimensional consistency. Hence the slope given must be multiplied by 2000 cm^3 g^{-1} to give the true slope to which the analysis applies. Since the true slope equals $2B$ according to Equation (68), we obtain

$B = (1/2)(2000\ \text{cm}^3\,\text{g}^{-1})(5.70 \cdot 10^{-7}\ \text{mole g}^{-1}) = 5.70 \cdot 10^{-4}\ \text{cm}^3\ \text{mole g}^{-2}$

The slope to intercept ratio for the $c = 0$ line equals $(6.78 \cdot 10^{-6})/(7.87 \cdot 10^{-7}) = 8.61$; this must be multiplied by $3\lambda^2/16\pi^2$ to obtain $\overline{R_g^2}$. Remember to convert $\lambda_0 = 436$ nm to the wavelength in acetone, which is the value used in this calculation: $436/1.359 = 321$ nm. Therefore $\overline{R_g^2} = 3(321)^2(8.61)/16\pi^2 = 1.69 \cdot 10^4$ nm^2 and the root-mean-square value is $R_g = 130$ nm. ∎

* * *

It is important to remember that the radius of gyration provides an unambiguous measure of a particle's extension in space. This quantity is evaluated from experimental data—as illustrated in Example 5.3—with no assumptions as to particle geometry. If we happen to know the shape of a particle as well, R_g can be translated into a geometrical dimension of the particle. We examine this further in the next section.

5.5c The Dissymmetry Ratio

If the intensity of light scattered by a colloidal dispersion is measured as a function of c and θ, the Zimm method enables us to convert this information into several parameters that characterize the colloid: M, B, and R_g. In some situations this is more information than is actually needed. If spatial extension is the only information sought, a simpler method for evaluating it employs the so-called dissymmetry ratio.

In this technique, one measures i/I_0 at $\theta = 45°$ and $\theta = 135°$ for dispersions at several different concentrations. It should be noted that the factor $(1 + \cos^2 \theta)$ in Equation (24) has the same value for these two angles of observation. Therefore any deviation of the ratio of the intensities z from unity must measure the ratio of the $P(\theta)$ values at these two angles [See Equation (47)]:

$$z = \frac{i_{45°}}{i_{135°}} = \frac{P(45°)}{P(135°)} \tag{73}$$

For example, we may examine the approximate form for $P(\theta)$ provided by Equation (66) to obtain

$$z = \frac{1 + (16\,\pi^2/3)\,(R_g/\lambda)^2 \sin 67.5°}{1 + (16\,\pi^2/3)\,(R_g/\lambda)^2 \sin 22.5°} \tag{74}$$

With this result—or its equivalent made with less restrictive approximations—values of z can be evaluated as a function of R_g/λ. From tables or plots of such calculated results, experimental dissymmetry ratios—extrapolated to $c = 0$ to eliminate the effects of solution nonideality—can be directly interpreted in terms of R_g.

We have noted previously that R_g is related to the geometrical dimensions of a body through expressions that are specific for the particle shape. Table 5.4 lists some of these relationships for shapes pertinent to colloidal systems. For a selected shape, one of the tabulated relationships can be used to replace R_g in Equation (74). What results is an expression that interprets z in terms of actual particle dimensions for the geometry chosen.

Figure 5.10 shows a plot of curves such as this drawn for spheres, coils, and rods. The characteristic dimension L_{ch} used in the abscissa of the figure is R_s for a sphere, r_{rms} for a coil, and L for a rod; the pertinent wavelength is λ_0/n. It will be noted from the ordinate values in Figure 5.10 that the scattering is always larger in the forward direction for particles showing interference effects. This is one reason why the presence of dust particles raises such havoc in a scattering experiment on particles that lie in the Rayleigh region.

TABLE 5.4 Relationships Between the Radius of Gyration and the Geometrical Dimensions of Some Bodies Having Shapes Pertinent to Colloid Chemistry

Geometry	Definition of parameters used as L_{ch}	Radius of gyration through the center of gravity
Random coil	$\overline{r^2}$: mean square end-to-end distance ($\propto n$)	$\overline{R_g^2} = \dfrac{\overline{r^2}}{6}$
Sphere	R_s: radius of sphere	$\overline{R_g^2} = \dfrac{3\,R_s^2}{5}$
Thin rod	L: length of rod (approximation for prolate ellipsoids for which $a/b \gg 1$)	$\overline{R_g^2} = \dfrac{L^2}{12}$
Cylindrical disk	R_d: radius of disk (approximation for oblate ellipsoids for which $a/b \ll 1$)	$\overline{R_g^2} = \dfrac{R_d^2}{2}$

Source: P. C. Hiemenz, *Polymer Chemistry: The Basic Concepts*, Marcel Dekker, New York, 1984.

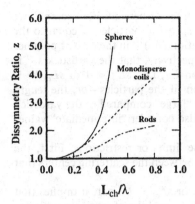

FIG. 5.10 Values of the dissymmetry ratio z versus the size parameter L_{ch}/λ for spheres, random coils, and rods. (Data from Stacey 1956.)

The development leading to Equation (66) in this section related the form factor $P(\theta)$ to the radius of gyration R_g, which is *one* measure of the "structure" of a particle. We can actually get much more information from the form factor. In the following section, we discuss this and illustrate the use of $P(\theta)$ for measuring the fractal dimension (defined in Chapter 1, Section 1.5b.1) of an aggregate.

5.6 INTERFERENCE EFFECTS AND STRUCTURE OF PARTICLES

The facts that we have explicitly included the intraparticle interference function $P(\theta)$ in the analysis of scattering intensities and that it is accessible experimentally allow us to characterize colloidal dispersions structurally in more detail than we have been able to so far. In order to understand this, we need to understand clearly what we mean by "small" or "large" values of θ or s and how they affect the behavior of $P(\theta)$. This will also help us to understand how (and why) it is possible to combine light scattering with x-ray or neutron scattering to study structures of particles and their aggregates.

5.6a The Form Factor and Particle Structure

The function $P(\theta)$ contains information on the *internal* structural details of the scatterers; this is the reason for its name "*form* factor" or "*intra*particle" structure factor, as defined in Section 5.5a.1. The length scales of the structural details contained in $P(\theta)$ depends roughly on the range of s^{-1} over which $P(\theta)$ is measured since, as we emphasized above, s^{-1} is the "yardstick" we use in the measurements. Notice that s^{-1} has units of length. We have so far alluded to the range of experimental measurements in terms of the magnitude of θ or s (i.e., as θ is "small" or "large" or as s is "small" or "large"). Before we discuss how $P(\theta)$ is used to obtain information about the geometrical structure of particles and in order to appreciate the physical significance of what is meant by "small" or "large," we need to make these notions more quantitative. For this, it is convenient to work with s rather than θ. The use of s is more general, as we see at the end of Section 5.6a.1.

5.6a.1 Physical Significance of the Magnitude of the Scattering Vector s

Since s^{-1} is a *dimensional* quantity, when we characterize it as "small" or "large" we implicitly do so relative to a suitable reference dimension L_{ch} (which is the characteristic dimension that represents the size of the scatterers). Otherwise, we can choose the units of s^{-1} to be angstrom or kilometer (or something even larger) and change the numerical value of s to something arbitrarily large or small! Let us define a new, dimensionless quantity Q by

$$Q = L_{ch}/s^{-1} = sL_{ch} \tag{75}$$

where L_{ch} is a characteristic length suitable for the particles in the dispersion. For example, we may choose L_{ch} to be equal to the radius R_s in the case of spherical particles or equal to the radius of gyration R_g or the root mean square radius of gyration $(\overline{R_g^2})^{1/2}$ in the case of polymer coils. With this definition, "small" or "large" s implies, respectively, that the yardstick s^{-1} is "large" or "small" compared to the characteristic dimension L_{ch}, i.e., $sL_{ch} \ll 1$ or $sL_{ch} \gg 1$. (Equivalently, we can say that the characteristic dimension of the particles — or, the length scale of the structural details of the particles — is "small" or "large" compared to the yardstick s^{-1}.) In addition to "small" or "large" values of s, we can also consider "intermediate" values of s, which are represented by $Q = sL_{ch} \sim 1$.

Let us now examine the physical significance of these limits or restrictions. First, the magnitude of s^{-1} is a measure of the distance of separation between the scattering centers that contribute to the intensity observed at that value of s.

1.   When s^{-1} is large compared to L_{ch}, i.e., $s^{-1} \gg L_{ch}$ or $sL_{ch} = Q \ll 1$, it implies that the scattering centers are separated by a distance larger than L_{ch} and therefore lie on different particles. As a result, there is no intraparticle interference in the intensity at that value of s, i.e., $P(\theta)$, or more accurately, $P(Q)$, tends to unity for $Q \ll 1$. We already noted this in Section 5.5a.1, in which we compared the wavelength of light λ (which is proportional to s^{-1}; see Equation (57)) directly to particle dimension ("L_{ch}") and noted that $P(\theta) \rightarrow 1$ for $\lambda \gg L_{ch}$.

2.   In the opposite limit, when s^{-1} is "small" or, equivalently, Q is large (typically, $Q \sim 1$ or larger), the measured $P(Q)$ consists of interference between closely lying scattering centers *within the same particle* (since the "yardstick" is small) and, therefore, contains information on the shape and intraparticle structure. We have more to say about this in Section 5.6b.

In general, the information contained in $P(Q)$ reflects the existence (or the lack) of interference effects *in the range of Q considered*. When the intensity measurements are obtained over a range of s^{-1}, we are, in effect, using a range of yardsticks to obtain the structural details of the particles over a number of length scales. However, comments (1) and (2) above highlight the fact that it is Q, the ratio of the reference dimension L_{ch} to the yardstick s^{-1}, that determines how we should interpret the information contained in $P(Q)$ (or, equivalently, in the intensity measured). This is, in fact, the rationale for defining the ratio L_{ch}/s^{-1} in Equation (75), which combines the two independent linear dimensions L_{ch} and s^{-1} into one variable.

The above discussion also draws our attention to another important way of looking at the measurements. Sometimes, we have more than one option for the characteristic dimension L_{ch}. One such example is a dispersion of colloidal aggregates (of diameter, say, d_{agg}) of smaller, primary particles (of diameter d_p). We can pick either d_{agg} or d_p as the characteristic dimension L_{ch}. Then, whether we call the $P(Q)$ measurements *inter*particle structure or *intra*particle structure depends on the choice of L_{ch}, i.e., our definition of "particle." For example, if we choose L_{ch} to be the diameter of the aggregate d_{agg}, then, for $Q (= d_{agg}/s^{-1})$ of the order of 1 or larger, $P(Q)$ describes the structure of the *aggregates*; i.e., the *intra*particle structure. We illustrate this in Example 5.4. If, on the other hand, we choose the diameter of the *primary* particle d_p as L_{ch}, the same range of s^{-1} corresponds to $Q \ll 1$, and $P(Q)$ is more appropriately referred to as *inter*particle structure (since our definition of "particle" now is the *primary* particle). In both cases, the information we have is the same, but our reference or perspective has changed. We return to this point in Section 5.6b.

As mentioned above, s (or s^{-1} or its dimensionless analog Q) is a more appropriate variable than θ to use in discussing the intraparticle (and interparticle) interference effects since θ is not the only variable that determines the range over which $P(Q)$ is measured. One can change the range by varying λ also (in fact, over a much larger range than possible by changing θ alone). This is the reason why sometimes other forms of radiation such as x-ray or neutron are used. We come back to this point in Section 5.6c.

5.6a.2   Different Regions of Interference Effects

Analytical expressions for $P(Q)$ can be derived for simple geometries of interest in colloid science (see, for example, Schmitz 1990), but it is sufficient for us to consider only the qualitative behavior of $P(Q)$ here. Such a picture of variation of $P(Q)$ with Q for aggregates of (primary) spherical particles of radius R_s is illustrated in Figure 5.11. One can identify four

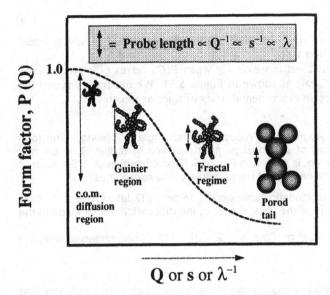

FIG. 5.11 Schematic representation of $P(Q)$ versus Q for fractal objects. The different parts of the curve corresponding to (a) the center-of-mass region, (b) the Guinier region, (c) the fractal region, and (d) the Porod region are indicated. The "probe length" is the "yardstick" corresponding to the measurement. (Adapted from Schmitz 1990.)

regions in this figure depending on the magnitude of Q, i.e., depending on the relative magnitudes of s^{-1} (suggested by the lengths of the arrows in the figure) and L_{ch}, which is taken to be the dimension of the aggregate (sketched in the figure):

1. As discussed above, in the limit of very small Q ($Q \ll 1$), $P(Q)$ approaches unity (see also Equation (63)), and there is no intraparticle interference since the "probe length" s^{-1} is much larger than R_s and the dimensions of the aggregate. This region is often called the "center-of-mass" region since the scattering centers act roughly as "centers of masses" that are uncorrelated with each other (i.e., $P(Q) \rightarrow 1$).

2. The next region, for Q's larger than the above limit but still small, signifies the onset of intraparticle interference. As we have already seen from Equations (63) and (65), the function $P(Q)$ here is of the form

$$P(Q) \approx 1 - \frac{1}{3}Q^2 \tag{76}$$

This region is known as the *Guinier region*.

3. Let us skip the next region temporarily and consider the range $Q \gg 1$. As mentioned above, intraparticle interference within the *primary* particles determines the function $P(Q)$ in this case, and the functional form of $P(Q)$ is determined by the shape of the primary particle (assumed here to be spherical). For spheres, one can show that

$$P(Q) \sim Q^{-4} \tag{77}$$

in this last region, which is known as the *Porod region*.

4. The third region is particularly important for dispersions of aggregates. As illustrated in the figure, Q in this case corresponds to the probe length s^{-1} larger than R_s, the radius of the primary particle, but comparable to the dimensions of the aggregates. The functional form of $P(Q)$ in this range depends on the detailed structure of the *aggregates* (i.e., distribution of the primary particles within the aggregates). If we think of the aggregate as the basic unit, $P(Q)$ describes the *intraaggregate* structure. For *fractal* aggregates (see Chapter 1, Section 1.5b), $P(Q)$ varies as

$$P(Q) \sim Q^{-d_f} \tag{78}$$

where d_f is known as the *fractal dimension* of the aggregate and is a measure of how compact the aggregate is, as discussed in Section 1.5b.

The above regions of $P(Q)$ stand out more clearly when $P(Q)$ versus Q is plotted on a *log-log* scale instead of on a linear scale as shown in Figure 5.11. We illustrate this and the above concepts in Example 5.4 using an experimental study of silica aggregates.

*   *   *

EXAMPLE 5.4   *Structure of Silica Aggregates Probed by Small-Angle Scattering.*   The form factor $P(s)$ for a silica sol consisting of spherical primary particles of radius $R_s = 2.7$ nm, obtained from light and x-ray scattering, is plotted in Figure 5.12 as a function of s (Schaefer et al. 1984). What can you infer from the data about the structure of the aggregates?

Solution: Instead of choosing a characteristic dimension L_{ch} to define Q, let us look at the data in terms of s directly. An examination of the data in light of the discussion above leads to the following conclusions:

1.   In the range of roughly $1 \cdot 10^{-3}$ nm^{-1} $\leq s \leq 2 \cdot 10^{-1}$ nm^{-1}, one observes from the figure that $P(s)$ is of the form

$$P(s) \sim s^{-2.1}$$

(remember the slope of a straight line in a log-log plot gives the exponent). This indicates that the dispersion consists of aggregates with a fractal dimension equal to 2.1 (a fairly loose aggregate). The lower limit of s in the fractal region is of the order of d_{agg}^{-1} (where, again, d_{agg} is the diameter of the aggregates), and the upper limit is related to the size of the primary particle, as we discuss below using the schematic representation of $i(s)$ versus s presented in Figure 5.13.

2.   For s^{-1} smaller than the radius of the silica particles (i.e., for sR_s of the order of unity or larger), $P(s)$ appears to decrease as s^{-4} with increasing s, thereby confirming that the primary particles are spherical as given.

3.   The data shown provide an example of how two different forms of radiation can be used effectively to cover a broad range of s^{-1}.

Note that the term *particle* in the intraparticle structure discussed here really refers to the aggregates. It is the internal structure of the aggregates that we have examined through the

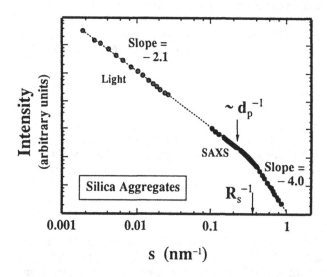

FIG. 5.12   Light scattering and small-angle x-ray scattering (SAXS) data for a dispersion of aggregates. The primary particles in the aggregates are monosize, spherical silica particles. The upper limit of s in the fractal region is roughly 0.2 nm^{-1}. Note log·log axes. (Data from Schaefer et al., 1984.)

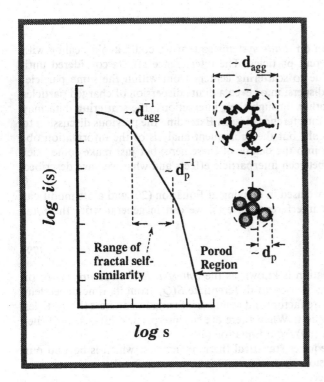

FIG. 5.13 A schematic illustration of the physical significance of the end points of the fractal region (see the text for the details). (Adapted from J. D. F. Ramsay and M. Scanlon, *Colloids and Surfaces*, **18**, 207 (1986).)

fractal region in the scattering curve. The "particles" with structures described by the Porod region are the primary particles, which are spherical silica particles in this case.   ∎

* * *

Comment (2) in the solution to Example 5.4 may appear to be superfluous since the example already states that the primary particles are spherical. However, that point is emphasized here in order to illustrate that scattering data can be used to follow the changes in the shapes of the primary particles if the physical situation may cause such changes. For example, if the primary particles are such that they coalesce and change their sizes or shapes, such changes may be followed using scattering experiments.

Moreover, as already pointed out in item (1) of the solution to Example 5.4, additional information about the aggregates and the primary particles may be obtained from plots such as Figure 5.12. This is illustrated schematically in Figure 5.13. This figure notes that the lower limit of the range of s in the fractal region is related to size of the aggregate; i.e., this limit is roughly d_{agg}^{-1}. The upper limit, on the other hand, is of the order of d_p^{-1}, where d_p is the diameter of the primary particle. As we noted in Chapter 1 (see Section 1.5b.1), fractal objects are *self-similar* objects consisting of similar structural features over a range of length scales. In the sketch shown in Figure 5.13, the aggregates look "similar" when we view them on length scales anywhere from roughly d_p to d_{agg}.

The above observations lead to the conclusion that the size of the aggregates in the dispersion represented by the data in Figure 5.12 is about 1000 nm, as already pointed out in item (1) in the solution to Example 5.4. The upper limit of s in the fractal region in Figure 5.12, i.e., $s = d_p^{-1} \approx 0.2$ nm^{-1}, is consistent with the fact that the primary particles in the example are spheres of diameter 5.5 nm.

5.6b Interparticle Structure

We have so far focused our attention on *dilute* systems so that we could avoid dealing with interference of scattering from different particles. The interference effects considered until now are restricted to interference due to scattering centers from within the same particle. When we have a fairly concentrated dispersion or even a dilute dispersion of charged particles that influence the position of each other through their interactions, the scattering data may have to be corrected for interparticle interference effects. Extending the previous discussion to *inter*particle interference is not difficult, but the subsequent analysis of the information obtained is not trivial. We shall not go into the details of these here, but just make some brief remarks to establish the connection between interparticle effects and what we have described so far for dilute systems.

Following the line of arguments we used in arriving at Equation (20) and at its modification, Equation (60), to account for interference effects, we can in general write the total intensity $i(Q)$ of the scattered light as

$$i(Q) \propto P(Q)S(Q) \tag{79}$$

The function $S(Q)$ in the above equation is known as the (*interparticle*) *structure factor* or *static structure factor* (the label *static* is used to differentiate $S(Q)$ from its time-dependent version, known as the *dynamic* structure factor) and contains information on how the particles are spatially distributed in the dispersion. When there are no interparticle effects, $S(Q)$ becomes unity and we recover Equation (60) from Equation (79).

A detailed discussion of $S(Q)$ requires statistical thermodynamics, which is beyond our scope here. We conclude this discussion by just noting three reasons for the importance of interparticle structure in colloid science:

1. In cases such as micellar solutions (discussed in Chapter 8) and in certain types of biological dispersions, interparticle interactions cannot be avoided. Surfactants in solutions form self-assembled structures (see Vignette I.3 on liposomes in Chapter 1) such as micelles, vesicles, and others at quite low surfactant concentrations. Vignette I.6 on polymer composites discusses another example of self-assembled colloidal structures. One can use scattering methods (particularly x-ray or neutron) to study the shape, size, and structure of such colloidal systems. In the case of micelles (especially those formed with ionic surfactants), the "particles" show significant interparticle interactions even at low concentrations. One cannot circumvent (as in the Zimm method) the interparticle interactions by diluting these dispersions for scattering experiments since these particles are thermodynamic entities and will dissociate (i.e., disintegrate) when diluted. Therefore, one is forced to account for the contribution of the interparticle interference to the measured intensities. If $S(Q)$'s can be established independently using suitable statistical mechanical models, then the form factor can be extracted for further analysis of shape and structure of the individual units. More commonly, one assumes models for both $P(Q)$ and $S(Q)$ and fits the observed intensities with the models in order to obtain structural information on the dispersion.

2. Measurement of the static structure factor using scattering techniques also provides us with a *nonintrusive* method to probe the structure of dispersions and the nature of interaction forces in colloids. Structural changes in colloids are particularly of interest in colloid-based techniques for fabrication of structural and special-purpose ceramics.

3. We have focused on only the *static* (i.e., the *time-averaged*) measurements of intensities so far. However, one can also obtain the *dynamic* structure factor (i.e., as a function of time) from scattering experiments. The dynamic structure factor can then be used as a probe of the rheological behavior of the dispersions discussed in Chapter 4.

In discussing the measurement and uses of form factors and static structure factors, we have, without stating explicitly, combined a discussion of light scattering with the use of other forms of radiation. In the following section we comment briefly on some of the similarities and differences between light scattering and x-ray and neutron scattering.

TABLE 5.5 Comparison of the Range Covered by Various Radiation Scattering Methods

Method	Typical wavelength (nm)	Range of s (nm^{-1})
Laser light scattering	500	$1 \cdot 10^{-3} - 4 \cdot 10^{-2}$
Small-angle x-ray scattering	0.15	$2 \cdot 10^{-2} - 4 \cdot 10^{-1}$
Small-angle neutron scattering	0.4	$7 \cdot 10^{-3} - 9 \cdot 10^{-1}$
Wide-angle neutron scattering	0.4	$1 \cdot 10^{1} - 5 \cdot 10^{1}$

5.6c   X-Ray and Neutron Scattering as Complements to Light Scattering

This is a good place to draw attention to x-ray and neutron scattering since we have already introduced a combination of x-ray and light scattering to examine the fractal structure of silica aggregates in Example 5.4.

Much of the general formalism treated above remains valid for scattering of x-rays and neutrons by particles, although the mechanisms by which x-rays, neutrons, and light are scattered differ from each other. We have already seen that light scattering is a consequence of the interaction of photons with the electronic structure of atoms. X-rays are scattered by electron clouds, whereas neutrons are scattered by the nuclei of the atoms and by the magnetic moments of the atoms. A comparison of the wavelengths and the magnitudes of Q's accessible through x-ray, neutron, and light scattering is given in Table 5.5. From the point of view of colloid science, x-ray and neutron scattering techniques have two major advantages:

1. The particles and the fluid are effectively transparent to x-rays and neutrons; i.e., their effective refractive indices are nearly the same. Therefore, the criterion we specified in Equation (61) is easily satisfied, and we can avoid the need for the more complicated Mie theory (see Section 5.7b) and use the Rayleigh-Debye theories.
2. The very small values of the wavelengths of x-rays and neutrons allow one to reach large values of the parameter s. Since s^{-1} is the yardstick for the range of details accessible through scattering techniques, x-rays and neutrons allow us to probe structure at much shorter length scales than possible with light. Thus, x-ray scattering and neutron scattering complement light scattering measurements. Light scattering is indispensable for studying structure of particles and clusters in the micrometer size range. X-rays and neutrons help us to probe smaller particles.

Finally, it is also important to note that the term *small-angle scattering* does not necessarily specify the same range of Q; small-angle light scattering probes a region of Q different from the one probed by, for example, small-angle x-ray scattering because of the differences in the wavelengths of light and x-rays (see Example 5.4 and Table 5.5).

5.7   SCATTERING BY LARGE, ABSORBING PARTICLES

All the applications of *light* scattering that have been discussed so far have been restricted to very small particles and fairly small indices of refraction or to fairly small particles and very small indices of refraction. We have finally reached the point at which it seems appropriate to relax all these restrictions and consider the scattering by a particle of arbitrary size and index of refraction.

5.7a   Scattering and Absorption

The general problem here is to solve Maxwell's electromagnetic equations both inside and outside a particle, for which the indices of refraction are different in the two regions. The problem has been solved for both spherical and cylindrical particles; we limit our considerations to the former.

5.7a.1 Accounting for Absorption Through Complex Refractive Index

In the most general case some provision must be made for the possibility that the particle absorbs as well as scatters light. This contingency is introduced by defining the refractive index of an absorbing material as a complex number $(n - ik)$ for which $i = \sqrt{-1}$. For nonabsorbing materials, $k = 0$. Both n and k are wavelength-dependent characteristics of the material; k obviously increases as the wavelength of an absorption peak is approached.

The idea of representing the refractive index of an absorbing material by a complex number may seem strange, so the following analysis will be helpful. Suppose we consider the passage of a beam of light through a layer (in the yz plane) of unspecified material of thickness Δx. If the layer contains a vacuum ($n = 1$), then the electric field transmitted through the layer will be given by Equation (7):

$$\mathbf{E}_{n=1} = \mathbf{E}_0 \cos \left[2\pi(\nu t - x/\lambda)\right] \tag{80}$$

On the other hand, suppose the layer consists of a material of refractive index n. The light will now take an increment of time Δt longer to pass through the layer owing to the delaying effect of the medium. In this case the emerging field would be given by

$$\mathbf{E}_n = \mathbf{E}_0 \cos \left[2\pi(\nu(t + \Delta t) - x/\lambda)\right] \tag{81}$$

The delay may be related to the thickness of the layer and the refractive index as follows:

$$\Delta t = t_n - t_{n=1} = \frac{\Delta x}{c/n} - \frac{\Delta x}{c} = (n - 1)\frac{\Delta x}{c} \tag{82}$$

where c is the speed of light in vacuum. This means Equation (81) may be written as

$$\mathbf{E}_n = \mathbf{E}_0 \cos \left[2\pi\left(\nu t + (n - 1)\frac{\Delta x}{\lambda} - \frac{x}{\lambda}\right)\right] \tag{83}$$

which shows that, compared to Equation (80), the field experiences a phase shift in the medium.

A very important relationship involving complex numbers is

$$e^{i\theta} = \cos \theta + i \sin \theta \tag{84}$$

Cosine trigonometric functions, in other words, are given by the real part of the function $e^{i\theta}$. This means that Equations (80) and (83) may be written

$$\mathbf{E}_{n=1} = \mathbf{E}_0 \exp \left[2\pi i(\nu t - x/\lambda)\right] \tag{85}$$

and

$$\mathbf{E}_n = \mathbf{E}_0 \exp \left[2\pi i\left(\nu t + (n - 1)\frac{\Delta x}{\lambda} - \frac{x}{\lambda}\right)\right] \tag{86}$$

It is only the real part of the complex number that is of interest to us. Equation (86) may be written as

$$\mathbf{E}_n = \mathbf{E}_0 \exp[2\pi i(\nu t - x/\lambda)] \exp \left(2\pi i(n - 1)\frac{\Delta x}{\lambda}\right) \tag{87}$$

$$= \mathbf{E}_{n=1} \exp \left(2\pi i(n - 1)\frac{\Delta x}{\lambda}\right)$$

As we have seen repeatedly in this chapter, the intensity of light is proportional to the field amplitude squared; therefore, we write

$$\frac{I_n}{I_{n=1}} = \exp \left(4\pi i(n - 1)\frac{\Delta x}{\lambda}\right) \tag{88}$$

Now let us consider the case in which the refractive index of the layer is a complex number. In that case, Equation (88) becomes

$$\frac{I_n}{I_{n=1}} = \exp\left(4\pi i(n + ik - 1)\frac{\Delta x}{\lambda}\right) = \exp\left(4\pi i(n - 1)\frac{\Delta x}{\lambda}\right)\exp\left(-4\pi k\frac{\Delta x}{\lambda}\right) \qquad (89)$$

The first factor of Equation (89) is exactly what the ratio $(I_n/I_{n=1})$ would be if the material were nonabsorbing. It is modified, however, by a second term, $\exp(-4\pi k\Delta x/\lambda)$, which is real but contains the imaginary part of the index of refraction.

5.7a.2 The Beer-Lambert Formalism

The *Beer-Lambert equation* is another formalism that might describe the arrangement we have been discussing. In the Beer-Lambert equation, the intensity of the transmitted light I_t relative to the intensity of the incident light I_0 is given by Equation (30), which may be written

$$\left(\frac{I_n}{I_0}\right)_{abs} = \exp(-\varepsilon_{abs}\Delta x) \qquad (90)$$

where ε_{abs} is the absorbance of the material. Because of its usefulness in analytical chemistry, chemistry students are more likely to be familiar with this formalism than with the one that leads to Equation (89). Comparison of Equations (89) and (90) reveals that both describe the attenuation of light due to absorption in terms of a negative exponential that is proportional to the path length through the absorbing material. Since the two approaches describe the same situation in the same functional form, the two proportionality factors must also be equal. Therefore the imaginary part of the complex refractive index and the absorbance must be related as follows:

$$\varepsilon_{abs} = 4\pi k/\lambda \qquad (91)$$

We observed in Section 5.4c that, for nonabsorbing systems, the turbidity is a concept analogous to the absorbance; that is,

$$(I_t/I_0)_{sca} = \exp(-\tau\Delta x) \qquad (92)$$

where the subscript *sca* stands for scattering. Examination of Equations (90) and (92) shows how the two complement each other: The first describes absorption without scattering; the second describes scattering without absorption. For a system that displays these two optical effects simultaneously, the following composite relationship applies:

$$(I_t/I_0) = \exp[-(\varepsilon_{abs} + \tau)\Delta x] \qquad (93)$$

The experimental extinction in a system that displays these two effects equals the sum of ε_{abs} plus τ.

5.7a.3 Absorption Cross Section and Efficiency

The literature on scattering with absorption often factors absorbance and turbidity into a product of several terms. Both the nomenclature and notation used in this area vary from author to author, but the principal breakdown of both ε_{abs} and τ is as follows:

$$\varepsilon_{abs} = c_N\pi R_s^2\chi_{abs} \qquad (94)$$

and

$$\tau = c_N\pi R_s^2\chi_{sca} \qquad (95)$$

where c_N is concentration of the particles (in terms of the number of dispersed particles per unit volume), R_s is their radius, and the χ terms are known as the *efficiency factors* for absorption and scattering. The quantity πR_s^2 in these equations gives the geometrical cross section of the dispersed spheres. This area times the efficiency factor defines a quantity known as the *cross section* C_{abs} and C_{sca} for absorption and scattering, respectively:

$$C_{abs} = \pi R_s^2\chi_{abs} \qquad (96)$$
$$C_{sca} = \pi R_s^2\chi_{sca} \qquad (97)$$

Dimensionally, ε_{abs} and τ describe the attenuation of light per unit optical path and have the units length^{-1}. The area πR_s^2 has units of length^2 per particle, and c_N has units of particles length $^{-3}$. The efficiency factors are dimensionless. The cross sections have units of area per particle. These should not be taken literally as cross-sectional areas but, rather, as the "blocking power" of a particle as far as the transmission of incident light is concerned.

As stated above, Maxwell's equations have been solved for spherical particles of arbitrary size and arbitrary refractive index. Furthermore, the refractive index may be complex, so the general theory applies to both nonabsorbing and absorbing particles. The solutions to Maxwell's equations are generally given in terms of the efficiency factors χ, the magnitudes of which depend on the wavelength of light, the size of the particles, the real and imaginary parts of the refractive index, and the angle of observation:

$$\chi = f(\lambda, R_s, n, k, \theta) \tag{98}$$

For nonscattering particles, χ_{sca} is zero, and for nonabsorbing particles χ_{abs} is zero. The results of such an analysis are generally reported by giving values of χ_{sca} and χ_{abs} as a function of a size parameter β, where β is defined as

$$\beta = 2\pi \frac{R_s}{\lambda} \tag{99}$$

The calculations involved in computing these results are formidable, and the results available prior to the widespread utilization of computers were very limited. As noted in the introduction to this chapter, the advent of the computer has broadened the applicability of light scattering enormously.

It is difficult to make any generalizations about the variation of the efficiencies with β since the functions are so complicated and vary greatly with the refractive index. About the best that can be done along these lines are the following:

1. At any given angle of observation, χ tends to be an oscillating function of β.
2. For nonabsorbing particles (n real), the amount of oscillation in the χ_{sca} versus β curve is more pronounced the larger n is—if the particles absorb, the amount of oscillation in the curves decreases with increasing k.
3. For any given value of β, χ varies with the angle of observation θ. The number of oscillations in this curve is greater for larger values of β and n.

These generalizations are consistent with the approximations discussed above in the chapter: Small particles with low values of n are the simplest to describe.

In the following two sections we consider two specific systems that have been studied extensively, aqueous dispersions of colloidal gold (Section 5.7b) and aqueous dispersions of colloidal sulfur (Section 5.7c). These illustrate some of the statements made above, as well as show the sort of information that may be obtained from light scattering experiments in this very general case. No attempt has been made to be comprehensive in this presentation. Much more has been done with the systems to be discussed, and cross sections for many other systems have been calculated. The book by Kerker (1969) contains a good bibliography of scattering functions published up to 1969.

5.7b The Mie Theory: Gold Sols

The preceding section indicated some of the complications that arise when the optical properties of dispersions are calculated without placing narrow limitations on the size and refractive index of the particles. The difficulties are still large but somewhat more manageable if only the refractive index is given full range, the particle dimensions being somewhat restricted. This is the situation that was treated by Mie in 1908.

Mie wrote the scattering and absorption cross sections as power series in the size parameter β, restricting the series to the first few terms. This truncation of the series restricts the Mie theory to particles with dimensions less than the wavelength of light but, unlike the Rayleigh and Debye approximations, applies to absorbing and nonabsorbing particles.

The following equations give some indication of the nature of these expansions:

$$\chi_{abs} = A_1\beta + A_2\beta^3 + A_3\beta^4 + \cdots \tag{100}$$

and

$$\chi_{sca} = A_4\beta^4 + \cdots \tag{101}$$

where the values of the coefficients A_1, A_2, A_3, and A_4 are listed in Table 5.6. In this presentation, we have intentionally limited the series to include no terms higher than fourth order in β. Thus only the leading term in χ_{sca} is represented, even though the first three terms in χ_{abs} are included. The neglect of higher order terms permits us to apply this discussion rigorously only to particles that are sufficiently small that terms in β^5 or higher would make a negligible contribution. Even with only these terms retained, it is possible to draw several informative conclusions about χ_{abs} and χ_{sca}.

1. The absorption and scattering efficiencies do not show the same dependence on the particle size parameter β.
2. Both numerical coefficients of the complex refractive index, n and k, appear in both χ_{abs} and χ_{sca} (Table 5.6).
3. The coefficients A_1, A_2, and A_3 equal zero, and A_4 reduces to Rayleigh's law if $k = 0$.
4. The efficiencies are functions of the dimensionless variable alone.
5. For dispersions of uniform spheres, the entire wavelength dependence of the extinction is given by Equations (100) and (101).
6. The "wavelength dependence of the extinction" is simply the spectrum of the dispersion, which is therefore predicted theoretically by the general equations.

In the remainder of this section we see how the theoretical calculations of Mie account for the observed spectrum of colloidal gold. In the next section we consider the inverse problem for a simpler system: how to interpret the experimental spectrum of sulfur sols in terms of the size and concentration of the particles. Both of these example systems consist of relatively monodisperse particles. Polydispersity complicates the spectrum of a colloid since the same χ value will occur at different λ values for spheres of different radii according to Equations (99)–(101).

Colloidal gold is of considerable historic importance in colloid chemistry since many of the scientists who led the early development of the field conducted experiments on this system.

TABLE 5.6 Values for the Constants A_1 to A_4 in Equations (100) and (101)

Coefficient	General case	Special case of $k = 0$
A_1	$\dfrac{24nk}{(n^2 + k^2)^2 + 4(n^2 - k^2) + 4}$	0
A_2	$\dfrac{4nk}{15} + \dfrac{20nk}{3[4(n^2 + k^2)^2 + 12(n^2 - k^2) + 9]}$ $+ \dfrac{4.8nk[7(n^2 + k^2)^2 + 4(n^2 - k^2 - 5)]^2}{[(n^2 + k^2)^2 + 4(n^2) - k^2) + 4]^2}$	0
A_3	$\dfrac{-192n^2k^2}{[(n^2 + k^2)^2 + 4(n^2 - k^2) + 4]^2}$	0
A_4	$\dfrac{\frac{8}{3}[(n^2 + k^2)^2 + n^2 - k^2 - 2]^2 + 36n^2k^2}{[(n^2 + k^2)^2 + 4(n^2 - k^2) + 4]^2}$	$\dfrac{\frac{8}{3}(n^2 - 1)^2}{(n^2 + 2)^2}$

Source: From P. B. Penndorf, *J. Opt. Soc. Am.*, **52**, 896 (1962).

Mie set out specifically to account for the brilliant colors displayed by sols of gold and other metals.

Chloroauric acid, $HAuCl_4$, is easily reduced to metallic gold by a wide variety of reducing agents. However, characteristics of the resulting gold are widely different for different reducing agents. Thus, if phosphorus is used, a polydisperse system containing very small particles forms rapidly. If the resulting colloid is used to seed a reaction in which hydrogen peroxide is the reducing agent, the following reduction takes place slowly without additional nucleation:

$$2AuCl_4^- + 3H_2O_2 \rightarrow 2Au + 8Cl^- + 6H^+ + 3O_2 \tag{A}$$

The gold particles grow to a larger size by this process, with considerable sharpening of the particle size distribution. The particle size may be regulated to some extent by varying the amount of reagents used. Thus approximately monodisperse colloids of several different particle sizes may be prepared and compared. They are found to display different colors, depending on the particle size: The smaller particles produce a red dispersion; somewhat larger particles impart a blue color to the dispersion.

In calculating efficiency factors for absorption and scattering, the wavelength dependence of both the real and imaginary parts of the refractive index must be considered.

Figure 5.14 shows the real and imaginary parts of the complex refractive index of gold plotted against the wavelength of light in air and in water. These values of the refractive index were used to calculate the values of C_{abs} and C_{sca} for three different size spherical gold particles. The results are shown in Figure 5.15 as a function of the wavelength in air. The figure also includes the sum of the two cross sections C_{ext}, which describes the total extinction.

The magnitude and location of the maximum for each of the cross section curves are informative. It will be observed that the height and the location of the absorption curve in Figure 5.15 change relatively little with particle size (note the different scales for the ordinates), as we might expect. The cross section for scattering, on the other hand, is practically negligible for the smallest of the particles and increases to be roughly 50 and 100% larger than

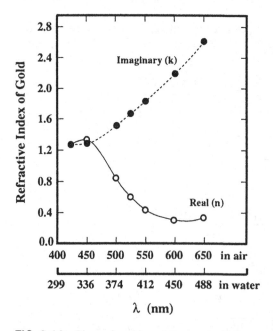

FIG. 5.14 The real and imaginary parts of the complex refractive index of gold versus wavelength in air and in water. (Data from Van de Hulst 1957.)

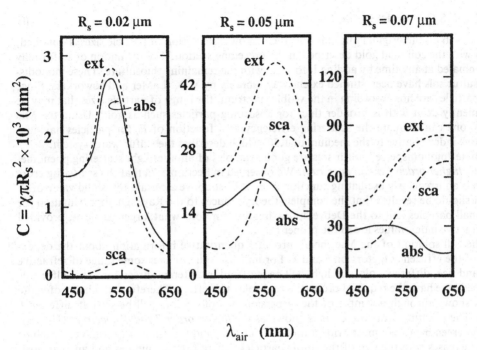

FIG. 5.15 Scattering coefficients versus wavelength for spheres of colloidal gold having three different radii. Note the different scales for the ordinates in each figure. (Data from Van de Hulst 1957.)

absorption for the larger particles. Also noteworthy is the fact that the wavelength location of the scattering maximum shifts to longer wavelengths as the particle size increases. This is a consequence of the fact that the efficiency depends on β rather than the separate values of R_s and λ. The wavelengths at which maximum total extinction occurs lie in the green, yellow, and red portions of the spectrum, respectively, for the particles with $R_s = 0.02$, 0.05, and 0.07 μm. On the basis of color complementarity, these would appear red, violet, and blue, respectively. These are the hues displayed by actual dispersions. Thus the Mie theory not only accounts for the color of the dispersions but also shows how the color displayed may be used to characterize the particle size of the dispersed phase.

Note that the index of refraction of the continuous phase and that of the dispersed particles enter the evaluation of the various efficiencies. The light that actually strikes the particles is used in the determination of χ. This light differs from that under vacuum by the refractive index of the medium. This effect enters the calculation of the χ values in that it is the ratio of the refractive index of the particle relative to that of the medium that determines the extinction.

5.7c Higher Order Tyndall Spectra: Monodisperse Sulfur Sols

Another colloidal system with light scattering characteristics that have been widely studied is the so-called monodisperse sulfur sol. Although not actually monodisperse, the particle size distribution in this preparation is narrow enough to make it an ideal system for the study of optical phenomena.

The colloid is prepared by rapidly mixing dilute solutions of sodium thiosulfate and hydrochloric acid so that the final concentration of each is about 0.002 M. The following reaction then occurs so slowly that the sulfur precipitates only on those particles that nucleate first:

$$H^+ + S_2O_3^{2-} \rightarrow HSO_3^- + S \tag{B}$$

Slow growth on the original nuclei is how the narrow distribution of particle sizes is obtained, just as with the colloidal gold described in the preceding section. The formation of sulfur may be terminated at any time by adding I_2 to react with the remaining thiosulfate. These monodisperse sulfur sols have been studied extensively, notably by V. K. LaMer and coworkers. Since these particles are nonabsorbing in the visible spectrum, the range of particle sizes that may be conveniently dealt with is broader than for absorbing particles such as gold. Using the Mie theory, one can evaluate the scattering efficiency as a function of R_s for particles having a refractive index relative to the medium of 1.50, which describes the sulfur-water system.

These "monodisperse" sulfur sols are good examples of another light scattering phenomenon: the *higher order Tyndall spectrum*. We observed in Section 5.7a that the scattering cross section is an irregularly oscillating function of θ, at least above a certain threshold value of β. Here it should be recalled that the complete theory reduces to the Rayleigh approximation for very small particles and to the Debye approximation for somewhat larger particles, provided the refractive index values are in the proper range.

The full solution of the Mie theory provides quantitative information about the dependence of the efficiency factors on θ and λ. For uniform spheres over some range of refractive index and size, different colors of light will be scattered in different directions. The sulfur sols described here have the required properties to display this effect. Therefore, if a beam of white light is shown through a sample of the dispersion, various colors will be seen at different θ values. The resulting array of colors is known as the *higher order Tyndall spectrum* (HOTS). Red and green bands are most evident in the sulfur sols, and the number of times these bands repeat increases with the size of the sulfur particles. Therefore the number and angular positions of the colored bands provide a unique characterization of the particle size. In the monodisperse sulfur sols, for example, particles having a radius of 0.30 μm are expected to show red bands at about 60, 100, and 140°. Particles with a radius of 0.40 μm, on the other hand, show red bands at about 42, 66, 105, 132, and 160°. Particle size determinations based on observations of this sort agree well with those determined from electron microscopy. These sulfur sols are quite easy to prepare, and it is interesting to observe the development of higher orders in the Tyndall spectrum as the thiosulfate decomposition reaction progresses.

It was once thought that the appearance of HOTS was evidence in itself for the presence of a monodisperse system. The argument was that one particle size would scatter, say, red light, at a particular angle, whereas another particle size would have the same R_s value and therefore the same scattering behavior for light of a complementary color. The resultant would be the obliteration of any distinct color: The scattered light would appear white. Although there may indeed be fortuitous cancellations of this sort at certain angles, it is also possible for certain bands to reinforce each other. In general, then, it is best to say that polydisperse systems may show HOTS, but in this case the angular distribution of bands is a characteristic of the particle size distribution. The angular location and number of bands as determined theoretically for uniform particles may not be used to interpret the HOTS of a polydisperse system correctly.

5.8 DYNAMIC LIGHT SCATTERING

As mentioned in Section 5.1, so far we have focused on what are known as static scattering experiments; that is, the intensities used in the methods discussed until now are time-averaged intensities at any given angle θ. In general, however, the intensity accessible in a scattering experiment depends on time t as well, since the scattering centers are in constant random motion due to their kinetic energy. The variation of the intensity with time, therefore, contains information on the random motion of the particles and can be used to measure the diffusion coefficient of the particles. The measured diffusion coefficient can then be used to determine the size of the particle. The class of light scattering methods based on the time dependence of the scattered light intensity is known as *dynamic light scattering* (DLS), and a large body of work on various aspects of this technique has appeared in the literature in recent years (Brown

1993; Chu 1991; Schmitz 1990). For example, the dynamic version of the diffusing wave spectroscopy described in Vignette V is a form of DLS, although in diffusing wave spectroscopy the method of analysis is different in view of multiple scattering. Most of the advanced developments are beyond the scope of this book. However, DLS is currently a routine laboratory technique for measuring diffusion coefficients, particle size, and particle size distributions in colloidal dispersions, and our objective in this section is to present the most essential ideas behind the method and show how they are used for particle size and size distribution measurements.

5.8a Intensity Fluctuations and the Siegert Relation

In a typical scattering experiment, a detector measures the intensity of the scattered radiation over a period of time, say, t_n, in discrete steps of Δt (see Figure 5.16a). As shown in the figure, the intensity $i(s,t)$ fluctuates around an *average* value because of the random motions of the scatterers. Until now, we have denoted the average by simply $i(s)$ for convenience. To be precise, however, the average should be denoted by, say, $\bar{i}(s)$, and this is an average *over time*, t, defined as

$$\bar{i}(s) = \lim_{t_n \to \infty} \frac{1}{t_n} \int_0^{t_n} i(s,t)\,dt \approx \lim_{n \to \infty} \frac{1}{n} \sum_{j=1}^{n} i(s,j\Delta t) \tag{102}$$

where the limit $t_n \to \infty$ reminds us that the measurement should be made over a sufficiently large time for the average to be accurate.

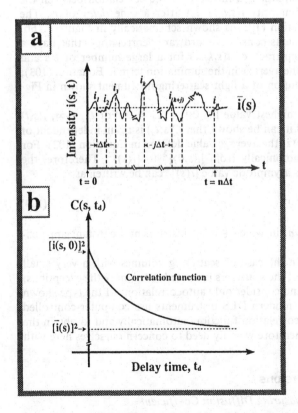

FIG. 5.16 Schematic illustration of intensity measurement and the corresponding autocorrelation function in dynamic light scattering: (a) variation of the intensity of the scattered light with time; (b) the variation of the autocorrelation function $C(s,t_d)$ with the delay time t_d.

Equation (102) also shows how the above time average is measured experimentally. Typically, the intensity is measured in a set of discrete time intervals, $t = \Delta t, 2\Delta t, 3\Delta t, \ldots$, etc., as illustrated in Figure 5.16a, and the arithmetic average shown in Equation (102) is an approximation of the average intensity over time $t_n = n\Delta t$.

In order to be able to use the *fluctuation* of the intensity around the average value, we need to find a way to represent the fluctuations in a convenient manner. In Section 5.3b in our discussion of Rayleigh scattering applied to solutions, we came across the concept of fluctuations of polarizabilities and concentration of scatterers and the role they play in light scattering experiments. In the present section, what we are interested in is the *time dependence* of such fluctuations. In general, it is not convenient to deal with detailed records of the fluctuations of a measured quantity as a function of time. Instead, one reduces the details of the fluctuations to what is known as the *autocorrelation function* $C(s,t_d)$, as defined below:

$$C(s,t_d) = \lim_{t_n \to \infty} \frac{1}{t_n} \int_0^{t_n} i(s,t)i(s,t + t_d)dt = \overline{i(s,0)i(s,t_d)}$$

$$\approx \lim_{n \to \infty} \frac{1}{n} \sum_{k=0}^{n} i(s,k\Delta t)i(s,(k + j)\Delta t) \tag{103}$$

where $t_d = j\Delta t$. The last part of the equation shows how the autocorrelation function is calculated experimentally when the intensity is measured in discrete time steps as illustrated in Figure 5.16a. The time t_d is known as the *delay time* since it represents the delay in time between the two signals $i(s,k\Delta t)$ and $i(s,(k + j)\Delta t)$ and is equal to $j\Delta t$ (see Figure 5.16a). The function $C(s,t_d)$ is obtained for a series of values of t_d by taking $j = 0, 1, 2, 3, \ldots$, etc. The autocorrelation function, as the name implies, is a measure of the correlation between the intensity $i(s,t_1)$ at any time t_1 and the intensity $i(s,t_1 + t_d)$ after a time *delay* of t_d. The correlation function obtained from Equation (103) is shown schematically in Figure 5.16b. Modern dynamic light scattering instruments consist of hardware "correlators" that have a number of channels or registers that keep track of $i(s,k\Delta t)$ for a large number of k's and automatically compute the products and the average in the summation term in Equation (103); see, for example, the schematic representation of a light scattering instrument shown in Figure 5.5.

The autocorrelation function has its highest value $\overline{[i(s,0)]^2}$ at $t_d = 0$. For $t_d \to \infty$, $i(s,t)$ and $i(s,t + t_d)$ become uncorrelated, and it can be shown that $C(s,t_d)$ is again independent of t_d and that it is given by $[\bar{i}(s)]^2$, where $\bar{i}(s)$ is the average value defined in Equation (102). For nonperiodic $i(s,t)$, $C(s,t_d)$ decreases monotonically from $\overline{[i(s,0)]^2}$ to $[\bar{i}(s)]^2$. Therefore, the ratio of the autocorrelation function to its asymptotic value $[\bar{i}(s)]^2$ can be written as

$$\frac{C(s, t_d)}{[\bar{i}(s)]^2} = g_2(s, t_d) = 1 + \xi|g_1(s, t_d)|^2 \tag{104}$$

an equation known as the *Siegert relation*, in which ξ is an instrumental constant approximately equal to unity.

The Siegert relation is valid except in the case of scattering volumes with a very small number of scatterers or when the motion of the scatterers is limited. We ignore the exceptions, which are rare in common uses of DLS, and consider only autocorrelations of the type shown in Equation (104). As mentioned above, modern DLS instruments use computer-controlled correlators to calculate the intensity autocorrelation function automatically and to obtain the results in terms of the function $g_1(s,t_d)$; therefore we only need to concern ourselves here with the interpretation of $g_1(s,t_d)$.

5.8b Dilute (Noninteracting) Dispersions

5.8b.1 *Monosize Spherical Particles: Measuring Diffusion Coefficient and Particle Size*

For dilute dispersions, i.e., those in which the interparticle spacing is so large that there are no particle-particle interactions, DLS simply measures the intensity fluctuations due to single-particle motion. For monosize, spherical particles, one can show rigorously that $g_1(s,t_d)$ decays exponentially as follows:

$$g_1(s,t_d) = \exp(-s^2 D t_d) \tag{105}$$

Here, D, which is the quantity we seek from $g_1(s,t_d)$, is the diffusion coefficient of the particle (and s is the magnitude of the "scattering vector" defined in Equation (57)). We can now use the *Stokes-Einstein equation* (see Equation (2.32) and the accompanying comment) to obtain the particle radius R_H from D:

$$D = \frac{k_B T}{6 \pi \eta R_H} \tag{106}$$

where η is the viscosity of the fluid, k_B is the Boltzmann constant, and T is the absolute temperature of the dispersion. The radius R_H measured in this manner is usually known as the *hydrodynamic radius* (hence the subscript H) since it relies on the Stokes coefficient, $6\pi\eta R_H$— a result from fluid (or, hydro) dynamics.

The measurement of the diffusion coefficient (and the hydrodynamic radius from D) is one of the most common uses of DLS, but the method can also be used as a nonintrusive technique for measuring the viscosity of a fluid. In this case, one uses "probe" particles with a known radius so that the unknown quantity in the Stokes-Einstein equation is η. More sophisticated uses of the DLS technique using essentially the above concept are discussed in specialized monographs (Brown 1993; Pecora 1985; Schmitz 1990). The diffusing wave spectroscopy mentioned in Vignette V also measures $g_1(t_d)$, but $g_1(t_d)$ in DWS is no longer a function of the angle θ since multiple scattering smears out the angle dependence of the intensity. As a result, the theoretical formalism needed for the analysis of the correlation function differs from what we have presented above and in Section 5.8b.2.

Example 5.5 illustrates one use of the DLS data.

* * *

EXAMPLE 5.5 *Determination of the Effective Diameter of an Enzyme Using Dynamic Light Scattering.* DLS analysis of a dilute solution of the enzyme phosphofructokinase in water at $T = 293\text{K}$ leads to the following data for the correlation function $g_2(s,t_d)$:

$s^2 t_d \times 10^{-10}$ (m^{-2} s)	0.4	0.8	1.2	1.6	2.0	2.4
$g_2(s,t_d)$	1.75	1.6	1.47	1.375	1.298	1.236
$s^2 t_d \times 10^{-10}$ (m^{-2} s)	2.8	3.2	3.6	4.0	5.0	10.0
$g_2(s,t_d)$	1.188	1.148	1.119	1.093	1.052	1.003

Assume that the enzyme is roughly spherical and that the instrument constant ξ in the Siegert relation is unity and determine the hydrodynamic radius R_H of the enzyme. Given that the partial molar volume $\overline{V}$ of the enzyme is $0.74 \cdot 10^{-3}$ m^3/kg and the molecular weight M is $4.78 \cdot 10^2$ kg/mol, determine the "dry radius" R_s for the enzyme and obtain the ratio (R_H/R_s). Can the difference between R_H and R_s be attributed to the bound water on the enzyme? The viscosity η of water at 293K may be taken as 0.001 kg/m s.

Solution: The given DLS data can be used to obtain the intensity autocorrelation function $g_1(s,t_d)$ by rewriting the Siegert relation as follows:

$$\ln g_1(s,t_d) = \ln [g_2(s,t_d) - 1]^{1/2} - 1/2 \ln \xi$$

This leads to

$s^2 t_d \times 10^{-10}$ (m^{-2} s)	0.4	0.8	1.2	1.6	2.0	2.4
$\ln g_1(s,t_d)$	−0.118	−0.23	−0.35	−0.464	−0.58	−0.696
$s^2 t_d \times 10^{-10}$ (m^{-2} s)	2.8	3.2	3.6	4.0	5.0	10.0
$\ln g_1(s,t_d)$	−0.81	−0.93	−1.04	−1.16	−2.45	−2.9

Equation (105) shows that the plot of $\ln g_1(s,t_d)$ versus $s^2 t_d$ should give a straight line with a slope equal to the negative of the diffusion coefficient D of the enzyme. A plot based on the above data gives a straight line with a slope of $-2.88 \cdot 10^{-11}$ m^2/s and an intercept of -0.026. Therefore, the diffusion coefficient based on the data is

$D = 2.88 \cdot 10^{-11} \, \text{m}^2/\text{s}$

(The magnitude of the intercept implies that the instrumental constant ξ is roughly 0.95.)
From the Stokes-Einstein relation, Equation (106), we then get

$$R_H = k_B T/(6\pi\eta D)$$
$$= [1.38 \cdot 10^{-23} \, (\text{J/K}) \cdot 293\text{K}] \div [(6\pi)0.001 \, (\text{kg/m s}) \, 2.88 \cdot 10^{-11}(\text{m}^2/\text{s})]$$
$$= 74.4 \cdot 10^{-9} \text{m}$$

The "dry radius" R_s can be calculated from

$$R_s = [(3/4\pi)\overline{V}M/N_A]^{1/3}$$
$$= [3/4\pi][0.74 \cdot 10^{-3}(\text{m}^3/\text{g})4.78 \cdot 10^2(\text{kg/mol})/6.02 \cdot 10^{23}\text{mol}^{-1}]^{1/3}$$
$$= 52 \cdot 10^{-9} \, \text{m}$$

The ratio (R_H/R_s) is therefore 1.43.

The source of this 43% difference between the "dry radius" and the hydrodynamic radius is unlikely to be the increase in diameter due to bound water. It is more likely that the shape asymmetry of the enzyme (i.e., the approximation that the enzyme is effectively spherical) is the source of the above difference. ∎

* * *

5.8b.2 Effect of Polydispersity: Measuring Size Distribution

The DLS measurements can also be used in more complicated situations, for example, (a) when interparticle interactions are important, (b) for dispersions with particles of other shapes, (c) for monitoring coagulation, and (d) when the dispersion is polydisperse. In all these cases, a significant amount of modeling is often necessary to interpret the measured autocorrelation function, and we do not consider them here. Instead, we restrict our attention to a brief discussion of item (d) above, namely, polydisperse systems, since it is concerned with a problem of more routine interest.

In the case of a polydisperse system, the overall decay of the function $g_1(s,t_d)$ is determined collectively by the decay rate (i.e., s^2D) corresponding to *each particle* (notice that s^2D varies with the particle size as evident from the Stokes-Einstein relation). In principle, the decay function in this case can be written formally in a simple manner as a weighted average of all possible decays:

$$g_1(s,t_d) = \lim_{n \to \infty} \frac{1}{n} \sum_{j=1}^{n} w_j(s^2D_j)\exp(-s^2D_jt_d) \tag{107}$$

where $w_j(s^2D_j)$ is a weighting function determined by the amount of particles in size range j; that is, the decay of $g_1(s,t_d)$ for a polydisperse system is an appropriately averaged function of the monodisperse case given in Equation (105). A number of methods are available for determining the size distribution from the experimentally determined $g_1(s,t_d)$. One of the simplest is based on what is known as the cumulant expansion, i.e., a series expansion of $\ln g_1(s,t_d)$, given by

$$\ln g_1(s,t_d) = \sum_{n=1}^{\infty} k_n(s) \frac{(-t_d)^n}{n!} \quad \text{in the limit } \tau_d \to 0 \tag{108}$$

$$= -\overline{D} s^2 t_d + \sigma^2 s^4 \frac{t_d^2}{2!} + \text{higher order terms}$$

where k_n is known as the *nth cumulant*. Equation (108) also shows that the first-order cumulant is related to the average diffusion coefficients of particles of all sizes (denoted here by $\overline{D}$) and the second-order cumulant to the standard deviation σ of the distribution of diffusion coefficients (see Chapter 1 and Appendix C for a discussion of standard deviation and some of the related statistical concepts).

Equation (108) is accurate only for small delay times and, in fact, the higher order terms obtained experimentally are not usually very reliable because of the "noise" in the data.

However, it does illustrate how one can determine an average particle size and a measure of the breadth of the distribution function from experimental data. A plot of $[\ln g_1(s,t_d)/(s^2t_d)]$ against (s^2t_d) will lead to a straight line for small t_d's, and $\overline{D}$ and s^2 can be obtained from the intercept and the slope of the straight line.

The logic of the above form of $g_1(s,t_d)$ and additional details are available in advanced books on DLS, and the above description is meant only to illustrate the basic ideas and one data-analysis approach. The cumulant analysis is often used as a first step before more advanced analytical procedures (each of which has its own advantages and disadvantages) are attempted. Most DLS instruments come with computer programs for the analysis of the size distribution, but we should bear in mind that each analysis technique has specific, and often restrictive, assumptions and none is "exact." As a consequence, the results of size distributions from DLS are best interpreted as semiquantitative indicators of polydispersity rather than a true representation of the distribution.

* * *

EXAMPLE 5.6   *Cumulant Analysis of Dynamic Light Scattering Data.*   A polystyrene latex dispersion supplied by a manufacturer is claimed to have a "very narrow" size distribution with an average particle diameter of 62 nm. An analysis of the dispersion using DLS leads to the following data for $\ln g_1(s,t_d)$. The DLS experiments are conducted at 20°C using a dispersion in water at a particle volume fraction of 0.005. The wavelength of the laser used and the angle at which the experiments are conducted correspond to $6.5345 \cdot 10^6$ m^{-1} for the magnitude of the scattering vector s. The viscosity of water at 20°C may be taken as 0.001 kg/m s.

$t_d \times 10^3$ (s)	0.05	0.1	0.15	0.2	0.25
$\ln g_1(s,t_d)$	−0.015	−0.0305	−0.0457	−0.061	−0.076
$t_d \times 10^3$ (s)	0.3	0.35	0.4	0.45	0.5
$\ln g_1(s,t_d)$	−0.091	−0.107	−0.122	−0.137	−0.152

Check if the specifications supplied by the manufacturer are correct. State any assumptions you make in your evaluation of the data.

Solution: Assume that the interparticle forces are negligible. Further, since the volume fraction of the dispersion used in the DLS experiments is very low, we may assume that the dispersion is sufficiently dilute and that multiple scattering is negligible.

Equation (108) shows that the cumulant expansion for $\ln g_1(s,t_d)$ may be rearranged to give

$$y = \frac{\ln g_1(s, t_d)}{x} \approx -\overline{D} + \frac{\sigma^2}{2}x \quad \text{with } x = s^2t_d$$

A plot of y versus x can now be used to obtain $\overline{D}$ from the intercept and σ^2 of the diffusion coefficient from the slope. Using the given data and the given value of s, we prepare a table of y versus x:

$x \times 10^{-10}$ (m^{-2} s)	0.2135	0.427	0.6405	0.854	1.0675
$y \times 10^{12}$ (m^2/s)	−7.03	−7.14	−7.135	−7.14	−7.15
$x \times 10^{-10}$ (m^{-2} s)	1.281	1.4945	1.708	1.9125	2.135
$y \times 10^{12}$ (m^2 s)	−7.10	−7.16	−7.14	−7.13	−7.12

It is clear from the table that the value of y is essentially constant over the entire range of given delay time. The slope, hence σ, is clearly negligible, and the intercept is approximately

$$\overline{D} = 7.14 \cdot 10^{-12} \text{ m}^2/\text{s}$$

The use of the Stokes-Einstein relation with the above value of the average diffusion coefficient leads to a hydrodynamic radius of roughly 30 nm, which is consistent with the specification of the manufacturer.   ∎

* * *

5.8c   Dispersions of Interacting Particles: Mutual and Self-Diffusion Coefficients

Even in the case of monodisperse systems, the observed decay rate of $g_1(s,t_d)$ (and hence the diffusion coefficient) in general depends on the angle at which the decay is measured if interparticle interference effects exist. In the case of *dilute* dispersions, in which interactions of all sorts among the particles may be neglected, the interference is negligible, and the diffusion coefficient measured is independent of the angle and is given by the Stokes-Einstein equation (for spherical particles). This diffusion coefficient is often called the *self-diffusion coefficient* (or *probe diffusion coefficient*) since it represents the unhindered Brownian motion of a typical particle.

Although the analysis becomes complex for more concentrated dispersions (or even for dilute dispersions of charged particles, which can interact over very large distances), some general observations on two limiting cases are useful:

1.   Measurements made at large enough values of Q ($= sR_s$) and for $t_d \rightarrow 0$: For $t_d \rightarrow 0$, the particles have very little time to wander far from their positions, i.e., the encounters with neighboring particles are negligible. Moreover, for large values of Q (e.g., large s) the interparticle interference is negligible since the range of interference in the observed intensity, represented by the magnitude of s^{-1}, is small. As a result, the measurements correspond again to the self-diffusion of the particles.

2.   Measurements at low Q's: At low Q's, because of the large magnitudes of s^{-1}, the measured intensity and its autocorrelation function are dominated by the *cumulative* diffusion of the particles. The measured decay rate thus represents the *cumulative* or *mutual diffusion coefficient* D_m given by

$$D_m = \frac{1}{6 \pi \eta R_s} \left(\frac{\partial \pi_{osm}}{\partial c_N} \right)_T \tag{109}$$

where π_{osm} is the osmotic pressure of the dispersion and c_N is the concentration of the particles in "number of particles/volume of dispersion." It is the cumulative diffusion coefficient that appears in the Fick's laws discussed in Chapter 2, and the diffusion experiments described in Chapter 2 measure this diffusion coefficient. Thus we have identified another method for measuring mutual diffusion coefficients for (at least spherical) solutes. For dilute dispersions, D_m in Equation (109) reduces to the self-diffusion coefficient. (Note that, for dilute systems, $\pi_{osm} = c_N k_B T$ and $(\partial \pi_{osm}/\partial c_N)_T = k_B T$.)

Moreover, the influence of the motions of the particles on each other (i.e., when the motion of a particle affects those of the others because of communication of stress through the suspending fluid) can also influence the measured diffusion coefficients. Such effects are called "hydrodynamic interactions" and must be accounted for in dispersions deviating from the dilute limit. Corrections need to be applied to the above expressions for D and D_m when particles interact hydrodynamically. These are beyond the scope of this book, but are discussed in Pecora (1985), Schmitz (1990), and Brown (1993).

We have made a note of the hydrodynamic interactions and other interactions to draw attention to an important fact. That is, the analysis of the DLS data is often quite complex, and a simple use of the single-exponential decay function and the Stokes-Einstein relation is not always sufficient, although many instruments available on the market use such an analysis and report an "effective size" for the particles in the dispersion.

REVIEW QUESTIONS

1.   Describe briefly what is meant by light *scattering* and the mechanism by which molecules scatter light.
2.   Explain what is meant by each of the following terms: (a) electric field, (b) intensity of light, (c) polarization of light.
3.   What is *Coulomb's law* and what are the units of the quantities that appear in Coulomb's law?
4.   Why is light scattering an important tool in colloid science? What is the range of dimensions of colloidal particles that can be probed by light scattering? Why?
5.   What is the difference between *static* light scattering and *dynamic* light scattering?
6.   How does light scattering differ from *x-ray* scattering and *neutron* scattering in terms of mechanisms as well as the range of interactions and structure that can be probed by each?
7.   What is meant by *Rayleigh scattering*? What are the important assumptions and limitations of the Rayleigh theory?

8. What is the *Rayleigh ratio*? How is it related to the intensity of the scattered light?
9. What is the relation between the Rayleigh ratio and the molecular weight and second virial coefficient of a dispersion of colloids or macromolecules?
10. What type of molecular weight average is measured by light scattering in the case of polydisperse solution?
11. What is meant by *turbidity* of a dispersion? How is the turbidity related to the Rayleigh ratio?
12. Explain what is meant by *interference of scattered light*. When is it important?
13. Why do liquids scatter light? Why isn't there a destructive influence of the lights scattered by all the molecules?
14. At what angle is the intensity of scattered light largest for large particles? Why?
15. How does the interference of scattered light affect the Rayleigh ratio?
16. What is the *Debye theory* of light scattering? What are its assumptions and limitations?
17. What is the definition of the *form factor* (or *intraparticle structure factor*)?
18. What is the magnitude of the scattering vector? How is it useful?
19. Discuss some of the limiting forms of the form factor. Explain what is meant by (a) the *center-of-mass region*, (b) the *Guinier region*, (c) the *fractal region*, and (d) the *Porod region*.
20. Discuss what is meant by *small-angle scattering*. How small is small?
21. What is the *Zimm plot*? Describe its use.
22. What is the *dissymmetry ratio*, and how is it used?
23. What is meant by *static structure factor* or *inter*particle structure factor? Describe its physical significance and use.
24. How is the light scattering theory formulated for large, absorbing particles? What is meant by "absorbing" particles?
25. What are absorbing and scattering cross sections?
26. What is the *Mie theory* of light scattering? What are its assumptions and limitations?
27. What is meant by *Tyndall spectra*?
28. What is meant by *dynamic light scattering* (DLS)? What is measured in DLS?
29. Why does the intensity of the light scattered by a dispersion fluctuate?
30. What is an *autocorrelation function*? Sketch qualitatively the autocorrelation function of the intensity of scattered light from a dispersion for a number of angles.
31. What is the *Siegert relation*? How is it related to the decay of intensity fluctuations measured in DLS experiments?
32. What is the relation between the diffusion coefficient of a monodispersed suspension of spherical particles and the decay of the intensity correlation function?
33. How is the particle size measured in DLS experiments? How is the size distribution measured in such experiments in the case of polydispersed colloids? Comment on the method(s) critically. How do interparticle interactions affect the above measurements?
34. What is the difference between the *self-diffusion coefficient* and the *mutual diffusion coefficient*? Which of these two is described by the Stokes-Einstein equation?

REFERENCES

General References (with Annotations)

Chu, B., *Laser Light Scattering: Basic Principles and Practice*, 2nd ed., Academic Press, Boston, 1991. (Advanced level. A detailed "manual" of light scattering. Covers basic theory as well as hardware and data-analysis procedures.)

Johnson, C. S., Jr., and Gabriel, D. A., *Laser Light Scattering*, Dover, New York, 1981. (Advanced level. An affordable, paperback edition of a long review article originally published as Chapter 5 in *Spectroscopy in Biochemistry*, Vol. 2 (T. E. Bell, Ed.), CRC Press, Boca Raton, FL, 1981. Covers classical as well as dynamic light scattering, including directed-flow light scattering [e.g., electrophoretic light scattering]. Illustrative applications use biological examples.)

Kerker, M., *The Scattering of Light and Other Electromagnetic Radiation*, Academic Press, New York, 1969. (Advanced level. A classic on the theory of light scattering.)

Schmitz, K. S., *An Introduction to Dynamic Light Scattering by Macromolecules*, Academic Press, Boston, 1990. (Advanced level. A textbook on DLS. Numerous examples from the literature.)

Zare, R. N., Spencer, B. H., Springer, D. S., and Jacobson, M. P., *Laser: Experiments for*

Beginners, University Science Books, Sausalito, CA, 1995. (Undergraduate level. A collection of experiments, most at the undergraduate level, on the use of lasers, light scattering, diffraction, and refraction. Chapter 2 describes a number of experiments highly relevant to the focus of the present book. The examples include particle size determination, studying aggregation, formation of microemulsion from a milky macroemulsion, and vesicle aggregation, among others. Chapter 3 describes diffraction and study of colloid crystals, discussed in Chapter 13 of the present book. Highly recommended.)

Other References

Abramowitz, M., and Stegun, I. A., Eds., *Handbook of Mathematical Functions with Formulas, Graphs, and Mathematical Tables*, U.S. Government Printing Office, Washington, DC, 1964.

Atkins, P. W., *Physical Chemistry*, 5th ed., W. H. Freeman, New York, 1994.

Brown, W. (Ed.), *Dynamic Light Scattering: The Method and Some Applications*, Clarendon Press, Oxford, England, 1993.

Huglin, M. B. (Ed.), *Light Scattering from Polymer Solutions*, Academic Press, New York, 1972.

Jackson, C., Nilsson, L. M., and Wyatt, P. J., *J. Appl. Polym. Sci.*, **43**, 99 (1989).

McIntyre, D., and Gormick, F. (Eds.), *Light Scattering from Dilute Polymer Solutions*, Gordon and Breach, New York, 1964.

Pecora, R. (Ed.), *Dynamic Light Scattering: Applications of Photon Correlation Spectroscopy*, Plenum, New York, 1985.

Schaefer, D. W., Martin, J. E., Wiltzius, P., and Cannel, D. S., *Phys. Rev. Lett.*, **52**, 2371 (1984).

Stacey, K. A., *Light Scattering in Physical Chemistry*, Butterworths, London, 1956.

Tanford, C., *Physical Chemistry of Macromolecules*, Wiley, New York, 1961.

Van de Hulst, H. C., *Light Scattering by Small Particles*, Wiley, New York, 1957.

Weitz, D. A., and Pine, D. J., Diffusing-Wave Spectroscopy. In *Dynamic Light Scattering: The Method and Some Applications* (W. Brown, Ed.), Clarendon Press, Oxford, England, 1993.

PROBLEMS

1. Use Figure 5.4a to show that $\sin^2 \phi_y + \sin^2 \phi_z = 1 + \cos^2 \phi_x$. Verify that the angle ϕ is one-half the tetrahedral angle when $\phi_x = \phi_y = \phi_z = \phi$.

2. The intensity of the light scattered by air was measured in the atmosphere at Mt. Wilson, California, in 1913. The following is a selection of the results obtained:*

λ_0 (nm)	350	360	371	384	397
$i_s/I_{o,u}$	0.459	0.423	0.377	0.338	0.285
λ_0 (nm)	413	431	452	475	
$i_s/I_{o,u}$	0.245	0.213	0.174	0.147	

   Show that these results are in agreement with the predictions of the Rayleigh theory, given that $(n - 1)$ is constant at $2.97 \cdot 10^{-4}$ over this range of wavelengths. Instead of $(N_A\rho/M)$, we could have multiplied the equation in Step 6 in Table 5.1 by N_0, the number of molecules per unit volume under appropriate conditions—taken to be STP in this example for simplicity— and use light scattering to measure this quantity. Under the conditions of this research, the numerical and geometrical factors work out to give $i_s/I_{o,u} = 0.1911\lambda_0^4 N_0$ in SI units. Use a representative observation to evaluate N_0 and, from this, Avogadro's number.

3. The turbidity of solutions of sodium silicate containing SiO_2 and Na_2O in a 3.75 ratio has been studied as a function of time.† When the total solute content is 0.02 g cm^{-3}, the following data are obtained:

Time (days)	0	1	3	15	45	88	166	199	351	455
$\tau \times 10^4$ (cm^{-1})	0.81	2.32	3.80	5.47	6.70	8.06	9.53	10.47	13.00	14.43

   (During the same time, the pH of the solution changes from 10.85 to 11.02.) Suggest a

---

*Fowle, F. E., *Astrophys. J.*, **40**, 435 (1914).

†Debye, P., and Hauman, R. V., *J. Phys. Chem.*, **65**, 5 (1961).

qualitative explanation for these observations in terms of the chemical behavior of silicates (be sure to include references to whatever sources you consult). Why is it important to the study of light scattering to realize that solutions such as these show a variation of turbidity with time?

4. Equation (2.63) suggests that the quantity $\overline{\delta c^2}$ in Table 5.2 is evaluated by solving

$$\overline{\delta c^2} = \int_0^\infty (\delta c)^2 P(\delta c) d(\delta c)$$

Argue that the appropriate form for $P(\delta c)$ is

$$P(\delta c) = A \exp\left(-\frac{G - G_0}{k_B T} \right)$$

Use the value of $G - G_0$ from Table 5.2 $[G - G_0 = \frac{1}{2}(\partial^2 G/\partial c^2)_0(\delta c)^2]$ to verify the value for $\overline{\delta c^2}$.

5. The refractive indices of NaCl solutions have been measured at 20°C as a function of concentration by means of a differential refractometer* with the following results:

$c \times 10^3$ (g NaCl cm^{-3})	3.749	5.468	7.498	7.920
$(n - n_0) \times 10^4$	6.73	9.83	13.18	14.06

Use these data to evaluate the quantity dn/dc for NaCl solutions in this concentration range. On the basis of these data, what is the apparent uncertainty introduced in a light scattering experiment through this quantity?

6. Above a certain concentration, the turbidity of sodium dodecyl sulfate (molecular weight = 288) solutions increases with concentration as if particles in the colloidal size range were present. Use the following data† to evaluate the apparent molecular weight of the species responsible for the scattering. For this system $H = 3.99 \times 10^{-6}$.

$c \times 10^3$ (g cm^{-3})	2.7	4.2	7.7	9.7	13.2	17.7	22.2
$\tau \times 10^4$ (cm^{-1})	1.10	1.29	1.71	1.98	2.02	2.14	2.33

Assuming the scattering centers are aggregates of sodium dodecyl sulfate molecules, estimate the number of these units in the aggregate.

7. The turbidity of Ludox (a colloidal silica manufactured by DuPont) has been studied as a function of concentration with the following results:‡

$c \times 10^2$ (g cm^{-3})	0.57	1.14	1.70	2.30
$\tau \times 10^2$ (cm^{-1})	1.56	2.97	4.25	5.36

Evaluate the molecular weight of the Ludox particles, using a value of $H = 4.08 \times 10^{-7}$ for the system. Calculate the characteristic diameter for these particles, assuming the particles to be uniform spheres of density 2.2 g cm^{-3}.

8. Criticize or defend the following proposition: The accompanying data for the turbidity of dodecylamine hydrochloride solutions§ suggest that at concentrations exceeding about 0.003 g cm^{-3} the solute associates into aggregates of colloidal dimensions.

$c \times 10^3$ (g cm^{-3})	0.77	1.73	3.10	3.31	6.15	8.31
$\tau \times 10^4$ (cm^{-1})	0	0	0	0.95	1.91	2.55

For this system H is approximately 7.7×10^{-6}.

---

*Debye, P., and Hauman, R. V., *J. Phys. Chem.,* **65**, 8 (1961).
†Tartar, H. V., and Lelong, A. L. M., *J. Phys. Chem.,* **59**, 1185 (1955).
‡Deželic, G., and Kratohvil, J. P., *J. Phys. Chem.,* **66**, 1377 (1962).
§Debye, P., *J. Phys. Chem.,* **53**, 1 (1949).

9. Krasna[*] has measured the turbidity of calf thymus DNA in aqueous solutions. The accompanying table gives $R_\theta \times 10^5$ (in cm^{-1}) for this system, measured at 546 nm:

	$c \times 10^6$ (g cm^{-3})			
θ (deg)	20.6	41.4	62.0	82.5
26	2.47	4.80	6.94	8.75
30	2.06	4.02	5.83	7.67
34	1.77	3.39	4.84	6.44
38	1.75	2.98	4.28	5.61
42	1.38	2.57	3.84	4.87
50	1.01	1.90	2.91	3.71
60	0.76	1.45	2.23	2.89

Prepare a Zimm plot of these results (using $K = 3.63 \times 10^{-7}$) and evaluate M, B, and the radius of gyration of the DNA in this preparation.

10. The following table gives $R_\theta/K \times 10^{-3}$ for different values of c and θ in the system polystyrene-decalin at 30°C, measured with the mercury 435.8-nm line:[†]

	$c \times 10^3$ (g cm^{-3})		
θ (deg)	0.50	0.99	1.49
30	0.735	1.34	—
45	0.685	1.27	2.04
60	0.625	1.17	1.62
75	0.562	1.05	1.49
90	0.510	0.96	1.37
105	0.467	0.88	1.25

Prepare a Zimm plot of these date and evaluate M, B, and the radius of gyration of the polymer under these conditions.

11. The following table shows the angular location of the green bands in the HOTS of various size spheres of relative refractive index 1.46:[‡]

Number of green bands	Radius (μm)					
	0.2	0.3	0.4	0.5	0.6	0.8
First	25°	7.5°				
Second	140°	77.5°	57.5°	42.5°	32.5°	17.5°
Third		150°	95°	72.5°	60°	40°
Fourth			140°	117.5°	85°	62.5°
Fifth				115°	110°	87.5°
Sixth				(blue)	130°	110°
Seventh					157.5°	127.5°
Eighth						142.5°

Formulate a generalization correlating the observed number of green bands in the HOTS for this system with the approximate particle size. Briefly describe how this information could be used to "grow" a monodisperse sulfur sol with a dimension that corresponds aproximately to a predetermined size.

[*]Krasna, A. I., *J. Colloid Interface Sci.,* **39**, 632 (1972).

[†]Lechner, M. D., and Schulz, G. V., *J. Colloid Interface Sci.,* **39**, 469 (1972).

[‡]Kerker, M., *The Scattering of Light and Other Electromagnetic Radiation,* Academic Press, New York, 1969, p. 409.

12. General solutions of light scattering equations generate oscillating curves when χ is plotted versus β for a particular value of the relative refractive index of the dispersed phase compared to the continuous phase. Such a curve can be described by the expression $\chi = K\beta^{-n}$, where $-n$ is the local value of the slope of the $\ln \chi$ versus $\ln \beta$ curve at specific values of β. Suppose this parameter n is known as a function of β for an experimental system. Describe the kind of experimental data and the analysis required thereof to yield a size parameter for the dispersed particles. What are the limitations of this method?

13. Suppose your employer intends to develop a new laboratory to characterize particles in the colloidal size range. Your assignment is to prepare a list of the equipment that should be purchased for such a facility. A brief justification for each major item should be included along with a priority ranking based on the versatility of the method. Assume that your laboratory is already well stocked with such nonspecialized items as laboratory glassware, balances, and the like.

14. Show that the average intensity $\bar{i}(s)$ given in Equation (102) is simply $(1/t_n)$ times the area under the function $i(s,t)$ in Figure 5.16a between $t = 0$ and $t = t_n$.

15. The DLS data for a dilute biological dispersion are given below. The experiment was conducted at 293K in water ($\eta = 0.001$ kg/m s). The scattering intensity was measured at a fixed angle of 30° using a laser with $\lambda = 480$ nm. Determine the average diffusion coefficient and the diameter of the particles based on this diffusion coefficient. What can you say about the polydispersity of the sample studied?

$t_d \times 10^3$(s)	0.05	0.1	0.15	0.2	0.25
$\ln [g_2(s,t_d) - 1]^{1/2}$	-0.25	-0.499	-0.747	-0.995	-1.242
$t_d \times 10^3$(s)	0.3	0.35	0.4	0.45	0.5
$\ln [g_2(s,t_d) - 1]^{1/2}$	-1.489	-1.735	-1.98	-2.225	-2.469

6

Surface Tension and Contact Angle

Application to Pure Substances

I don't think I said anything about the Third Dimension; and I am sure I did not say one word about "Upward, not Northward," for that would be nonsense, you know. How could a thing move Upward, and not Northward? Upward and not Northward! . . . How silly it is!

From Abbott's *Flatland*

6.1 INTRODUCTION

6.1a What Are Surface Tension and Contact Angle?

Why is it that insects like beetles can walk on water? Why do the bristles of a brush immersed in water cling together as the brush is pulled out? Phenomena such as these arise because of a special property of interfaces that separate two phases. Let us consider another example first. Everyone has had the experience of pouring more beverage into a cup or glass than that container could hold. In addition to the spills this causes, such an experience provides an opportunity to observe surface tension. Most liquids can be added to a vessel until the liquid surface bulges above the rim of the container. The liquid behaves as if it had a "skin" that prevents it—up to a point—from overflowing. Stated technically, a contractile force, which tends to shrink the surface, operates around the perimeter of the surface. This is what we mean when we talk about the *surface tension* of a liquid. All phase boundaries behave this way, not just liquid surfaces; however, the evidence for this is more apparent for deformable liquid surfaces.

The liquid "skin" described above is anchored to the solid walls of the container around the edges of the surface. The angle the liquid surface makes with the solid support is called the *contact angle*. The tendency of most liquids to climb walls—think of a meniscus in a capillary—is a manifestation of the existence of these angles.

Surface tension and contact angle are bulk properties, but they are both a consequence of intermolecular interactions that are short range in macroscopic terms. In the case of water, for example, the molecules like to be in the interior where they can form up to four hydrogen bonds; that is, the hydrogen atoms in a water molecule are weakly bonded to the oxygens of the neighboring molecules. At a water-air interface, however, the molecules have fewer neighbors, and water tries to minimize the number of "broken" bonds by minimizing the surface area. The bristles of the brush try to cling together as a consequence. The beetle does not sink because that would require the stretching of the surface. Unless an insect is heavy enough to counteract the force of surface tension, the insect will not sink and can walk on the surface. We have more to say about the molecular origin of both surface tension and contact angle in this chapter and in Chapter 10.

Surface tension and contact angle, however, are two different things, although they are closely related. Surface tension is a property of the *interface* between two phases, whereas the contact angle describes the edge of the two-phase boundary where it ends at a third phase. Two phases must be specified to describe surface tension; three are needed to describe contact angle. (In this chapter, we designate solid, liquid, and vapor phases by S, L, and V, respectively; if two liquids are involved, we call them L_1 and L_2 or A and B.)

6.1b Why Are They Important?

Surface tension and contact angle phenomena play a major role in many practical things in life. Whether a liquid will spread on a surface or will break up into small droplets depends on the above properties of interfaces and determines well-known operations such as detergency and coating processes and others that are, perhaps, not so well known, for example, preparation of thin films for resist lithography in microelectronic applications. The challenge for the colloid scientist is to relate the macroscopic effects to the interfacial properties of the materials involved and to learn how to manipulate the latter to achieve the desired effects. Vignette VI provides an example.

VIGNETTE VI   SURFACE CHEMISTRY AND MATERIALS SCIENCE:
Wettability of Surfaces and Fabrication of
Microstructured Materials

Why is understanding surface tension, contact angle, and their role in wettability of surfaces important? Waterproof fabrics, detergency, and coating processes (e.g., preparation of coated surfaces using spin coating) are standard examples that, perhaps, readily come to mind, but wettability of surfaces has a strong influence on many other phenomena of engineering significance as well. For example, the rate of evaporative heat transfer from surfaces (and, as a result, the effectiveness of heat-transfer equipment in which condensation and evaporation play a major role in the overall heat transfer) is influenced strongly by whether the liquid phase wets the surfaces or not.

Instead of considering wettability from the above traditional perspective, let us look at it from the opposite side: What if we can control wettability of a surface at microscopic length scales, perhaps *actively*? New developments in surface chemistry and fabrication of self-assembled monolayers (SAMs; see Chapter 7) have opened up precisely such a possibility. How can we take advantage of such opportunities?

As we see in this chapter, chemical affinities between a surface and a liquid at the molecular level (i.e., preferential molecular interactions) determine the wettability of a surface and the resulting shapes of the liquid drops. The chemical affinities can be controlled by allowing the formation of monolayers of surfactant molecules with desired properties (lyophilicity or lyophobicity) on a surface. Such a procedure, in combination with micromachining (for example, using a scanning tunneling microscope; see Vignette I.8 in Chapter 1) can be used to prepare surfaces with prechosen geometric patterns of wettability.

Figure 6.1 illustrates such an example (Abbott et al. 1992). Here, a gold surface is first coated with a hydrophilic SAM of alkanethiolate. The desired geometric pattern is then formed on the surface through micromachining (in this case, using a surgical scalpel), and the resulting features are covered with a hydrophobic SAM (of dialkyl disulfide) (see Fig. 6.1a). This procedure can be used to construct micrometer-scale hydrophobic lines on the surface so that the resulting shapes and distribution of liquid drops can be controlled (see Fig. 6.1b).

Similar techniques can also be used to control the wettability of a surface *actively* (in time scales of the order of seconds), for example, by electrochemically controlling the interconversion of less polar (less wettable) groups in the SAMs to more polar (more wettable) groups (Abbott and Whitesides 1994).

What benefits do we derive from such control? Control of wettability on small length scales using simple "wet chemistry" provides interesting opportunities, e.g., a simpler and more easily accessible alternative to optical lithography and new methods for the fabrication

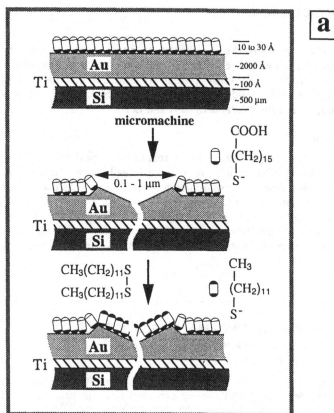

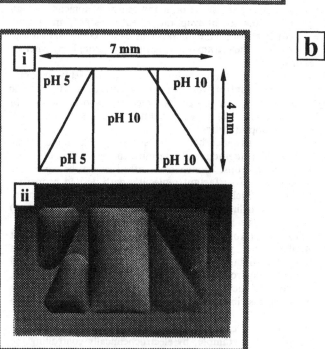

FIG. 6.1 Control of wettability of surfaces through chemistry: (a) schematic illustration of formation of hydrophobic lines on a hydrophilic surface with self-assembled monolayers (SAMs) and micromachining; (b) top view of the shapes and confinement of water drops on an engineered surface. (Reproduced with permission of Abbott et al. 1992.)

of optical switches, electrochemical valves, and pumps for the preparation and operation of diagnostic assays. The technique described above can also be used to generate droplets of controlled shapes so that analyses of such shapes can be used to measure surface excess free energies (Abbott et al. 1994).

Active control of wettability is a subject of current research activity and is beyond our scope here. However, the first step in this process is a study of interfacial tension and contact angle, their basis in thermodynamics, and methods to measure these properties. This is the objective of this chapter.

6.1b   Focus of This Chapter

Both this chapter and Chapter 7 are primarily concerned with the equilibrium behavior of surfaces. In this chapter the emphasis is on the surfaces between "pure" phases, whereas in Chapter 7 we consider the effects of solutes on the behavior of surfaces. Most of the phenomena we describe in this chapter will continue to apply to solutions and carry over into the next chapter.

Our major objectives in this chapter are to define surface tension and contact angle and describe how they are measured and what effects they have on the equilibrium behavior of materials.

1.   We begin with the definitions and a preliminary look at some measurement techniques in Section 6.2 and show that surface tension can also be thought of as the energy needed to create an interface (Section 6.3).

2.   The presence of surface tension has an important implication for the pressures across a curved interface and, as a consequence, for phase equilibria involving curved interphase boundaries. The equation that relates the pressure difference across an interface to the radii of curvature, known as the *Laplace equation*, is derived in Section 6.4, and the implications for phase equilibria are considered for some specific cases in Section 6.5.

3.   As mentioned in Section 6.1a, surface tension and contact angle determine wetting phenomena; we examine this in Section 6.6. We take a closer look at the definition of contact angle and some complications associated with it in Section 6.7.

4.   Now we are better equipped to consider measurements of surface tension in detail. Section 6.8 describes the use of shapes of menisci, drops, and bubbles for such measurements, and Section 6.9 considers the practically important case of contact of liquids with porous solids and powders.

5.   Finally, we close the chapter with a discussion of the relation between molecular interactions and surface tension and contact angle.

6.2   SURFACE TENSION AND CONTACT ANGLE: A FIRST LOOK

6.2a   Surface Tension as a Force

As the name implies, surface tension (denoted by γ) is a force that operates on a surface and acts perpendicular and inward from the boundaries of the surface, tending to decrease the area of the interface. In order to illustrate this, we use a simple apparatus based on this notion that might be used to measure surface tension (see Fig. 6.2a). In fact, other methods are used, but the arrangement of Figure 6.2a has the advantage of simplicity and serves to illustrate how the tension in the liquid surface is indeed measured by γ. The figure represents a loop of wire with one movable side on which a film could be formed by dipping the frame into a liquid. The surface tension of a stretched film in the loop will cause the slide wire to move in the direction of decreasing film area unless an opposing force F is applied. In an actual apparatus, the friction of the slide wire might be sufficient for this. In an idealized, frictionless apparatus like that in Figure 6.2a, the force opposing γ could be measured. The force evidently operates along the entire edge of the film and will vary with the length ℓ of the slide wire. Therefore it

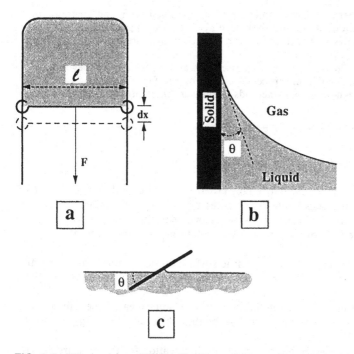

FIG. 6.2   Illustrations of liquid film formation, contact angle, and measurement of contact angle: (a) a wire loop with a slide wire on which a liquid film might be formed and stretched by an applied force **F**. (b) profile of a three-phase (solid, liquid, gas) boundary that defines the contact angle θ. (c) the tilted plate method for measuring contact angles.

is the force per unit length of edge that is the intrinsic property of the liquid surface. Since the film in Figure 6.2a has two sides, the surface tension as measured by this apparatus equals

$$\gamma = F/2\ell \tag{1}$$

Several points should be noted before proceeding any further:

1. Equation (1) defines the units of surface tension to be those of force per length or Nm^{-1} in SI or dynes cm^{-1} in the cgs system. We see presently that these are not the only units used for γ.
2. The apparatus shown in Figure 6.2a resembles a two-dimensional cylinder/piston arrangement. With this similarity in mind, the suggestion that surface tension is analogous to a two-dimensional pressure seems plausible. With certain refinements, this notion will prove very useful in Chapter 7.
3. A gas in the frictionless, three-dimensional equivalent to the apparatus of the figure would tend to expand spontaneously. For a film, however, the direction of spontaneous change is contraction.

6.2b   Contact Angle

A quantity that is closely related to surface tension is the *contact angle*. The contact angle θ is defined as the angle (measured in the liquid) that is formed at the junction of three phases, for example, at the solid-liquid-gas junction as shown in Figure 6.2b. Although the surface tension is a property of the two phases that form the interface, θ requires that three phases be specified for its characterization, as mentioned above. The above definition of contact angle is, however, highly simplified, and we take a more in-depth look at the concept later in this chapter.

6.2c   Measuring Surface Tension and Contact Angle: Round One

In this section we take an initial look at the experimental determination of γ and θ. The methods we discuss show the complementarity between these two parameters while introducing some important phenomena. As we proceed through the chapter, some additional complications are encountered. We return to the topic of measuring γ and θ in Section 6.8, in which some refinements can be discussed with more meaning.

Contact angle seems like a more straightforward quantity to measure than surface tension. Junctions such as that illustrated in Figure 6.2b are easy to observe and even photograph. This being the case, it seems easy enough to construct a tangent to the liquid surface at the point where it contacts the support and measure the enclosed angle. Actually, this is considerably easier to describe than to carry out. Although the foregoing is impractical as an experimental strategy, it does make the contact angle very accessible conceptually.

Figure 6.2c shows a variation of Figure 6.2b that has been used extensively to measure contact angles. This technique, known as the *tilted plate method*, varies the angle of inclination between a smooth solid penetrating the liquid surface until a position is found at which the liquid makes a horizontal contact with the solid. The horizontal is easier to judge than the tangent to a curve, so this method is preferable to that based on direct measurement of a junction like the one shown Figure 6.2b.

The situation shown in Figure 6.2b is one in which surface tension and contact angle considerations pull a liquid upward in opposition to gravity. A mass of liquid is drawn up as if it were suspended by the surface from the supporting walls. At equilibrium the upward pull of the surface and the downward pull of gravity on the elevated mass must balance. This elementary statement of force balance applies to two techniques by which γ can be measured if θ is known: the *Wilhelmy plate* and *capillary rise*.

6.2c.1   The Wilhelmy Plate Technique

Figure 6.3a represents a thin vertical plate suspended at a liquid surface from the arm of a tared balance. For simplicity, the plate is positioned so that the lower edge is in the same plane as the horizontal surface of the liquid away from the plate as shown in the figure. The manifestation of surface tension and contact angle in this situation is the entrainment of a

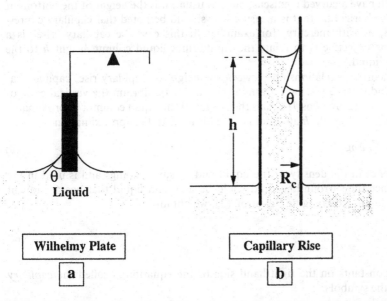

FIG. 6.3   Surface tension and capillary rise: (a) the Wilhelmy plate method for measuring γ; (b) schematic illustration of capillary rise in a cylindrical tube of radius R_c.

meniscus around the perimeter of the suspended plate. Assuming the apparatus is balanced before the liquid surface is raised to the contact position, the imbalance that occurs on contact is due to the weight of the entrained meniscus. Since the meniscus is held up by the tension on the liquid surface, the weight measured by the apparatus can be analyzed to yield a value for γ.

The observed weight w of the meniscus must equal the upward force provided by the surface. This, in turn, equals the vertical component of $\gamma - \gamma \cos \theta$, where θ is the contact angle — times the perimeter of the plate P since surface tension is a force per unit length of surface edge. Therefore, at equilibrium,

$$w = P\gamma \cos \theta \tag{2}$$

For a plate of rectangular cross section having length ℓ and thickness t, $P = 2(\ell + t)$; these dimensions can be accurately measured. By suspending the plate from a sensitive balance, we can also measure w with considerable accuracy. The apparatus is called a *Wilhelmy balance*, and the technique the *Wilhelmy plate method*. Thus, if the contact angle is known from an independent determination by, say, the tilted-plate method, then γ can be evaluated by Equation (2).

Strictly speaking, Equation (2) allows the vertical component of surface tension to be measured. Since this equals $\gamma \cos \theta$, we are actually making a single measurement that involves two parameters. If γ were independently known, the Wilhelmy plate method could also be used to determine θ. Whether we seek to evaluate γ, θ, or both, two experiments are needed, and these may not both involve the factor $\cos \theta$. In Section 6.8a we discuss a second type of measurement that can be made with the Wilhelmy apparatus that supplies a complementary observation so both γ and θ can be determined on a single instrument.

6.2c.2   The Capillary Rise Technique

Capillary rise is also a measure of the vertical component of surface tension, so it is an alternative to the Wilhelmy balance and not its complement. Figure 6.3b shows how a capillary rise experiment is carried out. Conventionally, the height of a liquid column in a capillary above the reference level in a large dish is measured. It is important that the dish be large enough in diameter so that the reference level has a well-defined horizontal surface. In the capillary the liquid will have a curved meniscus, and, as usual, it is the height of the bottom of the meniscus above the horizontal that is measured. It should be noted that capillary depression is also observed, as with mercury, for example. In this case the capillary "rise" is a negative quantity. Our objective is to relate the equilibrium liquid column height h to the surface tension of the liquid.

A simple — but incorrect — relationship between the height of capillary rise, capillary radius, contact angle, and surface tension is easily derived. At equilibrium the vertical component of the surface tension ($2\pi R_c \gamma \cos \theta$) equals the weight of the liquid column, approximated as the weight of a cylinder of height h and radius R_c. This leads to the approximation

$$2\pi R_C \gamma \cos \theta \approx \pi R_c^2 h \Delta \rho g \tag{3}$$

Here $\Delta \rho$ is the difference in the density of the liquid and its surroundings and is used in the above formulation since the surrounding fluid could be a second liquid that has a buoyant effect on the liquid column. Rearranging Equation (3), we obtain

$$\frac{R_c h}{\cos \theta} \approx \frac{2\gamma}{\Delta \rho g} \tag{4}$$

where the cluster of constants on the right-hand side of the equation is called the *capillary constant* and is given the symbol a^2:

$$a^2 = \frac{2\gamma}{\Delta \rho g} \tag{5}$$

Since $a^2 \propto R_c h$, a has units of length. Note that the height to which a liquid climbs in a capillary (assuming $\theta < 90°$) increases as R_c decreases. As would be expected, h is larger for large γ and small $\Delta \rho$. By measuring h for a capillary of known radius, Equation (5) permits an approximate value of γ to be determined if θ is known.

The approximation that limits this analysis of capillary rise originates from neglecting the weight of the liquid in the "crown" of the curved meniscus. We see in Section 6.8b that the height of capillary rise can be related to surface tension without making this approximation, although the connection is somewhat unwieldy. A more detailed description of the experimental aspects of the capillary rise method can be obtained from advanced textbooks (e.g., Adamson 1990).

There are numerous other methods for measuring surface tension that we do not discuss here. These include (a) the measurement of the maximum pressure beyond which an inert gas bubble formed at the tip of a capillary immersed in a liquid breaks away from the tip (the so-called *maximum bubble-pressure method*); (b) the so-called *drop-weight method*, in which drops of a liquid (in a gas or in another liquid) formed at the tip of a capillary are collected and weighed; and (c) the *ring method*, in which the force required to detach a ring or a loop of wire is measured. In all these cases, the measured quantities can be related to the surface tension of the liquid through simple equations. The basic concepts involved in these methods do not differ significantly from what we cover in this chapter. The experimental details may be obtained from Adamson (1990).

It is impossible to complete a discussion of the measurement of surface tension without saying something about the need for extreme cleanliness in any determination of γ. Any precision chemical measurement requires attention to this consideration, but surfaces are exceptionally sensitive to impurities. It is often noted that touching the surface of 100 cm^2 of water with a fingertip deposits enough contamination on the water to introduce a 10% error in the value of γ. Not only must all pieces of equipment be clean, but also the experiments must be performed within enclosures or in very clean environments to prevent outside contamination. In addition, both surface tension and contact angle should be measured under constant temperature conditions.

Both the Wilhelmy and capillary rise methods for determining γ have been based on the concept of surface tension as a force. While this point of view is useful for describing the experimental methods we have discussed, it is only one way of interpreting γ. An energetic interpretation is also possible that makes surface tension amenable to the powerful methods of thermodynamics.

6.3 THERMODYNAMICS OF SURFACES: SURFACE TENSION AS SURFACE FREE ENERGY

Application of a force infinitesimally larger than the equilibrium force to the slide wire in Figure 6.2a will displace the wire through a distance dx. The product of force and distance equals energy, in this case the energy spent in increasing the area of the film by the amount $dA = 2\ell\,dx$. Therefore the work done on the system is given by

$$\text{Work} = Fdx = \gamma 2\ell dx = \gamma dA \tag{6}$$

This supplies a second definition of surface tension: It equals the work per unit area required to produce a new surface. In terms of this definition, the units of γ are energy per area—J m^{-2} in SI or erg cm^{-2} in the cgs system.

We see, therefore, that there are two equivalent interpretations of γ: force per unit length of boundary of the surface and energy per unit area of the surface. The dimensional equivalency of the two is evident if the numerator and denominator of force length^{-1} are multiplied by length.

Equation (6) relates γ to the work required to increase the area of a surface. From thermodynamics, it will be recalled that work is a path-dependent process: How much work is done depends on how it is done. Based on this realization, then, it seems desirable to examine

Equation (6) a little more fully. As already noted, there is a tendency for mobile surfaces to decrease spontaneously in area. Therefore it is convenient to shift our emphasis from work done on the system to work done by the system in such a reduction of area. If the quantity $\delta w'$ is defined as the work done by the system when its area is changed, then Equation (6) becomes

$$\delta w' = -\gamma dA \tag{7}$$

According to Equation (7), a decrease in area (negative dA) corresponds to work done by the system, whereas an increase in area requires work to be done on the system (positive dA and negative $\delta w'$). This sign convention is consistent with the idea expressed in Chapter 1 that energy is stored in surfaces.

We are now in a position to relate the quantity $\delta w'$ to other thermodynamic variables. To do this a brief review of some basic thermodynamics is useful.

According to the first law of thermodynamics, the change in the energy E of a system equals

$$dE = \delta q - \delta w \tag{8}$$

in which δw is the work done by the system and δq is the heat absorbed by the system. The quantity δw is conveniently divided into a pressure-volume term and a non-pressure-volume term:

$$\delta w = \delta w_{pV} + \delta w_{non\text{-}pV} = pdV + \delta w_{non\text{-}pV} \tag{9}$$

It will be recalled from physical chemistry that chemical work is the usual substitution for $\delta w_{non\text{-}pV}$. However, the work defined by Equation (7) may also be classified as non-pressure-volume work.

The second law of thermodynamics tells us that for reversible processes

$$\delta q_{rev} = TdS \tag{10}$$

Substituting Equations (9) and (10) into Equation (8), with the stipulation of reversibility as required by Equation (10), enables us to write

$$dE_{rev} = TdS - pdV - \delta w_{non\text{-}pV} \tag{11}$$

Next we recall the definition of the Gibbs free energy G:

$$G = H - TS = E + pV - TS \tag{12}$$

which may be differentiated to give

$$dG = dE + pdV + Vdp - TdS - SdT \tag{13}$$

Substituting Equation (11) into (13) gives

$$dG_{rev} = TdS - pdV - \delta w_{non\text{-}pV} + pdV + Vdp - TdS - SdT \tag{14}$$

This is a fundamental equation of physical chemistry because it enables us to assign a physical significance to G as defined by Equation (12). Equation (14) shows that for a constant temperature, constant pressure, and reversible process

$$dG = -\delta w_{non\text{-}pV} \tag{15}$$

that is, dG equals the maximum non-pressure-volume work derivable from such a process since maximum work is associated with reversible processes.

We have already seen by Equation (7) that changes in surface area entail non-pressure-volume work. Therefore we identify $\delta w'$ from Equation (7) with $\delta w_{non\text{-}pV}$ in Equation (15) and write

$$dG = \gamma dA \tag{16}$$

Even better, in view of the stipulations made going from Equation (14) to Equation (15), we write

TABLE 6.1 Several Representative Values of γ, S^s, and H^s for a Variety of Liquids Near Room Temperature

Substance	γ at 20°C (mJ m^{-2})	$d\gamma/dT = -S^s$ (mJ m^{-2} K^{-1})	$H^s = \gamma - T(d\gamma/dT)$ (mJ m^{-2})
n-Hexane	18.4	−0.105	49.2
Ethyl ether	17.0	−0.116	51.0
n-Octane	21.8	−0.096	49.9
Carbon tetrachloride	26.9	−0.092	53.9
m-Xylene	28.9	−0.077	51.4
Toluene	28.5	−0.081	52.2
Benzene	29.0	−0.099	58.0
Chloroform	28.5	−0.135	68.3
1,2-Dichloroethane	32.2	−0.139	72.9
Carbon disulfide	32.3	−0.138	72.7
Water	72.8	−0.152	117.3
Mercury	484	−0.220	548

Source: D. H. Kaelbe, *Physical Chemistry of Adhesion*, Wiley, New York, 1971.

$$\gamma = \left(\frac{\partial G}{\partial A}\right)_{T,p} \tag{17}$$

Several things should be noted about Equation (17). This relationship identifies the surface tension as the increment in Gibbs free energy per unit increment in area. *The path-dependent variable $\delta w'$ is replaced by a state variable as a result of this analysis.* Another notation that is often encountered that emphasizes the fact that γ is identical to the excess Gibbs free energy per unit area arising from the surface is to write it as G^s. The energy interpretation of γ, then, has been carried to the point at which it has been identified with a specific thermodynamic function. Many of the general relationships that apply to G apply equally to γ. For example,

$$G^s = \gamma = H^s - TS^s \tag{18}$$

and

$$\left(\frac{\partial G^s}{\partial T}\right)_p = \left(\frac{\partial \gamma}{\partial T}\right)_p = -S^s \tag{19}$$

Equations (18) and (19) may be combined to give

$$\gamma = H^s + T\left(\frac{\partial \gamma}{\partial T}\right)_p \tag{20}$$

For water at 20°C, $\gamma = 7.28 \cdot 10^{-2}$ J m^{-2}; it is convenient to write this in millijoules since 72.8 mJ m^{-2} is numerically equal to the cgs value. For water $d\gamma/dT = -0.152$ mJ m^{-2} deg^{-1}; therefore $H^s = 117$ mJ m^{-2} by Equation (20). Additional values of G^s, S^s, and H^s for various substances are listed in Table 6.1.

6.4 SURFACE TENSION: IMPLICATIONS FOR CURVED INTERFACES AND CAPILLARITY

6.4a Pressure Difference across a Curved Interface: The Laplace Equation

In Section 6.2 we discussed the Wilhelmy and capillary rise experiments as if the supported liquid were hanging from a surface skin. While this is a convenient device for generating formulas, it is not an adequate description of the physical situation. Now that we have established the connection between surfaces and thermodynamics, we can remedy this situation. The insight that is central to this development is the realization that a pressure difference operates across a curved interface. The pressure difference is such that the greater pressure is on the concave side. Our objective in this section is to relate this pressure difference to the curvature of the surface.

Figure 6.4 shows a portion $ABCD$ of a curved surface. The surface has been cut by two planes perpendicular to one another. Each of the planes therefore contains a portion of arc where it intersects the curved surface. In the figure the radii of curvature are designated R_1 and R_2, and the lengths are designated x and y, respectively, for these two intercepted arcs. Now suppose the curved surface is moved outward by a small amount dz to a new position $A'B'C'D'$. Since the corners of the surface continue to lie along extensions of the diverging radial lines, this move increases the arc lengths to $x + dx$ and $y + dy$. Obviously the area of surface must also increase. The work required to accomplish this must be supplied by a pressure difference Δp across the element of surface area.

The increase in area when the surface is displaced is given by

$$dA = (x + dx)(y + dy) - xy = xdy + ydx + dxdy \approx xdy + ydx \qquad (21)$$

where the approximation arises from neglecting second-order differential quantities. The increase in free energy associated with this increase in area is given by γdA:

$$dG = \gamma(xdy + ydx) \qquad (22)$$

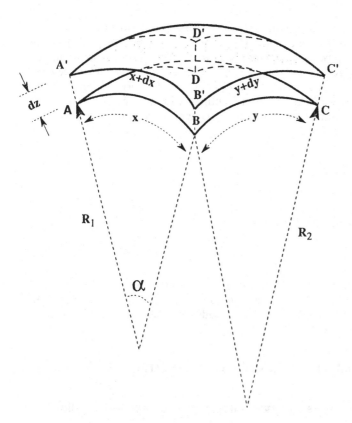

FIG. 6.4 Definition of coordinates describing the displacement of an element of curved surface $ABCD$ to $A'B'C'D'$.

If ordinary pressure-volume work is responsible for the expansion of this surface, then the work equals $\Delta p dV$, where dV is the volume swept by the moving surface. In terms of Figure 6.4, this equals

$$dw = \Delta p xy dz \tag{23}$$

Setting Equations (22) and (23) equal to one another gives

$$\gamma(xdy + ydx) = \Delta p xy dz \tag{24}$$

Notice that the arc lengths x and $(x + \Delta x)$ in Figure 6.4 are related to the angle α by $x = R_1\alpha$ and $(x + \Delta x) = (R_1 + dz)\alpha$. Therefore, we may set up the following proportions:

$$\frac{x + dx}{R_1 + dz} = \frac{x}{R_1} \tag{25}$$

Similarly,

$$\frac{y + dy}{R_2 + dz} = \frac{y}{R_2} \tag{26}$$

Equations (25) and (26) simplify to

$$\frac{dx}{xdz} = \frac{1}{R_1} \tag{27}$$

and

$$\frac{dy}{ydz} = \frac{1}{R_2} \tag{28}$$

Substituting Equations (27) and (28) into Equation (24) enables us to write the relationship of Δp to R_1, R_2, and γ:

$$\Delta p = \gamma\left(\frac{1}{R_1} + \frac{1}{R_2}\right) \tag{29}$$

This expression is known as the *Laplace equation* and was derived in 1805.

Until now we have not been specific as to the location of the curved surface under consideration. Since this is the case, Equation (29) is general and applies equally well to geometrical bodies with radii of curvature that are constant over the entire surface or to more intricate shapes for which the R_i's change from place to place on the surface. For the former category, there are several special cases of Equation (29) that are worthy of note:

1. For a spherical surface, $R_1 = R_2 = R_s$; therefore

$$\Delta p = 2\gamma/R_s \tag{30}$$

2. For a cylindrical surface, $R_1 \to \infty$; therefore

$$\Delta p = \gamma/R_2 \tag{31}$$

3. For a planar surface, $R_1 = R_2 \to \infty$; therefore

$$\Delta p = 0 \tag{32}$$

It is also possible for a portion of a surface to be locally saddle shaped; in such a case the two radii of curvature lie on opposite sides of the surface and have different signs. It is possible for p to be zero in this situation also.

The Laplace equation applied specifically to spherical surfaces can be derived in a variety of ways. Example 6.1 considers an alternative derivation that points out the thermodynamic character of the result quite clearly.

* * *

EXAMPLE 6.1 *Laplace Equation for Spherical Surfaces: A Thermodynamic Derivation.* The Maxwell relations play an important role in thermodynamics. By including the term dA in the usual differential form for dG, show that $(\partial V/\partial A)_{p,T} = (\partial\gamma/\partial p)_{A,T}$. Evaluate $(\partial V/\partial A)_{p,T}$ assuming a spherical surface and, from this, derive the Laplace equation for this geometry.

Solution: The derivation of the Maxwell relations treats dG as an exact differential and expands it as

$$dG = (\partial G/\partial p)_T dp + (\partial G/\partial T)_p dT$$

The coefficients are then matched with their counterparts in an alternative expression for dG: $dG = Vdp - SdT$. When surface effects are included, these two expressions become

$$dG = (\partial G/\partial p)_{T,A} dp + (\partial G/\partial T)_{p,A} dT + (\partial G/\partial A)_{p,T} dA$$

and

$$dG = Vdp - SdT + \gamma dA$$

or, at constant temperature,

$$dG = (\partial G/\partial p)_A dp + (\partial G/\partial A)_p dA \qquad \text{and} \qquad dG = Vdp + \gamma dA$$

Since the order of differentiation is immaterial for exact differentials, it follows that

$$\partial/\partial A[(\partial G/\partial p)_A]_p = \partial/\partial p[(\partial G/\partial A)_p]_A$$

or

$$(\partial V/\partial A)_p = (\partial\gamma/\partial p)_A$$

Since dV/dA can be written $(dV/dR_s)(dR_s/dA)$, which is easily evaluated for a sphere of radius R_s as $(4\pi R_s^2)(8\pi R_s)^{-1} = R_s/2$. Therefore $(\partial\gamma/\partial p)_{A,T} = R_s/2$ or $d\gamma = (1/2) R_s dp$. This result may be integrated to give $\gamma = (1/2) R_s\Delta p$ where the constant of integration has been set equal to zero since there would be no pressure difference if $\gamma = 0$. This is the Laplace equation for spherical particles.   ∎

* * *

6.4b   Laplace Equation and Capillary Rise

As noted above, it is possible that a different pair of radii of curvature applies at different locations on a surface. In this case the Laplace equation shows that Δp also varies with location. This is the reason for the variation of the pressure with z in the meniscus shown in the capillary in Figure 6.3b. As is often true of pressures, it is convenient to define pressure variations relative to some reference plane.

With this idea in mind, the horizontal surface in Figure 6.3b can be taken as a reference level at which $\Delta p = 0$. Just under the meniscus in the capillary the pressure is less than it would be on the other side of the surface owing to the curvature of the surface. The fact that the pressure is less in the liquid in the capillary just under the curved surface than it is at the reference plane causes the liquid to rise in the capillary until the liquid column generates a compensating hydrostatic pressure. The capillary possesses an axis of symmetry; therefore at the bottom of the meniscus the radius of curvature is the same in the two perpendicular planes that include the axis. If we identify this radius of curvature by b, then the Laplace equation applied to the meniscus is $\Delta p = 2\gamma/b$. Equating this to the hydrostatic pressure gives

$$2\gamma/b = \Delta\rho gh \tag{33}$$

or

$$bh = 2\gamma/\Delta\rho g = a^2 \tag{34}$$

Comparison of this result with Equation (4) shows the two relations to be of similar form, with the approximate form becoming exact under specific conditions. The condition under which the approximate result is exact is when the meniscus is a perfectly hemispherical depression at the liquid surface. When this is the case, $\theta = 0$ and $R_{cap} = R_{men} = b$. Since the meniscus usually only approximates a hemisphere, the use of the capillary radius in Equation (4) remains an approximate relationship. Although Equation (34) is exact, it is not particularly useful since b is not readily measurable. Note that if b were available, Equation (34) would permit the evaluation of γ without requiring that θ be known.

6.4c Arbitrary Variations in Radii of Curvature: Generalization of the Laplace Equation

A variety of drop, bubble, and meniscus shapes have axial symmetry. As is the case in the capillary in Figure 6.3b, p varies with z and, in general, the two radii of curvature may vary from position to position on the surface also. With these ideas in mind, the Laplace equation becomes

$$\Delta p(z) = \gamma[R_1^{-1}(x,y,z) + R_2^{-1}(x,y,z)] \tag{35}$$

In Equation (35) we have added the notion that Δp, R_1^{-1}, and R_2^{-1} may be functions of location in space for any given surface.

The following expressions from analytical geometry are general functions for R_1^{-1} and R_2^{-1} for surfaces with an axis of symmetry:

$$R_1^{-1} = \frac{d^2z/dx^2}{[1 + (dz/dx)^2]^{3/2}} \tag{36}$$

and

$$R_2^{-1} = \frac{dz/dx}{x[1 + (dz/dx)^2]^{1/2}} \tag{37}$$

Substitution of Equations (36) and (37) into Equation (35) generates a complicated differential equation with a solution that relates the shape of an axially symmetrical interface to γ. In principle, then, Equation (35) permits us to understand the shapes assumed by mobile interfaces and suggests that γ might be measurable through a study of these shapes. We do not pursue this any further at this point, but return to the question of the shape of deformable surfaces in Section 6.8b. In the next section we examine another consequence of the fact that curved surfaces experience an extra pressure because of the tension in the surface. We know from experience that many thermodynamic phenomena are pressure sensitive. Next we examine the effect of the increment in pressure small particles experience due to surface curvature on their thermodynamic properties.

6.5 EFFECTS OF CURVED INTERFACES ON PHASE EQUILIBRIA AND NUCLEATION: THE KELVIN EQUATION

In addition to capillarity, another important consequence of the pressure associated with surface curvature is the effect it has on the thermodynamic activity of substances. As a consequence, phase equilibria (including dissolution of chemical species in the different phases) are affected by the presence of interfaces. In this section we consider a few such cases.

6.5a Effect on Vapor-Liquid Equilibria

The influence of curvature on phase equilibria is most readily understood for liquids, for which the activity is measured by the vapor pressure of the liquid. Accordingly, suppose we consider the process of transferring molecules of a liquid from a bulk phase with a vast horizontal surface to a small spherical drop of radius R_s.

According to Equation (32), no pressure difference exists across a plane surface; the pressure is simply p_0, the normal vapor pressure. However, a pressure difference given by Equation (30) exists across a spherical surface. Therefore for liquid-vapor equilibrium at a spherical surface, both the liquid and the vapor must be brought to the same pressure $p_0 + \Delta p$. If we assume the liquid to be incompressible and the vapor to be ideal, ΔG for the process of increasing the pressure from p_0 to $p_0 + \Delta p$ is given by the following:

1. For the liquid,

$$\Delta G = \int_{p_0}^{p_0 + \Delta p} \overline{V}_L \, dp = \overline{V}_L \Delta p = \frac{2 \overline{V}_L \gamma}{R_s} \tag{38}$$

where $\overline{V}_L$ is the molar volume of the liquid.

2. For the vapor, since $\overline{V} = RT/p$ (in view of the ideal gas assumption),

$$\Delta G = RT \ln\left(\frac{p_0 + \Delta p}{p_0}\right) = RT \ln\left(\frac{p}{p_0}\right) \tag{39}$$

When liquid and vapor are at equilibrium, these two values of ΔG are equal:

$$RT \ln\left(\frac{p}{p_0}\right) = \frac{2 \overline{V}_L \gamma}{R_s} = \frac{2 M \gamma}{\rho R_s} \tag{40}$$

since the volume per mole equals M/ρ, where M and ρ are the molecular weight and density of the liquid, respectively. In either of these forms, this expression is known as the *Kelvin equation*. The Kelvin equation enables us to evaluate the actual pressure above a spherical surface and not just the pressure difference across the interface, as is the case with the Laplace equation.

In our derivation of the Kelvin equation, the radius is measured in the liquid. For a gas bubble in a liquid, the same equilibrium is involved, but the bubble radius is measured on the opposite side of the surface. As a consequence, a minus sign enters Equation (40) when it is applied to bubbles. Since γ is on the order of millijoules while RT is on the order of joules, Equation (40) predicts that ln (p/p_0) is very small. It is important to realize, however, that γ/R is also divided by the radius of the spherical particle and therefore becomes more important as R_s decreases. For water at 20°C, the Kelvin equation predicts values of p/p_0 equal to 1.0011, 1.0184, 1.1139, and 2.94 for drops of radius 10^{-6}, 10^{-7}, 10^{-8}, and 10^{-9} m, respectively. For bubbles of the same sizes in liquid water, p/p_0 equals 0.9989, 0.9893, 0.8976, and 0.339. These calculations show that the effect of surface curvature, while relatively unimportant even for particles in the micrometer range, becomes appreciable for very small particles.

The Kelvin effect is not limited to spheres. For example, the "neck" of liquid between two supports is described by a paraboloid of revolution. In this case the radius of curvature of the concave surface is outside the liquid and therefore introduces a minus sign into Equation (40). In addition, the factor of 2 is not required to describe the Kelvin effect for this geometry. Example 6.2 illustrates a test of the Kelvin equation based on this kind of liquid neck.

* * *

EXAMPLE 6.2   *Use of the Kelvin Equation for Determining Surface Tension.*   Figure 6.5 shows a plot of experimental data that demonstrates the validity of the Kelvin effect. Necks of liquid cyclohexane were formed between mica surfaces at 20°C, and the radius of curvature was measured by interferometry. Vapor pressures were measured for surfaces with different curvature. Use these data to evaluate γ for cyclohexane. Comment on the significance of the fact that the linearity of Figure 6.5 extends all the way to a p/p_0 value of 0.77.

Solution: For the paraboloid of revolution, the Kelvin equation becomes ln $(p/p_0) = -(M\gamma/\rho RTR_s)$. Therefore a plot of ln (p/p_0) versus $-1/R_s$ is expected to be a straight line of slope $M\gamma/\rho RT$. The slope of the line in Figure 6.5 is $1.13 \cdot 10^{-9}$ m. Equating this with $M\gamma/\rho RT$ and solving for γ, we obtain $\gamma = (1.13 \cdot 10^{-9})(779)(8.314)(293)/0.084 = 0.0255$ J m^{-2} = 25.5 mJ m^{-2}, which agrees with the value determined by conventional methods.

In addition to proving the Kelvin effect, the data in Figure 6.5 show that the surface tension of cyclohexane may be regarded as constant down to a radius of curvature given by the reciprocal of the abscissa corresponding to $p/p_0 = 0.77$, which is 4 nm. It is remarkable that the

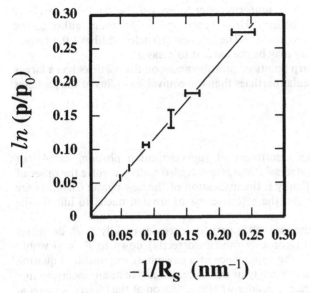

FIG. 6.5 Verification of the Kelvin equation for cyclohexane "necks" of different curvature between mica surfaces (discussed in Example 6.2). (Redrawn from L. R. Fisher and J. N. Israelachvili, *Nature*, **277**, 548 (1979).)

same value of γ as determined on macroscopic surfaces continues to apply to within an order of magnitude of molecular dimensions. ∎

* * *

6.5b Effect on the Solubility of Solids in Liquids

The Kelvin equation may also be applied to the equilibrium solubility of a solid in a liquid. In this case the ratio p/p_0 in Equation (40) is replaced by the ratio a/a_0, where a_0 is the activity of dissolved solute in equilibrium with a flat surface, and a is the analogous quantity for a spherical surface. For an ionic compound having the general formula $M_m X_n$, the activity of a dilute solution is related to the molar solubility S as follows:

$$a = (mS)^m (nS)^n \qquad (41)$$

Therefore for a solid sphere

$$\frac{2 M \gamma}{\rho R_s} = RT \ln\left(\frac{a}{a_0}\right) = (m + n)RT \ln\left(\frac{S}{S_0}\right) \qquad (42)$$

where S and S_0 are the solubilities of the spherical and flat particles, respectively. In principle, Equation (42) provides a thermodynamically valid way to determine γ for an interface involving a solid. The thermodynamic approach makes it clear that curvature has an effect on activity for any curved surface. The surface free energy interpretation of γ is more plausible for solids than the surface tension interpretation, which is so useful for liquid surfaces. Either interpretation is valid in both cases, and there are situations in which both are useful. From solubility studies on a particle of known size, γ_S can be determined by the method of Example 6.2.

Although the increase in solubility of small particles is unquestionably a real effect, using it quantitatively as a means of evaluating γ_{SL} is fraught with difficulties:

1. The difference in solubility between a small particle and a larger one will probably be less than 10%. Since a phase boundary exists at all, the solubility is probably low to begin, so there may be some difficulty in determining the experimental solubilities accurately.

2. Solid particles are not likely to be uniform spheres, even if the sample is carefully fractionated; rather, they will be irregularly shaped and polydisperse, although the particle size distribution may be narrow. The smallest particles will have the largest effect on the solubility, but they may be the hardest to measure.
3. The radius of curvature of sharp points or protuberances on the particles has a larger effect on the solubility of irregular particles than the equivalent radius of the particles themselves.

6.5c Effect on Nucleation

The Kelvin equation helps explain an assortment of supersaturation phenomena. All of these—supercooled vapors, supersaturated solutions, supercooled melts—involve the onset of phase separation. In each case the difficulty is the nucleation of the new phase: Chemists are familiar with the use of seed crystals and the effectiveness of foreign nuclei to initiate the formation of the second phase.

The Kelvin equation shows that the ratio S/S_0 or p/p_0 increases rapidly as R_s decreases toward zero. Applying Equation (40) rigorously (and incorrectly) down to $R_s = 0$ would imply infinite supersaturation and make the appearance of a new phase impossible. Equation (40) is derived on the basis of two phases already in existence. To arrive at an understanding of the emergence of a new phase, we must consider what is going on at the molecular level at the threshold of phase separation. A highly purified vapor, for example, may remain entirely as a gas even though its pressure exceeds the normal vapor pressure of the liquid for the temperature in question. At such a point the vapor state is thermodynamically unstable with respect to the formation of a liquid phase with flat surfaces. Whatever stability the vapor has is kinetic stability, arising from the high activation energy required to start the formation of the second phase. A crude kinetic picture of the processes occurring at the molecular level may be informative.

Although a supersaturated vapor is still a gas, it is a very nonideal gas indeed! Clusters of molecules are continually forming and disintegrating; these are the embryonic nuclei of the new phase. Some of these clusters will be dimers, some trimers, and in general n-mers. Each will have its own characteristic radius $R_{s,n}$. An abbreviated derivation of the rate of formation of n-mers proceeds as follows. The rate law will contain both a frequency factor and a Boltzmann factor. The energy term in the latter may be estimated by Equation (38) with volume of the cluster taken to be $(4/3)\pi R_{s,n}^3$. The frequency factor will involve the probability of additional molecules adding by collision; that is, it will depend on the surface area of the cluster $(4\pi R_{s,n}^2)$ and the frequency of collisions with a wall as given by kinetic molecular theory. Therefore the rate law may be approximated as

$$\text{Rate} \approx Z(4\pi R_{s,n}^2)\exp\left(-\frac{2\gamma}{R_{s,n}}\frac{4\pi R_{s,n}^3}{3 k_B T}\right) = c_1 p R_{s,n}^2 \exp(-c_2 R_{s,n}^2) \tag{43}$$

where Z is the collision frequency. Since the collision frequency certainly increases with p, the pressure has been factored out of the second expression in Equation (43), where c_1 and c_2 are constants.

Two aspects of Equation (43) are especially informative. First, we observe that the pre-exponential term increases with increasing $R_{s,n}$, whereas the exponential term decreases. This means that there exists some critical radius for which the rate law shows a maximum. Clusters with this critical radius may be compared with reaction intermediates or transition states in ordinary chemical reactions. Those clusters that manage to overcome the energy barrier associated with this critical size are capable of further growth, leading to the appearance of the new phase, and smaller clusters disintegrate. In addition, Equation (43) also shows that the rate of cluster growth increases as the pressure increases. All clusters, including those of the critical size, form more rapidly at higher pressures.

The preceding considerations suggest that a point is ultimately reached in the course of increasing supersaturation at which the liquefaction process becomes kinetically as well as thermodynamically favorable. It must be remembered here that the initial state of the system is one of instability, or, more correctly, metastability. Once liquefaction begins, the pressure drops until the radius-pressure combination that satisfies Equation (40) is reached.

A great deal of work has been done on the kinetics of phase formation; the arguments presented here are intended merely to suggest the direction taken in more detailed treatments. Many aspects of nucleation are of extreme interest in colloid and surface chemistry. The monodisperse colloids are formed by carefully controlling the formation of the solid phase. In the case of the monodisperse sulfur sols, for example, the decomposition of $S_2O_3^{2-}$ (Chapter 5, Section 5.7c) proceeds slowly to quite a high level of supersaturation (the solution must be free of foreign nuclei). Ultimately, clusters of sulfur atoms exceeding the critical size form, nucleate precipitation, and grow until the supersaturation is relaxed. Any additional sulfur that is formed beyond this point will deposit on these particles without forming a new "crop" of nuclei. Relatively monodisperse colloids formed by such condensation processes depend on the fact that supersaturation, and therefore nucleation, occur only once during the formation of the colloid. If the rate of crystallization were too slow, then a second stage of supersaturation and nucleation might be reached, and a polydisperse colloid would result.

Of course, nucleation may also be accomplished by seeding or adding externally formed nuclei. The monodisperse gold sols described in Chapter 5, Section 5.7b, are prepared by this method. The use of AgI crystals and other materials as nuclei in cloud seeding has also been studied extensively. This is especially interesting in view of possible applications to weather modification.

6.6   SURFACE TENSION AND CONTACT ANGLE: THEIR RELATION TO WETTING AND SPREADING PHENOMENA

In the sections above, a variety of important related concepts have been introduced. Among these are the equivalence of surface tension and surface free energy, the applicability of these concepts to solids as well as liquids, and the notion of the contact angle. All these concepts are brought together in discussing the physical situation sketched in Figure 6.6, which is important for understanding wetting, spreading, and related phenomena.

6.6a   Relationship Between Surface Tension and Contact Angle: The Young Equation

Suppose a drop of liquid is placed on a perfectly smooth solid surface, and these phases are allowed to come to equilibrium with the surrounding vapor phase. Viewing the surface tensions as forces acting along the perimeter of the drop enables us to write immediately an equation that describes the equilibrium force balance in the horizontal direction:

$$\gamma_{LV} \cos \theta = \gamma_{SV} - \gamma_{SL} \qquad\qquad (44)$$

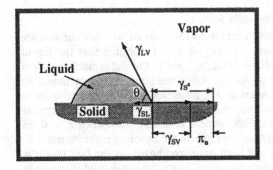

FIG. 6.6   Components of interfacial tension needed to derive Young's equation.

This result was qualitatively proposed by Thomas Young in 1805 and is generally known as *Young's equation*. It is also known as the *Young-Dupré equation* or the *Dupré equation* (MacRitchie 1990).

Young's equation is a plausible, widely used result, but its apparent simplicity is highly deceptive. The two terms that involve the interface between the solid and other phases cannot be measured independently, so experimental verification is difficult, although a variety of experiments have been directed along these lines. There are also a number of objections to Young's equation. These objections may be classified into two categories:

1. Those based on the noncompliance of the experimental system to the assumptions in the derivation

2. Those critical of the assumption of thermodynamic equilibrium in the solid

We discuss these two classes of objections separately.

6.6a.1   Effects of Surface Heterogeneities

Real solid surfaces may be quite different from the idealized one in the above derivation. Actual solid surfaces are apt to be rough and even chemically heterogeneous. This statement is true on a fine scale, even for carefully prepared surfaces. We discuss these complications in more detail in Section 6.7. In Chapter 9 we examine metal surfaces specifically to see the extreme conditions that must exist for these surfaces to be uniform down to an atomic scale.

The model surface of the derivation is thus the exception rather than the rule for solid surfaces. In principle, both roughness and heterogeneity can be incorporated into Young's equation in the form of empirical corrections. For example, if a surface is rough, a correction factor β is traditionally introduced as a weighting factor for $\cos \theta$, for which $\beta > 1$. The logic underlying this correction goes as follows. The factor $\cos \theta$ enters Equation (44) because of the projection of γ_{LV} onto the solid surface in the force balance. If the solid surface is rough, the corresponding surface area will be larger, but much of it will be "overshadowed" (i.e., the surface area will be underestimated) by the projection since Equation (44) assumes that the surface is smooth. The *roughness factor* β corrects for this effect (that is why $\beta > 1$). With the empirical correction factor for roughness included, Young's equation becomes

$$\beta \gamma_{LV} \cos \theta = \gamma_{SV} - \gamma_{SL} \tag{45}$$

A surface may also be chemically heterogeneous. Assuming, for simplicity, that the surface is divided into fractions f_1 and f_2 of chemical types 1 and 2, we may write

$$\gamma_{LV} \cos \theta = f_1(\gamma_{S_1V} - \gamma_{S_1L}) + f_2(\gamma_{S_2V} - \gamma_{S_2L}) \tag{46}$$

where $f_1 + f_2 = 1$.

Both roughness and heterogeneity may be present in real surfaces. In such a case, the correction factors defined by Equations (45) and (46) are both present. Although such modifications adapt Young's equation to nonideal surfaces, they introduce additional terms that are difficult to evaluate independently. Therefore the validity of Equation (44) continues to be questioned.

6.6a.2   Effect of the State of Equilibrium at the Solid Surface

More fundamental objections to Young's equation center on the issue of whether the surface is in a true state of thermodynamic equilibrium. In short, it may be argued that the liquid surface exerts a force perpendicular to the solid surface, $\gamma_{LV} \sin \theta$. On deformable solids a ridge is produced at the perimeter of a drop; on harder solids the stress is not sufficient to cause deformation of the surface. This is the heart of the objection. Is it correct to assume that a surface under this stress is thermodynamically the same as the idealized surface that is free from stress? Clearly, the troublesome stress component is absent only when $\theta = 0$, in which case the liquid spreads freely over the surface, and Figure 6.6 becomes meaningless.

In answer to this, the following argument has been suggested based on the fact that it is the difference $(\gamma_{SV} - \gamma_{SL})$ that appears in Young's equation. Since the same solid is common to both terms, it is really only the local difference at the surface between an adjacent phase that is liquid and one that is vapor that is being measured. According to this point of view, the nonequilibrium state of the solid is immaterial: The same solid is involved at both interfaces.

Another approach to resolving this theoretical objection to Young's equation is to eliminate the difference $(\gamma_{SV} - \gamma_{SL})$ from the equation entirely, replacing it by some equivalent quantity, thereby shifting attention away from the notion of solid surface tension. An example of this is the use of the heat of immersion to test Young's equation. We return to this in Section 6.6c.

In summary, then, Young's equation is still controversial despite the fact that it has been in existence since the beginning of the 19th century. The reader will appreciate that any relationship that has been around so long and has eluded definitive empirical verification has been the center of much research. Accordingly, the relationship is very widely encountered in the literature of surface chemistry.

6.6b Young's Equation and Equilibrium Film Pressure

To explore Young's equation still further, suppose we distinguish between γ_{SV} and γ_{S°, where the former describes the surface of a solid in equilibrium with the vapor of a liquid and the latter a solid in equilibrium with its own vapor. Since Young's equation describes the three-phase equilibrium, it is proper to use γ_{SV} in Equation (44). The question arises, however, what difference, if any, exists between these two γ's. In order to account for the difference between the two, we must introduce the notion of adsorption. In the present context adsorption describes the attachment of molecules from the vapor phase onto the solid surface. All of Chapter 9 is devoted to this topic, so it is unnecessary to go into much detail at this point. The extent of this attachment depends on the nature of the molecules in the vapor phase, the nature of the solid, and the temperature and the pressure.

For now we may anticipate a result from Chapter 7 to note that adsorption always leads to a decrease in γ. In the present context, therefore, we write

$$\gamma_{S^\circ} \geq \gamma_{SV} \tag{47}$$

We shall use the symbol π_e to signify the difference

$$\gamma_{S^\circ} - \gamma_{SV} = \pi_e \tag{48}$$

and call this quantity the *equilibrium film pressure*. The word *equilibrium* in this designation refers explicitly to the fact that the adsorbed molecules are in equilibrium with a drop of bulk liquid. The molecules adsorbed at an interface may be regarded as repelling one another or as rebounding off one another, thereby relieving some of the tension in the surface. This interpretation makes it sensible to call the reduction of γ due to adsorption π_e a "pressure." Note that π_e is a two-dimensional pressure, measuring the force exerted per unit length of perimeter (Newtons per meter) by the adsorbed molecules. We shall have a good deal more to say about this quantity in the following chapter. With these ideas in mind, Equations (44) and (48) may be combined to give

$$\gamma_{LV} \cos \theta = \gamma_{S^\circ} - \pi_e - \gamma_{SL} \tag{49}$$

Figure 6.6 shows the relationship among γ_{S°, γ_{SV}, and π_e. It is apparent that the value of θ might be quite different between equilibrium and nonequilibrium situations, depending on the value of π_e. There are several concepts that will assist us in anticipating the range of π_e values:

1. Spontaneously occurring processes are characterized by negative values of ΔG.
2. Surface tension is the surface excess free energy; therefore the lowering of γ with adsorption is consistent with the fact that adsorption occurs spontaneously.
3. Surfaces that initially possess the higher free energies have the most to gain in terms of decreasing the free energy of their surface by adsorption.
4. High-energy surfaces bind enough adsorbed molecules to make π_e significant. On the other hand, π_e is negligible for a solid that possesses a low-energy surface.

5. A surface energy in the neighborhood of 100 mJ m^{-2} is generally considered the cutoff value between high- and low-energy surfaces.

6. Silica, glass, metals, metal oxides, metal sulfides, and inorganic salts are examples of high-energy surfaces. Most solid organic compounds, including polymers, have low-energy surfaces.

7. Hard and soft solids are generally classified as having high- and low-energy surfaces, respectively. These criteria must be used cautiously since the mechanical properties of a solid depend on the concentration of defects and dislocations in the bulk.

Because of the simplification that results from $\pi_e = 0$ for low-energy surfaces, they are often chosen as model systems in fundamental research. Even when neglecting π_e is of questionable validity, γ_{SV} and γ_{S° are often used interchangeably for lack of suitable data. We see in Chapter 9 — Equation (9.7), for example — how π_e may be determined from experimental adsorption data. Otherwise, we generally assume $\pi_e = 0$.

6.6c Young's Equation and Heat of Immersion

6.6c.1 Enthalpy of Wetting

Small quantities of heat are generally evolved when a dry solid is immersed in a liquid. This can be measured calorimetrically and is called the *heat of immersion*. The physical process with which this heat is associated may be represented by the following equation:

$$\text{Dry solid} \xrightarrow{\ \text{liquid}\ } \text{wet solid} \tag{A}$$

Following the usual thermochemical convention, the heat of immersion ΔH_{im} may be written $H_{wet} - H_{dry}$, with these enthalpies expressed per unit area. Since heat is released by this process, ΔH_{im} is negative. The "wet" surface clearly describes the SL interface. For the "dry" surface, we ignore π_e and describe the surface by the S° notation. Therefore the heat of immersion may also be written

$$\Delta H_{im} = H^s_{SL} - H^s_{S^\circ} \tag{50}$$

Next we recall Equation (20), which gives us an expression for H^s:

$$H^s = \gamma - T\frac{\partial \gamma}{\partial T} \tag{51}$$

Applying Equation (51) to the right-hand side of Equation (44) — with $\gamma_{SV} = \gamma_{S^\circ}$ — gives

$$\Delta H_{im} = \left(\gamma_{SL} - T\frac{\partial \gamma_{SL}}{\partial T}\right) - \left(\gamma_{S^\circ} - T\frac{\partial \gamma_{S^\circ}}{\partial T}\right) \tag{52}$$

Using $\gamma_{LV}\cos\theta$ as a replacement for $\gamma_{S^\circ} - \gamma_{SL}$ we obtain

$$-\Delta H_{im} = \gamma_{LV}\cos\theta - T\frac{d}{dT}(\gamma_{LV}\cos\theta) \tag{53}$$

Carrying out the indicated differentiation yields

$$-\Delta H_{im} = \gamma_{LV}\cos\theta - T\cos\theta\frac{d\gamma_{LV}}{dT} - T\gamma_{LV}\frac{d\cos\theta}{dT} \tag{54}$$

which shows how Young's equation may be tested by comparing experimental heats of immersion with calculated values. Example 6.3 gives an idea of the magnitude of some of these quantities.

* * *

EXAMPLE 6.3 *Determination of Heat of Immersion from Surface Tension and Contact Angle.* Estimate the heat of immersion for the system for which γ and θ are 22 mJ m^{-2} and 30°, respectively, at 20°C. The temperature variations of γ and $\cos\theta$ are -0.10 mJ m^{-2} K^{-1} and 0.0010 K^{-1}, respectively. These values are close to those observed for liquid alkanes on Teflon. Comment on the implications of this estimate for the ease or difficulty in measuring ΔH_{im}.

Solution: We can estimate ΔH_{im} by substitution into Equation (54):

$$\Delta H_{im} = -\gamma_{LV}\cos\theta + T\cos\theta\,(d\gamma_{LV}/dT) + T\gamma_{LV}\,(d\cos\theta/dT)$$
$$= -(22)\cos 30 + 293\cos 30\,(-0.10) + 293\,(22)\,(0.0010)$$
$$= -38\,\text{mJ m}^{-2}$$

To get an idea of the problems associated with this kind of experiment, we estimate the temperature change in the liquid as a result of absorbing this heat. Using 2.4 J g^{-1} K^{-1} for the heat capacity (the value for *n*-octane) and taking $T = 1.6$ K as an arbitrary but convenient temperature change, we calculate

$$2.4\,\text{J K}^{-1}/\text{g} \cdot 1.6\,\text{K} \cdot (\text{m}^2/0.038\,\text{J}) \approx 100\,\text{m}^2\,\text{g}^{-1}$$

That is, for each gram of liquid, about 100 m^2 of surface must be wet. This overestimates the temperature change since it ignores the fact that the solid will absorb some heat. If the heat capacity of the solid is the same as the liquid, 100 m^2 of surface must be wet for each gram of solid-liquid mixture. Using a smaller temperature change would decrease the required area proportionately, but the fact remains that large areas are required to obtain measurable effects. Practically, this means work with powdered solids of small particle size. ∎

* * *

6.6c.2 Comments on the Measurement of Heats of Immersion

Although heats of immersion are small, this quantity is measurable. For systems in which both the heat of immersion and the necessary information about γ and θ have been measurable, the prediction of Equation (54) has been verified. Figure 6.7 shows some experimental results for *n*-alkanes wetting Teflon (polytetrafluoroethylene) surfaces. The open circles were determined calorimetrically; the closed ones were calculated from Equation (54). Even though the two sets of values diverge for alkanes larger than *n*-decane, the overall picture is quite acceptable. Incidentally, the value calculated in the example is close to the actual values, even though the numbers used in Example 6.3 were rounded off.

Several aspects of Figure 6.7 and Example 6.3 deserve further comment:

1. Calorimetry requires large areas of interface, which virtually demands powdered solids. We have not considered (yet) the problem of measuring θ and $d\theta/dT$ for powders. Solids that are available as large specimens with smooth surfaces (suitable,

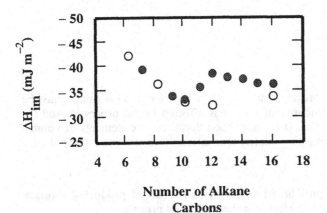

FIG. 6.7 Heats of immersion determined by calorimetry and calculated by Equation (54) for Teflon in various alkanes. (Redrawn from A. W. Neumann, *Adv. Colloid Interface Sci.*, **4**, 105 (1974).)

say, for the tilted plate determination of θ) can be pulverized to increase their surface. The reverse process is often not possible for powders, so a method for measuring powders is important.

2. Results like those shown in Figure 6.7 may be considered an experimental verification of Young's equation. In light of the first item, however, there may still be objections that the surfaces used for calorimetric studies and those used to study γ and θ are not identical even though they are nominally the same.

3. Accepting Equation (54) and Young's equation, on which it is based, suggests calorimetry as a method for measuring contact angles. At this time this is not practical, but the implication that contact angle is a thermodynamic property is a very important realization.

Although we established the thermodynamic significance of γ early in the chapter, θ has been allowed to drift. Its role is clear when we think of surface tension as a force: We use θ to project γ in a specified direction. In thermodynamic terms, contact angle has been an outsider in our presentation. Young's equation is the remedy to this. Rewriting Equation (44), we observe

$$\cos \theta = \frac{\gamma_{SV} - \gamma_{SL}}{\gamma_{LV}} \tag{55}$$

and $\cos \theta$ is fully described by various free energy terms. In view of this new-found (for us) significance, we return shortly to some additional experimental methods for the determination of θ. First, however, we consider the notions of adhesion, cohesion, and spreading, which will enhance our ideas of γ and θ as thermodynamic quantities.

6.6d Surface Tension and Cohesion, Adhesion, and Spreading

In this section we consider some hypothetical processes that provide us with additional ways of thinking about γ and θ. They are related to process (A) stated in Section 6.6c.1 but are not studied calorimetrically as is the case with immersion. In defining adhesion, cohesion, and spreading, we designate the phases A and B without specifying their physical state. Their surface with each other is designated AB; their individual surfaces with their own vapor or air (we make no distinction here) are designated by either A or B. With this notation in mind, we consider the following processes as they affect a unit area:

1. Cohesion:

No surface $\rightarrow$ $2A$ (or B) surfaces (B)

2. Adhesion:

$1AB$ surface $\rightarrow$ $1A + 1B$ surface (C)

3. Spreading (B on A):

$1A$ surface $\rightarrow$ $1AB + 1B$ surface (D)

Schematic illustrations of these processes are shown in Figure 6.8. Two things must be remembered about these sketches: One unit of surface is affected by the processes, and the shape of the affected area is immaterial. It is understood that these are elements of volume and area that are portions of macroscopic samples. Our interest is in the free energy change accompanying each process.

6.6d.1 Work of Cohesion

In Figure 6.8a—which applies to a pure liquid—the process consists of producing two new interfaces, each of unit cross section. Therefore, for the separation process,

$$\Delta G = 2\gamma_A = W_{AA} \tag{56}$$

The quantity W_{AA} is known as the *work of cohesion* since it equals the work required to pull apart a column of liquid A apart. It measures the attraction between the molecules of the two

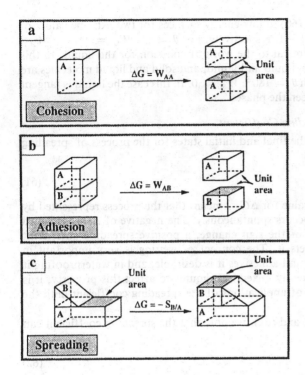

FIG. 6.8 Work of cohesion, adhesion, and spreading. Schematic illustrations of the processes for which ΔG equals (a) the work of cohesion; (b) the work of adhesion; and (c) the work of spreading.

portions. Recall the concept of cohesive energy density in terms of which we discussed in Section 3.4b interactions between pairs of identical molecules. Interpreting γ as half the work of cohesion shows that *surface tension measures the free energy change involved when molecules from the bulk of a sample are moved to the surface.*

6.6d.2 Work of Adhesion

Now let us consider the value of ΔG for the separation of A and B as represented in process (C) and Figure 6.8b. Taking the difference between the final and the initial free energies for this process yields

$$\Delta G = W_{AB} = \gamma_{final} - \gamma_{initial} = \gamma_A + \gamma_B - \gamma_{AB} \tag{57}$$

This quantity is known as the *work of adhesion* and measures the attraction between the two different phases.

The work of adhesion between a solid and a liquid phase may be defined by analogy with Equation (57):

$$W_{SL} = \gamma_{S^\circ} + \gamma_{LV} - \gamma_{SL} \tag{58}$$

By means of Equation (48), γ_{S° may be eliminated from this expression to give

$$W_{SL} = \gamma_{SV} + \pi_e + \gamma_{LV} - \gamma_{SL} \tag{59}$$

Finally, Young's equation may be used to eliminate the difference $(\gamma_{SV} - \gamma_{SL})$:

$$W_{SL} = \gamma_{SL}(1 + \cos\theta) + \pi_e \tag{60}$$

Neglecting π_e as we did in the last section, Equation (59) shows how γ and θ combined measure the work of adhesion between a solid and a liquid.

It is informative to apply Equation (60) to low-energy surfaces for two extreme values of θ, $0°$ and $180°$, for which $\cos \theta$ is 1 and -1, respectively. For $\theta = 0°$, $W_{SL} = 2\gamma_{LV} = W_{AA}$; the work of solid-liquid adhesion is identical to the work of cohesion for the liquid. In this case interactions between solid and solid, liquid and liquid, and solid and liquid molecules are all equivalent. At the other extreme, with $\theta = 180°$, $W_{SL} = 0$. In this case the liquid is tangent to the solid; there is no interaction between the phases.

6.6d.3   Spreading Revisited: The Spreading Coefficient

Last, if we take the difference between the final and initial states for the process of spreading B over A (symbolized B/A) we obtain

$$\Delta G = \gamma_{AB} + \gamma_B - \gamma_A \tag{61}$$

As usual with free energies, a negative value for $\Delta G_{B/A}$ means that the process represented by process (D) and shown in Figure 6.8c occurs spontaneously. The negative of $\Delta G_{B/A}$ is called the *spreading coefficient* $S_{B/A}$; because of the sign change, a positive spreading coefficient means B spreads freely over A and wets it. The concept of wetting is very important in numerous applications: With lubricants and adhesives it is desirable, and in waterproofing it is undesirable. Since additives are frequently mixed into liquids to affect this property, it is appropriate to postpone any discussion of applications of the spreading coefficient until the next chapter.

By combining Equations (56), (57), and (61), we note that the spreading coefficient can also be written

$$S_{B/A} = W_{AB} - W_{BB} \tag{62}$$

If $W_{AB} > W_{BB}$, the A-B interaction is sufficiently strong to promote the wetting of A by B. This is the significance of a positive spreading coefficient. Conversely, no wetting occurs if $W_{BB} > W_{AB}$ since the work required to overcome the attraction between two B molecules is not compensated by the attraction between A and B. Thus a negative spreading coefficient means that B will not spread over A.

6.7   CONTACT ANGLES: SOME COMPLICATIONS

6.7a   Advancing and Receding Contact Angles

In Section 6.2 we saw how the tilted plate method could be used to measure θ; we also noted that it could be determined by the Wilhelmy method if γ were measured independently. For that matter, the three-phase junction might be examined and θ determined by direct observation. These and many other methods—some of which we discuss later—have been used to measure θ. Even if the most careful experimental techniques are employed on carefully prepared surfaces, contact angle data are frequently confusing. The situation is best introduced by referring to Figure 6.9, which shows a drop on a tilted plane. "Teardrop" shapes such as this are familiar to everyone—just look at a raindrop on a window pane. The problem, of course, is that the contact angle is different at different points of contact with the support. It is conventional to call the larger value the *advancing angle* θ_a, and the smaller one the *receding angle* θ_r. The two may be quite different. The presence of contamination is definitely a contributing factor, but it is by no means the only one. Therefore, even with carefully purified materials, both advancing and receding contact angles should be measured.

All the techniques described here are easily conducted, so that both θ_a and θ_r may be observed. When the tilted plate method is used to evaluate the contact angle, θ_r values are obtained if the plate has been pulled out (emersion) from the liquid; θ_a results if the plate is pushed into the liquid (immersion). Likewise, both values of θ may be obtained from the Wilhelmy method, depending on whether the liquid is making an initial contact (θ_a) with the plate or is draining from it (θ_r).

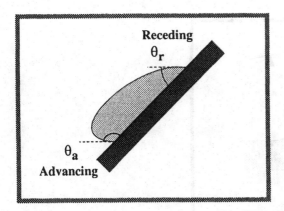

Receding
θ_r

θ_a
Advancing

FIG. 6.9 A drop on a tilted plane, showing advancing and receding contact angles. (Adapted from Johnson and Dettree 1969.)

6.7b   Contact Angle Hysteresis

A rather interesting example of the difference between advancing and receding contact angles is obtained from the Wilhelmy plate method when the contact angle has a nonzero value. Suppose the Wilhelmy plate is allowed to dip beneath the horizontal liquid surface, as shown in Figure 6.10a. In this case the weight of the meniscus as given by Equation (2) will be decreased by a term w', the buoyant force on the submerged plate. This buoyant force will clearly be proportional to the depth of immersion d. Therefore we write

$$w = \gamma P \cos \theta - w' = \gamma P \cos \theta - kd \tag{63}$$

where k is a suitable proportionality constant. This equation shows that the apparent weight w of the meniscus should give a straight line when plotted against the depth of immersion, and that the intercept should be proportional to $\cos \theta$. If a single value of θ applies to both the immersion and emersion steps, then the line shown in Figure 6.10b would result. Because of the difference between θ_a and θ_r, a curve like that shown in Figure 6.10c is obtained instead. When the immersion-emersion cycle is repeated, a hysteresis loop is obtained that may be as reproducible as the analogous loops observed in magnetization-demagnetization cycles. The difference $(\theta_a - \theta_r)$ is called the *hysteresis of a contact angle*.

The general requirement for hysteresis is the existence of a large number of metastable states that differ slightly in energy and are separated from each other by small energy barriers. The situation is shown schematically in Figure 6.11. The equilibrium contact angle corresponds to the free energy minimum. However, systems may be "frozen" in metastable states of somewhat higher energy by lacking sufficient energy to overcome the energy barrier separating them from equilibrium. An interesting experimental observation is that advancing and receding values of θ converge to a common value when the surface is vibrated. Presumably, the mechanical energy imparted to the liquid by the vibration assists it in passing over the energy barrier and reaching equilibrium. In this sense maximum vibration rather than vibration-free conditions may appear to be the ideal conditions for measuring θ, but "vibration" is a difficult parameter to control reproducibly, so this concept is of little practical help.

Now let us briefly consider the origin of these metastable states. If we exclude the effect of impurities, the metastable states are generally attributed to the roughness of the solid surface, its chemical heterogeneity, or both. Of course, a well-prepared laboratory sample will be fabricated and cleaned as effectively as possible to eliminate gross roughness and chemical heterogeneity. What we are talking about are microscopic irregularities that cannot be eliminated. In size and distribution, these will follow a random pattern on actual surfaces.

An informative model for contact angle hysteresis is obtained by postulating the surface

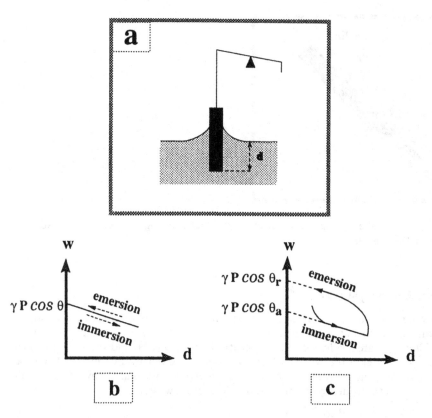

FIG. 6.10 Contact angle hysteresis: (a) weighing a meniscus in a Wilhelmy plate experiment versus the depth of immersion of the plate; (b) both the advancing and receding contact angles are equal; (c) $\theta_a > \theta_r$.

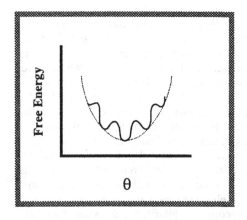

FIG. 6.11 Schematic energy diagram for metastable states corresponding to different contact angles.

to contain a set of concentric grooves on which a drop rests. Figure 6.12 represents the profile of two different drops on such a surface. In both of the profiles shown, the angles of contact between the liquid-vapor interface and the solid are identical, θ_0. With respect to the horizontal, however, two very different apparent contact angles are observed. The two extremes are identified as θ_a and θ_r in this model.

The two drop configurations in Figure 6.12 differ in surface area and in the elevation of their centers of gravity, thus they possess different energies. The change from one configuration to the other involves the distortion of the shape of the drop, which accounts for the energy barrier between the two configurations. Thus the model qualitatively accounts for the kind of metastable states shown in Figure 6.11 that are required for hysteresis. According to this model, contact angle hysteresis arises when a three-phase boundary becomes trapped in transit, lacking sufficient energy to surmount the energy barrier to a lower energy state. The teardrop profile of Figure 6.9 corresponds to the situation in which one edge of the drop has one configuration while the other edge has the second configuration.

This model can also be applied to hysteresis that arises from chemical heterogeneity; however, this time the surface is assumed to be smooth and to contain concentric rings of different chemical composition and hence different θ's. Actual heterogeneity may arise from impurities concentrated at the surface, from crystal imperfections, or from differences in the properties of different crystal faces. The distribution of such heterogeneities on an actual surface will obviously be more complex than the model considers, but the qualitative features of hysteresis are explained by the model nevertheless. Johnson and Dettree (1969) present additional details of model experiments of this sort.

If a surface consists of a patchwork array of high- and low-energy sites, it is generally assumed that the larger, advancing contact angle measures the contact angle that would be obtained for a smooth, homogeneous sample of the low-energy component:

$$\theta_a \approx \theta_{low\,E} \tag{64}$$

Conversely, the receding angle is taken as a measure of θ appropriate to the high-energy component:

$$\theta_r \approx \theta_{high\,E} \tag{65}$$

Both of these conclusions are consistent with Young's equation. All other things being equal in Equation (55), the larger $\gamma_{SV} \approx \gamma_{S^o}$ is, the larger $\cos \theta$ will be. Since the cosine increases as θ decreases, large values of γ_{S^o} result in smaller contact angles, and vice versa. We noted

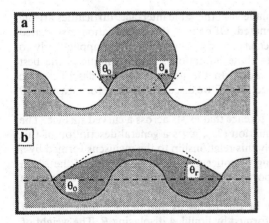

FIG. 6.12 Cross section of a drop resting on a surface containing a set of concentric grooves. For both profiles the contact angles are microscopically identical although macroscopically different. (Adapted from Johnson and Dettree 1969.)

previously that high-energy surfaces have more to gain by adsorption than low-energy surfaces. This helps explain why receding contact angles are less reproducible than advancing angles; adsorption by impurities affects this measurement the most. Correlations with surface roughness reveal that receding angles are also more sensitive to roughness than advancing angles. For these reasons θ_a is more reliable than θ_r as a characteristic of the solid. Obviously, the lower the hysteresis is, the more we can be sure that this quantity truly represents the solid.

In addition to roughness and heterogeneity, time-dependent effects also influence contact angle hysteresis. Dynamic studies have been conducted with liquid fronts moving across a solid substrate at variable speeds. Even the extrapolation to zero rate is unsatisfactory since the limiting dynamic values differ from equilibrium values. This suggests that insufficient mobility at the molecular level may also contribute to hysteresis. Viscous effects operating in the interface may contribute other forces besides those shown in Figure 6.6 that influence the value of the contact angle. This type of effect is harder to analyze than roughness and heterogeneity, but future research may help clarify this third cause of hysteresis.

With this background, let us now return to experiments that yield values for γ and θ.

6.8 MEASURING SURFACE TENSION AND CONTACT ANGLE: ROUND TWO

We have now established that both γ and θ have thermodynamic significance and have seen that their values as well as their temperature coefficients are of interest. In addition, we have seen that the measurement of contact angles presents some complications of its own. All this adds up to a need for more reliable and more accurate methods for the measurement of these parameters than those presented in Section 6.2. One of the most powerful strategies for this involves a second measurement made with the Wilhelmy plate.

6.8a Height of a Meniscus at a Wall

We saw in Section 6.2 that the Wilhelmy plate offers an accurate method for measuring $\gamma \cos \theta$. We thus have one experiment with two unknowns. The Wilhelmy balance measures the weight of the meniscus; in this section we examine the height to which the meniscus climbs on the same surface. We shall see that this distance—which may be accurately measured with a traveling microscope or cathetometer—also depends on γ and θ. The functional relationship between these parameters and the experimental variables is different from the case of the meniscus weight. Therefore we have two experiments with two unknowns that can be solved for γ and θ.

Since both of these measurements can be made with the same interface, difficulties arising from the pairing of mismatched data are eliminated. Of course, the complications associated with surface roughness and chemical heterogeneity persist, and the method applies only to solids that can be prepared as plates. Despite these limitations, this technique is the best general method available for determining γ and θ. In the next section we describe some additional ways of obtaining γ values, and in Section 6.9 we discuss a method for measuring θ on powdered solids.

In Section 6.4 we discussed the pressure difference that exists across a curved surface. The Laplace equation, in the form provided by Equation (35), gives a general description of this pressure difference Δp. Our objective is to apply this relationship to the meniscus formed by a liquid surface at a flat solid wall. The first thing to notice about this is that one of the R's in Equation (35) becomes infinite since the support is planar; hence this R^{-1} term disappears from the Laplace equation.

Figure 6.13a shows a cross-sectional view of this meniscus and defines some pertinent quantities for the experiment. The meniscus is formed by liquid A displacing B. The weight of the displaced B exerts a buoyant force on the meniscus; therefore the density difference must be used. The pressure difference is zero at the flat surface well removed from the wall. We define this to be the $z = 0$ plane. The meniscus makes contact with the wall at $z = h$. Our

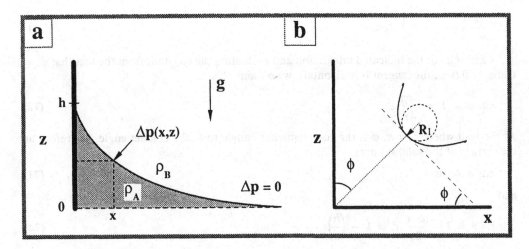

FIG. 6.13 Definitions of variables used to describe a meniscus formed by liquid A displacing liquid B in terms of (a) x and z and (b) ϕ.

interest is in some general point on the surface having the coordinates (x,z). Just beneath the surface at this point the pressure is less by $\Delta p(x,z)$ than in the reference plane. The liquid is elevated at this point by an amount sufficient to produce a compensating hydrostatic pressure. This pressure is given by $[(\rho_A - \rho_B)gz]$, which may be equated to the pertinent form of Equation (35):

$$\Delta\rho gz = \gamma/R_1 \tag{66}$$

where R_1 is the radius of curvature in the plane of Figure 6.13a. Substitution of Equation (36) for $1/R_1$ generates a second-order differential equation with a solution that gives $z = f(x)$. The mathematics are simplified considerably by using some trigonometric relations based on the tangent and the normal to the curve constructed at the point under consideration. Figure 6.13b shows the local radius of curvature of a general curve. The angle ϕ is the angle made by the extension of the normal with the z axis; ϕ also gives the angle between the tangent and the x axis. Inspection of Figure 6.13b reveals that the local slope of the tangent $dz/dx = -\tan\phi = -\sin\phi/\cos\phi$. Substituting this ratio into Equation (36) gives

$$\frac{1}{R_1} = \left[\frac{d}{dx}\left(-\frac{\sin\phi}{\cos\phi}\right)\right]\left[1 + \left(\frac{\sin\phi}{\cos\phi}\right)^2\right]^{-3/2} \tag{67}$$

When the indicated differentiations are carried out and the result is simplified, Equation (67) becomes

$$\frac{1}{R_1} = -\cos\phi\frac{d\phi}{dx} \tag{68}$$

which is the same as

$$\frac{1}{R_1} = -\frac{d\sin\phi}{dx} \tag{69}$$

Since $x \propto \sin\phi$ and $z \propto \cos\phi$, Equation (69) is equivalent to

$$\frac{1}{R_1} = -\frac{d\cos\phi}{dz} \tag{70}$$

Substituting this result into Equation (66) gives a relationship that is easily integrated:

$$\frac{\Delta \rho g}{\gamma} \int z \, dz = - \int d \cos \phi \tag{71}$$

Carrying out the indicated integration and evaluating the constant from the fact that $\phi = 0$ at $z = 0$ (i.e., the tangent is horizontal), we obtain

$$\cos \phi = 1 - \frac{\Delta \rho g}{2\gamma} z^2 \tag{72}$$

At the wall where $z = h$, ϕ is the complementary angle to θ, the contact angle. Therefore at the surface of the solid support

$$\cos \phi = \sin \theta \tag{73}$$

and

$$\sin \theta = 1 - \frac{\Delta \rho g}{2\gamma} h^2 = 1 - \left(\frac{h}{a}\right)^2 \tag{74}$$

where a^2 is given by Equation (5). This result shows that θ may be evaluated by measuring the height to which the meniscus climbs against a wall, provided that γ is known. Example 6.4 gives us an indication of the magnitude of the effect to be determined.

* * *

EXAMPLE 6.4 *Surface Tension and the Height of a Meniscus at a Wall.* Calculate the height to which an *n*-octane surface will climb on a Teflon wall (this is the same system used in Example 6.3) if γ is 22 mJ m^{-2}, $\theta = 30°$, and $\rho = 0.70$ g cm^{-3}. Comment on the ease or difficulty of making this measurement.

Solution: The density of air is insignificant compared to the liquid; hence ρ can be used for $\Delta\rho$. Because the density is given in cgs units, it is convenient to use this system of units throughout; remember, millijoule meter^{-2} and erg cm^{-2} are numerically identical. Substituting numerical values into Equation (74) gives

$$\sin 30 = 1 - [(0.70)(980) \, h^2/2(22)]$$

from which $h^2 = 0.032$ cm^2 and $h = 0.18$ cm.

Although h is not a large number, it is readily measurable using a cathetometer or low-magnification traveling microscope to measure the vertical distance between the level surface of the liquid and the top of the meniscus. The sighting is done along the liquid surface through a window in the container holding the liquid. Therefore the junction of interest must be viewed through the meniscus at the window. Special care must be taken to establish the right reference plane. If the contact angle is very small, the top of the meniscus might be difficult to see, but good illumination should remedy this. ■

* * *

6.8a.1 Simultaneous Measurement of Surface Tension and Contact Angle

Equation (74) provides a second relationship in addition to Equation (2), which expresses the connection between γ and θ and experimental quantities. The two simultaneous equations can be solved for γ and θ by the following steps:

1. Square Equation (74):

$$\sin^2 \theta = [1 - (\Delta\rho g h^2/2\gamma)]^2$$

2. Square Equation (2):

$$\cos^2 \theta = (w/P\gamma)^2$$

3. Recognize that $\sin^2 \theta + \cos^2 \theta$ equals unity:

$$1 = (w/P\gamma)^2 + (1 - \Delta\rho g h^2/2\gamma)^2$$

By multiplying out this last result and simplifying, we obtain

$$\gamma = \frac{\Delta\rho\, g\, h^2}{4} + \frac{w^2}{\Delta\rho\, g\, h^2\, P^2} \tag{75}$$

Substituting Equation (75) into Equation (2) gives

$$\cos\theta = \frac{4\,\Delta\rho\, g\, h^2\, P\, w}{(\Delta\rho)^2 g^2\, h^4\, P^2 + 4\, w^2} \tag{76}$$

Equations (75) and (76) allow γ and θ to be evaluated in a perfectly straightforward way from experimental quantities measured on a single system with one apparatus. For systems that can be fabricated into the necessary plates, this is an excellent way to measure these important parameters.

In the next section we examine the shapes of liquid surfaces possessing an axis of symmetry.

6.8b Surface Tension and Shapes of Drops, Bubbles, and Menisci

In the last section we saw that the Laplace equation describes the meniscus formed by a liquid surface at a flat wall. The situation is not essentially different — only more complicated — if the supporting wall is wrapped into a cylinder to generate a capillary meniscus with an axis of symmetry. As a matter of fact, any liquid interface with axial symmetry is described by the same mathematical formalism. For simplicity, we consider a liquid drop resting on a smooth horizontal surface as shown in Figure 6.14. Such a drop, incidentally, is called a *sessile* (sitting) drop. In the absence of gravity, a drop like this would be spherical since this geometry encloses the maximum volume within a minimum surface area. Any departure from a spherical shape increases the area and the surface free energy associated with it. A drop of liquid A in air — or, for that matter, any less dense medium B — is ordinarily acted on by gravitational as well as surface forces. Gravity forces the drop to lower its center of mass and thus flatten. Flattening increases the surface area; the equilibrium shape depends on the balance of the two forces. The same situation holds for the shapes of bubbles and menisci.

6.8b.1 Shape of a Sessile Drop: The Bashforth-Adams Equation

With this physical picture in mind, let us consider how the Laplace equation can be written to describe the profile of a sessile drop. Figure 6.14 represents the profile of a sessile drop; the

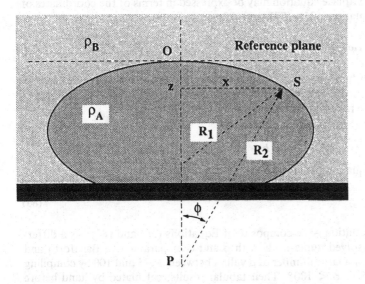

FIG. 6.14 Definition of coordinates for describing surfaces with an axis of symmetry (OP).

actual surface is generated by rotating the profile around the axis of symmetry, the z axis in this representation.

In Figure 6.14 the origin of the coordinate system O is situated at the apex of the surface. The two radii of curvature at point S are defined as follows. The one designated R_1 lies in the plane of the figure and describes the curvature of the profile shown. The radius of rotation of point S around the z axis equals x. The radius R_2 is the radius of curvature of the surface in the plane parallel to the z plane (normal to the plane of the figure) at S. The relationship between x and R_2 is given by

$$x = R_2 \sin \phi \tag{77}$$

since R_2 is defined as a vector originating at the z axis at P and making an angle ϕ with the axis of symmetry as shown in the figure.

Because of the symmetry of the surface, both R_1 and R_2 must be equal at the apex of the drop. The value of the radius of curvature at this location is symbolized b. Therefore at the apex (subscript 0)

$$(\Delta p)_0 = 2\gamma/b \tag{78}$$

according to Equation (30).

Next let us calculate the pressure at point S. At S the value of Δp equals the difference between the pressure at S in each of the phases. These may be expressed relative to the pressure at the reference plane through the apex (subscript 0) as follows:

1. In phase A,

$$p_A = (p_A)_0 + \rho_A g z \tag{79}$$

2. In phase B,

$$p_B = (p_B)_0 + \rho_B g z \tag{80}$$

Therefore Δp at S equals

$$(\Delta p)_S = p_A - p_B = (p_A)_0 - (p_B)_0 + (\rho_A - \rho_B)g z = (\Delta p)_0 + \Delta \rho g z \tag{81}$$

where $\Delta \rho = \rho_A - \rho_B$. Now Equation (78) may be substituted for Δp at the apex to give

$$(\Delta p)_S = 2\gamma/b + \Delta \rho g z \tag{82}$$

The general form of the Laplace equation may be expressed in terms of the coordinates of Figure 6.14 by combining Equations (35), (77), and (82):

$$\gamma \left(\frac{\sin \phi}{x} + \frac{1}{R_1} \right) = \frac{2\gamma}{b} + \Delta \rho \, g \, z \tag{83}$$

In Equation (83) R_1^{-1} is given by Equation (36). Expression (83) is known as the *Bashforth-Adams equation*. It is conventional to express this equation in dimensionless form by expressing all distances relative to the radius at the apex b:

$$\frac{\sin \phi}{x/b} + \frac{1}{R_1/b} = 2 + \frac{\Delta \rho \, g \, b^2}{\gamma} \frac{z}{b} = 2 + \beta \frac{z}{b} \tag{84}$$

The cluster of constants in Equation (84) is defined by the symbol β:

$$\beta = \frac{\Delta \rho \, g \, b^2}{\gamma} \tag{85}$$

The Bashforth-Adams equation — the composite of Equations (36) and (84) — is a differential equation that may be solved numerically with β and ϕ as parameters. Bashforth and Adams solved this equation for a large number of β values between 0.125 and 100 by compiling values of x/b and z/b for $0° < \phi < 180°$. Their tabular results, calculated by hand before the days of computers, were published in 1883. Other workers subsequently extended these tables. A very useful compilation of these results is found in Padday (1969). Table 6.2 shows a typical result from these tables for $\beta = 25$. The surface profiles sketched in Figure 6.15 were drawn from these tabulated results for (a) $\beta = +10.0$ and (b) $\beta = -0.45$.

It is apparent from inspection of Figure 6.15 that different values of β characterize different drop shapes. This is to be expected since β varies with $\Delta\rho$ compared to γ. According to the convention established in Figure 6.14, g and z are measured in the same direction. Therefore, when the Bashforth-Adams tables are applied to sessile drops, the sign of $\Delta\rho$ determines the sign of β. If $\rho_A > \rho_B$, β will be positive and the drop will be oblate in shape since the weight of the fluid tends to flatten the surface. If $\rho_A < \rho_B$ a prolate drop is formed since the larger buoyant force leads to a surface with much greater vertical elongation. In this case β is negative. A β value of zero corresponds to a spherical drop and, in a gravitational field, is expected only when $\Delta\rho = 0$. Positive values of β correspond to sessile drops of liquid in a gaseous environment. Negative β values correspond to sessile bubbles extending into a liquid. These statements imply that the drop is resting on a supporting surface. If, instead, the drop is suspended from a support (called pendant drops or bubbles), g and z are measured in opposite directions and have different signs; that is, for a pendant drop, z is measured upward from the apex, while g continues to define the downward direction. Because of this sign reversal, the liquid drop will be a prolate ($\beta < 0$) shape, and the gas bubble will be oblate ($\beta > 0$).

6.8b.2 Surface Tension from the Shapes of Sessile Drops

The Bashforth-Adams tables provide an alternate way of evaluating γ by observing the profile of a sessile drop of the liquid under investigation. If, after all, the drop profiles of Figure 6.15 can be drawn using β as a parameter, then it should also be possible to match an experimental drop profile with the β value that characterizes it. Equation (85) then relates γ to β and other measurable quantities. This method is claimed to have an error of only 0.1%, but it is slow and tedious and hence not often the method of choice in practice.

In order to obtain the profile of one of these drops experimentally, it is best to photograph

TABLE 6.2 x/b and z/b for $\beta = 25$ and $0° < \phi \le 180°$

ϕ (deg)	x/b	z/b	ϕ (deg)	x/b	z/b
5	0.08521	0.00368	95	0.48092	0.28243
10	0.16035	0.01348	100	0.47934	0.29435
15	0.22230	0.02712	105	0.47682	0.30594
20	0.27250	0.04288	110	0.47345	0.31665
25	0.31333	0.05974	115	0.46931	0.32662
30	0.34684	0.07713	120	0.46452	0.33585
35	0.37455	0.09475	125	0.45911	0.34433
40	0.39755	0.11236	130	0.45319	0.35204
45	0.41666	0.12985	135	0.44682	0.35901
50	0.43249	0.14711	140	0.44008	0.36519
55	0.44551	0.16405	145	0.43302	0.37061
60	0.45609	0.18063	150	0.42571	0.37526
65	0.46451	0.19678	155	0.41823	0.37915
70	0.47101	0.21246	160	0.41062	0.38231
75	0.47579	0.22761	165	0.40296	0.38472
80	0.47905	0.24221	170	0.39528	0.38643
85	0.48089	0.25626	175	0.38766	0.38744
90	0.48148	0.26966	180	0.38014	0.38776

Source: F. Bashforth, and J. C. Adams, *An Attempt to Test the Theory of Capillary Action*, Cambridge University Press, London, 1883.

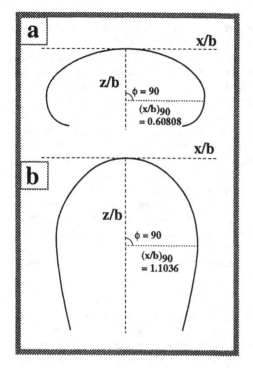

FIG. 6.15 Sessile drop profiles drawn from results of Bashforth and Adams's tables: (a) $\beta = 10.0$; and (b) $\beta = -0.45$. (Data from Padday 1969.)

the silhouette of the surface. Then experimental and theoretical profiles are compared in an effort to identify the β value that characterizes the experimental surface. With care, it is possible to interpolate between theoretical profiles and arrive at the β value that describes the surface under consideration. Equation (85) shows that knowledge of β alone is not sufficient to permit the evaluation of γ; b and $\Delta\rho$ must also be known. Evaluating the density difference poses no special difficulty. Let us next consider how b is measured.

Once β is known for a particular profile, the Bashforth-Adams tables may be used further to evaluate b:

1. For the appropriate β, the value of x/b at $\phi = 90°$ is read from the tables. This gives the maximum radius of the drop in units of b.
2. From the photographic image of the drop, this radius may be measured since the magnification of the photograph is known.
3. Comparing the actual maximum radius with the value of $(x/b)_{90°}$ permits the evaluation of b.

Figure 6.15a may be used as a numerical example to illustrate this procedure. Suppose an actual experimental drop profile is matched with theoretical profiles and is shown to correspond to a β value of 10.0. Then b is evaluated as follows:

1. The value of $(x/b)_{90°}$ for $\beta = 10$ is found to be 0.60808 from the tables.
2. Assume the radius of the actual drop is 5.00 mm at its widest point.
3. The first two items describe the same point; therefore $b = 5.00/0.608 = 8.22$ mm. This would be the radius at the apex of the drop shown in Figure 6.15a if the maximum radius were 5.00 mm.
4. This numerical example may be completed by using the definition of β to evaluate γ

for the liquid of Figure 6.15a. Assuming $\Delta\rho$ to be $+1.00$ g cm^{-3} and taking $g = 9.8$ m s^{-2} gives

$$\gamma = \Delta\rho g b^2/\beta = (10^3)(9.8)(8.22 \cdot 10^{-3})^2/10.0$$
$$= 0.0662 \text{ J m}^{-2} = 66.2 \text{ mJ m}^{-2}$$

Several additional points might be noted about the use of the Bashforth-Adams tables to evaluate γ. If interpolation is necessary to arrive at the proper β value, then interpolation will also be necessary to determine $(x/b)_{90°}$. This results in some loss of accuracy. With pendant drops or sessile bubbles (i.e., negative β values), it is difficult to measure the maximum radius since the curvature is least along the equator of such drops (see Figure 6.15b). The Bashforth-Adams tables have been rearranged to facilitate their use for pendant drops. The interested reader will find tables adapted for pendant drops in the material by Padday (1969). The pendant drop method utilizes an equilibrium drop attached to a support and should not be confused with the drop weight method, which involves drop detachment.

6.8b.3 Other Methods for Determining Surface Tension

Several other methods for determining γ—notably, the maximum bubble pressure, the drop weight, and the DuNouy ring methods (see Section 6.2)—all involve measurements on surfaces with axial symmetry. Although the Bashforth-Adams tables are pertinent to all of these, the data are generally tabulated in more practical forms that deemphasize the surface profile.

A method related to sessile drop and sessile bubble methods is a technique based on the measurement of the profile of a spinning drop. In this method a drop of a liquid is injected into a higher density liquid, and the whole container is rotated at a known angular velocity. The drop elongates because of the imposed centrifugal acceleration against the opposing effect of interfacial tension, which tries to reduce the area. The analysis of the shape of the drop is similar to the sessile drop method in principle, and the centrifugal acceleration can be related to the interfacial tension and the density difference from the liquids through a simple equation (see, for example, Miller and Neogi 1985). In contrast to the sessile drop and sessile bubble methods, the spinning drop technique requires no contact with a solid surface. Moreover, since the drop and the suspending liquid can be made quite thin, the method can be used even when the suspending liquid is moderately turbid. The spinning drop technique is particularly useful when the interfacial tension is very low.

In addition to the methods discussed here and in Section 6.2, there are a few other methods for measuring surface tension that are classified as *dynamic methods* as they involve the *flow* of the liquids involved (e.g., methods based on the dimensions of an oscillating liquid jet or of the ripples on a liquid film). As one might expect, the dynamic methods have their advantages as well as disadvantages. For example, the oscillating jet technique is ill-suited for air-liquid interfaces, but has been found quite useful in the case of surfactant solutions. A discussion of these methods, however, will require advanced fluid dynamics concepts that are beyond our scope here. As our primary objective in this chapter is simply to provide a basic introduction to surface tension and contact angle phenomena, we shall not consider dynamic methods here. Brief discussions of these methods and a comparison of the data obtained from different techniques are available elsewhere (e.g., see Adamson 1990 and references therein).

6.9 CONTACT OF LIQUIDS WITH POROUS SOLIDS AND POWDERS

Any solid can be pulverized into particles of small size; not all can be fabricated into the smooth supports we have discussed in this chapter. This consideration alone—not to mention the many practical applications of powdered solids—encourages us to look for a relationship that describes the junction of a liquid interface with such solids. One thought might be to compress the powder into a pellet and treat the surface of the pellet in the same way as we have treated other solid surfaces. On a fine scale the surface of such a pellet is rough, however, so hysteresis effects should be severe. Therefore we look for some alternative approach.

6.9a Penetration of Liquids Through Pores: Measurement of Contact Angles

Instead of the exterior surface of a powder pellet, let us consider the network of irregular channels that permeate it—the hole instead of the doughnut! Figure 6.16a represents a section through such a pellet for irregularly shaped particles. It is easy to imagine an arrangement such as that shown in Figure 6.16b in which the powder pellet is positioned as a plug in a cylinder such that liquid can be forced through it by a movable piston. For now our main requirement for such an apparatus is that we are able to measure the applied pressure p needed to force the liquid into the plug. In a further development we will also be interested in measuring the depth h into the plug that the liquid penetrates under a pressure p.

Commercial instruments are available that carry out these operations under the control of a computer, which also analyzes the results. Let us consider the penetration of the liquid into the pores of the plug to see the basis for this analysis.

Instead of the actual network of irregular channels, the interpretation of this experiment is based on a model that imagines the plug to consist of a bundle of cylindrical pores of radius R_c. The model is represented by Figure 6.16c. The intrusion of the liquid into the cylindrical pores in response to the applied pressure follows the same mathematical description as the rise of a liquid in a capillary. In view of the approximate nature of the model, it is adequate to use the Laplace equation in the form given by Equation (3) to describe this situation:

$$\frac{2\gamma_{LV}\cos\theta}{R_c} = p \tag{86}$$

In the capillary rise experiment $p = \Delta\rho g h$, whereas p is the applied pressure in the case discussed in the present section.

Now suppose a liquid is pushed through the plug that forms a contact angle of zero with the walls of the channels. Using the subscript 0 to represent this situation, Equation (86) becomes

$$p_0 = \frac{2\gamma_{LV,0}}{R_c} \tag{87}$$

since $\cos\theta = 1.0$ in this case. Equation (87) can be used to evaluate R_c for the plug. The value of R_c thus obtained is a parameter that is characteristic of the experiment; however, it might be very different from any physical distance within the plug. It might be called the radius of an equivalent cylinder, although some researchers feel that this name attaches too much significance to the R_c value thus obtained. The term *tortuosity* is sometimes applied to this

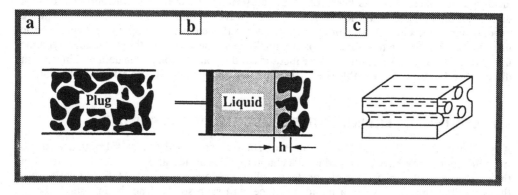

FIG. 6.16 Contact of liquids with pores and powders: (a) schematic view of pores through a plug of particles; (b) liquid intrusion into a plug under pressure; and (c) an idealized plug of cylindrical pores.

empirical R_c value. Whatever terminology is used, this evaluation of R_c amounts to a calibration with a known (with respect to θ) liquid so that R_c can be eliminated from Equation (86) to give

$$\frac{p_0 \gamma_{LV} \cos \theta}{\gamma_{LV,0}} = p \tag{88}$$

or

$$\cos \theta = \frac{p}{p_0} \frac{\gamma_{LV,0}}{\gamma_{LV}} \tag{89}$$

Note that values of γ are assumed to be available when this equation is used. Since numerous methods are available for measuring γ, there is no loss of applicability in assuming this. Since all of the quantities on the right-hand side of Equation (89) are measurable, this approach provides a method for determining θ for powdered solids. The method is not highly reliable, but is preferable to any technique based on the exterior surface of the plug as a liquid support.

6.9b Porosimetry: Structure of Porous Materials

Our approach until now has been to focus attention on the liquid-solid junction and to eliminate the parameter that characterizes the plug. Numerous porous solids exist for which there may be considerable interest in knowing the pore dimensions. The foregoing analysis can also be applied to this problem by assuming that both γ and θ are known. For this application the liquid is chosen for convenience, not because it is part of what is being measured. Mercury is generally used as the penetrating fluid, and the general technique is called *mercury porosimetry*. Contact angles of 130–140° are generally assumed, although it would clearly be preferable to use the θ value that is appropriate for the specific system.

When the pore structure of a solid is being investigated, it is desirable to learn as much as possible from porosimetry. In general, solids will possess a distribution of pore sizes, so the volume (or surface area) associated with pores of various radii is of interest. The significance of this sort of information is apparent for such solids as zeolite catalysts. These are aluminasilicates that can be synthesized with variable Si/Al ratios and variable pore sizes. Pore dimensions in the range of 0.8 to 2.0 nm can be obtained for solids with controllable acidity associated with Al^{3+} in the silica matrix. Zeolites are used on a large scale in the petroleum industry as catalysts for the cracking and isomerization of hydrocarbons (see Vignette I.9). Part of the specificity of these catalysts arises from the pore size distribution, so porosimetry becomes a valuable tool for the characterization of these materials.

In an apparatus based on Figure 6.16b, the volume of mercury forced into the pores of the solid can be measured as a function of the applied pressure. Equation (86) shows that higher pressures are required for smaller pores. Therefore incremental increases in p will result in the filling of pores of progressively smaller radii. The volume V that has intruded a porous solid at a pressure p gives the cumulative volume of all pores larger than the size associated with p. Since plots of V versus p give information about the cumulative pore distribution, it is the derivative of such data that measures the increment in pore volume associated with an increment in R_c. Written as a formula, $dV/dp \propto dV/(-dR_c)$ since V increases as R_c decreases.

To develop this last relation further, note that Equation (86) can be written as $pR_c =$ constant since γ and θ are independent of pore structure. Therefore $(p\,dR_c + R_c dp) = 0$, or

$$\frac{dp}{dR_c} = -\frac{p}{R_c} \tag{90}$$

With Equation (90), the slope of the experimental V-p plot can be developed as follows:

$$\frac{dV}{dR_c} = -\frac{dV}{dp}\frac{dp}{dR_c} = \frac{dV}{dp}\frac{p}{R_c} \tag{91}$$

Combining Equations (86) and (91) gives

$$\frac{dV}{dR_c} = \frac{dV}{dp} \frac{p^2}{2\,\gamma\cos\theta} \tag{92}$$

which shows how the local slope of the experimental data can be converted into the desired information about increments of volume for pores of different radius.

As noted above, much of the data manipulation of commercial porosimeters is computerized, so a pore size distribution is produced automatically by these instruments.

6.9c Contact Angle Measurement and Porosimetry Using the Poiseuille Equation

The rate at which a liquid penetrates a porous solid can also be measured and interpreted at the same level of approximation as used in the above section. The Poiseuille equation, Equation (4.20), gives an expression for the rate at which a liquid flows through cylindrical pores of radius R_c under a pressure p. We shall neglect the gravitational contribution to the driving force, which means that the following is better suited to horizontal than to vertical arrangements. If we write the volume rate of flow $dV/dt = d(\pi R_c^2 h)/dt$ and simplify, Equation (4.20) can be rewritten as

$$\frac{dh}{dt} = \frac{p\,R_c^2}{8\,\eta\,h} \tag{93}$$

where h is the depth of penetration of the intruding liquid and η is its viscosity.

If Equation (86) is used to eliminate p, Equation (93) becomes

$$\frac{dh}{dt} = \frac{2\,\gamma\cos\theta}{R_c} \frac{R_c^2}{8\,\eta\,h} = \frac{\gamma\,R_c\cos\theta}{4\,\eta\,h} \tag{94}$$

As with Equation (86), R_c can be eliminated from Equation (94) by calibration with a liquid for which $\theta = 0$. In that case Equation (94) can be used to determine θ for powders or porous solids. Alternatively, a value of θ may be assumed, and R_c can be evaluated from the rate of mercury intrusion. Equation (94) is called the *Washburn equation*.

Since advancing and receding contact angles are likely to be different in these experiments, mercury permeation curves are expected to be different, depending on whether the mercury is being pushed in or out of the plugs. This type of hysteresis is indeed observed. We encounter another type of hysteresis associated with pore filling in Chapter 9 (Section 9.7a).

6.10 MOLECULAR INTERPRETATION OF SURFACE TENSION

Throughout this chapter we have dealt with surface tension from a phenomenological point of view almost exclusively. From fundamental perspective, however, descriptions from a molecular perspective are often more illuminating than descriptions of phenomena alone. In condensed phases, in which interactions involve many molecules, rigorous derivations based on the cumulative behavior of individual molecules are extremely difficult. We shall not attempt to review any of the efforts directed along these lines for surface tension. Instead, we consider the various types of intermolecular forces that exist and interpret γ for any interface as the summation of contributions arising from the various types of interactions that operate in the materials forming the interface.

6.10a Some Molecular Interactions Important in Interpreting Interfacial Phenomena

To develop the above idea, several broad categories of molecular interactions should be considered.

1. *Hydrogen bonding*: Hydrogen atoms serve as bridges linking together two atoms of high electronegativity. In the present context these atoms are in separate molecules so the molecules themselves are mutually "attracted" by these bonds.

2. *Metallic bonding*: A sea of mobile electrons shared by the atoms of a metal contributes to the attraction between metal atoms in bulk samples.

3. *Permanent dipole interactions*: Polar molecules have relatively positive and negative regions. Regions of opposite charge on different molecules result in an attraction between these molecules. Molecules must possess a permanent dipole moment to display this effect.

4. *London forces*: Deformable electron clouds in adjoining molecules distort one another, resulting in an instantaneous polarity with accompanying attraction between the molecules involved. The polarizability of a molecule (see Chapter 5, Section 5.2b) is a measure of its tendency to display this effect.

Although additional entries could be included in this list, the foregoing entries are sufficient for our purposes. The main thing to recognize about these interactions is that all but London forces are highly specific. London forces require only the presence of electrons and therefore operate between all molecules. The other types of interactions, by contrast, require some specific feature: the metallic state, high electronegativity, or certain molecular geometry. Whether any of the other types of interactions operates or not, London forces are always present.

In Chapter 10 we discuss in considerable detail the attractions between molecules arising from their polarity and polarizability. In that chapter, we also see how surface tension data can be used to obtain information on London forces between macroscopic objects and vice versa, thus providing a direct link between the London force and surface tension. For now, however, it is sufficient to recognize that London forces attract all molecules together, regardless of their specific chemical nature. The other interactions itemized above operate only for systems possessing the requisite special features. What this means for surface tension is best illustrated by considering a specific surface. At the water-mercury interface, for example, the metallic bonds of the mercury end at the interface. Similarly, the hydrogen bonds of the water end at the surface. London attraction exists between water and mercury particles, however, just as it exists for water-water and mercury-mercury pairs. These various London attractions may differ in magnitude as a result of different molecular properties, but they share a common origin.

6.10b   Work Needed to Bring a Molecule to an Interface: The Girifalco-Good Equation

Next we apply these concepts to a molecule of A in two different environments: in the interior of a bulk phase and near an interface between two phases. A molecule in the interior of a bulk phase is surrounded on all sides by a homogeneous molecular environment of A. Any movement that would increase its separation from some neighbors would automatically decrease the separation from others, so large deviations from the equilibrium separation are improbable.

For a molecule at an interface between a condensed phase and the gas phase, the environment is quite asymmetrical. Movement toward the bulk phase is impeded by the excluded volume of the A molecules. Movement away from the bulk phase meets no resistance of this sort, although the prevailing attraction between molecules in the condensed phase opposes the A molecules from escaping the condensed phase altogether. Because of this, the equilibrium separation between the molecules at the surface will be larger than between those in the interior. The intermolecular separation has been "stretched" in bringing a molecule to the surface. The contractile force in the interface of the substance is simply the restoring force attempting to return the molecules to their bulk spacing. From an energetic point of view, the difference between the energy of the bulk and surface minimum separations gives the work needed to bring a molecule from the interior to the surface.

Next let us consider the situation in which a second condensed phase B adjoins the reference phase A. The new consideration in this case is the London component of attraction of the molecules in condensed phase B for the A molecules in the interface. This A-B attraction partially overcomes the A-A attraction that opposes the movement of an A molecule to any interface. As a consequence, there is a difference in the energy that must be expended to bring an A molecule to an interface with a gas and to an interface with another condensed phase. We call this difference ΔE^s.

It is argued that only the London component of intermolecular attractions operates across the interface to decrease the work required to bring a particle to the surface. Suppose we define ϕ as that fraction of the surface tension due to London forces. Next we use the sort of geometric mixing rule that was employed in Chapter 3, Section 3.4b, to estimate ΔE^s as follows:

$$\Delta E^s = \sqrt{(\phi_A \gamma_A)(\phi_B \gamma_B)} \tag{95}$$

The assignment of a geometric mean rather than an arithmetic mean or some other function of the two γ terms is justified primarily on the basis of the successful use of this type of averaging in the theory of nonelectrolyte solubility. Only the London component of γ is used since it is the part of γ that crosses phase boundaries.

According to these ideas, the work required to bring an A molecule to the AB interface is given by

$$(\text{Work})_A = \gamma_A - (\Delta E^s)_A = \gamma_A - \sqrt{(\phi_A \gamma_A)(\phi_B \gamma_B)} \tag{96}$$

when A and B are both in condensed phases. A similar expression applies to the work required to bring a B molecule to the interface. The total work of forming the AB interface is the sum of these contributions:

$$\gamma_{AB} = \left[\gamma_A - \sqrt{(\phi_A \gamma_A)(\phi_B \gamma_B)} \right] + \left[\gamma_B - \sqrt{(\phi_A \gamma_A)(\phi_B \gamma_B)} \right] \tag{97}$$

$$= \left[\gamma_A + \gamma_B - 2\sqrt{(\phi_A \gamma_A)(\phi_B \gamma_B)} \right]$$

Combining the two fractions into the parameter Φ, Equation (97) becomes

$$\gamma_{AB} = \gamma_A + \gamma_B - 2\Phi(\gamma_A \gamma_B)^{1/2} \tag{98}$$

with $\Phi = (\phi_A \phi_B)^{1/2}$. The relationship given by Equation (98) is called the *Girifalco-Good equation*. The parameter Φ may be viewed from two different points of view:

1. We may regard Φ as an empirical constant evaluated by fitting data to Equation (98) or some other form of the basic relationship.
2. We may estimate Φ from first principles by using molecular parameters and relationships from Chapter 10.

With a few additional refinements, the two approaches are found to give satisfactory agreement. The Φ values often lie in the range 0.5 to 1.0, which shows that, in general, half or more of the surface tension can be attributed to London forces.

Equation (98) may be combined with several other relationships of this chapter to generate expressions by which the contributions of London forces can be investigated. For example, if one of the condensed phases is a solid and the other a liquid, Equations (98) and (49) may be combined to give

$$\gamma_{LV} \cos \theta = -\gamma_L + 2\Phi(\gamma_S \gamma_L)^{1/2} - \pi_e \tag{99}$$

Equation (99) can be solved for any one of the factors, depending on the data available. Neglecting π_e as usual (although we might solve for this also), we obtain the following:

1. If γ_S, γ_L, and θ are known,

$$\Phi_{SL} = \gamma_L(1 + \cos \theta)/2(\gamma_S \gamma_L)^{1/2}$$

2. If γ_L, γ_S, and Φ are known,

$$\cos \theta = -1 + 2\Phi(\gamma_S/\gamma_L)^{1/2}$$

3. If γ_L, θ, and Φ are known,

$$\gamma_S = \gamma_L(1 + \cos\theta)^2/4\Phi^2$$

The third item provides a valuable way for evaluating γ_S, a quantity that is otherwise difficult to determine.

6.10c The Fowkes Approximation to the Girifalco-Good Equation

Rather than pursue the fractions ϕ or their composite Φ any further, we turn next to an empirical approach due to Fowkes to estimate the product $\phi\gamma$ for various substances. We use the symbol γ^d to represent the London contributions to γ estimated by his method. The notation is derived from the fact that London forces are also called *dispersion* (superscript d) *forces* (see, for example, Chapter 10, Section 10.7b). In this notation Equation (98) can be written

$$\gamma_{AB} = \gamma_A + \gamma_B - 2(\gamma_A^d\gamma_B^d)^{1/2} \tag{100}$$

In this version the relationship is called the *Girifalco-Good-Fowkes equation*. (We use a similar approach again in Chapter 10, e.g., see Equations (10.77) and (10.78), to determine the Hamaker constant for the van der Waals interaction forces.) Although we use $\phi\gamma$ and γ^d interchangeably, it is important to recognize that γ^d values are determined by a particular strategy as illustrated in Example 6.5.

* * *

EXAMPLE 6.5 *Estimation of Interfacial Tensions Using the Girifalco-Good-Fowkes Equation.* The following are the interfacial tensions for the various two-phase surfaces formed by n-octane (O), water (W), and mercury (Hg): for n-octane-water, $\gamma = 50.8$ mJ m^{-2}; for n-octane-mercury, $\gamma = 375$ mJ m^{-2}; and for water-mercury, $\gamma = 426$ mJ m^{-2}. Assuming that only London forces operate between molecules of the hydrocarbon, use Equation (100) to estimate γ^d for water and mercury. Do the values thus obtained make sense? Take γ values from Table 6.1 for the interfaces with air of these liquids.

Solution: In general, Equation (100) contains two unknowns: the dispersion components of γ for A and B. If γ^d for one of these is known, the other γ^d value can be calculated from experimental results. If it is assumed that $\gamma^d = \gamma$ for the hydrocarbon, then the experimental γ for an interface involving octane can be interpreted by Equation (100) to give the other γ^d value. Thus $375 = 484 + 21.8 - 2(\gamma_{Hg}^d \cdot 21.8)^{1/2}$; therefore $\gamma_{Hg}^d = 196$ mJ m^{-2}. Also $50.8 = 72.8 + 21.8 - 2(\gamma_w^d \cdot 21.8)^{1/2}$; therefore $\gamma_w^d = 22.0$ mJ m^{-2}. The two γ^d values are 30 and 40%, respectively, of the γ's for water and mercury. These fractions are somewhat less than expected in terms of the preceding discussion, but are not totally out of line. A more meaningful test is to use these values to estimate γ for the water-mercury interface. Repeating the above procedure, we obtain

$$\gamma_{Hg\text{-}W} = 484 + 72.8 - 2[(196)(22)]^{1/2} = 425 \text{ mJ m}^{-2}$$

which compares quite favorably with the experimental value of 426 mJ m^{-2}.   ■

* * *

Table 6.3 lists the data and calculated results for a number of hydrocarbons forming interfaces with water and mercury. The relative constancy of the calculated γ^d values adds to their plausibility.

The average values of γ^d for water and mercury are clearly successful in their ability to calculate γ correctly for the water-mercury interface. Individually, however, they are the averages of slightly divergent values measured for several different interfaces with hydrocarbons. In this sense the values of γ^d we have considered are analogous to mean bond energies in physical chemistry, which also are averages obtained from a variety of compounds. Although mean bond energies are very useful, they are by nature insensitive to unique, specific effects. With both mean bond energies and values of γ^d, the user must be careful that no such special interactions are present, otherwise quite serious errors could arise. Furthermore, it is important to realize that any errors are perpetuated and compounded by this scheme for evaluating γ^d.

TABLE 6.3 Experimental Values for γ_H, γ_{H-W}, and γ_{H-Hg}, with H = Hydrocarbon, at 20°C (All Values in mJ m^{-2})[a]

| Hydrocarbon | γ | Mercury (γ_{Hg} = 484) | | Water (γ_W = 72.8) | |
		γ_{H-Hg}	γ_{Hg}^d	γ_{H-W}	γ_{H-w}^d
n-Hexane	18.4	378	210	51.1	21.8
n-Heptane	18.4	–	–	50.2	22.6
n-Octane	21.8	375	199	50.8	22.0
n-Nonane	22.8	372	199	–	–
n-Decane	23.9	–	–	51.2	21.6
n-Tetradecane	25.6	–	–	52.2	20.8
Cyclohexane	25.5	–	–	50.2	22.7
Decalin	29.9	–	–	51.4	22.0
Benzene	28.85	363	194	–	–
Toluene	28.5	359	208	–	–
o-Xylene	30.1	359	200	–	–
m-Xylene	28.9	357	211	–	–
p-Xylene	28.4	361	203	–	–
n-Propylbenzene	29.0	363	194	–	–
n-Butylbenzene	29.2	363	193	–	–
Average			≈ 201		≈ 21.9

Source: Data from F. M. Fowkes, *Ind. Eng. Chem.*, **56**, 40 (1964).
[a]γ_{Hg}^d and γ_{Hw}^d are calculated as in Example 6.5.

6.10d　London Components of Interfacial Tensions of Solid Surfaces

It is not difficult to apply the concept of the dispersion component of γ to solid surfaces. In doing this, it is necessary to treat high- and low-energy surfaces differently. We shall not consider solid interfaces in detail; our treatment is limited to the following observations:

1. For low-energy surfaces, $\pi_e \approx 0$. Manipulation of Young's equation (Equation (44)) generates a relationship that expresses γ_s^d in terms of θ and other experimental quantities.

2. For high-energy surfaces, $\pi_e > 0$ owing to adsorption. Relationships have been derived that express γ_s^d in terms of gas adsorption.

Values of γ_s^d that have been determined for high- and low-energy surfaces by these two methods are listed in Table 6.4.

Examination of the values of γ^d for high- and low-energy solids in Table 6.4 reveals several interesting points. To begin, there is a slight overlap in the data inasmuch as polypropylene was studied by the gas adsorption technique, even though it might reasonably be expected to resemble paraffin wax or polyethylene in γ^d value. As the table shows, the value of γ^d for this substance lies in the same neighborhood as the values for these other compounds, even though very widely different procedures were used to arrive at the different values. The multiple values of γ^d listed for some high-energy solids correspond to evaluations based on the adsorption of different gases at widely different temperatures. For example, the γ^d values for TiO$_2$ are obtained from adsorption studies conducted with butane at 0°C, heptane at 25°C, and N$_2$ at −195°C. With these ideas in mind, the observed variation between ostensibly duplicate values becomes more acceptable. The evaluation of γ_s^d also depends on the accuracy of the value of γ for a substance other than the solid. Again we see that errors may be propagated in the analyses that led to the results presented in Table 6.4.

TABLE 6.4 Values of γ_s^d (in mJ m^{-2}) for a Variety of Solids as Determined From Contact Angles and From Gas Adsorption

	γ_s^d	
Material	By measurements of θ	By measurements of π_e
Dodecanoic acid on Pt	10.4, 13.1	—
Polyhexafluoropropylene	11.7,* 18.0	—
Polytetrafluoropropylene	19.5	—
n-C$_{36}$H$_{74}$	21.0	—
n-Octadecylamine on Pt	22.0,* 22.1	—
Paraffin wax	23.2,* 25.5	—
Polypropylene	—	26, 28.5
Polytrifluoromonochloroethylene (Kel F)	30.8	—
Nylon-6,6	33.6*	—
Polyethylene	31.3,* 35.0	—
Polyethyleneterephthalate	36.6*	—
Polystyrene	38.4,* 44.0	—
BaSO$_4$	—	76
Silica	—	78
Anatase (TiO$_2$)	—	89, 92, 141
Iron	—	89, 106, 108
Graphite	—	115, 120, 123, 132

Source: Most data from F. M. Fowkes, *Ind. Eng. Chem.*, **56**, 40 (1964); the data with an asterisk are from D. H. Kaelbe, *Physical Chemistry of Adhesion*, Wiley, New York, 1971.

REVIEW QUESTIONS

1. What is *surface tension*? What are its *units*? Explain the physical significance of surface tension.
2. What is *contact angle*? What is the *range* of magnitudes it can take?
3. Describe a few methods to measure surface tension and contact angle of a liquid on a solid substrate and discuss the assumptions and approximations involved in the measurement procedure.
4. Why are the surface tension and contact angle thermodynamic properties of a substance? What does it mean?
5. What is the *Laplace equation*? What is the *Kelvin equation*? What are the differences between the two?
6. What assumptions have been made in the derivation of the Kelvin equation in the text? How restrictive are these assumptions?
7. What is the *Young-Dupré equation*? What are the approximations made in its derivation? Discuss the merits of those approximations.
8. Describe the terms *wetting, cohesion, adhesion,* and *spreading*.
9. Define the following terms and their relation to surface energies: (a) work of adhesion, (b) work of cohesion, and (c) spreading coefficient.
10. What is the physical significance of the spreading coefficient?
11. What is *equilibrium film pressure*? What is its physical significance?
12. What is the *Bashforth-Adams equation*? How would you use it to determine the surface tension of a liquid and the contact angle of a liquid with a surface?
13. List a few methods for the measurement of surface tension and contact angle. Discuss the basic principles involved in each method. What are the experimental advantages and disadvantages in each case?
14. What are some of the *dynamic methods* for measuring surface tension? What are the differences between these and the *static methods*?

15. Explain the basic principle behind *porosimetry*. How would you use a porosimetry experiment to measure the contact angle of a liquid with a powdered solid? What are the limitations of this approach?
16. Present a molecular interpretation of surface tension.
17. What is the *Girifalco-Good equation*? What is the *Fowkes approximation* to the Girifalco-Good equation?

REFERENCES

General References (with Annotations)

Adamson, A. W., *Physical Chemistry of Surfaces*, 5th ed., Wiley, New York, 1990. (Primarily graduate level. More details on surface tension measurement may be found in this volume.)
Israelachvili, J. N., *Intermolecular and Surface Forces*, 2d ed., Academic Press, London, 1991. (Undergraduate and graduate levels. Contains good discussions and illustrations of the relation among surface tension, wetting, and contact angle phenomena and van der Waals forces.)
Miller, C. A., and Neogi, P., *Interfacial Phenomena: Equilibrium and Dynamic Effects*, Marcel Dekker, New York, 1985. (Graduate level. An advanced-level textbook with a focus on interfaces between liquids, interfacial tension, and dynamics of interfaces.)

Other References

Abbott, N. L., Folkers, J. P., and Whitesides, G. M., *Science*, **257**, 1380 (1992).
Abbott, N. L., and Whitesides, G. M., *Langmuir*, **10**, 1493 (1994).
Abbott, N. L., Whitesides, G. M., Racz, L. M. and Szekely, J., *J. Am. Chem. Soc.*, **116**, 290 (1994).
Adam, N. K., *The Physics and Chemistry of Surfaces*, Dover, New York, 1968.
Good, R. J. In *Surface and Colloid Science*, Vol. 11 (R. J. Good and R. R. Stromberg, Eds.), Plenum, New York, 1979.
Johnson, R. E., Jr., and Dettree, R. H. In *Surface and Colloid Science*, Vol. 2 (E. Matijević, Ed.), Wiley, New York, 1969.
MacRitchie, F., *Chemistry at Interfaces*, Academic Press, San Diego, CA, 1990.
Neumann, A. W., and Good, R. J. In *Surface and Colloid Science*, Vol. 11 (R. J. Good and R. R. Stromberg, Eds.), Plenum, New York, 1979.
Padday, J. F. In *Surface and Colloid Science*, Vol. 1 (E. Matijević, Ed.), Wiley, New York, 1969.

PROBLEMS

1. The frictional force was measured on a strand of viscous rayon fiber moving through a wad of identical fibers as a function of the water content of the wadded fiber.* It was found that the friction incresed from 59 to 133 mN cm^{-1} when the water content decreased to the point at which capillary "necking" between the fibers occurred. A model for this situation may be visualized by considering two parallel, tangent cylinders connected by a "neck" of water held in the neighborhood of the contact by capillary forces. Sketch the situation represented by this model and explain why the force of fiber-fiber attraction increases with decreasing water content. Use this model to discuss (a) the behavior of a wet paintbrush, (b) the practice of wetting the tip of a thread before threading a needle, and (c) the dewatering of cellulose fibers to form paper.

2. The accompanying data give experimental values of the capillary rise for various liquids:†

Liquid	$\Delta\rho$ (g cm^{-3})	h (cm)	R_c (cm)
Water	0.9972	1.4343	0.100990
Benzene	0.8775	1.5425	0.043135
$CHCl_3$	1.4869	1.9210	0.019320

---

*Skelton, J., *Science*, **190**, 15 (1975).
†Richards, T. W., and Carver, E. K., *J. Am. Chem. Soc.*, **43**, 827 (1921).

Use values from the following table (interpolated from Padday 1969) to evaluate a^2 (and γ) by successive approximation:

β	0	0.02	0.04	0.06	0.08	0.10	0.12	0.14	0.16	0.18	0.20
$(x/b)_{90°}$	1.00	0.997	0.994	0.991	0.978	0.984	0.981	0.978	0.975	0.972	0.970

Compare the values of γ calculated by this procedure with those obtained by the approximation given by Equation (4).

3. Finely dispersed sodium chloride particles were prepared, their specific area was measured, and their solubility in ethanol at $25°C$ was studied.* It was found that a preparation with a specific area of 4.25×10^5 cm^2 g^{-1} showed a supersaturation of 6.71%. Estimate the radius of the NaCl ($\rho = 2.17$ gm cm^{-3}) particles, assuming uniform spheres. Calculate γ for the NaCl-alcohol interface from the solubility behavior of this sample.

4. Enüstün and Turkevich† prepared SrSO$_4$ ($\rho = 3.96$ g cm^{-3}) precipitates under conditions that resulted in different particle sizes. Particle sizes were characterized by electron microscopy, and solubilities were determined at $25°C$ by a radiotracer technique. In the following data the supersaturation ratios are presented for different preparations, each of which is characterized by an average particle width and a minimum particle width:

x_{mean} (Å)	x_{min} (Å)	S/S_0
247	96	1.43
269	130	1.35
388	155	1.28
541	168	1.29
629	252	1.16
1260	378	1.10
1660	500	1.07

Which size parameter gives the best agreement with the Kelvin equation? Explain. Use the best fitting data to evaluate γ for the SrSO$_4$-H$_2$O interface.

5. Bartell and Osterhof‡ describe an experimental procedure for measuring the work of adhesion between liquids and solids. With carbon (lampblack) as the solid, the following values for the work of adhesion were obtained:

Liquid	Benzene	Toluene	CCl$_4$	CS$_2$	Ethyl ether	H$_2$O
W_{AB} (erg cm^{-2})	109.3	110.2	112.4	122.1	76.4	126.8

Use these data together with the surface tensions of the pure liquids from Table 6.1 to calculate the spreading coefficients for the various liquids on carbon black. Use your results to interpret the authors' observations: "About equal quantities of water and organic liquid were put into a test tube with a small amount of the finely divided solid and shaken. It was noted that the carbon went exclusively to the organic liquid phase."

6. The effect of mutual saturation on the L-V and L-L interfacial tensions is effectively illustrated by considering the spreading coefficient of one liquid on another using both the initial (unsaturated) and equilibrium values of γ. Use the following data§ to calculate $S_{B/A}$ (equilibrium) and $S'_{B/A}$ (nonequilibrium):

*Van Zeggeren, F., and Benson, G. C., *Can. J. Chem.,* **35,** 1150 (1957).

†Enüstün, B. V., and Turkevich, J., *J. Am. Chem. Soc.,* **82,** 4502 (1960).

‡Bartell, F. E., and Osterhof, H. J., *J. Phys. Chem.,* **37,** 543 (1933).

§Harkins, W. D., *The Physical Chemistry of Surface Films,* Van Nostrand-Reinhold, Princeton, NJ, 1952.

	H_2O/air	H_2O/IAA	IAA/air	H_2O/air	H_2O/CS_2	CS_2/air
γ' (erg cm^{-2})	72.8	5.0	23.7	72.8	47.4	32.4
γ (erg cm^{-2})	25.9	5.0	23.6	70.3	48.4	31.8

Describe what happens when a drop of pure isoamyl alcohol (IAA) is placed on the surface of pure water. What happens with the passage of time? Repeat this description for the case of pure CS_2 on pure water.

7. Water drops were formed at the mercury-benzene interface by means of a syringe, and the contact angle (measured in the water) was recorded as a function of time.* For the interface between Hg and benzene saturated with water, γ was measured independently as a function of time. The following table summarizes these data (all measured at 25°C):

Time (hr)	$\gamma_{Hg\text{-}benzene}$ (erg cm^{-2})	θ_{obs} (deg)
0.10	363.0	118
0.42	359.5	119
1.0	358.0	122
2.5	354.5	138
5.0	350.0	144
13	336.0	—
23	—	180

Using 379.5 and 34.0 erg cm^{-2}, respectively, as the values of γ for the mercury-water and benzene-water interfaces, compare the observed contact angles with the predictions of Young's equation. Comment on the fact that constant values are used for $\gamma_{Hg\text{-}W}$ and $\gamma_{benzene\text{-}W}$.

8. A cylindrical rod may be used instead of a rectangular plate in a slight variation of the Wilhelmy method. Derive an expression equivalent to Equation (63) for a suspended solid of cylindrical geometry. Prepare semiquantitative plots analogous to Figure 6.10c based on the following data, assuming cylindrical rods 1.0 mm in diameter at the air-water interface:†

Metal	ρ_s (g cm^{-3})	θ_a (deg)	θ_r (deg)
Au	19.3	70	40
Pt	21.5	63	28

In both cases the metal surfaces were carefully polished, washed, steamed, and then heated in an oven for 1 hr at 100°C.

9. The vertical rod method of the preceding problem was used to study the contact angle of water at the gold-water-air junction at 25°C. The following data show how the value of θ_a depends on the prior history of the metal surface.‡ For a gold surface polished, washed, and heat treated for 1 hr at T°C, then allowed to stand in air for t hr we obtain

T (°C)	100	200	300	400	500	600	600	600	600	600	600
t (h)	< ¼	< ¼	< ¼	< ¼	< ¼	¼	1	5	10	24	120
θ_a (deg)	68	57	45	36	25	13	22	38	47	53	55

Calculate the work of adhesion between water and gold for each of these cases on the assumption that $\pi_e = 0$. Is the variation in W_{SL} consistent (qualitatively? quantitatively?) with the expected validity of the assumption concerning π_e?

10. The tendency of spilled mercury to disperse as small drops that roll freely on most surfaces is a well-known characteristic of this liquid. Discuss this behavior in terms of (a) the work of adhesion and the spreading coefficient for mercury on various substrates, (b) surface tension

*Bartell, F. E., and Bjorkland, C. W., *J. Phys. Chem.,* **56**, 453 (1952).
†Bartell, F. E., Culbertson, J. A., and Miller, M. A., *J. Phys. Chem.,* **40**, 881 (1936).
‡Bartell, F. E., and Miller, M. A., *J. Phys. Chem.,* **40**, 889 (1936).

versus contact angle (measured in Hg) as causes of this behavior, and (c) the implications of the Kelvin equation on the health hazards associated with mercury spills.

11. Use the data of Table 6.2 to plot the profile of a drop with $\beta = 25$. Measure (in cm) the radius of the drop you have drawn at its widest point. By comparing this value with the value of $(x/b)_{90°}$ from the table, evaluate b (in cm) for the drop as you have drawn it. Suppose an actual drop is characterized by this value of β. If the actual radius at the widest point is 0.25 cm and $\Delta\rho = 0.50$ g cm^{-3}, what is γ for the interface of the drop?

12. Suppose the drop profile shown in Figure 6.15b describes an actual drop for which the radius at the widest point equals 0.135 cm. Use the value of $(x/b)_{90°}$ from the figure to calculate γ for each of the following situations:

System	$|\Delta\rho|$ (g cm^{-3})
(a) Oil in water	0.20
(b) Water in oil	0.20
(c) Oil in air	0.80
(d) Air in water	1.00
(e) Water in air	1.00

State whether the drop is pendant or sessile in each case.

13. Sometimes it is difficult to locate the bottom of the meniscus of a colorless solution in a buret. If this is the case, it is surely that much more difficult to try to estimate contact angles by direct observation of a meniscus. In view of this, criticize or defend the following proposition: Before attempting to read contact angles directly by viewing the surface through a low-power microscope, a new worker should practice with simulated drops (recommended by Neumann and Good 1979). By masking off the lower part of a sphere such as a ball bearing and looking at its silhouette, a practice junction is obtained. The angle estimated can be compared with the true value by calculating the true value from the height of the apex and the width of the base. In fact, values for actual drops can be determined by this method if the profile of the entire drop is visible in the microscope.

14. The following combinations of θ, γ_L, and p values were reported by Bartell and Whitney* for the wetting of silica plugs by various liquids:

Liquid	p (g cm^{-2})	γ_L (dyne cm^{-1})	θ
Nitrobenzene	125	25.3	41°25′
Chloroform	192	31.6	22°11′
Benzene	215	34.7	19°16′
Toluene	221	36.5	—
CCl$_4$	265	44.5	—
Hexane	333	51.0	—

Use the first three sets of data to determine a value of R_c for the plug; then use this value to determine θ for the remaining liquids against silica. Are large or small contact angles more sensitive to, say, a 5% error in R_c?

15. Drake† has reported the accompanying data for the porosimetry analysis of a catalyst preparation; the cumulative pore volume occupied by mercury is given for the applied pressures indicated:

Pressure (psi)	Volume (cm^3 g^{-1})	Pressure (psi)	Volume (cm^3 g^{-1})
1,000	0.082	20,000	0.228
1,500	0.115	25,000	0.249

---

*Bartell, F. E., and Whitney, C. E., *J. Phys. Chem.*, **36**, 3115 (1932).
†Drake, L. C., *Ind. Eng. Chem.*, **41**, 781 (1949).

2,000	0.132	30,000	0.276
3,000	0.150	35,000	0.307
4,000	0.160	40,000	0.336
6,000	0.177	45,000	0.358
10,000	0.190	50,000	0.363
15,000	0.213		

Plot these data and from the tangents to the curve estimate dV/dR_c at 2000, 10,000, 30,000, and 45,000 psi. To what radii do these pressures correspond? Use $\gamma = 484$ mJ m^{-2} and $\theta = 140°$ for these calculations.

16. By a suitable combination of Equations (49) and (97) show that

$$\gamma_S^d = \frac{\gamma_{LV}^2}{4\gamma_L^d}(1 + \cos\theta)^2$$

for low-energy surfaces. Use the following data (see Table 6.4 for reference) to evaluate either γ_S^d or γ_L^d as appropriate:

S	L	γ_{LV} (erg cm^{-2})	γ^d (erg cm^{-2})	θ (deg) in liquid
Dodecanoic acid on Pt	α-Bromonaphthalene	44.6	10.4 for S	92
Kel F	α-Bromonaphthalene	44.6	30.8 for S	48
Paraffin wax	Glycerol	63.4	36 for L	97
Paraffin wax	Fluorolube	20.2	13.5 for L	31

17. The equation derived in the preceding problem suggests that a plot of $\cos\theta$ (as ordinate) versus $\sqrt{\gamma_L^d}/\gamma_L$ (as abscissa) should be linear with a slope of $2\sqrt{\gamma_S^d}$ and an intercept of -1. Describe how this result can be used to evaluate γ_S^d when contact angle measurements are made on a particular solid with a variety of liquids for which γ_L and γ_L^d are known. Describe how the same graphing procedure can be used to evaluate $\sqrt{\gamma_L^d}/\gamma_L$ when contact angles are measured in a liquid of unknown γ_L^d on different solids of known γ_S^d. Use the data of the preceding problem to illustrate these two graphical interpretations. Include the additional datum that α-bromonaphthalene forms a contact angle of 58.5° with paraffin (see Table 6.4 for reference).

18. If γ_L and θ are measured for a homologous series of liquids on a given low-energy solid, a plot of $\cos\theta$ versus γ_L results in a straight line. Verify that this is the case for the following data, determined for alkanes on Teflon at 20°C:*

Compound	γ_L (mJ m^{-2})	θ (deg)	Compound	γ_L (mJ m^{-2})	θ (deg)
Hexadecane	27.6	46	Nonane	22.9	32
Tetradecane	26.7	44	Hexane	21.8	26
Dodecane	25.4	42	Heptane	20.3	21
Undecane	24.7	39	Octane	18.4	12
Decane	23.9	35	Pentane	16.0	spreads

The intercept at $\theta = 0$ is viewed as a kind of critical state for these systems, and the corresponding surface tension is represented by γ_c. What is the value of γ_c for Teflon?

19. Use Equation (99) to show that γ_c as defined in the last problem is equal to γ_S^d for those systems where $\gamma_L = \gamma_L^d$. What else must be assumed to prove this?

*Fox, H. W., and Zisman, W. A., J. Colloid Sci., 5, 514 (1950).

7

Adsorption from Solution and Monolayer Formation

You are living on a plane. What you style Flatland is the vast level surface of what I may call a fluid, on, or in, the top of which you and your countrymen move about, without rising above it or falling below it.

From Abbott's *Flatland*

7.1 INTRODUCTION

The interest of scientists on monolayers of oil on a water surface is frequently traced to Benjamin Franklin's now-famous experiment on stilling the waves with oil in the pond at Clapham Common in England on a windy day,* but the striking properties of such monolayers are brought to light by an equally simple experiment most students have seen at home or high school. Shake some pepper or flour on a pan of water and touch the surface of the water with a bar of soap; you can see the pepper or flour race to the edges. Why does an oil film dampen the surface? Why does the pepper retreat so rapidly to the edge of the water's surface? Investigations of such questions are what lead us to the material we discuss in this chapter.

7.1a Surfactant Layers: Langmuir and Gibbs Layers and Langmuir-Blodgett Films

Monolayers formed by a substance that is insoluble in the liquid subphase are labeled *Langmuir layers*, whereas those formed by substances soluble in the bulk but adsorbed preferentially at the interface are known as *Gibbs layers*. Langmuir layers transferred to solid substrates are known as *Langmuir-Blodgett* (LB) layers (see Vignette VII below). Our objective in this chapter is to lay the basic foundations necessary for understanding adsorption and formation of such layers and their structures. The Langmuir-Blodgett films usually consist of multiple layers; however, we restrict ourselves to monolayers.

Until now, we have avoided considering the above topics by intentionally excluding solutes of variable concentration from our consideration of surfaces. Now, the effects of such solutes are our specific interest. We are especially concerned with a particular class of solutes that show dramatic effects on surface tension. These are said to be *surface active* and are often simply called *surfactants*. Our primary emphasis here is on the relationship between adsorp-

---

*An engaging discussion of the history of Benjamin Franklin's experiment and a relatively nontechnical treatment of monolayers and bilayers of surfactants and their implications to biochemistry and biology are presented by Tanford, a pioneer of what is known as the *hydrophobic effect* and the biological applications of mono- and multilayers (Tanford 1989). Almost all of the material discussed in this highly readable volume is relevant to the focus of this chapter.

tion phenomena and surface tension/surface energy. In the process of describing experimental methods, results, and interpretations, however, a variety of related concepts enters the picture, as explained in Section 7.1c.

7.1b Why Are Monolayers and Multilayers Important?

Traditionally, monolayer and multilayer adsorption have been used in detergency, mineral processing, flotation, stability of food and pharmaceutical emulsions, and the like, and, as a consequence, the topics of this chapter have been a central part of colloid science. In recent years, however, research on monolayer and multilayer deposition has mushroomed rapidly because of significant new opportunities.

At the most fundamental level, monolayers of surfactants at an air-liquid interface serve as model systems to examine condensed matter phenomena. As we see briefly in Section 7.4, a rich variety of phases and structures occurs in such films, and phenomena such as nucleation, dendritic growth, and crystallization can be studied by a number of methods. Moreover, monolayers and bilayers of lipids can be used to model biological membranes and to produce vesicles and liposomes for potential applications in artificial blood substitutes and drug delivery systems (see, for example, Vignette I.3 on liposomes in Chapter 1).

Equally important are the new developments in electronic and optical materials and other "advanced materials." Developments in chemical synthesis and in experimental techniques such as atomic force microscopy, scanning tunneling microscopy, and optical imaging techniques have opened the way for manipulating the structure and properties of ultrathin films. As a result, it might be possible to "engineer" surface coatings with special optical, electrical, and magnetic properties, and the monolayer and multilayer structures engineered in this manner can be used for fabricating microelectronic devices, chemical and biochemical sensors, and optical switches and storage devices, to name a few.

Vignette VII presents a historical snapshot of the early experiments of Agnes Pockels that led to the above exciting possibilities. A couple of specific examples along these lines and some additional information follow in Sections 7.10b and 7.10c.

VIGNETTE VII MOLECULAR MONOLAYERS AND MULTILAYERS: Langmuir-Blodgett Films

When Agnes Pockels (1862–1935) of Brunswick, Germany, sent a registered letter on January 12, 1891, to Lord Rayleigh (John William Strutt, Third Baron Rayleigh, 1842–1919) describing her simple apparatus for studying properties of oil films and oil-contaminated water, little did she realize that her work and the essential design of her apparatus would pave the way for fundamental research on molecular films for possible use in molecular electronics, bio- and chemical sensors, tribology, and catalysis, among others.

Pockels described in the letter her design of a rectangular tin trough with a thin tin strip laid across it. The trough was filled to the brim with water, with a thin layer of oil covering the surface of the water on one side of the tin strip and clean water on the other side. The tin strip served to vary the area of the oil-contaminated surface, and a balance measured the force necessary to lift a small disk (a button) from the surface. Pockels used this setup to study the surface tension of the oil-contaminated layer.

This groundbreaking work from a woman with a "domestic background" impressed Lord Rayleigh immensely, and he sent it to *Nature*, a prestigious British journal, with his strong recommendation for publication (Pockels 1891). Additional papers from Pockels and Lord Rayleigh followed, and Lord Rayleigh proposed that layers a single molecule in thickness could be studied using the Pockels technique.

While the experiments of Pockels and Rayleigh generated some activity on the properties of thin films, the area was essentially dormant until the work of Irving Langmuir (1881–1957; Nobel Prize, 1932, for surface chemistry) at the General Electric Corporate Research Laboratories in Schenectady, New York. Langmuir developed a number of new techniques (including the well-known surface film balance named after him) to study the chemistry and physics of monolayers on water surfaces. Such "floating" monolayers are now known as *Langmuir films*, in honor of Langmuir's pioneering work.

The scientific and industrial importance of monolayers and multilayers of organic and polymeric materials would not have been possible without the work of another pioneer, Katherine Blodgett (1898–1979), the first woman scientist to join G.E. Research Laboratories, and the first woman to obtain a doctorate from the famed Cavendish Laboratory in Cambridge, England. Blodgett devised an apparatus to transfer fatty acid monolayers to solid substrates from water surfaces. The films thus produced are now known as *Langmuir-Blodgett films* and can be fabricated using a number of different substances (see Fig. 7.1).

The Langmuir-Blodgett films are believed to hold considerable promise in numerous high-technology applications such as molecular electronics, piezoelectric organic films, waveguides, nonlinear optics, and optical information storage, in addition to classical applications such as adhesion, prevention of corrosion, catalysis, and solubilization (see Section 7.10 and Roberts 1990). The challenge to the surface chemists, physicists, and engineers is to devise and fabricate films with special chemical and physical properties (i.e., use molecular engineering to manipulate the architecture of the films to provide the desired properties). This field is thus a very active area of current research and development in colloid and surface science.

7.1c   Focus of This Chapter

The material in this chapter is organized broadly in two segments. The topics on monolayers (e.g., basic definitions, experimental techniques for measurement of surface tension and surface-pressure-versus-area isotherms, phase equilibria and morphology of the monolayers, formulation of equation of state, interfacial viscosity, and some standard applications of monolayers) are presented first in Sections 7.2–7.6. This is followed by the theories and experimental aspects of adsorption (adsorption from solution and Gibbs equation for the relation between

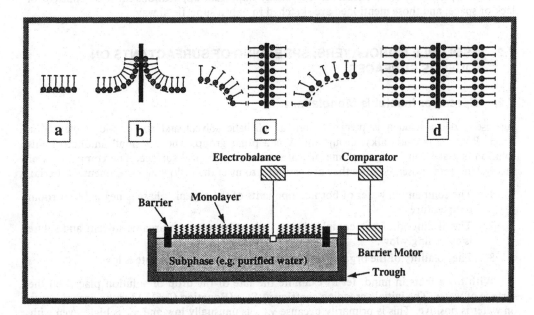

FIG. 7.1 Deposition of multilayers using the Langmuir-Blodgett technique: (a and b) monolayer deposition; and (c and d) multilayer deposition.

surface tension and surface excess concentration, experimental aspects of Gibbs equation, adsorption on solid substrates, the Langmuir equation and its application, and examples of traditional and some modern applications of adsorbed layers) in Sections 7.7–7.10. We close the chapter with a brief discussion of adsorption in the presence of applied potentials as this is an important topic in many aspects of colloid science.

Despite the above organization, in studying the material of this chapter it may be helpful to realize that the topics covered may be grouped in several different ways. Let us enumerate what these various ways of looking at the material are.

First, we may focus our attention on the solubility of the adsorbed species in one or both of the adjacent phases. In this way two broad categories of phenomena emerge: insoluble and soluble surface layers.

A second way of classifying the material is on the basis of the experimental methods involved. For mobile interfaces, surface tension is easily measured. For these it is easiest to examine the surface tension-adsorption relationship starting with surface tension data. When insoluble surface films are involved, we shall see how the difference in γ between a clean surface and one with an adsorbed film may be measured directly. For solid surfaces, surface tension is not readily available from experiments. In this case adsorption may be measurable directly, and the relationship between adsorption and surface tension may be examined from the reverse perspective.

Third, the material of this chapter is a mix of descriptive and theoretical concepts. The most important descriptive observation is the existence of two-dimensional phases. The theoretical content of the chapter is mostly thermodynamic in origin. Three major results are the equations named after Gibbs, Langmuir, and Lippmann. We are mostly concerned with uncharged surfaces, except for a brief discussion of electrolyte adsorption at a polarizable mercury electrode in Section 7.11.

Finally, the material may also be regarded as a mixture of fundamentals and applications. Although the entire book stresses principles, applications are considered from time to time as examples of more abstract ideas. This is also the intent of the sections on applications in this chapter. In addition, however, many applications of adsorption phenomena are the basis of large and important areas of technology. To omit mention of them would lead to a very incomplete picture of these fields. As it is, many important applications must be omitted for lack of space, and those mentioned are sketched in only a superficial way.

7.2   INSOLUBLE MONOLAYERS: SPREADING OF SURFACTANTS ON AQUEOUS SURFACES

7.2a   Spread or Insoluble Monolayers

Suppose a dilute solution is prepared from an aliphatic solvent and an organic solute RX in which R is a long-chain alkyl group and X is a polar group. Then, a small amount of this solution is placed on a large volume of water with a horizontal surface. The components of this system were chosen because they are assumed to meet the following experimental criteria:

1. The solubility in water of both components of the organic phase is negligible at room temperature.
2. The likelihood of any complex being formed between the organic solvent and solute is exceedingly low.
3. The volatility of the organic solvent is high and that of the solute is low.

With these facts in mind, let us examine the fate of the drop of solution placed on the surface of water. The initial spreading coefficient $S_{O/W}$ (Equation (6.61)) for the organic layer on water is positive. This is primarily because $\gamma_{O/W}$ is unusually low and γ_W is high, even with an adsorbed layer of the organic solvent. After spreading, we allow sufficient time to elapse for all the solvent to evaporate from the spread layer. At this point the surface will contain a layer of the organic solute similar to that which would result from the spreading of a sessile drop of pure liquid solute or from the adsorption of vapors of the solute component from the gas phase. Using a solution with a volatile solvent to form such a layer is a very common technique and has the advantage of permitting very small amounts of solute to be quantitatively deposited on a surface.

The nature of the layer that remains after the solvent has evaporated depends on the amount of solute deposited and the area available to it. It is convenient to distinguish among three situations in this regard. If the amount of added material and the area are such that the water surface is covered uniformly to a depth of one molecule with the solute, the resulting film is called a *monolayer*. On the other hand, submonolayer coverage and multilayer coverage result when the amount of added material per area is less or more, respectively, than that which produces the monolayer. In this chapter we are concerned mostly with degrees of coverage up to and including the monolayer. If a large excess of spread material (beyond the amount needed for monolayer coverage) is used, the excess collects into droplets of a bulk phase. The equilibrium situation is then identical to what would be produced by the spreading of a sessile drop of the solute material. Films of the sort described here are called either *spread monolayers* (when the method of their preparation is stressed) or *insoluble monolayers* (when the chemical nature of the solute is emphasized). We use these terms interchangeably.

7.2b   Some Properties of Spread Monolayers

Now let us examine some of the properties of the spread monolayer that we have described. It was seen in the preceding chapter (e.g., Equation (6.48)) that the presence of an adsorbed layer lowers the surface tension of an interface. The phenomenon is quite general, so we redefine π (no subscript) in the following symbols:

$$\pi = \gamma_0 - \gamma \tag{1}$$

where γ_0 refers to the surface tension of any phase in the absence of an adsorbed layer, and γ refers to the tension of the same surface with an adsorbed layer. Specific subscripts are used only when the problem clearly involves more than one interface.

The spread monolayer just described may be discussed from two points of view. First, there are those aspects of the film that pertain explicitly to the chemical nature of the components: water and the organic solute. Second, there are certain properties of the monolayer that depend on physical variables such as temperature, area of the water surface, and number of molecules of RX present. Let us briefly discuss both of these viewpoints.

7.2b.1   Structural Aspects of Monolayers

The organic solute RX is a prototype of an important array of surface-active materials. Many surface-active substances are composed of what are known as *amphipathic* molecules. This term simply means that the molecule consists of two parts, each of which has an affinity for a different phase. We are concerned mostly with surfaces in which one of the phases is aqueous, so the surfactants that we consider will contain polar or ionic groups, or "heads," and nonpolar organic residues, or "tails." In the compound RX, for example, R is an alkyl group, generally containing 10 or more carbon atoms. The literature of surface chemistry contains many references to these organic groups by both their International Union of Pure and Applied Chemistry (IUPAC) and their common names. Table 7.1 lists some of the more commonly encountered examples. In RX, the polar X group may be $-OH$, $-COOH$, $-CN$, $-CONH_2$, or $-COOR'$, or an ionic group such as $-SO_3^-$, $-OSO_3^-$, or $-NR_3^+$.

With the foregoing ideas in mind, one characteristic of the adsorbed monolayer becomes apparent: molecular orientation at surfaces. For a film of RX on water, the picture that emerges is one in which the polar groups are incorporated into the aqueous phase with the hydrocarbon part of the molecule oriented away from the water. Such details as the depth of immersion of the tail and the configuration of the alkyl group are best approached by considering how the properties of the monolayer depend on the physical variables.

TABLE 7.1   IUPAC and Common Names for a Variety of Normal
Saturated and Unsaturated Surface Active Compounds

	Normal, saturated compounds		
	Carboxylic acids		Alcohols, amines, sulfates, etc.
n (number of C atoms)	IUPAC name	Common name	Common name
12	Dodecanoic	Lauric	Lauryl
14	Tetradecanoic	Myristic	Myristyl
16	Hexadecanoic	Palmitic	Cetyl
17	Heptadecanoic	Magaric	Heptadecyl
18	Octadecanoic	Stearic	Stearyl
20	Eicosanoic	Arachidic	Eicosyl, arachic
22	Docosanoic	Behenic	Docosyl

	Normal, unsaturated carboxylic acids	
	IUPAC name	Common name
18	cis-9-Octadecenoic	Oleic
18	cis,cis-6,9-Octadecenoic	Linoleic
18	cis,cis,cis-3,6,9-Octadecenoic	Linolenic
18	trans-9-Octadecenoic	Elaidic
22	cis-9-Docosenoic	Erucic
22	trans-9-Docosenoic	Brassidic

7.2b.2   Surface Pressure Versus Area Isotherms

Next let us consider some of the physical properties of the spread monolayer we have described. Equation (1) states that the surface tension of the covered surface will be less than that of pure water. It is quite clear, however, that the magnitude of γ must depend on both the amount of material adsorbed and the area over which it is distributed. The spreading technique already described enables us to control the quantity of solute added, but so far we have been vague about the area over which it spreads. Fortunately, once the material is deposited on the surface, it stays there—it has been specified as insoluble and nonvolatile for precisely this reason. This means that some sort of barrier resting on the surface of the water may be used to "corral" the adsorbed molecules. Furthermore, moving such a barrier permits the area accessible to the surface film to be varied systematically. In the laboratory this adjustment of area is quite easy to do in principle. As we see below, the actual experiments must be performed with great care to prevent contamination.

Suppose that the initial film is spread on water that fills to the brim a shallow tray made of some inert material. Rods with low-energy surfaces may then be drawn across the water to adjust the area accessible to the molecules of the monolayer. Figure 7.2a indicates schematically how such an arrangement might appear. In practice, several barriers would be used, first to sweep the surface free of insoluble contaminants and then to confine the monolayer.

7.2b.3   Wilhelmy-Plate Technique for Surface Tension Measurement

Next, an experiment such as that shown in Figure 7.2b could be conducted. The apparatus consists of a pair of *Wilhelmy plates* attached to two arms of a balance. One plate contacts the clean surface and the other the surface with the monolayer. Note that the barrier separates the two portions of surface. The surface tension will be different in the two regions, and the weight (and volume) of the meniscus entrained by the plate will be larger for the clean surface

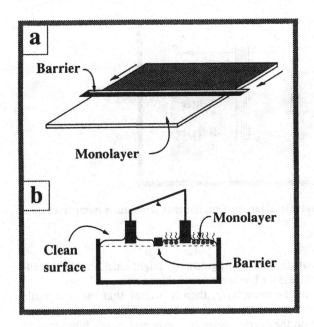

FIG. 7.2 Schematic illustrations of a monolayer and a Wilhelmy plate arrangement for surface tension measurement: (a) schematic illustration of a barrier delineating the area of a monolayer; and (b) a Wilhelmy plate arrangement for measuring the difference in γ on opposite sides of barrier.

because of its higher surface tension (Equation (6.2)). Ideally, the two plates are identical in weight, perimeter, and contact angle, although the last may be difficult to achieve in practice. If these conditions are met, however, the additional weight needed to bring the apparatus to balance measures the difference in γ for the two surfaces. By Equation (6.2), this is given by

$$w_{clean} - w_{film} = P \cos \theta \, (\gamma_0 - \gamma) \tag{2}$$

Of course, there is no necessity to measure both γ_0 and γ on the same apparatus; they may be determined independently by any of the methods of Chapter 6. The experiment represented by Figure 7.2b is intended mainly to emphasize that the surface tension of the two areas will be different. Furthermore, as the barrier is moved in such a way as to compress the area of the spread monolayer, the value of π will increase.

Although π and the area A of the surface vary inversely, the precise functional form by which they are related is more difficult to describe. For very large areas, π and A show a simple inverse proportionality such as pressure and volume for an ideal gas. As the area is decreased, a more complex relationship is needed to connect these variables, just as the equation of state becomes more complex for nonideal gases and condensed phases. This analogy of π and A with p and V turns out to be a very profitable way of thinking about insoluble monolayers. For one thing, it suggests an alternative to the difference between two values of surface tension as a means of measuring π. In addition, the analogy to three-dimensional states suggests models for understanding monolayers. In succeeding sections each of these points is developed in greater detail.

Identification of area as the two-dimensional equivalent of volume is a straightforward geometrical concept. That π should be interpreted as the two-dimensional equivalent of pressure is not so evident, however, even though the notion was introduced without discussion in Chapter 6, Section 6.6. Figure 7.3 helps to clarify this equivalency as well as suggest how to compare quantitatively two- and three-dimensional pressures. The figure sketches a possible profile of the air-water surface with an adsorbed layer of amphipathic molecules present. In

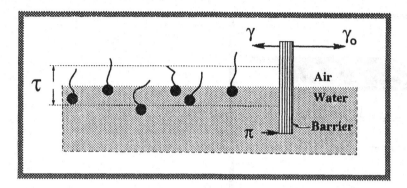

FIG. 7.3 Schematic profile of the air-water interface at a barrier that separates a monolayer from the clean surface.

general, we must allow for the fact that different configurations might exist among the adsorbed molecules; nevertheless, the surface layer has some mean thickness τ.

If the barrier represents the limit of the monolayer, then it is clear that the contractile force exerted by the surface is different on opposite sides of the barrier. Since γ is less than γ_0, it is as if the film were exerting a force on the barrier along the perimeter of the film equal to π. Force per unit length—the units of γ—is the two-dimensional equivalent of force per unit area, the units of pressure in the bulk.

The surface layer does not have zero thickness, of course, even though it is conceptually convenient to think of it as two-dimensional matter. If we assume that the film pressure π extends over the entire thickness of the film, then it is an easy problem to convert the two-dimensional pressure to its three-dimensional equivalent. Taking 10 mN m^{-1} as a typical value for π and 1.0 nm as a typical value for τ enables us to write

$$p = \frac{\pi}{\tau} = \frac{10^{-2}\,\text{N m}^{-1}}{10^{-9}\,\text{m}} = 10^7\,\text{N m}^{-2} \tag{3}$$

or

$$p = 10^7\,\text{N m}^{-2} \cdot \frac{1\,\text{atm}}{1.013 \cdot 10^5\,\text{N m}^{-2}} \approx 100\,\text{atm} \tag{4}$$

In view of this calculation, it is not too surprising that insoluble monolayers do not usually display a simple inverse proportionality between π and A. At pressures this high, three-dimensional matter is not likely to obey the ideal gas law either.

7.3 EXPERIMENTAL MEASUREMENT OF FILM PRESSURE

7.3a LANGMUIR FILM BALANCE

The considerations of the preceding section suggest a second way to study spread monolayers. This technique involves measuring the film pressure directly rather than calculating it from surface tension differences by Equation (1). Figure 7.4a is a schematic representation of an apparatus called the Langmuir film balance after Irving Langmuir, a pioneer in this field as mentioned in Vignette VII. Its base is a shallow tray or trough of some inert material. As was the case in Figure 7.2a, the surface must be swept by barriers both to clean the surface and to compress monolayers. In the Langmuir balance, however, one of the barriers is attached to a pivoted arm, arranged in such a way that a torque balance around point P can be measured, for example, by adding weights to the pan. As the figure shows, an insoluble monolayer may be confined to a portion of the surface adjacent to the pressure-sensing float. The shaded area

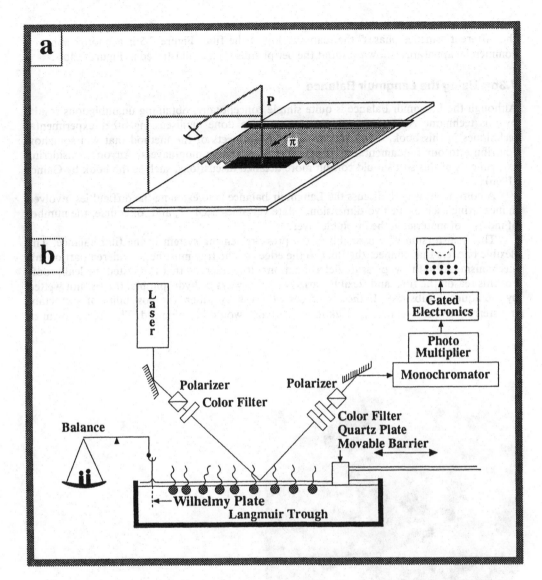

FIG. 7.4 Langmuir film balance: (a) a schematic representation of a Langmuir balance; and (b) a Langmuir trough with a laser optics instrument to measure the orientations of the hydrocarbon tails of the surfactant molecules. The apparatus shown monitors the orientation of the tails through the second harmonic signals generated at various angles of incident light beam. (Redrawn with permission of G. A. Somarjai, *Introduction to Surface Chemistry and Catalysis*, Wiley, New York, 1994.)

in the figure corresponds to the area of the film. It is obviously adjustable by moving the other barrier. To prevent the film from leaking past the edges of the float, flexible barriers connect the ends of the float to the edges of the tray.

By means of this apparatus, it is possible to vary the area of a spread monolayer and measure the corresponding film pressure directly. Many different variations of the film balance are available, and a number of instrumentational techniques can be combined with the Langmuir balance to obtain information on the microstructure of the films and the properties of the films. Figure 7.4b illustrates, for example, a laser optics arrangement to monitor the molecular orientation of the hydrocarbon tails of the surfactant molecules. Below in this chapter we come across another technique, based on fluorescence microscopy, for observing the different "surface phases" that can develop in the film. Figure 7.5 is a photograph of a commercial apparatus, shown without the peripherals such as illustrated in Figure 7.4b.

7.3b Using the Langmuir Balance

Although the Langmuir balance is quite simple conceptually, obtaining unambiguous results by this technique is far from simple. As we have done in discussing other experimental techniques in this book, we only touch on those aspects of the method that will somehow contribute to our fundamental understanding of insoluble monolayers. Anyone considering experiments of this sort should consult more detailed discussions, such as the book by Gaines (1966).

A convenient way to discuss the Langmuir balance is to examine the difficulties involved in measuring each of the two-dimensional state variables π, A, T, and, of course, the number of moles n of material in the insoluble layer.

The float-torsion wire assembly is the pressure-sensing system in the film balance. The flexible barriers that connect the float to the edges of the tray must be considered part of this mechanism. As with gas pressure determinations, it is essential that the system be leakproof. For this reason the float and flexible barriers must always be hydrophobic, that is, not wetted by the aqueous substrate. If these surfaces were wet by water, the possibility of surfactant transferring to them — that is, a leak in the system — would be enhanced. Thin pieces of mica

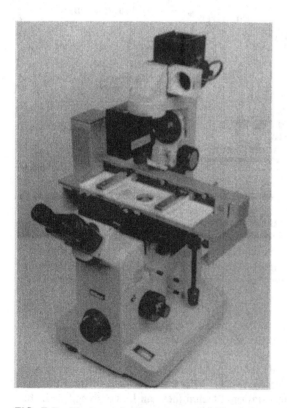

FIG. 7.5 Photograph of a commercial film balance. The photograph shows a "minitrough" with microscopy setup. (Courtesy of KSV Instruments, Ltd., Höyläämötie 11B, SF-00380 Helsinki, Finland.)

are commonly used as float material, and platinum ribbons or threads of silk or nylon are often used for flexible barriers. All of these are waxed to give them suitably hydrophobic characteristics. The sweeping barriers and the tray must also be hydrophobic for the same reasons. These are usually waxed metal, although Teflon is also quite popular because of its inertness. The barriers must make intimate contact with the edges of the tray, also to prevent leaks. Therefore both tray and barriers must be carefully machined to assure good contact. Plastics are generally unsuitable as barrier materials because they are too light to make good contacts.

To convert the measured torque into a two-dimensional pressure, it is necessary to know both the length of the float and the distance between the float and the torsion wire. The distance measurement requires that the water level be controlled quite accurately. As far as the length of the float is concerned, the flexible connectors must be included in this figure. Since they are anchored at one end, only part of their length may be considered a part of the pressure-sensing system. Some approximations are required here, but if the length of the connector is small compared to the total length of the float, the error is negligible.

The float is effectively a two-dimensional manometer, and, like its open-ended counterpart, it measures the film pressure difference between the two sides of the float. This is another reason why it is imperative that no leakage occur past the float assembly: Leakage would increase the pressure on the reference side of the float. For the same reason, the side of the float opposite the monolayer must be carefully checked for any possible source of contamination, not just misplaced surfactant. One way of doing this is to slide a barrier toward the float from that side to verify that no displacement of the float occurs. In all aspects of film pressure measurement, the torque must be measured with sufficient sensitivity to yield meaningful results.

Measuring the area of the film is less troublesome. If the edges of the tray are parallel and the barriers perpendicular to them, the area of the rectangular surface is easily determined. The curvature of the surface at the hydrophobic boundaries introduces a small error but—since the total area is of the order of magnitude of 10^{-2} m^2—this is generally negligible.

The results obtained in π versus A experiments may be sensitive to the rate at which the film area is changed. We shall not discuss the factors responsible for this, but merely note that the same film pressures should be obtained on compression and expansion if true equilibrium values are being measured.

Temperature is an important variable in any equation of state. The experiments we are describing are isothermal; therefore it is important that both the water substrate and the adjoining vapor be thermostated.

Next let us consider those difficulties associated with the determination of the amount of material deposited on the surface. We have already noted that the method of depositing insoluble monolayers by spreading permits the accurate determination of n. Since the spreading technique requires solvent volatility, care must be exercised to prevent the stock solutions from changing concentration due to evaporation prior to their application to the surface. Also, precise microvolumetric methods must be used to dispense the solution on the aqueous surface since the quantity used is small. The solvent (as well as the solute) must be free from contaminants. There is also the possibility that the solvent will extract spreadable contaminants from the waxed surfaces of the float, barriers, and tray. Some workers advocate addition and evaporation of one drop at a time to minimize this. Oily contaminants may also reach the water surface from the fingers and from the atmosphere. These last sources are particularly hard to control: Tests for reproducibility and blank compressions (i.e., moving the barrier toward the float on a "clean" surface) are the best evidence of their absence.

Not all solvents are equally suitable for spreading monolayers. The requirement that the solvent evaporate completely is self-evident. It has been suggested that if the organic solvent dissolves much water, the properties of the monolayer will be different from those in which no water is trapped. Verification that no artifacts are entering the observations from the solvent may be accomplished by conducting duplicate experiments with different solvents.

Until now we have concentrated on those difficulties in using the Langmuir balance that arise from the determination of π, A, T, and n. A few remarks are also in order about considerations that may affect the monolayer and originate in the adjacent phases. We have already discussed contaminants originating in the gaseous phase. For some monolayer materials, air oxidation may also be a problem. The aqueous substrate is the source of a wide assortment of contaminants in addition to spreadable oily matter. Ionic impurities, including those that affect the pH, are quite troublesome. The charge state of many amphipathic molecules, for example, amines and carboxylic acids, is obviously pH dependent. Salts or complexes formed between amphipathic molecules and ions in the aqueous phase will have different monolayer properties from those of the unreacted surfactant molecule.

An extensive discussion of the Langmuir balance technique and a comparison with the Wilhelmy plate method are given by MacRitchie (1990). This book also discusses modifications that are possible to the techniques and other experimental details.

7.4 RESULTS OF FILM BALANCE STUDIES

The preceding section shows that it is possible to determine π-A isotherms for surfaces just as p-V isotherms may be measured for bulk matter. The results that are obtained for surfaces are analogous to bulk observations also, although some caution must be expressed about an overly literal correlation between bulk and surface phenomena. We return to a discussion of these reservations below. There can be no doubt, however, that analogies with bulk behavior supply a familiar framework within which to consider π-A isotherms.

The curve sketched in Figure 7.6, which is drawn with grossly distorted coordinates to encompass all features, contains several similarities to p-V isotherms. Not all the features

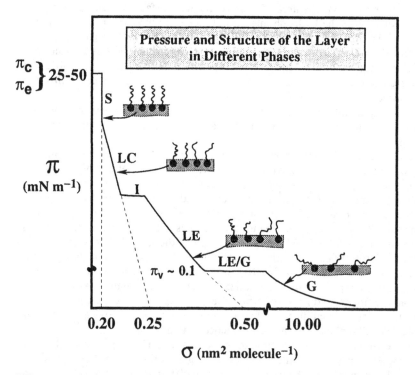

FIG. 7.6 Composite two-dimensional pressure π versus area σ isotherm, which includes a wide assortment of monolayer phenomena. Note that the scale of the figure is not uniform so that all features may be included on one set of coordinates. The sketches of the surfactants show the orientations of the molecules in each phase at various stages of compression.

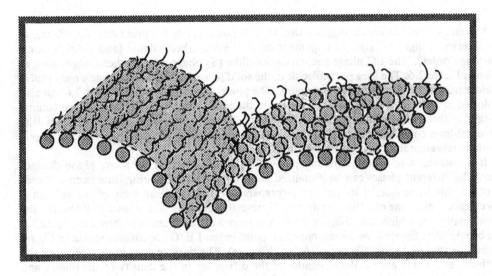

FIG. 7.7 Schematic illustration showing the collapse of the film.

shown here are always observed, nor are all known idiosyncrasies of π-A isotherms represented. The presence or absence of various features and their π-A coordinates vary with temperature for a particular amphipathic molecule and from one amphipathic substance to another. Last, there is some diversity in the terminology used to describe various monolayer phenomena. In short, Figure 7.6 is a composite isotherm that will introduce and summarize a variety of observations.

In this section we discuss in turn the various two-dimensional phases and phase equilibria represented in Figure 7.6, progressing from low values of π to high ones. The existence of these two-dimensional states and the properties they possess are presumably unfamiliar to most readers. Therefore it is important to keep the following ideas in mind in reading this section:

1. We are concerned with two-dimensional matter situated at the boundary between two bulk phases.
2. The properties of the two-dimensional phases are relatively independent of the properties of the bulk phases of the same material.
3. Many surface states are two-dimensional analogs of three-dimensional states. As with any analogy, however, there are points of similarity and points of difference between the surface and bulk states. For most of the states we discuss, we consider the phenomenological behavior as represented by Figure 7.6 and Figure 7.7.

7.4a Microstructural Phases in Monolayers

Let us first take a look at the types of microstructural phases one typically finds in a monolayer. For this consider the isotherm shown in Figure 7.6. The Langmuir layers have a (two-dimensional) "gaslike" distribution of surfactants when the area per molecule is large compared to the dimensions of the molecule. In this phase, denoted as G in Figure 7.6, the hydrocarbon tails of the molecules make significant contact with water. As the concentration of the surfactant at the surface increases, the gas phase begins condensing to a liquidlike phase known as the *liquid-expanded* (LE) phase. This two-phase region corresponds to the plateau region in Figure 7.6 (analogous to the three-dimensional gas-liquid coexistence region on the pressure-volume isotherm). At higher surfactant concentrations the gas phase completely condenses to the LE phase. In the LE phase, the hydrocarbon tails of the surfactants lift from the surface of the water but remain disordered. At further increases in surfactant concentration, the LE phase is transformed to another "liquid" phase known as the *liquid-condensed* (LC) phase. Current research suggests that the LC phase is not a liquid (Knobler 1990b), as one observes a higher degree of alignment of the hydrocarbon chains (and possibly some long-range order). The LC phase becomes a solidlike (S) phase with further compression as shown in Figure 7.6. The area per molecule in the solid phase is close to what one would expect for close packing of the hydrccarbon chains. The phase diagram shown in Figure 7.6 can also be plotted in terms of a temperature-versus-area diagram (similar to the temperature-volume diagram in three dimensions) as shown in Figure 7.8, in which some sample tie lines (the horizontal lines connecting coexisting phases) are also marked. The diagram shown is only a schematic representation, as emphasized in the legend.

It is instructive to examine the morphology of a monolayer as the above phase changes occur. The different phases can be visualized to a limited extent by using fluorescence micro-scopy. In this technique, a monolayer is prepared with a small amount of solute with a fluorescence label. One can then excite the fluorescence markers with a laser and observe the morphology using a high-sensitivity camera. A sequence of such results is shown in Figure 7.9 (Knobler 1990b). Figure 7.9a corresponds to a point in the LE/G coexistence region in Figure 7.8 (at a density of 1 molecule per 61 Å^2 in this case). The dark regions in the figure are the gas "bubbles", which appear dark because of the difference in the density of the phases and (possibly) because of the quenching of the fluorescence of the labeled molecules in the gas phase. The white regions are the LE phase.

As the layer is compressed at constant temperature (i.e., as one moves horizontally from right to left in Fig. 7.8), one sees the sequence of morphologies shown in Figures 7.9b–7.9e. The layer in Figure 7.9b is still the LE/G coexistence region, but the amount of gas phase has clearly decreased. On further increase in density one enters the single-phase LE region (Fig. 7.9c), as indicated by the completely white layer. Figure 7.9d may be misleading as it appears similar to Figures 7.9a and b, but in this case the dark circles correspond to the LC phase in which the (fluorescence-labeled) probe molecules have low solubility (hence the contrast). Figure 7.9e shows that the LC phase increases in quantity at a higher solute concentration. Although the fluorescence technique described here provides only black-and-white pictures and it may appear that the technique is limited to two-phase regions, detection of more than two phases can be achieved by analyzing the evolution of the features in the morphology.

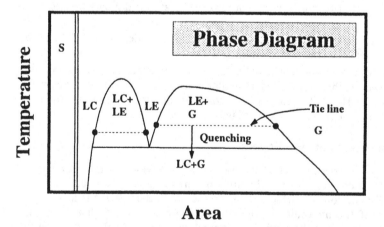

Area

FIG. 7.8 A schematic representation of the temperature-versus-area diagram for a Langmuir layer (a two-dimensional phase diagram). The coexistence regions are exaggerated for clarity. The horizontal lines shown are the tie lines. The arrow marked "quenching" starts at the LE + G coexistence region and is used in the text to illustrate the morphological changes when a two-phase liquid-expanded/gaslike (LE + G) mixture is quenched. (Redrawn with permission of Knobler 1990b.)

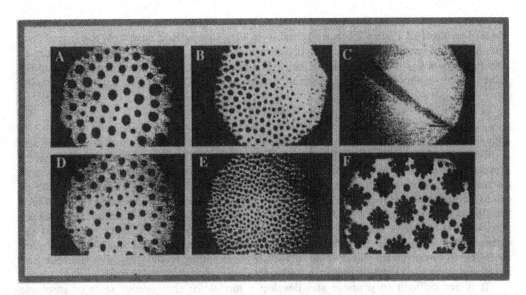

FIG. 7.9 Fluorescence microscope pictures of a monolayer of pentadeconoic acid (PDA) containing 1% fluorescent probe (4-(hexadecylamino)-7-nitrobenz-2-oxa-1,3-diazole, i.e., NBD-hexadecylamine): (A) 1 molecule per 61 Å² at 25°C with G (dark) and LE (white) phases; (B) 1 molecule per 50 Å² at 25°C with G (dark) and LE (white) phases; (C) 1 molecule per 36 Å² at 25°C with a single LE phase; (D) 1 molecule per 27 Å² at 25°C with LC (dark) and LE (white) phases; (E) 1 molecule per 24 Å² at 25°C with LC (dark) and LE (white) phases; (F) temperature quench starting at the LE/G coexistence region (overall density = 1 molecule per 51 Å²). The final point is a three-phase region consisting of LC, LE, and G phases. See the text for details. (Redrawn with permission of Knobler 1990b.)

One such case is shown in Figure 7.9f, which is obtained by quenching a state in the LE/G coexistence region to a point in the LC/LE/G three-phase region (the arrow marked "quenching" in Fig. 7.8). The large circles in the centers of the flower-shaped domains correspond to the LC phase. Circling the LC "drops" are small circular domains of the gas phase. These grow at the expense of the LE phase, and with further quenches one moves into an LC/G mixture (see the arrow marked "quenching" in Fig. 7.8). A considerable number of studies have appeared in the literature in recent years on the morphological changes in Langmuir monolayers (see Knobler 1990b) because of the potential and the projected uses of such monolayers in advanced technological applications (see Sections 7.6 and 7.10).

7.4b The "Gas" Phase and Its Models

As evident from the above discussion, if measurements can be made at sufficiently low pressures, all monolayers will display gaseous behavior, represented by region G in Figure 7.6. The gaseous region is characterized by an asymptotic limit as $n \rightarrow 0$. In the limit of very low film pressures, a two-dimensional equivalent to the ideal gas law applies:

$$\pi A = nRT \tag{5}$$

where R is the gas constant, usually in SI units. This is a convenient place to define another quantity, the area occupied per molecule in the interface σ. Since R equals Avogadro's number times the Boltzmann constant, and nN_A equals the total number of surface molecules, we may write

$$\pi\left(\frac{A}{nN_A}\right) = \pi\sigma = k_B T \tag{6}$$

As a model for this highly expanded state we may choose a situation like that depicted in one of the insets of Figure 7.6 corresponding to very low surface coverage; the inset corresponding to the gas phase shows the hydrocarbon chain lying flat on the surface, blocking an area $\pi\ell^2$, where ℓ is the length of the "tail." We may then use Equation (6) to calculate the value of π corresponding to this area per molecule. At 25°C and with $\ell = 1.0$ nm, we obtain

$$\pi = \frac{k_B T}{\sigma} = \frac{(1.38 \cdot 10^{-23})(298)}{(3.14)(10^{-9})^2} = 1.31 \text{ mN m}^{-1} \tag{7}$$

Allowing the above amount of area per molecule is equivalent to taking the distance of closest approach for the typical distance between the molecules. Therefore ideal behavior will be expected only if the areas per molecule are larger than this, on the order of 10 nm^2 perhaps. Correspondingly lower film pressures will be involved also. If we recall that the pressure of this example is equivalent to a bulk gas pressure of about 13 atm, it is less surprising that such low film pressures are needed to observe gaseous monolayer behavior. The repulsion between particles in a charged monolayer increases the effective area these molecules occupy at the surface. The effect of this is to increase the pressure of charged films, making them more accessible to measurement. Since ionic surfactants are soluble, techniques other than the film balance must be used to study their π-σ isotherms.

It is not difficult to propose and develop a model for the gaseous state of insoluble monolayers. The arguments parallel those developed in kinetic molecular theory for three-dimensional gases and lead to equally appealing results. The problem, however, is that many assumptions of the model are far less plausible for monolayers than for bulk gases. To see this, a brief review of the derivation seems necessary.

7.4b.1 Equation of State from Kinetic Theory

Suppose we imagine a single molecule bouncing back and forth across a surface between two restraining barriers. If we define the direction of this motion to be the x direction and the velocity of the molecule to be v_x, then the change in momentum at each collision (if we assume they are elastic) is

$$\Delta(\text{momentum})/\text{collision} = mv_x - (-mv_x) = 2 m v_x \tag{8}$$

where m is the mass of the molecule. The time interval between two successive collisions at the same wall is given by

$$\text{elapsed time/collision} = 2\ell/v_x \tag{9}$$

if the distance between barriers is ℓ since the molecule must cross the distance between the barriers twice before returning to the same spot. The force exerted by the molecule on impact equals the rate of change in momentum, which, in turn, equals the ratio of Equation (8) to Equation (9):

$$\Delta(\text{momentum})/\Delta(\text{time}) = F_x = m v_x^2/\ell \tag{10}$$

This force is converted to two-dimensional pressure by dividing it by the length of the edge to which the force is applied. Assuming the accessible surface area to be a square means that the length of this edge is also ℓ; the pressure contribution of this one collision equals

$$\pi = F_x/\ell = mv_x^2/\ell^2 \tag{11}$$

The quantity ℓ^2 in this equation clearly describes the area accessible to the molecule.

Since pressure is isotropic, we assume that the forces on the perpendicular barriers are identical. Therefore, if the surface contains N molecules, they behave as if $N/2$ were exerting a pressure given by Equation (11) on the barriers perpendicular to the x direction, with the other $N/2$ exerting an identical pressure in the other direction. That is, for a surface containing N molecules

$$\pi A = \frac{N}{2} m\overline{v^2} = N(\overline{KE}) \tag{12}$$

The average value of the square velocity has been used in Equation (12) to allow for the fact that a distribution of molecular velocities exists. The nature of the averaging procedure to be used in this case is well established from physical chemistry. We also know from physical chemistry that the average kinetic energy per molecule $(\overline{KE})$ per degree of freedom is

$$\frac{\overline{KE}}{\text{degree of freedom}} = \frac{1}{2} k_B T \tag{13}$$

Since the molecules on the surface have two translational degrees of freedom, Equations (12) and (13) may be combined to give

$$\pi \frac{A}{N} = \pi \sigma = k_B T \tag{14}$$

which is identical to Equation (6).

The foregoing derivation is a straightforward two-dimensional analog of the three-dimensional case and leads to a result that describes the experimental facts. From a pragmatic point of view, it is a great success. One of the theoretical assumptions underlying Equation (13), however, is that translational quantum states are sufficiently close together to justify treating them as continuous rather than discrete. This is unquestionably true for gases. For an amphipathic molecule with a polar head that contacts—and interacts with—the aqueous substrate, it is somewhat harder to justify. We see below that there is a totally different way of looking at Equation (14) that is free from this objection.

Like its three-dimensional counterpart, Equation (5) is a limiting law, which means that deviations may be expected at higher pressures, lower temperatures, or with more strongly interacting molecules. Figure 7.10 is a plot of $\pi\sigma/k_B T$ versus π for several members of the carboxylic acid homologous series. The film balance was used to collect the data only for the C_{12} acid. Shorter chain compounds are soluble and were investigated by surface tension measurements and interpreted by the Gibbs equation, which we discuss in Section 7.7.

The main point to note about Figure 7.10 is the strong resemblance it bears to similar plots for three-dimensional gases. Negative deviations occur at low pressures, becoming more pronounced as the length of the alkyl chain increases. As with gases, this may be attributed to attraction between molecules, an effect that increases with chain length. At higher pressures, deviations tend to be positive. This is analogous to the excluded volume effect for gases, except that it becomes an excluded area in two dimensions.

* * * *

EXAMPLE 7.1 *Determination of the Molecular Weight of a Solute from π Versus A Isotherms.* The molecular weight of a gas may be determined by measuring the mass of the sample as well as a set of p, V, and T values for the same sample. Use this fact to criticize or defend the

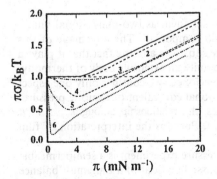

FIG. 7.10 Plots of $\pi\sigma/k_B T$ versus π for *n*-alkyl carboxylic acids: (1) C_4, (2) C_5, (3) C_6, (4) C_8, (5) C_{10}, and (6) C_{12}. (Data from N. K. Adam, *Chem. Rev.*, 3, 172 (1926)).

following proposition: The molecular weight of a solute in a monolayer may be calculated by the formula $M = mRT/\pi A$, where A/m gives the area mass $^{-1}$ of the monolayer under a pressure π and at a temperature T.

Solution: The formula given is the two-dimensional analog of the ideal gas law; therefore two conditions must be met to justify its use. First, the monolayer must be in the gaseous state, and, second, the gas pressure must approach zero. The first point may appear trivial since no one would apply $pV = nRT$ to a bulk liquid sample for which p, V, and T had been measured. The two-dimensional state of a monolayer is not directly perceptible, however, and π-σ data must be evaluated to verify that the monolayer is indeed in the G state. With π-A data measured over a range of (low) π's, Figure 7.10 shows that the limiting value of $\pi A/RT$ is the number of moles in the sample. Dividing the mass of the sample by this value of n yields the correct molecular weight. The proposition needs to be qualified by adding "in the limit of $\pi \rightarrow 0$." ∎

* * *

7.4b.2   Van der Waals Equation of State for Monolayers

We noted above that the applicability of Equation (14) to insoluble monolayers is severely restricted to very low values of π. Figure 7.10 shows that the deviations from Equation (14) with increases in π are very similar to what is observed for nonideal gases. Specifically, the positive deviations associated with excluded volume effects in bulk gases and the negative deviations associated with intermolecular attractions are observed. It is tempting to try to correct Equation (14) for these two causes of nonideality in a manner analogous to that used in the van der Waals equation:

$$\left(\pi + \frac{a}{\sigma^2}\right)(\sigma - b) = k_B T \tag{15}$$

where a and b are the two-dimensional analogs of the van der Waals constants. Note that b, the excluded area per molecule, is conceptually equivalent to σ^0, although which of the values $(\sigma_{LE}^0, \sigma_{LC}^0, \sigma_S^0)$ best fits the data cannot be predicted a priori.

The temperature at which the van der Waals equation goes from one having three real roots to one having one real root is generally identified with the critical temperature. In the same way, Equation (15) may be considered to connect both the gaseous (G) and liquid-expanded (LE) states in monolayers, the transition between which also displays a critical point. Statistical mechanics shows, however, that the van der Waals constant a explicitly ignores orientation effects as contributing anything to the energy in gases; it is hard to imagine the properties of insoluble monolayers as being independent of orientation. Therefore any attempt to correct Equation (14) in such a way as to extend its range encounters difficulties. Ultimately, all objections to two-dimensional equations of state seem to center on their neglect or unsatisfactory inclusion of the substrate.

7.4b.3   Monolayers as Two-Dimensional Binary Solutions

An alternative way of looking at monolayers is to consider them as two-dimensional binary solutions rather than as two-dimensional phases of a single component. The advantage of this approach is that it does acknowledge the presence of the substrate and the fact that it plays a role in the overall properties of the monolayer. Although quite an extensive body of thermodynamics applied to two-dimensional solutions has been developed, we consider only one aspect of this. We examine the film pressure as the two-dimensional equivalent of osmotic pressure. It will be recalled that, at least for low osmotic pressures, the relationship among π_{osm}, V, n, and T is identical to the ideal gas law (Equation (3.25)). Perhaps the interpretation of film pressure in these terms is not too farfetched after all!

As we saw in Chapter 3, the heart of any osmotic pressure experiment is a semipermeable membrane that allows the solvent but not the solute to pass. The float of a Langmuir balance accomplishes this. That portion of surface with the monolayer is considered to be the two-dimensional solution; the clean surface is the two-dimensional solvent. The solvent can certainly pass from one region to another (remember that the mechanism of the partitioning has nothing to do with the equilibrium osmotic pressure) through the bulk substrate. However, the insoluble solute is restrained by the float.

For osmotic equilibrium, the chemical potential of the solvent must be the same on both sides of the membrane. In the two-dimensional analog also the chemical potential must be the same for the water on both sides of the float. The presence of the solute lowers the chemical potential of the solvent, but the excess pressure compensates for this. Therefore, by analogy with Equation (3.19), we write

$$\mu_{1s}^0 = \mu_{1s}^0 + RT \ln a_{1s} + \int_0^\pi \overline{A}_1 \, dp \tag{16}$$

where the subscript s indicates the surface and the subscript 1 identifies the solvent. If the partial molal area of the solvent $\overline{A}_1$ is assumed to be independent of π, Equation (16) may be integrated to give

$$-RT \ln a_{1s} = \pi \overline{A}_1 \tag{17}$$

For ideal (dilute) solutions the activity is replaced by the mole fraction x_{1s}, which, for a two-component surface solution, equals $(1 - x_{2s})$. With the customary expansion of the logarithm as a power series (see Appendix A), these substitutions yield

$$RTx_{2s} = \pi \overline{A}_1 \tag{18}$$

the two-dimensional equivalent of Equation (3.23). Finally, it is necessary to relate the surface mole fraction and the molar area of the solvent to more familiar variables.

The surface mole fraction is entirely analogous to the bulk value of this quantity,

$$x_{2s} = \frac{n_{2s}}{n_{1s} + n_{2s}} = \frac{N_{2s}}{N_{1s} + N_{2s}} \tag{19}$$

where the n terms are the numbers of moles, and the N terms are the numbers of molecules. For dilute surface solutions—that is, expanded monolayers—$N_{1s} \gg N_{2s}$; therefore

$$x_{2s} \cong N_{2s}/N_{1s} \tag{20}$$

The total area of the surface may be written

$$A_T = n_{1s}\overline{A}_1 + n_{2s}\overline{A}_2 = N_{1s}\sigma_1^0 + N_{2s}\sigma_2^0 \tag{21}$$

Applying these various relationships to Equation (18) leads to the following result for low film osmotic pressures:

$$\pi(A_T - N_{2s}\sigma_2^0) = n_{2s}RT = N_{2s}k_BT \tag{22}$$

Dividing through by N_{2s} to express the total area as area per solute molecule gives

$$\pi(\sigma - \sigma_2^0) = k_BT \tag{23}$$

Equation (23) obviously gives the two-dimensional ideal gas law when $\sigma \gg \sigma_2^0$ and with the σ_2^0 term included represents part of the correction included in Equation (15). This model for surfaces is, of course, no more successful than the one-component gas model used in the kinetic approach; however, it does call attention to the role of the substrate as part of the entire picture of monolayers. We saw in Chapter 3 that solution nonideality may also be considered in osmotic equilibrium. Pursuing this approach still further results in the concept of phase separation to form two immiscible surface solutions, which returns us to the phase transitions described above.

* * *

EXAMPLE 7.2 *Use of the van't Hoff Equation for Monolayers.* A monolayer of egg albumin was spread on a concentrated aqueous solution of ammonium sulfate and π-A data were collected at 25°C by Bull (1945). Use the two-dimensional van't Hoff equation to evaluate the molecular weight of the albumin if $(\pi/c')_0 = 5.54 \cdot 10^5$ erg g^{-1}. In this expression c' is the two-dimensional concentration in practical units, g cm^{-2}. How does this interpretation of π-A data compare with the interpretation given in Example 7.1?

Solution: According to the van't Hoff equation,

$$1/M = (1/RT)(\pi/c')_0 = (5.54 \cdot 10^5)/(8.314 \cdot 10^7)(298)$$
$$= 2.24 \cdot 10^{-5} \text{ mole g}^{-1}$$

therefore $M = 44,700$ g mole^{-1}. This is completely equivalent to the interpretation given in Example 7.1 since $\pi \to 0$ as $c' \to 0$. Therefore the same limiting value is obtained whether the limit is taken in terms of π or c'.   ■

*   *   *

In summary, we see that insoluble monolayers may be viewed either as examples of two-dimensional phases of one component or as two-dimensional solutions with two components. The former model is somewhat simpler and is often adequate. The latter, although more complex, is more realistic. In spite of our interest in the monolayer, we must not neglect the fact that none of the monolayer phenomena would exist without the aqueous phase as the substrate.

7.4c   The Liquid-Expanded Phase

Next, let us return to Figure 7.6, discussing the features labeled LE and LE/G. As mentioned above, the horizontal line of the LE/G region in Figure 7.6 is analogous in every way to the corresponding feature in bulk matter. At a given temperature there is a constant-pressure region over which a significant compression occurs. The film pressures at which LE/G equilibrium occurs are known as film vapor pressures π_v. Like the gaseous state itself, the LE/G equilibrium occurs at very low pressures. Tetradecanol, for example, has a two-dimensional vapor pressure of $1.1 \cdot 10^{-4}$ N m^{-1} at 15°C.

It is important to remember the significance of π_v. It refers specifically to the equilibrium between two surface states. There is a danger of confusing π_v with the equilibrium spreading pressure π_e, introduced in Chapter 6. The latter is the pressure of the equilibrium film that exists in the presence of excess bulk material on the surface. It is the equilibrium spreading pressure that is involved in the modification of Young's equation (Equation (6.49)), for which a bulk phase is present on the substrate. For tetradecanol at 15°C, the equilibrium spreading pressure is about $4.5 \cdot 10^{-2}$ N m^{-1}, so π_e and π_v are very different from one another.

The temperature variation of π_v may be analyzed by a relationship analogous to the Clapeyron equation to yield the two-dimensional equivalent to the heat of vaporization. The numerical values obtained for this quantity more nearly resemble the bulk values for hydrocarbons than those for polar molecules. This suggests that most of the change in the surface transition involves the hydrocarbon tail of the molecule rather than the polar head.

Finally, note that the LE/G equilibrium region disappears above a certain temperature that is the two-dimensional equivalent of the critical temperature for liquid-vapor equilibrium (see Fig. 7.8).

As described in Section 7.4a, the liquid-expanded state is the first of several condensed states. Because it is bound on the low-pressure side by a two-phase region with a critical temperature, the LE state is easily compared to a bulk liquid state. Since gaseous behavior is observed only at very low pressures, it is easy to extrapolate the isotherm for the LE state to the $\pi = 0$ value, as the dashed line in Figure 7.6 shows. For amphipathic molecules with saturated unbranched R groups, this intercept is in the range 0.45 to 0.55 nm^2. We shall identify the limiting area per molecule (superscript zero) for this state (subscript LE) by the symbol σ_{LE}^0. The presence of branched chains or double bonds—particularly in the cis configuration—increases the value of this limiting area.

The precise structural details of the liquid-expanded state at the molecular level are not fully understood, but several generalizations do appear to be justified. The value of σ_{LE}^0 is several times the actual cross-sectional area of the amphipathic molecule, when oriented perpendicular to the surface. At the same time the area per molecule is considerably less than could be permitted if the entire tail were free to move in the surface. The inset corresponding to the LE phase in Figure 7.6 represents a model of the surface at the molecular level. Here, part of the hydrocarbon chain lies in the surface, and some has been lifted out of the surface plane. That portion of the tail in the surface defines the effective area per molecule. Neither this area nor the length of the chain out of the surface will be the same for all molecules, so experimental values of σ_{LE}^0 correspond to average values. Furthermore, there will be considerable lateral interaction between those segments that are not in contact with the substrate.

The compressibility of the LE state is expected to be far less than that of the gaseous state, but not yet incompressible, because the average area per molecule may be altered by squeezing additional CH_2 groups out of contact with the water.

7.4d The Intermediate Liquid Phase

With sufficient compression the isotherm of the LE state shows a sharp break to enter a situation variously known as the intermediate or transition state, indicated by I in Figure 7.6. Recent results suggest that the region I does represent a first-order phase transition (Knobler 1990a).

For a number of years, no satisfactory molecular interpretation existed for the I state. As its name implies, the I state was thought of as an intermediate transition state between the LE and LC states. If, however, the I state is a transition state that somehow becomes trapped in transit, it should be described as a metastable rather than a stable state. We saw in Chapter 6, Section 6.7 that metastable states are associated with nonreversible behavior, and I states seem to be reversible. A considerable number of studies have emerged in recent years on the evolution of the LC phase from the LE phase. It is known that very complex patterns (e.g., fractal dendritic structures) develop during this transition. In some cases, additional phases have also been identified (see Knobler 1990a). We shall not go into these here and merely emphasize that the isotherm shown in Figure 7.6 and the phase diagram Figure 7.8 are highly simplified.

7.4e The Liquid-Condensed and the Solid Phases

If the area of an insoluble monolayer is isothermally reduced still further, the compressibility eventually becomes very low. Because of the low compressibility, the states observed at these low values of σ are called *condensed* states. In general, the isotherm is essentially linear, although it may display a well-defined change in slope as π is increased, as shown in Figure 7.6. As mentioned above, the (relatively) more expanded of these two linear portions is the *liquid-condensed* state LC, and the less expanded is the *solid* state S. It is clear from the low compressibility of these states that both the LC and S states are held together by strong intermolecular forces so as to be relatively independent of the film pressure.

Because of the near linearity of these portions of the isotherm, it is easy to extrapolate both regions to their value at $\pi = 0$. The intercepts for the solid and liquid-condensed regions, σ_S^0 and σ_{LC}^0, respectively, differ only slightly. Values of σ_{LC}^0 for alcohols are about 0.22 nm^2, and for carboxylic acids about 0.25 nm^2, more or less independent of the length of the hydrocarbon chain. The intercept σ_S^0 has a value of about 0.20 nm^2, independent of both the length of the chain and the nature of the head. The film pressures in the condensed states (LC or S) are of the same magnitude as the equilibrium spreading pressure for amphipathic molecules.

For the condensed LC and S states a molecular interpretation is again possible. In both the values of σ^0 are close to actual molecular cross sections when the molecules are oriented perpendicular to the surface. The difference between these two regions seems to involve the polar part of the molecule more than the hydrocarbon chain, which was more important for the more expanded states. The difference between σ_S^0 and σ_{LC}^0 may involve a more efficient packing of the heads or the formation of fairly specific lateral interactions through hydrogen bonds, for example. The values of σ^0 that are observed for monolayers of saturated *n*-alkyl compounds are only slightly larger than the close-packed cross sections obtained for these compounds in the bulk solid state by x-ray diffraction.

Additional compression eventually leads to the collapse of the film. The pressure π_c at which this occurs is somewhere in the vicinity of the equilibrium spreading pressure. Figure 7.7 represents schematically how this film collapse may occur. The mode of film buckling shown in Figure 7.7 is not the only possibility: head-to-head as well as tail-to-tail configurations can be imagined. The second structure strongly resembles that of cell membranes, which we discuss in the next chapter.

Film collapse represents the squeezing out of the surface of highly ordered aggregates that may quite plausibly be regarded as nuclei to bulk phase particles. If collapse marked the appearance of the amphipathic material in a bulk phase, then the collapse pressure and the equilibrium spreading pressure should be identical. Here a significant complication appears. The collapse pressure is highly sensitive to the rate at which the film is compressed. This indicates nonequilibrium conditions, showing that the two-dimensional solid phase resembles three-dimensional solids in this respect also, a difficulty in the attainment of thermodynamic equilibrium.

In summary, it must be emphasized again that there are wide variations in the properties of insoluble monolayers. Extensive compilations of π-A data for a large number of surfactants are available, although not over the full range of densities (Mingotaud et al. 1993). Some of the phenomena reported in the literature are probably artifacts due to impurities or nonequilibrium conditions. Others are probably unique effects that apply only to a very specific system. In the descriptive material of this section both phenomenological and modelistic information were provided for various stages along the isotherm. At the very least, the models serve the pedagogical function of assisting the student in remembering an assortment of probably unfamiliar facts. At best, the models provide the basis for quantitatively understanding these phenomena. In the next section we take a more quantitative look at the model for the gaseous state.

7.5 VISCOUS BEHAVIOR OF TWO-DIMENSIONAL PHASES

Our discussion of two-dimensional phases has drawn heavily on the analogy between bulk and surface behavior. This analogous behavior is not restricted to thermodynamic observations, but extends to other areas also. The viscosity of surface monolayers is an excellent example of this. To illustrate the parallel between bulk and surface viscosity, let us retrace some of the introductory notions of Chapter 4, restricting the flow to the surface region.

7.5a Coefficient of Surface Viscosity

We begin by defining the coefficient of surface viscosity η^s. Equation (4.1) serves to define the *bulk* viscosity; for surfaces we ignore the area extending in the z direction and consider the force per unit edge of the surface ℓ. Assuming a velocity gradient dv/dy exists between two edges of an element of area, we can write the two-dimensional analog of Equation (4.1) as

$$\frac{F}{\ell} = \eta^s \frac{dv}{dy} \tag{24}$$

This gives η^s units of mass time^{-1} in contrast to bulk viscosity, which has dimensions mass length^{-1} time^{-1}. In discussing Figure 7.3 we noted that—from a chemical if not geometrical point of view—surfaces extend over a thickness τ. By analogy with Equation (3), we might expect surface and bulk viscosities to be related by

$$\eta = \eta^s/\tau \tag{25}$$

7.5b Measurement of Surface Viscosity

Next let us consider how surface viscosities can be measured. A variety of methods for measuring η^s exist, including a method based on concentric rings, the two-dimensional equivalent of the concentric cylinder viscometer. We limit our discussion to the analog of the capillary viscometer. Figure 7.11a shows an arrangement by which a monolayer can be pushed through a narrow channel by a moving barrier. Following a derivation that parallels the development of the Poiseuille equation (Equation (4.20)), the area rate at which the monolayer emerges from the channel under an applied pressure $\gamma^0 - \gamma = \Delta\gamma$ can be obtained. As in Section 4.4, we continue to describe the locations within the channel in terms of the distance r from its center line, with h the distance of the wall from the center line (see Figure 7.11b). The following steps are highlights of the derivation and parallel the presentation in Section 4.4:

1. Viscous and surface pressure forces balance under stationary-state conditions along the edge of the film a distance r from the center line. In terms of Figure 7.11b, this force balance is given by

$$\eta^s \ell \frac{dv}{dr} + \Delta\gamma r = 0 \tag{26}$$

2. Integrating Equation (26) and using the nonslip condition at the wall ($v = 0$ at $r = h$) to evaluate the constant of integration yields

$$\frac{\eta^s \ell}{\Delta\gamma} v = \frac{1}{2} (h^2 - r^2) \tag{27}$$

3. The rate of emergence of area, A/t, of the monolayer from the channel is twice the integral of $v dr$ between $r = 0$ and $r = h$:

$$\frac{A}{t} = \frac{2}{3} \frac{\Delta\gamma}{\eta^s \ell} h^3 \tag{28}$$

This is the two-dimensional equivalent of Poiseuille's equation. All of the other quantities besides η^s in Equation (28) are measurable, so η^s can be evaluated by measuring the rate at which the monolayer flows through the channel. In practice, a second barrier is moved along in front of the advancing interface to maintain a constant film pressure for an insoluble monolayer.

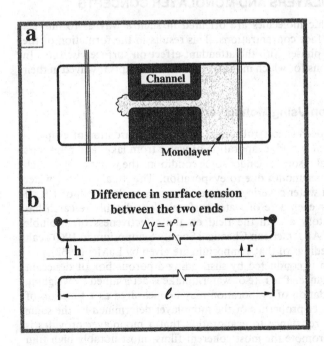

FIG. 7.11 Schematic representation of a surface viscometer: (a) a monolayer is pushed through a narrow channel; and (b) definition of variables for analysis.

In the two-dimensional gaseous state, surface viscosities can be as low as 10^{-8} kg s^{-1}, while for condensed states values range from 10^{-7} to 10^{-5} kg s^{-1}. While it makes sense that the surface viscosity increases as we move from G to L to S states, the numbers themselves mean little to us. To remedy this we repeat the sort of calculation done for two-dimensional pressure and examine what Equation (25) tells us about the equivalent bulk viscosity. As with Equation (3), we assume $\tau = 1.0$ nm; therefore a surface viscosity of 10^{-7} kg s^{-1} is equivalent to a bulk viscosity of

$$\eta = \frac{\eta^s}{\tau} = \frac{10^{-7}\,\text{kg s}^{-1}}{10^{-9}\,\text{m}} = 10^2\,\text{kg s}^{-1}\,\text{m}^{-1} = 10^3\,\text{P} \tag{29}$$

a value that suggests a consistency like that of butter for the molecules of the monolayer in condensed states. This is consistent with a high degree of lateral interaction between the vertically oriented chains. Once again, we see that a highly localized phenomenon translates into a dramatic effect when scaled up to macroscopic dimensions. In the next section we see some applications of monolayers or monolayer concepts that take advantage of the properties we have discussed.

Surface viscosities have been measured for soluble and insoluble monolayers, for charged and uncharged molecules, and at water-air and water-oil interfaces. We will not consider all these possibilities, but note instead that all share as a common feature the presence of polar head groups in the aqueous phase. In general, this means hydrogen bonding will occur between the amphipathic molecules at the surface and the substrate. A certain amount of water is dragged along with the surface molecules as a consequence. This effect was not taken into account in the derivation of Equation (28), although it has been investigated extensively. The effect of the entrained substrate may be incorporated into Equation (28) by multiplying the equation by a correction factor of the form $(1 + 2h\eta/\pi\eta^s)^{-1}$. This shows that the resistance to the motion of the monolayer due to this effect increases as the ratio η/η^s increases. This factor is incorporated into Equation (28) when η^s is determined experimentally by this method.

7.6 APPLICATIONS OF MONOLAYERS AND MONOLAYER CONCEPTS

At the surface of water, amphipathic molecules are oriented in such a way as to interact extensively, at least for ordinary surface concentrations. This results in the formation of the various two-dimensional condensed phases with the attendant effect on surface viscosity. In this section we consider some situations for which monolayers or the concepts involved in their discussion find application.

7.6a Retardation of Evaporation Using Monolayers

One area in which monolayers have been successfully employed is the retardation of evaporation. Particularly in arid regions of the world, evaporation of water from lakes and reservoirs constitutes an enormous loss of a vital resource. Under some conditions the water level of such bodies may change as much as 1 ft per month due to evaporation. The usual unit for water reserves is the acre-foot, a volume of water covering an acre of surface to the depth of 1 ft. It equals about 1/3 million gallons for each acre of water surface. Considerable research has been conducted both in the laboratory and in the field on the effectiveness of insoluble monolayers in reducing evaporation. An American Chemical Society Symposium in 1960 dealt exclusively with this topic; the proceedings of that symposium are given by LaMer (1962).

Laboratory research in this area is conducted by suspending a porous box of desiccant very close to the surface of a film balance. The rate of water uptake is determined by weighing at various times. This way the retardation of evaporation may be measured as a function of film pressure and correlated with other properties of the monolayer determined by the same method. As might be expected, the resistance to evaporation that a monolayer provides is enhanced by those conditions that promote the most coherent films, most notably high film pressures and straight-chain compounds. To see how this is quantified, consider the Example 7.3.

* * *

EXAMPLE 7.3 *Suppression of Evaporation by Monolayers.* The rate of evaporation is quantified by a parameter called the transport resistance r. For water with octadecanol monolayers at surface pressures of 10, 20, 30, and 40 mN m^{-1}, r is about 1, 2, 3, and 4 s cm^{-1}, respectively. This resistance drops off rapidly at lower pressures and approaches $2 \cdot 10^{-3}$ s cm^{-1} for pure water. By considering the rate of water uptake as a diffusion problem, suggest how these r values are calculated from data collected in an experiment like that described above. Use the fact that $1/r$ is dimensionally equivalent to the diffusion coefficient D divided by a length.

Solution: To be collected by the desiccant, molecules evaporating from a surface of area A must diffuse across a gap of width Δx between the water surface and the desiccant. The gap contains the monolayer as well as the air space, so the diffusion coefficient used is an effective value rather than the actual D value for a homogeneous region. According to Equations (2.20) and (2.22), the rate at which the desiccant increases in weight is given by

$$dQ/dt = AD(\Delta c/\Delta x)$$

In this expression Δc describes the difference between the concentration of water vapor at the water surface and that at the desiccant surface: $\Delta c = c_w - c_{des}$. Since $c_{des} \ll c_w$, Δc may be replaced by c_w, which in turn equals pM/RT, where p and M are the vapor pressure and molecular weight, respectively, of water. With this substitution the expression for dQ/dt becomes $A(D/\Delta x)(pM/RT)$. The ratio $D/\Delta x$ has units length time^{-1}, so we identify it as the reciprocal of the transport resistance. Note that r increases as the effective value of Δx increases and the effective value of D decreases. Thus l/r is the only unknown in the expression $dQ/dt = A(1/r)(pM/RT)$ and can be calculated from the measured rate of weight increase with different monolayers present. ∎

* * *

To be acceptable for use in the field, the monolayer material must have the following properties:

1. It must spread easily, probably as bulk material, so a high value of π_e is desirable.
2. It must be self-healing since surface ripples will disrupt the monolayer. This implies viscous rather than rigid monolayers.
3. It must be inexpensive, which means, effectively, capable of forming good films from naturally occurring mixtures.
4. It must be nontoxic and free from other deleterious effects on aquatic life.

Hexadecyl and octadecyl alcohol have been extensively studied and shown to be highly effective in evaporation retardation. Scattering powdered samples of commercial-grade alcohols by boat on lake surfaces or the continuous addition of alcohol slurries from floating dispensers are two of the methods that have been employed to apply these monolayers. Wind conditions and the activity of aquatic birds have a considerable effect on the stability of the monolayer and therefore on the rate at which the monolayer chemicals must be reapplied. Rates of application rarely exceed 0.5 lb acre^{-1} day^{-1}, however, so that the cost of the materials used is not excessive.

An indication of the effectiveness of such field treatment is seen in Figure 7.12, which compares the amount of water lost by evaporation from two small adjacent lakes in Illinois. Treatment of North Lake (ordinate) with commercial hexadecanol was begun in late June 1957; untreated South Lake was the control (abscissa). Prior to treatment the evaporation losses from the two lakes were identical, as shown by the 45° line in the figure. After treatment was begun, however, the loss of water from North Lake fell considerably behind that from South Lake. By the end of the summer a difference of about 40% in the water loss was observed. This was equivalent to about 7600 gallons of water saved per pound of hexadecanol used. For areas where water is scarce—the southwestern United States, Israel, and western Australia, for example—such conservation of water is highly valued, and research continues to look for methods to improve the efficiency of this technique.

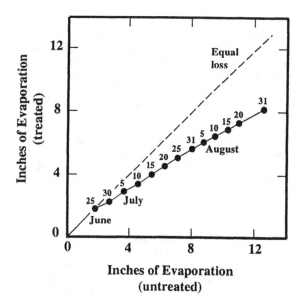

FIG. 7.12 Comparison of the water level in two adjacent lakes during the summer, 1957. The ordinate shows the level in the lake with the monolayer; the abscissa is the level in the untreated lake. (Redrawn with permission of LaMer 1962.)

7.6b Damping of Waves

Surface viscosity also has observable effects on macroscopic bodies of water. As mentioned in the opening paragraph of this chapter, the calming effect on surface turbulence of pouring oil on the sea has been known from antiquity. In terms of the concepts of this chapter, the increase in surface viscosity produced by the film has a damping effect on waves. Such damping has been studied both theoretically and in laboratory situations; the book by Davies and Rideal (1961) contains some interesting photographs of the ripples on a pond before and after the application of hexadecanol to the surface.

7.6c Stabilization of Emulsions and Foams

Emulsions and foams are two other areas in which dynamic and equilibrium film properties play a considerable role. Emulsions are colloidal dispersions in which two immiscible liquids constitute the dispersed and continuous phases. Water is almost always one of the liquids, and amphipathic molecules are usually present as emulsifying agents, components that impart some degree of durability to the preparation. Although we have focused attention on the air-water surface in this chapter, amphipathic molecules behave similarly at oil-water interfaces as well. By their adsorption, such molecules lower the interfacial tension and increase the interfacial viscosity. Emulsifying agents may also be ionic compounds, in which case they impart a charge to the surface, which in turn establishes an ion atmosphere of counterions in the adjacent aqueous phase. These concepts affect the formation and stability of emulsions in various ways:

1. Most emulsions are formed by some sort of comminution process in which large blobs of the dispersed phase are eventually ground down to small drops. This is a complex process, but basically consists of drops becoming elongated under shearing forces, necking, and finally separating into smaller drops. The adsorption of a surface film with the attendant lowering of γ and increase in η^s clearly enters the picture.

2. The first step in the "breaking" of an emulsion is the coming together of the individual drops. If water is the continuous phase and the emulsifier is ionic, then it is the ion atmo-

spheres of the approaching particles that make the first contact. We have already seen in Chapter 4 that the overlap of the electrical double layers affects the viscosity of the dispersion. We see in Chapter 13 that the details of this first encounter can determine whether the drops form an aggregate or go their separate ways.

3. If droplets can aggregate into a single kinetic unit, they might also coalesce into a single geometrical unit. This involves rupture of the thin film of continuous phase that separates them in an aggregate. Again, surface tension and surface viscosity are certainly pertinent to the coalescence process.

The huge variety of emulsions used as food, medicinal, cosmetic, and other industrial products make these colloids important practical systems in which the surface monolayers exert considerable influence. We have already discussed the use of lecithin to control the viscosity and the texture of chocolate in Vignette IV in Chapter 4.

Foams are colloidal systems in which a gas is the dispersed phase. Although a whole range of concentrations is possible, we shall focus on those foams that consist of volume-filling, distorted polyhedra separated by liquid films. For aqueous foams the high area of air-water interface requires adsorption to lower the surface tension sufficiently to make the foam in the first place. Foams drain by losing liquid through the channels that occur at the junction of the planar film surfaces. Such surfaces meet with curved menisci between them. This means that the pressure is lower in the junction than in the flat faces of the film according to the Laplace equation, Equation (6.29). As a consequence, liquid flows from the planar regions into these junctions — called plateau borders — through which the drainage occurs. As the film thickness of the continuous phase decreases, the probability of rupture due to thermal or mechanical fluctuations increases. Surface energetics and viscosity are important here also. As with emulsions, foams occur in many systems familiar to consumers, such as fire-fighting foams, whipped cream, shaving lather, and the head on a glass of beer!

7.6d Preparation of Langmuir-Blodgett Films

As we pointed out in the vignette at the beginning of this chapter, many potential applications emerge when the Langmuir layers are transferred to a solid substrate. We have some more to say about Langmuir-Blodgett films in Section 7.10c, but it is clear, based on what we have discussed so far, that understanding the different structural features of Langmuir layers and how to control the stability of the layers is the first prerequisite for depositing Langmuir-Blodgett layers on solid substrates.

Finally, it is worth noting that the monolayers of the type discussed in the previous sections may serve as good model systems for examining some of the theories of condensed matter physics.

Although we started out this chapter by discussing insoluble monolayers, it is evident that we have slipped into examples for which soluble amphipathics are being considered. In the next section we examine the thermodynamics of adsorption from solution.

7.7 ADSORPTION FROM SOLUTION: THERMODYAMICS

Until now we have discussed only insoluble monolayers. Although their behavior is complex, they have the conceptual simplicity of being localized in the interface. It has been noted, however, that even in the case of insoluble monolayers, the substrate should not be overlooked. The importance of the adjoining bulk phases is thrust into even more prominent view when soluble monolayers are discussed. In this case the adsorbed material has appreciable solubility in one or both of the bulk phases that define the interface.

7.7a The Gibbs Equation: Multicomponent Systems

Gibbs treated this situation as part of his investigations into phase equilibria. Suppose we consider two phases α and β in equilibrium with a surface s dividing them. For the system so constituted, we may write

$$G = G^{\alpha} + G^{\beta} + G^{s} \tag{30}$$

where the superscripts indicate the contribution from each category. For the bulk phases

$$G = E + pV - TS + \sum_i \mu_i n_i \tag{31}$$

where the chemical potential terms are summed for all components i. The superscript has been omitted for convenience. The volume term is replaced by an area term in the corresponding expression for G^s:

$$G^s = E^s + \gamma A - TS^s + \sum_i \mu_i n_i \tag{32}$$

where μ_i's and n_i's here are for the surface phase. Substituting Equations (31) and (32) into Equation (30) and taking the total derivative yields

$$dG = \sum_{\alpha,\beta,s} \left(dE + pdV + Vdp - TdS - SdT + \sum_i \mu_i\, dn_i + \sum_i n_i\, d\mu_i \right) + Ad\gamma \tag{33}$$
$$+ \gamma dA$$

For a reversible process

$$dE = \delta q - \delta w = \sum_{\alpha,\beta,s} dE = \sum_{\alpha,\beta,s} [TdS - (pdV + \delta w_{non\text{-}pV})] \tag{34}$$

Substituting this result into Equation (33) gives

$$dG = \sum_{\alpha,\beta,s} \left(Vdp - SdT + \sum_i \mu_i\, dn_i + \sum_i n_i d\mu_i - \delta w_{non\text{-}pV} \right) + Ad\gamma + \gamma dA \tag{35}$$

As we saw in Chapter 6 (Equation (6.16)), the quantity γdA may be equated to non-pressure-volume work when surface energy is being considered. With this consideration, Equation (35) simplifies still further to become

$$dG = \sum_{\alpha,\beta,s} \left(Vdp - SdT + \sum_i \mu_i\, dn_i + \sum_i n_i d\mu_i \right) + Ad\gamma \tag{36}$$

Another well-known relationship from thermodynamics may be introduced at this point:

$$dG = Vdp - SdT + \sum_i \mu_i\, dn_i \tag{37}$$

Applying Equation (37) to the bulk phases and the surface and subtracting the result from Equation (36) gives

$$\sum_i n_i^{\alpha} d\mu_i + \sum_i n_i^{\beta} d\mu_i + \sum_i n_i^{s} d\mu_i + Ad\gamma = 0 \tag{38}$$

When only one phase is under consideration, only one of the terms in Equation (30) is required, and only one of the bulk phase summations in Equation (38) survives. The result in this case is the famous *Gibbs-Dühem equation*:

$$\sum_i n_i\, d\mu_i = 0 \tag{39}$$

It will be recalled from physical chemistry that this relationship permits the evaluation of the activity of one component from measurements made on the other in binary solutions.

By means of the Gibbs-Dühem equation, we may eliminate the terms in Equation (38) that apply to bulk phases and write

$$\sum_i n_i^{s} d\mu_i + Ad\gamma = 0 \tag{40}$$

This is the *Gibbs adsorption equation* that relates γ to the number of moles and the chemical potentials of the components in the interface.

7.7b Two-Component Systems

In subsequent developments we consider only two-component systems and identify the solvent (usually water) as component 1 and the solute as component 2. In terms of this stipulation, Equation (40) becomes

$$n_1^s\, d\mu_1 + n_2^s\, d\mu_2 + A\, d\gamma = 0 \tag{41}$$

It is conventional to divide Equation (41) through by A to give

$$d\gamma = -\frac{n_1^s}{A}\, d\mu_1 - \frac{n_2^s}{A}\, d\mu_2 \tag{42}$$

The quantity n_i^s/A is called the *surface excess* of component i and is given the symbol Γ_i

$$\Gamma_i = n_i^s/A \tag{43}$$

In this notation, Equation (42) becomes

$$-d\gamma = \Gamma_1\, d\mu_1 + \Gamma_2\, d\mu_2 \tag{44}$$

We return to further simplifications of this equation after a brief discussion of how to define the position of a surface and how to define surface excess quantities.

7.7c Location of the Surface and the Meaning of Surface Excess Properties

Before proceeding any further, it is necessary to examine just what the concept of a surface excess means. To do this it is convenient to consider the changes that occur in some general property P as we move from phase α to phase β. The situation is represented schematically by Figure 7.13, in which x is the distance measured perpendicular to the interface. The scale of this figure is such that variations at the molecular level are shown. The interface is not a surface in the mathematical sense, but rather a zone of thickness τ across which the properties of the system vary from values that characterize phase α to those characteristic of β. In spite of this, we generally do not assign any volume to the surface, but treat it as if the properties of α and β applied right up to some dividing plane situated at some specific value of x. What is this position x_0 at which we draw such a boundary?

Suppose the solid line in Figure 7.13 represents the actual variation of property P. The

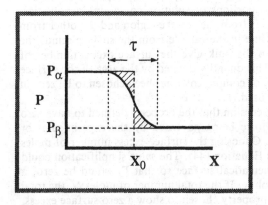

FIG. 7.13 Variation of some general property P with perpendicular distance from the surface in the vicinity of an interface between two phases α and β.

squared-off extensions of the bulk values of this property represent the approximation made in assuming the surface to have zero thickness. Then the shaded area to the left of x_0 shows the amount by which the value of P for the system as a whole has been overestimated by extending P_α. Likewise, the shaded area to the right of x_0 shows how the extension of P_β leads to an underestimation of P for the system as a whole. In principle, the "surface" may be located at an x value such that these two areas compensate for one another; that is, x_0 may be chosen so that the two shaded areas in the figure are equal.

This is where the trouble begins! Generally speaking, the kind of profile sketched in Figure 7.13 will be different for each property considered. Therefore we may choose x_0 to accomplish the compensation discussed herein for one property, but this same line will divide the profiles of other properties differently. The difference between the "overestimated" property and the "underestimated" one accounts for the "surface excess" of this property.

From the point of view of thermodynamics—which is oblivious to details at the molecular level—the dividing boundary may be placed at any value of x in the range τ. The actual placement of x_0 is governed by consideration of which properties of the system are most amenable to thermodynamic evaluation. More accurately, that property that is least convenient to handle mathematically may be eliminated by choosing x_0 so that the difficult quantity has a surface excess of zero.

For example, if the property in Figure 7.13 was G and the dividing surface was placed so that the two shaded regions would be equal, then there would be no surface excess G: The last term in Equation (30) would be zero. The Gibbs free energy is convenient to work with, however, so such a choice for x_0 would not be particularly helpful. Until now we have not had any reason to identify the surface of physical phases with any specific mathematical surface. We had not, that is, until Equation (44) was reached. Now things are somewhat different.

Suppose the property represented in Figure 7.13 is the number of moles of solvent per unit area in a slice of solution at some value of x. This quantity will clearly undergo a transition in the vicinity of an interface. We choose x_0 so that the shaded areas are equal when this is the quantity of interest. This placement of the dividing surface means

$$\Gamma_1 = 0 \tag{45}$$

With this situation, Equation (44) becomes

$$d\gamma = -\Gamma_2 \, d\mu_2 \tag{46}$$

The physical significance of Γ_2 is determined by the arbitrary placement of the mathematical surface that made $\Gamma_1 = 0$; that is, Γ_2 equals the algebraic difference between the "overestimated" and "underestimated" areas of the curve describing moles of solute when this curve is divided at a location x_0 that makes the surface excess of the solvent zero.

It is important to realize that the mathematical dividing surface just discussed is a reference level rather than an actual physical boundary. What is physically represented by this situation may be summarized as follows. Two portions of solution containing an identical number of moles of solvent are compared. One is from the surface region and the other from the bulk solution. The number of moles of solute in the sample from the surface minus the number of moles of solute in the sample from the bulk give the surface excess number of moles of solute according to this convention. This quantity divided by the area of the surface equals Γ_2. To emphasize that the surface excess of component 1 has been chosen to be zero in this determination, the notation Γ_2^1 is generally used.

It should be evident from the foregoing discussion that the property defined to have zero surface excess may be chosen at will, the choice being governed by the experimental or mathematical features of the problem at hand. Choosing the surface excess number of moles of one component to be zero clearly simplifies Equation (44). The same simplification could have been accomplished by defining the mathematical surface so that Γ_2 would be zero, a choice that would obviously deemphasize the solute. If the total number of moles N, the total volume V, or the total weight W had been the property chosen to show a zero surface excess, then in each case both Γ_1 and Γ_2 (which would be identified as Γ^N, Γ^V, or Γ^W for these three conventions) would have nonzero values. Last, note that the surface "excess" is an algebraic quantity that may be either positive or negative depending on the convention chosen for Γ. A variety of different experimental methods are encountered in the literature to measure "surface excess" quantities; one must be careful to understand clearly what conventions are used in the definition of these quantities.

7.7d Relation Between Surface Tension and Surface Excess Concentration

Equation (46), one form of the Gibbs equation, is an important result because it supplies the connection between the surface excess of solute and the surface tension of an interface. For systems in which γ can be determined, this measurement provides a method for evaluating the surface excess. It might be noted that the finite time required to establish equilibrium adsorption is why dynamic methods (e.g., drop detachment) are not favored for the determination of γ for solutions. At solid interfaces, γ is not directly measurable; however, if the amount of adsorbed material can be determined, this may be related to the reduction of surface free energy through Equation (46). To understand and apply this equation, therefore, it is imperative that the significance of Γ_2 be appreciated.

Now let us return to the development of Equation (46). The chemical potential depends on the activity according to the equation

$$\mu_2 = \mu_2^0 + RT \ln a_2 \tag{47}$$

In applying these results to adsorption from solution, the activity equals the pressure or concentration multiplied by the activity coefficient f. Differentiation of Equation (47) at constant temperature yields

$$d\mu_2 = RT \frac{da_2}{a_2} = RT\, d \ln (fc) \tag{48}$$

This relationship may also be applied to the adsorption of gases by replacing concentration by gas pressure and continuing to use the appropriate activity coefficient. We return to the application of this result to the adsorption of gases in Chapter 9.

For adsorption from dilute solutions the activity coefficient approaches unity, in which case the combination of Equations (46) and (48) leads to the result

$$\Gamma_2^1 = -\frac{c}{RT}\left(\frac{d\gamma}{dc}\right)_T = -\frac{1}{RT}\left(\frac{d\gamma}{d \ln c}\right)_T \tag{49}$$

This form of the Gibbs equation shows that the slope of a plot of γ versus the logarithm of concentration (or activity if the solution is nonideal) measures the surface excess of the solute. It might also be noted that the choice of units for concentration is immaterial at this point.

7.8 THE GIBBS EQUATION: EXPERIMENTAL RESULTS

7.8a Typical Variations of Surface Tension in Aqueous Solutions

Surface tensions for the interface between air and aqueous solutions generally display one of the three forms indicated schematically in Figure 7.14. The type of behavior indicated by curves 1 and 3 indicates positive adsorption of the solute. Since $d\gamma/dc$ and therefore $d\gamma/d \ln c$ are negative, Γ_2^1 must be positive. On the other hand, the positive slope for curve 2 indicates a negative surface excess, or a surface depletion of the solute. Note that the magnitude of negative adsorption is also less than that of positive adsorption.

Curve 1 in Figure 7.14 is the type of behavior characteristic of most un-ionized organic compounds. Curve 2 is typical of inorganic electrolytes and highly hydrated organic compounds. The type of behavior indicated by curve 3 is shown by soluble amphipathic species, especially ionic ones. The break in curve 3 is typical of these compounds; however, this degree of sharpness is observed only for highly purified compounds. If impurities are present, the

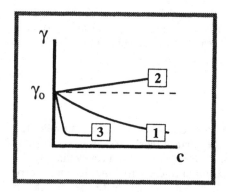

FIG. 7.14   Three types of variation of γ with c for aqueous solutions: (1) simple organic solutes, (2) simple electrolytes, and (3) amphipathic solutes.

curve will display a slight dip at this point. All three of these curves correspond to relatively dilute solutions. At higher concentrations effects other than adsorption may lead to departures from these basic forms. We say a bit more about adsorption from binary solutions over the full range of compositions in Section 7.9c.4.

7.8a.1   Simple Organic Solutes

For the limit as $c \rightarrow 0$, curve 1 may be presented by the equation of a straight line:

$$\gamma = \gamma_0 - mc \tag{50}$$

where m is the initial slope of the line. This is the same as

$$\pi = mc \tag{51}$$

From Equation (50) $d\gamma/dc = -m$, and from Equation (51) $c = \pi/m$; therefore Equation (49) may be written

$$\Gamma_2^1 = \pi/RT \tag{52}$$

Recalling the definition of Γ_2^1 provided by Equation (43), we see that Equation (52) may also be written

$$\pi A = n_2^s RT \tag{53}$$

the two-dimensional ideal gas law again! Those carboxylic acids containing less than 12 carbons in the alkyl chain for which results were presented in Figure 7.10 were investigated by this method. This same analysis also applies to the branch of curve 3 in Figure 7.14 as $c \rightarrow 0$.

7.8a.2   Simple Electrolytes

Curve 2 in Figure 7.14 indicates a negative surface excess of simple electrolytes. This means that portions of solution from both the surface and bulk regions that contain the same number of moles of solvent will have more solute in the bulk region than at the surface. Obviously, the surface is enriched over the bulk in solvent, a fact that is easily understood when the hydration of the ions is considered. Water molecules interact extensively with ions, a fact that accounts in part for the excellent solvent properties of water for ionic compounds. To move an ion directly to the air-water interface would require considerable energy to partially dehydrate the ion. Accordingly, the first couple of molecular diameters into the solution will be a layer of essentially pure water, the ions being effectively excluded from this region. The surface tension is not that of pure water, but is increased slightly due to the small surface deficiency of solute. Other highly solvated solutes such as sucrose also show this effect.

7.8a.3 Amphipathic Solutes

Curve 3 in Figure 7.14 applies primarily to amphipathic species. Most long-chain amphipathic molecules are insoluble unless the hydrophobic alkyl part of the molecule is offset by an ionic head or some other suitably polar head such as a polyethylene oxide chain, $-(CH_2CH_2O)_n-$. Like their insoluble counterparts, these substances form an oriented monolayer even at low concentrations. Figure 7.15 shows some actual experimental plots of type 3 for the ether that consists of a dodecyl chain and a hexaethylene oxide chain ($n = 6$) in the general formula just given. Example 7.4 illustrates the application of the Gibbs equation to these data.

* * *

EXAMPLE 7.4 *Determination of Surface Excess Concentration from Surface Tension Data.* The slope of the 25°C line in Figure 7.15 on the low-concentration side of the break is about -16.7 mN m^{-1}. Calculate the surface excess and the area per molecule for the range of concentrations shown. How would Figure 7.15 be different if accurate measurements could be made over several more decades of concentration in the direction of higher dilution? Could the data still be interpreted by Equation (49) in this case?

Solution: The surface excess is constant over this range of concentrations as indicated by the linearity of Figure 7.15. Equation (49) can be used directly to evaluate Γ_2^1. Since base 10 logarithms are used in the figure, we write

$$\Gamma_2^1 = -(0.0167)/(2.303)(8.314)(298) = 2.93 \cdot 10^{-6} \text{ mole m}^{-2}$$

The reciprocal of this gives the area of surface occupied by a mole of adsorbed molecules. Division by Avogadro's number converts the reciprocal of Γ_2^1 into a value for σ:

$$\sigma = (1 \text{ m}^2/2.93 \cdot 10^{-6} \text{ mole}) \cdot (1 \text{ mole}/6.02 \cdot 10^{23} \text{ molecules}) \cdot (10^{-9} \text{ nm}/1 \text{ m})^2$$
$$= 0.56 \text{ nm}^2$$

If accurate measurements could be made to increasingly lower concentrations, the surface excess would gradually decrease toward zero. This means that the lines in Figure 7.15 must eventually curve until they show a slope of zero at infinite dilution. Curved lines on a semilogarithmic plot of γ versus c are interpreted by drawing tangents at the concentrations of interest and applying the Gibbs equation to the slopes of the tangents to give the corresponding surface excesses. ∎

* * *

The polar heads of the solute molecules in Figure 7.15 are much bulkier than those of the simple amphipathic molecules with insoluble monolayers that we discussed above. This is

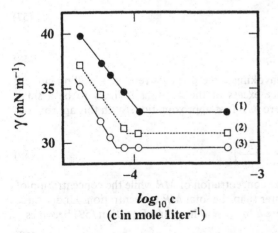

FIG. 7.15 Plot of γ versus $\log_{10} c$ for the dodecyl ether of hexaethylene oxide at three temperatures: (1) 15°C, (2) 25°C, and (3) 35°C. (Redrawn with permission of J. M. Corkill, J. F. Goodman, and R. H. Ottewill, *Trans. Faraday Soc.*, **57**, 1927 (1961).)

especially true when the hydration of the ether oxygens is considered. In view of this, the value calculated in Example 7.4 is probably about as small a value as these molecules can achieve. This suggests that the amphipathic molecules are in a highly condensed surface state at the concentrations investigated in Figure 7.15.

The break in curve 3 in Figure 7.14 is characteristic of this type of plot for soluble amphipathic molecules. Note that it appears in the experimental curves of Figure 7.15 also. The break is understood to indicate the threshold of *micelle* formation (see Chapter 1, Section 1.3a), known as the *critical micelle concentration* (see Chapter 8). We do not discuss this phenomenon any further since the next chapter is devoted entirely to micelles and related structures.

7.8b   Effect of Ionic Dissociation on Adsorption

It is instructive to consider the effect of dissociation on the adsorption of amphipathic substances since many of the compounds that behave according to curve 3 are electrolytes. We consider only the case of strong $1:1$ electrolytes; for weak electrolytes the equilibrium constant for dissociation must be considered.

If an ionic solute is totally dissociated into positive and negative ions, then its activity is given by

$$a_{MR} = a_M a_R \cong c_M c_R \tag{54}$$

where the subscripts M and R refer to the cation and amphipathic anion, respectively. Analogous results would be obtained if the cation were the amphipathic species. The approximation included in Equation (54) applies to the case in which the activity coefficient equals unity. Substituting this result into Equation (49) gives

$$\Gamma^1_{MR} = -\frac{1}{2RT}\left(\frac{d\gamma}{d\ln c}\right)_T \tag{55}$$

The assumption that no other electrolyte is present is implicit in this result. Now let us consider what happens when the system also contains a nonamphipathic electrolyte with a common ion to the surface-active electrolyte.

If a second electrolyte MX is present in addition to MR, then Equation (44) must be written

$$-d\gamma = \Gamma^1_M d\mu_M + \Gamma^1_R d\mu_R + \Gamma^1_X d\mu_X \tag{56}$$

Now the condition of surface neutrality becomes

$$\Gamma^1_M = \Gamma^1_R + \Gamma^1_X \tag{57}$$

so Equation (56) may be written

$$-d\gamma = \Gamma^1_R(d\mu_M + d\mu_R) + \Gamma^1_X(d\mu_M + d\mu_X) \tag{58}$$

This result may now be simplified by invoking some previous results. Recalling curve 2 from Figure 7.14, we know that the surface excess of the X^- ions is likely to be a small negative number that we shall set equal to zero as a first approximation. With this approximation, Equation (58) becomes

$$-d\gamma \approx \Gamma^1_R(d\mu_M + d\mu_R) = \Gamma^1_R RT\left(\frac{dc_M}{c_M} + \frac{dc_R}{c_R}\right) \tag{59}$$

Now let us consider a small change in the concentration of MR while the concentration of MX remains constant and considerably greater than the total MR concentration. Under these conditions, $dc_M = dc_R$ and $c_M \gg c_R$; therefore $d\ln c_M \ll d\ln c_R$ and Equation (59) becomes

$$-d\gamma = \Gamma^1_R(RT\,d\ln c) \tag{60}$$

Equations (55) and (60) are thus seen to describe the adsorption of MR in the absence of electrolyte and in the presence of swamping amounts of electrolyte, respectively. It is clear from the difference between these two results that extreme care must be taken in the study of charged monolayers if the effect of the charge on the state of the monolayer is to be properly considered in the interpretation of experimental results.

The difference between Equations (55) and (60) may be qualitatively understood by comparing the results with the Donnan equilibrium discussed in Chapter 3. The amphipathic ions may be regarded as restrained at the interface by a hypothetical membrane, which is of course permeable to simple ions. Both the Donnan equilibrium (Equation (3.85)) and the electroneutrality condition (Equation (3.87)) may be combined to give the distribution of simple ions between the bulk and surface regions. As we saw in Chapter 3 (e.g., see Table 3.2), the restrained species behaves more and more as if it was uncharged as the concentration of the simple electrolyte is increased. In Chapter 11 we examine the distribution of ions near a charged surface from a statistical rather than a phenomenological point of view.

7.8c Measuring Surface Excess Concentrations

We have noted previously that measuring γ as a function of concentration is a convenient means of determining the surface excess of a substance at a mobile interface. In view of the complications arising from charge considerations, the need for an independent method for measuring surface excess becomes apparent. Some elaborate techniques have been developed that involve skimming a thin layer off the surface of a solution and comparing its concentration with that of the bulk solution.

A simpler method for verifying the Gibbs equation involves the use of isotopically labeled surfactants. If the isotope emits a low-energy β particle, the range of β in water will be very low. Thus a detector placed just above the surface will count primarily those emissions originating from the surface region. Tritium (^3H), for example, emits a 0.0186-MeV β particle with a range in water of only about 17 μm, which means that only a negligible fraction of the β particles can travel farther than this in water. In fact, most are absorbed in an even shorter distance, so any ^3H β particles detected above an aqueous solution of tritiated surfactant probably originate within approximately 3 μm of the surface. The contribution of the bulk solution to the "background" of the former measurement is made using the same isotope in a compound that is known not to be adsorbed. By such studies the kinds of effects just described have been investigated and verified.

The surface-active substances we have discussed have been purified, research-quality materials. In practical situations the cost of synthesizing and purifying such surfactants is prohibitive. The materials commercially used, therefore, are inevitably mixtures. Commercial surfactants originate, for example, from the esterification of sugars or the sulfonation of alkyl-aryl mixtures. Such mixtures are marketed under a bewildering variety of trade names, and often as members of number- or letter-designated series that correspond, roughly, to a set of homologs. Table 7.2 lists examples of several specific members of such series, along with a brief description of the general nature of the family to which they belong.

7.9 ADSORPTION ON SOLID SURFACES

7.9a The Langmuir Equation: Theory

Throughout most of this chapter we have been concerned with adsorption at mobile surfaces. In these systems the surface excess may be determined directly from the experimentally accessible surface tension. At solid surfaces this experimental advantage is missing. All we can obtain from the Gibbs equation in reference to adsorption at solid surfaces is a thermodynamic explanation for the driving force underlying adsorption. Whatever information we require about the surface excess must be obtained from other sources.

If a dilute solution of a surface-active substance is brought in contact with a large adsorbing surface, then extensive adsorption will occur with an attendant reduction in the concentration of the solution. To meet the requirement of a large surface available for adsorption, the solid—which is called the *adsorbent*—must be finely subdivided. From the analytical data

TABLE 7.2 Some Familes of Commercial Surfactants and Specific Examples from Each

Name of series	General chemical nature	Specific designation ("____") and chemical nature of example	Example
Igepon "____"	Fatty acid amide of methyltaurine	"TN" R = palmityl	$RCON(CH_3)C_2H_4SO_3^-Na^+$
Aerosol "____"	Alkyl ester of sulfosuccinic acid	"OT" R = octyl	CH_2-COOR | CH_2-COOR | $SO_3^-Na^+$
Span "____"	Fatty acid esters of anhydrosorbitols	"60" R = stearyl	HO$-$CH$-$C HOH | \\ CH$_2$ CH$-$CHOH$-$COOR \\ / O
Tween "____"	Fatty acid ester and ethylene oxide esters of anhydro-sorbitols	"21" n = 4, R = lauryl	ROOCCH$-$ CHOH | | CH$_2$ CH(OC$_2$H$_4$)$_n$OH \\ / O
Triton "____"	Ethylene oxide ethers of alkyl benzene	"X-45" n = 5, R = octyl	$R-\langle\bigcirc\rangle-(OC_2H_4)_nOH$
Hyamine "____"	Alkylbenzyl dimethyl ammonium salts	"3500" R = $C_{12} - C_{16}$	$R-\langle\bigcirc\rangle-\overset{+}{N}(CH_3)_2Cl^-$ | $CH_2\phi$

describing the concentration change in the solution as well as a knowledge of the total amount of solid and solution equilibrated, it is possible to determine the amount of solute adsorbed—which is called the *adsorbate*—per unit weight of adsorbing solid. If the specific area of the adsorbing solid is known, then the results may be expressed as amount adsorbed per unit area. These studies are generally conducted at constant temperature, and the results—which relate the amount of material adsorbed to the equilibrium concentration of the solution—describe what is known as the *adsorption isotherm*.

One isotherm that is both easy to understand theoretically and widely applicable to experimental data is due to Langmuir and is known as the *Langmuir isotherm*. In Chapter 9, we see that the same function often describes the adsorption of gases at low pressures, with pressure substituted for concentration as the independent variable. We discuss the derivation of Langmuir's equation again in Chapter 9 specifically as it applies to gas adsorption. Now, however, adsorption from solution is our concern. In this section we consider only adsorption from dilute solutions. In Section 7.9c.4 adsorption over the full range of binary solution concentrations is also mentioned.

Suppose we imagine a dilute solution in which both the solvent (component 1) and the solute (component 2) have molecules that occupy the same area when they are adsorbed on a surface. The adsorption of solute may then be schematically represented by the equation

$$\text{Adsorbed solvent} + \text{Solute in solution} \rightarrow \text{Adsorbed solute} + \text{Solvent in solution} \quad (61)$$

The equilibrium constant for this reaction may be written as

$$K' = \frac{a_2^s a_1^b}{a_1^s a_2^b} \tag{62}$$

where a stands for the activity of the species and the superscripts s and b signify surface and bulk values, respectively. Next let us assume that the two-dimensional surface solution is ideal, an assumption that enables us to replace the activity at the surface by the mole fraction at the surface x^s:

$$K' = \frac{x_2^s a_1^b}{x_1^s a_2^b} \tag{63}$$

Since the surface contains only two components, $x_1^s + x_2^s = 1$ and Equation (63) becomes

$$K' = \frac{x_2^s a_1^b}{(1 - x_2^s) a_2^b} \tag{64}$$

Equation (64) may be rearranged to give

$$x_2^s = \frac{K' a_2^b / a_1^b}{K' a_2^b / a_1^b + 1} \tag{65}$$

In dilute solutions the activity of the solvent is essentially constant, so the ratio K'/a_1^b may be defined to equal a new constant K, in terms of which Equation (65) becomes

$$x_2^s = \frac{K a_2^b}{K a_2^b + 1} \tag{66}$$

This is one form of the Langmuir adsorption isotherm.

An equivalent form of the Langmuir equation expressed in slightly different variables is obtained by writing both x_1^s and x_2^s in Equation (63) in terms of area fractions occupied by '1' and '2'. We have already postulated that both the solvent and solute molecules occupy equal areas on the surface. Therefore, x_i^s equals the fraction of the surface occupied by component i, θ_i. Since $\theta_1 + \theta_2 = 1$, we have

$$\theta_2 = \frac{K a_2^b}{K a_2^b + 1} \tag{67}$$

In this form the Langmuir equation shows how the fraction of surface adsorption sites occupied by solute increases as the solute activity in solution increases. From now on we drop the subscript 2 and the superscript b. Since Equation (67) is written solely in terms of the solute, these designations are redundant.

Two limiting cases are of special interest:

1. At infinite dilution $a \to 0$ and Equation (67) becomes

$$\theta = Ka \tag{68}$$

2. If $Ka \gg 1$, Equation (67) becomes

$$\theta = 1 \tag{69}$$

Equation (68) shows that θ increases linearly with an initial slope that equals K. This slope will be larger the farther to the right the equilibrium represented by Equation (61) lies. At higher concentrations, Equation (69) indicates that saturation of the surface with adsorbed solute is achieved. Figure 7.16a shows how these two limiting conditions affect the appearance of the isotherm.

Experimentally, one does not measure the fraction of sites containing adsorbed solute directly; instead, either the number of moles of solute adsorbed per unit weight of adsorbent n_2^s/w or the number of moles per unit area of adsorbent n_2^s/A is measured. These quantities are related by the equation

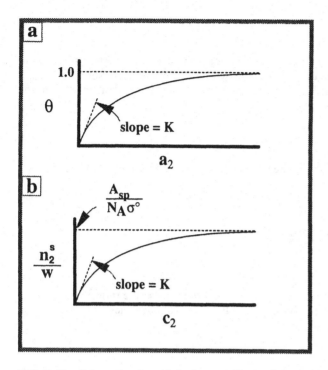

FIG. 7.16 Schematic plots of the Langmuir equation showing the significance of the initial slope and the saturation value of the ordinate: (a) the fraction covered versus solute activity; and (b) the number of moles of solute adsorbed per unit weight of adsorbent versus concentration.

$$\frac{n_2^s}{A} A_{sp} = \frac{n_2^s}{w} \tag{70}$$

where A_{sp} is the *specific area* of the adsorbent (see Chapter 1, Section 1.2). The fraction covered is related to these quantities as follows:

$$\theta = \frac{n_2^s}{A} N_A \sigma^0 = \frac{n_2^s N_A \sigma^0}{w A_{sp}} \tag{71}$$

where N_A is Avogadro's number and σ^0 is the area occupied per molecule. The level at which saturation adsorption occurs may be identified with $\theta = 1$. Therefore Equation (71) shows the saturation values of the usual ordinates to be either

$$\left(\frac{n_2^s}{w}\right)_{sat} = \frac{A_{sp}}{N_A \sigma^0} \tag{72}$$

or

$$\left(\frac{n_2^s}{A}\right)_{sat} = \frac{1}{N_A \sigma^0} \tag{73}$$

Since the entire derivation of the Langmuir isotherm assumes dilute solutions, the concentration c_2 of the solute (denoted by c, for simplicity) rather than the activity is generally used in presenting experimental results. Figure 7.16b shows how actual experimental data might appear.

7.9b The Langmuir Equation: Application

Many systems that definitely do not conform to the Langmuir assumptions — the adsorption of polymers, for example — nevertheless display experimental isotherms that resemble Figure 7.16. Although these can be fitted to Equation (67), the significance of the constants is dubious. Therefore the Langmuir equation is often written as

$$m\left(\frac{n_2^s}{w}\right) = \frac{(m/b)c}{(m/b)c + 1} \tag{74}$$

where m and m/b are regarded simply as empirical constants. A method for obtaining the numerical values for these constants from experimental data is easily seen by rearranging Equation (74) to the form

$$\frac{c}{n_2^s/w} = mc + b \tag{75}$$

This form suggests that a plot of $c/(n_2^s/w)$ versus c will be a straight line of slope m and intercept b.

If the experimental system matches the model, then the values of m and b can be assigned a physical significance by comparing Equation (74) with Equations (67) and (71):

$$m = \frac{N_A \sigma^0}{A_{sp}} \tag{76}$$

and

$$m/b = K \tag{77}$$

If the model does not apply, these constants are treated merely as empirical parameters that describe the adsorption isotherm.

When the model does apply, the experimental value of m permits A_{sp} to be evaluated if σ^0 is known, or σ^0 to be evaluated if A_{sp} is known. It is often difficult to decide what value of σ^0 best characterizes the adsorbed molecules at a solid surface. Sometimes, therefore, this method for determining A_{sp} is calibrated by measuring σ^0 for the adsorbed molecules on a solid of known area, rather than relying on some assumed model for molecular orientation and cross section.

Example 7.5 illustrates how adsorption data can be interpreted if the data conform to the Langmuir model.

*   *   *

EXAMPLE 7.5 *Use of Langmuir Adsorption Isotherm.* The moles of solute B adsorbed per gram of solid C were determined by measuring concentration changes in the solution. The accompanying results report the adsorption versus the concentration of the equilibrium solution:

c (mole B liter^{-1})   0.75 1.40 2.25 3.00 3.50 4.25
n/w · 10^4 [mole B
 (g C)$^{-1}$]   6.00 8.00 9.57 10.0 10.4 10.8

Plot these data in the form suggested by Equation (75) and evaluate the slope and intercept. If A_{sp} for solid C is known by independent study to be 325 m^2 g^{-1}, what is σ^0 for the adsorbate? Alternatively, suppose σ^0 for the adsorbate is known to be 0.25 nm^2 on this surface. What value of A_{sp} is consistent with the adsorption data?

Solution: The ratio $c/(n/w)$ is evaluated as required to test Equation (75):

c(n/w)$^{-1}$ · 10^{-3} (g C liter^{-1})   1.25 1.75 2.35 3.00 3.35 3.95

A plot of these values against the equilibrium concentrations is shown in Figure 7.17. The slope and intercept of the line drawn are 769 g C (mole B)$^{-1}$ and 0.0700 g C liter^{-1}, respectively.
 Equation (76) permits the interpretation of the slope m.

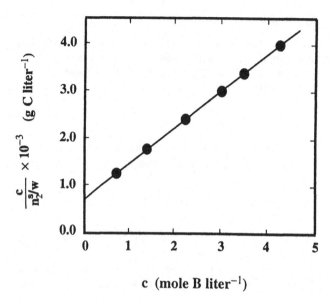

FIG. 7.17 Plot of the Langmuir equation in the form given by Equation (75) for data in Example 7.5.

If A_{sp} is known,

$$\sigma = mA_{sp}/N_A = (769)(325)(10^9)^2/6.02 \cdot 10^{23} = 0.42 \text{ nm}^2$$

and if σ^0 is known,

$$A_{sp} = N_A\sigma^0/m = (6.02 \cdot 10^{23})(0.25)(10^{-9})^2/769 = 196 \text{ m}^2 \text{g}^{-1}$$

The value of K is given by the ratio m/b according to Equation (77). For this (hypothetical) system, $K = (769)/(0.0700) = 1.10 \cdot 10^4$ liter (mole B)$^{-1}$. These reciprocal concentration units are appropriate for K of Equation (66) since the activity of the bulk solvent has been absorbed into the definition of K.   ∎

* * *

The method of Example 7.5 applied to the adsorption of benzene, naphthalene, and anthracene on carbon black from heptane solutions gives values of 0.42, 0.67, and 0.83 nm^2, respectively, for σ^0. The progression of sizes indicates that the molecules lie flat on the surface of the carbon.

It might also be noted that K' (Equation (62)) may be related to ΔG^0 for the adsorption process if the model applies to the experimental system. Therefore, from studies of adsorption at different temperatures, values of ΔH^0 and ΔS^0 may be determined for the process described by Equation (61). It must be emphasized that compliance with the form predicted by the Langmuir isotherm is not a sensitive test of the model; therefore interpretations of this kind must be used cautiously.

In summary, adsorption from dilute solutions frequently displays the qualitative form required by the Langmuir equation. If this form is observed, it may be quantitatively described by Equation (75), in which m and b are empirical constants. Sometimes there may be a justification for further interpretation of these parameters in terms of the theoretical model.

7.9c Limitations of the Langmuir Equation

We should not be too surprised that the Langmuir equation often yields only an empirical isotherm. There are several reasons why real systems are likely to deviate from the theoretical model:

1. The adsorption process described by Equation (61) is a complex one involving several different kinds of interactions: solvent-solute, solvent-adsorbent, and solute-adsorbent.

2. Few solid surfaces are homogeneous at the molecular level.

3. Few monolayers are ideal.

4. Our interest often extends beyond the region of dilute concentrations.

We briefly comment on each of these limitations.

7.9c.1 Multiplicity of Interactions

In discussing adsorption from solution, there is nothing that can be done about the multiplicity of possible interactions, except possibly to avoid systems in which highly specific interactions are to be expected. In Chapter 9 we again discuss the Langmuir isotherm as it applies to the adsorption of gases. In that case there are considerably fewer interactions involved in the adsorption process, making it more amenable to analysis.

7.9c.2 Surface Heterogeneity

The assumption of surface homogeneity is one that was not explicitly stated in deriving the Langmuir equation. It is essential, however; otherwise a different value of K would apply to Equation (61) at various places on the surface. Attempts to deal with surface heterogeneity have been undertaken, but this enterprise seems more likely to be successful for gas adsorption rather than for adsorption from solution because the variety of interactions that must be considered is less in the former than in the latter. There is an equation — known as the *Freundlich isotherm* — that may be derived by assuming a certain distribution function for sites having different ΔG^0 values for the process represented by Equation (61) and assuming Langmuir adsorption at each type of site. The Freundlich isotherm is given by the expression

$$\theta = ac^{1/n} \tag{78}$$

in which a and n are constants with $n > 1$. This equation was in use long before the interpretation of a certain distribution of sites was assigned to it. Therefore it is best regarded as an empirical isotherm, the constants for which may be evaluated from the slope and intercept of a log-log plot of θ versus c. The Freundlich isotherm is no cure-all for surface heterogeneity: Its theoretical derivation depends on a highly specific distribution of site energies. In addition, the Langmuir equation gives adequate results in many cases in which surface heterogeneity is known to be present. Note that with $n = 1$ the Freundlich isotherm is identical to the low-concentration limit of the Langmuir isotherm (Equation (68)), and with $n = \infty$ to the high-concentration limit (Equation (69)).

7.9c.3 Multilayer Adsorption

In the Langmuir derivation the adsorbed molecules are allowed to interact with the adsorbent but not with each other: The adsorbed layer is assumed to be ideal. This necessarily limits adsorption to a monolayer. Once the surface is covered with adsorbed molecules, it has no further influence on the system. The assumption that adsorption is limited to monolayer formation was explicitly made in writing Equations (72) and (73) for the saturation value of the ordinate. It is an experimental fact, however, that adsorption frequently proceeds to an extent that exceeds the monolayer capacity of the surface for any plausible molecular orientation at the surface. That is, if monolayer coverage is postulated, the apparent area per molecule is only a small fraction of any likely projected area of the actual molecules. In this case the assumption that adsorption is limited to the monolayer fails to apply. A model based on multilayer adsorption is indicated in this situation. This is easier to handle in the case of gas adsorption, so we defer until Chapter 9 a discussion of multilayer adsorption.

7.9c.4 Adsorption from Concentrated Solutions

Next let us consider adsorption from solutions that are not infinitely dilute. Suppose, for example, that adsorption is studied over the full range of binary liquid concentrations. Figure 7.18 is an example of such results for the benzene-ethanol system adsorbed on carbon. At

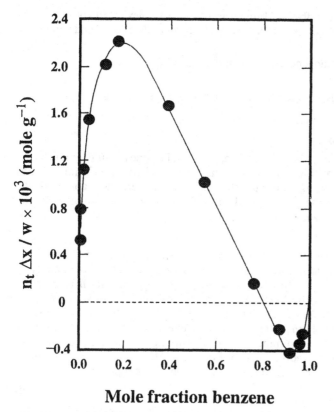

FIG. 7.18 Adsorption on carbon from the ethanol-benzene system. The ordinate equals the total number of moles of solution times the change in solution mole fraction per unit weight of carbon. (Data from F. E. Bartell and C. K. Sloan, *J. Am. Chem. Soc.*, **51**, 1643 (1929).)

first these results appear quite bewildering, displaying maximum, minimum, and negative adsorption. Recall, however, that what is actually measured is an isotherm of concentration change. The observed change in concentration is then expressed as moles of solute adsorbed. In a totally different range of solution concentrations, the solvent rather than the solute may adsorb. The associated change in the solution would then be an increase in solute concentration or the apparent negative adsorption of solute. A curve like that shown in Figure 7.18 should therefore be understood as a composite of two distinctly different isotherms. A good deal of work has been done with composite isotherms, particularly toward separating them into individual isotherms. A summary of this kind of research can be found in Kipling (1965).

7.10 APPLICATIONS OF ADSORPTION FROM SOLUTION

No discussion of adsorption from solution is anywhere near complete unless it includes some indication of its enormous practical applicability. As a matter of fact, the examples we briefly consider—detergency and flotation—encompass a wide variety of concepts from almost all areas of surface and colloid chemistry. We have chosen to stress principles rather than applications, however, so these subjects will receive an amount of attention that belies their actual importance. Following these traditional applications, two examples of new applications that are envisioned for surfactant layers deposited on solid substrates are discussed.

7.10a   Detergency and Flotation: Similarities

It is impossible to do justice to the complex phenomena of *detergency* and *flotation* in a few paragraphs. All we can do is point out some of the ways in which the principles of colloid and surface chemistry apply in these areas. There are several ways in which detergency and flotation phenomena resemble one another:

1.   Both terms give simple names to processes involving many different steps. The more familiar of the two, detergency, may be defined as the process by which some unwanted foreign matter is removed from a substrate by a combination of chemical treatment, temperature, and mechanical agitation. Flotation is the process by which a specific mineral component of an ore mixture is separated from other components (called "gangue") by being concentrated in the froth of an aerated slurry. Chemical additives and mechanical forces are involved here also.

2.   In actual practice, both detergency and flotation deal with systems that are terribly difficult to idealize by any sort of model. In a laundering operation, for example, there will be present a variety of different fabric surfaces (cotton, polyester, etc.), different kinds of foreign matter (particulate, oily, etc.), and different chemical additives (detergent, inorganic phosphate, fluorescent whitening agents, as well as the solvent, water). In flotation, all three states of matter—solid, liquid, and gas—are involved and each of these involves several chemical components. The ore is a complex mixture of minerals (assumed to be crushed to such an extent that each particle is a different phase), the air is a mixture of gases (including chemically reactive oxygen), and the liquid contains at least three deliberately added reagents (known as regulator, collector, and frother), in addition to whatever dissolved minerals are present in the water.

3.   A third point of resemblance between detergency and flotation (perhaps redundant in view of what has already been said) is that both have developed largely by empirical research with (partially satisfactory) explanations trailing far behind the actual practice.

7.10a.1   Detergency

With this much general background, let us now consider these two processes separately. In discussing detergency we must first examine the availability of the surfactant. Weak acid soaps form insoluble compounds with Ca^{2+}, for example, and are converted to insoluble molecular acids at low pH levels. One of the reasons for the addition of inorganic phosphate to laundry products is to prevent or minimize these reactions. In this discussion we assume that the impurity has not been imbibed into the interior of the fiber (soaking might help if it has) and that it is a semiliquid soiled spot rather than a solid contaminant with which we are dealing. One advantage of washing this type of soiled material at high temperatures is that the viscosity of the oily spot is lowered so that the shape of these drops is more readily altered.

The process of removing an oily drop from a solid substrate may be described in terms of the work of adhesion, given by Equation (6.57). Applying this idea to the separation of oil (*O*) from solid (*S*) gives

$$W_{adhesion} = \gamma_{OW} + \gamma_{SW} - \gamma_{OS} \tag{79}$$

where the subscript *W* describes the aqueous solution. For the separation to be spontaneous in the thermodynamic sense, this quantity (ΔG) must be negative. Positive surface excesses of surfactant molecules at the interface between the aqueous phase and the oil and/or the solid will lower γ for these surfaces. This change is a favorable one for the process of removing foreign matter.

In addition, the contact angle between an oil spot and a solid surface to be cleaned may be a contributing factor in detergency. For example, Figures 7.19b and 7.19d illustrate schematically two different situations for an oily drop being lifted off a substrate by currents in the adjacent phase. The contact angles θ_l between the drop and the substrate are assumed to be the same at "lift off" (Figs. 7.19b and 7.19d) as in the quiescent state (Figs. 7.19a and 7.19c). It is evident from the figure, however, the necking of the drop for $\theta_1 < 90°$ is likely to leave a residue, whereas $\theta_1 > 90°$ would lead to a clean detachment. Young's equation (Equation (6.44)) may be applied to this situation:

$$\gamma_{OW} \cos \theta_1 = \gamma_{SW} - \gamma_{OS} \tag{80}$$

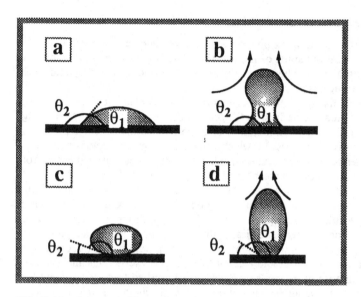

FIG. 7.19 Schematic illustration of several configurations of three phases useful in the discussion of detergency and flotation. The shaded region represents the soiled spot in detergency and θ_1 is the relevant contact angle; the shaded region is an air bubble in flotation, and θ_2 is the appropriate contact angle. The arrows in (b) and (d) indicate flow in the adjacent phase.

where θ_1 is measured in the oil drop as shown in Figure 7.19. Equation (80) shows that $\theta_1 >$ 90° and $\theta_1 <$ 90° correspond to $\gamma_{SW} < \gamma_{OS}$ and $\gamma_{SW} > \gamma_{OS}$, respectively. Any adsorption at the solid-water interface will lower γ_{SW} and therefore be conducive to a contact angle that favors the complete "rollback" of the oily spot.

Once the dirty spot is removed from the substrate being laundered, it is important that it not be redeposited. Solubilization of the detached material in micelles of surfactant has been proposed as one mechanism that contributes to preventing the redeposition of foreign matter. Any process that promotes the stability of the detached dirt particles in the dispersed form will also facilitate this. We see in Chapter 11 how electrostatic effects promote colloidal stability. The adsorption of ions—especially amphipathic surfactant ions—onto the detached matter assists in blocking redeposition by stabilizing the dispersed particles. Materials such as carboxymethylcellulose are often added to washing preparations since these molecules also adsorb on the detached dirt particles and interfere with their redeposition.

7.10a.2 Flotation

Now let us turn to a brief examination of *flotation*. Virtually all nonferrous metallic ores are concentrated by the flotation process. Sulfide ores have been studied particularly extensively, although the method has been used with oxides and carbonates as well as such nonmetallic materials as coal, graphite, sulfur, silica, and clay. Something on the order of a billion tons of ore a year are processed in this way.

We assume that the ore has been pulverized and mixed with water so that our involvement begins with a slurry known as *pulp*. We consider, in turn, the chemical nature and the effects of each of the three broad classes of chemicals added in the flotation process.

The first class of chemical additives to be considered is the regulator, a compound that affects the adsorption of the collectors. Regulators, like catalysts, may be positive or negative in their role. For the case in which the collector adsorption is enhanced, the regulator is called an activator; when the effect is negative, it is called a depressant. Regulators are frequently compounds that control the pH and sequester metallic cations that would otherwise compete with the mineral particle surfaces for the surface-active collectors. The pH affects not only the availability of certain collectors, but also the charge of the mineral particles (see Chapter 11,

Section 11.2 for a discussion of potential-determining ions). This last consideration plays a role in determining whether the mineral particles will be dispersed as small units (easier to lift by flotation) or whether they will be aggregated. Ammonia, lime, and sources of CN^- and HS^- are commonly used as regulating agents.

Collectors are surface-active additives that adsorb onto the mineral surface and prepare the surface for attachment to an air bubble so that it will float to the surface. Therefore collectors must adsorb selectively if flotation is to result in any fractionation of the crude ore. In addition, the adsorbed collector must impart a hydrophobic character to the particle surface so that an air bubble will attach to the mineral or vice versa.

Amphipathic substances such as we have discussed throughout this chapter are used as collectors. Alkyl compounds with C_8 to C_{18} chains are widely used with carboxylate, sulfate, or amine polar heads. For sulfide minerals, sulfur-containing compounds such as mercaptans, monothiocarbonates, and dithiophosphates are used as collectors. The most important collectors for sulfides are xanthates, the general formula for which is

$$R-O-\overset{\overset{\displaystyle S}{\|}}{C}\diagdown_{S-CH_3}$$

In the collectors used, R is generally in the C_2 to C_6 range. Xanthates are readily oxidized to dixanthogens, and the extent of this reaction may have a big effect on the efficiency of the collector.

The fundamental role of the collector is to produce a solid surface that is sufficiently hydrophobic so that it will attach to an air bubble when the pulp is aerated. Figure 7.19 may also be used to represent this situation, except that for flotation the shaded region is an air bubble. Since contact angles are measured in the liquid phase, the contact angle in the flotation case will be θ_2. For good bubble adhesion contact angles greater than 90° are preferred. Unlike the parallel situation in detergency, the adhesion of the bubble rather than its detachment is required for the success of the process.

Once again, we may use Young's equation to decide what adsorption situation is most conducive to values of $\theta_2 > 90°$

$$\gamma_{WA}\cos\theta_2 = \gamma_{AS} - \gamma_{SW} \tag{81}$$

From this, we see the optimum condition corresponds to $\gamma_{SW} \gg \gamma_{AS}$, or extensive adsorption at the air-solid surface and minimum adsorption at the solid-water interface. The hydrophobic nature of the collectors and their chemical affinity for specific solids promote this situation.

The formation of a large bubble that facilitates flotation requires a large area of attachment or, more specifically, a large perimeter of attachment since the three-phase contact boundary occurs along the perimeter. Increasing this perimeter is favored by a positive value of the spreading coefficient (Equation (6.61)). In the notation of this problem (air spreading on solid in water), the spreading coefficient equals

$$S_{A/S} = \gamma_{SW} - (\gamma_{AW} + \gamma_{SA}) \tag{82}$$

The collector lowers γ_{SA}, an effect favorable to a positive spreading coefficient.

Finally, the frothing agents are intended to stabilize the mineral-laden foam at the surface of the aeration tank until it can be scooped off. Alkyl or aryl alcohols in the C_5 to C_{12} range are typical frothers. We have already seen how the long-chain members of this series form monolayers at the air-water surface. This lowers γ_{AW}, which is beneficial to the stability of the foam and also favors a large contact angle (if $\theta_2 > 90°$), positive spreading, and large bubbles for flotation. Neither the collector nor the frother is adsorbed exclusively at the solid-air or the water-air surface where their respective effects would be greatest. To a certain extent these two classes of additives compete with each other for adsorption sites; therefore conditions

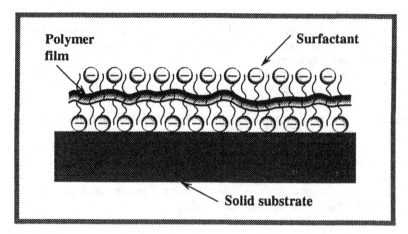

FIG. 7.20 An admicelle (a bilayer adsorbed on a solid substrate) as a two-dimensional solvent for a polymerization reaction. (Redrawn with permission of J. Wu, J. H. Harwell, and E. A. O'Rear, *J. Phys. Chem.*, **91**, 623 (1987).)

under which each produces the maximum effect are difficult to achieve, so compromise conditions in which the net effect is optimized are sought.

Another aspect of the frothers used is the fact that they form fairly condensed and therefore relatively viscous slow-draining films. In addition to thermodynamic considerations, then, kinetic factors are also important in stabilizing the froth.

A number of additional applications of the ideas of this chapter could be profitably considered if space permitted. Included among these are adhesives, lubricants, waterproofing, and the recovery of oil from the pores of rocks. Like detergency and flotation, these topics involve a variety of surface and colloid phenomena. The interested reader will find an introduction to these fields in some of the references listed at the end of this chapter, especially Adamson (1990), Davies and Rideal (1961), and Osipow (1962).

7.10b   Surfactant Films as Two-Dimensional Solvents and Reactors

Surfactants adsorbed on solid substrates can also be used as "two-dimensional solvents" or "reactors" (see Fig. 7.20). The objective is to form thin polymer layers on a solid surface for potential use in devices that require ultrathin polymer films with specific properties (e.g., ultrathin photoresists and waveguides for integrated optical systems). The basic idea is to form a layer of surfactants on a suitably treated solid surface so that the hydrocarbon tails of the surfactants project out from the surface. An additional layer of surfactants can then be deposited with the hydrocarbon chains of the second layer in contact with the exposed tails from the first layer (see Fig. 7.20). The resulting film, a bilayer adsorbed on a solid substrate, is sometimes called an *admicelle* (*ad*sorbed *micelle*). This bilayer serves as a two-dimensional solvent for the chosen monomer. The monomer solution is exposed to the surfactant film and, when the appropriate amount of monomers is dissolved in the film, the polymerization reaction is initiated by chemical, thermal, or photochemical methods. The surfactant film serves to localize the polymerization reaction and allows a thin polymer film to form on the solid surface. In principle, such a procedure can be used for fabricating single and multilayer polymer coatings for use in a wide variety of applications such as corrosion protection, solid lubricants, conducting or semiconducting films, and films for controlled release of drugs, in addition to the ones mentioned above.

7.10c   Langmuir-Blodgett Films

As mentioned above (Vignette VII and Section 7.6d), Langmuir-Blodgett (LB) films have received considerable attention in recent years as potentially capable of providing a number of applications. Excellent reviews of such applications are available in the compendium on LB films by Roberts (1990). The projected applications include the use of LB films in molecular electronics, passive thin-film applications such as electron-beam microlithography and lubrication, piezoelectric films, optical devices (e.g., waveguides), semiconductor devices (e.g., metal-insulator-semiconductor [MIS] diodes), and chemical and biological sensors. Some of these applications are speculative at this time, whereas others are closer to reality. Detailed discussions of any of these will require concepts that are outside the area of colloid science, and we restrict ourselves to one brief example (Ball 1994), which will illustrate how the structures and properties of LB films are taken advantage of in devising novel applications. Specialized monographs such as MacRitchie (1990), Roberts (1990), Ulman (1991), and Tredgold (1994) contain other examples.

For instance, optical memories or switches may be fabricated using LB films such as the one illustrated in Figure 7.21. The LB film in this case is made of molecules consisting of 7,7,8,8-tetracyano-p-quinodimethane (TCNQ, for short) at one end and a long hydrocarbon chain on the other. Each hydrocarbon chain contains an azobenzene unit, which can switch between two isomeric forms when exposed to visible and ultraviolet light. The stacking of the TCNQ units is affected by the isomeric state of the azobenzene unit, and this in turn changes the conductivity of the film. One can thus use photoisomerization to change the film from conducting to semiconducting, and it might be possible to use such an LB film in electronic devices such as MIS sandwich structures.

7.11 ADSORPTION IN THE PRESENCE OF AN APPLIED POTENTIAL

We conclude this chapter with a discussion of adsorption at the interface between mercury and a solution (usually aqueous) under the influence of an applied potential. The γ value for this interface is easily measured, and potential and electrolyte concentration can be studied as variables. The Nernst equation provides a familiar reminder that potentials can be dealt with by thermodynamics.

Moving a charge of q coulombs to a surface at which the potential is ψ volts involves a quantity of work (in joules)

$$\delta w = q\psi \tag{83}$$

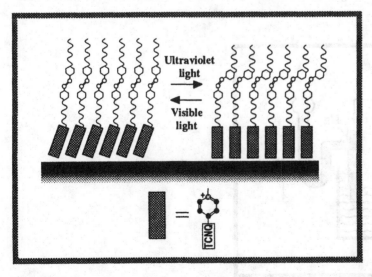

FIG. 7.21 An optical switching device based on Langmuir-Blodgett films. (Redrawn with permission of P. Ball 1994.)

Since this is non-pressure-volume work, it can be identified with a change in Gibbs free energy. In addition, the charge q carried by n moles of ions having a relative charge of $\pm z$ is given by $q = \pm nN_A ze$, where e is the proton charge, $1.60 \cdot 10^{-19}$ C. The product $N_A e$ is called the Faraday constant $\mathfrak{F}$ and equals 96,480 C. In view of these ideas, we can write

$$\Delta G = \pm nN_A ze\psi = \pm zn\mathfrak{F}\psi \qquad (84)$$

where the sign depends on whether the charge is moved with or against the potential.

7.11a Electrocapillarity

Figure 7.22 illustrates an apparatus by which the relationship between interfacial tension, applied potential, and electrolyte concentration can be investigated using what is known as the *electrocapillary effect*. Since mercury has a contact angle (measured in the mercury) that is greater than 90°, the mercury-solution interface is depressed (again in reference to the mercury). If an etched mark is placed on the capillary, it is possible to bring the mercury-solution meniscus to that mark by varying the height of the mercury reservoir. It turns out that this is quite sensitive to the potential E between the electrodes. The height of the capillary depression (a negative capillary "rise") can be readily converted to the interfacial tension through Equation (6.4). We see, therefore, that γ for the Hg-solution interface depends on the electrical potential across the system. In addition, the detailed shape of the so-called electrocapillary curve—a plot of γ versus E—depends on the concentration and nature of the electrolyte present. Figures 7.23a and 7.23b show examples of typical electrocapillary curves.

Several generalizations are evident from an inspection of Figure 7.23:

1. The general shape of the curves is roughly parabolic. The coordinates of the maximum in the electrocapillary curve depend on the electrolyte content of the system. Since γ decreases on both sides of the electrocapillary maximum and since reductions in γ are associated with adsorption, we conclude that adsorption increases as we move in either direction from the maximum; that is, the electrocapillary maximum seems to be a point of minimum adsorption.

2. The left-hand branch of the electrocapillary curves (also called the rising or ascending branch) is sensitive to the chemical nature of the anion present. In Figure 7.23a, for example, different potassium salts are used as the electrolytes. Although the ascending branches of these curves differ, the descending branches (to the right of the maximum) lie on a common curve. Figure 7.23b shows that the right-hand branch of the curves is sensitive to the nature (and

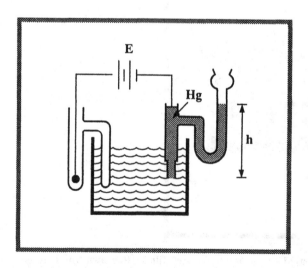

FIG. 7.22 Schematic illustration of an apparatus to measure the electrocapillary effect.

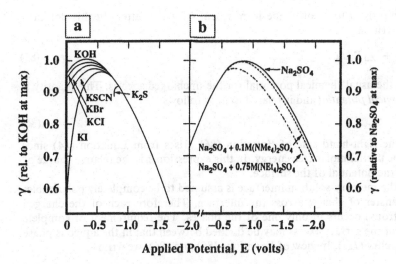

FIG. 7.23 Typical electrocapillary curves: (a) anions are adsorbed; and (b) cations are adsorbed. (Redrawn with permission of N. K. Adam, *The Physics and Chemistry of Surfaces*, Dover, New York, 1968.)

concentration) of the cation. These observations, coupled with the above remarks about adsorption, suggest that the anion is preferentially adsorbed on the ascending branch of the curve, the cation is preferentially adsorbed on the descending branch, and that neither ion is preferentially adsorbed at the maximum, although both may be adsorbed in equivalent amounts.

3. The charge carriers in the system — ions in the solution and electrons in the mercury — come to an equilibrium distribution that is consistent with the applied potential. To the left of the maximum, the mercury surface is a positively charged anode (this branch is also called the anodic branch), toward which anions are attracted. Electrons in the mercury are repelled from the surface by the anions in the water, so the potential on the mercury must become progressively more positive (relative to the maximum) for increased anion adsorption to occur. To the right of the maximum, the mercury is negative (the cathodic branch), attracts cations, and must become progressively more negative (relative to the maximum) for increased cation adsorption.

4. The potential at which the maximum occurs is different for different ions. In light of the third item, the differences in the voltage coordinates of the maxima must reflect differences in the chemical (as opposed to purely electrostatic) affinities of the ions for the interface. Nonionic solutes have also been investigated extensively, but we do not go into this aspect of the subject.

Electrocapillary phenomena have been studied for a long time; the apparatus shown in Figure 7.22 is essentially that used by G. Lippmann in 1875 in his comprehensive studies of electrocapillarity. We do not examine either the experimental or the theoretical aspects of this system in great detail; however, an interpretation of the results that is more quantitative than that just outlined qualitatively is possible with relatively little additional effort.

7.11b   The Lippmann Equation

As a starting point, it is convenient to return to the kind of argument that leads to the Gibbs-Dühem equation (Equation (39)) for one-phase systems and to the Gibbs adsorption equation (Equation (44)) for systems containing an interface. In the present context, we are interested not only in the interface, but also in possible charge effects at that interface. Accordingly, it is convenient to distinguish between charged and uncharged components in the system; we use the subscript i to identify the former and j for the latter. In this notation, Equation (44) may be written

$$-d\gamma = \sum_i \Gamma_i d\bar{\mu}_i + \sum_j \Gamma_j d\mu_j \tag{85}$$

In this expression μ_j is the usual chemical potential for the uncharged species. The quantity $\bar{\mu}_i$ is called the *electrochemical potential* and is related to μ_j as follows:

$$\bar{\mu}_i = \mu_i + z_i e \psi \tag{86}$$

The second term on the right-hand side of Equation (86) arises from Equation (84) since chemical potential is the partial molar free energy. In this expression z_i is the relative charge of the ith species and ψ is the potential of the surface.

In this discussion the mercury-solution interface is assumed to be completely polarizable; that is, there is no transfer of charge across the interface. Therefore each of the charged species (including electrons) occurs in only one of the phases. The requirement of complete polarizability means that the surface excess may be divided between that in the aqueous phase (W) and that in the mercury (Hg). In view of Equations (85) and (86), we write

$$
\begin{aligned}
&-d\gamma \\
&= \sum_i (\Gamma_i d\mu_i)_W + \sum_i (z_i \Gamma_i e \psi)_W + \sum_i (\Gamma_i d\mu_i)_{Hg} + \sum_i (z_i \Gamma_i e \psi)_{Hg} + \sum_j \Gamma_j d\mu_j \quad (87)
\end{aligned}
$$

Since the system as a whole is electrically neutral, the charge density on each of the phases must be equal and opposite; that is,

$$\sum_i (z_i e \Gamma_i)_W = -\sum_i (z_i e \Gamma_i)_{Hg} \tag{88}$$

Substituting Equation (88) into Equation (87) yields

$$-d\gamma = \sum_i (\Gamma_i d\mu_i)_W + \sum_i (\Gamma_i d\mu_i)_{Hg} + \sum_j \Gamma_j d\mu_j + \sum_i (z_i \Gamma_i e)_W (\psi_W - \psi_{Hg}) \tag{89}$$

The difference in potential between the two phases varies as dE, the externally applied potential difference. Therefore Equation (89) may also be written

$$-d\gamma = \sum_i (\Gamma_i d\mu_i)_W + \sum_i (\Gamma_i d\mu_i)_{Hg} + \sum_j \Gamma_j d\mu_j + \sum_i (z_i \Gamma_i e)_W dE \tag{90}$$

For a completely polarizable electrode, the concentration of each of the components in both solutions is constant and therefore so is the chemical potential for each. Therefore we obtain

$$-\left(\frac{\partial \gamma}{\partial E}\right)_\mu = \sum_i (z_i e \Gamma_i)_W \tag{91}$$

This result is known as the *Lippmann equation*. The charge density on the right-hand side of this equation refers exclusively to the solution phase; therefore the subscript is no longer retained.

For monovalent ions, Equation (91) is simply

$$-\left(\frac{\partial \gamma}{\partial E}\right)_\mu = e(\Gamma_+ - \Gamma_-) \tag{92}$$

7.11c   Use of Electrocapillary Curves

When the surface excess of anions exceeds that of cations, $d\gamma/dE$ is positive, as is observed in Figure 7.23a. Negative values of $d\gamma/dE$ correspond to larger surface excesses of cations, as shown by Figure 7.23b. Finally, the condition $d\gamma/dE = 0$ corresponds to equal amounts of positive and negative adsorbed charge, that is, surface neutrality. Note that this is not the same as saying no ions are adsorbed. The slope of the electrocapillary curve measures the difference between the cation and anion surface excesses. This is explored further in Example 7.6.

* * *

EXAMPLE 7.6 *Interpretation of Adsorption Using Electrocapillary Curves.* In electrocapillary curves like those shown in Figure 7.23a, the ascending branches cross for NaCl and $Ca(NO_3)_2$ solutions, with the steeper NaCl curve cutting across the flatter $Ca(NO_3)_2$ curve. Use this observation to criticize or defend the following proposition: Both solutes must show equal adsorption at the crossover point since the lowering of γ depends on the surface excess. Since both show the same value for γ, both must have the same surface excess.

Solution: According to the Lippmann equation, it is not γ per se, but rather $d\gamma/dE$ that is pertinent to these data. Even though the two interfaces show the same value of γ at the point in question, they do not have the same slopes. In fact, $(d\gamma/dE)_{NaCl} > (d\gamma/dE)_{Ca(NO_3)_2}$, which means, according to Equation (91),

$$-e[(+1)\Gamma_{Na^+} + (-1)\Gamma_{Cl^-}] > -e[(+2)\Gamma_{Ca^{2+}} + (-1)\Gamma_{NO_3^-}]$$

or

$$(\Gamma_{Cl^-} - \Gamma_{Na^+}) > (\Gamma_{NO_3^-} - 2\Gamma_{Ca^{2+}})$$

This does not mean that $\Gamma_{Cl^-} > \Gamma_{NO_3^-}$, but rather that the surface excess of the anion compared to the cation is greater for NaCl than for $Ca(NO_3)_2$. ■

* * *

Another interpretation of the electrocapillary curve is easily obtained from Equation (89). We wish to investigate the effect of changes in the concentration of the aqueous phase on the interfacial tension at constant applied potential. Several assumptions are made at this point to simplify the desired result. More comprehensive treatments of this subject may be consulted for additional details (e.g., Overbeek 1952). We assume that (a) the aqueous phase contains only 1 : 1 electrolyte, (b) the solution is sufficiently dilute to neglect activity coefficients, (c) the composition of the metallic phase (and therefore μ_i^{Hg}) is constant, (d) only the potential drop at the mercury-solution interface is affected by the composition of the solution, and (e) the Gibbs dividing surface can be located in such a way as to make the surface excess equal to zero for all uncharged components ($\Gamma_j = 0$). With these assumptions, Equation (89) becomes

$$-d\gamma = RT (\Gamma_+ + \Gamma_-)_W d\ln c + e(\Gamma_+ - \Gamma_-)_W dE \qquad (93)$$

where c is the concentration of the electrolyte. If we specify constant applied potential, Equation (93) becomes

$$-\left(\frac{\partial \gamma}{\partial \ln c}\right)_E = RT (\Gamma_+ + \Gamma_-)_W \qquad (94)$$

This result shows that the vertical displacements (at fixed potential) of the electrocapillary curve with changes in electrolyte concentration measure the sum of the surface excesses at the solution surface. Curves such as those in Figure 7.23b may be interpreted by this result. We have already seen that $\Gamma_+ = \Gamma_-$ at the electrocapillary maximum (where $E = E_{max}$); therefore

$$-\left(\frac{\partial \gamma}{\partial \ln c}\right)_{E_{max}} = 2 RT\Gamma_{+,W} \qquad (95)$$

A final result that can be extracted from the Lippmann equation is readily obtained by differentiating Equation (91) with respect to E at constant μ:

$$-\left(\frac{\partial^2 \gamma}{\partial E^2}\right)_\mu = \left(\frac{\partial (e\Sigma_i z_i \Gamma_i)}{\partial E}\right)_\mu = C \qquad (96)$$

The quantity $(e\Sigma_i z_i \Gamma_i)$ gives the charge density of the surface, and d(charge density)$/dE$ is called the *differential capacitance C* of an interface. Although experimental values of surface capacitance provide valuable information about the distribution of charge at an interface, we shall not pursue this topic. Our interest in Equation (96) is primarily in the suggestion it offers that the distribution of charge at an interface can be modeled as a capacitor. When we discuss the ion atmosphere at a charged surface in Chapter 11, we begin by using the parallel plate capacitor as a preliminary model.

REVIEW QUESTIONS

1. Provide a simple mechanistic description of adsorption.
2. What is your conceptual picture of an interface between two liquids or a liquid and a gas? For example, in the case of a gas-liquid interface, can you imagine an infinitesimally thin surface that separates the gas from the liquid? If not, how would you define a surface that serves as the interface between the gas and the liquid?
3. What is a *surface-active agent* (or a *surfactant*)? Why is it often referred to as an *amphipathic* molecule?
4. What are *Langmuir* and *Gibbs* layers? What is a *Langmuir-Blodgett* film? How do they differ from each other?
5. Describe the formation and the possible structures of a surfactant monolayer at an air-water interface.
6. What is the meaning of "two-dimensional" phases?
7. What types of two-dimensional phases can occur in a monolayer when the surface concentration of the surfactant and the temperature are changed individually?
8. Describe a technique to visualize the structures that form in a Langmuir film.
9. Describe the *Langmuir film balance* and how it is used to measure surface tension.
10. What is the meaning of *surface pressure*? Describe one or two models of pressure-area isotherms for a monolayer.
11. List a few applications of monolayers. List a few applications in which the formation of monolayer at a fluid-fluid interface is used in industry.
12. What is a *Gibbs surface*? What are the definition and meaning of *surface excess* properties?
13. Explain why the air-water interfacial tension decreases with adsorption of a surfactant at the interface.
14. What is the *Gibbs equation*, and how is it derived?
15. Sketch the surface tension of water as a function of solute concentration in the bulk for different types of solutes such as an ionizable surfactant, a nonionic surfactant, and an inorganic electrolyte. Why is there a sharp change in such a plot in the case of ionic surfactants?
16. What is the *Langmuir adsorption isotherm*, and when is it applicable? Describe its limitations.
17. What is a *Freundlich isotherm*?
18. List a few (traditional as well as modern) applications of adsorbed layers of surfactants.
19. What is the *electrocapillary effect*? How is it used to study adsorption of solutes under an applied potential?

REFERENCES

General References (with Annotations)

Adamson, A. W., *Physical Chemistry of Surfaces*, 5th ed., Wiley, New York, 1990. (Graduate level. A more extended and somewhat more advanced treatment of adsorption at liquid interfaces in Chapter 4 and adsorption at solid-liquid interfaces in Chapter 11.)

Ball, P., *Designing the Molecular World: Chemistry at the Frontier*, Princeton University Press, Princeton, NJ, 1994. (Undergraduate level. A *Scientific-American*-style discussion of the role of chemistry in the molecular design of materials. Chapter 7 has a discussion of Langmuir and Langmuir-Blodgett layers. Recommended as an introduction.)

Davies, J. T., and Rideal, E. K., *Interfacial Phenomena*, Academic Press, New York, 1961. (Undergraduate level. A standard reference on interfacial phenomena. Somewhat outdated, but has good discussions on many of the conventional applications of adsorption of surfactants at interfaces, such as detergency, lubrication, etc. Also contains some interesting photographs of the ripples on a pond before and after the application of hexadecanol to the surface.)

Gaines, G. L., *Insoluble Monolayers at Liquid-Gas Interface*, Wiley, New York, 1966. (Undergraduate level. A classic on the subject.)

MacRitchie, F., *Chemistry at Interfaces*, Academic Press, San Diego, CA, 1990. (Graduate and undergraduate levels. This book is devoted almost entirely to the contents of the present chapter and to surface tension and contact angle phenomena discussed in Chapter 6. A number of applications of monolayers and multilayers, including Langmuir-Blodgett films, are summarized in Chapter 11. Chapter 10 is devoted to the implications of adsorption and monolayer formation in biology and medicine.)

Tanford, C., *Ben Franklin Stilled the Waves*, Duke University Press, Durham, NC, 1989. (A nontechnical treatment. A historical perspective of Langmuir layers is intertwined nicely with a readable, popular introduction to surfactant films.)

Other References

Adam, N. K., *The Physics and Chemistry of Surfaces*, Dover, New York, 1968.

Bull, H. B., *J. Am. Chem. Soc.*, **67**, 4 (1945).

Kipling, J. J., *Adsorption from Solutions of Nonelectrolytes*, Academic, New York, 1965.

Klassen, V. I., and Mokrousov, V. A., *An Introduction to the Theory of Flotation*, Butterworths, London, 1963.

Knobler, C. M., *Adv. Chem. Phys.*, **77**, 397 (1990a).

Knobler, C. M., *Science*, **249**, 870 (1990b).

LaMer, V. K. (Ed.), *Retardation of Evaporation by Monolayers*, Academic Press, New York, 1962.

Mingotaud, A.-F., Mingotaud, C., and Patterson, L. K., *Handbook of Monolayers*, Vols. 1 and 2, Academic Press, San Diego, CA, 1993.

Osipow, L. I., *Surface Chemistry*, Van Nostrand-Reinhold, Princeton, NJ, 1962.

Overbeek, J. Th. G. In *Colloid Science*, Vol. 1 (H. R. Kruyt, Ed.), Elsevier, Amsterdam, Netherlands, 1952.

Pockels, A., *Nature*, **43**, 437 (1891).

Roberts, G. (Ed.), *Langmuir-Blodgett Films*, Plenum Press, New York, 1990.

Tredgold, R. H., *Order in Thin Organic Films*, Cambridge University Press, Cambridge, England, 1994.

Ulman, A., *Ultrathin Organic Films*, Academic Press, Boston, MA, 1991.

PROBLEMS

1. By analogy with the behavior of gases, monolayers are expected to expand to cover an entire surface. When the underlying liquid is flowing, the motion supplies a natural barrier to spreading at the edge of the monolayer. Use this concept to interpret the following bucolic scene:*

   On a calm day an observer who finds the right place on a stream or a river will see an unobtrusive yet startling phenomenon, a line on the surface of the water. The line may lie still, or it may contort itself, one way and another, in response to eddies. Very likely he will think a spider thread has fallen onto the water and try to cut it with his canoe paddle. As the disturbance caused by the cutting fades, the line reappears, mended and whole.

   Use the concepts of this chapter to discuss the existence of the bulge or line at the edge of the monolayer.

2. In some general chemistry laboratory courses the following experiment is done to evaluate Avogadro's number. A watch glass is filled to the brim with water, and then a solution of stearic acid in benzene is slowly deposited dropwise on the surface until such

---

*McCutchen, C. W., *Science,* **170,** 61 (1970).

time that a drop is added that "will not spread out, but will instead form a thick, lens-shaped layer."* What is the name of the film pressure at the "end point" of this experiment? What would be a reasonable estimate for σ at this pressure? Estimate the number of 0.005-ml drops of stearic acid solution ($c = 0.200$ g liter^{-1}) needed to form a monolayer on a watch glass with a 14-cm diameter. Outline how data such as these could be interpreted to lead to a value for N_A. Discuss some of the sources of error in this experiment.

3. Isotherms of π versus σ at both 15 and 25°C were studied for oleic acid on a substrate of pH 2.0 using a high-sensitivity film balance.† The following results were obtained:

σ (Å^2 molecule^{-1})	π (dyne cm^{-1})	
	At 15°C	At 25°C
10,000	0.028	0.037
8,000	0.035	0.040
6,000	0.044	0.052
4,000	0.055	0.071
3,000	0.058	0.086
2,000	0.076	0.095
1,000	0.075	0.095
500	0.074	0.095

Prepare a plot of π versus σ from these data. What is the apparent significance of the discontinuity in the curves? What quantity can be evaluated from the temperature variation of this discontinuity? Estimate this quantity from the available data.

4. The accompanying data give σ values corresponding to different film pressures for monolayers of various lecithins spread on 0.1 M NaCl at 22°C:‡

π (dyne cm^{-1})	σ (Å^2 molecule^{-1})				
	Dibehenoyl	Distearoyl	Dipalmitoyl	Dimyristoyl	Dicapryl
2	51.7	53.3	96.7	96.7	99.2
4	50.0	52.5	88.3	90.0	93.8
6	49.2	50.8	82.2	85.0	86.7
8	48.3	50.0	76.7	81.7	82.5
10	47.9	49.5	66.7	77.5	78.3
15	46.7	48.0	53.3	70.8	71.7
20	45.5	46.7	50.0	65.8	65.8
30	45.0	45.0	46.3	58.3	58.3
40	44.7	44.7	45.0	53.8	53.8

Plot π versus σ for these isotherms and label the apparent two-dimensional phase present for various parts of the curves. Write the structural formulas for each of the lecithins and discuss the features of the curves in terms of the structure of the molecules.

5. If gas densities can be measured as a function of pressure, the molecular weight of a gas may be calculated from the expression

---

*Ifft, J. B., and Roberts, J. L., Jr., *Frantz/Malm's Essentials of Chemistry in the Laboratory*, 3d ed., W. H. Freeman, San Francisco, CA, 1975.
†Pagano, R. E., and Gershfeld, N. L., *J. Colloid Interface Sci.*, **41**, 311 (1972).
‡Phillips, M. C., and Chapman, D., *Biochim. Biophys. Acta*, **163**, 301 (1968).

$$M = RT\left(\frac{d}{p}\right)_{lim\ p\to 0}$$

Likewise, if surface concentrations (in weight per area) are measured as a function of π, the molecular weight of a solute that forms an insoluble monolayer may be determined. Romeo and Rosano* obtained the following data for a monolayer of acetyl lipopolysaccharide on 0.2 M NaCl at 20° C:

c (mg m^{-2})	0.06	0.09	0.11	0.14	0.17	0.23
π (millidyne cm^{-1})	10.3	16.4	20.4	25.9	34.3	50.0

What is the molecular weight of the acetyl lipopolysaccharide?

6. Cockbain† measured the interfacial tension of the water–decane surface at various concentrations of sodium dodecyl sulfate (NaDS). The experiments were done at 20°C both in the presence and absence of NaCl. Use the suitable form of the Gibbs equation in each case to calculate Γ_R^1 and σ at γ values of 10 and 20 dyne cm^{-1} from the following data:

Pure H$_2$O		0.1 M NaCl (swamping)	
c_{NaDS} (mole liter^{-1})	γ (dyne cm^{-1})	c_{NaDS} (mole liter^{-1})	γ (dyne cm^{-1})
0.0079	8.5	0.0014	5.2
0.00694	10.8	0.000694	11.7
0.00521	15.3	0.000347	17.4
0.00347	20.8	0.000173	22.7
0.001735	28.3	0.0000867	27.5

Is the variation of σ with interfacial film pressure qualitatively consistent with the expected behavior of monolayers in general? Of charged monolayers in particular?

7. Blank‡ has reported the permeability (in cm^3 of gas s^{-1} cm^{-2} surface) of various spread monolayers to water vapor at 25°C. For several different RX-type compounds at different π values, the permeabilities are as follows:

R	X	π (dyne cm^{-1})	Permeability $\times$ 10^3 (cm^3 s^{-1} cm^{-2})
C_{16}	OH	44	380
C_{18}	OH	44	300
C_{17}	COOH	24	430

Discuss the observed differences in permeability between (a) the two alcohols at the same film pressure and (b) the two 18-carbon surfactants at different pressures. In your comments include comparisons of the molecular structure of the surfactants and the efficiencies of these monolayers in retarding evaporation.

8. The pendant drop method has been used§ to measure the interfacial tension at the surface between mercury and cyclohexane solutions of stearic acid at 30 and 50°C.

---

*Romeo, D., and Rosano, H. L., J. Colloid Interface Sci., 33, 84 (1970).

†Cockbain, E. G., Trans. Faraday Soc., 50, 874 (1954).

‡Blank, M. In Retardation of Evaporation by Monolayers (V. K. LaMer, Ed.), Academic Press, New York, 1962.

§Ambwani, D. S., Jao, R. A., and Fort, T., Jr., J. Colloid Interface Sci., 42, 8 (1973).

Interpret the accompanying data by means of the Gibbs equation to evaluate Γ^1 and σ for stearic acid at these two temperatures when the equilibrium bulk concentrations are 10^{-3}, 10^{-4}, and 10^{-5} M:

$T = 30°C$		$T = 50°C$	
γ (dyne cm^{-1})	c_{Eq} (mole liter^{-1})	γ (dyne cm^{-1})	c_{Eq} (mole liter^{-1})
362	4.8×10^{-6}	364	4.8×10^{-6}
355	8.5×10^{-6}	362	9.6×10^{-6}
334	6.6×10^{-5}	354	7.4×10^{-5}
307	2.7×10^{-4}	334	4.4×10^{-4}
286	1.0×10^{-3}	314	1.6×10^{-3}

Estimate the concentrations at which the stearic acid film at the interface reaches a condensed packing at the two temperatures.

9. A scintillation counter is used to measure tritium β particles adjacent to the surfaces of tritiated sodium dodecyl sulfate in 0.115 M aqueous NaCl solution and tritiated dodecanol in dodecanol. The former system is surface active and the latter is not, so the difference between the measured radioactivity above the two indicates the surface excess of sodium dodecyl sulfate. The number of counts per minute arising from the surface excess A_s is related to the surface excess in moles per square centimeter Γ^1 by the relationship $A_s = 4.7 \times 10^{12} \Gamma^1$.* Use the following data (25°C) to construct the adsorption isotherm for sodium dodecyl sulfate on 0.115 M NaCl:

	Activity $\times 10^{-3}$ (cpm)		Surfactant concentration $\times 10^3$
^{3}H μC (g solution)$^{-1}$	Tritiated nonsurfactant	Tritiated surfactant	(mole kg^{-1})
2	–	1.9	0.17
4	–	2.1	0.34
6	–	2.3	0.50
8	0.50	2.5	0.67
10	–	2.6	0.84
15	0.95	2.9	1.20
20	–	3.2	1.65
30	1.85	3.8	2.45

Briefly outline how the proportionality constant between A_s and Γ^1 might be determined experimentally.

10. A quantity called the HLB (for hydrophile–lipophile balance) number has proved to be a useful way to match a surfactant to a particular application. For example, surfactants with HLB numbers in the range 4 to 6 produce water-in-oil emulsions; those in the range 7 to 9 are useful as wetting agents; those ranging between 8 and 18 produce oil-in-water emulsions; and those with values in the range 13 to 15 make good detergents. Use these considerations plus the following specific examples to formulate a generalization about the dependence of the HLB number on the molecular structure of the surfactant:†

---

*Muramatsu, M., Tajima, K., Iwahashi, M., and Nukina, K., *J. Colloid Interface Sci.*, **43**, 499 (1973).

†Osipow, L. I., *Surface Chemistry*, Van Nostrand-Reinhold, Princeton, NJ, 1962.

Surfactant	HLB number
Sodium dodecyl sulfate	40
Sodium oleate	18
Tween 80 ($n = 20$, $R =$ oleate)	15
Sorbitan monolaurate	8.6
Span 60	4.7
Sorbitan tristearate	2.1

11. The adsorption of straight-chain fatty acids from n-heptane on Fe_2O_3 has been studied.* In all cases studied the adsorption isotherms conform with the Langmuir equation. The following are values of the amount adsorbed at saturation:

Fatty acid	$\left(\dfrac{n_2^s}{w}\right)_{sat} \times 10^5$ (mole g^{-1})
Acetic	3.00
Propionic	2.36
n-Butyric	2.11
n-Hexanoic	1.78
n-Heptanoic	1.40
n-Octanoic	1.30
Lauric	1.04
Myristic	0.97
Palmitic	0.91
Magaric	0.82
Stearic	0.81

Calculate the area per molecule (σ) of each on the saturated surface if the Fe_2O_3 is known to have a specific surface of 3.45 m^2 g^{-1}. Do the adsorbed molecules appear to be in the same two-dimensional phase in each of these systems?

12. The adsorption of various aliphatic alcohols from benzene solutions onto silicic acid surfaces has been studied.† The experimental isotherms have an appearance consistent with the Langmuir isotherm. Both the initial slopes of an n/w versus c plot and the saturation value of n/w decrease in the order methanol > ethanol > propanol > butanol. Discuss this order in terms of the molecular structure of the alcohols and the physical significance of the initial slope and the saturation intercept. Which of these two quantities would you expect to be most sensitive to the structure of the adsorbed alcohol molecules? Explain.

13. The Michaelis–Menton equation is an important biochemical rate law. It relates the rate of the reaction v to a substrate concentration $[S]$ in terms of two constants v_{max} and K_M:

$$v = \frac{v_{max}[S]}{K_M + [S]}$$

It will be noted that this equation follows the same functional form as the Langmuir equation, specifically $v \to v_{max}$ as $[S]$ increases. The biochemical literature contains three different graphical procedures to evaluate the constants v_{max} and K_M from kinetic data:

---

*Allen, T., and Patel, R. M., *J. Colloid Interface Sci.*, **35**, 647 (1971).
†Hoffman, R. L., McConnell, D. G., List, G. R., Evans, C. D., *Science*, **157**, 550 (1967).

Name of method	Plotted on ordinate	Plotted on abscissa
Lineweaver–Burk	$1/v$	$1/[S]$
Hanes	$[S]/v$	$[S]$
Eadie	v	$v/[S]$

Describe how the three equivalent variations of the Langmuir equation would be plotted. Give the interpretation of the slope and intercept in each case.

14. In laboratory tests of flotation, fluorite (CaF_2) particles (range of particle diameters, 0.074–0.147 mm) with oleic acid as collector were aerated under three different conditions. These conditions and the percent CaF_2 recovery after 10 min are as follows:[*]

Aeration conditions	Percent recovery after 10 min of aeration
Bubbles precipitated from solution	5
Bubbles produced by mechanical means	20
Combination of both means of bubble production	70

Suggest an interpretation for this variation in the efficiency of fluorite recovery.

15. Grahame[†] gives the following data for γ_{max} versus log c (corrected for activity) for the interface between mercury and aqueous KI at 18°C:

γ_{max} (dyne cm^{-1})	390	398	407	414	419	422
log c (c in mole liter^{-1})	0.5	0	−0.5	−1.0	−1.5	−2.0

Use these results to estimate Γ, the surface excess of KI at the electrocapillary maximum, for 1.0 and 0.1 M KI. Express your results as moles of KI adsorbed per square centimeter and as total charge q_T per square meter.

16. Show by the double integration of Equation (96), that is $-(\partial^2\gamma/\partial E^2)_\mu = C$, that a parabolic relationship between $\gamma_{max} - \gamma$ and $E - E_{max}$ is expected if the capacitance of the double layer is constant. Use the equation you derive to test the assumed constancy of C for the interface between mercury and 1 M NaCl from the following data:

γ (dyne cm^{-1})	340	376	396	410	418	423^a	421
E (V)	−0.02	−0.08	−0.18	−0.28	−0.38	−0.56^a	−0.68
γ (dyne cm^{-1})	415	405	396	384	373	358	340
E(V)	−0.78	−0.88	−0.98	−1.08	−1.18	−1.28	−1.38

^{a}Maximum.

Double-layer capacitance values are usually expressed as microfarads per square centimeter; remember that practical electrical units, including the farad, are consistent with SI units. Comment on these results in terms of anion and cation adsorption.

---

*Klassen, V. I., and Mokrousov, V. A., *An Introduction to the Theory of Flotation*, Butterworth, London, England, 1963.
†Grahame, D. C., *Chem. Rev.*, **41**, 441 (1947).

8

Colloidal Structures in Surfactant Solutions
Association Colloids

I for my part have never known an Irregular who was not also what Nature evidently intended him to be— ... up to the limits of his power, a perpetrator of all manner of mischief.

From Abbott's *Flatland*

8.1 INTRODUCTION

8.1a What Is Self-assembly and What Are Association Colloids?

In the last chapter we examined the tendency of surfactant molecules in aqueous solutions to adsorb at a surface in the form of a monolayer. In this chapter we continue to study surfactant solutions, this time considering a few of the many possible modes of organization they can adopt within a *bulk* phase. This process of organization is thermodynamically driven and is spontaneous, as in the case of Langmuir layers we discussed in Chapter 7. It is therefore often called *self-assembly*, and the resulting aggregates are known as *association colloids*. Both chapters share an interest in amphipathic solutes: Chapter 7 focuses on surface activity, while this chapter is concerned with structures in the colloidal size range formed by these molecules. In both, the head-to-head/tail-to-tail ordering of the amphipathic molecules is observed. The difference is that in Chapter 7 the ordering occurs at the surface, under the influence of the surface, while it occurs in the bulk in this chapter. Quite an assortment of bulk surfactant structures is known, including ordinary and reverse micelles, liquid crystals, bilayers, vesicles, and microemulsions. This chapter is concerned primarily with micelles (Sections 8.2–8.8) and microemulsions (Sections 8.9 and 8.10), although some additional structures are mentioned in passing.

The colloidal structures we examine in this chapter are formed as a result of physical interactions among amphipathic molecules, rather than by covalent bonding. This sort of physical association has been recognized for a long time, although contemporary students may be relatively (or totally!) unaware of it. It is interesting to note that in the early days of polymer chemistry, macromolecules were believed to be physically associated rather than covalently bonded structures. The birth of modern polymer chemistry can be traced to the acceptance of the covalent character of these substances. Associated structures do exist, however, and we see by the end of this chapter that their investigation is a very lively area of chemical research.

8.1b Why Are Association Colloids Important?

In the last couple of decades one important insight has triggered a tremendous upsurge of interest in surfactant structures. This is the recognition that these structures may mimic biological structures in some ways. Enzymes, for example, are protein molecules into which a reactant molecule somehow "fits" to form a reactive intermediate. The highly efficient and specific catalytic effect of enzymes makes their investigation one of the most fruitful areas of biochemical research. Likewise, cell membranes not only compartmentalize biological systems, but also play a variety of functions in the life of the cell. Surfactant structures can be used as model systems to mimic both enzymes and membranes. A whole new field of mimetic (exhibiting mimicry) chemistry has grown around this concept, and colloidal structures formed by surfactants are at the center of the entire subject. Surfactant aggregates known as liposomes are common in physiological systems, and the use of specially designed liposomes as drug-delivery vehicles was discussed in Chapter 1 in Vignette I.3. Self-assembled structures such as micelles and reverse micelles also play an increasingly important role in separation processes in engineering and environmental science and technology; one such example is described in Vignette VIII.

**VIGNETTE VIII   SEPARATION PROCESSES BASED ON
MICELLAR ENCAPSULATION:
Micellar-Enhanced Ultrafiltration**

Removal of dissolved organics and polyvalent cations from contaminated groundwater, industrial wastewater, and oil-field-produced water in a cost-effective manner is one of the most important engineering challenges of this decade. Hazardous waste sites contain groundwaters polluted by leaking underground organic storage tanks, and the leaching of the hazardous materials and accidental waste releases contribute further to the water pollution problems. The use of water in chemical industries in separation operations is common, and the water thus used picks up at least small amounts of dissolved organics and other hazardous chemicals.

The current technologies popular for removing such wastes include adsorption on activated carbon, ion-exchange membranes, and bioremediation. Activated carbon adsorption is effective but expensive since regeneration of the carbon is often difficult and expensive. Ion-exchange membranes are restricted to specific ions. An alternative technology that is currently investigated relies on the ability of surfactants to self-assemble in water as *micelles* and trap hydrocarbon contaminants. We have already seen an example of encapsulation of chemical species by surfactant aggregates known as liposomes in Chapter 1 and have seen the use of surfactants in detergency in Chapter 7. Micellar encapsulation works on similar principles and may turn out to be simultaneously effective for both dissolved organics and polyvalent ions since an ionic micelle with an appropriate head group can adsorb ionic contaminants.

"Biologically friendly" ionic surfactants can be added to the wastewater at concentrations above the threshold value beyond which the surfactants self-assemble to form micelles. The resulting micelles can trap the hydrocarbon wastes since the hydrocarbon solutes prefer the hydrocarbon interior of the micelle over the aqueous environment outside. In addition, ionic wastes in the water adsorb to the polar heads of the surfactants (see Fig. 8.1). The resulting waste-laden micelles can then be removed more easily using ultrafiltration methods. Such a process, known as *micellar-enhanced ultrafiltration* (MEUF), can be made continuous, scalable, cost effective, and environmentally friendly (through the use of biodegradable surfactants).

Some of the topics we discuss in this chapter are essential for understanding processes such as MEUF. The same ideas can also be used for other separation processes (e.g., protein separation in reverse micelles) and in genetic engineering, as mentioned in Vignette I.3 in Chapter 1. We also see in this chapter other applications such as using micelles as "microreactors," i.e., using the unique environment inside micelles for catalysis and material synthesis.

8.1c   Focus of This Chapter

As mentioned above, the bulk of this chapter is concerned with the simplest of the surfactant aggregates that form in solution, namely, micelles. There is plenty to focus on in this respect, and the material is already very complex.

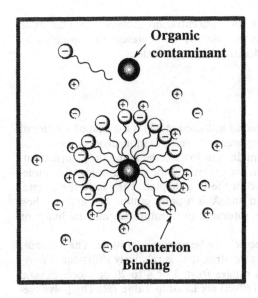

FIG. 8.1   Use of micelles to trap hydrocarbon and ionic wastes.

1.   In Section 8.2 we are concerned with the threshold concentration of surfactants at which micellization occurs. This concentration, known as the *critical micelle concentration* (CMC for short), is one of the most important properties of surfactant solutions. We look at two different ways of modeling micellization and discuss briefly when they are appropriate.

2.   In Sections 8.3 and 8.4, we examine the structure of the different parts of a micelle and illustrate how the shape of the micelle is determined by the relative sizes of the head group and the tail of the surfactant. This relative measure provides a geometric rationale for the structures of aggregates that are possible, and is simple to use.

3.   Following this, the thermodynamic arguments needed for determining CMC are discussed (Section 8.5). Here, we describe two approaches, namely, the *mass action model* (based on treating micellization as a "chemical reaction") and the *phase equilibrium model* (which treats micellization as a phase separation phenomenon). The entropy change due to micellization and the concept of *hydrophobic effect* are also described, along with the definition of thermodynamic standard states.

4.   The next two sections focus on solubilization and catalysis in micelles, two important applications of micelles.

5.   An overview of other forms of micellar systems follows in the next three sections. Formation of reverse micelles, in nonaqueous media, is discussed briefly in Section 8.8. Sections 8.9 and 8.10 present an introduction to microemulsions (oil, or water, droplets stabilized in water or oil, respectively) and their applications.

6.   Finally, a discussion of surfactant self-assembly will not be complete without a mention of surfactant assemblies in biological systems. Although they are outside the scope of our book, we have already drawn attention to such biological applications of colloid science in Chapters 1 and 7 and above in this chapter. Some additional discussion is provided in the last section of this chapter (Section 8.11).

8.2   SURFACTANTS IN SOLUTION: EXPERIMENTAL OBSERVATIONS AND MODELS

8.2a   Self-Assembly into Micelles: Representation as Reactions

Curve 3 in Figure 7.14 was presented as a typical illustration of the way surface tension γ varies with concentration for an amphipathic solute in aqueous solution. In discussing that figure, we noted that the break in the curve marks the concentration above which these solute molecules aggregate to form clusters known as *micelles*. If we represent the amphipathic species by S, then this clustering process can be described by the reaction

$$nS \leftrightarrow S_n$$

(A)

in which S_n is the micelle with a *degree of aggregation n*. Reaction (A) is the first of a series of reactions of increasing complexity we use to represent the micellization process. Note that Reaction (A) is reversible: Dilution shifts the equilibrium toward the monomeric surfactant. The surfactant species that undergo micellization are essentially the same amphipathic molecules with adsorption behavior that we discussed in the last chapter. They have the general formula RX, in which R is a hydrocarbon chain and X is a polar group. The hydrocarbon chains in R are ordinarily C_8 or greater, may be saturated or unsaturated, may be linear or branched, and may contain an aromatic ring.

The polar X group in the amphipathic molecule may be nonionic or ionic. The nonionic X group most widely used in the study of surfactant structures is relatively short-chain poly-oxyethylene moieties, $-(C_2H_4O)_x-$, in which x ranges from 3 or 4 to 20 or more. Several commercial surfactants of this sort (Tween and Triton) are listed in Table 7.2. The polyoxye-thylene "heads" of these molecules are essentially short-chain polymers themselves and, as such, are generally polydisperse. The x values that characterize these preparations are average values, and a distribution of chain lengths around the average is typical of commercial non-ionic surfactants.

Among numerous possible ionic groups, sulfate ($-SO_4^-$), sulfonate ($-SO_3^-$), and car-boxylate ($-CO_2^-$) are the most common anionic groups; various quaternary ammonium groups ($-NR_3^+$) are the most common X substituents in cationic surfactants. Zwitterionic surfactants, which combine both positively and negatively charged groups, have also been investigated. With ionic surfactants, the amphipathic ion is accompanied by a counterion. Although other possibilities exist, we consider only univalent counterions. In general, we write ionic surfactants as M^+S^- and assume these are 100% dissociated into M^+ and S^-. A common surfactant of this type is sodium dodecyl sulfate, often written SDS, in which $M^+ = Na^+$ and $S^- = C_{12}H_{25}SO_4^-$.

The clustering of low molecular weight anionic surfactant molecules to form micelles can also be represented by the following reaction:

$$nS^- + mM^+ \leftrightarrow (S_n^- M_m^+)^{z-}$$

(B)

in which n is the degree of aggregation and the net charge z of the micelle is given by $z = n - m$. Note that the charge of the micelle can also be expressed as the fraction ionized, $\alpha = (n - m)/n$. This formalism is readily extended to micelles formed from cationic surfactants; for nonionics, m and z are zero.

Reaction (B) is clearly an extension of Reaction (A), with the former admitting the possibility of counterion binding to the micelle. Additional refinements can be introduced into this reaction. The micelle still carries a net charge of $-z$, which means that zM^+ ions must be present in solution to assure electroneutrality. This may be included in the representation of micellization by writing

$$nS^- + mM^+ \leftrightarrow (S_n^- M_m^+)^{z-} + zM^+$$

(C)

Unless noted otherwise, we use anionic micelles as prototypes; that is, we use Reaction (B) to represent the formation of micelles.

8.2b Critical Micelle Concentration

The threshold concentration at which micellization begins is known as the *critical micelle concentration*. Surface tension is by no means the only property of the solution that displays a discontinuity in slope when plotted against increasing concentration. Figure 8.2 shows schematically how several different experimental quantities vary with concentration. The figure idealizes things somewhat since the various experimental quantities tend to "see" slightly different thresholds for micellization. However, the differences in CMC values by various methods of determination are not so diverse as to invalidate the notion of a "critical concentration." Below the breaks in the curves in the figure, anionic surfactants behave as expected for strong electrolytes; above the CMC, however, the behavior observed is consistent with the presence of particles in the colloidal size range. That this is true for the curves shown in Figure 8.2 is seen by the following arguments:

1. In the case of variation of osmotic pressure π with concentration c, to a first approximation we know that π/c is proportional to $1/M$ (Equation (3.34)). This is roughly equivalent to saying that the slope of a plot of π versus c is proportional to $1/M$, where M is the molecular weight. (Osmotic pressure measurements generally give the number-average molecular weight $\overline{M}_n$ as illustrated in Chapter 3, Section 3.3c. However, as also noted there, in the case of micellar solutions, osmometry can lead to *weight average* molecular weight $\overline{M}_w$; see Puvvada and Blankschtein 1989.) The decrease in the slope of the osmotic pressure plot at the CMC indicates an increase in the average molecular weight of the solute at this point.

2. In the case of turbidity τ, to a first approximation (Hc/τ) is proportional to $1/M$ (Equation (5.39)). This means that the slope of a plot of τ versus c is roughly proportional to M. The break in the curve again corresponds to an increase in the molecular weight of the solute.

3. The conductivity κ of the solution decreases at the CMC owing to the lower mobility of the larger micelles. Dividing by concentration to convert to equivalent conductivity Λ leads to a sharp reduction in the last quantity at the CMC.

These data can be used quantitatively as well as qualitatively to determine the molecular weight of the micelle. For example, the colloidal particles may be analyzed by light scattering by modifying Equation (5.39) as follows:

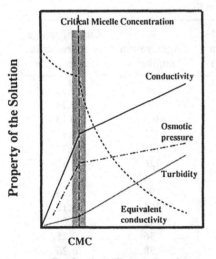

FIG. 8.2 Schematic illustration of variation of properties of a surfactant solution with surfactant concentration. Note the change in behavior at the critical micelle concentration (CMC). The shading emphasizes the fact that the CMC is not necessarily sharply defined.

$$\frac{H(c - c_{CMC})}{\tau - \tau_{CMC}} = \frac{1}{M} + 2B(c - c_{CMC}) \tag{1}$$

since the micelles form in a solution that already has a turbidity τ_{CMC} at the CMC. Turbidity data can then be plotted as in Figure 5.7 to give the molecular weight and the second virial coefficient of the micelles. The second virial coefficient can be analyzed further to give the net charge of the micelle if the amphipathic species are ionic as assumed. Most of the techniques we considered in Chapters 2–5, along with numerous other experimental procedures, have been used to characterize micelles. Of particular interest to us at this point are the average molecular weight $\overline{M}$ and the net charge z. Also of interest are the particle shape and hydration (available from sedimentation and diffusion measurements) and polydispersity (available by combining determinations of average $\overline{M}_n$ and $\overline{M}_w$). For nonionic micelles the degree of association is given by the ratios of the molecular weights of the micelle and of the individual surfactant molecules. For ionic micelles the measured molecular weight includes the bound counterions, but n can be determined if z is known.

Table 8.1 lists experimental CMC values for several ionic surfactants in water and aqueous NaCl solutions, as well as n and α values measured at the CMC. The data in Table 8.1, supplemented by numerous additional studies, allow us to make several generalizations about the state of aggregation and charge of micelles at the CMC:

1. All other things being equal, the aggregation number n increases as the length of the hydrocarbon chain increases.

2. The aggregation number for sodium dodecyl sulfate increases with increasing amounts of the indifferent electrolyte NaCl.

3. There is a general tendency for α to decrease owing to the addition of electrolyte. More extensive data show that α is generally in the range 0.1 to 0.4; a value of about 0.25 is a good choice for an "average" fraction of ionization.

TABLE 8.1 Critical Micelle Concentration, Degree of Aggregation, and Effective Fractional Ionization for Several Surfactants With and Without Added Salt

Surfactant	Solution	Critical micelle concentration (mole liter^{-1})	Aggregation number n	Ratio of charge to aggregation number, z/n
Sodium dodecyl	Water	0.00810	80	0.18
sulfate	0.02 M NaCl	0.00382	94	0.14
	0.03 M NaCl	0.00309	100	0.13
	0.10 M NaCl	0.00139	112	0.12
	0.20 M NaCl	0.00083	118	0.14
	0.40 M NaCl	0.00052	126	0.13
Dodecylamine	Water	0.01310	56	0.14
hydrochloride	0.0157 M NaCl	0.01040	93	0.13
	0.0237 M NaCl	0.00925	101	0.12
	0.0460 M NaCl	0.00723	142	0.09
Decyl trimethyl	Water	0.06800	36	0.25
ammonium bromide	0.013 M NaCl	0.06340	38	0.26
Dodecyl trimethyl	Water	0.01530	50	0.21
ammonium bromide	0.013 M NaCl	0.01070	56	0.17
Tetradecyl trimethyl	Water	0.00302	75	0.14
ammonium bromide	0.013 M NaCl	0.00180	96	0.13

Source: J. N. Phillips, *Trans. Faraday Soc.*, **51**, 561 (1955).

4. As a further refinement of the previous item, additional studies suggest that the order of ion binding to negative micelles is $Cs^+ > Rb^+ > Na^+ > Li^+$ and for positive micelles $I^- > Br^- > Cl^-$; that is, larger ions that are more polarizable and that tend to be less hydrated bind more effectively.

5. For nonionics, increasing the polyoxyethylene chain length for a constant R group decreases n.

6. Those factors that increase n tend to lower the CMC.

7. An assortment of additional evidence suggests that, at the CMC, many micelles are roughly spherical and relatively monodisperse.

The explanations for some of these may be seen from the relation between surfactant packing and the resulting shapes of the micelles described in Section 8.4. Here we merely note that the variables in terms of which micelles have been examined are by no means exhausted by this list. We encounter some additional experimental observations as we proceed. At this point, however, these results indicate that the tendency toward aggregation is increased by those variations that increase the hydrophobic character of the surfactant: increasing the hydrocarbon chain length, decreasing the polyoxyethylene chain length, or increasing counterion binding.

8.2c Micellization: Chemical Reaction or Phase Equilibrium?

The last point we consider in this section is the question of whether micellization should be viewed in terms of chemical reaction equilibrium or phase equilibrium. If we think of micellization as a chemical reaction, then Reaction (A) should surely be written as a sequence of stepwise additions:

$$2S \overset{s}{\leftrightarrow} S_2 \leftrightarrow S_3 \leftrightarrow S_4 \leftrightarrow \cdots \leftrightarrow S_n$$

(D)

The law of mass action implies that a continuous distribution of species — monomers, dimers, trimers, etc. — should be present. Increasing the surfactant concentration should shift the equilibria in Reaction (D) to the right, but the observed transition from monomers to n-mers with $n > 50$ is not expected. Experimentally, large micelles do form sharply at the CMC in aqueous solutions, and — to a good approximation — the concentration of monomeric surfactant in solution varies little above the CMC. *This type of behavior is typical of phase equilibria*; examples include the (constant) concentration of a saturated solution and the (constant) vapor pressure of a liquid. Energetically, some minimum value of n is apparently necessary before the exclusion of hydrophobic tails from the aqueous medium is effective. Once the solution is concentrated enough for aggregates with this critical n value to be formed, any additional surfactant added to the solution goes into the micelle.

We see in Section 8.8 that surfactants undergo aggregation in nonaqueous solvents also, but the degree of aggregation is very much less ($n < 10$), and the threshold for aggregation is far less sharp than in water. The mass action model for micellization seems preferable for nonaqueous systems.

In summary, whether a reaction equilibrium or a phase equilibrium approach is adopted depends on the size of the micelles formed. In aqueous systems the phase equilibrium model is generally used. In Section 8.5 we see that thermodynamic analyses based on either model merge as n increases. Since a degree of approximation is introduced by using the phase equilibrium model to describe micellization, micelles are sometimes called *pseudophases*.

An illustration of how both the reaction equilibrium and phase equilibrium models can be applied to micellization is provided by Example 8.1.

* * *

EXAMPLE 8.1 *Reaction Equilibrium and Phase Equilibrium Models of Micellization.* Research in which the CMC of an ionic surfactant M^+S^- is studied as a function of added salt, say M^+X^-, shows that a plot of log CMC versus log ($[MX]$ + CMC) yields a straight line of slope (α − 1). Derive this relationship by considering Reaction (B) as describing both a chemical equilibrium and a phase equilibrium.

Solution: As a chemical equilibrium, Reaction (B) is described by an equilibrium constant K, which may be written

$$K = [\text{micelle}]/[S^-]^n[M^+]^m$$

where the brackets signify molar concentrations of the indicated species, activity effects being neglected. Taking logarithms and rearranging gives

$$n \log [S^-] + m \log [M^+] = \log ([\text{micelle}]/K)$$

As a phase equilibrium, the concentration of surfactant in equilibrium with the micellar pseudophase equals the CMC regardless of the apparent concentration as long as micelles are present. Indicating the CMC value by a subscript, we write

$$\log [S^-]_{\text{CMC}} = (1/n) \log ([\text{micelle}]/K) - (m/n) \log [M^+]$$

At the CMC the total cation concentration equals the CMC value plus the concentration of the added 1 : 1 electrolyte: $[M^+] = [M^+]_{\text{salt}} + [M^+]_{\text{CMC}}$. Since $m/n = 1 - \alpha$, these results can be combined to give

$$\log [S^-]_{\text{CMC}} = (1/n) \log ([\text{micelle}]/K) - (1 - \alpha) \log \{[M^+]_{\text{salt}} + [M^+]_{\text{CMC}}\}$$

which is the desired result. The picture of micellar charge as developed here is in reasonable agreement with that determined by other methods.   ■

* * *

8.3   STRUCTURE OF MICELLES

In considering the structure of micelles, we continue to base our discussion on aqueous, anionic surfactant solutions as prototypes of amphipathic systems. Cationic micelles are structured no differently from anionics, and nonionics are described parenthetically at appropriate places in the discussion. We summarize present thinking about the structure of micelles at surfactant concentrations equal to or only slightly above the CMC. We see that in nonaqueous systems (Section 8.8) and in concentrated aqueous systems (Section 8.6), the surfactant molecules are organized quite differently from the structure we describe here.

Although McBain suggested over 80 years ago that soap molecules form micellar structures of lamellar and spherical shape (McBain 1913), most of the subsequent work focused on spherical micelles. The earliest concrete model for spherical micelles is attributed to Hartley (1936), whose picture of a liquidlike hydrocarbon core surrounded by a hydrophilic surface layer formed by the head groups, has been essentially verified by modern techniques, and the "Hartley model" still dominates our thinking. We present an overview of the structure of the micelle first and then go on to examine the details a little bit more closely.

8.3a   Internal Structure of Micelles: An Overview

We saw in the last section that at the CMC surfactant molecules cluster into roughly spherical aggregates containing 50–100 amphipathic units. For purposes of orientation, let us take an imaginary journey radially outward from the center of a micelle into its aqueous surroundings, identifying some distinctly different regions for subsequent discussion. This first inventory is incomplete; we add refinements as we proceed. Moving outward from the center of the micelle, we find the following:

1.   The central core is predominantly hydrocarbon. The expulsion of the hydrophobic tails of the surfactant molecules from the polar medium is an important driving force behind micellization. The amphipathic molecules aggregate with their hydrocarbon tails pointing together toward the center of the sphere and their polar heads in the water at its surface.

For the surfactant this mode of organization competes with monolayer adsorption, which it somewhat resembles.

2. In ionic micelles the hydrocarbon core is surrounded by a shell that more nearly resembles a concentrated electrolyte solution. This consists of ionic surfactant heads and bound counterions in a region called the *Stern layer* (see Chapter 11, Section 11.8). Water is also present in this region, both as free molecules and as water of hydration.

3. In typical nonionic micelles the shell surrounding the hydrocarbon core also resembles a concentrated aqueous solution, this time a solution of polyoxyethylene. The ether oxygens in these chains are heavily hydrated, and the chains are jumbled into coils to the extent that their length and hydration allow.

4. Beyond the Stern layer, the remaining z counterions exist in solution. These ions experience two kinds of force: an electrostatic attraction drawing them toward the micelle and thermal jostling, which tends to disperse them. The equilibrium resultant of these opposing forces is a diffuse ion atmosphere, the second half of a double layer of charge at the surface of the colloid. Chapter 11 provides a more detailed look at the diffuse part of the double layer.

As the first step in refining our ideas about the various regions in and around micelles, it is important to realize that in reality these domains are very different from the static, well-delineated regions described. The various steps represented by Reaction (D) take place very rapidly, so the micelle exists in a state of *dynamic equilibrium*. Furthermore, by simple rotations around carbon-carbon bonds in the hydrocarbon chains, the spatial extension of that chain varies, either pulling the head deeper into the core or allowing it to protrude into the Stern layer. This, coupled with the comings and goings of ions and water molecules, makes the surface of the micelle quite turbulent at the molecular level. These considerations alone argue that the regions cited above are not sharply defined on a molecular scale. We see presently that uncertainties as to the extent of water penetration into the core also blur the distinction between these regions.

8.3b Structure of Micelles: Some Additional Details

Now let us retrace our way, in reverse order, through the various regions of the micelle enumerated above.

8.3b.1 The Exterior of the Micelle

The diffuse part of the double layer is of little concern to us at this point. Chapters 11 and 12 explore in detail various models and phenomena associated with the ion atmosphere. At present it is sufficient for us to note that the extension in space of the ion atmosphere may be considerable, decreasing as the electrolyte content of the solution increases. As micelles approach one another in solution, the diffuse parts of their respective double layers make the first contact. This is the origin of part of the nonideality of the micellar dispersion and is reflected in the second virial coefficient B as measured by osmometry or light scattering. It is through this connection that z can be evaluated from experimental B values.

Ionic micelles will migrate in an electric field, and the ion atmosphere of the colloidal particle is dragged along with it. Interpretation of micellar mobility (conductivity experiments) must take this into account. The same is true, however, of the mobility of simple ions, but the situation is more involved here since the micelle and the ion atmosphere have comparable dimensions. We see in Chapter 12 how particle and double-layer dimensions affect the interpretation of mobility experiments.

In the second item above, the presence of bound and free water molecules was noted. Both bound ions and ionic surfactant groups are hydrated to about the same extent in the micelle as would be observed for the independent ions. The dehydration of these ionic species is an endothermic process, and this would contribute significantly to the ΔH of micellization if ion dehydration occurred. In the next section we discuss the thermodynamics of micellization, but it can be noted for now that there is no evidence of a dehydration contribution to the ΔH of micelle formation. The extent of micellar hydration can be estimated from viscosity data, assuming spherical particles. From this approach, sodium dodecyl sulfate micelles are estimated to be hydrated to about 39% by weight. For this system, 24 wt% hydration can be explained by assuming that sulfate groups and sodium ions are hydrated with 1 and 4 water molecules, respectively, at $\alpha = 0.3$. Although charge solvation accounts for a considerable part of the hydration of a micelle, these estimates show that an additional 15 wt% hydration remains unaccounted.

One important point to recognize about the Stern layer in ionic micelles is that the bound counterions help overcome the electrostatic repulsion between the charged heads of the surfactant molecules. For nonionics no such repulsion exists. It is incorrect to think that ionic micelles form and then adsorb counterions. The Stern layer is part of the micelle, and the energetics of its formation are part of the thermodynamics of micellization.

8.3b.2   The Core

Next let us return to the hydrocarbon core of the micelle. Of particular interest are the geometrical constraints imposed on the micellar structure by the length of the hydrocarbon tail and the location of the yet unexplained water of hydration. As a first approximation, the core of the micelle can be viewed as a drop of liquid hydrocarbon, the maximum radius of which equals the length of the fully extended hydrocarbon tail. To get an idea of the size of this region, let us consider a numerical example.

* * *

Example 8.2   *Calculating the Geometric Parameters of the Core of a Micelle.*   Calculate the radius, volume, and surface area of the core of a micelle formed by the aggregation of dodecyl groups. The fully extended chain has a zigzag structure with angles of 109.5°, and the carbon-carbon bond length is 0.154 nm. The van der Waals radius of the terminal methyl group equals 0.21 nm, and 0.06 nm may be taken as half the length of the bond to the polar head.

Solution: The radius of the spherical core equals the length of the fully extended hydrocarbon tail (see Example 8.3 below). The 12 carbon atoms are connected by 11 bonds, each of length 0.154 nm. What must be added together, however, are the projections of these bond lengths along the direction of the chain. The distance between every other carbon in the fully extended chain—the base of a triangle opposite the tetrahedral angle—is given by the law of cosines:

$$a = (b^2 + c^2 - 2bc \cos \theta)^{1/2} = [(2)(0.154)^2(1 - \cos 109.5°)]^{1/2}$$
$$= 0.252 \text{ nm}$$

Half of this is the projection of each bond along the chain length; the sum of these projections is $(11)(0.252/2) = 1.39$ nm. Adding the contributions of the two ends to this gives the radius of the sphere: $1.39 + 0.21 + 0.06 = 1.66$ nm. (Compare this with the length obtained from Equation (4) for a chain with 12 carbon atoms.)

The volume and surface area of the core are now readily calculated:

$$V = (4/3)\pi(1.66)^3 = 19.1 \text{ nm}^3, \quad \text{and}$$
$$A = 4\pi(1.66)^2 = 34.5 \text{ nm}^2 \quad\quad\quad \blacksquare$$

* * *

Dividing the surface area calculated in the example by typical aggregation numbers gives the area per head group at the surface of the core. Using n values of 50 and 100 gives 0.69 and 0.35 nm^2 per group, respectively, quite plausible numbers in view of the behavior of surfactants in monolayers.

The alkyl tails of surfactant molecules are not lines without thickness that can radiate outward from a common center in unlimited number. Instead, the chains themselves occupy a certain volume, represented by the bars in Figure 8.3. In this figure the circles represent head groups and the shaded region is water. Figure 8.3a shows schematically that close packing of the head groups requires unacceptable overlapping of the chain ends if the radius of the core is equal to the length of the fully extended chain. Figure 8.3b continues to use bars for the radius of the sphere, but fans them out in such a way as to avoid overlapping. The hole in the center is an artifact of this pictorial representation, but the wedges that allow water to pene-

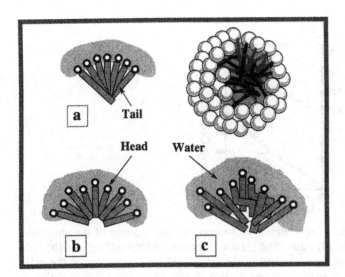

FIG. 8.3 A schematic representation of the structure of an aqueous micelle. Three possibilities are illustrated: (a) tails overlap at the center; (b) water penetrates core; and (c) chain protrusion and bending correct deficiencies of (a) and (b).

trate deeply into the core are a real feature of this model. The representation in Figure 8.3c differs from the preceding models as follows:

1. Some surfactant molecules protrude from the surface of the core farther than others, thereby alleviating crowding at the center of the micelle.
2. Chain flexibility is explicitly acknowledged, allowing some chains to twist and bend in such a way as to fill wedges that would otherwise contain water.
3. These modifications, taken separately or combined, overcome objections to the hypothetical structures shown in Figure 8.3a and 8.3b, which incidentally are whimsically known as *reef* and *fjord* structures, respectively.
4. Because of the protrusion of portions of the hydrocarbon chains into the Stern layer, the core acquires a rough surface. This model seems like a reasonable "snapshot" of what we have already described as a rapidly changing surface region.

One consequence of roughness at the surface of the micellar core is an increased contact between water and hydrocarbon. Figure 8.3b seems unrealistic because the water-hydrocarbon contact is scarcely less than in the bulk solution, a situation that apparently undermines an important part of the driving force for micellization. Figure 8.3c minimizes this effect without eliminating it. At the same time it allows for some water entrapment, which accounts for that part of the micellar hydration that was unexplained by the hydration of ions and charged groups.

8.3b.3   The Palisade Layer

The rough water-hydrocarbon surface of the core introduced in Figure 8.3c suggests that the core of the micelle should really be considered as two distinct regions: an inner core that is essentially water-free and a hydrated shell between the inner core and the polar heads. This partly aqueous shell is sometimes called the *palisade layer*. The extent to which the hydrocarbon chains protrude into the water is problematic, but we can get an idea of the volume of the palisade layer as follows.

Suppose a section of chain three methylenes long defines the thickness of the palisade layer. For a dodecyl chain this corresponds to the outer 3/12 of the radius, meaning that the radius of the inner, anhydrous core is only 3/4 of what we have been using. Cubing this fraction shows that $(3/4)^3 = 0.42$ is the fraction of the original core that is anhydrous, while

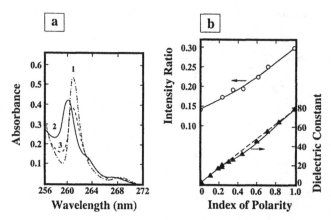

FIG. 8.4 Determination of the microenvironment of a molecule: (a) a portion of the ultraviolet spectrum of benzene in (1) heptane, (2) water, and (3) 0.4 M sodium dodecyl sulfate; and (b) ratio of the intensity of the solvent-induced peak to that of the major peak for benzene versus the index of solvent polarity. The relative dielectric constant is also shown versus the index of polarity. (Redrawn, with permission, from P. Mukerjee, J. R. Cardinal, and N. R. Desai, In *Micellization, Solubilization and Microemulsions*, Vols. 1 and 2 (K. L. Mittal, Ed.), Plenum, New York, 1976.)

58% is hydrated. Other fractions can be calculated by assuming other extents of protrusion. What is significant about this calculation is the fact that hydration of a relatively small fraction of the chain results in the presence of water in a significantly larger fraction of the volume of the micelle. After these idealized calculations, we should ask: Is there any experimental evidence for the presence of water in micelles?

We see below in this chapter that there is a great deal of interest in molecules that are "guests" in micelles. In the present context water molecules are the guests and their abundance and location have been and continue to be the basis for considerable research and controversy. We merely describe one set of pertinent experiments and their interpretation.

Figure 8.4a shows a portion of the ultraviolet spectrum of benzene in three media: (1) heptane, (2) water, and (3) 0.4 M aqueous sodium dodecyl sulfate. It is not the prominent peaks in these spectra that interest us, but rather the small bands located 3.6 nm on the long-wavelength side of the major features. This band is absent in benzene vapor, but is present with variable intensity in solutions. Accordingly, it is described as a solvent-induced band with an intensity that depends on the polarity of the solvent.

This intensity is shown in Figure 8.4b, in which the characteristics of spectra measured in different reference liquids and liquid mixtures are plotted. The abscissa in Figure 8.4b is an index of solvent polarity, specifically the molar concentration of $-OH$ groups in the reference liquids relative to the concentration of such groups in water, namely, 55.5 mole liter^{-1}. Thus abscissa values of 0.0 and 1.0 correspond to hydrocarbon and water, respectively, as solvents; intermediate values describe solvents of intermediate polarity.

The ordinate in Figure 8.4b gives the ratio of the intensity of the solvent-induced band relative to the intensity of the major peak at a 3.6-nm shorter wavelength. With this as a calibration curve, the behavior of the benzene in the sodium dodecyl sulfate micelles can be interpreted in terms of the micellar microenvironment of the benzene molecules. In this case the microenvironment is about 62% of the way between hydrocarbon and water.

Figure 8.4b also shows the relative dielectric constant of the reference liquids plotted against the same polarity index. The spectrum of the benzene suggests that the benzene molecules are located in a microenvironment of relative dielectric constant 46 (ϵ_r values for alkanes and water are about 2 and 78, respectively; $\epsilon_r = 43$ for glycerol).

This experiment introduces the use of a probe molecule to explore the microenvironment within a micelle. The results in this case show the environment to be quite "wet," but this observation alone does not tell us where either the benzene or the water is. Any one of the models in Figure 8.3 has regions of water-hydrocarbon contact. We examine some additional data in Section 8.6 that support the palisade layer as the location of benzene in these micelles.

8.4 MOLECULAR ARCHITECTURE OF SURFACTANTS, PACKING CONSIDERATIONS, AND SHAPES OF MICELLES

Although one expects the *details* of the molecular architecture of a surfactant molecule to play a prominent role in the shapes and structures of association colloids the surfactant forms in solution, a remarkably useful picture of the shapes of the resulting aggregates can be obtained using packing considerations based on some of the crude, general geometric features of the surfactant. This is somewhat analogous to the simple van der Waals picture of gas-liquid phase transition in atomic and molecular fluids in which one uses two parameters (one for the "size" or "excluded-volume" effects of repulsion and another for attraction) to deduce the general features of phase transition. In the case of surfactant solutions, one can predict with a reasonable accuracy the shapes of the association colloids resulting from self-assembly using three effective geometric parameters of the surfactant: (a) the optimal head group area a_h, (b) the volume v_t of the tail, and (c) the critical chain length $l_{c,t}$ of the tail (see Fig. 8.5).

In addition to its simplicity, such a geometric argument can also be used to predict the changes in the structure of the aggregates as variables such as pH, charge, electrolyte concentration, and chain length of the tail are varied. The importance of the relative effects of the head group area and the size of the tails was first emphasized by Tatar (1955), and the details were developed subsequently by Tanford (1980) and others (see Wennerström and

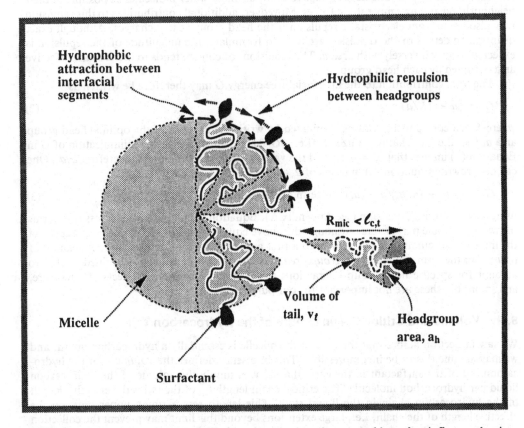

FIG. 8.5 Illustration of the geometric parameters for a surfactant and how they influence the size of the micelle. (Redrawn from Israelachvili 1991.)

Lindman 1979 and Chapter 17 of Israelachvili 1991). In what follows we restrict ourselves to the features essential for illustrating how packing considerations influence the resulting shapes of the surfactant aggregates and ignore the presence of water inside the Stern layer.

We first define the basic geometric parameters of the surfactants and follow this up with an illustration and a discussion of packing considerations that allow us to predict the shapes of the aggregates.

8.4a Optimal Head Group Area

The surface area a taken up by the head group of each surfactant molecule on the surface of a micelle depends on a number of factors, some of which tend to increase the area per head group while others tend to decrease it. This idea of two mutually opposing forces, suggested by Tanford (1980), refers to (a) an attractive force caused by the hydrophobic attraction of the hydrocarbon chain units at the hydrocarbon-water interface, and (b) a repulsive force between adjacent head groups arising from hydrophilic, steric, and ionic (in the case of charged head groups) repulsion. These opposing forces together determine the *optimal* area occupied by the head group, as illustrated in Figure 8.5.

The hydrophobic attraction is due to the preference of the hydrocarbon units near the surface to be close to its counterparts in the adjacent surfactant molecule. This attraction tends to decrease the effective area occupied by the head group. The attractive free energy contribution arising from this force is proportional to a and can be written as γa, where γ is about 20–50 mJ m^{-2}. The repulsive force mentioned in the previous paragraph is more complex and is not well understood presently. We know that this repulsion is caused by the tendency of the hydrophilic head groups to allow as many water molecules as possible in their neighborhood and by simple steric forces. Moreover, additional contributions to this repulsion may also come from electrostatic repulsion if the head groups carry charges. Although exact quantitative details of the repulsion are hard to formulate, the magnitude of the repulsion is expected to vary inversely with area a. The repulsion, of course, tends to increase the effective area occupied by each head group.

The total contribution to the interfacial free energy G may therefore be written as

$$G = \gamma a + (K/a) \tag{2}$$

where K is a constant that can be eliminated by writing it in terms of the optimal head group area a_h i.e., the area that minimizes G (i.e., $\partial G/\partial a = 0$ at $a = a_h$). The minimization of G in Equation (2) implies that $K = \gamma a_h^2$, and the value of G at the minimum is therefore $2\gamma a_h$. One can now rewrite Equation (2) in terms of a_h as

$$G = 2\gamma a_h + (\gamma/a_h)(a - a_h)^2 + \text{term of the order of } (a - a_h)^3 \tag{3}$$

using Taylor's series expansion (see Appendix A). Equation (3) allows us to express G in terms of two measurable parameters. Although the above equations are approximate, they represent the interactions among the surfactants to a first approximation and illustrate the meaning of *optimal* in the term *optimal head group area*. The above arguments can be refined further to account for specific head group interactions (e.g., ionic interactions), effects of curvature, and so on, but these are not important for our purpose here.

8.4b Volume and Critical Chain Length of the Hydrocarbon Tail

We saw in Section 8.3b.2 that the core of the micelle is essentially a hydrocarbon liquid, and we may assume this to be incompressible. This in essence defines the *volume* v_t of the hydrocarbon tail of the surfactant in the core; that is, v_t is simply the volume of the hydrocarbon liquid per hydrocarbon molecule. The critical chain length $\ell_{c,t}$ of the tail is the effective length of the hydrocarbon chain in the liquid state. This length sets a rough upper limit on the effective length of the chain, i.e., large extensions beyond this limit may prevent the collection of hydrocarbon chains from being considered a liquid. The chain length thus defined is a semiempirical parameter, although it is expected to be of the same order as the length of the fully extended hydrocarbon molecule ℓ_{max}. Following arguments similar to Example 8.2, Tanford (1980) has given the following expressions for $\ell_{c,t}$ (and v_t) for saturated hydrocarbon chains of n carbon atoms:

$$\ell_{c,t} \leq \ell_{max} = (0.154 + 0.1265\,n)\,\text{nm} \tag{4}$$

and

$$v_t = (27.4 + 26.9n) \cdot 10^{-3}\,\text{nm}^3 \tag{5}$$

Since both expressions are linear in n, for large values of n (in fact, even for n larger than 5), the ratio $(v_t/\ell_{c,t})$ approaches $0.21\,\text{nm}^2$, which defines the minimum cross-sectional area a hydrocarbon can have.

Once the estimates for a_h, v_t, and $\ell_{c,t}$ are available, one can determine the preferred shapes of the surfactant aggregates using geometric packing considerations, as discussed in the following subsection and illustrated in Example 8.3.

8.4c Packing Considerations and Shapes of Aggregates

The basic packing considerations that restrict the shape of an aggregate are rather straightforward and simply serve to reconcile the volume-to-surface-area ratio of an aggregate of any shape to the requirements imposed by the optimal head group area and the liquidlike structure of the core. The optimal head group area determines the number of surfactants that can be accommodated in an aggregate of any specified shape and, therefore, the volume of the corresponding hydrocarbon tails. The packing considerations demand that the shape and the size of the core of the aggregate for the volume of the tails thus obtained be such that the aggregate has a *liquidlike* hydrocarbon core. For instance, for a spherical micelle to have a liquidlike core, the ratio $(R_s/\ell_{c,t})$, where R_s is the radius of the micelle, has to be less than or equal to unity. For larger R_s, the chains will extend further or will have more space than needed for a liquidlike core. This implies that the volume-to-surface-area ratio of the micelle specified by the radius R_s must be consistent with the ratio of the volume of the chain of the surfactant in question to its optimal head group area; otherwise, aggregation into a spherical micelle is not possible for the given surfactant (see Example 8.3).

These considerations imply that a dimensionless group, known as the *packing parameter* $\mathcal{P}$ given by

$$\mathcal{P} = v_t/(a_h\ell_{c,t}) \tag{6}$$

can be defined and used as an indicator of the shapes one can expect for the aggregate. We illustrate this using the example below and consider subsequently what this implies for predicting and controlling the aggregate shapes using variables such as pH, electrolyte concentration, and the like.

* * *

Example 8.3 *Packing Parameter for Spherical Micelles.* Show that the packing parameter $\mathcal{P}$ of a surfactant has to be less than 1/3 for it to form spherical micelles.

Solution: For a spherical micelle of radius R_s and aggregation number n_a, one has

$$n_a = [(4/3)\pi R_s^3]/v_t = 4\pi R_s^2/a_h$$

This implies that

$$(v_t/a_h) = (R_s/3)$$

that is,

$$\mathcal{P} = v_t/(a_h\ell_{c,t}) = (R_s/\ell_{c,t})(1/3)$$

Since R_s has to be less than or equal to $\ell_{c,t}$ for the core of the micelle to be liquidlike, $\mathcal{P}$ has to be less than or equal to 1/3 for the micelle to be spherical. ∎

* * *

One can extend the above analysis to come up with the values of the packing parameter for which different shapes of aggregates are favored (Israelachvili 1991). These are summarized in Table 8.2. The results shown also serve as "rules of thumb" for what one can expect as one changes the chemical conditions of the solvent or the structure of the surfactant or for controlling the shape or aggregation number. For example,

1. One can change the optimal head group area by (a) decreasing or increasing the electrolyte concentration in the case of ionic surfactants or (b) changing the pH to effect an increase or decrease in the dissociation of the head group. Similarly, in the case of nonionic surfactants (such as polyoxyethylene surfactants), changing the degree of ethoxylation of the head group, or decreasing the temperature (which increases the extent of hydration), will increase the optimal head group area.
2. One can also change the ratio $(v_t/\ell_{c,t})$ by increasing the number of alkyl chains, introducing branching in the tails, using unsaturated hydrocarbon chains, and so on.

Some of the observations summarized at the end of Section 8.2b follow from the above. Although the above considerations provide some guidelines for predicting or controlling the aggregate structures based on the architecture of the surfactant molecules and the packing details, we must remember that the simplicity of the above development could be misleading. For example, the delineation points in the packing parameter do not indicate an abrupt change in the shape of the aggregates. For instance, a value of $\mathcal{P} = 0.35$ does not imply that the micelle is cylindrical. It simply means that the surfactants cannot pack themselves "neatly" into spheres; the shapes are simply expected to be slightly nonspherical. Further, we have not accounted for a number of "second-order" considerations such as the effect of curvature on the interfacial energy, effects of specific head group interactions such as ion bridging, and so on. These and other factors may be found in advanced or more specialized books such as those by Israelachvili (1991) and Clint (1992).

Now we turn our attention to thermodynamic considerations of micellization.

8.5 CRITICAL MICELLE CONCENTRATION AND THE THERMODYNAMICS OF MICELLIZATION

In this section we consider the thermodynamics of micellization from two points of view: the *law of mass action* and *phase equilibrium*. This will reveal the equivalency of the two approaches and the conditions under which this equivalence applies. In addition, we define the thermodynamic standard state, which must be understood if derived parameters are to be meaningful.

8.5a Mass Action Model

In the mass action approach we use Reaction (B) as a prototype for the process of micellization. The equilibrium constant for this reaction is given by

$$K = a_{mic}/a_S^n a_M^m \tag{7}$$

in which the a's are the activities of the indicated species. The well-known thermodynamic result $\Delta G^0 = -RT \ln K$ can be applied to Equation (7) to give ΔG^0 for Reaction (B), that is, the ΔG^0 value for micelle formation:

$$\Delta G^0 = -RT(\ln a_{mic} - n \ln a_S - m \ln a_M) \tag{8}$$

Dividing both sides of Equation (8) by n expresses this free energy change per mole of surfactant; we shall label this ΔG^0_{mic}. At the CMC, $a_M \approx a_S = a_{CMC}$, so per mole of surfactant, Equation (8) becomes

$$\Delta G^0_{mic} = RT\left[\left(1 + \frac{m}{n}\right) \ln a_{CMC} - \frac{1}{n} \ln a_{mic}\right] \tag{9}$$

TABLE 8.2 Packing Parameter and Its Relation to Shapes of Aggregates

Lipid	Critical packing parameter	Critical packing shape	Structures formed
Single-chained lipids (surfactants) with large head group areas: SDS in low salt	$<1/3$	Cone	Spherical micelles
Single-chained lipids with small head group areas: SDS and CTAB in high salt, nonionic lipids	$1/3$–$1/2$	Truncated cone	Cylindrical micelles
Double-chained lipids with large head group areas, fluid chains: phosphatidyl choline (lecithin), phosphatidyl serine, phosphatidyl glycerol, phosphatidyl inositol, phosphatidic acid, sphingomyelin, DGDG,[a] dihexadecyl phosphate, dialkyl dimethyl ammonium salts	$1/2$–1	Truncated cone	Flexible bilayers, vesicles
Double-chained lipids with small head group areas, anionic lipids in high salt, saturated frozen chains: phosphatidyl ethanolamine phosphatidyl serine $+$ Ca^{2+}	~ 1	Cylinder	Planar bilayers
Double-chained lipids with small head group areas, nonionic lipids, poly (cis) unsaturated chains, high T: unsaturated phosphatidyl ethanolamine, cardiolipin $+$ Ca^{2+} phosphatidic acid $+$ Ca^{2+} cholesterol, MGDG[b]	>1	Inverted truncated cone or wedge	Inverted micelles

Source: Adapted with permission from Israelachvili 1991.
[a]DGDG: digalactosyl diglyceride, diglucosyl diglyceride.
[b]MGDG: monogalactosyl diglyceride, monoglucosyl diglyceride.

Since n is large, the second term on the right is smaller than the first at the CMC and can be neglected. Introducing this approximation yields

$$\Delta G^0_{mic} \approx RT\left(1 + \frac{m}{n}\right)\ln a_{CMC} \tag{10}$$

Critical micelle concentrations occur in dilute solutions, so activity may be replaced by the concentration of the surfactant at the CMC:

$$\Delta G^0_{mic\cdot} \approx RT\left(1 + \frac{m}{n}\right)\ln c_{CMC} \tag{11}$$

Equation (11) can be used to evaluate ΔG^0_{mic} from readily available CMC values. Note that setting $m = 0$ for ionic micelles is equivalent to reverting from Reaction (B) to (A) for a description of micellization. The ΔG^0_{mic} value calculated in this case describes the contribution of the surfactant alone without including the contribution of counterion binding. Since $m = 0$ for nonionics, the surfactant contribution alone is useful when comparisons between ionic and nonionic micelles are desired.

8.5b   Defining the Standard State: The Phase Equilibrium Approach

It is apparent that CMC values can be expressed in a variety of different concentration units. The measured value of c_{CMC} and hence of ΔG^0_{mic} for a particular system depends on the units chosen, so some uniformity must be established. The issue is ultimately a question of defining the standard state to which the superscript on ΔG^0_{mic} refers. When mole fractions are used for concentrations, ΔG^0_{mic} directly measures the free energy difference per mole between surfactant molecules in micelles and in water. To see how this comes about, it is instructive to examine Reaction (A) – this focuses attention on the surfactant and ignores bound counterions – from the point of view of a phase equilibrium. The thermodynamic criterion for a phase equilibrium is that the chemical potential of the surfactant (subscript S) be the same in the micelle (superscript mic) and in water (superscript W): $\mu_S^{mic} = \mu_S^W$. In general, $\mu_i = \mu_i^0 + RT\ln a_i$, in which μ_i^0 is the standard state for the chemical potential. We write the activity as the product of the mole fraction and an appropriate *activity coefficient* f_i. We recognize that some of the surfactants are in the water and some are in the micelle and use the labels W and mic to differentiate between the two. If x is the mole fraction of surfactant, then

$$\mu_S^W = \mu_S^{0,W} + RT\ln x_W + RT\ln f_W \tag{12}$$

In this equation the standard state corresponds to the state that results from letting $f_W \to 1$ and $x_W \to 1$, in which case $\mu_S^W = \mu_S^{0,W}$. Letting $f_W \to 1$ is equivalent to saying that the surfactant behaves ideally, and letting $x_W \to 1$ is equivalent to having "pure" surfactant possessing the kind of interactions it has when surrounded by water. Physically, this corresponds to an infinitely dilute solution of surfactant in water. Using the primed symbol to represent the chemical potential of surfactant in micelles per mole of micelles, we write

$$(\mu_S^{mic})' = (\mu_S^{0,mic})' + RT\ln\left(\frac{x_{mic}}{n}\right) + RT\ln f_{mic} \tag{13}$$

The mole fraction of micelles is given by x_{mic}/n for particles of aggregation n. By the same logic as used above, $(\mu_S^{0,mic})'$ describes the surfactant in a standard state of pure micelle. We can write Equation (13) per surfactant molecule by dividing it by n:

$$\mu_S^{mic} = \mu_S^{0,mic} + \frac{RT}{n}\ln\left(\frac{x_{mic}}{n}\right) + \frac{RT}{n}\ln f_{mic} \tag{14}$$

The condition for phase equilibrium is given by equating Equations (12) and (14):

$$\mu_S^{0,mic} + \frac{RT}{n}\ln\left(\frac{x_{mic}}{n}\right) + \frac{RT}{n}\ln f_{mic} = \mu_S^{0,W} + RT\ln x_W + RT\ln f_W \tag{15}$$

This last result is seen to be equivalent to Equation (11) with $m = 0$ under the following conditions:

1. Both activity coefficients are set equal to unity, causing the $\ln f$ terms in Equation (15) to go to zero.
2. For large n,

$$\frac{RT}{n}\ln\left(\frac{x_{mic}}{n}\right) \ll \frac{RT}{n}\ln x_W$$

so the smaller term can be neglected.
3. Since the concentration of surfactant in water is essentially constant and roughly equal to the CMC value after micellization, $x_W = c_{CMC}$.
4. This shows that $(\mu_S^{0,mic} - \mu_S^{0,W})$ is identical to ΔG_{mic}^0 since it describes the difference per mole between the free energy of surfactant in a micelle and in water.

Although there are some aspects of micellization that we have not taken into account in this analysis—the fact that n actually has a distribution of values rather than a single value, for example—the above discussion shows that CMC values expressed as mole fractions provide an experimentally accessible way to determine the free energy change accompanying the aggregation of surfactant molecules in water. For computational purposes, remember Equation (3.24), which states that $x_2 \approx n_2/n_1$ for dilute solutions. This means that CMC values expressed in molarity units, [CMC], can be converted to mole fractions by dividing [CMC] by the molar concentration of the solvent, [solvent]: $x_2 \approx$ [CMC]/[solvent]; for water, [solvent] = 55.5 mole liter^{-1}.

The Gibbs-Helmholtz equation provides another familiar thermodynamic relationship that is useful in the present context:

$$\Delta H_{mic}^0 = \left(\frac{\partial(\Delta G^0/T)}{\partial(1/T)}\right)_p = -T^2\left(\frac{\partial(\Delta G^0/T)}{\partial T}\right)_p \tag{16}$$

Using Equation (11) as the expression for ΔG^0, Equation (16) becomes

$$\Delta H_{mic}^0 = R\left(1 + \frac{m}{n}\right)\left(\frac{\partial \ln c_{CMC}}{\partial(1/T)}\right)_p = -RT^2\left(1 + \frac{m}{n}\right)\left(\frac{\partial \ln c_{CMC}}{\partial T}\right)_p \tag{17}$$

The first version of Equation (17) suggests that a plot of $\ln c_{CMC}$ versus $1/T$ is a straight line of slope $\Delta H_{mic}^0/[R(1 + m/n)]$ if ΔH_{mic}^0 and m/n are independent of T. Backtracking to Equation (9) or (15) shows that n must also be constant with respect to temperature for this to be valid. In fact, n increases with temperature for polyoxyethylene nonionics; any temperature dependence of n is generally assumed to be absent in ionic systems. Even if the $\ln c_{CMC}$ versus $1/T$ plot is nonlinear, the second form of Equation (17) allows a value of ΔH_{mic}^0 to be evaluated from the slope of a tangent to a plot of $\ln c_{CMC}$ versus T at a particular temperature. Once ΔG_{mic}^0 and ΔH_{mic}^0 are known, the entropy of micellization is readily obtained from $\Delta G = \Delta H - T\Delta S$. Example 8.4 illustrates the use of these relationships.

* * *

Example 8.4 *Gibbs Energy and Entropy Changes Due to Micellization.* At 25°C the CMC for sodium dodecyl sulfate is $8.1 \cdot 10^{-3}$ M. An analysis of the temperature variation of the CMC according to Equation (17) (with $m = 0$) gives $\Delta H_{mic}^0 = 2.51$ kJ mole^{-1}. Evaluate ΔG_{mic}^0 and ΔS_{mic}^0 from these data. Omit counterion binding by taking $m = 0$.

Solution: Convert molarity to mole fraction concentration units:

$x = 8.1 \cdot 10^{-3}$ mole liter^{-1}/55.5 mole liter^{-1} = $1.46 \cdot 10^{-4}$

Then $\Delta G_{mic}^0 = (8.314)(298) \ln(1.46 \cdot 10^{-4}) = -21.9$ kJ mole^{-1};

$\Delta S_{mic}^0 = (\Delta H_{mic}^0 - \Delta G_{mic}^0)/T = 2.51 - (-21.9) = 81.9$ J K^{-1} mole^{-1}

Note that the quantity ($T\Delta S_{mic}^0$) represents about 90% of the value of ΔG_{mic}^0. ∎

* * *

Some values of ΔG_{mic}^0, ΔH_{mic}^0, and ΔS_{mic}^0 as determined by this method are listed in Table 8.3. Several observations can be made concerning these results:

1. By taking $m = 0$, we have focused on the aggregation of surfactant only, making it easier to compare ionic and nonionic micelles.
2. The ΔG_{mic}^0 values are negative, indicating spontaneous micellization.
3. The ΔH_{mic}^0 values are both positive and negative. Values of ΔH_{mic}^0 calculated by this method generally show poor agreement with those determined calorimetrically, at least for ionic surfactants.
4. The ΔS_{mic}^0 values are positive and make a far larger contribution to ΔG_{mic}^0 than ΔS_{mic}^0.

Despite controversies and uncertainties as to the best values for ΔH_{mic}^0, there is no question that the principal driving force behind the aggregation of surfactant molecules is a large, positive value of ΔS_{mic}^0. We return to this below.

The variation of ΔG_{mic}^0 with changing molecular parameters has been extensively investigated. It is an experimental fact that $\ln c_{CMC}$ varies linearly with the number of carbon atoms in the alkyl chain of the surfactant molecules, the CMC decreasing with increasing chain length. This variation is readily explained by breaking ΔG_{mic}^0 into contributions from the terminal methyl group (subscript CH$_3$), chain methylene groups (subscript CH$_2$), and the polar head group (subscript PH):

$$\Delta G_{mic}^0 = \Delta G_{CH_3} + \nu \Delta G_{CH_2} + \Delta G_{PH} \qquad (18)$$

In Equation (18), ν is one less than the number of carbons in the alkyl chain. If we assume that neither ΔG_{CH_3} nor ΔG_{CH_2} is affected by the length of the tail, then Equation (18) can be combined with Equation (11) to give

$$\ln c_{CMC} = \frac{\nu \Delta G_{CH_2}}{RT} + \text{constant} \qquad (19)$$

which is the equation of a straight line ($\ln c_{CMC}$ versus ν) and explains the observed behavior. The slope of such a plot for an assortment of n-alkyl ionic surfactants averages about -0.69, which corresponds to a ΔG_{CH_2} of -1.72 kJ mole^{-1} at 25°C. Since 0.693 equals $\ln 2$, it follows that the addition of a methylene group to a hydrocarbon chain decreases the CMC of these

TABLE 8.3 Some Thermodynamic Properties for the Micellization Process at or Near 25°C for Various Surfactants

Surfactant	ΔG_{mic}^0 (kJ mole^{-1})	ΔH_{mic}^0 (kJ mole^{-1})	ΔS_{mic}^0 (J K^{-1} mole^{-1})
Dodecyl pyridinium bromide	-21.0	-4.06	$+56.9$
Sodium dodecyl sulfate[a]	-21.9	$+2.51$	$+81.9$
N-Dodecyl-N,N-dimethyl glycine	-25.6	-5.86	$+64.9$
Polyoxyethylene(6) decanol	-27.3	$+15.1$	$+142.0$
N,N-Dimethyl dodecyl amine oxide	-25.4	$+7.11$	$+109.0$

Source: Data from J. H. Fendler and E. J. Fendler, *Catalysis in Micellar and Macromolecular Systems*, Academic Press, New York, 1975.
[a]Calculated in Example 8.4.

surfactants by almost exactly a factor of 2. In the presence of added indifferent electrolyte, $\Delta G_{CH_2} \approx -2.9$ kJ mole^{-1}, the aggregation being facilitated by the additional ions. With this type of data, it is possible to back-calculate the contribution to ΔG^0_{mic} of various polar head groups and, with these, to estimate the ΔG^0_{mic} and CMC values of other surfactants as the sum of group contributions. Extensive tabulations of these contributions are available.

8.5c Entropy Change During Micellization and the Hydrophobic Effect

We conclude this section with a brief discussion of the relatively large, *positive* values of ΔS^0_{mic}, which we have seen are primarily responsible for the spontaneous formation of micelles. At first glance it may be surprising that ΔS for Reaction (A) is positive; after all, the number of independent kinetic units decreases in this representation of the micellization process. Since such a decrease results in a negative ΔS value, it is apparent that Reaction (A) is incomplete as a description of micelle formation. What is not shown in Reaction (A) is the aqueous medium and what happens to the water as micelles form. The water must experience an increase in entropy to account for the observed positive values for ΔS^0_{mic}.

The key to understanding this entropy increase is the extensive hydrogen bonding that occurs in water. To a first approximation, the water molecule has a tetrahedral shape with the oxygen atom at the center, hydrogen atoms at two of the apex positions, and two lone pairs of electrons at the remaining apex positions. Hydrogen bonds form between the hydrogen atoms on one molecule and the oxygen lone pairs on another, effectively building a loose network of tetrahedra bound at the corners. Because of thermal fluctuations, various parts of this network continuously break and reform in liquid water, but at equilibrium a high average level of hydrogen bonding prevails. Many properties of water are due to this hydrogen bonding.

Next suppose a hydrocarbon moiety such as the tail of a surfactant is embedded in the water. Since water forms no hydrogen bonds with the alkyl group, the alkyl group merely occupies a hole in the liquid water structure. Our first thought might be that such cavities require the breaking of hydrogen bonds, and that their formation should involve a positive enthalpy contribution. As we have seen, a positive ΔS^0_{mic} is the dominant driving force behind surfactant clustering. We are therefore forced to conclude that water molecules at the surface of the cavity regenerate the hydrogen-bonded water network, as if the hydrocarbon chains were nucleation sites for network formation. As a result, water molecules become more ordered around the hydrocarbon with an attendant decrease in entropy. The standard state we have used means that the various Δ's measure the difference between surfactant in micelles and surfactant in water. Removing the surfactant from the water and placing it in a micellar environment allows the cavity to revert to the structure of pure liquid water. The highly organized cavity walls return to normal, hydrogen-bonded liquid with an increase in entropy. Incidentally, enhanced hydrogen bonding at the walls of the cavity largely compensates for the breaking of hydrogen bonds to form the cavity, so the enthalpy change is slight.

This explanation for the entropy-dominated association of surfactant molecules is called the *hydrophobic effect* or, less precisely, *hydrophobic bonding*. Note that relatively little is said of any direct "affinity" between the associating species. It is more accurate to say that they are expelled from the water and—as far as the water is concerned—the effect is primarily entropic. The same hydrophobic effect is responsible for the adsorption behavior of amphipathic molecules and plays an important role in stabilizing a variety of other structures formed by surfactants in aqueous solutions.

8.6 SOLUBILIZATION

Above the CMC, a number of solutes that would normally be insoluble or only slightly soluble in water dissolve extensively in surfactant solutions. The process is called *solubilization*, the substance dissolved is called the *solubilizate*, and—in this context—the surfactant is called the *solubilizer*. The result is a thermodynamically stable, isotropic solution in which the solubilizate is somehow taken up by micelles since the enhancement of solubility begins at the CMC. This observation, in fact, provides one method for determining the CMC of a surfactant; it must be used cautiously, however, since the solubilizate may change the CMC from what would be observed in its absence. There is an upper limit to the amount of material that can be solubilized in a given surfactant solution; beyond this limit the excess solubilizate displays normal phase separation.

8.6a   Location of the Solubilizate

There is considerable interest in establishing the location within a micelle of the solubilized component. As we have seen, the environment changes from polar water to nonpolar hydrocarbon as we move radially toward the center of a micelle. While the detailed structure of the various zones is disputed, there is no doubt that this gradient of polarity exists. Accordingly, any experimental property that is sensitive to the molecular environment can be used to monitor the whereabouts of the solubilizate in the micelle. Spectroscopic measurements are ideally suited for determining the microenvironment of solubilizate molecules. This is the same principle used in Section 8.3, in which the ultraviolet spectrum of solubilized benzene was used to explore the solvation of micelles. Here we take the hydration for granted and use similar methods to locate the solubilizate.

In addition to ultraviolet spectroscopy, resonance methods—such as electron spin resonance and nuclear magnetic resonance (NMR)—have been widely used to establish the site of solubilization. We briefly consider the use of proton NMR for such an investigation. Like the electron, the proton may have spin quantum numbers of $\pm 1/2$, resulting in two different nuclear quantum states. In a magnetic field these states differ in energy, the difference corresponding to the nuclear magnetic moment aligning with or against the magnetic field. If energy of the appropriate frequency is supplied, transitions from one state to the other are possible, and the radiation is absorbed. The absorbed frequency measures the separation of the two states by the familiar formula $\Delta E = h\nu$, where h is Planck's constant. The energy separation of the affected states depends on the strength of the applied magnetic field as modified by the local environment. Thus even protons in different parts of a molecule absorb at slightly different frequencies since the magnetic field each experiences is somewhat different from the applied field owing to the shielding effect of nearby electrons. Furthermore, neighboring protons cause a peak in an absorption spectrum to be split into a multiplet with characteristic relative intensities. This aids the identification of peaks in an NMR spectrum.

In addition to one part of a molecule influencing the NMR spectrum of another part, the medium in which the molecule is embedded also has an effect. Therefore the NMR spectrum of a solubilized molecule has the potential not only to reveal the location of a solubilized molecule in a micelle, but also to give information about its orientation.

Figure 8.6 shows some data measured for benzene solubilized in hexadecyl pyridinium chloride. The abscissa in the figure shows the extent of solubilization expressed as moles of solubilizate per mole of surfactant. The ordinate values show shifts in the resonance frequencies from water taken as an internal standard. Shifts of peaks arising from protons in the pyridinium ring, in benzene, and in methylene groups in the alkyl tail are shown versus the extent of solubilization.

The following points summarize these observations:

1. All peaks are shifted toward higher fields because of the diamagnetic effect of the solubilized benzene.
2. The slope of the shift versus extent of solubilization curve is much steeper for the protons on the benzene and pyridinium rings than for protons in methylenes in the tail. In fact, the first two increase in roughly parallel fashion.
3. Since the charged pyridinium ring must be at the surface, the benzene that parallels it in chemical shift must have a similar location.
4. The far less sensitive response of the chemical shift of the methylene protons to the extent of solubilization suggests that the core of the micelle is mainly composed of alkyl chains and is relatively unaffected by the benzene.

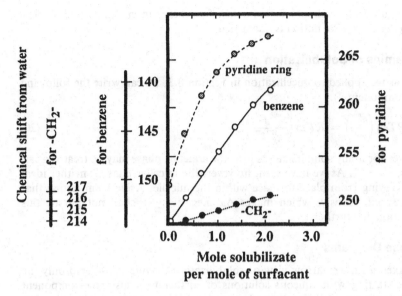

FIG. 8.6 Chemical shifts of protons in benzene, pyridium rings, and methylene groups versus the ratio of moles of solubilized benzene to moles of hexadecyl pyridinium chloride. (Redrawn, with permission, from J. C. Eriksson, *Acta Chem. Scand.*, **17**, 1478 (1963).)

In another related study (using hexadecyltrimethyl ammonium bromide micelles), isopropyl benzene was solubilized, and the chemical shifts of aromatic and alkyl protons were observed. The results suggest that the isopropyl benzene molecules are oriented such that the isopropyl groups are buried more deeply in the core of the micelle, while the benzene ring is in the more hydrated palisade layer. This plus the conclusion of Item 3 is consistent with the description presented in Section 8.3, which located the benzene in a relatively polar portion of the micelle.

8.6b Extent of Solubilization and Its Relation to Location of the Solubilizate

The extent of solubilization and the location of solubilizate in the micelle are related to one another. In the most hydrophobic inner core of the micelle, there simply is not as much room for the solubilizate compared to the hydrated palisade layer. In the case of polyoxyethylene nonionics, the core is surrounded by a mantle of aqueous hydrophilic chains, and solubilization may occur in both the core and the mantle. The relative amount of solubilization in these two regions of the nonionic micelle depends on the polarity of the solubilizate. Nonionics appear relatively more hydrophobic at higher temperatures, presumably owing to an equilibrium shift that favors dehydration of the ether oxygens. At some elevated temperature called the *cloud point*, solutions of these surfactants undergo phase separation. As the cloud point is approached, the solubilization of nonpolar solubilizates increases, probably because of an increase in the aggregation number of the micelles. For polar solubilizates, solubilization decreases owing to dehydration of the polyoxyethylene chains accompanied by their coiling more tightly. These observations demonstrate that nonpolar compounds are solubilized in the core of the micelle, while polar solubilizates are located in the mantle. Both of the temperature effects cited here are consistent with variations in the space available for the solubilized molecules in the micelle.

We saw in the last section how the removal of a surfactant molecule from water into a micellar environment has a negative ΔG° value. Qualitatively, it is not surprising that other organic solutes — the solubilizates — should decrease their free energy by entering micelles. Next we briefly consider how this general notion is quantified.

8.6c Thermodynamics of Solubilization

Using the same logic as we applied to micellization in Section 8.3, we can write the following for the solubilizate (component 3):

$$\mu_3^{0,mic} - \mu_3^{0,W} = RT\ln\left(\frac{a_3^{mic}}{a_3^{W}}\right) = RT\ln\left(\frac{x_3^{mic} f_3^{mic}}{x_3^{W} f_3^{W}}\right) \tag{20}$$

Since solubility in water for many solubilizates is low, the aqueous phase may be treated as an ideal solution, that is, $f_3^{W} = 1$. As we have seen, however, the micellar phase is neither ideal nor simple, with f_3^{mic} varying from place to place within the micelle. Aside from noting that solubilization occurs spontaneously, which makes $\Delta\mu^0$ negative, we shall not pursue this approach to solubilization any further.

8.6d Micellar Phase Diagrams

Until now we have taken a rather narrow view of solubilization, having considered only the uptake of solubilizate starting with aqueous solutions of surfactant. This three-component system can clearly take on a much wider range of proportions than we have discussed so far. A ternary phase diagram describes the full range of possibilities. For surfactant systems these diagrams can be quite complex, even though we shall limit ourselves to relatively simple cases. We begin with a brief review of the method for reading a ternary phase diagram:

1. The apex points of an equilateral triangle are identified with the three pure components, say, A, B, and C.

2. Each of the sides represents one of the possible binary combinations, say AB, BC, and CA. The fraction of the distance along the side of the triangular diagram measures the fractional composition of the binary mixture.

3. The perpendicular distance from any point within the triangle to one of the bases is proportional to the amount in the mixture of the component opposite the base in question. Since the sum of the three perpendiculars from any point to the sides equals the height of the triangle, expressing these lengths as fractions of the height permits any point in the triangle to be described as being some fraction toward A, some fraction toward B, and some fraction toward C. These same fractions can be used to describe either the weight fraction or mole fraction compositions of a ternary mixture.

4. Since ternary phase diagrams are drawn at constant temperature and pressure, the phase rule allows one, two, or three phases to be present. In two-phase regions, tie lines connect points having the compositions of the equilibrium phases. Three-phase regions have triangular shapes; the coordinates of the corners of the triangular zone give the compositions of the equilibrium phases.

Readers desiring a more detailed review of ternary phase diagrams will find the topic discussed in most physical chemistry textbooks or in the concise but comprehensive monograph on ternary equilibrium diagrams by West (1982).

Figure 8.7 shows the ternary phase diagram for water, hexanoic acid, and sodium dodecyl sulfate at 25°C. Seven different areas are shown in the figure, which has been used to describe the solubilization of polar dirt by surfactant solutions in detergency applications. The following comments refer to these seven different regions and explain the labeling used in Figure 8.7:

1. *Two liquids*: This is a two-phase region in which two liquid solutions — each containing three components — are in equilibrium. The two solutions could exist as distinct layers (e.g., in a separatory funnel) or as an emulsion in which droplets of one phase are dispersed in the second.

2. L_1 *and* L_2: Each of these phases is a homogeneous, isotropic liquid solution containing three components. The sort of system we have focused on in this section is the L_1 region in which the acid is solubilized in aqueous micelles. In L_2 the reverse is true: The acid is the

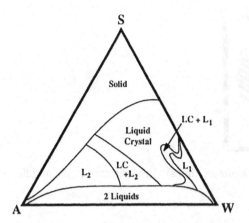

FIG. 8.7 Ternary phase diagram for water (W), hexanoic acid (A), sodium dodecyl sulfate (S) at 25°C. See text for a description of the various regions. (Redrawn, with permission, from A. S. C. Lawrence, *Chem. Ind.*, **44**, 1764 (1961).)

solvent, and water is solubilized in reverse micelles. We have more to say about the solubilization of water in reverse micelles in Section 8.8.

3. *Liquid crystal*: As the name implies, this is an ordered yet fluid phase in which water, surfactant, and solubilizate combine to form anisotropic, organized structures. These are called *lyotropic mesomorphic phases*, as opposed to *thermotropic mesomorphs*, which form when certain organic crystals are heated.

4. L_1 or L_2 plus liquid crystal: Each of these is a two-phase region in which the liquid crystalline phase exists in equilibrium with one of the isotropic liquid phases. Tie lines must be determined experimentally to give the precise compositions of the phases in equilibrium.

5. *Solid*: Eventually the system becomes saturated with surfactant and solid sodium dodecyl sulfate precipitates out.

The phase diagram shown in Figure 8.7 is simpler than many other ternary phase diagrams inasmuch as only one liquid crystal phase exists, and there are no three-phase triangles. By contrast, the water-decanol-sodium-caprylate phase diagram at 20°C shows 5 different liquid crystal phases and 10 triangular regions, in each of which 3 different sets of phases are in equilibrium.

8.6e Liquid-Crystalline Phases

Liquid crystals are a fascinating topic of study in their own right, but we limit our discussion to a brief description of the ordering in some of the possible structures. In all cases the amphipathic molecules are oriented in such a way as to minimize the contact between water and the alkyl chains. Whether the polar head points outward or not depends on which component dominates the continuous phase; the minor component is solubilized inside the micellar structures.

At relatively low concentrations of surfactant, the micelles are essentially the spherical structures we discussed above in this chapter. As the amount of surfactant and the extent of solubilization increase, these spheres become distorted into prolate or oblate ellipsoids and, eventually, into cylindrical rods or lamellar disks. Figure 8.8 schematically shows (a) spherical, (b) cylindrical, and (c) lamellar micelle structures. The structures shown in the three parts of the figure are called (a) the *viscous isotropic phase*, (b) the *middle phase*, and (c) the *neat phase*. Again, we emphasize that the orientation of the amphipathic molecules in these structures depends on the nature of the continuous and the solubilized components.

The "crystallinity" of liquid crystal phases refers to the large assortment of ways these micellar structures can be organized within a bulk phase. For example, spherical micelles of

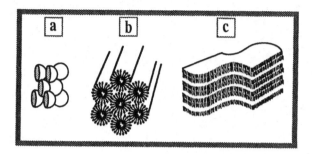

FIG. 8.8 Schematic representations of surfactant structures in (a) viscous isotropic, (b) middle, and (c) neat liquid crystal phases.

either type can form an ordered bulk phase by arranging themselves into either face-centered or body-centered cubic arrangements. Likewise, rodlike micelles of either type can be arranged into square or hexagonal packing. When all of the shape and packing possibilities are taken into account, it is no wonder that surfactant phase diagrams are complicated!

8.7 CATALYSIS BY MICELLES

For about four decades—between 1920 and 1960—the study of micelles was conducted mostly by researchers who would be classified as colloid scientists. Since about 1960 there has been an immense upswing of interest in micelles, with biochemists, organic chemists, inorganic chemists, and even physicists joining the ranks of those engaged in micellar research. The discovery that micellar media can affect the rates of chemical reactions has triggered this surge of interest in other branches of chemistry, and the notion that micelles might serve as models for enzymes has further stimulated research in this area. Studies have been conducted with a variety of objectives, including elucidation of reaction mechanisms, clarification of enzyme catalysis, investigations of micelles themselves, and applications to synthetic chemistry.

8.7a An Example: Enzyme Catalysis

Enzymes and micelles resemble each other with respect to both structure (e.g., globular proteins and spherical aggregates) and catalytic activity. Probably the most common form of enzyme catalysis follows the mechanism known in biochemistry as *Michaelis-Menton kinetics*. In this the rate of the reaction increases with increasing substrate concentration, eventually leveling off. According to this mechanism, enzyme E and substrate A first react reversibly to form a complex EA, which then dissociates to form product P and regenerate the enzyme:

$$E + A \underset{k_{-1}}{\overset{k_1}{\rightleftharpoons}} EA \overset{k_2}{\rightarrow} P + E$$

$$(E)$$

The various k's are the rate constants for the specific reactions shown. Standard kinetic analysis of this mechanism predicts that the rate of product formation is given by

$$\text{Rate} = \frac{k_1 k_2 [E]_0 [A]}{k_2 + k_{-1} + k_1 [A]} = \frac{k_2 [E]_0 [A]}{K_M + [A]} \tag{21}$$

in which $[E]_0$ is the total concentration of enzyme, and $K_M = (k_2 + k_{-1})/k_1$ is called the *Michaelis constant*. Note that as $[A]$ increases, K_M can be neglected in the denominator, allowing cancellation of $[A]$ and explaining the plateau in rate. The Michaelis constant inversely measures the affinity between enzyme and substrate: A small value of K_M means the enzyme binds the substrate tightly.

Figure 8.9 illustrates that micellar catalysts product similar effects. The chemical reaction under investigation in the figure is that between the crystal violet carbonium ion and the hydroxide ion to form the corresponding alcohol:

$$(CH_3)_2N-\text{⟨⟩}-C^+ + OH^- \longrightarrow (CH_3)_2N-\text{⟨⟩}-C-OH$$

(F)

In Figure 8.9 pseudo-first-order rate constants at 30°C for the rate of this reaction in 0.003 M NaOH are plotted versus the concentration of various alkyl trimethyl ammonium bromides. Several things should be noted about these data:

1. The catalytic role of the surfactant begins more or less sharply at the CMC. The C_{10} and C_{12} surfactants also show catalytic activity above their respective CMCs at higher concentrations.

2. The enhancement of rate qualitatively follows Michaelis-Menton kinetics, with both the initial slope and the final plateau increasing with increasing length of the alkyl tails of the surfactant.

3. The initial (starting at the CMC) slope of these curves is inversely proportional to K_M and increases with increasing binding effectiveness between the micelle and the substrate (since K_M itself inversely measures affinity).

These results can be rationalized by picturing the crystal violet carbonium ion as solubilized and oriented in the micelle, followed by attack by the aqueous hydroxide ion. The catalytic effect of the micellar solution has a twofold origin: a concentration effect and an

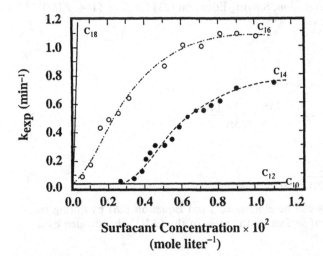

FIG. 8.9 Experimental rate constants for Reaction (F) versus concentration for alkyl trimethyl ammonium bromides with the indicated chain length. (Redrawn, with permission, from J. Albrizzio, J. Archila, T. Rodulfo, and E. H. Cordes, *J. Org. Chem.*, **37**, 871 (1972).)

effect on the transition state of the reaction. For the system shown in Figure 8.9 the micelles are positively charged, and the OH^- reactant is concentrated in the Stern layer of such micelles. In addition, the cationic crystal violet should be less stable in the cationic micelles than the zwitterionic transition state. Thus the transition state is electrostatically favored in the micelle compared to a nonsolubilized species. That electrostatic as well as hydrophobic effects are involved in the enhancement of the rate of Reaction (F) is evident from the fact that the relative rates with and without surfactant are 241/17 for cationic micelles and 1/17 for anionic micelles. It has also been observed that other anions inhibit the catalysis of Reaction (F), presumably by competing with OH^- for adsorption sites in the Stern layer.

8.7b Quantitative Analysis of Micellar Catalysis

Having looked at an example of micellar catalysis, let us next consider how such results are analyzed quantitatively. By analogy with Reaction (E), we visualize the micelle M and the substrate A entering a solubilization equilibrium characterized by an equilibrium constant K:

$$M + A \underset{k_0}{\overset{K}{\rightleftharpoons}} MA \overset{k_m}{\longrightarrow} P \qquad (G)$$

The solubilized substrate MA is the analog of the complex ES in Reaction (E), and the product P is formed in the micelle with the rate constant k_m. The product can also form from the substrate without involving the micelle; k_0 is the rate constant for the last process. The experimental (subscript exp) rate constant for Reaction (G) is then a weighted sum of the two constants k_m and k_0:

$$k_{exp} = f_0 k_0 + f_m k_m = f_0 k_0 + (1 - f_0) k_m \qquad (22)$$

where f_0 and f_m are the fractions of substrate in the bulk solution and solubilized in the micelles, respectively. The second version of Equation (22) arises from the recognition that $f_0 + f_m = 1$. The equilibrium constant in Reaction (G) can be written as

$$K = \frac{[MA]}{[M][A]} = \frac{f_m [A]_0}{[M] f_0 [A]_0} = \frac{f_m}{(1 - f_m)[M]} \qquad (23)$$

where $[A]_0$ is the total substrate concentration. Solving Equation (23) for $f_0 = (1 + K[M])^{-1}$ and substituting into Equation (22) gives

$$k_{exp} = \frac{k_0 + K k_m [M]}{1 + K[M]} \qquad (24)$$

Subtracting k_0 from both sides of Equation (24) yields

$$k_{exp} - k_0 = \frac{k_0 + K k_m [M]}{1 + K[M]} - k_0 = \frac{K k_m [M] - K k_0 [M]}{1 + K[M]} \qquad (25)$$

which, on inversion, gives

$$\frac{1}{k_{exp} - k_0} = \frac{1}{k_m - k_0} + \frac{1}{(k_m - k_0) K[M]} \qquad (26)$$

Finally, the concentration of micelles can be eliminated from Equation (26) by noting that $[M]$ is given by the number of moles of surfactant in excess of the CMC value divided by the degree of aggregation of the micelle, or

$$[M] = (c - c_{CMC})/n \qquad (27)$$

where c is the total concentration of surfactant. Substituting Equation (27) into Equation (26) gives

$$\frac{1}{k_{exp} - k_0} = \frac{1}{k_m - k_0} + \frac{n}{K(k_m - k_0)} \frac{1}{c - c_{CMC}} \qquad (28)$$

Recall that k_{exp} and k_0 are rate constants with and without surfactant, respectively, for the reaction in question, and that c and c_{CMC} are surfactant concentrations in the reaction mixture and at the CMC, respectively. Therefore k_{exp}, k_0, c, and c_{CMC} are experimentally accessible. Equation (28) predicts that a plot of $(k_{exp} - k_0)^{-1}$ versus $(c - c_{CMC})^{-1}$ is a straight line with a slope and intercept that have the following significance:

Intercept $= 1/(k_m - k_0)$ (29)

Slope $= n/[K(k_m - k_0)]$ (30)

Slope/intercept $= n/K$ (31)

Since k_0 is known, Equation (29) allows k_m to be evaluated. Likewise, the equilibrium constant for the binding of the substrate to the micelle can be evaluated from Equation (31) if n is known from a separate experiment. This method of analysis of catalyzed reactions is called a *Lineweaver-Burke plot* after the corresponding technique in biochemistry. Example 8.5 illustrates the use of these relationships.

*   *   *

Example 8.5 *Estimation of Equilibrium and Rate Constants in Micellar Catalysis.* Using $k_0 = 0.050$ min^{-1} and $c_{CMC} = 1.0 \cdot 10^{-3}$ M, estimate k_m and K/n for the C_{16} data in Figure 8.9. Because of the scatter in the data, take points at regular intervals from the drawn curve in the figure as the basis for this estimate.

Solution: The following points are read from the C_{16} curve in Figure 8.9:

$c \cdot 10^3$ (M)	1.5	2.0	2.5	3.0	4.0	5.0	6.0	7.0	8.0	9.0
k_{exp} (min^{-1})	0.25	0.40	0.52	0.66	0.80	0.90	1.0	1.1	1.1	1.1

Calculate $(c - c_{CMC})^{-1}$ and $(k_{exp} - k_0)^{-1}$ using $c_{CMC} = 1.0 \cdot 10^{-3}$ M and $k_0 = 0.05$ min^{-1}:

$(c - c_{CMC})$ 10^3 (M)	$(k_{exp} - k_0)$ (min^{-1})	$(c - c_{CMC})^{-1}$ (liter mole^{-1})	$(k_{exp} - k_0)^{-1}$ (min)
0.5	0.20	2000	5.00
1.0	0.35	1000	2.90
1.5	0.47	670	2.10
2.0	0.61	500	1.60
3.0	0.75	330	1.30
4.0	0.85	250	1.20
5.0	0.95	200	1.10
6.0	1.05	170	0.95
7.0	1.05	140	0.95
8.0	1.05	130	0.95

These data produce a reasonably linear Lineweaver-Burke plot of slope $2.2 \cdot 10^{-3}$ min mole liter^{-1} and intercept 0.62 min.

According to Equation (29), $k_m - k_0 = $ intercept$^{-1} = 1/0.62 = 1.61$; therefore $k_m = 1.7$ min^{-1}, a value 34 times larger than the estimated k_0 value.

According to Equation (31), $K/n = $ intercept/slope $= 0.62/2.2 \cdot 10^{-3} = 280$ liter mole^{-1}. ∎

*   *   *

The derivation of Equation (24) involves a number of assumptions, some of which are questionable in many cases. For example, in the above derivation, we assume the following:

1. The association between substrate and micelle shown in Reaction (G) follows 1 : 1 stoichiometry; that is, only one substrate molecule is taken up per micelle. If this is not the case, n is not the actual aggregation number of the micelle, but rather the number of surfactant molecules per solubilizate molecule.

2. The substrate does not complex with monomeric surfactant, and rate enhancement begins sharply at the CMC. Reactant molecules, like other solubilizates, may change the CMC of a surfactant, and premicellar clusters could influence the rate.

3. There are no competitive processes that inhibit the reaction.

8.7c   Electrolyte Inhibition in Micellar Catalysis

In discussing Reaction (F), we remarked that other anions are observed to compete with OH^- in the Stern layer. This sort of electrolyte inhibition is widely observed, and the dependence of the inhibition on both the size and charge of the ions generally corresponds to expectations. For example, in the base-catalyzed hydrolysis of carboxylic esters in the cationic micelles, anions inhibit the reaction in the order $NO_3^- > Br^- > Cl^- > F^-$. For acid-catalyzed ester hydrolysis in anionic micelles, the cation of added salt inhibits the reaction in the order $R_4N^+ > Cs^+ > Rb^+ > Na^+ > Li^+$.

Inhibition may be incorporated into the mechanism of micellar catalysis in the same way it is handled in enzyme kinetics. Representing the inhibitor by I, we can revise Reaction (G) as follows:

$$
\begin{array}{c}
M + A \underset{K}{\rightleftharpoons} MA \xrightarrow{k_m} P \\
+ \qquad\qquad\qquad \nearrow \\
I \qquad\quad k_0 \\
\updownarrow K_1 \\
MI
\end{array}
$$

(H)

where K_I is the binding constant for the inhibitor. Assuming that solubilization of one inhibitor molecule into the micelle blocks the addition of substrate leads to

$$
k_{exp} = \frac{k_0 + k_0 K_I[I] + k_m K[M]}{1 + K_I[I] + K[M]}
$$

(32)

which reduces to Equation (24) when $[I]$ or K_I equals zero. Equation (32) is also an oversimplification, but nevertheless it satisfactorily accounts for many observations.

Figure 8.10a is an example of some data in which the effect of added salt is more than a competitive ion-binding phenomenon. The reaction involved is the decarboxylation of 6-nitrobenzisoxazole-3-carboxylate, catalyzed by hexadecyl trimethyl ammonium bromide micelles:

(I)

In the presence of sodium tosylate (NaTos), both the rate constant k_{exp} and the viscosity of the solution show maxima at the same electrolyte concentration. The dramatic variation in η shown in Figure 8.10a suggests that the sodium tosylate alters the shape of the micelles, first producing rodlike structures that subsequently break up into more compact structures. The complicated phase diagrams of surfactants make this a plausible explanation. Effects such as this clearly complicate the picture not only of inhibition, but also of micellar catalysis in general.

Equilibrium constants for the binding between substrates and micelles — Reaction (G) — generally range from 10^3 to 10^6 for hydrophobic organic substrates. Furthermore, they are expected to increase as the hydrophobic character of the substrate increases. Figure 8.10b shows that this effect sometimes overshoots optimum solubilization. The figure shows, on a

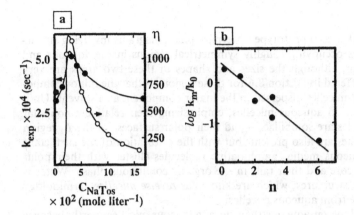

FIG. 8.10 Micellar catalysis: (a) the micelle catalyzed rate constant for Reaction (I) versus the concentration of sodium tosylate. The viscosity of the system versus sodium tosylate concentration is also shown (redrawn, with permission, from C. A. Bunton, In *Reaction Kinetics in Micelles* (E. H. Cordes, Ed.), Plenum, New York, 1973)); (b) ratio of k_m to k_0 for Reaction (J) versus the number of carbon atoms in the acyl group (redrawn, with permission, from A. K. Yatsimirski, K. Martinek, and I. V. Berezin, *Tetrahedron*, **27**, 2855 (1971)).

logarithmic scale, the ratio k_m/k_0 versus the number of carbon atoms in acyl groups for the following reaction:

$$\text{(J)}$$

The reaction is catalyzed by anionic micelles. In this case increasing the carbon content of the acyl group inhibits the reaction. Apparently, increasingly hydrophobic substrates are buried more and more deeply in the micelle such that they are inaccessible to the aqueous aldoxime anion.

8.7d Micellar and Enzyme Catalysis: Differences

Exploring the parallel between catalysis by enzymes and by micelles has been beneficial to both areas of research, but has also revealed two major differences between the two systems. First, the effectiveness of micelles in enhancing the rates of reactions is generally less than that of enzymes; for micelles, the effect rarely exceeds 100-fold. Second, micellar catalysts are less specific than enzymes. The lock-and-key model for enzyme activity implies specific binding sites in a protein molecule with a configuration maintained by covalent bonds. The corresponding features in micelles are zones of a particular polarity in rapidly changing equilibrium structures. Although the parallels between enzymes and micellar catalysts will undoubtedly continue to be profitably explored, it seems likely that future developments of micellar catalysts will probably emphasize synthetic chemistry.

8.8   REVERSE MICELLES

Examination of Figure 8.7—taken as a prototype of ternary phase diagrams involving surfactants—shows L_1 and L_2 phases occupying roughly symmetrical regions in the aqueous and
organic corners of the diagram, although the sizes and shapes of these two areas are very
different. The interpretation offered in Section 8.6 for these regions is that the minor component is solubilized in surfactant micelles dispersed in the major component as a solvent. Until
now we have focused attention on aqueous micelles, emphasizing their relatively nonpolar
cores—where nonpolar molecules are solubilized—and their polar surfaces. In the L_2 region
of Figure 8.7, solubilized micelles are also present, but with the orientation of the surfactant
molecules reversed. In nonaqueous media, amphipathic molecules cluster with their polar
heads together in the micellar core and their tails in the organic continuous phase. Water is
solubilized in the core of these structures, which are known as *reverse micelles*, terminology
that emphasizes their difference from aqueous micelles.

Far more work has been done on aqueous than on nonaqueous micelles, partly because
the number of amphipathic species that dissolve in nonpolar solvents is considerably fewer
than in water-soluble surfactants. Aerosol OT, Triton X-100, and various Spans—the structures of which are given in Table 7.2—are examples of substances that have been widely used
in the study of reverse micelles. Fundamental understanding of these structures is most likely
to come from research involving pure, well-defined chemicals. Commercial products—petroleum sulfonates and the like—are widely studied from an applications point of view, but the
interpretation of such experiments is often clouded by the uncertain effect of impurities. Even
though we have approached this discussion from the perspective of water-containing systems,
the behavior of anhydrous systems is obviously of interest. Through solvation of polar groups
and hydrogen bonding, water can promote aggregation in nonpolar media; complete removal
of water as an impurity in both the solvent and the surfactant can be very difficult. Since
two-component systems are simpler than those with three components, let us consider some
anhydrous solutions of surfactant in nonaqueous solvents.

8.8a   Surfactant Aggregation in Nonaqueous Media

Surfactant aggregation in an anhydrous, nonpolar medium differs in several important respects from aggregation in water. The most apparent of these differences is that the hydrophobic effect plays no role in the formation of reverse micelles. The amphipathic species are
relatively passive in aqueous micellization, being squeezed out of solution by the water. In
contrast, surfactant molecules play an active role in the formation of reverse micelles, which
are held together by specific interactions between head groups in the micellar core.

The differences in the solubility parameters (Chapter 3, Section 3.4b) of the hydrocarbon
tail of the surfactant and the solvent have also been examined as contributing to reverse
micelle formation. It is interesting to note that for the micellization of potassium benzene
sulfonate in heptane, $\Delta H = -79.5 \text{ kJ mole}^{-1}$ and $\Delta S = -62.8 \text{ J K}^{-1} \text{ mole}^{-1}$. In contrast to
aqueous systems, spontaneous micellization is largely due to a large *negative enthalpy change*
with an unfavorable entropy change opposing micellization.

Another striking difference between aqueous and anhydrous, nonaqueous systems is the
size of the aggregates that are first formed. As we have seen, n is about 50 or larger for
aqueous micelles, while for many reverse micelles n is about 10 or smaller. A corollary of the
small size of nonaqueous micelles and closely related to the matter of size is the blurring of the
CMC and the breakdown of the phase model for micellization. Instead, the stepwise buildup
of small clusters as suggested by Reaction (D) is probably a better way of describing micellization in anhydrous systems. When the clusters are extremely small, the whole picture of a polar
core shielded from a nonaqueous medium by a mantle of tail groups breaks down.

It is sometimes argued that the "reverse micelle" terminology is an inappropriate comparison to aqueous micelles. Since water can be solubilized by these micelles, causing an increase
in n, the reverse micelle model and vocabulary do seem useful for ternary systems.

Proton NMR has been used to measure both the onset of micellization and the tendency toward solubilization in nonaqueous surfactant solutions. In an NMR spectrum the observed displacement of a resonance frequency—the chemical shift δ—depends on the environment of the molecule. Also, protons in the methylene groups of surfactants absorb at a different point in the spectrum than, say aromatic protons, so a surfactant can be readily monitored in, say, a benzene solution. If δ for a proton in a surfactant molecule is plotted against the concentration of the amphipathic species, a graph showing two linear regions of different slope results. In contrast to the properties of aqueous systems (Fig. 8.2), the transition between the two linear regions is gradual, but extrapolation to their point of intersection defines the CMC. If a solubilized molecule has an identifiable feature in an NMR spectrum, then solubilization can also be monitored by NMR. Example 8.6 explores this further.

* * *

Example 8.6 *A Nuclear Magnetic Resonance Study of Solubilization in a Reverse Micelle.* Let δ be the chemical shift produced by the proton of a solubilized molecule and differentiate between the experimental shift (subscript *exp*) and the δ's produced by molecules in bulk (subscript 0) or micellar (subscript *m*) environments. Arguing by analogy with the derivation of Equation (24) shows that

$$\delta_{exp} = (\delta_0 + \delta_m K[M])/(1 + K[M])$$

where K is the binding constant for the interaction of the solubilizate and the micelle. Suggest how experimental NMR data can be interpreted to give a quantitative value for K.

Solution: Incorporation of solubilizate B into micelle M can be represented as

$$B + M \overset{K}{\rightleftarrows} MB$$

for which K is the equilibrium constant. Define f_0 and f_m as the fraction of the solubilizate in the bulk solution and in the micelle, respectively. It follows by analogy with Equation (22) that $\delta_{exp} = f_0\delta_0 + f_m\delta_m = f_0\delta_0 + (1 - f_0)\delta_m$ since $f_m + f_0 = 1$.
By analogy with Equation (23),

$$K = f_m[B]_0/[M]f_0[B]_0 = (1 - f_0)/f_0[M]$$

with $[B]_0$ = total solubilizate concentration.
Finally,

$$\delta_{exp} = (\delta_0 + \delta_m K[M])/(1 + K[M])$$

with $[M] = (c - c_{CMC})/n$ by analogy with Equation (27).
This can be rearranged by analogy with Equation (28) to give

$$1/(\delta_{exp} - \delta_m) = 1/(\delta_m - \delta_0) + [n/K(\delta_m + \delta_0)][1/(c - c_{CMC})]$$

If δ is measured for different concentrations of surfactant and δ_0 and c_{CMC} are known, then this result can be plotted in a linear form and K/n evaluated from the intercept/slope ratio (compare Equation (31)). ∎

* * *

8.8b Some Uses of Reverse Micelles

Some of the amphipathic species that have been used in the investigation of reverse micelles are capable of dissociation under suitable conditions. Metal carboxylates, alkyl aryl sulfonates, sulfosuccinates, and alkyl ammonium salts are examples of compounds with a high degree of ionic character. Coulomb's law describes the force between two charges q_1 and q_2 separated by a distance r in a medium of relative dielectric constant ε_r,

$$F_{Coul} \propto \frac{q_1 q_2}{\varepsilon_r r^2} \tag{33}$$

Electrolyte dissociation is expected to decrease as this force increases. This dissociation— as measured by pK for the dissociation process—should be inversely proportional to ε_r when a given electrolyte is studied in a series of solvents of various polarities. Since r in Equation (33) is the sum of anion and cation radii, the above statement is true only if the state of solvation of the ions remains the same as the solvent varies. Alkali metal salts of dinonyl naphthalene sulfonic acid show decreasing dissociation as the solvent changes from ethanol ($\varepsilon_r = 24.3$) to acetone ($\varepsilon_r = 20.7$) to ethyl acetate ($\varepsilon_r = 6.02$). The quantitative magnitude of the difference between solvents is not the same for all cations, however, indicating that solvation of the cation also enters the picture. In solvents of very low polarity — like hydrocarbons ($\varepsilon_r = 2$-3) — dissociation is negligible.

The situation with respect to dissociation makes the small core of the reverse micelle a unique microenvironment, approaching the properties of ionic crystals, but at the same time readily accessible to solubilizates and reactants. This is why the interior of the reverse micelle is so effective in solubilizing water and explains why anhydrous systems are so difficult to obtain. Under nearly anhydrous conditions, as few as one solubilized water molecule per reverse micelle might be obtained. This degree of solubilization corresponds to 10 moles of solute per 0.018 kg water, or about 550 molal — an intriguing microenvironment for the investigation of catalytic effects! In addition, for amphipathic molecules with, say, weakly basic polar groups like sulfonates, cations with various strengths as Lewis acids, can be used as counterions.

One investigation of the catalytic activity of such a system used the reverse micelles formed in decane by sulfosuccinates with various cations. This medium was used to study the reaction

(K)

in which the benzylchloroformate reactant is basic and should therefore react strongly with acidic cations. The results were qualitatively similar to those shown in Figure 8.9 and could be analyzed similarly. The rate constants were evaluated at different temperatures, and activation energies were determined by a standard Arrhenius plot of $\ln k$ versus $1/T$. Table 8.4 shows the kinetic parameters so determined when the cations in the reverse micelle were Na^+, Al^{3+}, Ce^{3+}, and Zn^{2+}. Since the slope in this kind of plot is proportional to the activation energy and since the activation energy is very different for various cations, the lines in the Arrhenius plots can cross at accessible temperatures. This makes the relative effectiveness of the various cations in catalyzing Reaction (K) a matter of temperature.

TABLE 8.4 Activation Energies and Arrhenius Preexponential Factors for Reaction (K) Catalyzed by Reverse Micelles Containing the Indicated Cation

Cation	E_a (kJ mole^{-1})	$\ln A$
Na^+	129	25.3
Al^{3+}	148	32.2
Ce^{3+}	84	11.7
Zn^{2+}	267	71.4

Source: F. M. Fowkes, D. Z. Becher, M. Marmo, C. Silebi, and C. C. Chao, In *Micellization, Solubilization and Microemulsions*, Vols. 1 and 2, (K. L. Mittal, Ed.), Plenum, New York, 1976.

Use of reverse micelles in synthetic chemistry to improve the rate and the yield of reactions seems likely to be a fruitful area of research in the future. In addition to catalysis, several other applications of reverse micelles can be cited. Just as nonpolar dirt is solubilized in aqueous micelles, so, too, polar dirt that would be unaffected by nonpolar solvents may be solubilized into reverse micelles. This plays an important role in the dry cleaning of clothing. Motor oils are also formulated to contain reverse micelles to solubilize oxidation products in the oil that might be corrosive to engine parts.

8.9 EMULSIONS AND MICROEMULSIONS

The term *microemulsion* was coined in 1958 to describe a fairly specific class of colloidal systems. Before we discuss these, a brief discussion of the broader term *emulsion* seems in order. If two immiscible liquids are shaken together, they will ordinarily separate rapidly into two distinct layers that can be divided in, say, a separatory funnel. Although any immiscible liquids might be considered, in this discussion we refer to the two as oil (abbreviation O) and water (abbreviation W). Next, instead of merely shaking the two liquids together, suppose we add a surfactant — often called an *emulsifying agent* in this context — and then vigorously mix the components in a blender or homogenizer of some sort. The milling together of the constituents causes one to be dispersed (the inner phase) in the other (the continuous phase): An emulsion is produced. There can be many variations in this procedure — in the nature of the components, their proportions, the milling process, the temperature, and so on; however, a few broad generalizations are possible.

1.   The dispersed particles are spheres of great polydispersity. As noted in Appendix C (Section C.3b), this is generally the case in dispersions prepared by comminution such as this.

2.   The average particle size is at the upper end of the colloidal size range (on the order of micrometers), and the particles are usually visible in a light microscope. We shall describe these as *coarse emulsions* when we want to emphasize their size range. Because the particles are relatively large and polydisperse, coarse emulsions look white when examined visually.

3.   Emulsions are two-phase systems and — because of the free energy associated with the oil-water interface — are thermodynamically unstable with respect to separation into oil and water layers.

4.   Oil may be the dispersed phase and water the continuous phase — designated an O/W emulsion — or water may be dispersed in oil (W/O). The form obtained depends on the specifics of the system, including the temperature. Compatibility with either oil or water on dilution is an easy way of establishing which phase is continuous.

5.   The surfactant is adsorbed at the oil-water interface in the oriented fashion of monolayers. Judging from monolayer studies at the air-water interface, saturating the surface with surfactant lowers the surface tension γ by 25–50 mN m^{-1} (Fig. 7.6).

6.   If the surfactant is ionic and imparts a charge to the interface, then the dispersed particle will be surrounded by an ion atmosphere. We see in Chapters 11 and 13 how an ion atmosphere surrounding a particle may slow down the rate at which such particles come together. This is one of the ways by which an emulsion may achieve some degree of kinetic stability.

Many of these concepts were introduced in Chapter 1 and seem fairly straightforward as abstract propositions. Things are not always so clear in concrete instances, however. The situation of microemulsions is a case in point.

8.9a Microemulsions

Historically, the term *microemulsion* was applied to systems prepared by emulsifying an oil in aqueous surfactant and then adding a fourth component, called a *cosurfactant*, generally an alcohol of intermediate chain length. Benzene, water, potassium oleate, and hexanol might be the components of a typical microemulsion formulation. What is observed experimentally is that the usual milky emulsion becomes transparent on addition of the alcohol. Light scattering and an assortment of other techniques reveal that the resulting system consists of either O/W or W/O dispersions with particles having diameters in the 10- to 100-nm size range. Whether oil or water is continuous, the extent of uptake of the other component may be appreciable. In summary, the following differences between microemulsions and coarse emulsions should be noted:

1. Microemulsions contain particles at least an order of magnitude smaller than those in coarse emulsions.
2. Microemulsions are clear, and coarse emulsions are cloudy.
3. Microemulsions form spontaneously; coarse emulsions ordinarily require vigorous stirring.
4. Microemulsions are stable with respect to separation into their components; coarse emulsions may have a degree of kinetic stability, but ultimately separate.

The term *microemulsion* seems quite firmly established as the name for the sort of system described above. Still, there has been and continues to be a great deal of controversy as to the exact nature of these systems and the suitability of this vocabulary. The word *emulsion* implies the presence of two phases with an interfacial free energy associated with the phase boundary. Their small size makes the specific area large for microemulsions with a large free energy contribution from γ. Both the spontaneous formation of microemulsions and their stability with respect to separation are hard to reconcile with these considerations. It has even been suggested that the mixed film of surfactant and cosurfactant make the interfacial free energy negative. Alternatively, the increase in overall free energy with decreasing particle size may be offset by a favorable and hence negative $T\Delta S$ term in which ΔS describes the entropy of mixing the microemulsion particles with molecules of the dispersion medium. Since the number of microemulsion particles increases with decreasing particle size, the $T\Delta S$ term becomes more favorable with decreasing size. This idea is hard to apply quantitatively because of uncertainty as to the value—or meaning—of γ in these systems.

8.9b Microemulsions Viewed As "Swollen Micelles"

A totally different way of looking at microemulsions—and one that connects this topic with previous sections of the chapter—is to view them as complicated examples of micellar solubilization. From this perspective, there is no problem with spontaneous formation or stability with respect to separation. Furthermore, ordinary and reverse micelles provide the basis for both O/W and W/O microemulsions. From the micellar point of view, it is the phase diagram for the four-component system rather than γ that holds the key to understanding microemulsions.

The difference in perspective between the emulsion and micellar points of view is suggested by Figure 8.11. The small aggregate on the left represents a micelle with little or no solubilization. From left to right, the "particles" increase in size owing to increasing solubilization. The circle on the right represents an emulsion particle: an oil drop with a monolayer of surfactant on the surface. An actual continuum of states such as that suggested in Figure

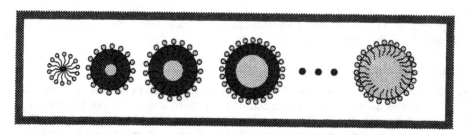

FIG. 8.11 Schematic progression from micelle (at left) to emulsion droplet (at right). Various degrees of solubilization, including microemulsions, lie between the two extremes.

8.11 is not physically attainable—most emulsions are prepared by comminution rather than condensation—but microemulsions do lie between the extremes. The conflicting schools of thought concerning microemulsions arise from the difference in perspective: One side looks from the emulsion point of view, and the other from the micellar point of view. *Swollen micelles* is another term for microemulsions; this terminology clearly reflects the different perspective of its originators.

8.9c Phase Diagrams for Microemulsions

We saw in Section 8.6 that phase diagrams are an effective way of representing the complex behavior of surfactant systems. Let us take a look at microemulsions in terms of phase diagrams. It turns out that nonionic surfactants form microemulsions at certain temperatures without requiring cosurfactants. Since only three components are present, these have somewhat simpler phase diagrams; this kind of system offers a convenient place to begin.

Figure 8.12 is a composite of both experimental and schematic, interpretive portions. The rectangular diagram shows the experimental behavior of the system water-cyclohexane-polyoxyethylene-(8.6)-nonyl phenol ether. All systems studied contained 5% surfactant with variable proportions of cyclohexane and water. Temperature was the experimental variable, and the nature of the phases present was recorded as the temperature was changed. The presence of different equilibrium phases is represented by different areas on the rectangular diagram, which is divided diagonally by a pair of lines that trace a ribbonlike pattern. This pair of lines cross twice, so the ribbon is divided into three areas. These, plus the regions

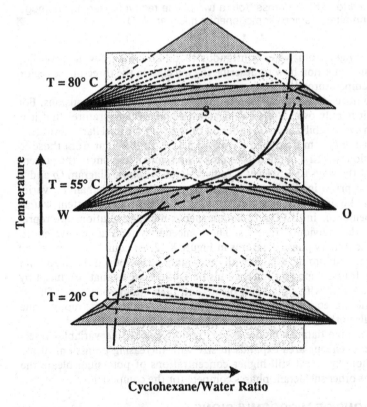

FIG. 8.12 Rectangular figure shows the phase diagram for water-cyclohexane (at 5% surfactant) versus temperature. Superimposed ternary phase diagrams offer an interpretation of the phases present. (Redrawn, with permission, from M. L. Robbins, In *Solution Chemistry of Surfactants*, Vols. 1 and 2 (K. L. Mittal, Ed.), Plenum, New York, 1979.)

above and below the ribbon, define five different phase situations encountered at various temperatures and compositions.

The idealized ternary phase diagrams that have been superimposed on the experimental plot in Figure 8.12 help us understand the five different regions of the rectangular diagram. The stacked triangles represent the oil-water-surfactant (abbreviation S) phase diagrams at different (constant) temperatures. Since the experimental data were collected at 5% surfactant, the rectangular plot slices through the triangles 5% of the distance toward S from the O-W base of the triangle. Example 8.7 offers practice in reading triangular phase diagrams and helps identify the phases present in the five different regions of the rectangular composition-temperature diagram.

* * *

Example 8.7 *Interpreting Phase Diagrams of Microemulsions.* At each of the three temperatures for which a ternary phase diagram is provided, describe the phases present when the diagram is crossed from the W-S side to the O-S side in the slice at 5% S. Describe the homogeneous phases in terms of the apparent solubilization.

Solution: Recall that on ternary diagrams the radial lines are tie lines representing two-phase regions. The triangular regions are three-phase regions, and the remaining area toward the S apex of the triangles is a homogeneous phase.

At 20°C, moving from left to right, we cross from a one-phase region (homogeneous micellar, O/W) into a two-phase region (oil plus homogeneous micellar, O/W).

At 55°C, moving from left to right, we cross from a two-phase region (water plus homogeneous micellar, O/W) into a three-phase region (oil plus water plus homogeneous micellar). Next we enter a different two-phase region (oil plus homogeneous micellar, W/O).

At 80°C, moving from left to right, we cross from a two-phase region (water plus homogeneous micellar, W/O) to a one-phase region (homogeneous micellar, W/O). ∎

* * *

Since the micellar phase changes from water-continuous to oil-continuous with increasing temperature, it is an intriguing question how to describe the micelles in the three-phase region that exists at intermediate temperatures.

In this study it is the homogeneous micellar phases that comprise the microemulsions. For this 5% surfactant system it is only over a relatively narrow range of temperatures that it is possible to have fairly extensive solubilization of either oil or water in the micellar solutions.

It could be argued that the system described in Figure 8.12 is so different from those to which the name microemulsion was first applied as to make the figure irrelevant to the present discussion. However, one of the ways to represent a four-component phase diagram (p and T constant) is to use a trigonal prism in which one of the components—rather than T—varies along the rectangular faces. Figure 8.13a is a partial phase diagram for the system water-benzene-potassium oleate-pentanol. In this figure homogeneous micellar solutions are represented by unshaded areas in the individual triangles. These micellar systems are equivalent to the homogeneous areas in the ternary phase diagrams in Figure 8.12. At equilibrium, up to 50 wt% benzene can be incorporated into the system with only minor variations in the maximum water uptake. The resulting four-component microemulsion is the same as that produced by first emulsifying the oil and water with potassium oleate and then adding pentanol. This is true because we are considering equilibrium phase diagrams. As with all thermodynamic conclusions, the diagram tells nothing about the rate at which equilibrium is achieved.

Figure 8.13b shows the same data as Figure 8.13a, replotted with the variables interchanged. Note how the homogeneous area expands in size with increasing potassium oleate concentration up to 0.44 mole kg^{-1}. At still higher concentrations of potassium oleate the homogeneous area shrinks as other surfactant phases compete for the components.

8.10 SOME APPLICATIONS OF MICROEMULSIONS

Systems in which one liquid phase is finely dispersed in another under the stabilizing influence of one or more additional components find applications in countless areas. As consumers, we encounter many of these every day. Floor waxes, shaving lotions, beverage concentrates,

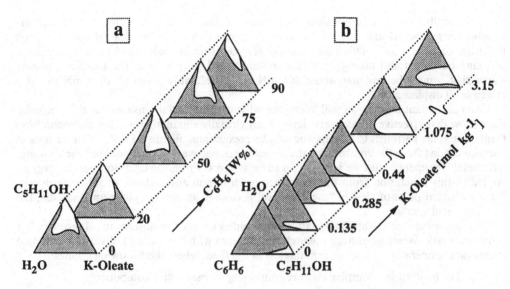

FIG. 8.13 Two representations of a portion of the phase diagram for the water-benzene-potassium oleate-pentanol system. The unshaded regions represent homogenous solutions. (a) Redrawn, with permission, from S. Friberg and I. Burasczeska, *Prog. Colloid Polym. Sci.*, **63**, 1 (1978). (b) Redrawn, with permission, from C. U. Herrmann, U. Wurz, and M. Kahlweit, In *Solution Chemistry of Surfactants*, Vols. 1 and 2 (K. L. Mittal, Ed.), Plenum, New York, 1979.

pesticide preparations, cold creams, and pharmaceutical products are a few of the more common examples. In recent years a great deal of research in this area has been directed toward the problem of *tertiary* oil recovery.

8.10a Tertiary Oil Recovery

First, the recovery of oil from natural reservoirs occurs in three stages. During the *primary* recovery stage the pressure of natural gases in the reservoir pushes the oil out. When the gas pressure is no longer adequate, water is pumped into the reservoir to force the oil out. This is called water flooding and represents the *second stage* of oil recovery. Primary and secondary oil recovery leave about 70% of the total oil in place, much of it trapped in the pore structure of the reservoir by capillary and viscous forces. It is estimated that in U.S. oil fields alone 300 billion barrels of oil are not recoverable by primary or secondary processes, and that of this 25 billion to 60 billion barrels are potentially recoverable through some tertiary process.

Numerous methods have been explored to recover at least some of this vast resource. Injection of oil-miscible fluids, gases under high pressure, and steam — either separately or in combination — have all been tried with various degrees of success. This is where microemulsions enter the picture. Under optimum conditions an aqueous surfactant solution — which may also contain cosurfactants, electrolytes, polymers, and so on — injected into an oil reservoir has the potential to solubilize the oil, effectively dispersing it as a microemulsion.

Any attempt to represent the trapped oil by a manageable model is bound to be an oversimplification. To see qualitatively, however, how capillary forces trap the oil and how surfactant solutions offer a potential for freeing it, imagine a cylindrical pore containing a slug of oil. Furthermore, assume the oil is in contact with water and that the interface is hemispherical. This assumption about the shape of the interface makes the water-oil-rock contact angle zero and is equivalent to neglecting θ ($\cos 0 = 1.0$). Although a primitive picture of oil in a rocky reservoir, this model can be described by a single size parameter r, the radius of the pore and the radius of curvature of the interface. The Laplace equation (Equation (6.29)) may then be used to describe the pressure across the oil-water interface, a pressure that must be exceeded to displace the oil. According to the Laplace equation, $\Delta p \propto \gamma/r$ and, since r is small, Δp will be large unless r is offset by a small value of γ. Using r values that are sensible for geological structures, it has been estimated that $\Delta p \approx 500$ psi per foot if γ is on the order of 10 mN m^{-1}. This kind of pressure drop is unattainable under field conditions. Working backward and taking Δp as an attainable 1–2 psi per foot, the Laplace equation shows that γ must be less than about 0.1 mN m^{-1}, preferably closer to 10^{-3} mN m^{-1} for effective oil displacement.

Any surfactant adsorption will lower the oil-water interfacial tension, but these calculations show that effective oil recovery depends on virtually eliminating γ. That microemulsion formulations are pertinent to this may be seen by reexamining Figure 8.11. Whether we look at microemulsions from the emulsion or the micellar perspective, we conclude that the oil-water interfacial free energy must be very low in these systems. From the emulsion perspective, we are led to this conclusion from the spontaneous formation and stability of microemulsions. From a micellar point of view, a "pseudophase" is close to an embryo phase and, as such, has no meaningful γ value.

It is apparent that extrapolating laboratory studies on microemulsions to oil recovery is a formidable task. While laboratory research is conducted with pure solutes, distilled water, and at constant temperature, these are meaningless in the field, where the following applies:

1. The oil itself is a complex mixture containing surface-active components.
2. Commercial petroleum sulfonates, *a mixture of compounds*, are the most widely used surfactants.
3. Groundwater contains dissolved minerals, and, in practice, brine is used as the aqueous component.

In addition, viscosity considerations may be as important or more important than capillarity; fortunately, microemulsions also have relatively low viscosities!

Despite the obvious difficulty of the tertiary oil recovery problem, this is a major area of surfactant research since the potential rewards for success are very great.

8.10b Polymer Synthesis

Another area of microemulsion application is in the synthesis of certain polymers. The process is called *emulsion polymerization*, a misnomer since micelles rather than emulsion drops are the site of the polymerization reaction. Because of the commercial importance of polymers, this process has been extensively researched and is quite well understood. We only consider some highlights of the process.

Emulsion polymerization is applicable only to monomers that are relatively insoluble in water, such as styrene. A coarse emulsion of monomer in aqueous surfactant is prepared with a water-soluble initiator, say, H_2O_2 in the solution. The surfactant concentration is above the CMC, so surfactant molecules are present as monomers, micelles, and emulsifiers at the oil-water interface. Even an insoluble liquid like styrene dissolves in water to some extent. Therefore the monomer is present in coarse emulsion drops, solubilized in micelles, and as dissolved molecules in water. A schematic illustration of the distribution of surfactant, monomer, and polymer in an emulsion polymerization process is shown in Figure 8.14.

The H_2O_2 molecules undergo thermal decomposition to form hydroxyl free radicals $\cdot OH$ that initiate the polymerization. The overall reaction for the polymerization of styrene can be represented as

(L)

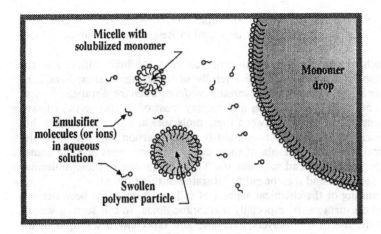

FIG. 8.14 Schematic representation of the distribution of surfactant and monomer in an emulsion polymerization. (Redrawn, with permission, from J. W. Vanderhoff, E. B. Bradford, H. L. Tarkowski, J. B. Shaffer, and R. M. Wiley, *Adv. Chem.*, **34**, 32 (1962).)

When x is large, the uniqueness of the end group(s) can be ignored; familiar polystyrene is the product.

In emulsion polymerization the first step in Reaction (L) takes place in water between dissolved monomer and initiator fragments. The resulting free radical is solubilized in micelles, in which it quickly reacts with solubilized monomer to form polymer. The low concentration of monomer in the aqueous phase prevents this from occurring to any appreciable extent in the water, although, by diffusion, there continues to be a flux of monomer from emulsion drops into micelles. Likewise, any polymerization that occurs in the coarse emulsion drops themselves is insignificant because of the much greater numerical abundance of micelles than the emulsified styrene droplets. The polymer chains grow in micelles until the process is terminated by reaction with another radical. Thus polymer growth is either propagating or terminating in micelles at any time; therefore half the micelles in a reaction mixture contain growing chains under stationary-state conditions. Both the rate of polymerization and the average molecular weight of the polymer depend on the surfactant concentration—via the concentration of micelles—in emulsion polymerization, while this has no effect on polymerizations conducted in nonmicellar solutions.

The aqueous polymer dispersion that results from emulsion polymerization is called a *latex*. In applications, the polymer may be separated, or the latex may be used directly as in paints and floor coatings.

As the conversion to polymer proceeds, the micelles become progressively more and more swollen by the polymer-monomer mixture. As with other microemulsions, it eventually becomes problematic as to whether the resulting dispersed particles should be called micelles or swollen polymer particles with adsorbed surfactant.

8.11 BIOLOGICAL MEMBRANES

It is neither feasible nor appropriate in a book like this to give a detailed presentation of biological membranes, which compartmentalize living matter and perform numerous cell functions as well. However, because of the impetus to the study of surfactants that the membrane-mimetic properties of surfactant structures have provided, it would be a mistake to exclude some mention of membranes in this chapter. We have already noted in connection with Figure 7.7 that a monolayer may collapse into a bilayer that leaves the surfactant in a tail-to-tail configuration. This is exactly the arrangement of molecules in the lipid portion of a cell membrane. Protein molecules are arrayed in or on this lipid bilayer. Vignette I.2 presented in Chapter 1 discusses this briefly. More details may be found in Bergethon and Simons (1990) and Goodsell (1993).

Figure 8.15 is a sketch of one possible relationship between the lipid bilayer and the membrane proteins. Molecules are free to move laterally in these membranes; hence the structure pictured in Figure 8.15 is called the *fluid mosaic model* of a cell membrane.

Cell membrane lipids are natural surfactants and display most of the properties of synthetic surfactants. The principal difference between these molecules and the surfactants that we discussed above in the chapter is that lipids contain two hydrocarbon tails per molecule. Table 8.5 shows the general structural formula of these cell membrane lipids and the names and formulas for some specific polar head substituents. The alkyl groups in these molecules are usually in the C_{16}–C_{24} size range and may be either saturated or unsaturated.

Much of our understanding of the chemical aspects of cell membranes has been derived from model systems based on surfactants, especially membrane lipids. In this section we are primarily concerned with the use of monolayers, bilayers, and especially black lipid membranes and vesicles as cell membrane models.

Fundamental membrane research has benefited greatly from the study of monolayers. One of the most important discoveries from this sort of research is the very existence of *two-dimensional phases and phase transitions*. Generally, studies of the sort that can be carried out with monolayers and bilayers cannot be directly extended to living cells, but some exceptional cases have shown that the extrapolation is valid. For example, it is known from monolayer studies that the presence of unsaturated hydrocarbon chains in lipid monolayers prevents some phase transitions from occurring as the temperature is lowered. Certain mutants of *Escherichia coli* are unable to synthesize fatty acids and hence can be manipulated through the compounds they are provided as nutrients. Abnormal levels of saturated hydrocarbon can

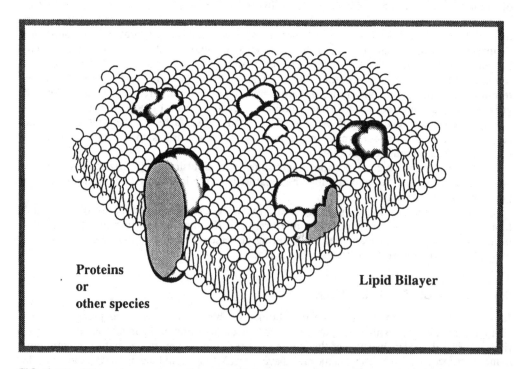

FIG. 8.15 Schematic representation of a biological membrane. The amphipathic phospholipid molecules form a bilayer with protein molecules embedded in it. (Redrawn, with permission, from S. J. Singer and G. L. Nicolson, *Science*, **175**, 720 (1972).)

TABLE 8.5 Names and Structures of Some Typical
Phospholipid Surfactants

$$
\begin{array}{l}
\text{O} \\
\| \\
\text{R}-\text{C}-\text{O}-\text{CH}_2 \\
\text{R}-\text{C}-\text{O}-\text{CH} \\
\| \\
\text{O} \qquad\qquad \text{O} \\
\qquad\qquad\quad \| \\
\qquad \text{CH}_2-\text{O}-\text{P}-X \\
\qquad\qquad\qquad\quad | \\
\qquad\qquad\qquad \text{O}-
\end{array}
$$

Name	X
Phosphatidic acid	$-\text{OH}$
Phosphatidylcholine	$-\text{OCH}_2\text{CH}_2\overset{+}{\text{N}}(\text{CH}_3)_3$
Phosphatidylethanolamine	$\text{OCH}_2\text{CH}_2\text{NH}_2$
Phosphatidylserine	$-\text{OCH}_2\text{CCHNH}_2$ $\qquad\quad\;\,\text{COOH}$
Phosphatidylthreonine	$-\text{OCH}-\text{CH}-\text{NH}_2$ $\qquad\quad\,\text{CH}_3\text{COOH}$
Phosphatidylglycerol	$-\text{OCH}_2\text{CHCH}_2\text{OH}$ $\qquad\qquad\;\text{OH}$

thereby be introduced into the cell membranes of these organisms. Under appropriate conditions the same phase transitions occur in cell membranes as in model monolayers; still, the organisms survive. This is probably because patches of different two-dimensional phases segregate on the cell walls, with the unchanged regions carrying out normal biological functions.

Studies with monolayers and bilayers as models show that modifications of chain conformation can alter the effective thickness of the hydrocarbon portion of a bilayer; this, in turn, can have profound effects on its permeability. Suppose, for example, that a two-dimensional phase transition involved cooperatively inducing just one kink per molecule in otherwise fully extended chains. Such a kink shortens the extension of the chain by 0.127 nm, or thins a bilayer by 0.25 nm. We saw in Example 8.2 that this distance is about the length projected by two carbon-carbon bonds in the direction of the chain. Therefore, cooperative induction of such kinks has the same effect on the thickness of the hydrocarbon part of the bilayer as removing two carbon atoms from the tail groups.

Even closer to cell membranes than monolayers and bilayers are organized surfactant structures called *black lipid membranes* (BLMs). Their formation is very much like that of an ordinary soap bubble, except that different phases are involved. In a bubble, a thin film of water — stabilized by surfactants — separates two air masses. In BLMs an organic solution of lipid forms a thin film between two portions of aqueous solution. As the film drains and thins, it first shows interference colors but eventually appears black when it reaches bilayer thickness. The actual thickness of the BLM can be monitored optically as a function of experimental conditions. Since these films are relatively unstable, they are generally small in area and may be formed by simply brushing the lipid solution across a pinhole in a partition separating two portions of aqueous solution.

Because they are binary lipid films sandwiched in water, BLMs are excellent models for membranes in terms of electrical and permeability properties. Membrane potential, conductivity, and capacitance measurements have all been made on BLMs, and the effects on these quantities of electrolytes, proteins, and organic additives have all been studied. For example, 2,4-dinitrophenol, a known hydrogen ion carrier, has been observed to decrease the resistance significantly between electrodes on opposite sides of a BLM. Similarly, high applied voltages thin the BLM, with a simultaneous reversible increase in its permeability and conductance. The rate of permeation of water through BLMs is greater than that of nonpolar compounds; however, ions permeate BLMs more slowly than biological membranes.

The final surfactant structures we consider as models for biological membranes are *vesicles*. These are spherical or ellipsoidal particles formed by enclosing a volume of aqueous solution in a surfactant bilayer. When phospholipids are the surfactant, these are also known as *liposomes*, as we have already seen in Vignette I.3 in Chapter 1. Vesicles may be formed from synthetic surfactants as well. Depending on the conditions of preparation, vesicle diameters may range from 20 nm to 10 μm, and they may contain one or more enclosed compartments. A multicompartment vesicle has an onionlike structure with concentric bilayer surfaces enclosing smaller vesicles in larger aqueous compartments.

Phospholipid vesicles form spontaneously when distilled water is swirled with dried phospholipids. This method of preparation results in a highly polydisperse array of multicompartment vesicles of various shapes. Extrusion through polymeric membranes decreases both the size and polydispersity of the vesicles. Ultrasonic agitation is the most widely used method for converting the lipid dispersion into single-compartment vesicles of small size.

Phase transitions, electrical properties, and permeability have all been investigated for vesicle bilayers. Especially important are the molecular dynamics of transverse motion through the bilayer. Use of isotopic labels shows that lipid molecules can flip-flop from one surface to another in the bilayer. Techniques have been developed for incorporating protein and other molecules into liposomes, and it has been observed that proteins facilitate lipid flip-flop in the bilayer. Liposomes shrink and swell osmotically as the activity of water in the surrounding aqueous phase is changed by additives. Depending on the surface phase state, the addition of cholesterol to the liposome may decrease the water permeability of the bilayer in vesicles. Transport of ions through the vesicle walls is thought to occur either through channels or via carriers. Since these surfaces are orders of magnitude more permeable to H^+ and OH^- compared to other univalent ions, transmission through the vesicle wall by a hydrogen-bonded network of water molecules is proposed just as the high mobility of these ions in water is explained.

Only a few of the observed properties of vesicles have been enumerated here. The examples cited are sufficient, however, to illustrate how these "synthetic cells" are ideally suited as models for research into the structure and functioning of cell membranes.

REVIEW QUESTIONS

1. Explain why surfactant aggregates form in solution.
2. What is the *critical micelle concentration* (CMC), and how is it measured?
3. Do different types of measurements of CMC lead to the same value for CMC? Explain.
4. In aqueous systems, the enthalpy change due to micellization is usually positive, and micellization is driven by entropy change. Explain the reason for the positive entropy change.
5. What is meant by the *optimal head group area* of a surfactant? What is the *packing parameter* $\mathcal{P}$? Explain how packing considerations can be used to determine the possible shapes of the micellar aggregates.
6. Consider a spherical micelle with charged head groups. What changes in shape would you expect as you increase the concentration of added electrolytes?
7. Describe the details of the structure of a micelle.
8. What is a *palisade layer*? How does one determine if there is water penetration in a micelle and, if there is any, where the water molecules are?
9. Under what conditions can the formation of micelles be described in terms of *reaction-equilibrium formalism*? When is the *phase equilibrium model* appropriate?

10. What does *solubilization in a micelle* mean?
11. Draw a typical phase diagram for an oil-water-surfactant system and identify at least some of the phases that can be found.
12. Why is catalysis using micelles as "microreactors" advantageous sometimes? When is it advantageous?
13. What are *reverse micelles*? When would you expect them?
14. What are *microemulsions*? How do they differ from coarse emulsions?
15. What is a *swollen micelle*?
16. Give some examples of lipids that form biological membranes.
17. Why do biological membranes and bilayers usually consist of double-tailed surfactants?

REFERENCES

General References (with Annotations)

Ball, P., *Designing the Molecular World—Chemistry at the Frontier*, Princeton University Press, Princeton, NJ, 1994. (Undergraduate level. A *Scientific-American*-style tour of chemistry at the molecular level and its implications to science and technology. Chapter 7, entitled "A Soft and Sticky World: The Self-Organizing Magic of Colloid Chemistry," is an excellent and engaging overview of surfactant-and polymer-based colloidal systems of interest in technology, biology, and medicine.)

Clint, J. H., *Surfactant Aggregation*, Chapman and Hall, New York, 1992. (Graduate level. An advanced monograph on surfactant solutions and on self-assembly ["aggregation"] in surfactant solutions. A research reference at the graduate level.)

Evans, D. F., and Wennerström, H., *The Colloidal Domain: Where Physics, Chemistry, Biology, and Technology Meet*, VCH Publishers, New York, 1994. (Undergraduate and graduate levels. A textbook on colloids covering many of the topics discussed in the present book, but presented almost exclusively from the point of view of surfactant systems and self-assembly.)

Israelachvili, J. N., *Intermolecular and Surface Forces*, 2d ed., Academic Press, New York, 1991. (Undergraduate and graduate levels. This book is an excellent starting point for advanced concepts on modeling micellization and on molecular and surface forces that drive self-assembly. Graduate level, but portions of the material should be accessible to undergraduate students.)

Myers, D., *Surfaces, Interfaces, and Colloids: Principles and Applications*, VCH Publishers, New York, 1991. (Undergraduate level. A qualitative overview of micelles, microemulsions, and their applications.)

Rosen, M. J., *Surfactants and Interfacial Phenomena*, 2d ed., Wiley, New York, 1989. (Reference. A good source of reference for properties of surfactants and surfactant solutions. Brings a strong industrial perspective.)

Tanford, C., *The Hydrophobic Effect. The Formation of Micelles and Biological Membranes*, 2d ed., Wiley, New York, 1980. (Undergraduate level. A classic reference by a pioneer on the hydrophobic effect on the relevance of surfactants to biological membranes.)

West, D. R. F., *Ternary Equilibrium Diagrams*, 2d ed., Chapman and Hall, New York, 1982. (A "manual" of ternary phase diagrams. Concise.)

Other References

Bergethon, P. R., and Simons, E. R., *Biophysical Chemistry: Molecules to Membranes*, Springer-Verlag, New York, 1990.

Goodsell, S., *The Machinery of Life*, Springer-Verlag, New York, 1993.

Hartley, C. S., *Aqueous Solutions of Paraffin Chain Salts*, Hermann and Cie, Paris, 1936.

McBain, J. W., *Trans. Faraday Soc.*, 9, 99 (1913).

Puvvada, S., and Blankschtein, D., *J. Phys. Chem.*, 93, 7753 (1989).

Tatar, H. V., *J. Phys. Chem.*, 59, 1195 (1955).

Wennerström, H., and Lindman, B., *Physics Reports*, 52, 1 (1979).

PROBLEMS

1. Use the data for sodium dodecyl sulfate in Table 8.1 to test the equation developed in Example 8.1 and to evaluate α for these micelles. Criticize or defend the following proposition: The α

values in the table are more accurate than that evaluated here because they are based on individual light scattering experiments and reflect the variation of α with changing salt concentration; by contrast, the α value calculated here is an average value based on the assumption that m/n, and hence α, is independent of salt concentration.

2. Shinoda* examined the variation of the CMC of various potassium alkyl malonates, $RCH(COOK)_2$, with the addition of univalent salts. For $R = C_8$, C_{12}, C_{14}, and C_{16}, the log-log plots of CMC verus counterion concentration produce parallel straight lines of slope -1.12. Criticize or defend the following proposition: According to the analysis presented in Example 8.1, the slope of this type of plot equals $-(1 - \alpha)$, meaning that $\alpha = -0.12$; this negative function apparently means that the micelle binds an excess of counterions and has the opposite charge from that expected.

3. The spectra of substituted pyridinium iodides are characterized by charge transfer bands involving the interaction of pyridinium and iodide ions. Mukerjee and Ray† showed that this band is shifted about 90 nm toward the red for dodecyl pyridinium iodide, which forms micelles, compared to methyl pyridinium iodide, which does not. They measured λ_{max} for the micelles in mixed solvents of variable relative dielectric constant and obtained the following results:

λ_{max} (nm)	281	284	288	298
ε_r	43	38	33	34

Estimate the effective dielectric constant at the surface of the micelle from the fact that λ_{max} occurs at 286 nm for dodecyl pyridinium iodide micelles in water. In light of the value estimated in Section 8.3 for the dielectric constant in the vicinity of solubilized benzene, does it seem likely that the value of ε_r for bulk water applies in the Stern layer?

4. Ionescu et al.‡ measured the CMC of hexadecyl trimethyl ammonium bromide in water-dimethyl sulfoxide (DMSO) mixtures at 25 and 40°C:

Mole fraction of DMSO	CMC $\times$ 10^3 (mole liter^{-1})	
	$T = 25°C$	$T = 40°C$
0.000	0.92	1.00
0.027	1.48	1.51
0.060	2.24	2.51
0.098	3.60	3.98
0.144	5.62	6.30
0.201	8.91	10.00
0.275	14.00	22.00
0.366	None	None

Use these data to evaluate ΔG_{mic}^0 for hexadecyl trimethyl ammonium bromide and to estimate (remember, only two temperatures were measured) ΔH_{mic}^0 and ΔS_{mic}^0. Discuss the values obtained in light of infrared and NMR experiments, which indicate formation of the stoichiometric compound $DMSO \cdot 2H_2O$ at $x_{DMSO} = 0.33$.

5. Both adsorption from solution and micellization occur as a result of the hydrophobic effect. To test the correspondence between these two effects, Rosen§ assembled ΔG^0 values for adsorption at the air-water interface and for micellization of a number of linear and branched surfactants. The following is a selection of these data:

*Shinoda, K., *J. Phys. Chem.*, **59**, 432 (1955).

†Mukerjee, P., and Ray, A., *J. Phys. Chem.*, **70**, 2144 (1966).

‡Ionescu, L. G., Tokuhiro, T., Czerniawski, B. J., and Smith, E. S., In *Micellization, Solubilization and Microemulsions*, Vols. 1 and 2 (K. Mittal, Ed.), Plenum, New York, 1979.

§Rosen, M. J., In *Micellization, Solubilization and Microemulsions*, Vols. 1 and 2 (K. Mittal, Ed.), Plenum, New York, 1979.

Compound	Temperature (°C)	ΔG^0 (kJ mole^{-1})	
		Adsorption	Micellization
$n\text{-}C_9SO_4Na$	25	−22.4	−17.3
$n\text{-}C_{14}SO_4Na$	25	−30.0	−25.2
$p\text{-}n\text{-}C_8\phi SO_3Na$	70	−27.7	−23.4
$p\text{-}n\text{-}C_{14}\phi SO_3Na$	70	−38.5	−34.1
$n\text{-}C_{10}SO_4Na$	50[a]	−27.8	−19.8
$n\text{-}C_{16}SO_4Na$	50[a]	−38.6	−30.4
$p\text{-}n\text{-}C_{12}\phi SO_3Na$	75	−35.6	−32.4
$p\text{-}C_6CHCH_2\phi SO_3Na$	75	−34.8	−28.7
$\quad\quad\; \overset{\mid}{\underset{}{C_4}}$			
$p\text{-}(C\!-\!\overset{\mid}{\underset{\mid C}{C}}\!-\!)_4\phi SO_3Na$	75	−34.4	−27.5

[a]Measured at the hexane-water inferface.

Use these data to criticize or defend the following propositions: (a) The CMC is a good indicator of a surfactant's adsorption effectiveness since the ΔG^0 values for adsorption and micellization both show parallel changes with increasing chain length. (b) The dodecyl benzene sulfonates have some of the most favorable ΔG^0 values among the data shown; this shows that branching has no adverse effect on either adsorption or micellization.

6. Danbrow and Rhodes* used potentiometric titration data to determine the distribution of benzoic acid (HB) between water and nonionic micelles. The commercial surfactant used has the average formula $C_{16}(OC_2H_4)_{24}$. They obtained the following results in 4% surfactant solutions:

$[HB]_0$ (mmole liter^{-1})	4.055	8.804	13.63	18.18	24.30	28.21
HB_m (mmole)	0.1859	0.3497	0.5067	0.6660	0.7790	0.9460

These authors consider two postulates: (a) If the benzoic acid is solubilized in the micellar core, then dimerization should occur, in which case $HB_m/[HB]^2 =$ const. (b) If the solubilization occurs at the micellar surface, then surface saturation analogous to Langmuir adsorption should occur. Test each postulate with the data provided and decide which gives the better fit.

7. Tokiwa and Aigami† used proton NMR to study the solubilization of benzyl alcohol, 2-phenyl ethanol, and 3-phenyl propanol in sodium dodecyl sulfate micelles. The upfield chemical shift of the aromatic protons in the micelle from their location in water was measured as a function of solubilization; it tends to increase with solubilization much like the results in Fig. 8.6. The following are some of the results obtained:

Compound	Chemical shift at 0.05 mole of solubilizate (100 g surfactant solution)$^{-1}$ (cps)	(alcohol)$_m$/ (alcohol)$_w$	mole alcohol/ mole surfactant
ϕCH_2OH	3.0	2.28	4.48
ϕCH_2CH_2OH	8.5	4.35	4.08
$\phi CH_2CH_2CH_2OH$	17.0	11.80	2.81

Criticize or defend the following proposition: The more carbon atoms there are in the alcohols, the more hydrophobic these compounds become and the more enriched the micelles become relative to the aqueous phase; the magnitude of the chemical shift increases as the extent of solubiliation in the micelles increases owing to the diamagnetic effect of the phenyl groups.

---

*Danbrow, M., and Rhodes, C. T., J. Chem. Soc., Supplement II and Indexes 6166 (1964).
†Tokiwa, F., and Aigami, K., Kolloid Z. Z. Polym., 246, 688 (1971).

8. Duynstee and Grunwald* present some experimental data for Reaction (F) in the presence of hexadecyl trimethyl ammonium bromide (CTABr, C = cetyl) and sodium dodecyl sulfate (NaLS, L = lauryl). Sodium hydroxide was the source of OH^- in all cases. A pseudo-first-order rate constant of 2.40×10^{-2} s^{-1} is observed for k_{CTABr}. Use the following absorbance data to evaluate k_{NaLS} for this reaction:

t (min)	0	4	7	10	13	17	25	35	41	51
Absorbance	0.734	0.652	0.618	0.577	0.537	0.489	0.408	0.327	0.293	0.244
	60	68	85	124	$\cdots$					
	0.211	0.189	0.160	0.129	0.155					

Recall that for first-order kinetics a plot of ln (fraction unreacted) versus time has a slope $-k$. Also note that the reaction reaches an equilibrium characterized by an absorbance 0.115; the data must be corrected for this. For both the anionic and cationic micelles, qualitatively sketch, emphasizing the charge state, the micelle, the solubilized substrate, and the approaching OH^- reactant. Indicate how these pictures are consistent with the experimental rate constants.

9. In the same research described in Problem 8, the authors also examined the rate of Reaction (F) in pure water and in NaCl solution to guarantee that the ionic surfactants were not displaying an electrolyte activity effect as is ordinarily observed with ion combination reactions. For 10^{-5} M crystal violet and 0.01 M NaOH, they observed the following:

Solution	Pure water	0.01 M NaCl	0.01 M CTABr
$k \times 10^4$ (s^{-1})	17.1	16.4	240

Use these data and the Debye–Hückel theory of electrolyte nonideality to criticize or defend the following proposition: Indifferent electrolytes always inhibit the rates of ion combination reactions because the activity coefficients are fractions. The data for CTABr show an enhancement of rate so this cannot be due to an activity effect. In these data, the k's for pure water and aqueous NaCl are essentially identical, so no activity effects operate in the absence of micelles either.

10. Bunton and Robinson† studied the effect of sodium dodecyl sulfate micelles on the rate of the reaction between OH^- and 2,4-dinitrochlorobenzene. These negative micelles have an inhibiting effect on the reaction, yet the kinetic data can be analyzed according to Equation (24). Use the following data and the authors' CMC value of 0.0064 M to estimate K/n, where K is the binding constant between the dodecyl sulfate micelles and the 2,4-nitrochlorobenzene:

$c_{NaC_{12}SO_4} \times 10^2$ (M)	0.0	1.41	1.85	2.80	3.90	5.60
$k \times 10^5$ (liter mole^{-1} s^{-1})	14.2	9.40	7.70	6.20	5.10	3.10

From the increased solubility of 2,4-dinitrochlorobenzene in sodium dodecyl sulfate solutions (without NaOH), the authors of this research estimate a K/n of 44. Criticize or defend the following proposition: The presence of OH^- could change the n value for sodium dodecyl sulfate micelles, so exact agreement between K/n values determined by the two methods is not necessarily expected.

11. The reaction of Problem 10 was studied at two different temperatures, and, from the temperature dependence of the rate constants, the authors† determined $\Delta H\ddagger$ and $\Delta S\ddagger$, the enthalpy and entropy of activation, respectively. The following values of these parameters were obtained in pure water and in 0.01 M sodium dodecyl sulfate (NaLS) and 0.01 M hexadecyl trimethyl ammonium bromide (CTABr):

#Duynstee, E. F. J., and Grunwald, E., *J. Am. Chem. Soc.,* **81**, 4542 (1959).
†Bunton, C. A., and Robinson, L., *J. Am. Chem. Soc.,* **90**, 5972 (1968).

Solvent	$\Delta H\ddagger$ (kcal mole^{-1})	$\Delta S\ddagger$ (cal K^{-1} mole^{-1})
Water	21.3	−5.2
CTABr	16.4	−13.4
NaLS	21.3	−5.6

Use these values to criticize or defend the following proposition: The smaller endothermic value for $\Delta H\ddagger$ in CTABr means the product molecules must be more readily expelled from these micelles, making the enthalpy contribution more favorable to the reaction in this case. The solubilized substrate has a higher entropy, so the decrease in entropy for the micellar reaction is larger. The last problem shows that the reaction occurs about twice as fast in water as in 0.01 M NaLS. The rate in water determines the kinetic parameters in the last case.

12. Chemistry students are certainly familiar with the regular tetrahedron (methane, sp^3 hybrids, etc.), but may not have considered this geometry as the basis for a quaternary phase diagram. McCarthy# discusses these as extensions of triangular phase diagrams: Each of the four faces of a regular tetrahedron is the ternary phase diagram that results as the concentration of component X in a system goes to zero. The opposite apex represents pure component X, and planes slicing through the tetrahedron parallel to any face have a constant percentage of X, the magnitude of which depends on their placement. Resketch one of the versions of Figure 8.13 using this kind of tetrahedral representation. This sould be done qualitatively and on a large enough scale to separate the various slices. Criticize or defend the following proposition: Both the tetrahedral and prismatic quaternary diagrams show essentially the same thing; in the tetrahedral diagram the advantage of having all four ternary diagrams contained therein is offset by the shrinking size of the slices as an apex is approached.

13. Proteins are polyamids formed from amino acids having the formula $H_2N-CHR-COOH$; therefore different R groups occur along the polymer backbone according to the amino acid sequence in the protein. Write structural formulas for the R groups in the following amino acids:

Alanine (Ala)*	Glycine (Gly)	Serine (Ser)*
Arginine (Arg)	Isoleucine (Ile)*	Threonine (Thr)
Asparagine (Asn)	Leucine (Leu)*	Tyrosine (Tyr)*
Aspartic acid (Asp)	Lysine (Lys)	Valine (Val)*
Glutamine (Gln)	Phenylalanine (Phe)*	

Kaiser and Kezdy† have suggested that helical structures for certain amino acid sequences in proteins may be stabilized by a hydrophobic interaction with cell membrane lipids. In the list above, R groups marked with an asterisk may be considered hydrophobic. Construct a model for residues 1–22 of human growth hormone releasing factor in the following way. Roll a piece of paper into a cylinder and sketch a helix on the surface. Evenly space the names of five consecutive amino acids along each turn of the helix. On a second cylinder, mark off portions of helix two turns long and carefully enter seven amino acids along each two-turn length. The models with 5 and 3.5 amino acid residues per turn are approximations of the π and α helical structures shown in protein crystals. Which of the two appears more effective in concentrating hydrophobic groups along one edge of the helix? The first 22 amino acids in this protein occur in the following order:

1. Tyr	9. Ser	16. Gln
2. Ala	10. Tyr	17. Leu
3. Asp	11. Arg	18. Ser
4. Ala	12. Lys	19. Ala
5. Ile	13. Val	20. Arg
6. Phe	14. Leu	21. Lys
7. Thr	15. Gly	22. Leu
8. Asn		

#McCarthy, P., *J. Chem. Educ.*, **60**, 922 (1983).
†Kaiser, E. T., and Kezdy, F. J., *Science*, **223**, 249 (1984).

14. On the basis of the model produced in Problem 13, criticize or defend the following proposition: A roughly 20-residue α-helix has a length approximately equal to the thickness of a membrane bilayer. A bundle of three, four, or more of these helices — with hydrophobic and hydrophilic sides — could orient themselves in such a way as to form a hydrophilic channel through the membrane (see Vignette I.2 and Figure 1.2). While the individual helices may be stable, their aggregation into this sort of channel involves bringing several membrane proteins together and is therefore entropically unfavorable.

9

Adsorption at Gas–Solid Interfaces

> When I cut through your plane as I am now doing, I make your plane a section which you, very rightly, call a Circle. For even a sphere—which is my proper name in my own country—if he manifest himself at all to an inhabitant of Flatland—must needs manifest himself as a Circle.
>
> From Abbott's *Flatland*

9.1 INTRODUCTION

9.1a Physisorption and Chemisorption

Adsorption at the solid-gas interface is traditionally subdivided into two broad classes: *chemisorption* and *physisorption* (i.e., physical adsorption).

- As the name implies, *chemisorption* comes very close to the formation of chemical bonds between the adsorbent (e.g., the solid) and the adsorbate (gas). In this case electron exchange between the adsorbent and the adsorbate occurs. Two consequences of this are that the associated heat effects are comparable to those that accompany ordinary chemical reactions and that the process is not always reversible. It is possible, for example, to adsorb (chemisorb) oxygen on carbon and desorb CO or CO_2.

- In *physical adsorption*, on the other hand, the energy effects are comparable to those that accompany physical changes such as liquefaction and are completely reversible for nonporous solids. In contrast to chemisorption, the adsorbent and the adsorbate in this case interact relatively weakly through van der Waals forces (see Chapter 10). Physical adsorption is the easier of the two types of adsorption and provides much background needed for an understanding of chemisorption.

A major portion of this chapter is concerned with physical adsorption, particularly from a global thermodynamic point of view. This is followed by a molecular-scale examination of crystalline surfaces and a brief discussion of chemisorption and its relevance to heterogeneous catalysis.

9.1b Focus of This Chapter

As should be evident from the discussions in Chapters 6 and 7, adsorption phenomena play a major role in colloid and surface chemistry. We also come across other examples in Chapters 11 and 13. Adsorption, especially at solid-gas interfaces, is very important in heterogeneous catalysis, as highlighted in Vignette IX. In this chapter, the focus is the introduction of quantitative measurement and the description of adsorption at solid-gas interfaces.

9.1b.1 Physisorption

There are several different ways in which the topics pertaining to physical adsorption can be subdivided. We are primarily concerned with nonporous solids, briefly discussing porous materials only in Section 9.7.

1.   It is convenient to divide the extent of adsorption into three categories: submonolayer, monolayer, and multilayer. We discuss them in this order. The thermodynamics of adsorption may be developed around experimental isotherms or around calorimetric data. We begin with the definition of adsorption isotherms and how they are determined experimentally (Section 9.2).

2.   Adsorption isotherms may be derived from a consideration of two-dimensional equations of state, from partition functions by statistical thermodynamics, or from kinetic arguments. Even though these methods are not fundamentally different, they differ in ease of visualization. We consider examples of each method in Sections 9.3 and 9.4.

3.   Multilayer adsorption and the popular Brunauer-Emmett-Teller (BET) method of analysis are described in Section 9.5. This section also describes the determination of specific areas by gas adsorption. Low-temperature N_2 adsorption and the BET method of analysis are so widely used for this purpose that these topics will receive special attention.

4.   A brief description of calorimetric analysis of adsorption follows in Section 9.6.

5.   Adsorption in porous solids, an important topic in catalysis and other areas, is presented in Section 9.7, in which the adsorption hysteresis and capillary condensation are introduced.

9.1b.2 Chemisorption

The global thermodynamic approach used in the above sections is insensitive to details at the atomic level and can only yield a gross characterization of the surface. Properties such as the specific surface area and the presence or absence of pores can be determined using the above approach since only the average surface—not atomic details—is involved. The existence of a distribution of surface energy sites can also be inferred from adsorption data, but the method falls short when it comes to specifics about this distribution. Observations on an atomic scale are needed to learn more about the details of the surface structure. Such observations comprise the subject matter of the last two sections of the chapter.

1.   A great assortment of high-technology (and high-cost!) methods exist for examining surfaces on a fine scale, and new methods are being developed continually, as we saw in Vignette I.8. The techniques available permit us to characterize solid surfaces in a variety of ways: pictorially, chemically, and crystallographically. Rather than attempt to catalog all of these and say a few things about each, it seems preferable to single out one example and develop it in a bit more detail. Accordingly, in Section 9.8 we discuss one such method—*low-energy electron diffraction* (LEED)—for studying surfaces.

2.   There are numerous other techniques that we shall not discuss but that also tend to be known by their initials, as evident from Vignette I.8 on STM (scanning tunneling microscopy), SPM (scanning probe microscopy), and AFM (atomic force microscopy) and related discussions in Chapter 1. The literature on this subject contains what Adamson (1990) calls "a veritable alphabet soup of designations, many of which are contrived acronyms." Table 8.1 of Adamson (1990) (which is eight pages long) provides an excellent summary of the numerous techniques that are currently available for surface analysis. These include a wide variety of *spectroscopic* methods for analyzing the composition of a surface both qualitatively and quantitatively. An example is *Auger electron spectroscopy* (AES). What makes AES so suitable for chemical analysis of surfaces is the shallow escape depth of Auger electrons from solid surfaces. Only those electrons originating within two or three atomic layers of the surface are able to escape without suffering the sort of inelastic interactions that obscures the features of the spectrum. Several additional techniques, including the ones based on the scanning probe method highlighted in Vignette I.8, are described in Hubbard (1995).

3.   Just as an assortment of experimental techniques exists, so too are there numerous

types of solid surfaces that might be examined. Not all methods can be used for all surfaces, and the techniques we consider are particularly well suited for the study of metal surfaces. Therefore we also limit our examples to metal surfaces (Section 9.9). Although this obviously excludes a number of interesting and important solid surfaces, it also includes some topics of great significance, including friction and corrosion in engineering materials, semiconductors for electronic devices, and catalytic surfaces for many chemical reactions.

VIGNETTE IX HETEROGENEOUS CATALYSIS AND CHEMICAL VAPOR DEPOSITION: Adsorption—Setting the Stage for Catalysis and Thin-Film Growth

Adsorption of gases at a solid surface can be a bother or a boon! It is a nuisance when one needs a very clean surface, devoid of any adsorbed species, for research or chemical analysis of a solid. On the other hand, it is a convenient tool in vacuum technology for pumping gases out of vacuum chambers (as in cryogenic pumping; see Section 9.8a.2). And, adsorption can be a vehicle for molecular engineering—as in the case of catalysis or in layer-by-layer fabrication of ordered films on surfaces! Perhaps more striking is the role of adsorption in the last context, and we highlight this in this vignette.

We have already noted in Vignette I.9 that selectivity of a catalyst is of extreme importance in practice. But, how does the surface of a solid serve as a catalyst? How does adsorption enter the picture? What issues are of interest in this context? We already touched on the answers to the first two questions in Vignette I.9. A further elaboration of these is presented schematically in Figure 9.1. The adsorption of a molecule on a surface contributes to the disassembly of the molecule and allows the molecule to react with the other reactants in the vicinity. Adsorption can either promote or *hinder* a reaction or may contribute to "side" reactions that generate undesirable products. This is the reason selectivity of a surface is of utmost importance. Depending on the strength of the adsorbate-adsorbent bond, an adsorbed molecule (or a disassembled part of it) may become free to wander along the surface and to participate in chemical reactions and in the nucleation and growth of islands of two-dimensional surface phases. Therefore, quantifying the energetics of adsorption, the amount of adsorbed material, the chemical nature of products (if any), and their relation to the atomic-level structure of the surface itself become issues of paramount importance not only in catalysis, but also in the technology of thin-film growth on surfaces. These issues serve as the motivation for the types of theoretical and experimental procedures we develop in this chapter and for the interest in tools such as scanning probe microscopy and spectroscopy highlighted in Vignette I.8.

As is evident from the above comments, adsorption is also the embryonic event that precedes nucleation and growth of oriented films—*epitaxial films*—on surfaces (if the conditions are right; see Figure 9.2). Chemisorption is again the stage-setter in this case (as is often the case in heterogeneous catalysis). Epitaxial growth (which is not restricted to gas-solid systems) has significant industrial importance and depends on, among other things, interactions among adsorbed molecules, adsorbent/adsorbed-molecule interactions, structure of the surface (also known as the substrate), temperature, and surface diffusion of adsorbed molecules.

Many of the issues relevant to the above examples of adsorption (and the attendant) processes are beyond the scope of the present chapter. The topics we cover—such as adsorption isotherms, relating bulk phase conditions to the extent of adsorption and surface phases, and low-energy electron diffraction (LEED) for studying surface structure—are the essential first steps in understanding the problems and the prospects engendered by adsorption phenomena.

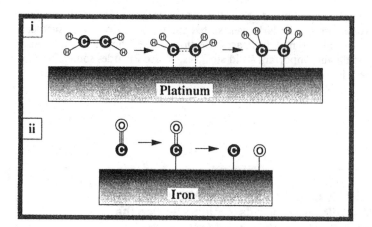

FIG. 9.1 A schematic illustration of the role of a substrate on adsorption and catalysis. Sketch (i), on chemisorption of ethylene on platinum, shows the breaking of C=C double bond so that the carbon atoms can form bonds to the surface. Sketch (ii) shows how iron substrate disassembles carbon monoxide into carbon and oxygen atoms. Breaking these bonds in gaseous phases requires a considerable amount of energy, whereas a catalytic surface allows these under moderate conditions. The broken fragments then may wander on the surface and react with other fragments or molecules in the gas. (Adapted and redrawn with permission from Ball 1994.)

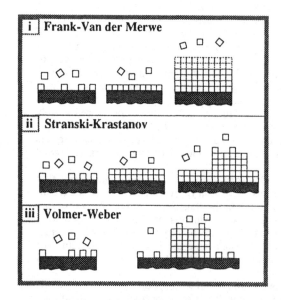

FIG. 9.2 A schematic representation of three topologically different epitaxial growth modes for crystallization on surfaces through adsorption. Sketch (i) illustrates layer-by-layer growth, known as *Frank-Van der Merwe growth*. Sketch (iii) illustrates just the opposite: Crystallites nucleate and grow on immediate contact between the substrate and the adsorbate. This is known as *Volmer-Weber growth*. Sketch (ii), known as *Stranski-Krastanov growth*, is an intermediate mode. (Adapted from A. Zangwill, *Physics at Surfaces*, Cambridge University Press, Cambridge, England, 1988.)

9.2 EXPERIMENTAL AND THEORETICAL TREATMENTS OF ADSORPTION: AN OVERVIEW

9.2a Adsorption Isotherms

Adsorption experiments are conducted at constant temperature, and an empirical or theoretical representation of the amount adsorbed as a function of the equilibrium gas pressure is called an *adsorption isotherm*. Adsorption isotherms are studied for a variety of reasons, some of which focus on the adsorbate while others are more concerned with the solid adsorbent. In Chapter 7 we saw that adsorbed molecules can be described as existing in an assortment of two-dimensional states. Although the discussion in that chapter was concerned with adsorption at liquid surfaces, there is no reason to doubt that similar two-dimensional states describe adsorption at solid surfaces also. Adsorption also provides some information about solid surfaces. The total area accessible to adsorption for a unit mass of solid — the specific area A_{sp} — is the most widely encountered result determined from adsorption studies. The energy of adsorbate-adsorbent interaction is also of considerable interest, as we see below.

9.2b Adsorption: Some Experimental Considerations

We saw in Chapter 6, Section 6.7, that solid surfaces are notoriously heterogeneous, particularly with respect to roughness and chemical composition. From the point of view of specific area determination, roughness is not too troublesome since the adsorbed gas molecules can generally cover the hills and valleys of the surface with ease. Pores with very small dimensions pose more of a problem. For now, we assume that such pores are absent; we take up the question of adsorption on porous solids in Section 9.7. Chemical heterogeneity affects the energetics of adsorption. For simplicity, we often assume that the surface is characterized by a single adsorption energy. Actually, a distribution of surface sites with differing adsorption energies may be present, and some indication of this may be extracted from adsorption data. As an approach to surface characterization, adsorption studies are indirect and give average rather than specific descriptions. We have already seen in Vignette I.8 that solid surfaces can be probed more directly for information on a molecular scale using scanning probe spectroscopy; we see below in this chapter (Section 9.8) how low-energy electron diffraction can be used to study the structure of surfaces on a molecular scale. For solids with high specific area, however, gas adsorption is the method of choice for quantifying this feature; we have more to say about it in Section 9.5.

9.2b.1 Experimental Determination of Adsorption Isotherms

Adsorption studies for the experimental determination of adsorption isotherms are conducted in a vacuum apparatus from which all gases can be removed prior to the addition of the adsorbate being studied. After the solid is introduced into the sample tube and the tube is attached to the vacuum line, it is generally pretreated by some sort of degassing procedure. This is a combination of heating and pumping to ensure the removal of physically adsorbed contaminants. Consideration must be given to the possibility of changing the solid when the temperature of degassing is selected. A heat treatment that is too vigorous may result in changes in any chemisorbed layer, which — from our point of view at least — amounts to a change in the adsorbent itself. If extensive enough, such changes may alter the surface area of a solid. Even less drastic changes are sufficient to alter the adsorption energy of a surface.

The range of pressures over which adsorption studies may be conducted is — in principle — from zero to p_0, the saturation pressure or the normal vapor pressure of the material at the temperature of the experiment. At the low-pressure end of this range adsorption will be slight, so the determination of the isotherm involves measuring small differences in pressure at low pressures. This is not easy to do experimentally, although relatively modern low-pressure techniques have greatly extended this region. As the pressure approaches p_0, adsorption often increases rapidly as if anticipating phase separation at the surface (resulting from multilayer adsorption in which most of the adsorbed molecules behave as if they were in the bulk liquid state). As a matter of fact, if the solid is porous, the vapor may actually condense in the small pores at $p < p_0$ (see Section 9.7).

Figure 9.3 is a sketch of an apparatus that can be used to determine the equilibrium extent of gas adsorption as a function of pressure. We outline how such an experiment is conducted at ambient temperature, even though adsorption studies are frequently conducted at low temperatures, particularly when determination of A_{sp} is the objective of the experiment. A known mass of adsorbent is introduced into the sample tube and degassed as described above. Then the following set of pressure-volume readings are made, described here in terms of Figure 9.3.

1. The sample tube and gas burette are evacuated and then a nonadsorbing gas—frequently helium—is introduced into the gas burette. The burette is graduated with respect to volume and also serves as one leg of a manometer so that both the volume and pressure of gas in the burette can be measured. Ambient temperature is assumed to apply throughout.

2. The three-way stopcock is opened to connect the gas burette with the sample tube. The new pressure and volume are read. From this, the volume of the dead space—the volume beyond the three-way stopcock that is not occupied by sample—can be determined (see Example 9.1).

3. The nonadsorbed gas is pumped out and replaced by the adsorbate. Its volume, pressure, and temperature are measured, and from these the number of moles of gas introduced (initially) into the apparatus can be determined.

4. The three-way stopcock is opened to connect the burette and sample tube, and volume and pressure are measured again. Since the dead space is known, the (final) number of moles of gas can be calculated. The difference between the initial and final number of moles gives the number of moles adsorbed.

5. This amount of adsorbed gas is in equilibrium with bulk gas at the pressure read in Step 4.

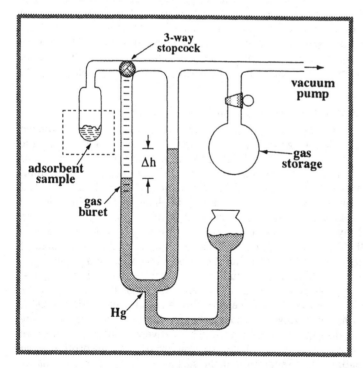

FIG. 9.3   Schematic illustration of a gas adsorption apparatus.

These steps describe the determination of a single point on an adsorption isotherm. By adjusting the mercury level, we can increase the pressure of the equilibrium gas with more adsorption occurring. Steps 3 and 4 are thus repeated until the full isotherm is mapped. Example 9.1 illustrates numerically how a point on the isotherm is established.

* * *

EXAMPLE 9.1 *Construction of Adsorption Isotherms*: The following pressure-volume (p-V) data were collected at a temperature of 22°C. The V's are volumes in the gas burette, and the numerical subscripts refer to the steps itemized above.

With helium, p_1 = 21.71 torr, V_1 = 12.90 cm^3, p_2 = 16.50 torr, V_2 = 10.90 cm^3. With adsorbate, p_3 = 12.85 torr, V_3 = 13.70 cm^3, p_4 = 3.24 torr, V_4 = 5.00 cm^3. What is the volume of the dead space? How many moles are adsorbed at the final equilibrium pressure, 3.24 torr?

Solution: Successive applications of the ideal gas law allow us to calculate the desired quantities.

The total volume to which the gas has access after the stopcock is opened is the sum of the burette volume and the dead space V_d. Therefore $V_d + V_2 = p_1V_1/p_2 = (21.71)(12.9)/(16.50) =$ 16.97, or V_d = 6.07 cm^3. *It is convenient to use R = 62,360 cm^3 torr K^{-1} mole^{-1}* as the value of the gas constant in these calculations. The initial moles of adsorbate are given by $n_i = p_3V_3/RT$ = (12.85)(13.70)/(62360)(295) = 9.57 · 10^{-6}. After adsorption equilibrium is established, n_f = (3.24)(5.00 + 6.07)/(62360)(295) = 1.95 · 10^{-6}. The difference $n_i - n_f$ = (9.57 − 1.95) · 10^{-6} = 7.62 · 10^{-6} mole gives the amount adsorbed at an equilibrium pressure of 3.24 torr. ∎

* * *

If the adsorption isotherm is to be determined at some temperature other than room temperature—liquid nitrogen temperature, for example—the sample tube is placed in a suitable thermostat. This is indicated by the dotted line in Figure 9.3. In this case two sets of readings are made with the nonadsorbed gas, one at room temperature and one with the thermostat in place. In this way the partitioning of the dead space between the two temperature regions can be determined. Several additional considerations should be cited that are important in actual practice:

1. The gases used for adsorption must be of high purity.
2. The gas burette should itself be thermostated unless the laboratory has very good temperature control.
3. The nonideality for the adsorbate at low temperatures must be taken into account unless the volume of the low-temperature dead space is minimized.
4. Sufficient time must be allowed for equilibrium to be established. A way of checking this experimentally is to lower the pressure and observe that desorption follows the same curve as adsorption.

9.2b.2 Classification of Adsorption Isotherms

Experimental gas adsorption isotherms are traditionally classified into one of the five types shown in Figure 9.4.

- The Type I isotherm is reminiscent of Figure 7.16, the *Langmuir isotherm*. The plateau is interpreted as indicating monolayer coverage. We see that this type of behavior implies a sufficiently specific interaction between adsorbate and adsorbent to be more typical of chemisorption than physical adsorption.
- Type II adsorption, by contrast, is widely observed with physical adsorption and is interpreted to mean multilayer adsorption.
- We see in Section 9.5b (Example 9.5) that Type III behavior occurs when the heat of liquefaction is more than the heat of adsorption.
- Types IV and V are analogs of Types II and III, except for the leveling off that occurs at pressures below p_0. These cases are associated with porous solids in which the adsorbate condenses in the small pores at $p < p_0$.

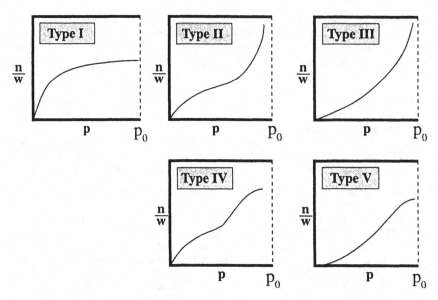

FIG. 9.4 Qualitative shapes of the five general types of gas adsorption isotherms. (See text for a discussion of their physical significance.)

If the specific area of the solid is the information sought, it is the amount of adsorption at monolayer coverage that must be measured. This is readily available in Type I adsorption, but requires considerable interpretation when multilayer adsorption takes place. Once determined, however, the number of molecules required to saturate a surface times the area occupied per molecule gives the surface area of a sample. This divided by the mass of the sample gives A_{sp}. In addition, once the adsorption at monolayer coverage is identified, all other extents of adsorption can be expressed as fractions or multiples of the monolayer.

9.2c   Adsorption Isotherms: Theoretical Considerations

This makes a convenient point of contact with theory since models for adsorption inevitably subdivide the surface into an array of adsorption sites that gradually fill as the pressure increases. If θ is defined as the fraction of sites filled, then $\theta = 1$ corresponds to monolayer coverage, with $\theta < 1$ or $\theta > 1$ to submonolayer and multilayer coverages, respectively. Theoretical isotherms predict how θ varies with p in terms of some particular model for adsorption. It turns out that a set of experimental points can often be fitted by more than one theoretical isotherm, at least over part of the range of the data; that is, theoretical isotherms are not highly sensitive to the model on which they are based. A comparison between theory and experiment with respect to the temperature dependence of adsorption is somewhat more discriminating than the isotherms themselves.

Since it is relatively easy to fit experimental adsorption data to a theoretical equation, there is some controversy as to what constitutes a satisfactory description of adsorption. From a practical point of view, any theory that permits the amount of material adsorbed to be related to the specific surface area of the adsorbent and that correctly predicts how this adsorption varies with temperature may be regarded as a success. From a theoretical point of view, what is desired is to describe adsorption in terms of molecular properties, particularly in terms of an equation of state for the adsorbed material, where the latter is regarded as a two-dimensional state of matter.

As we see in the course of the chapter, these two approaches frequently clash. The adsorption isotherm of Brunauer, Emmett, and Teller (BET), which is discussed in Section 9.5, is an excellent example of this. The model on which the BET isotherm is based has been criticized by many theoreticians. At the same time, the isotherm itself has become virtually the standard equation for determining specific areas from gas adsorption data. Ross (1971) summarizes the situation effectively by comparing it to a master chef who concocts a palatable dish out of an old shoe. For some, the end result is what matters: a palatable dish. For others, the starting material dominates their opinions: the old shoe. To present a relatively accurate picture of the current state of affairs in this area, it is necessary to present both of these viewpoints. We attempt not to take too one-sided a position, but to give some indication of each side.

9.3 THERMODYNAMICS OF ADSORPTION: PHENOMENOLOGICAL PERSPECTIVE

9.3a Relating Equations of State to Isotherms

To see how the equation of state of two-dimensional matter and the adsorption isotherm are related, we return to the Gibbs equation (Equation (7.46)):

$$-d\gamma = \Gamma_2 d\mu_2 \tag{1}$$

Since we are concerned here with adsorption from the gas phase, the chemical potential may be related to the pressure of the gas by

$$\mu_2 = \mu_2^0 + RT \ln (fp) \tag{2}$$

where f is the activity coefficient (see Equations (7.47) and (7.48)). In this discussion we assume that the gas behaves ideally, although in analyzing experimental results it may be necessary to include the correction required by the nonideality of the gas. Combining Equations (1) and (2) leads to

$$-d\gamma = RT\Gamma_2 \, d\ln p \tag{3}$$

In the present context it is convenient to use Equation (7.43) to eliminate Γ_2 from Equation (3) and express the surface excess as the number of moles adsorbed per unit area:

$$-d\gamma = \frac{nRT}{A} \, d\ln p \tag{4}$$

Next we substitute the product of sample weight times specific area for A in this equation to obtain

$$-d\gamma = \frac{RT}{A_{sp}} \frac{n}{w} \, d\ln p \tag{5}$$

For a given adsorbent and an isothermal experiment, T and A_{sp} are constants. The ratio n/w is the equilibrium amount adsorbed that will be a function of p. Therefore Equation (5) may be integrated as follows:

$$-\int d\gamma = \frac{RT}{A_{sp}} \int \frac{n}{w} \, d\ln p \tag{6}$$

The constant of integration may be evaluated by recognizing that n/w goes to zero as p approaches zero. Under these circumstances $\gamma \to \gamma_0$; therefore Equation (6) becomes

$$\gamma_0 - \gamma = \pi = \frac{RT}{A_{sp}} \int_0^p \frac{n}{w} \, d\ln p \tag{7}$$

If the experimental isotherm (n/w as a function of p) is known, then Equation (7) may be integrated either analytically or graphically to give the two-dimensional pressure as a function of coverage. This relationship therefore establishes the connection between the two-and three-dimensional pressures that characterize the surface and bulk phases. This is how adsorption data could be used to determine the film pressure in equilibrium with a drop of bulk liquid on a solid surface as discussed in Section 6.6b.

9.3b   Ideal Behavior: The Henry Law Limit

As an illustration of the kind of information obtainable from Equation (7), suppose we consider the situation in which the equilibrium adsorption of a gas is described by the isotherm

$$n/w = mp \tag{8}$$

where m is a constant. Equations (7) and (8) may be combined to give

$$\pi = \frac{RT}{A_{sp}} \int_0^p mp \, \frac{dp}{p} = \frac{RT}{A_{sp}} mp = \frac{N_A k_B Tn}{A_{sp} w} \tag{9}$$

Since the quantity $(A_{sp}w/nN_A)$ equals σ, area per molecule, Equation (9) may be written

$$\pi\sigma = k_B T \tag{10}$$

the two-dimensional ideal gas law for the surface phase!

The adsorption isotherm — Equation (8) — associated with this surface equation of state is called the *Henry law limit*, in analogy with the equation that describes the vapor pressure of dilute solutions. The constant m, then, is the adsorption equivalent of the Henry law constant. When adsorption is described by the Henry law limit, the adsorbed state behaves like a two-dimensional ideal gas.

9.3c   Deviations from Ideality

Equation (8) may also be written as

$$\theta = m'p \tag{11}$$

if the specific area of the adsorbent and the cross-sectional area of the adsorbate are known (Equation (7.71)). We see, therefore, that compliance with Henry's law implies that a log-log plot of θ versus p yields a straight line of unit slope. Figure 9.5 shows some experimental results for adsorption at 77.4K plotted in this way for pressures down to 10^{-10} torr. Note that even at these low pressures the Henry law limit is not yet reached (it would give a 45° line in Fig. 9.5), although argon on Pyrex appears to be approaching this limiting behavior. Note, further, that any errors introduced in evaluating θ would vertically shift the curves but would not change their slopes. We may conclude, therefore, that — just as with monolayers on aqueous substrates — the compliance of an adsorbed layer with the ideal gas law is a form of behavior that is extremely difficult to observe. The implication of this is that significant departures from two-dimensional ideality already set in at very low surface coverage.

9.3c.1   Adsorbent-Adsorbate Interactions

In general, there are two different types of interactions in which adsorbed molecules may participate. They are the interaction between the adsorbed molecules and the adsorbent and the interaction between the adsorbed molecules themselves. In the Henry law region, we have seen that the adsorbed layer behaves ideally. This is to be expected in view of the low surface concentration (θ very small) of adsorbate. The adsorbed molecules definitely do interact with the adsorbent, however, and at very low coverage the interaction energy might be very sensitive to surface heterogeneity. Any "hot spots" on the surface would adsorb first, less energetic patches next, then the normal surface sites. In Sections 9.4b and 9.6 we see how isotherms measured at several different temperatures may be interpreted to yield information on the energy of adsorption.

9.3c.2   Adsorbate-Adsorbate Interactions

The second type of interaction possible for adsorbed molecules is direct adsorbate-adsorbate interaction. Interactions of this sort are expected to lead to deviations from ideality in the two-dimensional phase just as they lead to deviations from ideal behavior for bulk gases. In this case surface equations of state, which are analogous to those applied to nonideal bulk gases, are suggested for the adsorbed molecules. The simplest of these allows for an excluded area correction (see Equation (7.23)):

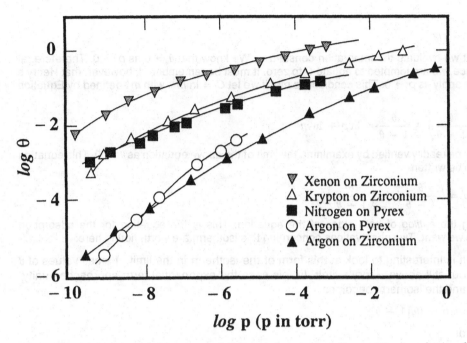

FIG. 9.5 A log-log plot of θ versus p for xenon, krypton, and argon on zirconium and nitrogen and argon on Pyrex. (Redrawn with permission from J. P. Hobson, "Physical Adsorption at Extremely Low Pressures." In *The Solid-Gas Interface*, Vol. 1 (E. A. Flood, Ed.), Marcel Dekker, New York, 1967.)

$$\pi(\sigma - \sigma^0) = k_B T \tag{12}$$

The use of the above equation of state to construct the corresponding adsorption isotherm is illustrated in Example 9.2.

* * *

EXAMPLE 9.2 *Evaluating Isotherms from Equations of State: Finite-Size Effects of Adsorbates.* Use Equation (12) along with Equation (7) to derive an adsorption isotherm that accounts for excluded-area adsorbate-adsorbate interactions.

Solution: According to Equation (7), $d\pi = -d\gamma$. We now recall that

$$\sigma = \frac{A_{sp} w}{n N_A}$$

Therefore Equation (5) may be written as

$$\frac{\sigma \, d\pi}{k_B T} = d \ln p$$

From Equation (12), $d\pi$ is given by

$$d\pi = -k_B T (\sigma - \sigma^0)^{-2} \, d\sigma$$

Combining the above two equations we get

$$-\frac{\sigma \, d\sigma}{(\sigma - \sigma^0)^2} = d \ln p$$

which on integration leads to

$$\ln\left(\frac{\theta}{1 - \theta}\right) + \frac{\theta}{1 - \theta} = \ln p + C$$

where

$$\sigma/\sigma^0 = 1/\theta$$

Next we evaluate the integration constant C. We know that $\theta \to 0$ as $p \to 0$. Therefore, at first glance, we are tempted to equate C to zero. It must be remembered, however, that Henry's law must apply as $p \to 0$. This condition is met if we let $C = \ln m'$, with m' defined by Equation (11):

$$\ln\left(\frac{\theta}{1-\theta}\right) + \frac{\theta}{1-\theta} = \ln p + \ln m'$$

This may be readily verified by examining the limit of the above equation as $\theta \to 0$. This equation may also be written

$$m'p = \left(\frac{\theta}{1-\theta}\right)\exp\left(\frac{\theta}{1-\theta}\right)$$

by taking the *antilog* of both sides of the equation. This is the equation for the adsorption isotherm we want. Two observations concerning this isotherm are worth noting here:

1. It is interesting to look at this form of the isotherm in the limit of small values of θ but still above Henry's limit. In this case the exponential term approaches unity, and the isotherm becomes

   $$m'p = (\theta/[1-\theta])$$

   or

   $$\theta = \frac{m'p}{1 + m'p}$$

   which is identical in form to the Langmuir equation (see Equation (7.67)).

2. The isotherm also reveals that for $m'p \geqslant 1$, θ approaches unity as an upper limit. Thus at both the upper and lower limits, the isotherm gives the same results as the Langmuir equation. At intermediate values the two functions differ slightly, but it would probably be difficult to distinguish between them in fitting experimental data.   ■

* * *

9.3c.2a   Van der Waals Isotherm and Surface Phases.   This approach can be extended to other forms of two-dimensional equations of state. For example, still greater nonideality in the two-dimensional equation of state might be represented by the *van der Waals* analog:

$$\left(\pi + \frac{a}{\sigma^2}\right)(\sigma - b) = k_B T \tag{13}$$

in which the van der Waals factor b and σ^0 from Equation (12) are identical in principle. Table 9.1 shows the isotherm resulting from this equation; the table also lists the result of Example 9.2 and one other example that leads to what is known as the *Harkins-Jura isotherm*. We return to the Harkins-Jura isotherm below in this section.

What makes the two-dimensional van der Waals equation of state especially interesting is the fact that there exists a temperature above which there is only one real root to Equation (13) and below which some values of π correspond to three values of θ. This situation is shown schematically in Figure 9.6a. With bulk gases and also insoluble monolayers on water, the three-root region is identified with a region of two-phase equilibrium. Is there any evidence for this type of phase equilibrium in the two-dimensional layer of adsorbed gas on a solid substrate? Figure 9.6b shows several data for the adsorption of krypton on specially treated graphite over a range of temperatures from about 77 to 91K. These plots are adsorption isotherms, not π versus σ diagrams, but nevertheless there is quite clear evidence of a two-phase region. A few additional assumptions lead to the prediction that the two-dimensional critical temperature should be one-half the value of the three-dimensional critical temperature

TABLE 9.1 Examples of Adsorption Isotherms Based on Some Two-Dimensional Equations of State

Two-dimensional equation of state	Isotherm	Comments
$\pi(\sigma - \sigma^0) = k_B T$	$m'p = \left(\dfrac{\theta}{1-\theta}\right)\exp\left(\dfrac{\theta}{1-\theta}\right)$	1. The equation of state accounts for only the excluded-area interaction. 2. Reduces to Equation (11) as $\theta \to 0$. 3. Reduces to the Langmuir equation for low θ's above the Henry limit.
$\left(\pi + \dfrac{a}{\sigma^2}\right)(\sigma - b) = k_B T$	$m'p = \dfrac{\theta}{1-\theta}\exp\left(\dfrac{\theta}{1-\theta} - \dfrac{2a\theta}{bk_B T}\right)$	1. The coefficients a and b are two-dimensional analogs of the three-dimensional van der Waals coefficients. 2. Reduces to the above case when $a = 0$.
$\pi = -C_1\sigma + C_2$	$\ln p = -\dfrac{C_1}{2N_A RT}A_{sp}^2\left(\dfrac{n}{w}\right)^{-2} + C_3$	1. Surface behaves like an incompressible bulk phase. 2. Known as the *Harkins-Jura Isotherm*.

according to this model. The two T_c values for Kr are 86K and 210K, and the corresponding values for argon are 65 and 151K.

There is another feature in Figure 9.6b that deserves additional comment. It is the existence of a second set of vertical segments in the isotherms at values of θ in the range of 0.65–0.78. This suggests a second phase equilibrium in the two-dimensional matter, perhaps something analogous to the transition between the *LE* and *LC* states of monolayers on water (see Chapter 7) or to a liquid-solid transition. Indeed, beyond the second two-phase region, the slope of the isotherm changes sharply in a way that corresponds to a much less compressible surface state.

Before exploring the consequences of this feature, it is first necessary to make certain that the result is not just an artifact arising from surface heterogeneities. We noted that the graphite substrate for which the data of Figure 9.6b were collected was "specially treated." A few details of this treatment should be mentioned. Graphite and anhydrous $FeCl_3$ are introduced into opposite ends of a Pyrex tube; the evacuated tube is then sealed and brought to about 300°C. A slight temperature gradient is maintained between opposite ends of the tube so the $FeCl_3$ distills to the graphite, where it reacts to form an interlamellar compound. Rapidly heating this compound to 900°C or higher results in the expulsion of the $FeCl_3$ and the attendant delamination of the graphite. The specific surface area of the final product—known as exfoliated graphite—is roughly two orders of magnitude greater than that of the initial graphite sample. The newly exposed graphite planes constitute a remarkably uniform solid adsorbent if all traces of $FeCl_3$ are removed. There is some evidence that the same vertical segments are observed in isotherms measured on the parent graphite as are observed in the exfoliated graphite. This argues that these features are not artifacts of the delamination process.

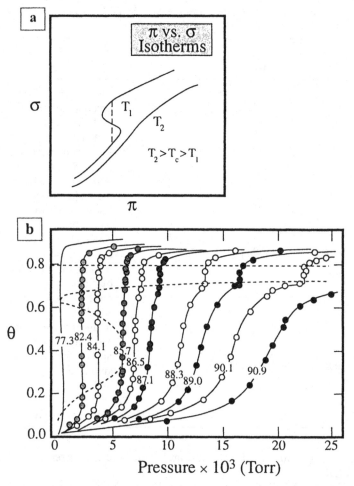

FIG. 9.6 Adsorption isotherms and surface phases. (a) schematic illustration of π versus σ isotherms in the vicinity of a two-dimensional critical temperature. (b) experimental data for the adsorption of krypton on exfoliated graphite showing similar features. (Data from A. Thomy and X. Duval, *J. Chem. Phys.*, **67**, 1101 (1970).)

9.3c.2b Harkins-Jura Isotherm. The Harkins-Jura equation (see Table 9.1) suggests that a plot of $\ln p$ versus $(n/w)^{-2}$ should be a straight line with a slope that is proportional to the square of the specific area. In experiments with solids of known area the linearity predicted by the Harkins-Jura equation has been observed. Furthermore, the proportionality constant relating the observed slopes to A_{sp}^2 is independent of the nature of the adsorbent to a first approximation. Thus solids for which A_{sp} is known may be used to "calibrate" this method for a particular adsorbed species; then the specific area of an unknown may be determined using the same adsorbate. The Harkins-Jura isotherm therefore introduces the possibility of determining specific areas by gas adsorption studies, although it is only one of many isotherms that permit the evaluation of A_{sp}.

Although it may give satisfactory values for A_{sp}, the Harkins-Jura equation leaves something to be desired at the molecular level. For example, the linear π versus σ equation of state—the starting point of the derivation of the Harkins-Jura isotherm—represents the relatively incompressible state of the surface phase (i.e., $\theta = 0.7$ in Fig. 9.6b). (This equation is obtained in analogy with the approximately linear π versus σ equation for insoluble monolayers discussed in Chapter 7.) However, in most instances of physical adsorption, no saturation limit of adsorption appears. As $p \rightarrow p_0$ the amount of material adsorbed increases asymptotically. Multilayer adsorption is the only reasonable model for this observation. Except for special cases, the neighborhood of $\theta = 1$ is obscured by the onset of multilayer adsorption. In other words, there is a mismatch between the situation described by the Harkins-Jura model and that suggested by macroscopic observations. We noted above that many isotherms are insensitive to the assumptions of their derivation. In line with that observation is the fact that the Harkins-Jura equation does fit a fairly wide range of experimental data and gives reasonable values of specific surfaces despite these objections. This is one example of an "old shoemaster chef" situation; we see presently that it is only one of several such cases.

9.3d Summary of Observations

Several points might be noted in summarizing the results of Section 9.3:

   1.   In principle, it is possible to correlate an adsorption isotherm and a two-dimensional equation of state by working from either direction; that is, we may start with an experimental isotherm and develop the associated equation of state (as in going from Equation (8) to Equation (10)), or we may proceed from the equation of state to the isotherm (as illustrated in Example 9.2).

   2.   Relatively small increases in complexity for the equation of state result in considerably more complex equations for the adsorption isotherms. The gross features of the more complex isotherms are also given by simpler isotherms. This means that it is very difficult to choose among various isotherms in terms of the goodness of fit to experimental data. Therefore it is difficult to conclude from an experimental isotherm what the two-dimensional surface phases are like.

   3.   Two-dimensional equations of state are a useful source of isotherms, however, even though the test of the isotherm must be made in terms of some criterion other than an ability to describe adsorption. For example, the ability of an isotherm to predict the temperature dependence of adsorption or the specific area of an adsorbent is a more sensitive test of an isotherm than merely describing the way n/w increases with p.

   Our approach until now has been to discuss adsorption isotherms on the basis of the equation of state of the corresponding two-dimensional matter. This procedure is easy to visualize and establishes a parallel with adsorption on liquid surfaces (Chapter 7); however, it is not the only way to proceed. In the following section we consider the use of statistical thermodynamics in the derivation of adsorption isotherms and examine some other approaches in subsequent sections.

9.4   THERMODYNAMICS OF ADSORPTION: A STATISTICAL PERSPECTIVE

The *partition function* is the central feature of statistical thermodynamics. From the partition function the various thermodynamic variables such as entropy, enthalpy, and free energy may be evaluated. It is also possible, in principle, to deduce the equation of state for a system from the partition function.

   It should be apparent—since an adsorption isotherm can be derived from a two-dimensional equation of state—that an isotherm can also be derived from the partition function since the equation of state is implicitly contained in the partition function. The use of partition functions is very general, but it is also rather abstract, and the mathematical difficulties are often formidable (note the cautious "in principle" in the preceding paragraph). We shall not attempt any comprehensive discussion of the adsorption isotherms that have been derived by the methods of statistical thermodynamics; instead, we derive only the Langmuir equation for adsorption from the gas phase by this method. The interested reader will find other examples of this approach discussed by Broeckhoff and van Dongen (1970).

9.4a   Preliminaries: Statistical Thermodynamics of Bulk Gases

A brief review of the statistical thermodynamics of ideal (bulk) gases will help us get started. In addition to reviewing some relevant physical chemistry, it will supply us with some expressions that may be useful since the two-dimensional ideal gas law applies to adsorbed molecules as a limiting case.

The partition function Q is defined by the equation

$$Q = \sum_i g_i \exp\left(- \frac{\varepsilon_i}{k_B T} \right) \tag{14}$$

in which g_i and ε_i represent the *degeneracy* and the *energy*, respectively, of the ith state. An important property of a partition function is its factorability into contributions arising from translation and internal degrees of freedom:

$$Q = Q_{trans} Q_{int} \tag{15}$$

For N *indistinguishable* molecules that do not interact with each other, the total partition function Q_N is given by $(1/N!)$ times the Nth power of $(Q_{trans} Q_{int})$, where we take the Q's without the subscript N as the partition functions for individual molecules. The translational portion of the partition function is relatively easy to evaluate for ideal gases (see, for example, Atkins 1994). Substituting its value into Equation (15) gives

$$Q_N = \frac{1}{N!} (Q_{trans} Q_{int})^N = \left[V \left(\frac{2 \pi m k_B T}{h^2} \right)^{3/2} \right]^N \left(\frac{e}{N} Q_{int} \right)^N \tag{16}$$

for N molecules of mass m in a volume V at temperature T. (The expression $(e/N)^N$ on the right-hand-side of Equation (16) comes from the use of Stirling's approximation for $\ln N!$ for large N, i.e., $\ln N! \approx N \ln N - N$, or $N! \approx N^N e^{-N} = (N/e)^N$.)

The easiest quantity to evaluate from this expression is the *Helmholtz free energy F*:

$$F = -k_B T \ln Q_N \tag{17}$$

A variety of other thermodynamic functions may be evaluated from this. For example, the chemical potential – the quantity equalized in equilibrium calculations – of species i in a multi-component system is given by

$$\mu_i = \left(\frac{\partial F}{\partial N_i} \right)_T = -k_B T \frac{\partial \ln Q_N}{\partial N_i} \tag{18}$$

Also, to calculate the equation of state, we recall

$$p = -\left(\frac{\partial F}{\partial V} \right)_T \tag{19}$$

If we apply Equations (17) and (19) to Equation (16), for example, we get the ideal gas law:

$$p = k_B T \left(\frac{\partial \ln Q_N}{\partial V} \right)_T = \frac{N k_B T}{V} \tag{20}$$

9.4b Statistical Thermodynamics of Adsorption

The statistical thermodynamic approach to the derivation of an adsorption isotherm goes as follows. First, suitable partition functions describing the bulk and surface phases are devised. The bulk phase is usually assumed to be that of an ideal gas. From the surface phase, the equation of state of the two-dimensional matter may be determined if desired, although this quantity ceases to be essential. The relationships just given are used to evaluate the chemical potential of the adsorbate in both the bulk and the surface. Equating the surface and bulk chemical potentials provides the equilibrium isotherm.

We apply this method to the derivation of the Langmuir isotherm both to illustrate the method and to see the assumed nature of the surface energy states on which it is based.

9.4b.1 Localized, Monolayer Adsorption: The Langmuir Isotherm

The Langmuir isotherm is based on the assumption of localized adsorption. This means that an adsorbed molecule has such a high statistical preference for a certain surface site as to possess a negligible translational entropy in the adsorbed state. Localized adsorption is thus seen to be very plausible for chemisorption, in which the adsorbed molecules and the adsorbent interact quite specifically. For nonspecific physical adsorption a nonlocalized or mobile layer seems to be a more plausible picture. We have already discussed the Henry law type of isotherm and the ideal gas equation of state that are associated with the simplest type of mobile adsorption.

The adsorption sites on the surface are assumed to be uniform and to bind the adsorbate with an energy ε per molecule or E per mole; that is, the potential energy of a molecule in the gaseous state is zero, and in the adsorbed state it is $-\varepsilon$. Note that this adsorption energy is a characteristic of the interaction between the adsorbed molecules and the adsorbent. As such, it is the same not only for all parts of the surface but also for all degrees of surface coverage. This is equivalent to saying that the adsorbed molecules do not interact with each other.

The surface is assumed to consist of M adsorption sites. Suppose we consider the case in which N of the sites are occupied; that is, N molecules are adsorbed. To write the partition function Q for the surface molecules, we must ask how these molecules differ from those in the gas phase (superscript g). Some of the internal degrees of freedom may be modified by the adsorption (Q^s_{int}), but the most notable difference will be in the translational degrees of freedom. From three equivalent translational degrees of freedom, the adsorbed molecule goes to two highly restrained translational degrees of freedom (remember the adsorption is localized) and one vibrational degree of freedom normal to the surface(s):

$$Q^g_{trans.\,3d} \rightarrow Q^s_{trans.\,2d} Q^s_{vib}$$ (21)

With these ideas in mind, we may assemble the partition function of the surface molecules as follows:

$$Q^s = g_N \left\{ Q^s_{trans.\,2d} Q^s_{vib} Q^s_{int} \exp\left[-\left(-\frac{\varepsilon}{k_B T} \right)\right] \right\}^N$$ (22)

In this expression, the degeneracy factor g_N represents the number of ways the N molecules may be placed on M sites, and is given by the combinatorial formula (see Equation (3.52) in Section 3.4a.2 or Equation (2.46) in Section 2.6a):

$$g_N = \frac{M!}{N!\,(M-N)!}$$ (23)

Combining Equations (22) and (23) gives the following expression for the partition function of the adsorbed molecules:

$$Q^s = \frac{M!}{N!\,(M-N)!} \left(Q^s_{trans.\,2d} Q^s_{vib} Q^s_{int} \right)^N \exp\left(\frac{N\varepsilon}{k_B T} \right)$$ (24)

Application of Equation (17) to Equation (24) gives the Helmholtz free energy of the adsorbed molecules:

$$F^s = -k_B T \left(\ln M! - \ln N! - \ln(M-N)! + \frac{N\varepsilon}{k_B T} + N \ln \left(Q^s_{trans.\,2d} Q^s_{vib} Q^s_{int} \right) \right)$$ (25)

Since N and M are large, the factorials may be expanded by Sterling's approximation ($\ln x! \approx x \ln x - x$) to give

$$F^s = -k_B T \left(M \ln M - N \ln N - (M-N) \ln(M-N) + \frac{N\varepsilon}{k_B T} \right.$$

$$\left. + N \ln \left(Q^s_{trans.2d} Q^s_{vib} Q^s_{int} \right) \right)$$ (26)

The chemical potential of the adsorbed molecules is given, according to Equation (18), by the derivative of F^s with respect to N:

$$\mu^s = k_B T \left[\ln \left(\frac{N}{M-N} \right) - \frac{\varepsilon}{k_B T} - \ln \left(Q^s_{trans.\,2d} Q^s_{vib} Q^s_{int} \right) \right]$$ (27)

The condition of equilibrium between the adsorbed molecules and molecules in the gas state requires that the chemical potential for the adsorbed species be the same in both the gas phase and the adsorbed state:

$$\mu^s = \mu^g \tag{28}$$

Applying Equations (17) and (18) to Equation (16) shows the chemical potential for an ideal gas to be

$$\mu^g = -k_B T \ln\left(\frac{k_B T}{p}\right)\left(\frac{2\pi m k_B T}{h^2}\right)^{3/2} Q_{int}^g \tag{29}$$

since $V = RT/p$.

Equating Equations (27) and (29) gives

$$\frac{N}{M - N} = \frac{p}{k_B T}\left(\frac{h^2}{2\pi m k_B T}\right)^{3/2}\frac{Q_{trans.2d}^s Q_{vib}^s Q_{int}^s}{Q_{int}^g}\exp\left(\frac{\varepsilon}{k_B T}\right) \tag{30}$$

We may group the following terms together to define a new quantity K:

$$K = \frac{1}{k_B T}\left(\frac{h^2}{2\pi m k_B T}\right)^{3/2}\frac{Q_{trans.2d}^s Q_{vib}^s Q_{int}^s}{Q_{int}^g}\exp\left(\frac{\varepsilon}{k_B T}\right) \tag{31}$$

in terms of which Equation (30) becomes

$$\frac{N}{M - N} = Kp \tag{32}$$

Now if we divide both the numerator and denominator of the left-hand side by M and recognize that $N/M = \theta$, we obtain

$$\frac{\theta}{1 - \theta} = Kp \tag{33}$$

or

$$\theta = \frac{Kp}{1 + Kp} \tag{34}$$

two forms of the *Langmuir adsorption isotherm* (see Equation (7.67)).

The quantity K defined by Equation (31) may easily be expanded somewhat further. First, we write the two-dimensional partition function by analogy with its three-dimensional counterpart, Equation (16). To do this, replace V by the area accessible to the adsorbed molecule σ and the exponent by 2/2 ($= 1$) in place of 3/2 since two rather than three degrees of freedom are involved. Therefore we obtain

$$Q_{trans.2d}^s = \sigma\frac{2\pi m k_B T}{h^2} \tag{35}$$

If the energy separating the vibrational quantum states is small relative to the thermal energy, the partition function for vibration is approximately given by

$$Q_{vib}^s = \frac{k_B T}{\varepsilon_{vib}} \tag{36}$$

For physical adsorption the approximation involved here is expected to be valid. Finally, the energy of vibration may be replaced by $h\nu$, where ν is the frequency with which the adsorbed molecules vibrate against the adsorbent:

$$Q_{vib}^s = k_B T/h\nu \tag{37}$$

Combining Equations (35) and (37) with Equation (31) gives

$$K = \frac{\sigma}{\nu}(2\pi m k_B T)^{-1/2}\frac{Q_{int}^s}{Q_{int}^g}\exp\left(\frac{\varepsilon}{k_B T}\right) \tag{38}$$

It should be noted that this quantity has the units length^2 force^{-1}, reciprocal pressure units, as required by Equation (34).

We saw from Equation (7.75) how to rearrange the Langmuir equation into a form that permits graphical evaluation of the parameters. For the adsorption of gases this becomes

$$p/(n/w) = mp + b \qquad (39)$$

and predicts a straight line when $[p/(n/w)]$ is plotted versus p, with

$$\text{Slope} = m = N_A\sigma/A_{sp} \qquad (40)$$

and

$$\text{Slope/Intercept} = m/b = K \qquad (41)$$

Figure 9.7 shows some data for the adsorption of ethyl chloride on charcoal. Since these data were collected at different temperatures, the ratio p/p_0 is used as the independent variable in fitting the data to Equation (39). That is, to compare the adsorption at different temperatures, the pressure is expressed as a fraction of the equilibrium vapor pressure at that temperature. The data for each of the three temperatures in Figure 9.7 give quite good straight lines when plotted according to the linear form of the Langmuir equation (Equation (39)); the interpretation of this analysis is discussed in Example 9.3.

* * *

EXAMPLE 9.3   *Analysis of Adsorption Data Using the Langmuir Isotherm.*   Slope and intercept values for the linearized plots of the data in Figure 9.7 are as follows:

Temperature (°C)	Slope (g g^{-1})	Intercept (g g^{-1})
−15.3	1.9	0.040
0.0	2.0	0.047
20.0	2.0	0.088

Note that since p/p_0 is used in the linearization, the abscissa is dimensionless and the slope and intercept have the same units. Figure 9.7 suggests that the data at −15.3 and 0.0°C converge to a common saturation value, but this is less clear for the data at 20°C. Do the linear plots clarify this situation? From the temperature dependence of the K's for these data, estimate the adsorption energy in this system.

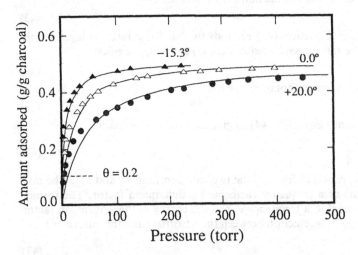

FIG. 9.7   Plot showing how the amount of ethyl chloride adsorbed on charcoal (in g g^{-1}) varies with pressure at −15.3, 0, and 20°C. (Data from F. Goldman, and M. Polanyi, *Z. Phys. Chem.*, **132**, 321 (1928).)

Solution: The slopes of these lines are identical (within experimental error), and the combination of Equations (40) and (7.72) shows that the reciprocal of the slope gives the adsorption at saturation. We conclude, therefore, that all three isotherms converge to the same saturation level of adsorption, namely, $(2.0)^{-1} = 0.50$ g g^{-1}.

The ratio of the slope to intercept values gives K for the adsorption at the three temperatures: i.e., $K = 48$, 43, and 23 at -15.3, 0.0, and 20.0°C, respectively. Multiplying Equation (38) through by $T^{1/2}$ and taking logarithms gives $\ln (T^{1/2}K) = (\epsilon/k_B)(1/T)$ + const. $= (E/R)(1/T)$ + const. While the quality of the data and the number of points scarcely justify the interpretation, a value for E/R of about 1700 can be estimated, suggesting an apparent value for E of about 14 kJ mole^{-1}. The word *apparent* is included until the effect of using p/p_0 instead of p in the linearization of the Langmuir equation is clarified. ∎

* * *

As noted above, the ability to predict correctly the temperature dependence of adsorption is a more stringent test of an isotherm than mere correlation of adsorption data. For this reason an independent measure of the energy of adsorption is clearly desirable. We return to this in Section 9.6.

Until now, we have focused our attention on those adsorption isotherms that show a saturation limit, an effect usually associated with monolayer coverage. We have seen two ways of arriving at equations that describe such adsorption: from the two-dimensional equation of state via the Gibbs equation or from the partition function via statistical thermodynamics. Before we turn our attention to multilayer adsorption, we introduce a third method for the derivation of isotherms, a *kinetic* approach, since this is the approach adopted in the derivation of the multilayer, BET adsorption isotherm discussed in Section 9.5. We introduce this approach using the Langmuir isotherm as this would be useful in appreciating the common features of (and the differences between) the Langmuir and BET isotherms.

9.4b.2 Langmuir Isotherm from Kinetic Theory

Suppose we consider a surface consisting of a total of M adsorption sites, M_1 of which are filled with adsorbed molecules to a depth of one molecule. Thus the number of bare sites M_0 is $(M - M_1)$. This situation is defined to be in equilibrium when the rate at which molecules attach to bare spots is the same as the rate at which they escape from monolayer regions. The rate of the adsorption process R_a is proportional to both the pressure and the number of available sites:

$$R_a = k_a p M_0 \tag{42}$$

The rate of desorption R_d is proportional to the number of sites occupied:

$$R_{d,1} = k_{d,1} M_1 \tag{43}$$

The subscript 1 for R_d and k_d in Equation (43) reminds us that these refer to layer 1 only (monolayer). At equilibrium the rates given by Equation (42) and (43) are equal:

$$k_a p M_0 = k_{d,1} M_1 \tag{44}$$

When adsorption is restricted to a monolayer, one has

$$M_0 + M_1 = M \tag{45}$$

a result that may be substituted into Equation (44) to give the Langmuir equation:

$$\frac{k_a}{k_{d,1}} p = \frac{M_1}{M - M_1} = \frac{\theta}{1 - \theta} \tag{46}$$

It is worthwhile to examine the expected form of the two rate constants k_a and $k_{d,1}$. The rate constant for desorption consists of a frequency factor and a Boltzmann factor. The former may be assumed to be proportional to a frequency ν, and the energy in the Boltzmann factor may be identified with the energy of interaction between the adsorbate and the adsorbent:

$$k_{d,1} \propto \nu \exp\left(-\frac{\epsilon}{k_B T}\right) \tag{47}$$

From kinetic theory of gases (see Atkins 1994), the number of collisions per unit area per unit time Z between gas molecules and a wall equals

$$Z = (2\pi m k_B T)^{-1/2} p \tag{48}$$

If this is multiplied by a surface area σ, the result is the rate of surface collisions. Then by analogy with Equation (42), we may take k_a as

$$k_a \propto (2\pi m k_B T)^{-1/2} \sigma \tag{49}$$

The ratio $k_a/k_{d,1}$ may be set equal to the Langmuir coefficient K. Therefore, from Equations (49) and (47) we get

$$K = \frac{k_a}{k_{d,1}} \propto (2\pi m k_B T)^{-1/2} \frac{\sigma}{\nu} \exp\left(\frac{\varepsilon}{k_B T}\right) \tag{50}$$

a result that is identical to Equation (38) as far as translational factors are concerned.

This kinetic-theory-based view of the Langmuir result provides no new information, but it does draw attention to the common starting assumptions of the Langmuir derivation and the BET derivation (Section 9.5a). This kinetic derivation of the Langmuir equation is especially convenient for obtaining an isotherm for the adsorption of two gases. This is illustrated in Example 9.4.

* * *

EXAMPLE 9.4 *Kinetic-Theory-Based Description of Binary Adsorption.* Assume that two gases A and B individually follow the Langmuir isotherm in their adsorption on a particular solid. Use the logic that results in Equation (46) to derive an expression for the fraction of surface sites covered by one of the species when a mixture of the two gases is allowed to come to adsorption equilibrium with that solid.

Solution: The rate of desorption of A is given by Equation (43) as $R_{d,A} = k_{d,A} M_A$ and the rate of adsorption of A by Equation (42). Allowing for the fact that both A and B occupy surface sites, we have $R_{a,A} = k_{a,A} p_A M_0 = k_{a,A} p_A (M - M_A - M_B)$. In this expression p_A is the partial pressure of A. At equilibrium the rates of adsorption and desorption are equal; therefore $k_{d,A} M_A = k_{a,A} p_A(M - M_A - M_B)$. Defining $\theta_i = M_i/M$ and $K_i = k_{a,i}/k_{d,i}$, we can write this last result as $\theta_A = K_A p_A(1 - \theta_A - \theta_B)$. A similar expression can be written for $\theta_B : \theta_B = K_B p_B(1 - \theta_A - \theta_B)$.

Next the ratio of θ_A/θ_B can be used to eliminate one of the θ's from the above expressions. Since $\theta_A/\theta_B = K_A p_A/K_B p_B$, $\theta_B = (K_B p_B/K_A p_A)\theta_A$. Using this to eliminate θ_B, we obtain $K_A p_A = \theta_A(1 + K_A p_A + K_B p_B)$, which rearranges to $\theta_A = K_A p_A/(1 + K_A p_A + K_B p_B)$.

A similar expression is obtained for $\theta_B : \theta_B = K_B p_B/(1 + K_A p_A + K_B p_B)$.

The fraction of surface sites covered by one gas or the other is given by $\theta_A + \theta_B = (K_A p_A + K_B p_B)/(1 + K_A P_A + K_B P_{BB})$. Note that the expressions for θ_A, θ_B, and $\theta_A + \theta_B$ all reduce to simple Langmuir expressions when one of the p or K values is zero. ∎

* * *

9.5 MULTILAYER ADSORPTION: THE BRUNAUER-EMMETT-TELLER EQUATION

9.5a Derivation of the Brunauer-Emmett-Teller Isotherm

As noted above, the range of pressures over which gas adsorption studies are conducted extends from zero to the normal vapor pressure of the adsorbed species p_0. An adsorbed layer on a small particle may readily be seen as a potential nucleation center for phase separation at p_0. Thus at the upper limit of the pressure range, adsorption and liquefaction appear to converge. At very low pressures it is plausible to restrict the adsorbed molecules to a monolayer. At the upper limit, however, the imminence of liquefaction suggests that the adsorbed molecules may be more than one layer thick. There is a good deal of evidence supporting the idea that multilayer adsorption is a very common form of physical adsorption on nonporous solids. In this section we are primarily concerned with an adsorption isotherm derived by Brunauer, Emmett, and Teller in 1938; the theory and final equation are invariably known by the initials of the authors: BET.

The BET isotherm has subsequently been derived by a variety of methods, but, as mentioned in the previous section, we follow the approach of the original derivation, namely, a kinetic description of the equilibrium state. Like the Langmuir isotherm, the BET theory begins with the assumption of localized adsorption. There is no limitation as to the number of layers of molecules that may be adsorbed, however; hence there is no saturation of the surface with increasing pressure. In general, the derivation assumes that the rates of adsorption and desorption from each layer are equal at equilibrium and that adsorption or desorption can occur from a particular layer only if that layer is exposed, that is, provided no additional adsorbed layers are stacked on top of it.

If we allow the possibility that a surface site may be covered to a depth of more than one molecule, then the following modifications to the derivation presented in Section 9.4b.2 are required. First, we define M_i to be the number of sites covered to a depth of i molecules. Second, Equation (45) is modified to

$$M = M_0 + M_1 + M_2 + \ldots + M_n = \sum_{i=0}^{n} M_i \tag{51}$$

where the summation includes all thicknesses of coverage from zero to n, the maximum.

Now let us consider the composition of the second layer ($i = 2$). By analogy with Equation (42), the rate at which molecules adsorb to form layer 2 is proportional to p and M_1. For simplicity, the proportionality constant is assumed to be the same as that for adsorption on the bare surface. Therefore we write

$$R_a = k_a p M_1 \tag{52}$$

In a similar fashion, we assume the rate of desorption from the second layer to be proportional to M_2:

$$R_{d,2} = k_{d,2} M_2 \tag{53}$$

The constant $k_{d,2}$ is assumed to be of the same form as that given by Equation (47) for the first layer, with one important modification. While desorption from the first layer involves detaching a molecule from the adsorbent as a substrate, desorption from the second layer involves detachment from another adsorbed molecule of the same kind. The adsorption energy ε was used in the Boltzmann factor in the first case; the energy of vaporization ε_v is a more appropriate value to use for the analogous quantity in the second case. We shall assume the frequency factor to be unchanged and write

$$k_{d,2} = \nu \exp\left(-\frac{\varepsilon_v}{k_B T}\right) \tag{54}$$

At equilibrium, the rate of adsorption and the rate of desorption from the second layer are also equal; therefore

$$k_a M_1 p = k_{d,2} M_2 \tag{55}$$

As a matter of fact, the same expression also applies to the third, fourth, . . . , ith levels. As a first approximation, the "activation energy" for desorption is the same for all layers after the first. This leads to the generalization

$$k_a M_{i-1} p = k_{d,i} M_i \tag{56}$$

for $2 \leq i < n$.

Equation (56) enables us to relate M_i to M_{i-1} just as Equation (44) relates M_1 to M_0. Moreover, through repeated application of Equation (56) we may relate M_i to M_1 or M_0 as follows:

$$M_i = \left(\frac{k_a}{k_{d,i}} p\right)^{i-1} M_{i-(i-1)} = \left(\frac{k_a}{k_{d,t}} p\right)^{i-1} M_1 = \left(\frac{k_a}{k_{d,i}}\right)^{i-1} \frac{k_a}{k_{d,1}} p^i M_0 \tag{57}$$

Since k_a is assumed to be the same for each layer and $k_{d,i}$ differs from k_d only in the value of the energy in the Boltzmann factor, this may be written

$$M_i = \frac{k_a^i p^i M_0}{v^i [\exp (-\varepsilon_v/k_B T)]^{i-1} \exp (-\varepsilon/k_B T)} \tag{58}$$

Multiplying the numerator and denominator of the right-hand side by $\exp (-\varepsilon_v/k_B T)$ we get

$$M_i = \frac{k_a^i p^i M_0}{[v \exp (-\varepsilon_v/k_B T)]^i} \frac{\exp (-\varepsilon_v/k_B T)}{\exp (-\varepsilon/k_B T)} \tag{59}$$

which may be written

$$M_i = x^i c M_0 \tag{60}$$

with c and x defined as follows:

$$x = \frac{k_a p}{v \exp (-\varepsilon_v/k_B T)} = \frac{k_a}{k_{d,i\geq 2}} p \tag{61}$$

and

$$c = \exp \left(\frac{\varepsilon - \varepsilon_v}{k_B T} \right) \tag{62}$$

Further, Equation (60) may now be substituted into Equation (51) to give

$$M = M_0 + M_1 + M_2 + \ldots + M_n = M_0 + \sum_{i=1}^{n} x^i c M_0 \tag{63}$$

Under some circumstances there may be a reason to restrict adsorption to a finite number of layers, that is, assign some specific value to n. In general, however, $n \to \infty$ as $p \to p_0$ is usually taken as the upper limit for this summation.

At this point we may define two other quantities in terms of the variables involved in Equation (59): the total volume of gas adsorbed and the volume adsorbed at monolayer coverage, V and V_m, respectively. The total volume is obviously the sum of the volume held in each type of site V_i, which is proportional to iM_i:

$$V = \sum_{i=1}^{n} V_i \propto \sum_{i=1}^{n} i M_i \tag{64}$$

The volume adsorbed at monolayer coverage is simply proportional to the total number of sites irrespective of the depth to which they are covered:

$$V_m \propto M = M_0 + \sum_{i=1}^{n} M_i \tag{65}$$

that is, V_m equals the volume of gas that would be adsorbed if a monolayer were formed. However, note that in writing Equation (65) it is not assumed that the monolayer is completely filled before other layers are formed. On the contrary, the picture allows for the coexistence of all types of patches, with adsorbed molecules stacked to various depths on each. There is no implication that a filled monolayer is required for multilayer formation.

Taking the ratio of Equation (64) to Equation (65) eliminates the unspecified proportionality constant and gives

$$\frac{V}{V_m} = \frac{\Sigma_i i M_i}{M_0 + \Sigma_i M_i} = \frac{c \Sigma_i i x^i}{1 + c \Sigma_i x^i} \tag{66}$$

The ratio V/V_m may be identified with θ, which may have values greater than unity in the case of multilayer adsorption. All that remains to be done to complete the BET derivation is to evaluate the summations in Equation (66).

To assist in the evaluation of the summations, suppose we consider the quantity $[x(1 - x)^{-1}]$. The factor in parentheses may be expanded as a power series (see Appendix A) to give

$$x(1 - x)^{-1} = x(1 + x + x^2 + \dots) = \sum_{i=1}^{n} x^i \tag{67}$$

Next we consider the quantity $\{x(d/dx)(\Sigma_i x^i)\}$. Carrying out the indicated differentiation gives

$$x \frac{d}{dx} \sum_i x^i = x \sum_i i x^{i-1} = \sum_i i x^i \tag{68}$$

Equation (67) may now be used as a substitution for $\Sigma_i x^i$ in Equation (68) to yield

$$\sum_i i x^i = x \frac{d}{dx} [x(1 - x)^{-1}] = \frac{x}{(1 - x)^2} \tag{69}$$

Substituting Equations (67) and (69) into Equation (66) gives

$$\frac{V}{V_m} = \frac{cx (1 - x)^{-2}}{1 + cx(1 - x)^{-1}} \tag{70}$$

This last result may be simplified further to become

$$\frac{V}{V_m} = \frac{cx}{(1 - x) [1 + (c - 1)x]} \tag{71}$$

Equation (71) is the result generally defined as the Brunauer-Emmett-Teller (BET) equation.

. The condition that $V \to \infty$ is seen to correspond to $x = 1$ by Equation (71). Recalling the definition of x given by Equation (61) and that $V \to \infty$ as $p \to p_0$ permits us to write

$$1 = \frac{k_a}{k_{d,i\geq 2}} p_0 \tag{72}$$

Therefore Equation (61) becomes

$$x = \frac{p}{p_0} \tag{73}$$

Thus the independent variable in the BET theory is the pressure relative to the saturation pressure. Therefore the BET equation describes the volume of gas adsorbed at different values of p/p_0 in terms of two parameters V_m and c. Furthermore, the model supplies a physical interpretation to these two parameters.

9.5b Testing the Brunauer-Emmett-Teller Theory

Basically there are three criteria against which the success of the BET theory may be evaluated: its ability to "fit" adsorption data, correct prediction of the temperature dependence of adsorption, and correct evaluation of specific area. We discuss these three issues in this section.

9.5b.1 Evaluation of Brunauer-Emmett-Teller Constants

The easiest way to evaluate the BET constants is to rearrange Equation (71) into the following linear form:

$$\frac{1}{V} \frac{x}{1 - x} = \frac{c - 1}{c V_m} x + \frac{1}{c V_m} \tag{74}$$

This form suggests that a plot of $(1/V)[x/(1 - x)]$ against x should yield a straight line, with

$$\text{Slope} = m = \frac{c - 1}{c V_m} \tag{75}$$

and

$$\text{Intercept} = b = 1/c V_m \tag{76}$$

Equations (75) and (76) may be solved to supply values of V_m and c from the experimental results:

$$V_m = \frac{1}{m + b} \tag{77}$$

and

$$c = \frac{m}{b} + 1 \tag{78}$$

In the following few paragraphs we first examine some of the general features of gas adsorption as predicted by the BET theory. Next we consider how well the theory actually fits experimental data. The use of experimental V_m values in the evaluation of A_{sp} is taken up in the following section.

9.5b.2 Predictions of Brunauer-Emmett-Teller Theory: Success in Fitting Experiments

Figure 9.8 enables us to observe some of the general features of the BET isotherm. The most apparent aspect concerning all the curves in this figure is their rapid increase as $p/p_0 \rightarrow 1$. It is also apparent, however, that the shape of the curve is sensitive to the value of c, especially for low values of c. Note particularly that the BET equation encompasses both Type II and Type III isotherms (Fig. 9.4). For values of c equal to 2 or less, the curves show no inflection point, whereas the inflection becomes increasingly pronounced as c increases above 2. In view of the wide diversity of curve shapes that are consistent with the BET equation and the relative insensitivity of adsorption data to the model underlying a particular equation, we might expect that the BET equation will fit experimental data rather successfully.

Figure 9.9a is a plot of some actual experimental data showing the volume of N_2—

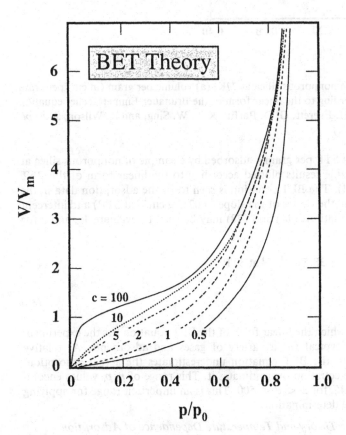

FIG. 9.8 Plots of V/V_m versus p/p_0 for several values of the parameter c, calculated according to the BET theory by Equation (71).

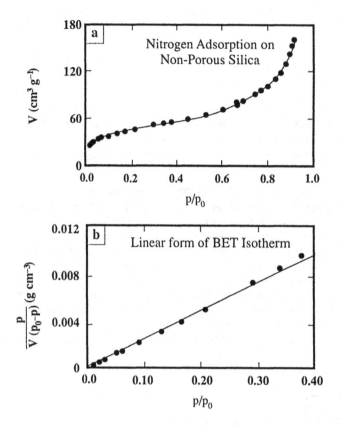

FIG. 9.9   Nitrogen adsorption on nonporous silica at 77K. (a) volume per gram (in cm³ per gram at STP) versus p/p_0; and (b) according to the linear form of the Brunauer-Emmett-Teller equation (Equation (74)). (Data from D. H. Everett, G. D. Parfitt, K. S. W. Sing, and R. Wilson, *J. Appl. Chem. Biotechnol.*, **24**, 199 (1974).)

expressed as cubic centimeters at STP per gram — adsorbed by a sample of nonporous silica at 77K. Figure 9.9b shows these same results plotted according to the linear form of the BET equation, given by Equation (74). The BET equation is seen to fit the adsorption data in the range $0.05 < p/p_0 < 0.30$. From the values of the slope (0.0257 g cm^{-3} at STP) and intercept ($2.85 \cdot 10^{-4}$ g cm^{-3} at STP), Equations (77) and (78) may be used to evaluate V_m and c for this system:

$$V_m = \frac{1}{(257 + 2.85) \cdot 10^{-4}} = 38.5 \text{ cm}^3 \text{ g}^{-1} \text{ at STP} \tag{79}$$

$$c = \frac{257 \cdot 10^{-4}}{2.85 \cdot 10^{-4}} + 1 = 91.2 \tag{80}$$

The range of pressures over which the linear form of the BET equation fits the experimental data in Figure 9.9 is fairly typical for a variety of gases and adsorbents. At relative pressures below the range of fit, the BET equation underestimates the actual adsorption, whereas above $p/p_0 = 0.35$, the equation overestimates it. This range of p/p_0 values encompasses the region in which $V = V_m$ for $2 < c < 500$. This is an important range for applying the BET equation to surface area determination.

9.5b.3   Brunauer-Emmett-Teller Theory and Temperature Dependence of Adsorption

The parameter c describes the temperature dependence of adsorption. In the derivation already presented, certain simplifying assumptions were made as to the constancy of k_a and ν_i in each of the layers. Various modifications of this assumption might be made, but they would involve minor temperature effects at best. To a first approximation, then, one should be able to evaluate $(\varepsilon - \varepsilon_v)$ from a knowledge of experimental c values, or vice versa. Proceeding in the first manner leads to values of $(\varepsilon - \varepsilon_v)$ that are too low, perhaps half their expected value. When we consider the region of fit on which the evaluation of c is based, it is not difficult to see why values of c are too low. The data are not fitted to the earliest stages of adsorption that are associated with larger values of ε; hence c is underestimated. The definition of c allows us to conclude some unfinished business pertaining to Example 9.3. This is examined in Example 9.5.

* * * *

EXAMPLE 9.5   *Calculating the Adsorption Energy from the Brunauer-Emmett-Teller Isotherm.* The BET analysis uses p/p_0 rather than p as a variable just as we used this pressure ratio to compare Langmuir adsorption at different temperatures in Example 9.3. What corrections, if any, are needed in the "apparent adsorption energy" of about 14 kJ mole^{-1} as calculated in Example 9.3?

Solution: The answer to this is obtained by comparing Equations (31) and (62). When p/p_0 is used instead of p, we obtain c, not K, from the linearized Langmuir equation via Equation (41). The exponential energy term evaluated from the analysis of $\ln K$ versus $1/T$ is seen by Equation (62) to be the difference between the adsorption energy and the energy of vaporization of the adsorbate. For ethyl chloride $E_v = 23$ kJ mole^{-1}; therefore $E = 23 + 14 = 37$ kJ mole^{-1}; this is the actual adsorption energy for the ethyl chloride-charcoal system discussed in Example 9.3.

Note also that small values of c—which account for Type III adsorption isotherms—result when $\varepsilon_v > \varepsilon$ in Equation (62). This was anticipated in our remarks about Figure 9.4.   ■

* * *

9.5b.4   Brunauer-Emmett-Teller Theory and Specific Surface Area

The final criterion for judging the success of a theoretical isotherm is its ability to measure specific surface areas. Since the BET equation has become a standard in this regard, we devote this subsection to the discussion of this topic.

With monolayer adsorption, we saw how the saturation limit could be related to the specific surface area of the adsorbent. The BET equation permits us to extract from multilayer adsorption data (by means of Equation (77)) the volume of adsorbed gas that would saturate the surface if the adsorption were limited to a monolayer. Therefore V_m may be interpreted in the same manner that the limiting value of the ordinate is handled in the case of monolayer adsorption. Since it is traditional to express both V and V_m in cubic centimeters at STP per gram, we write (see Equation (7.72))

$$V_m = \left(\frac{n}{w}\right)_{sat} (22{,}414 \text{ cm}^3 \text{ mole}^{-1}) = \frac{A_{sp} (22{,}414)}{N_A \sigma^0} \tag{81}$$

Note that it is assumed that V_m has been expressed on a "per gram" basis in writing Equation (81). If this is not the case, the area of the actual sample rather than its specific area is given by Equation (81). If the area occupied per molecule on the surface is known, the specific surface may be evaluated from Equation (81):

$$A_{sp} = \frac{V_m N_A \sigma^0}{22{,}414} \tag{82}$$

This last quantity is something that may be determined by independent methods; therefore the BET theory may be tested at this point.

It is clear that the same value of A_{sp} must be obtained for a particular adsorbent regardless of the nature of the adsorbate used. Studies of this sort lead to quite consistent values of A_{sp}, provided the adsorbed species all have access to the same surface. On a porous surface, for example, large adsorbate molecules may not be able to enter small cavities that are accessible to smaller molecules.

In the preceding discussion, it has been assumed that a value of σ^0 is known unambiguously. This quantity is obviously the "yardstick" by which moles of adsorbed gas are converted to areas. Any error in this quantity will invalidate the determination of A_{sp}. What is generally done is to assume the adsorbed material has the same density on the surface that it has in the bulk liquid at the same temperature and to assume the molecules are close packed on the surface. In view of the earlier discussion of surface phases, this is seen to be a somewhat risky procedure. The safest way to proceed would be to evaluate σ^0 for a particular adsorbate from independent measurements of V_m and A_{sp}.

In view of the difficulty in translating measured gas adsorption into absolute specific surface areas, it is not surprising that self-consistency is often normative in this matter. In this sense, at least, an area of 16.2 $\overset{\circ}{A}{}^2$ for nitrogen has become something of a standard. Values for other common adsorbed species may be found in the works of Adamson (1990), Broeckhoff and van Dongen (1970), and Kantro et al. (1967). It is probably not surprising that polar molecules such as water display values of σ^0 that are sensitive to the nature of the substrate. Using nitrogen adsorption to evaluate A_{sp} and then using the latter to evaluate σ^0 for water has led to values of 12.5, 10.4, and 11.4 $\overset{\circ}{A}{}^2$ for amorphous silica, calcium hydroxide, and calcium silicate hydrate, respectively.

In spite of a variety of objections to the BET theory, V_m values from N_2 adsorption studies have become a very common means for determining specific surface areas. As a matter of fact, an IUPAC commission was organized in 1969 to study N_2 adsorption with the objective of preparing reference standards for surface area determinations. In this project four silicas and four carbon blacks of different particle size were investigated independently in 13 different laboratories. Nitrogen adsorption data were analyzed by the BET method over the best fit region, and values of V_m were converted to A_{sp} using 16.2 $\overset{\circ}{A}{}^2$ as the value of σ^0. Table 9.2 shows the values of A_{sp} and c obtained in the 10 laboratories that studied the silica of Figure 9.9. The average value of the specific surface from these determinations is 163.4 ($\pm6\%$) m^2 g^{-1}. This value agrees with our analysis of Figure 9.9 obtained by combining Equations (79) and (82):

$$A_{sp} = \frac{(38.5)\ (6.02 \cdot 10^{23})\ (16.2 \cdot 10^{-20})}{22,400} = 168\ \text{m}^2\,\text{g}^{-1} \tag{83}$$

Samples of this material as well as others investigated are now available for calibration purposes as gas adsorption standards. Details may be found in the reference cited for the data of Table 9.2.

TABLE 9.2 Values of A_{sp} and c as Determined in 10 Different Laboratories by the Brunauer-Emmett-Teller Method for the Same Silica Sample Shown in Figure 9.9

Laboratory	A_{sp} (m^{2}g^{-1})	c
A	166.4	92
B	162.8	101
C	174.0	100
D	148.5	166
E	173.5	70
F	166.0	98
G	167.9	91
H	143.6	113
I	169.5	62
J	161.7	122
Average	163.4	102
Standard deviation	10.0 (6%)	29 (28%)

Source: D. H. Everett, G. D. Parfitt, K. S. W. Sing, and R. Wilson, *J. Appl. Chem. Biotechnol.*, **24**, 199 (1974).

At first glance, it may seem surprising that the BET method is as successful as it is in the evaluation of A_{sp}. After all, the values of c are not in particularly good agreement with expected values, nor, conversely, are $(\varepsilon - \varepsilon_v)$ values calculated from experimental c's in good agreement with expectations. Probably a significant part of the discrepancy between theory and experiment with respect to c arises from the heterogeneity of the surface. The BET equation—like the Langmuir equation to which it reduces in certain limits—assumes that a single adsorption energy applies to all surface sites. As we saw in Chapter 6, Section 6.7, a distribution of surface energy states is not uncommon with solids. In this regard, both the Langmuir and BET models are unrealistic. However, sites with high adsorption energies are apt to be covered by adsorbate molecules first. If these represent a relatively small fraction of the surface sites, this effect gets masked fairly quickly as the first layer fills. In addition, lateral interactions between adsorbed molecules start at zero and increase as the surface begins to fill. Even these interactions tend to level off after a certain degree of coverage is achieved. Therefore, even though a single adsorption energy is unrealistic, there are compensating effects that tend to minimize the variation.

Although we have often mentioned the adsorption energy in this chapter, we have not yet discussed any procedure by which this can be measured quantitatively except as some sort of an average quantity. Since most solid surfaces are heterogeneous, it is desirable to be able to examine adsorption energy in a more discriminating way, for example, as a function of coverage or pretreatment. In the following section we see how this can be done.

9.6 ENERGETICS OF ADSORPTION

As we have seen, an adsorption isotherm is one way of describing the thermodynamics of gas adsorption. However, it is by no means the only way. Calorimetric measurements can be made for the process of adsorption, and thermodynamic parameters may be evaluated from the results. To discuss all of these in detail would require another chapter. Rather than develop all the theoretical and experimental aspects of this subject, therefore, it seems preferable to continue focusing on adsorption isotherms, extracting as much thermodynamic insight from this topic as possible. Within this context, results from adsorption calorimetry may be cited for comparison without a full development of this latter topic.

9.6a Definitions of Heat of Adsorption

The approach we follow is essentially that used to derive the Clapeyron equation (Atkins 1994). Suppose we consider an infinitesimal temperature change for a system in which adsorbed gas and unadsorbed gas are in equilibrium. The criterion for equilibrium is that the free energy of both the adsorbed (subscript s) and unadsorbed (subscript g) gas change in the same way:

$$dG_s = dG_g \tag{84}$$

The following equations may be written for these two quantities if the temperature change is assumed to cause no change in the amount of adsorbed material:

$$dG_g = -S_g dT + V_g dp \tag{85}$$

and

$$dG_s = -S_s dT + V_s dp \tag{86}$$

Substituting Equations (85) and (86) into Equation (84) and rearranging gives

$$\left(\frac{\partial p}{\partial T} \right)_{n_s} = \frac{S_g - S_s}{V_g - V_s} \tag{87}$$

In writing this last result, it has been explicitly noted that the number of moles of adsorbed gas n_s is constant. If the process under consideration is carried out reversibly, $S_g - S_s$ may be replaced by q_{st}/T, where q_{st} is known as the *isosteric* (the same coverage) *heat of adsorption*:

$$S_g - S_s = q_{st}/T \tag{88}$$

Combining Equations (87) and (88) leads to the result

$$\left(\frac{\partial p}{\partial T}\right)_{n_s} = \frac{q_{st}}{T(V_g - V_s)} \tag{89}$$

Equation (89) may be integrated if the following assumptions are made: (a) $V_g \gg V_s$, so that V_s may be neglected, (b) the gas behaves ideally so that the substitution $V_g = RT/p$ may be used, and (c) q_{st} is independent of T. With these assumptions, Equation (89) integrates to

$$\ln\left(\frac{p_1}{p_2}\right) = -\frac{q_{st}}{R}\left(\frac{1}{T_1} - \frac{1}{T_2}\right) \tag{90}$$

Equation (90) shows that the isosteric heat of adsorption is evaluated by comparing the equilibrium pressure at different temperatures for samples showing the same amount of surface coverage. The data of Figure 9.7 may be used as an example to see how this relationship is applied.

For an arbitrarily chosen extent of adsorption, a horizontal line such as the dashed line in Figure 9.7 may be drawn that cuts the various isotherms at different pressures. The pressure coordinates of these intersections can be read off the plot. According to Equation (90), a graph of $\ln p$ versus $1/T$ should be linear with a slope of $(-q_{st}/R)$. From Figure 9.7, for example, when the adsorption is 0.10 g ethyl chloride (g charcoal)$^{-1}$ (which corresponds to $\theta = 0.2$), the equilibrium pressures are 0.20, 0.63, and 2.40 torr at -15.3, 0, and 20°C, respectively. When plotted in the manner just described, these data yield a line of slope -5330 K. Multiplication by R gives $q_{st} = 44.3$ kJ mole^{-1} as the isosteric heat of adsorption for this system at $\theta = 0.2$. Table 9.3 lists values of q_{st} for different θ values as calculated from the data in Figure 9.7.

9.6b Comparison of Adsorption Energies Measured Using Different Methods

Table 9.3 is based on the same data that were analyzed according to the Langmuir equation in Section 9.4b.1. Examples 9.3 and 9.5 show that these data are consistent with an adsorption energy of about 37 kJ mole^{-1} according to the Langmuir interpretation.

TABLE 9.3 Values of the Isosteric Heat of Adsorption at Different Values of θ for the Data Shown in Figure 9.7 as Evaluated by Equation (90)

θ	q_{st} (kJ mole^{-1})
0.06	56.9
0.08	47.3
0.10	46.4
0.20	44.3
0.30	41.4
0.40	40.2
0.50	40.2
0.60	41.0
0.70	38.1
0.80	37.2

In presenting this result initially, the remark was made that an independent determination of the adsorption energy would be desirable. Although the present reinterpretation of the same data is not exactly an "independent" determination, it does extract an energy quantity from the experimental results that is free of any assumed model for the mode of adsorption. Accordingly, it is informative to compare the two interpretations. The Langmuir model assumes that a single energy applies to all adsorption sites. Therefore any data analyzed according to this model cannot yield more than one energy. The isosteric heat of adsorption, on the other hand, is evaluated at different degrees of surface coverage. The data in Table 9.3 show that this quantity definitely varies with coverage, tending to level off as $\theta \rightarrow 1$. This is consistent with the picture of the more active "hot spots" being covered first. The average energy that the Langmuir analysis yields is approximately the same as the energy toward which q_{st} converges.

There are several additional thermal quantities besides q_{st} and E that may be generically called "heats of adsorption." One of these is the *integral heat of adsorption Q_n*, which is related to the isosteric heat of adsorption as follows:

$$Q_n = \int_0^n q_{st} \, dn \tag{91}$$

It applies to the process in which n moles of adsorbate are transferred from the bulk gas to the surface, starting from a bare surface. The integral heat of adsorption is determined from data such as those contained in Table 9.3 by graphical integration.

In addition, there are a number of different calorimetric methods to determine heats of adsorption. For example, we may distinguish between isothermal and adiabatic heats depending on the type of calorimeter involved. Of course, thermodynamic relationships exist among these various quantities. We shall not pursue these topics, but one should be aware of the differences and seek precise definitions if the need arises.

The data shown in Figure 9.10 indicate both the kind of data that may be obtained by direct calorimetric study of gas adsorption and some evidence of the effect of preheating on the properties of surfaces. The figure shows the calorimetric heat of adsorption of argon on carbon black. The broken line indicates the behavior of the untreated black, and the solid line is the "same" adsorbent after heating at 2000°C in an inert atmosphere, a process known as *graphitization*. The horizontal line indicates the heat of vaporization of argon.

There are several interesting aspects of this figure that are quite generally observed.

1. The untreated carbon black shows the effect of surface heterogeneity, an effect that becomes smeared out as the coverage increases. The surface of the untreated black contains a certain amount of oxygen in a variety of functional groups (e.g., ether, carbonyl, hydroxyl, and carboxyl). Graphitization results in both the reduction of these oxygen-containing groups and the sharpening of both the basal and prismatic crystallographic planes. Electron micrographs of carbon black before and after graphitization are shown in Figure 1.9.

2. After graphitization, the most notable feature is the sharp discontinuity at monolayer coverage. Beyond the monolayer the heat of adsorption is close to the heat of vaporization (as required by the BET theory) but does show some influence of the surface as well. At coverage below $V/V_m = 1$ the heat of adsorption increases with increasing coverage, probably due to lateral interactions between the adsorbed molecules.

Figure 9.10 is an extreme example of the effect of heating on the properties of an adsorbent. Degassing prior to measuring an isotherm is done under far less severe conditions; nevertheless, these conditions should always be reported when adsorption studies are conducted because of the possibility of surface modification on heat treatment.

Another calorimetric technique for measuring the heat of adsorption consists of comparing the heat of immersion (see Chapter 6, Section 6.6c) of bare solid with that of a solid preequilibrated with vapor to some level of coverage. Table 9.4 summarizes some results of this sort. The experiment consisted of measuring the heats of immersion of anatase (TiO_2) in benzene with the indicated amount of water vapor preadsorbed on the solid. Small quantities of adsorbed water increase the heat of immersion more than threefold so that it approaches the value for water itself. Most laboratory samples will be contaminated with adsorbed water.

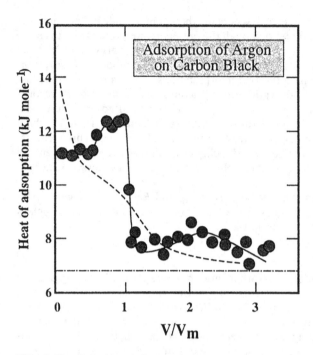

FIG. 9.10 Calorimetric heats of adsorption as a function of coverage for argon on carbon black at 78K. The dashed line represents untreated black; the solid line is after graphitization at 2000°C. The horizontal line is the heat of vaporization of argon. (Redrawn with permission from R. A. Beebe and D. M. Young, *J. Phys. Chem.*, **58**, 93 (1954).)

Unless the material has been carefully pretreated, the actual nature of the surface may be quite different from what would be expected nominally.

9.7   ADSORPTION IN POROUS SOLIDS

High-specific-area solids of the type studied by gas adsorption consist of small particles for which the radius of an equivalent sphere is given by Equation (1.2). This figure may be a fairly reasonable measure of a characteristic linear dimension even of irregularly shaped particles,

TABLE 9.4   Effect of Traces of Adsorbed Water on the Heat of Immersion of TiO_2 in Benzene

Amount of water adsorbed (mmol kg^{-1})	Heat of Immersion (mJ m^{-2})
0.0	150
2.0	250
4.0	320
10.0	450
17.0	506
Pure H_2O	520

Source: G. E. Boyd, and W. D. Harkins, *J. Am. Chem. Soc.*, **64**, 1195 (1942).

provided the gas adsorption is restricted to the exterior surface of the particles. Any cracks or pores in the solid particles will expose additional adsorbing surfaces and increase the total specific area of the material. Because of this complication, we have specified nonporous solids at several places in this chapter. The particles of high-specific-area solids are small at the beginning, so it follows immediately that any pores in these solids will necessarily have very small dimensions.

In Chapter 6, Section 6.9, we discussed the pressure required to force a liquid—most commonly mercury—into the pores of a solid. Our emphasis in that section was on the pores *between* powder particles when the particles are tightly pressed into a plug. In this section we are not concerned with these interstitial pores since powders are not densely packed in adsorption studies; instead, our interest is in the pores *within* the particles themselves. The difference is a matter of emphasis, however, and both liquid intrusion and gas adsorption complement one another for the study of porous solids.

9.7a Adsorption Hysteresis and Capillary Condensation

We restrict our attention to only one aspect of the adsorption behavior of porous solids: the *hysteresis* they display in their adsorption isotherms. A schematic illustration of the phenomenon is shown in Figure 9.11. Although the region enclosed by the hysteresis loop may have a variety of shapes, in all cases there are two quantities of adsorbed material for each equilibrium pressure in the hysteresis range. That branch of the loop that corresponds to adsorption (increasing pressure) inevitably displays less adsorption at any given pressure than the desorption branch (decreasing pressure). In many cases hysteresis loops such as this are reproducible, although they are not reversible in the thermodynamic sense. Since irreversible processes are involved, the substitution of q_{st}/T for ΔS in the derivation of Equation (90) is not valid. One can go through the motions of evaluating $\partial p/\partial T$ for a system that displays hysteresis, but the "apparent q_{st} values" so obtained (one for each branch) are not easily related to calorimetric adsorption energies.

Adsorption hysteresis is often associated with porous solids, so we must examine porosity for an understanding of the origin of this effect. As a first approximation, we may imagine a pore to be a cylindrical capillary of radius r. As just noted, r will be very small. The surface of any liquid condensed in this capillary will be described by a radius of curvature related to r. According to the Laplace equation (Equation (6.29)), the pressure difference across a curved interface increases as the radius of curvature decreases. This means that vapor will condense

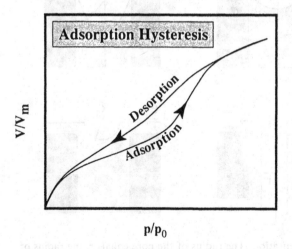

FIG. 9.11 A schematic illustration of hysteresis between the adsorption and desorption branches of an experimental isotherm.

in small capillaries at pressures less than the normal vapor pressure p_0, which is defined for flat surfaces. The condensation of vapors in small capillaries is an equilibrium phenomenon, however, so capillary condensation in itself does not account for hysteresis. It does point out the fact that a liquid-vapor surface is also involved in the adsorption on porous solids for $p < p_0$. In fact, the leveling off of the adsorption in Type IV and Type V isotherms before p_0 is reached is the result of liquid condensation in small pores at these pressures.

For simplicity, let us assume that the liquid condensed in a pore has a surface that is part of a sphere of radius R_s, with $R_s > r$. For a spherical surface we may use the Kelvin equation (Equation (6.40)) to calculate p/p_0

$$N_A k_B T \ln\left(\frac{p}{p_0}\right) = -\frac{2M\gamma}{\rho R_s} \tag{92}$$

A minus sign has been introduced in the Kelvin equation because the radius is measured outside the liquid in this application, whereas it was inside the liquid in the derivation of Chapter 6. In hysteresis, adsorption occurs at relative pressures that are higher than those for desorption. According to Equation (92), it is as if adsorption-condensation took place in larger pores than desorption-evaporation. Since the pore dimensions are presumably constant, we must seek some mechanism consistent with this observation to explain hysteresis.

9.7b Some Models of Capillary Condensation

Figure 9.12 contains sketches for several different models of pores that will be useful in our discussion of capillary condensation. Figure 9.12a is the simplest, attributing the entire effect just described to variations in pore radius with the depth of the pore. That is, when liquid first begins to condense in the pore, the larger radius R_a determines the pressure at which the adsorption-condensation occurs. Once the pore has been filled and the desorption-evaporation branch is being studied, the smaller radius R_d determines the equilibrium pressure. Although bottlenecked pores of this sort may exist in some cases, this model seems far too specialized to account for the widespread occurrence of hysteresis.

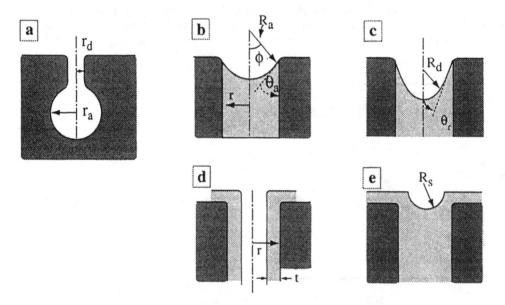

FIG. 9.12 Five models for capillary condensation. The radius of the pore equals r, the radius of curvature of the spherical meniscus is R_s, and t is the thickness of the adsorbed layer. The subscripts a and d refer to adsorption and desorption.

Figures 9.12b and 12c represent another model, based on contact angle hysteresis. These sketches represent the surface of liquid in a capillary during the adsorption and desorption stages of hysteresis, respectively. In Figure 9.12b the capillary is filling; in Figure 9.12c it is emptying. Accordingly, the advancing and receding values of the contact angle apply to adsorption and desorption, respectively. The radius R_s of the spherical surface and the radius r of the capillary are related through the contact angle θ and its complementary angle ϕ (see Fig. 9.12):

$$r = R_s \sin \phi = R_s \cos \theta \tag{93}$$

Substituting Equation (93) into Equation (92) gives

$$N_A k_B T \ln \left(\frac{p}{p_0} \right) = -\frac{2 M \gamma \cos \theta}{\rho r} \tag{94}$$

For a pore of constant radius the equilibrium pressure decreases as $\cos \theta$ increases or as θ decreases. It will be recalled from Chapter 6 that advancing contact angles are larger than receding ones. Therefore this model is consistent with the observation that desorption-evaporation occurs at lower relative pressures than adsorption-condensation. The only objection to this explanation of adsorption hysteresis is that it makes no reference whatsoever to adsorption!

Figures 9.12d and 12e illustrate another model for adsorption hysteresis that considers multilayer adsorption explicitly. During adsorption the capillary is viewed as a cylinder of radius $(r - t)$, with t the thickness of the adsorbed layer at that pressure. This is represented by Figure 9.12d. For such a surface the Kelvin equation becomes

$$N_A k_B T \ln \left(\frac{p}{p_0} \right) = -\frac{M \gamma}{\rho (r - t)} \tag{95}$$

since one of the radii of curvature is infinite. As multilayer adsorption proceeds, however, the thickness of the adsorbed layer equals r. Once the pore is thus filled, its surface becomes a meniscus that may be treated by Equation (94) as a portion of a sphere, as shown in Figure 9.12e. In the event that $\theta = 0$, Equation (92) applies with $R_s = r$. Comparison of adsorption and desorption in this situation is particularly easy if $t \ll r$ in Equation (95). In that case, the right-hand side of Equation (95) is proportional to $1/r$, and comparison with Equation (92) reveals that

$$\left(\frac{p}{p_0} \right)_d^{1/2} = \left(\frac{p}{p_0} \right)_a \tag{96}$$

Since both of these ratios are less than unity, $p_a > p_d$. This model is qualitatively consistent with the observed hysteresis but is difficult to apply quantitatively because it neglects differences between the adsorbed material on the surface and that in the bulk liquid.

Gas adsorption data may be analyzed for the distribution of pore sizes. What is generally done is to interpret one branch of the isotherm and use an appropriate equation to calculate the effective pore radius at a given pressure. The amount of material adsorbed or desorbed for each increment or decrement in pressure measures the volume of pores with that effective radius.

9.8 ADSORPTION ON CRYSTAL SURFACES

Until now, our treatment in this chapter of the solid-gas interface has been very one-sided, focusing almost entirely on the adsorbed layer(s). To be sure, we have extracted some information about the solid, e.g., the specific surface area and the presence or absence of pores. In addition, we have extracted values for the energy of interaction between the solid adsorbent and the adsorbate molecules but, other than this, the influence of the solid has been ignored. Likewise, the solids we have considered have been high-surface-area powders, presenting either a multitude of crystal faces or amorphous surfaces to the adsorbing gas molecules. Because of this approach, we have not asked whether the order of a crystalline surface carries over to monolayer or submonolayer adsorption. As a matter of fact, the notion that there may be order in a layer of adsorbed molecules actually contradicts the models on which the isotherm presented above were based.

In this section we discuss the adsorption on crystal surfaces. First, we begin with low-energy electron diffraction (LEED) — an experimental method for examining crystal surfaces — and introduce some basic crystallographic concepts needed to interpret the experimental measurements. Then we look at the implication of the adsorbate structure to adsorption and the structure of adsorbed layers using LEED measurements.

9.8a   Low-Energy Electron Diffraction

Diffractometry provides an excellent tool for examining structure so we turn now to *low-energy electron diffraction* to study the order at a specific face of a single crystal, with and without adsorbed molecules. For the remainder of the chapter, we focus attention on the faces of the metal crystals. There are several reasons for this choice:

1. Many metals crystallize in the relatively simple cubic (either primitive, face-centered, or body-centered) structures.

2. Metal surfaces have been studied extensively by LEED in the context of a number of practical applications such as corrosion, friction, and semiconductor devices.

3. Metal surfaces are important as catalysts, and this discussion of LEED enables us to introduce some ideas about chemisorption and catalysis.

9.8a.1   Basic Principle

In LEED a beam of low-energy electrons rather than x-rays is used to form the diffraction pattern, but otherwise many of the concepts, relationships, and vocabulary are based on x-ray diffraction (Van Hove and Tong 1979). Accordingly, our discussion of LEED includes a review of pertinent aspects of this topic. Since diffractometry can get quite involved, we tailor our review to those subjects most helpful in getting us started and leave more advanced concepts for further study.

The de Broglie concept of wave-particle duality enables us to calculate the wavelength of an electron. According to the *de Broglie equation*,

$$\lambda = \frac{h}{p} = \frac{h}{[2 \text{ m } (KE)]^{1/2}} \tag{97}$$

in which h is Planck's constant, and the momentum of the electron, $p = [2m(KE)]^{1/2}$, can be calculated in terms of its mass and the accelerating voltage that accounts for its kinetic energy. A straightforward application of this formula enables us to calculate that a voltage of 150 V gives electrons a wavelength of 0.1 nm, the order of magnitude of interatomic spacings in crystals:

$$\lambda = 6.63 \cdot 10^{-34} \text{ J s}/[2(9.11 \cdot 10^{-31} \text{ kg})(1.6 \cdot 10^{-19} \text{ C})(150 \text{ V})]^{1/2} = 10^{-10} \text{ m} = 0.10 \text{ nm}$$

Thus the electrons used in LEED are usually in the range 10 to 300 eV, lower in energy by one to two orders of magnitude than those used in scanning electron microscopy (SEM; Chapter 1) and Auger electron spectroscopy (AES; see Section 9.1b.2). Davisson (Nobel Prize, 1937), working with a series of collaborators, is generally credited with the first application of electrons to diffraction studies.

What makes LEED of particular interest in the study of surfaces is the fact that low-energy electrons such as these are only able to penetrate and reemerge by diffraction from a few atomic thicknesses of the target material. Even at that, the diffracted beam acquires a background of secondary electrons picked up through inelastic interactions with the solid. Figure 9.13 is a schematic of a LEED apparatus. Note that several grids are interposed between the specimen and the viewing screen. One of these is adjusted in potential so that only

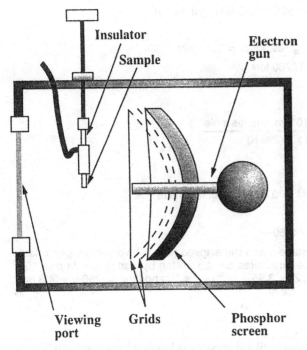

Insulator

Sample

Electron gun

Viewing port   **Grids**   **Phosphor screen**

FIG. 9.13 Schematic illustration of a low-energy electron diffraction (LEED) apparatus. (Redrawn with permission from Atkins 1994.)

electrons of the original energy—those elastically diffracted—are allowed to pass; secondary electrons of lower energy are blocked. The diffracted beam is accelerated by a final, positively charged grid to produce an image on the fluorescent screen.

We saw in connection with the discussion of Figure 9.5 that measurable gas adsorption occurs even at gas pressures as low as 10^{-10} torr. As a matter of fact, the two-dimensional density of the adsorbed molecules is not low enough to conform to the two-dimensional ideal gas law even when the pressure is on the order of 10^{-10} torr. A question of considerable practical importance, then, is how low the pressure must be for an initially clean surface to remain that way for a reasonable period of time. The above reference to adsorption cites equilibrium data that are not useful for answering questions of rate.

Instead, we must turn to the kinetic molecular theory of gases for an estimate of the frequency with which molecules collide with a solid surface. We shall not be misled, however, if we anticipate that this pressure is low. Example 9.6 is a numerical examination of gas collisions with walls.

* * *

EXAMPLE 9.6 *Rate of Atomic Collisions as a Function of Pressure.* Assuming 10^{19} atoms per square meter as a reasonable estimate of the density of atoms at a solid surface, estimate the time that elapses between collisions of gas molecules at 10^{-6} torr and 25°C with surface atoms. Use the kinetic molecular theory result that relates collision frequency to gas pressure through the relationship $Z = 1/4 \ \bar{v}N/V$, for which the mean velocity of the molecules $\bar{v} = (8RT/\pi M)^{1/2}$ and N/V is the number density of molecules in the gas phase and equals pN_A/RT. Repeat the calculation at 10^{-8} and 10^{-10} torr.

Solution: If 10^{19} collisions occurred per square meter per second, each surface atom would be hit an average of once each second. Therefore we must divide 10^{19} by Z at the required pressures to obtain the elapsed time. Assuming a molecular weight of 28 g mole^{-1} for the gas, we obtain

$(8RT/\pi M)^{1/2} = [8(8.314 \text{ J K}^{-1} \text{ mole}^{-1})(298) / \pi(0.028 \text{ kg mole}^{-1})]^{1/2}$

$= 475 \text{ m s}^{-1}$

Converting the gas pressure to SI units gives, for $p = 10^{-6}$ torr,

$p = [10^{-6} \text{ torr} \cdot 1.013 \cdot 10^5 \text{ N m}^{-2}] / 760 \text{ torr}$

$= 1.33 \cdot 10^{-4} \text{ N m}^{-2}$

and

$N/V = \dfrac{(1.33 \cdot 10^{-4} \text{ Nm}^{-2})(6.02 \cdot 10^{23} \text{ molecules mole}^{-1})}{(8.314 \text{ J K}^{-1} \text{ mole}^{-1})(298 \text{ K})}$

$= 3.24 \cdot 10^{16} \text{ molecules m}^{-3}$

Therefore

$Z = (1/4) \bar{v} N = (475)(3.24 \cdot 10^{16})/4 = 3.85 \cdot 10^{18} \text{ collisions m}^{-2} \text{ s}^{-1}$

and

Elapsed time $= 10^{19}/3.85 \cdot 10^{18} = 2.60 \text{ s}$

Since N/V and Z are directly proportional to p and the elapsed time is inversely proportional to p, the values of these quantities at other pressures can be written by inspection. At $p = 10^{-8}$ torr, $N/V = 3.24 \cdot 10^{14}$ molecules m^{-2}, $Z = 3.85 \cdot 10^{16}$ m^{-2} s^{-1}, and time $= 260 \text{ s} = 4.3 \text{ min}$; at $p = 10^{-10}$ torr, time $= 2.6 \cdot 10^4$ s $= 7.2 \text{ hr}$. ■

* * *

9.8a.2 Ultrahigh Vacuum Requirement

The calculation in Example 9.6 shows that until gas pressure is lowered below, say, 10^{-8} torr, the rate of surface bombardment is too great for the surface to remain clean long enough for an experiment to be carried out. Pressures on the order of 10^{-6} torr are achieved in ordinary vacuum lines with diffusion and mechanical pumping. Any pressure below this is considered ultrahigh vacuum (UHV); today devices are commercially available that routinely achieve pressures in the range 10^{-9} to 10^{-12} torr.

An assortment of different pumping arrangements is employed to achieve UHV, including ion pumps, cryopumps, and getter pumps. With each of these the pumped gas is stored within the system rather than being removed, as is the case with diffusion pumps. This means that caution must be exercised so that the stored gas is not re-released into the system during the course of an experiment. Ion pumping is achieved by ionizing with electrons the gas to be removed and then collecting these ions at a metallic cathode, where they remain adsorbed. Getter pumps selectively adsorb or dissolve the gas without using ionization to collect the molecules, and cryopumps use low temperatures to promote adsorption. These three modes of pumping are thus the highly specific reverse of outgassing procedures: Adsorption at a specific site (the pump) is promoted to lower the pressure.

It is not enough merely to reach a low pressure; it must also be possible to measure it. Ionization gauges are almost always used for this purpose. In these the residual gas is ionized, collected at an electrode, and the resulting current measured. The current varies linearly with the gas pressure down to about 10^{-11} torr. If the ions are separated by mass — making the gauge a mass spectrometer — then the partial pressures of various gases in the vacuum chamber can be determined.

Next let us consider the preparation of the sample itself. Above we referred to the desirability of keeping a surface clean; nothing was said about the problem of obtaining such a surface in the first place.

9.8a.3 Sample Preparation

Two preferred methods for preparing clean surfaces consist of generating the surfaces under high vacuum at the beginning. Using remote control manipulators to crush a sample under vacuum exposes fresh surface to an environment in which adsorption equilibrium is very slow. This technique produces a heterogeneous array of crystal faces, however. Far more suitable for the specific and localized examination that diffraction methods offer is crystal cleavage under vacuum. Again by use of remote manipulators, a sample may be cleaved along a predetermined crystal plane to expose a fresh surface of known character. Surface repair through annealing may be necessary after this relatively violent procedure. The need for annealing also means that UHV chambers must possess the capability for heating samples in a controlled and measurable way. Sample preparation by cleavage plus annealing gives the best defined surfaces for subsequent examination.

Prior to a discussion of the use of LEED for structural measurements, a review of a few concepts from crystallography is needed for interpreting data from LEED.

9.8b Some Basic Concepts from Crystallography

A crystal is an orderly array of atoms or molecules but, rather than focusing attention on these material units, it is helpful to consider some geometrical constructs that characterize its structure. It is possible to describe the geometry of a crystal in terms of what is called a *unit cell*: a parallelepiped of some characteristic shape that generates the crystal structure when a three-dimensional array of these cells is considered. We then speak of the lattice defined by the intersections of the unit cells on translation through space. Since we are interested in crystal surfaces, we need to consider only the two-dimensional faces of these solids. In two dimensions the equivalent of a unit cell is called a *unit mesh*, and a net is the two-dimensional equivalent of a lattice. Only four different two-dimensional unit meshes are possible.

Figure 9.14 shows a set of points in one plane of a cubic structure. Much of our discussion in this section is based on cubic unit cells for simplicity, but the results are quite general. It also turns out that many metals crystallize with cubic unit cells. We may regard this as representing the surface net of a cubic lattice. Several sets of parallel lines connecting various points in the net have been drawn in the figure. These rows of atoms — like lines etched on a diffraction grating — are the origin of the LEED diffraction.

The lines in Figure 9.14 are, of course, the edges of crystallographic planes that slice through the crystal and are identified by the so-called Miller indices just as the lines themselves are. Since we are working in two dimensions, only two indices are needed to characterize them. The sets of numbers labeling the lines in Figure 9.14 are called the *Miller indices* — represented as *hk* — for that line. The easiest way to remember the significance of these indices is to note that h and k count the number of lines (of type hk) crossed in moving from one net point to the next in the x and y directions, respectively, for cubic cells. Thus the 11 line in Figure 9.14 cuts across the surface in such a way that one of these lines is crossed in moving from one net point to the next in the x direction and one is crossed in moving from point to point in the y direction.

Before we leave Figure 9.14, there are two additional points to be made concerning the spacing and atom density of the various lines. The first thing to note for the simple array of points in Figure 9.14 is that the actual spacing of the net points is exactly the same in all directions, the edge length of the unit cell (a cube) δ. Each of the various sets of lines that can be drawn through this net has its own characteristic spacing, say, d_{hk} for the hk line. What makes this significant is that in LEED studies electrons are reflected off the various crystallographic planes and constructive interference occurs when the spacing between the diffracting surfaces d_{hk} and the angle of incidence satisfy the Bragg equation (see Equation (1.21)),

$$n\lambda = 2d_{hk} \sin \theta \tag{98}$$

where λ is the wavelength of the electrons and n ($= 1, 2, 3, \ldots$) is the order of the diffraction. The Bragg equation is discussed in connection with microscopy in Chapter 1, Section 1.6. The point here is this: First-order ($n = 1$) diffraction results in reinforcement at different angles for different reflecting planes, even when the actual net is characterized by a single distance δ. Our interest is in the relationship between δ and the spacing d_{hk} for planes with those indices. It is possible to show that there are only three area-filling, two-dimensional shapes: square, rectangular, and oblique parallelogram. Table 9.5 shows how the inter-row spacing d_{hk} is related to the Miller indices and the side lengths a and b for these different shapes of two-dimensional unit cells.

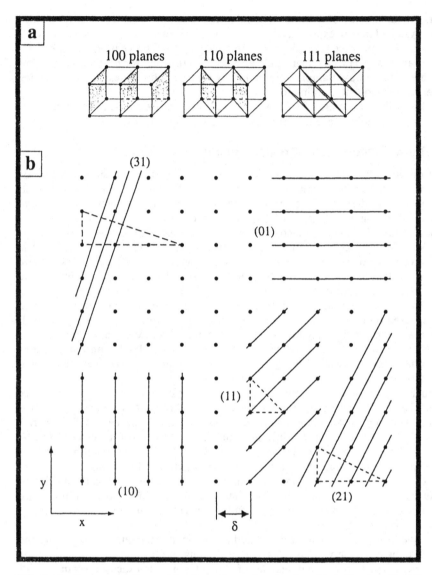

FIG. 9.14 Some crystallographic planes through a simple cubic lattice in which the atomic spacing is δ. The numbers in parentheses are the Miller indices of (a) planes and (b) lines defined by the edges of planes.

A second result that is evident from an inspection of Figure 9.14 is the fact that planes with lower Miller indices have higher atomic densities. Those planes characterized by high Miller indices strike out across the lattice at acute angles and cross several planes of atoms before actually intercepting an atom. This makes the average distance between atoms greater in planes that have high Miller indices. The surface tension is less in planes of low Miller index than in planes of higher index, in which the interatomic spacing is larger. If it were possible to deform the former surface into the latter, work would have to be done against the attractive forces between atoms to increase this separation. We may think of this work as an increment in surface free energy in the planes of high index that is not present when the Miller indices are lower. Surfaces of lower γ and hence lower Miller index are thermodynamically more stable. Surface studies conducted on the specific face of a crystal generally involve the 100, 110, or

TABLE 9.5 Relationship Between the Spacing d_{hk} and the Side Lengths δ and δ' in the Two-Dimensional Unit Cells of Square, Rectangular, and Oblique Parallelogram Nets

Geometry	Relationship
Square ($\delta = \delta'$)	$d_{hk}^{-2} = (h^2 + k^2)/\delta^2$
Rectangle ($\delta \neq \delta'$)	$d_{hk}^{-2} = (h/\delta)^2 + (k/\delta')^2$
Oblique, general ($\delta \neq \delta'$; $\alpha \neq 90°$)[a]	$d_{hk}^{-2} = (h^2/\delta^2 \sin^2 \alpha) + (k^2/\delta'^2 \sin^2 \alpha) - (2hk \cos \alpha/\delta\delta' \sin^2 \alpha)$
Oblique, hexagonal ($\delta = \delta'$; $\alpha = 120°$)[a]	$d_{hk}^{-2} = 4(h^2 + hk + k^2)/3\delta^2$

[a]The angle between sides is α in the oblique case and equals 120° for the hexagonal element.

111 plane for this reason. (Remember, three indices are necessary to describe the planar face of a crystal.)

9.8c Interpretation of Low-Energy Electron Diffraction Patterns

Generally, LEED experiments are conducted on specified faces of single crystals. When this is done, the diffraction pattern produced consists of a series of spots with a location, shape, and intensity that can be interpreted in terms of the surface structure. We focus attention on what can be learned from the location and shape of the spots since the study of intensity is beyond the scope of this book. It is generally assumed that the surface examined by LEED is an extension of an already-known bulk crystal structure. The correctness of this assumption can be tested, and results are often expressed in terms of modifications of the three-dimensional structure at the surface. Before we turn to the LEED patterns below, we must first figure out how they are read.

Figure 9.15a follows the path of the diffracted electron beam from the edge of the planes—separated by the distance d_{hk}—to the photographic surface where the LEED pattern is observed. The diffracting planes are shown in greatly exaggerated size in this version, and the angle of incidence is purposely drawn to be exceptionally small. Small angles like these are not typical of LEED, but the argument that follows will help index the LEED pattern.

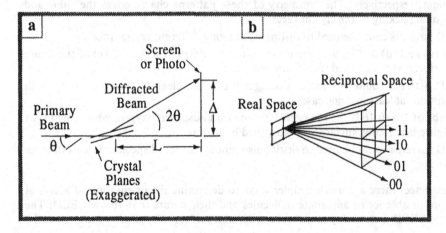

FIG. 9.15 Path of a diffracted electron beam in low-energy electron diffraction (LEED), and indexing of points in reciprocal space. (a) interaction of a diffracted beam with a photographic plate for small angles of incidence; and (b) illustration of the indexing of points in reciprocal space relative to the primary beam, labeled 00.

In Figure 9.15a the diffracted beam produces a spot on the photographic plate a distance Δ from the point at which the primary beam strikes. From trigonometry, it is evident that $\tan 2\theta = \Delta/L$. For the small angles considered here $\tan x \approx x$; therefore $2\theta \approx \Delta/L$, or $\theta \approx \Delta/2L$. For small angles, $\sin x \approx x$ also; therefore $\sin \theta \approx \Delta/2L$ for the situation shown in the figure. Combining this small-angle approximation with the Bragg condition for first-order diffraction,

$$2 d_{hk} \sin \theta = \lambda \tag{99}$$

we obtain

$$\lambda = d_{hk} \frac{\Delta}{L} \tag{100}$$

Since d_{hk} and L are constant, Equation (100) predicts that Δ is inversely proportional to d_{hk}, with Δ the distance of the diffraction spot from the spot produced by the primary. Figure 9.14 shows that d_{hk} is largest for planes of low Miller indices; since Δ varies inversely with d_{hk}, it follows that spots nearest the primary spot are due to low Miller index planes. Likewise, more distant spots are due to planes of higher index. There is a reciprocal relationship between the location of the spot on the photographic plate and the separation of the planes responsible for the spot.

Since every point in an LEED photograph is associated with a set of rows across a net, it is standard to identify the points by the same index numbers as their originating lines. Figure 9.15b shows how the points in a photograph are identified by a pair of indices hk. The primary beam is labeled 00, and the distance Δ in the approximation described above is measured outward from 00.

Now that the theory and vocabulary of LEED are established, let us look at some simple LEED patterns and see what they tell us about solid surfaces.

9.8d   Low-Energy Electron Diffraction Applied to Metal Surfaces

Figure 9.16 is a set of photographs of LEED patterns on clean metal surfaces and surfaces with adsorbed gases. Several variables are in effect to produce the LEED patterns shown:

1.   Parts (a) and (d) of Fig. 9.16 are the 100 surface of tungsten and the 111 surface of platinum, respectively. The symmetry of these patterns characterizes the cubic and hexagonal packing of the crystal faces.
2.   LEED patterns could be used to distinguish among different crystal faces.
3.   Parts (a) and (d) of Fig. 9.16 are clean surfaces, while parts (b) and (c) of the figure are surfaces with adsorbed species present.
4.   LEED patterns could be used to distinguish between clean surfaces and those with adsorption, at least in some cases.
5.   Part (b) of Fig. 9.16 is the tungsten 100 face with adsorbed oxygen, while part (c) of the figure is the same surface with adsorbed hydrogen.
6.   LEED patterns could be used to distinguish among adsorbed species, at least in some cases.

In most instances there are much simpler ways to determine the orientation of a crystal and the presence or absence of adsorbate molecules and their nature than to use LEED. The unique power of LEED is its ability to measure order at a surface. Therefore we may state the following:

1.   LEED enables us to determine the structure of the solid surface and to compare this with the bulk structure.

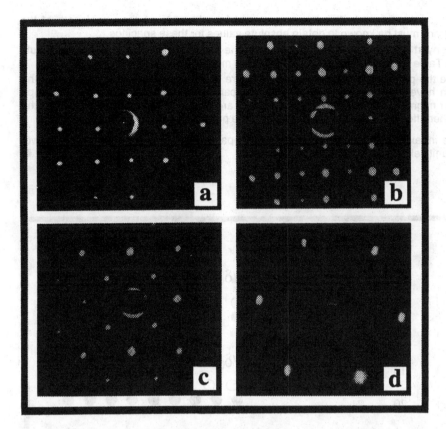

FIG. 9.16 Low-energy electron diffraction (LEED) patterns for clean metal surfaces and surfaces with adsorption: (a) clean W(100) surface; (b) W(100) with adsorbed oxygen; (c) W(100) with adsorbed hydrogen; and (d) clean Pt(111). (Figs. 19.16a–19.16c reprinted with permission from P. J. Estrup, In *Modern Diffraction and Imaging Techniques in Materials Science* (S. Amelincks, R. Gevers, G. Remaut, and J. Van Landuyt, Eds.), North Holland, Amsterdam, Netherlands, 1970; Fig. 9.16d reprinted with permission of Somorjai 1981.)

2. LEED allows us to observe and measure the order that exists in some adsorbed monolayers.

Item 2 in this last list is discussed somewhat further in Example 9.7.

* * *

EXAMPLE 9.7 *Comparison Between Bulk and Surface Structures Using Low-Energy Electron Diffraction Patterns.* If we accept that an LEED pattern has the same symmetry as the net of surface atoms responsible for its formation, what additional information is needed to complete the comparison between bulk and surface structures for the tungsten surface shown in Figure 9.16a?

Solution: We assume the crystal structure of the bulk metal, including the dimensions of the unit cell, are known. Comparison, then, depends on determining this quantity for the surface.

1. Taking photographic magnification into account, measure the distance of the diffraction spots from the origin, the spot produced by the primary beam.

2. These distances are multiples of the crystallographic spacing d_{hk}. The spacing may be calculated by evaluating a magnification factor from the experimental geometry in the manner suggested by Figure 9.15a.

3. The d_{hk} values so calculated are expressed in units of λ, so the wavelength of the electrons must be known to obtain absolute values for these spacings.
4. To convert d_{hk} values to the dimensions of the two-dimensional unit cell by a formula from Table 9.5, the spots in the LEED pattern must be indexed.
5. Since the pattern is a simple square in Figure 9.16a and since the spots nearest the origin have the lowest indices, the pattern can be indexed just as rectilinear graph paper might be marked off. Negative indices are possible; these are written above the number affected. Figure 9.17a shows how the points in Figure 9.16a are indexed.

From the index numbers and the experimental parameters, the dimensions of the two-dimensional unit cell can be determined.   ■

* * *

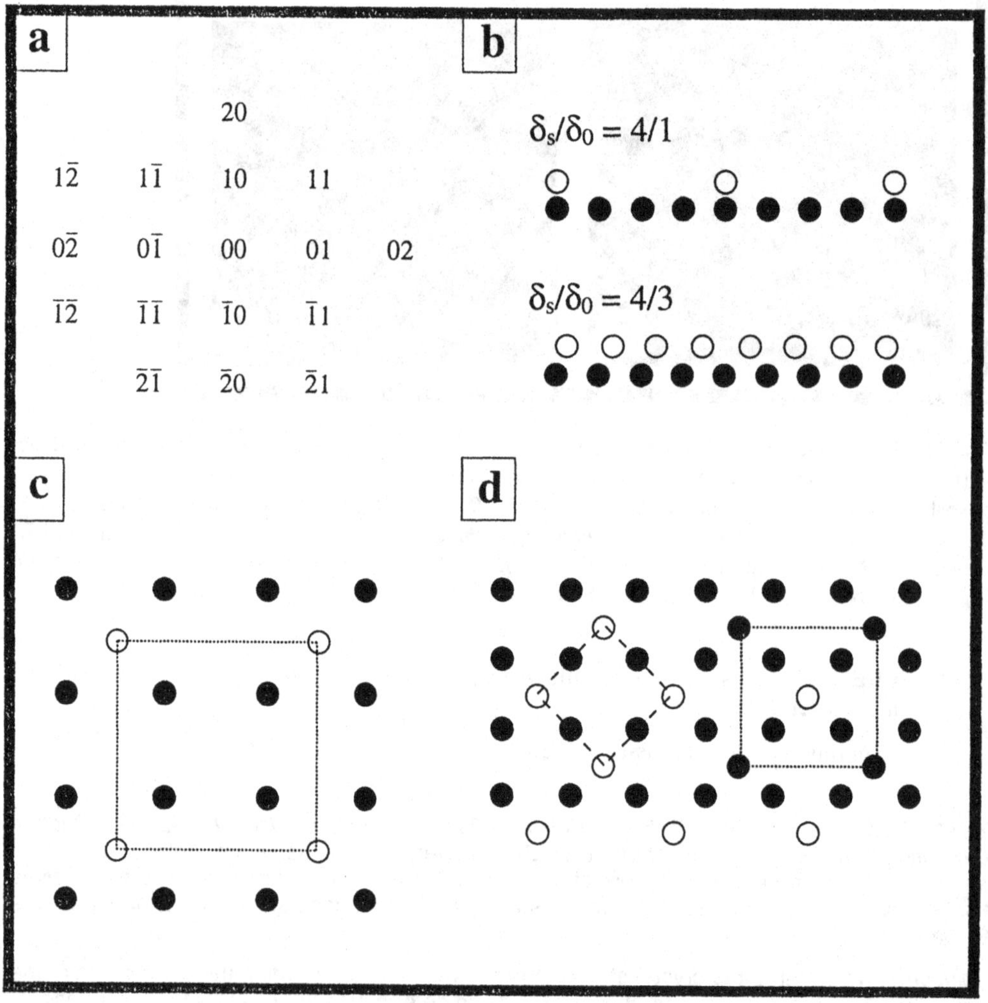

FIG. 9.17 Illustrations of LEED spots and coherent structures formed by adsorbed species. (a) indexing of the LEED spots of W(100) pattern in Figure 9.16a as described in Example 9.7; (b) side view of coherent structures formed by adsorbed (open circles) species on metal surface; (c) unit mesh for W(100) with adsorbed oxygen (open circles); (d) unit mesh for W(100) with adsorbed hydrogen (open circles).

For the 100 face of tungsten, written W(100), it has been found that the characteristic dimension of the unit cell in the two-dimensional lattice is 6% less at the surface than in the bulk crystal, although the packing geometry remains the same in both. The change in separation without a change in symmetry for surface atoms is called *relaxation* and is widely observed, particularly in crystal faces with relatively low packing efficiency. For example, the contraction of the Al(110) and Mo(100) surfaces are 5–15% and 11–12%, respectively. By contrast, the high atomic density 111 surfaces of silver and platinum show no contraction, and the Fe(111) surface contracts by 1.5%. This phenomenon may be explained as an attempt by surface atoms to compensate for their lower coordination number by pulling closer to their neighbors for more effective bonding. In some instances, it is observed that clean surfaces have totally different structures from what would be expected on the basis of bulk structures; this is called *surface reconstruction*.

Although UHV is required for LEED measurement, there is considerable interest in applying this technique to surfaces that carry adsorbed species. In view of our discussion of adsorption equilibrium above in the chapter, there is a difficulty here since adsorbed molecules imply an equilibrium gas phase. One way around this problem is to study surfaces at a sufficiently low coverage that the equilibrium gas pressure is compatible with the LEED technique. When higher pressures are desired, the surface is first equilibrated, then the excess gas is pumped out, and the surfaces before and after adsorption are compared through LEED. Chemisorption is better suited for study by this method than physical adsorption because the adsorbed layer remains intact when the equilibrium gas is removed.

Figures 9.16b and c are examples of this, and we discuss this figure in detail. First, let us consider the circumstances in which we might expect LEED to be sensitive to adsorbed species. From the perspective of approaching gas molecules, the surface of a single crystal presents an ordered array of adsorption sites (see, for example, Fig. 1.22 in Vignette I.8). We might anticipate that this order is carried over into the adsorbed layer, at least at low coverage, in which crowding and lateral interactions are not complications. Note also that this implies highly specific adsorption; this is another reason why LEED is especially valuable in the study of chemisorption. Since LEED responds to surface order, randomly adsorbed molecules such as those held by weak physical adsorption might go undetected. It was suggested at the beginning of this section that the LEED pattern of a clean surface might be taken as evidence that no adsorption had occurred. Actually, it may only indicate the absence of order in an adsorbed layer. A supplementary chemical analysis is a more definitive proof of surface cleanliness.

Figure 9.17b schematically represents a cross-sectional view of the surface of a solid and represents the topmost layer of atoms by shaded circles. The open circles represent molecules in an ordered pattern on the solid substrate. Since the adsorbed molecules are ordered, their structure on the surface is characterized by what is called a *supernet*. Suppose we define δ_0 as the characteristic spacing of the substrate and δ_s the equivalent quantity for the supernet. Then the two arrangements in Figure 9.17b are described by the ratios $\delta_s/\delta_0 = 4/1$ and $\delta_s/\delta_0 = 4/3$. Building on the notion of reciprocal distances as developed in the discussion of Figure 9.15, it follows that the adsorbed layer with $\delta_s/\delta_0 = 4/1$ should produce spots with a separation that is 1/4 that of the substrate. Likewise, for the case when $\delta_s/\delta_0 = 4/3$, a pattern of spots with a separation that is 3/4 that of the substrate is predicted. Thus, if the substrate produces spots at, say, 0 and 1, extra spots would be expected at 1/4, 2/4, and 3/4 for the $\delta_s/\delta_0 = 4$ case, and at 3/4, 6/4, and 9/4 when $\delta_s/\delta_0 = 4/3$. The cases illustrated here are called *coincident structures* since the two patterns coincide periodically. When there is no correlation between two structures, they are said to be *incoherent*.

With this background we are now in a position to make sense of the LEED pattern shown in Figure 9.16b. The substrate is the same W(100) surface shown in Figure 9.16a, but in Figure 9.16b the surface carries adsorbed oxygen. In the LEED pattern additional spots appear midway between the spots produced by the clean tungsten. The extra spot at the 1/2 position means that $\delta_s/\delta_0 = 2$ for the oxygen in this experiment; that is, the dimension of the oxygen unit mesh is twice that of the unit mesh of the substrate. Figure 9.17c is a schematic illustration of a possible supernet that is consistent with this description. It should be pointed out that spot location alone does not tell us where in the unit mesh of the supernet the adsorbed oxygens are located, only that they display a periodic structure with two times the repeat distance of the solid. The relative positions of the adsorbate and substrate atoms within their respective nets may be calculated from intensity data.

This example of a supernet makes it evident that an assortment of these supernets is possible with different values of δ_s/δ_0. This suggests that a system of designation is needed to distinguish among the various possibilities. One such system uses the notation $p(n \times m)$ in which the letter p (for primitive) indicates that the adsorbed net has the same primitive unit mesh as the substrate. In this system of notation n and m are integers, not necessarily the same, that express the expansion of the supermesh dimensions relative to the substrate along the mesh axes. Since LEED patterns of this type portray specific crystal faces under definite adsorption conditions, they are labeled to identify both the solid and adsorbate structures. Thus Figure 9.16b is a photograph of the $W(100)-p(2 \times 2)$—oxygen LEED pattern. This system of labeling is known as the *Wood notation*.

Figure 9.16c is another example of an LEED pattern of the $W(100)$ surface carrying an adsorbate, this time hydrogen. By their greater brightness, the spots originating from the substrate can be identified, and it is seen that the extra spots caused by the adsorbate once again lie midway between the tungsten spots, but this time with a different orientation from the array produced by the substrate. The midpoint positioning of these extra spots suggests that this adsorbate also shows a 2×2 supernet, but one that is rotated by 45° compared to the substrate net. A schematic illustration of the two nets is shown in Figure 9.17d, in which the substrate atoms are represented by filled circles and the adsorbed species by open circles. The dashed square in the figure shows the unit mesh of the supernet and emphasizes its 45° orientation relative to the substrate. The unit mesh dimensions in the dashed square are the same as those of the substrate, so the distances seem to be wrong even though the symmetry is correct. The solid square in the figure shows that there is a second way of looking at the identical supernet: as a 2×2 enlargement of the substrate net but possessing a centered rather than primitive packing. This is a preferable way of describing the supernet since it accounts for both the symmetry and the separation of the unit mesh. These two descriptions do not involve different supernets, but merely different ways of looking at the same net.

Since the version represented by the solid square in Figure 9.17d best characterizes the adsorption, it is described as a $c(2 \times 2)$ net, the c reminding us that this is a *centered* structure. The full description of the LEED pattern in Figure 9.16c in the Wood notation is therefore written $W(100)-c(2 \times 2)$—hydrogen.

The solid squares in Figures 9.17c and 9.17d represent the unit mesh of the adsorbed layers of oxygen and hydrogen, respectively, on the $W(100)$ surface. Both are seen to have the square symmetry of the underlying tungsten surface, but with oxygen showing a primitive net and hydrogen a centered structure. An additional detail about the surface with the adsorbed hydrogen is that the $c(2 \times 2)$ structure shown in Figures 9.16c and 9.17d persists until the surface coverage is half a monolayer. This seems too neat for an accidental circumstance. On the other hand, if we assume the H_2 is dissociated and the adsorbed species are hydrogen atoms, the same amount of adsorbed material is enough to form a monolayer. The chemisorbed hydrogen atoms form a surface layer characterized by a square, centered unit mesh. In Wood notation the surface layer is $W(100)-c(2 \times 2)-H$, indicating the fact that atomic hydrogen rather than H_2 is the adsorbate. Oxygen is also adsorbed in the monatomic state on the $W(100)$ surface.

In the adsorption studies we have discussed, the expansion of the unit mesh is the same in both directions, but this need not be the case. Examples in which the expansion along different axes of the mesh varies are $p(4 \times 2)-O$ for the adsorption of O_2 on $Mo(111)$, $p(3 \times 15)-O$ for O_2 on $Pt(111)$, $c(4 \times 2)-S$ for H_2S on $Au(100)$, and $c(9 \times 5)-CO$ for CO on $W(110)$. Somorjai (1981, 1994) has assembled extensive tables of this sort of information. Note that many but not all adsorbates are dissociated on the metal surfaces. Finally, it is not necessary for the supernet and the substrate to show the same angles between sides of their respective meshes. The Wood notation does not apply in these cases, but an alternative notation exists that we shall not pursue. Example 9.8 deals with a surface at which the adsorbate and the substrate display different nets.

* * *

EXAMPLE 9.8 *An Example of the Lattice Structures of the Adsorbate and Adsorbed Layers*
The unit mesh of the Pt(111) surface is a parallelogram with $\delta = \delta' = 0.277$ nm and having angles of 60° and 120°. This surface adsorbs *n*-butane with a unit mesh that is a parallelogram having $\delta = 0.480$ nm and $\delta' = 0.733$ nm and angles of 71° and 109°. Show that these two nets come into periodic register if the short side of the supermesh coincides with the diagonal of the substrate mesh.

Solution: Use the law of cosines to determine the length of the two diagonals of the substrate unit mesh:

$$L^2 = \delta^2 + \delta'^2 - 2\delta\delta' \cos 60° = 2(0.277)^2 - 2(0.277)^2(0.500)$$

$$L = 0.277 \text{ nm}$$

$$L' = [2(0.277)^2 - 2(0.277)^2(-0.500)]^{1/2} = 0.480 \text{ nm}$$

Since the short side of the supernet mesh has $\delta = 0.480$ nm, this mesh registers with the substrate along the diagonal of the latter. Since the long diagonal of the substrate and the short side of the super mesh register, this common line makes a convenient reference line to test the register in the perpendicular direction. Half the length of the short diagonal, $0.277/2 = 0.139$ nm, measures the perpendicular distance from this base to the corner atom. Also, the long side of the supernet mesh makes an angle of 71° with the base. Therefore the long side of the supernet mesh projects a length $(0.733 \sin 71) = 0.693$ nm onto the short diagonal of the substrate. The ratio $0.693/0.139 = 5.00$ shows that the edge of the supermesh and the diagonal of the substrate net will coincide again at the fifth row of atoms perpendicular to the diagonal that serves as the base for this calculation. Even though the two parallelograms have different side lengths, angles, and orientations, they do become coincident at periodic intervals. ■

* * *

Next, we see some additional examples of LEED applied to the study of metallic catalysts.

9.9 METAL SURFACES AND HETEROGENEOUS CATALYSIS

The topic of heterogeneous catalysis is the point at which the study of surfaces and the study of catalysts meet, as we have illustrated in Vignette IX (see also Somorjai 1994). It has been recognized for a long time that heterogeneous catalysts owe their activity to the properties of their surfaces. Until fairly recently, however, catalyst particle size and hence specific area were the major surface parameters that could be varied and monitored in the study of the surface effects. Much of the interest in characterizing high-surface-area solids through physical adsorption that we discussed at the beginning of this chapter originates from this application. With the advent of LEED and various types of surface spectroscopy, the role of surface science in the study of these catalysts is greatly expanded.

A wide variety of solid surfaces is used as catalysts in an even wider assortment of industrial processes (see, for example, Richardson 1989 and Somorjai 1994); we limit our discussion to metal catalysts. While these represent only a fraction of all catalytic systems, they do include a number of industrially important examples. Table 9.6 lists some metals used as commercial catalysts and indicates briefly the types of reactions for which they are employed. In this section we emphasize the effect on catalytic activity of the chemical and crystallographic properties of metal surfaces.

It should be emphasized here that there is a vast difference between the microenvironment of the catalyst surface as examined by the type of analytical techniques mentioned in Section 9.1 and the overall surface that influences commercial processes. Until the modern techniques became available, however, catalyst preparation was mostly a matter of trial and error; we have now entered an era in which science has a chance to catch up with technology. It seems fairly safe to predict that a greatly increased understanding of heterogeneous catalysis will emerge as modern surface chemistry matures.

TABLE 9.6   Examples of Some Metal Catalysts and the Reactions They Catalyze

Metal	Reaction
Cobalt	Fischer-Tropsch synthesis of hydrocarbons from CO and H_2
Iron	Haber synthesis of ammonia from N_2 and H_2
Platinum	Hydrogenation of vegetable oils
Platinum-palladium	Oxidation of hydrocarbons and CO and reduction of NO_x in automobile exhaust
Platinum-rhenium and platinum-tin	Reforming alkanes to aromatic hydrocarbon
Platinum-rhodium	Oxidation of NH_3 to HNO_3

9.9a   Catalysis at Surfaces

In our discussion of micellar catalysts in Chapter 8, we noted that effective catalysts have two features: the ability to accelerate the rate of a reaction and the ability to do so selectively. Chemistry students are familiar with the general notion that catalysts modify the mechanistic path of a reaction in such a way as to lower the activation energy and make the conversion of reactants to products more probable. One of the easiest places to see this is in reactions of diatomic gas molecules. In the gas phase the mechanism for the reaction of hydrogen and oxygen to form water involves the following steps, among others:

$$H_2 + O_2 \rightarrow HO_2^{\bullet} + H^{\bullet}$$
$$\text{(A)}$$

$$H_2 + HO_2^{\bullet} \rightarrow HO^{\bullet} + H_2O$$
$$\text{(B)}$$

$$H_2 + HO^{\bullet} \rightarrow H^{\bullet} + H_2O$$
$$\text{(C)}$$

How much simpler things would be if these were monatomic gases and there was no need for all the juggling between intermediate species to dispose of unused molecular fragments! We saw in our discussion of LEED that molecules of this sort are chemisorbed at metal surfaces in the dissociated state. The combination of chemisorbed hydrogen and oxygen atoms to form water clearly follows a different mechanism than in the gas phase. The fact that the reaction occurs rapidly in the presence of platinum and not at all when the reactants are mixed without the metal shows that the activation energy has been lowered tremendously by this modification.

A similar reaction of great commercial importance is the synthesis of ammonia from the diatomic elements. The catalysts that are used commercially in this reaction are mixtures of iron and iron oxide with the oxides of potassium and aluminum. Indicating the adsorbed species by the subscript (a), the mechanism for the surface-catalyzed synthesis of ammonia is thought to involve the steps

$$N_{2(g)} \leftrightarrow N_{2(a)}$$
$$\text{(D)}$$

$$H_{2(g)} \leftrightarrow H_{2(a)}$$
$$\text{(E)}$$

$$N_{2(a)} \leftrightarrow 2N_{(a)}$$
$$\text{(F)}$$

$$H_{2(a)} \Leftrightarrow 2H_{(a)}$$

(G)

$$N_{(a)} + H_{(a)} \Leftrightarrow NH_{(a)} \Leftrightarrow NH_{2(a)} \Leftrightarrow NH_{3(a)}$$

(H)

$$NH_{3(a)} \to NH_{3(g)}$$

(I)

The reaction between chemisorbed atoms dispenses with the problem of disrupting the triply bonded N_2 molecule in the gas phase, but Reaction (F) may also be more complicated than what is shown here.

The above are equilibrium reactions, and their successful exploitation requires that they be carried out under conditions in which the equilibrium favors the product. Specifically, this requires that the adsorbed species in Reactions (D)–(I) not be held so tightly on the catalyst surfaces as to inhibit the reaction. On the other hand, strong interaction between adsorbate and catalyst is important to break the bonds in the reactant species. Optimization involves finding a compromise between scission and residence time on the surface. Although we are especially interested in metal surfaces, those constituents known as promoters in catalyst mixtures are also important. It is known, for example, that the potassium in the catalyst used for the ammonia synthesis shifts Equilibrium (F) to the right and also increases the rate of Reaction (D) by lowering its activation energy from 12.5 kJ mole^{-1} to about zero.

In addition to affecting reaction energetics, other atoms at metal surfaces also influence the selectivity of the catalyst. Platinum catalysts are excellent examples of this since the platinum surface is capable of catalyzing a number of hydrocarbon reactions, including hydrogenation, dehydrogenation, isomerization, ring opening, and dehydrocyclization. Which of these processes is most favored by a particular catalyst is sensitive to the presence of foreign atoms at the platinum surface. Auger electron spectroscopy and other surface spectroscopies are clearly important techniques for the study of these effects. It is known, for example, that a partial monolayer of oxygen on platinum enhances reactions involving scission of the carbon-carbon bond and inhibits dehydrogenation reactions. Conversely, gold on platinum blocks $C-C$ scission without affecting dehydrogenation or isomerization. Sulfur, which often poisons catalysts, increases the selectivity of Pt-Re catalysts.

At low pressures hydrocarbon reactions on Pt are controlled by the clean metal surface. At higher pressures a carbonaceous layer controls the selectivity of the catalyst. Since the second condition describes how catalysts are generally used, the carbonaceous layer becomes an intrinsic part of the catalyst. At low temperatures hydrocarbons are physically adsorbed on Pt surfaces; at high temperatures a graphitic coating poisons the catalyst. Under intermediate conditions of temperature, various hydrocarbon fragments are present on the surface that determine its reactivity. It is noteworthy that the hydrogen atoms in these surface fragments are readily available for reaction, as evidenced by isotope exchange studies with deuterium. Carbonaceous matter, oxygen, and other atoms are thought to affect catalyst activity in several ways. For example, the electronic structure of the metal may be altered, surface restructuring may occur, and specific surface sites may be blocked.

9.9b Crystallographic Structure of Catalysts and Catalytic Behavior

The crystallographic character of surfaces has been shown to be of great importance in determining their catalytic behavior. Particularly interesting are those surfaces designed with well-characterized "roughness." To see how this is accomplished, consider slicing through a face-centered cubic platinum crystal at a small angle relative to, say, the 111 plane. This would result in a surface of high Miller index, low atomic density, and high surface free energy. Such a slice can be stabilized, however, by forming terraces of 111 planes separated by steps of, say, 100 planes. If the angle of the cut is small relative to the 111 plane, the steps are only one atom high; the width of the terrace depends on the angle of the cut, with narrower terraces resulting from steeper cuts. Application of crystallographic concepts shows that a slice through a face-centered cubic crystal along a plane with Miller indices 7,5,5 is equivalent to a series of 111 terraces five atoms wide separated by 100 steps one atom high.

The slice through a bulk crystal can differ from both the 111 plane and the 100 plane by small angles. This produces a kink in the face of the step. By an extension of the analysis that leads to step characterization, these kinks can also be characterized. For example, a plane with Miller indices 10,8,7 has 111 terraces seven atoms wide, 110 steps one atom high, and kinks of 100 orientation every two atoms. Because of the greater thermodynamic stability of the planes of low Miller index, these surfaces of ordered roughness are stable and can be prepared and studied. Since it is sensitive to periodicity over a domain about 20 nm in diameter, LEED "sees" the pattern associated with terraces of various widths and may be used to characterize these surfaces. Satisfactory LEED patterns do not require absolute uniformity of terrace width but may be obtained with experimental surfaces that display a distribution of widths.

The idea that catalyst surfaces possess a distribution of sites of different energies has been around since the 1920s, but it has not been possible until fairly recently to show that adsorption sites on terraces, steps, and kinks differ in energy. For example, hydrogen shows stronger bonding to steps and kinks on platinum than on the 111 terraces. In addition, the activation energy for H_2 dissociation is about zero on the step face and about 8.4 kJ mole^{-1} on the terrace plane. In addition, carbon monoxide is adsorbed with dissociation on the kinks of Pt, but in the molecular form on the steps and terraces.

The behavior of catalyst surfaces with respect to selectivity and poisoning by foreign atoms is now known to depend on the location of these atoms with respect to the terrace, step, and kink structure of the surface. By using the well-characterized faces of single crystals as catalytic surfaces, the effects of these variables have been examined for some reactions. Figure 9.18 is an example of the results of such a study. The hydrogenation of cyclohexene to cyclohexane is the reaction involved, and the figure shows how the ordered roughness and oxygen content of the platinum surface affect this reaction at 150°C. In Figure 9.18 the ordinate gives the number of molecules converted to product per second per Pt atom—the *turnover number* of the catalyst—and the abscissa gives the oxygen content of the surface as measured by Auger spectroscopy.

The three different lines in the figure correspond to platinum surfaces of different Miller

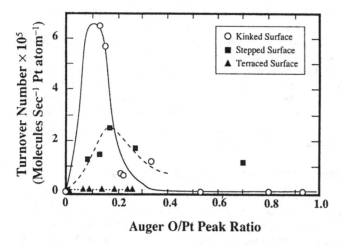

FIG. 9.18 Number of molecules of cyclohexene converted to cyclohexane per second per Pt atom versus oxygen content of platinum surface. The oxygen content is expressed as the ratio of O to Pt Auger peak heights. The data shown are for kinked (circles), stepped (squares), and terraced (triangles) platinum surfaces. (Redrawn with permission from Somorjai 1981.)

indices, corresponding to different degrees of ordered roughness. The data points marked by circles were measured on an 11,9,8 surface that consisted of terraces, steps, and kinks; the squares on a 7,5,5 surface that had terraces and steps; and the triangles on 111, which corresponds to terrace sites only.

The experiments described in Figure 9.18 were conducted under low-pressure conditions so the following generalizations apply:

1. In the absence of oxygen all three surfaces show undetectable catalytic activity.
2. The Pt(111) surface continues to show undetectable activity as the oxygen content increases, although the activity passes through a maximum for both the stepped and kinked surfaces with increasing oxygen content.
3. Compared at their respective maxima, the kinked surface is more active than the stepped surface.
4. The oxygen content at the maximum is slightly higher for the stepped surface than for the kinked surface.

These observations suggest that maximum catalytic activity for this reaction occurs at a kinked surface with about one-third of the surface covered by oxygen. For a competitive reaction, the dehydrogenation of cyclohexene to benzene, the kinked surface is also the most active. In this case, however, the optimum oxygen content is at a coverage of about one-half compared to one-third for the hydrogenation reaction on which Figure 9.18 is based. Differences such as these show that the selectivity of various catalysts is traceable to the chemical and physical nature of the surface. Even though the reasons for this behavior are not yet understood, these types of data hold out the promise that catalysts may someday be custom designed for maximum effectiveness. Polymer and pharmaceutical chemistry offer many examples of molecules that are designed with a specific function in mind. The methods and results of this chapter, augmented by some of the methods described in Chapter 1, suggest that this may become possible for catalysts in the not too distant future. Many of the modern developments in the area of heterogeneous catalysis may be found in Somorjai (1994).

REVIEW QUESTIONS

1. What is the difference between *chemisorption* and *physisorption*?
2. What is an *adsorption isotherm*?
3. Describe an experimental procedure for determining adsorption isotherms.
4. Describe the various types of isotherms observed in experiments. Describe the significance of the different forms of the isotherm.
5. How is an equation of state related to an adsorption isotherm? What is the basic thermodynamic principle that governs the equilibrium between the surface phase and bulk phase?
6. Discuss the implication of the *van der Waals equation of state* for a two-dimensional phase to the corresponding adsorption isotherm and to the analysis of adsorption data.
7. How is statistical thermodynamics used for deriving adsorption isotherms? What are the similarities and differences between this procedure and the one based on phenomenological thermodynamics? How is the kinetic theory of gases used for deriving adsorption isotherms?
8. What are the assumptions implicit in the *Langmuir adsorption isotherm*?
9. What is meant by the *Henry law limit* in the case of an adsorption isotherm?
10. What is the *Harkins-Jura isotherm*? What assumption does it make concerning the nature of the surface phase? Is the assumption consistent with the experimental observations?
11. What is the *BET isotherm*? How does it differ from the Langmuir isotherm?
12. How is the BET isotherm used to determine the surface area of powders?
13. What is meant by *calorimetric* analysis of adsorption?
14. What is *isosteric heat of adsorption*? How is it related to the pressure-versus-temperature relationship at constant surface coverage?
15. What is *adsorption hysteresis*?
16. What is *capillary condensation*? Describe some models of capillary condensation?
17. Describe the basic principles of operation of *low-energy electron diffraction (LEED)*.
18. Why are very low pressures needed for LEED experiments?

19. Describe how LEED patterns are interpreted.
20. What is meant by *heterogeneous catalysis*?
21. What is meant by *selectivity* of a catalyst?
22. Describe how adsorption assists in catalysis.
23. List a few examples of metal catalysts and what reactions they catalyze.

REFERENCES

General References (with Annotations)

Adamson, A. W., *Physical Chemistry of Surfaces*, 5th ed., Wiley, New York, 1990.   (Graduate level. A good source of extended discussions on the topics considered in this chapter. This volume includes more advanced material as well.)

Atkins, P., *Physical Chemistry*, 5th ed., W. H. Freeman, New York, 1994.   (Undergraduate level. An excellent, introductory-level book on physical chemistry for basic concepts not described in detail in this chapter.)

Ball, P., *Designing the Molecular World: Chemistry at the Frontier*, Princeton University Press, Princeton, NJ, 1994.   (Undergraduate level. A highly readable, *Scientific-American*-style, popular introduction to the many facets of molecular design using chemistry. Chapter 2 discusses chemical reactions and heterogeneous catalysis.)

Richardson, J. T., *Principles of Catalyst Development*, Plenum Press, New York, 1989.   (Undergraduate level. An introduction to catalysis and catalyst development from an engineering perspective. The emphasis throughout is on the practical aspect of the subject, rather than on theory.)

Somorjai, G. A., *Introduction to Surface Chemistry and Catalysis*, Wiley, New York, 1994.   (Undergraduate level. This in-depth treatment of surface chemistry and catalysis brings the experience and perspectives of a pioneer in the field to the general audience. The book is meant to be an introductory-level description of modern developments in the area for students at the junior level. However, it is also an excellent source of the current literature and contains numerous, extensive tables of data on kinetic parameters, surface structure of catalysts, and so on. Chapter 3, "Thermodynamics of Surfaces," and Chapter 7, "Catalysis by Surfaces," cover information relevant to the present chapter. Chapter 8 discusses applications in tribology and lubrication (not discussed in this chapter).)

Other References

Broeckhoff, J. C. P., and van Dongen, R. H., Mobility and Adsorption on Homogeneous Surfaces. In *Physical and Chemical Aspects of Adsorbents and Catalysts* (B. G. Linsen, Ed.), Academic Press, New York, 1970.

Hubbard, A. T. (Ed.), *The Handbook of Surface Imaging and Visualization*, CRC Press, Boca Raton, FL, 1995.

Kantro, D. L., Brunauer, S., and Copeland, L. E., BET Surface Areas: Methods and Interpretations. In *The Solid-Gas Interface*, Vol. 1 (E. A. Flood, Ed.), Marcel Dekker, New York, 1967.

Lowell, S., *Powder Surface Area and Porosity*, 3d ed., Chapman and Hall, New York, 1991.

Prutton, M., *Surface Physics*, 2d ed., Oxford University Press, Oxford, England, 1983.

Ross, S., Monolayer Adsorption on Crystalline Surfaces. In *Progress in Surface and Membrane Science*, Vol. 4 (J. F. Danelli, M. D. Rosenberg, and D. A. Cadenhead, Eds.), Academic Press, New York, 1971.

Somorjai, G. A., *Chemistry in Two Dimensions: Surfaces*, Cornell University Press, Ithaca, NY, 1981.

Van Hove, M. A., and Tong, S. Y., *Surface Crystallography by LEED*, Springer-Verlag, Berlin, 1979.

PROBLEMS

1. Consider the following linear equation of state appropriate for a relatively incompressible surface state (e.g., above $\theta \approx 0.70$ in Fig. 9.6b)

$$\pi = -C_1\sigma + C_2$$

where C_1 and C_2 are known constants. Derive the corresponding adsorption isotherm, known as the *Harkins-Jura isotherm* (see Table 9.1),

$$\ln p = -\frac{C_1}{2 N_A RT} A_{sp}^2 \left(\frac{n}{w}\right)^{-2} + C_3$$

(where C_3 is a constant of integration) for this linear equation of state.

2. An isotherm that is not too difficult to derive by the methods of statistical mechanics assumes an adsorbed layer that obeys the two-dimensional analog of the van der Waals equation. The result of such a derivation is the equation (see Table 9.1)

$$\ln p = \ln\left(\frac{\theta}{1-\theta}\right) + \frac{\theta}{1-\theta} - \frac{2a}{bk_B T}\theta + \ln K$$

where a, b, and K are constants, the first two being the two-dimensional van der Waals constants. Like its three-dimensional counterpart, this equation predicts that at the critical point for the two-dimensional matter

$$\left(\frac{\partial \ln p}{\partial \theta}\right)_{T_c} = 0 \quad \text{and} \quad \left(\frac{\partial^2 \ln p}{\partial \theta^2}\right)_{T_c} = 0$$

From the second of these derivatives, evaluate θ_c as predicted by this model. Use this value of θ_c and the first of these derivatives to evaluate the relationship between T_c and the two-dimensional a and b constants. How does this result compare with the three-dimensional case? The van der Waals constant b is four times the volume of a hard-*sphere* molecule. What is the relationship between the two dimensional b value and the area of a hard-*disk* molecule?

3. Use the linear form of the Langmuir equation to evaluate $(n/w)_{sat}$ and K for the adsorption of pentane on carbon black from the higher pressure values in the following data. Use the ratio p/p_0 rather than p only to normalize pressures relative to the equilibrium vapor pressure of pentane at different temperatures.* (All pressures are in torr.)

$T(°C)$	-63.7		0		5.24		20.5	
p_0	3.48		187.5		235.6		445.1	
	p	$g\,C_5\,g^{-1}$	p	$g\,C_5\,g^{-1}$	p	$g\,C_5\,g^{-1}$	p	$g\,C_5\,g^{-1}$
	0.024	0.2827	0.0533	0.1062	3.8	0.2299	0.284	0.1061
	0.028	0.2904	0.1405	0.1322	7.2	0.2535	0.675	0.1317
	0.067	0.3162	0.203	0.1427	19.6	0.2939	0.95	0.1421
	0.103	0.3276	0.89	0.1908	36.7	0.3147	3.62	0.1884
	0.233	0.3459	2.5	0.2305	53.6	0.3231	9.7	0.2262
	0.288	0.3507	5.0	0.2506	88.5	0.3313	15.9	0.2469
	0.671	0.3581	14.6	0.2964	199.0	0.3418	62.9	0.3002
	1.460	0.3647	29.2	0.3184			90.4	0.3107
			45.4	0.3272			155.2	0.3204
			80.7	0.3353			328.7	0.3321
			161.0	0.3433			428.7	0.3372

---

*Polanyi, M., and Goldmann, F., *Z. Phys. Chem.*, **132**, 321 (1928).

4. Examine the temperature variation of K values from the preceding problem by means of the procedure given in Example 9.3. (To decrease computational effort, various members of the class may be assigned different temperatures to analyze in Problem 3. The K values may then be pooled for this problem.)

5. The accompanying data* give the volume of N_2 at STP adsorbed on colloidal silica at the temperature of liquid nitrogen as a function of the ratio p/p_0. Plot these results according to the linear form of the BET equation. Evaluate c, V_m, and A_{sp} from these results, using 16.2 Å^2 as the value for σ^0.

V at STP (cm^3 g^{-1})	p/p_0	V at STP (cm^3 g^{-1})	p/p_0
44	0.008	117	0.558
52	0.025	122	0.592
57	0.034	130	0.633
61	0.067	148	0.692
64	0.075	165	0.733
65	0.083	194	0.775
70	0.142	204	0.792
77	0.183	248	0.825
78	0.208	296	0.850
85	0.275		
90	0.333		
96	0.375		
100	0.425		
109	0.505		

6. The following data give the volume at STP of nitrogen and argon adsorbed on the same nonporous silica at $-196°C$:†

	V at STP (cm^3 g^{-1})	
p/p_0	Nitrogen	Argon
0.05	34	23
0.10	38	29
0.15	43	32
0.20	46	38
0.25	48	41
0.30	51	43
0.35	54	45
0.40	58	50
0.45	58	54
0.50	61	55
0.60	68	62
0.70	77	69
0.80	89	79
0.90	118	93

Using 16.2 Å^2 as the N_2 cross section, calculate A_{sp} for the silica by the BET method. What value of σ^0 is required to give the same BET area for the argon data?

---

*Everett, D. H., Parfitt, G. D., Sing, K. S. W., and Wilson, R., *J. Appl. Chem. Biotechnol.*, **24**, 199 (1974).

†Payne, D. A., Sing, K. S. W., and Turk, D. H., *J. Colloid Interface Sci.*, **43**, 287 (1973).

7. Use the accompanying data* to criticize or defend the following proposition: Self-consistent A_{sp} values for nonporous solids are obtained at 77 and 90K by using values of σ^0 for N_2 equaling 16.2 and 17.0 Å^2, respectively. These are consistent with the density of liquid N_2 at these two temperatures. For the same self-consistency in A_{sp} using O_2 as the adsorbate, σ^0 values of 14.3 and 15.4 Å^2 must be used at these two temperatures. This suggests that O_2 is somewhat more loosely packed on the surface than in the liquid state at 90K compared to 77K.

	N_2		O_2	
T (K)	ρ (g cm^{-3})	V_{sp} (cm^3 g^{-1})	ρ (g cm^{-3})	V_{sp} (cm^3 g^{-1})
77	0.808	1.238	1.204	0.831
90	0.751	1.332	1.140	0.877

8. Use the data from Problem 3 to estimate the isosteric heat of adsorption of pentane on carbon black at $\theta \simeq 0.3$, 0.6, and 0.9. Under what conditions would greater variation of q_{st} be expected? What prevents these conditions from being examined in this problem? How does q_{st} compare with the energy of adsorption for this system as determined in Problem 4? How does the difference between q_{st} and E_{ads} compare with ΔH_v for pentane?

9. Colloidal carbon was formed by the slow pyroloysis of polyvinylidene chloride. Surface oxidation occurs during subsequent storage in air. By heating at elevated temperatures, the following gases are evolved† (heats of immersion were also measured after the different thermal pretreatments):

	Cumulative amount desorbed (mmol g^{-1})			Heat of immersion (cal g^{-1})
T (°C)	CO	CO$_2$	H$_2$	
100	0.03	0.02	–	11.0
200	0.05	0.14	–	9.8
300	0.15	0.33	–	9.2
400	0.35	0.44	–	8.7
500	0.63	0.51	–	8.2
600	0.95	0.55	–	7.7
700	1.300	0.55	–	7.0
800	1.580	0.55	–	6.2
900	1.800	0.55	0.03	6.0
1000	1.823	0.556	0.269	5.9

Using 7.9 and 9.1 Å^2 as the areas of oxide desorbing as CO and CO$_2$, respectively, estimate the area (in m^2 g^{-1}) occupied by each of these oxide types. If the specific area of the carbon is 1100 m^2 g^{-1}, what percentage of the surface is covered with oxide? Discuss the variation of the heat of immersion with the removal of the surface oxides.

---

*Hanna, K. M., Odler, I., Brunauer, S., Hagymassy, T., and Bodor, E. E., *J. Colloid Interface Sci.*, **45**, 27 (1973); Also see Osipow, L. I., in Ross, S., Monolayer Adsorption on Crystalline Surfaces, in *Progress in Surface and Membrane Science*, (Danell, T. E., Rosenberg, M.D. and Cadenhead, D. A., Eds.), Academic Press, New York, 1971.
†*Barton, S. S., Evans, M. J. B., and Harrison, B. H., *J. Colloid Interface Sci.*, **45**, 542 (1973).

10. Samples of rutile TiO_2 were outgassed at elevated temperatures; in some instances this was followed by subsequent exposure to O_2 at 150°C. The following results were obtained by the BET analysis of N_2 adsorption after various preliminary treatments:*

Pretreatment	A_{sp} (m² g⁻¹)	c
150°C + O_2	11.0	200
210°C + O_2	11.7	100
200°C, no O_2	11.9	520
240°C, no O_2	11.3	370
470°C + O_2	11.7	450
560°C + O_2	11.9	1210

Criticize or defend the following proposition: After the initial removal of physically adsorbed water, partial dehydroxylation of the surface occurs with increasing temperature. In the absence of O_2, Ti^{3+} cations are present that interact with N_2 in pretty much the same way as the isolated hydroxyl groups on the surface. At still higher temperatures surface diffusion permits the hydroxyl groups to migrate into patches, which show stronger interactions with N_2. The specific surface is not essentially changed throughout treatment.

11. Low-energy electron diffraction diagrams reveal that O_2 adsorbs on a Pd(110) surface through a series of ordered structures that interchange reversibly with increasing surface coverage obtained by varying the pressure and temperature:

$$\text{Clean Pd(110)} \rightleftharpoons (1 \times 3) \rightleftharpoons (1 \times 2) \rightleftharpoons c(2 \times 4) \rightleftharpoons c(2 \times 6)$$

The equilibrium pressure-temperature coordinates of the transitions between one LEED pattern and another were measured by Ertl and Rau† and were found to obey the two-dimensional Clausius-Clapeyron equation. When $\log_{10} p$ is plotted versus T^{-1}, straight lines of slope -1.75×10^4, -1.68×10^4, -1.35×10^4, and -1.05×10^4 K, respectively, are obtained for the four transitions above. Use these data to evaluate ΔH for each of the phase transitions of the adsorbed oxygen layer. Criticize or defend the following proposition: Since the only difference between these structures is the extent of oxygen coverage, the values of ΔH may be viewed as isosteric heats of adsorption for the surface at their respective degrees of coverage.

12. The object of this problem is to demonstrate that only certain shapes are possible for the unit mesh of a surface net. The following steps outline the proof: (a) Draw two parallel lines; one is the x axis, the other is at $y = $ const. (b) Draw a transverse cutting across the two lines and making an angle θ with the x axis. (c) Cutting the x axis at another location, draw another transverse, this time making an angle $-\theta$ with the axis. The two transverse lines should be mirror reflections of each other. (d) Imagine a surface atom at each of the four intersections of these lines. If the unit mesh is to be a regular polygon, then three of the lines must be equal to each other and equal to the interatomic spacing d. Make any necessary adjustments in the drawing so that the three shorter lengths — say, the base and the two transverses — are equal. (e) The longer side of the drawing has a length $d + 2d \cos \theta$ since each of the transverse lines project a length $d \cos \theta$ in the x direction. For this array of atoms to come into periodic register, the ratio $(d + 2d \cos \theta)/d$ must be an integer. What values of n and θ satisfy this condition, and what shapes do the corresponding meshes possess?

13. Firment and Somorjai‡ showed that the C_4–C_8 n-alkanes adsorb on the Pt(111) surface in ordered monolayers for which the unit mesh is a parallelogram with the following dimensions:

---

*Parfitt, G. D., Urwin, D., and Wiseman, T. J., *J. Colloid Interface Sci.,* **36**, 217 (1971).

†Ertl, G., and Rau, P., *Surface Sci.,* **15**, 443 (1969).

‡Firment, L. E., and Somorjai, G. A., *J. Chem. Phys.,* **66**, 2901 (1977).

Hydrocarbon	δ (Å)	δ' (Å)	Degree
n-Butane	4.80	7.33	71.0
n-Pentane	4.80	16.79	74.0
n-Hexane	4.80	9.99	75.6
n-Heptane	4.80	22.29	78.0
n-Octane	4.80	12.69	79.5
Pt(111)	2.77	2.77	60.0

Look up the formula for the area of a parallelogram and calculate the area per mole, assuming one molecule per unit mesh. Compare this area with the Pt(111) unit mesh for which the dimensions are also given. Do the data make any more sense if it is assumed that some of these alkanes form surface structures with two molecules per unit mesh? Prepare a plot of the area per molecule versus the number of carbon atoms in the chain. Criticize or defend the following proposition: The amount of free area per unit mesh in these packings is equivalent to one Pt mesh.

14. The authors of the research cited in Problem 13 point out that the width of the alkane molecule in the bulk crystals is 4.79, 4.78, and 4.70 Å for n-octane, n-heptane, and n-hexane, respectively. This corresponds almost exactly to one of the dimensions of the unit mesh of these molecules on the Pt(111) surface. The data tabulated in Problem 13 show that the longer dimensions of these cells tend to increase with the chain length. To establish whether this is a quantitative correlation, plot δ' versus the number of methylenes N in the alkane and determine the slope and intercept of the resulting graph. This gives an equation of the type $\delta' = mN_{CH_2} + b$, where m is the length contribution per methylene and $b/2$ is the length contribution per terminal methyl group. Compare these lengths with the dimensions of a fully extended hydrocarbon chain as presented in Example 8.2. What does this reveal about the configuration of these molecules on the Pt(111) surface?

10

van der Waals Forces

Already the difficulties of avoiding a collision in a crowd are enough to tax the sagacity of even a well-educated Square; but if no one could calculate the Regularity of a single figure . . . all would be chaos and confusion, and the slightest panic would cause serious injuries.

From Abbott's *Flatland*

10.1 INTRODUCTION

10.1a What Are van der Waals Forces?

This chapter is concerned with one of the most important forces in surface and colloid chemistry, namely, *van der Waals forces* between atoms, molecules, or particles. These forces have their origin in the dipole or induced-dipole interactions at the atomic level and are therefore of extreme importance in almost all aspects of the study of materials. The strength of van der Waals forces increases in the case of interaction between macroscopic objects such as colloidal particles since typically each particle has a large number of atoms or molecules. We see in this chapter that there are three major types of van der Waals forces, and that one of these, known as the *dispersion force*, is always present (like the gravitational force). This fact implies that it is not always possible to ignore van der Waals forces in colloid science, although, as we see in Chapter 13 on colloid stability, there are methods one can use to minimize or override the influence of these forces in order to promote the stability of colloids against coagulation. For example, in Chapter 11 we see how ionic atmospheres near charged surfaces can override van der Waals attraction and prevent coagulation. This is discussed further in Chapter 13, in which we also see how polymer layers adsorbed (or grafted) on particles are often used to "mask out" the influence of van der Waals attraction.

Our objectives in this chapter are to look into the origin of van der Waals forces, see how they affect macroscopic behavior and properties of materials, and establish relations for scaling up the molecular-level forces to forces between macroscopic bodies.

10.1b Why Are van der Waals Forces Important?

One of the most important things to bear in mind in studying van der Waals forces is that this topic has ramifications that extend far beyond our discussion here. Van der Waals interactions, for example, contribute to the nonideality of gases and, closer to home, gas adsorption. We also see how these forces are related to surface tension, thereby connecting this material with the contents of Chapter 6 (see Vignette X below). These connections also imply that certain macroscopic properties and measurements can be used to determine the strength of van der Waals forces between macroscopic objects. We elaborate on these ideas through illustrative examples in this chapter.

VIGNETTE X   WETTING AND SPREADING OF LIQUIDS:
Stability of Thin Liquid Films

Why does a drop of pentane spread into a thin film when placed on a water surface, whereas a larger hydrocarbon such as dodecane breaks up into smaller droplets? This is not an academic question, as should be evident from the importance of wetting and contact angle phenomena that we discussed in Chapter 6. Why is it that we can produce relatively stable bubbles with a soap solution but not with pure water? Water droplets on an oily surface, dewdrops on a blade of grass, and soap bubbles or foams are so common in our daily life that they rarely engage our attention, but to a scientist they are a constant reminder of the ubiquitous van der Waals forces!

The very fact that the vapor phase of many substances can condense to form a liquid is a consequence of the existence of *attractive* van der Waals forces between atoms or molecules. An attractive intermolecular force is *not* needed for a gas to condense into a solid; solidification can occur purely as a result of "excluded-volume" interactions among the molecules at sufficiently large densities. The pressure in a fluid, the cohesion between materials, and the existence of surface energy or surface tension all result, partially or wholly, from van der Waals forces.

Let us return to the differing behavior of pentane and dodecane on water. The energy of a film (of thickness d) of substance 1 spread between a planar interface of materials 2 and 3 can be traced to the energies of interaction between 2 and 3 through 1. In macroscopic terms, under certain conditions (see Section 10.6a), this energy is proportional to $(-A/(12\pi d^2))$, where the proportionality constant is a material property known as the *Hamaker constant*, which, in this case, depends on certain physical properties of all the three materials. This energy is the *macroscopic* van der Waals energy and results from the cumulative effects of individual, *intermolecular* van der Waals energies summed over all the molecules of the macroscopic objects. It turns out that in the case of pentane, A can become negative (i.e., the van der Waals energy is positive and the force is repulsive) so that pentane spreads on the water's surface thereby separating the air (material 3) from water (material 2) to the extent permitted by other forces such as gravity. In the case of dodecane, however, the Hamaker constant is positive, the van der Waals force is attractive, and the film continues to thin and breaks up into small, lens-shaped droplets. The reason for the lens shape is discussed in Chapter 6.

Now we can understand what happens in the case of soap bubbles. As discussed in Chapter 7, the soap molecules spread on both air-water interfaces enclosing a film of water and provide the repulsion necessary to maintain a water film, similar to the case we discuss in Example 11.3. In the absence of soap molecules, the (attractive) air-air van der Waals forces through the water film rupture the film so that the bubbles are unstable.

Phenomena such as the ones described above are usually (and conveniently) described in terms of macroscopic properties such as surface tension, contact angle, and so on. This is what was done in Chapters 6 and 7. In the present chapter, we probe the *molecular* origin of van der Waals forces, go into some of the details of how they scale up in the case of macroscopic bodies, and illustrate their importance in molecular as well as macroscopic phenomena.

10.1c   Focus of This Chapter

The focus of this chapter is to present a fairly comprehensive introduction to the van der Waals forces. In addition to giving an introduction to the origin of these forces, we also illustrate the prevalence of these forces through examples of their role in some of the common macroscopic phenomena that are not necessarily directly related to colloidal problems.

1.   However, we begin with a classical example of van der Waals forces in colloid science, namely, their role in coagulation and colloid stability, in Section 10.2. This discussion also sets the stage for a more detailed discussion of colloid stability in Chapter 13.

2.   Following this, we review briefly the so-called power law molecular interaction forces in Section 10.3 and develop the details of the different kinds of van der Waals forces in Section

10.4. Two illustrations of the implications of these forces to macroscopic behavior of materials are also provided here.

3.   The van der Waals forces scale up from atomic distances to colloidal distances undiminished. How the molecular forces scale up in the case of large objects, expressions for such forces, definition of the Hamaker constant, and theories based on bulk material properties follow in Sections 10.5–10.7.

4.   More often than not one deals with colloidal objects immersed in a liquid or other such media, and therefore interactions between similar or dissimilar materials in an arbitrary medium are of importance in colloid science. Moreover, it is very useful to relate such dissimilar interactions to those between identical particles in vacuum. In the last section (Section 10.8) we present what are known as "combining relations" for accomplishing this. The van der Waals forces between macroscopic objects are usually attractive, but under certain circumstances they (and, as a consequence, the Hamaker constant) can be negative, as noted in Vignette X. A brief discussion of this completes Section 10.8.

10.2 VAN DER WAALS FORCES AND THEIR IMPORTANCE IN COLLOID AND SURFACE CHEMISTRY

Almost all interfacial phenomena are influenced to various extents by forces that have their origin in atomic- and molecular-level interactions due to the induced or permanent polarities created in molecules by the electric fields of neighboring molecules or due to the instantaneous dipoles caused by the "positions" of the electrons around the nuclei. These forces consist of three major categories known as *Keesom* interactions (permanent dipole/permanent dipole interactions), *Debye* interactions (permanent dipole/induced dipole interactions), and *London* interactions (induced dipole/induced dipole interactions). The three are known collectively as the *van der Waals interactions* and play a major role in determining material properties and behavior important in colloid and surface chemistry. The purpose of the present chapter is to outline the basic ideas and equations behind these forces and to illustrate how they affect some of the material properties of interest to us.

Of the three forces mentioned above, the London force is always present (like the gravitational force) because it does not require the existence of permanent polarity or charge-induced polarity in the molecules. Even neutral atoms or molecules such as helium or hydrocarbons give rise to the London interaction. As a consequence, the London interaction plays a special role in colloid and surface chemistry. As mentioned above, it influences physical adsorption and surface tension; in addition, it is important in adhesion, wetting phenomena, structure of macromolecules such as proteins and other biological and nonbiological polymer molecules, and stability of foams and thin films. It also plays a very important part in determining the strengths of solids, properties of gases and liquids, heat of melting and vaporization of solids, and the like. We illustrate some of these using simple examples in subsequent sections.

Before we proceed to a molecular-level description of the van der Waals forces, it is useful to note a few other important points of information:

1.   The van der Waals forces are always attractive (although, as we see in Section 10.8b, the London forces between two macroscopic bodies immersed in a medium can be repulsive, depending on the material properties).

2.   They are relatively long ranged compared to other atomic- or molecular-level forces and can have an interval of influence ranging from about 0.2 nm to over 10 nm.

3.   The London force is also often called the *dispersion force*. The word *dispersion* here has nothing to do with the role of the London force in colloidal dispersions, but is the result of the role this type of interaction force plays in the dispersion of light in the visible and ultraviolet wavelengths.

4.   The dispersion force between two atoms, molecules, or large bodies is influenced by the presence of other nearby particles. Nevertheless, we consistently add pairwise interactions between the atoms in separate bodies as our procedure for scaling up

the interactions. This must be viewed as an approximation since perturbations by neighboring atoms limit the additivity of these forces.

More detailed and advanced information on these forces can be found in the book by Israelachvili (1991), which is devoted completely to intermolecular and surface forces. Here, we focus on the essential basic information and examples. Before we proceed to a physical explanation of these forces and the necessary equations, it is useful to explore the role played by the van der Waals forces in colloid stability since this theme reappears in our discussions of electrical double-layer forces in Chapter 11 and polymer-induced forces in Chapter 13.

10.2a Colloid Stability, van der Waals Attraction, and Potential Energy Curves

Colloid stability serves as a convenient example to illustrate the importance of the strength and range of van der Waals attraction between macroscopic bodies in a practical context and to introduce the idea of potential energy curves.

We introduced the concept of coagulation in Chapter 1, Section 1.4a, as that process by which two (or more) dispersed particles (primary or otherwise) cluster together to form an aggregate in which the individual units retain their identity but lose their kinetic independence. The fact that the primary particles are held together in these aggregates is evidence of the existence of attractive forces between the particles. The fact that some dispersions are stable with respect to the coagulation process is evidence that other forces that compete against attraction are also operative. In any specific system it is the relative magnitude of the attractive and repulsive forces between the particles that governs the coagulation behavior of the system.

Figure 10.1a is a schematic illustration of the kinds of interactions described above. The figure shows the potential energy of the interacting particles as a function of the distance of separation between them. It turns out that potential energy is a more useful quantity to deal with than force, although the force is given by the negative local slope of one of the curves in an illustration like Figure 10.1a. The figure shows two potential energy curves, one corresponding to attraction and the other to repulsion. By convention, the potential energy associated with repulsion is defined as positive, while the attraction is negative. The combination of repulsion and attraction allows us to speak of the "height" of energy "barriers" and the "depth" of energy minima.

An important aspect of Figure 10.1a is the fact that both the attraction and repulsion vary with the distance of separation between the bodies involved. Regardless of the specific shapes of the curves, both modes of interaction become weaker as the separation becomes larger. At sufficiently large distances the particles exert no influence on each other. The curves are deliberately interrupted at very small separations in view of the possibility of highly specific interactions at small separations. Our interest is in interactions of a more general type. In the schematic illustration shown, the two components vary in roughly the same way with distance, but this effective cancellation is only one of an assortment of possible attraction-repulsion combinations. One of our objectives in this section is to consider some of the combinations that are of interest.

The attractions of interest in colloid stability usually arise from the van der Waals forces, and we see below that they are scaled-up versions of the same intermolecular attractions discussed in the following section and that contribute to the nonideality and ultimately liquefaction of gases. In view of Figure 10.1a, we are interested in both the magnitude and the distance dependence of these attractive forces.

Our discussion of the repulsion between particles requires a more drawn out development. In Chapter 11 we consider the effects of ion atmosphere as the origin of repulsion. In Chapter 13, we discuss polymer-induced repulsive forces, which are often used to screen out the van der Waals forces. In these cases also, we are interested in both the magnitude and the distance dependence of these interactions.

Whether what we are discussing is attraction or repulsion, it is important to realize that the interaction between a pair of colloidal particles involves some fundamental physical

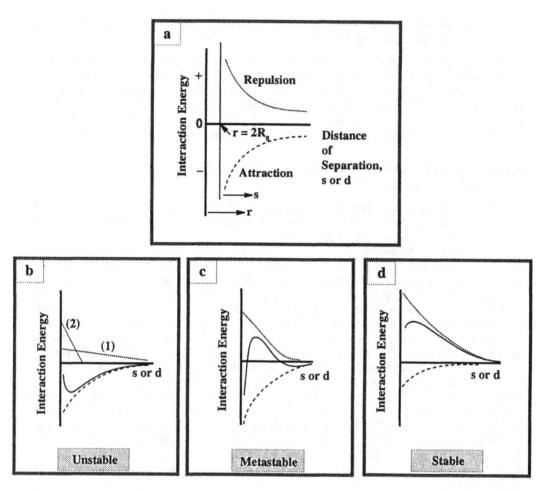

FIG. 10.1 Potential energy curves for the interaction of two colloidal particles, each of radius R_s. Negative values correspond to attraction and positive values to repulsion: (a) definition of variables; (b) repulsion less than attraction in magnitude and/or range; (c) repulsion and attraction comparable in magnitude and range; and (d) attraction less than repulsion.

phenomena and some geometrical considerations. The precise shapes of the curves in a plot like Figure 10.1a depend on both. For the quantitative interpretation of experimental results, it is important that both the physics and the geometry of the theory accurately (or at least approximately) describe the experimental system. From the point of view of pedagogy, however, we are more interested in physical phenomena than in particle geometry. For this reason we idealize dispersed particles as spheres and often discuss interactions between facing planar surfaces of bulk specimens of matter.

The axis representing the distance of separation in Figure 10.1a is defined in terms of the geometry of the interacting bodies. For spherical particles we define r as the variable that describes the separation of their centers. For spheres of (equal) radius R_s the separation of their surfaces along the line of centers is given by $(r - 2R_s)$, which we define as s. These are illustrated in Figure 10.1a. We define the distance of surface separation of two blocks of material by the symbol d.

The attraction and repulsion curves that describe the interaction of a pair of colloidal particles are usually not as evenly matched as those shown in Figure 10.1a. Next let us consider some other possible combinations of interest. In Figures 10.1b–10.1d the individual attraction and repulsion curves are shown as dashed or dotted lines, and the resultant of the two is indicated by a solid line. It is the net interaction that governs the behavior, and the components have been adjusted to span an assortment of possible behaviors.

In Figure 10.1b the attraction is much stronger than the repulsion at large distances of separation. It is easy to imagine this occurring as a result of two different situations with respect to the repulsion component. The repulsion might be relatively small at all distances, as indicated by line 1 in Figure 10.1b, or it may drop off over a much shorter range of distance than the attraction, as shown by line 2. In either case the attraction dominates at the larger separations, and the particles can achieve a lower state of energy by maintaining the distance of separation corresponding to the potential energy minimum. Stated somewhat differently, it would take an energy corresponding to the depth of the minimum to disrupt an aggregate that had formed as a result of the particles adopting this equilibrium separation. If we identify the potential energy axis with free energy, it is apparent that the initial dispersed state is unstable with respect to the final aggregated state in this case.

In Figure 10.1d the relative magnitudes of attraction and repulsion are just the opposite of those in Figure 10.1b. In Figure 10.1d the repulsion dominates. In this case the separated particles are lower in energy than the aggregate, and stability with respect to coagulation is indicated.

Figure 10.1c is an intermediate situation in which attraction and repulsion each have regions of dominance. In this case there is a shallow minimum at large separations, a maximum at somewhat smaller separations, and a deep minimum at small separations. These minima are known as the *secondary* and *primary* minima, respectively, in the order cited. The actual depths of the minima and height of the maximum depend on the particulars of the components. Except for admitting the possibility of its existence, we shall not consider the secondary minimum any further. What is of particular interest to us is the fact that a minimum in energy results from coagulation in the primary minimum; however, access to that minimum first requires that an energy barrier be overcome. Such a system may be viewed as metastable, possessing a degree of *kinetic* stability even though it lacks thermodynamic stability. That is, coagulation is predicted, but it is expected to occur slowly. In this sense the barrier serves as an obstacle along the path to coagulation; its height is analogous to the activation energy in ordinary reaction chemistry. This last observation suggests that the kinetics of coagulation may offer some clue as to the height of the maximum in this intermediate situation.

Potential energy curves of the type shown in Figure 10.1 are thus seen to be useful constructs for understanding and describing coagulation phenomena. They also illustrate the role of the attractive van der Waals forces on the stability of colloids.

In the next section we consider the molecular origins of attractions between colloidal particles.

10.3 Molecular Interactions and Power Laws

To understand the origin of the attraction between colloidal particles, it is necessary to back off a bit and consider the interactions between individual molecules. Macroscopic interactions — as we shall call the interactions between colloids since these particles are large compared to atomic dimensions — are the summation of the pairwise interactions of the constituent molecules in the individual particles. Therefore we begin by examining the interactions between a pair of isolated molecules.

Our primary interest in this section is to discuss the functional form that relates potential energy to the distance of separation x for various types of interactions. For many interactions an inverse power dependence on the separation describes the potential energy. Several examples of this are shown in Table 10.1. The main point to be observed now is that the value of the exponent in the inverse power dependence on the separation differs widely for the various types of interactions. An immediate consequence of this is that the range of the interactions is quite different also.

TABLE 10.1 Partial List of Interactions Between Pairs of Isolated Ions and/or Molecules, with a Listing of Functions That Describe the Potential Energy Versus Separation, Along with Appropriate Proportionality Constants

Description	Φ	Definitions and restrictions*	Attributed to	Value of n in $\Phi \propto x^{-n}$
Ion 1–ion 2	$\dfrac{(ze)_1(ze)_2}{4\pi\varepsilon_0 x}$	z = valence, e = electron charge under vacuum—otherwise ε_r in denominator (sign depends on the z value)	Coulomb	1
Ion 1–permanent dipole 2	$\dfrac{(ze)_1\mu_2\cos\theta}{4\pi\varepsilon_0 x^2}$	μ = dipole moment, θ = angle between line of centers and axis of dipole; length of dipole small compared to x (sign depends on z and orientation)	Coulomb	2
Permanent dipole 1–permanent dipole 2	$\dfrac{(\text{const.})\mu_1\mu_2}{4\pi\varepsilon_0 x^3}$	Numerical constant (including sign) depends on orientation: const. $=\sqrt{2}$ for average overall orientations; const. $= +2$ for parallel and -2 for antiparallel alignment	Coulomb	3
Permanent dipole 1–induced dipole 2	$-\dfrac{(\alpha_{0,1}\mu_2^2 + \alpha_{0,2}\mu_1^2)}{(4\pi\varepsilon_0)^2 x^6}$	α_0 = polarizability (always negative)	Debye	6
Permanent dipole 1–permanent dipole 2	$-\dfrac{2}{3}\dfrac{\mu_1^2\mu_2^2}{(4\pi\varepsilon_0)^2 k_B T x^6}$	Free rotation of dipoles (always negative)	Keesom	6
Induced dipole 1–induced dipole 2	$-\dfrac{3h}{2}\dfrac{\nu_1\nu_2}{\nu_1 + \nu_2}\dfrac{\alpha_{0,1}\alpha_{0,2}}{(4\pi\varepsilon_0)^2}\dfrac{1}{x^6}$	ν = characteristic vibrational frequency of electrons (always negative)	London	6
Induced dipole 1–induced dipole 2 (retarded)	$-\dfrac{23}{8\pi^2}hc\dfrac{\alpha_{0,1}\alpha_{0,2}}{(4\pi\varepsilon_0)^2 x^7}$	h = Planck's constant, c = speed of light; applies if $x > c/\nu$ (always negative)	Casimir and Polder	7
Repulsion	$+\dfrac{\xi}{x^{12}}$	Exponent in range 9 to 15; 12 mathematically convenient (always positive)		12

*The signs indicated are for the sign of Φ.

It is those functions with an inverse sixth-power dependence on the separation that are our main concern in this chapter. Those power laws with exponents greater or less than 6 are included in Table 10.1 mainly to emphasize the point that many types of interactions exist and that these are governed by different relationships. The interactions listed are by no means complete: Interactions of quadrupoles, octapoles, and so on might also be included, as well as those due to magnetic moments; however, all of these are less important than the interactions listed. Let us now examine Table 10.1 in greater detail.

The first three entries in Table 10.1 include Coulomb's law and two results that follow directly from it by treating stationary dipoles as a pair of charges and adding all pairwise interactions. What is important to note about these results is that the sign may be positive or negative—corresponding to repulsion or attraction—depending on the charge of ions, the orientation of the dipoles, or both.

By contrast, those results that involve an inverse sixth-power law are always negative; that is, attraction always results from interactions of the following types:

1. Permanent dipole/induced dipole interaction (*Debye equation*)
2. Permanent dipole/permanent dipole interaction (*Keesom equation*)
3. Induced dipole/induced dipole interaction (*London equation*)

The inverse seventh-power law is a special case of the induced dipole/induced dipole interaction that applies to the case of large separations; this is discussed in more detail in Section 10.5. As mentioned above, the three attractive interactions listed above are collectively known as van der Waals attraction. In Section 10.4 we discuss in greater detail the significance and the origin of the van der Waals attractions listed in Table 10.1. In the present section we are interested in only the exponent in the power law.

The last entry in Table 10.1 is the least well defined of those listed. This is of little importance to us, however, since our interest is in attraction, and the final entry in Table 10.1 always corresponds to repulsion. The reader may recall that so-called hard-sphere models for molecules involve a potential energy of repulsion that sets in and rises vertically when the distance of closest approach of the centers equals the diameter of the spheres. A more realistic potential energy function would have a finite (though steep) slope. An inverse power law with an exponent in the range 9 to 15 meets this requirement. For reasons of mathematical convenience, an inverse 12th-power dependence on the separation is frequently postulated for the repulsion between molecules.

In general, the combined effects of van der Waals attraction and interparticle repulsion at the molecular level may be represented by the equation

$$\Phi = \xi x^{-12} - \beta x^{-6} \tag{1}$$

in which the constant β has been used to represent the various constants in the Debye, Keesom, and London equations. Since the two terms on the right-hand side of Equation (1) correspond to opposing tendencies, the total potential energy function will display a minimum, the coordinates of which will describe an equilibrium situation. By differentiating Equation (1) with respect to x and setting the result equal to zero, the coordinates of the minimum can be evaluated. These are readily found to be

$$x_m = (2\xi/\beta)^{1/6} \tag{2}$$

and

$$\Phi_m = -(\beta/2)x_m^{-6} = -\xi x_m^{-12} \tag{3}$$

where the subscript reminds us that these are values at the minimum.

Equation (3) can be used to eliminate β and ξ from Equation (1) to obtain

$$\Phi = -\Phi_m[(x/x_m)^{-12} - 2(x/x_m)^{-6}] \tag{4}$$

Equations (1) and (4) or other variations of the 12–6 power law are often called the *Lennard-Jones potential*. The numerical values of the constants in the Lennard-Jones potential may be obtained from studies of the compressibility of condensed phases, the virial coefficients of gases, and by other methods. A summary of these methods and other expressions for the molecular interaction energy can be found in the book by Moelwyn-Hughes (1964).

Figure 10.2 is a plot of the Lennard-Jones function for methane, for which $\xi = 6.2 \cdot 10^{-134}$ J m^{12} and $\beta = 2.3 \cdot 10^{-77}$ J m^6 ($x_m = 0.42$ nm and $\Phi_m = 2.1 \cdot 10^{-21}$ J).

Curves of this sort occur in many places in physical chemistry, and it is important to realize that they are the result of two contributions: a very short-range repulsion and a relatively long-range attraction. It is the latter with which we are primarily concerned, so we turn next to an examination of the origins of these inverse sixth-power attractions.

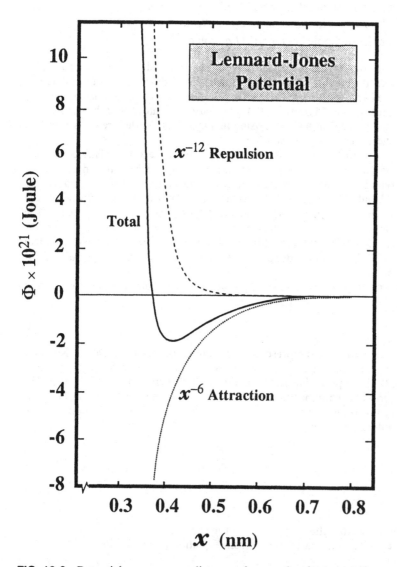

FIG. 10.2 Potential energy versus distance of separation for two methane molecules.

10.4 MOLECULAR ORIGINS AND THE MACROSCOPIC IMPLICATIONS OF VAN DER WAALS FORCES

In this section we outline the molecular origins of the Debye, Keesom, and London forces and discuss the strengths of these forces relative to each other. In addition, we also outline how macroscopic properties and behavior (such as the heat of vaporization of materials, nonideality of equations of state, and condensation of gases) can be traced to the influence of the above van der Waals forces and illustrate these through specific examples. Another example of the van der Waals forces, namely, the relation between the surface tension (or surface energy) of materials and the London force, is discussed in Section 10.7.

10.4a Molecular Origins of van der Waals Attractions

In Table 10.1 we saw that all random dipole-dipole interactions follow the inverse sixth-power law except the so-called *retarded* van der Waals attraction, which varies with the inverse seventh power of the separation. In this section we examine briefly the physical basis of the three different inverse sixth-power laws that describe intermolecular attractions. Space limitations prevent us from deriving the Debye, Keesom, and London expressions in detail. More complete derivations may be found in many physical chemistry textbooks, such as that by Moelwyn-Hughes (1964). The abbreviated discussion we present should be sufficient, however, to indicate the connection between these attractions and molecular parameters.

Interactions between dipoles, whether permanent or induced, are the result of the electric field produced by one dipole (subscript 1) acting on the second dipole (subscript 2). Therefore the first factor to consider in discussing such interactions is the field E produced by a dipole and measured a distance x from the dipole, with x large compared to the length of the dipole. Since we are dealing with the forces between charges, we must be attentive to the matter of units, as was the case in the similar discussion in Chapter 5, Section 5.2. The field has as units charge length^{-2}, but in SI this requires that we divide the charge by $4\pi\varepsilon_0 = 1.112 \cdot 10^{-10}$ J^{-1} C^2 m^{-1}, in which case the field has the units N C^{-1} (i.e., force per unit charge) or V m^{-1} (i.e., potential gradient). The factor $4\pi\varepsilon_0$ is not required when cgs units are used; and, since the factor 4π may cancel, some of these relationships look different when written for other systems of units.

The field is a vector quantity and may be resolved into components as shown in Figure 10.3. We define θ to be the angle between the axis of the dipole and the line that connects the point under consideration with the center of the dipole and along which x is measured. The field may be resolved into the following components:

1. Parallel to the line of centers:

$$E_{\parallel} = -(2\mu_1/4\pi\varepsilon_0)x^{-3}\cos\theta \tag{5}$$

2. Perpendicular to the line of centers:

$$E_{\perp} = -(\mu_1/4\pi\varepsilon_0)x^{-3}\sin\theta \tag{6}$$

The total field is the square root of the sum of the squares of Equations (5) and (6), or

$$
\begin{aligned}
E &= -(\mu_1/4\pi\varepsilon_0)x^{-3}(\sin^2\theta + 4\cos^2\theta)^{1/2} \\
&= -(\mu_1/4\pi\varepsilon_0)x^{-3}(1 + 3\cos^2\theta)^{1/2}
\end{aligned} \tag{7}
$$

10.4a.1 Debye Interaction

Now let us consider the effect of such a field on a particle with no permanent dipole of its own. The field will induce a dipole in the second molecule; the magnitude of the induced dipole moment μ_2 is proportional to the field with the polarizability of the second molecule α_2 the proportionality constant (see Equation (5.11)):

$$\mu_2 = \alpha_2 E \tag{8}$$

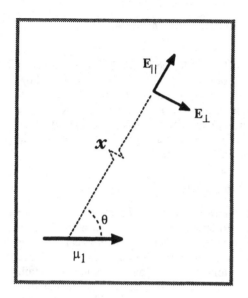

FIG. 10.3 The electric field a distance x from a dipole. The field is resolved into components parallel and perpendicular to the line of centers along which x is measured.

The potential energy of the second dipole due to this field is given by $-\mu_2 E$ or $-\alpha_2 E^2$. To this must be added the energy necessary to induce the dipole $(1/2)\alpha_2 E^2$ since the second is not a permanent dipole. Therefore the total potential energy Φ of the second dipole is

$$\Phi_2 = -(1/2)\alpha_2 E^2 \tag{9}$$

Substituting Equation (7) into this expression and averaging over all orientations yields

$$\Phi_2 = -[\alpha_2 \mu_1^2/(4\pi\epsilon_0)^2]x^{-6} \tag{10}$$

The second dipole acts on the original dipole in a similar fashion, giving a second contribution to the interaction energy that is identical to Equation (10) except that the subscripts are interchanged. The total potential energy of attraction is the sum of these two contributions:

$$\Phi_D = -[(\alpha_2\mu_1^2 + \alpha_1\mu_2^2)/(4\pi\epsilon_0)^2]x^{-6} \tag{11}$$

This is the *Debye equation* (subscript D) in Table 10.1 that describes the attraction between a permanent dipole and an induced dipole.

10.4a.2   Keesom Interaction

If this argument is applied to two *permanent* dipoles, the polarizability may be regarded as the sum of two contributions: one that is independent of the presence of the permanent dipole α_0 and a second that is the average effect of the rotation of the molecules in the electric field. The molar polarization P of a substance is given by

$$P = (N_A/3\epsilon_0)[\alpha_0 + (\mu^2/3k_B T)] \tag{12}$$

The second term inside the brackets thus gives the average value of the orientation contribution to the polarizability. Note that the magnitude of the permanent dipole contribution is expressed relative to thermal energy (k_B is the Boltzmann constant) since increased thermal jostling tends to scramble the permanent dipoles. It is the second term in the brackets that is used in Equation (11) for the interaction of two permanent dipoles:

$$\Phi_K = -(2/3)\{\mu_1^2\mu_2^2/[k_B T(4\pi\epsilon_0)^2]\}x^{-6} \tag{13}$$

This is the *Keesom equation* (subscript K) from Table 10.1; it applies to the interaction of two permanent dipoles.

10.4a.3 London (Dispersion) Interaction

Finally, we turn our attention to the third contribution to van der Waals attraction, *London* (or *dispersion*) forces between a pair of induced dipoles. It will be noted that (at least) one permanent dipole is needed for the preceding sources of attraction to be operative. No such restriction is present for the London component. Therefore this latter quantity is present between molecules of all substances, whether or not they have a permanent dipole. These are the same forces that we considered in Chapter 6 when we discussed the Girifalco-Good-Fowkes equation.

The London equation for the attraction between a pair of induced dipoles is a quantum mechanical result that represents one of the contributions to the "bond" between a pair of particles. Like other quantum mechanical results, the interaction energy emerges as part of the solution to the Schrödinger equation. We dispense with a rigorous examination of the situation and consider only the physical model and the final results.

Figure 10.4 represents the situation we wish to consider. It represents two dipoles with length ℓ_i that is negligible compared to the distance between their centers. The dipoles are formed by the symmetrical vibration of electrons in the two particles. According to Table 10.1, the potential energy of two dipoles in this arrangement is $\pm 2\mu^2(4\pi\varepsilon_0)^{-1}x^{-3}$ or $\pm 2(e\ell_1)(e\ell_2)(4\pi\varepsilon_0)^{-1}x^{-3}$ since $\mu_i = e\ell_i$. In addition, each of the vibrating dipoles may be regarded as a harmonic oscillator for which the potential energy is given by

$$\Phi = (1/2)K\ell_i^2 \tag{14}$$

in which

$$K = (e^2/\alpha_0) \tag{15}$$

Combining these various energy contributions gives the following expression to be used as the potential energy of this system:

$$\Phi_T = (1/2)K\ell_1^2 \pm 2(e\ell_1)(e\ell_2)(4\pi\varepsilon_0)^{-1}x^{-3} + (1/2)K\ell_2^2 \tag{16}$$

When this net potential energy function is substituted into the one-dimensional Schrödinger equation and the suitable mathematical operations are carried out, the allowed energies E are found to be

$$E = [n_i + (1/2)]h\nu_i + [n_j + (1/2)]h\nu_j \tag{17}$$

in which

$$\nu_i = \nu\{1 - [2\alpha_0/(4\pi\varepsilon_0 x^3)]\}^{1/2} \tag{18}$$

and

$$\nu_j = \nu\{1 + [2\alpha_0/(4\pi\varepsilon_0 x^3)]\}^{1/2} \tag{19}$$

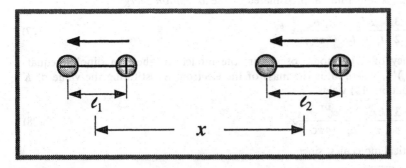

FIG. 10.4 A linear arrangement of two dipoles used to define the potential energy in the Schrödinger equation for the London interaction energy.

and the n_i and n_j are integers. Note that the terms in Equation (17) are formally identical to the energy of quantized harmonic oscillators of frequency v_i and v_j. In addition, we observe that v_i and v_j approach v as $x \to \infty$. Thus v is identified as the frequency of vibration for the system in the case in which the electrons vibrate independently.

The lowest energy for the two coupled oscillators is the situation in which $n_i = n_j = 0$, in which case Equation (17) becomes

$$E = (1/2)h(v_i + v_j) \tag{20}$$

On the other hand, the energy of two independent oscillators in their ground state is given by

$$E = 2[(1/2)hv] \tag{21}$$

The difference between Equations (20) and (21) gives the contribution of the interaction to the potential energy:

$$\Phi_L = (1/2)h[(v_i + v_j) - 2v] \tag{22}$$

This is one way of writing the London (subscript L) attraction between two induced dipoles.

If the expressions for v_i and v_j given by Equations (18) and (19) are substituted into Equation (22), the following result is obtained:

$$\Phi_L = (1/2)hv\left\{ \{1 - [2\alpha_0/(4\pi\varepsilon_0 x^3)]\}^{1/2} + \{1 + [2\alpha_0/(4\pi\varepsilon_0 x^3)]\}^{1/2} - 2\right\} \tag{23}$$

Expanding the square roots by the binomial expansion (see Appendix A) and retaining no terms higher than second order yields

$$\Phi_L = -\frac{h v \alpha_0^2}{2 (4\pi\varepsilon_0)^2} x^{-6} \tag{24}$$

Several modifications of Equation (24) are also important:

1. When the molecules are capable of vibration in all three dimensions, the numerical constant in Equation (24) becomes 4 rather than 2:

$$\Phi_L = -\frac{3 h v \alpha_0^2}{4 (4\pi\varepsilon_0)^2} x^{-6} \tag{25}$$

2. When unlike molecules are involved, their individual frequencies and polarizabilities are involved, and the expression equivalent to Equation (24) is

$$\Phi_L \approx -\frac{3}{2} h\left(\frac{v_1 v_2}{v_1 + v_2}\right) \frac{\alpha_{0,1}\alpha_{0,2}}{(4\pi\varepsilon_0)^2} x^{-6} \tag{26}$$

which is the result shown in Table 10.1. Note that this result becomes identical to the three-dimensional version of Equation (24) when the atoms are identical.

3. The quantity hv in Equation (25) may be regarded as some energy that characterizes the system and is sometimes approximated by the *ionization energy I*:

$$\Phi_L \approx -\frac{3}{2}\left(\frac{I_1 I_2}{I_1 + I_2}\right) \frac{\alpha_{0,1}\alpha_{0,2}}{(4\pi\varepsilon_0)^2} x^{-6} \tag{27}$$

4. The frequency of a harmonic oscillator, the model for the two dipoles, equals $(1/2\pi)(K/m_e)^{1/2}$, where m_e is the mass of the electron. Substituting the value of K given by Equation (15) yields

$$\Phi_L = -\frac{3}{8}\left(\frac{h e}{\pi}\right) \frac{\alpha_0^{3/2}}{m_e^{1/2} (4\pi\varepsilon_0)^2} x^{-6} \tag{28}$$

for two identical molecules, since

$$v = \frac{1}{2\pi}\sqrt{\frac{e^2}{\alpha_0 m_e}} \tag{29}$$

Equations (25)–(28) are widely encountered expressions for the London attraction between two molecules.

In examining the Debye, Keesom, and London equations we see that (a) they share as a common feature an inverse sixth-power dependence on the separation and (b) the molecular parameters that describe the polarization of a molecule, polarizability and dipole moment, serve as proportionality factors in these expressions. For a full discussion of the experimental determination of α_0 and μ, a textbook of physical chemistry should be consulted (Atkins 1994). For our purposes, it is sufficient to note that the molar polarization of a substance can be related to its relative dielectric constant ε_r by

$$P = \frac{M}{\rho} \frac{\varepsilon_r - 1}{\varepsilon_r + 2} \tag{30}$$

where M and ρ are the molecular weight and density, respectively (Atkins 1994, Chapter 22).

Combining Equations (12) and (30) gives the general result

$$\frac{M}{\rho} \frac{\varepsilon_r - 1}{\varepsilon_r + 2} = \frac{N_A}{3\,\varepsilon_0} \left(\alpha_0 + \frac{\mu^2}{3k_BT} \right) \tag{31}$$

Thus studies of ε_r as a function of T may be analyzed to yield values of both α_0 and μ. For substances with no permanent dipole moment, $\mu = 0$ and $\varepsilon_r = n^2$, where n is the refractive index at long wavelengths. For such a system Equation (31) becomes

$$\frac{M}{\rho} \frac{n^2 - 1}{n^2 + 2} = \frac{N_A}{3\,\varepsilon_0} \alpha_0 \tag{32}$$

To use Equation (32) it is necessary to extrapolate to infinite wavelength (or zero frequency) to obtain the unperturbed polarizability since the electric field of the light also alters the molecule. Failure to carry out such an extrapolation introduces far less error, however, than is introduced by an approximation such as Equation (27).

10.4a.4 Relative Magnitudes of the Individual Contributions to van der Waals Interaction

In general, we may think of any molecule as possessing a dipole moment and a polarizability. This means that each of the three types of interaction may operate between any pair of molecules. Of course, in nonpolar molecules for which $\mu = 0$, two of the three sources of attraction make no contribution.

As we have already noted, all molecules display the dispersion component of attraction since all are polarizable and that is the only requirement for the London interaction. Not only is the dispersion component the most ubiquitous of the attractions, but it is also the most important in almost all cases. Only in the case of highly polar molecules such as water is the dipole-dipole interaction greater than the dispersion component. Likewise, the mixed interaction described by the Debye equation is generally the smallest of the three.

For a pair of identical molecules, Equations (11), (13), and (25) may be combined to give the *net* van der Waals attraction (subscript A) Φ_A:

$$\Phi_A = -\frac{1}{(4\pi\,\varepsilon_0)^2} \left(2\alpha_{0,1}\mu_1^2 + \frac{2\mu_1^4}{3\,k_BT} + \frac{3}{4}\,h\nu_1\alpha_{0,1}^2 \right) x^{-6} = -\beta_{11}x^{-6} \tag{33}$$

The interaction parameter β_{11} is defined

$$\beta_{11} = \frac{1}{(4\pi\,\varepsilon_0)^2} \left(2\alpha_{0,1}\mu_1^2 + \frac{2\mu_1^4}{3\,k_BT} + \frac{3}{4}\,h\nu_1\alpha_{0,1}^2 \right) \tag{34}$$

where the subscript 11 has been added to β as a reminder that this result applies to a pair of identical molecules.

We have gone through a rather complicated list of equations without looking at the magnitude of the various energy contributions. This is the subject of Example 10.1.

* * *

EXAMPLE 10.1   *Relative Magnitudes of van der Waals Forces and Relation to Heat of Vaporiza-tion.*   The parameter β_{11} must have units energy length[6] in order to satisfy Equation (33). Verify these units as well as the dimensional consistency of each of the three terms in Equation (34). Taking $\mu = 1.0$ debye and $\alpha = 10^{-39}$ C^2 m^2 J^{-1}, calculate the amount of energy needed to separate a pair of molecules from 0.3 nm to ∞. Scaled up by Avogadro's number, how does this energy compare with typical enthalpies of vaporization?

Solution: First, we examine the units of each of the terms in Equation (34). Example 5.1 makes it easy to analyze the units of terms containing α. That example shows that $(\alpha/4\pi\varepsilon_0)$ has units length[3]. Polarizabilities are often tabulated as $(\alpha/4\pi\varepsilon_0)$, thereby having volume units.

The dipole moment has units of charge length, or C m in SI. The square of these units divided by the units of $(4\pi\varepsilon_0)$ therefore has units C^2 $m^2/(C^2$ J^{-1} $m^{-1})$ or J m^3.

The first term in Equation (34) involves the product of α and μ^2, each expressed in multiples of $(4\pi\varepsilon_0)$, and therefore has units J m^6 as required.

Since $k_B T$ has units of energy, the second term in Equation (34) is seen to have units $(J\ m^3)^2/J = J\ m^6$.

Since $h\nu$ has units of energy, the third term has units J $(m^3)^2 = J\ m^6$. The debye is widely used as a unit of dipole moment. It is equal to 10^{-18} esu cm. To convert this to SI we write

$$1\ \text{debye} = 10^{-18}\ \text{esu cm} \cdot (1\ \text{m}/100\ \text{cm}) \cdot (1.60 \cdot 10^{-19}\ \text{C}/4.80 \cdot 10^{-10}\ \text{esu})$$
$$= 3.336 \cdot 10^{-30}\ \text{C m}$$

In Equation (34) the Debye term is given by

$$[2\alpha\mu^2/(4\pi\varepsilon_0)^2] = [2 \cdot (10^{-39})(3.34 \cdot 10^{-30})^2]/(1.11 \cdot 10^{-10})^2$$
$$= 1.81 \cdot 10^{-78}\ \text{J m}^6$$

and the Keesom term by

$$[(2/3)\mu^4/k_B T(4\pi\varepsilon_0)^2] = [2(3.34 \cdot 10^{-30})^4]/[3(1.38 \cdot 10^{-23})(293)(1.11 \cdot 10^{-10})^2]$$
$$= 1.67 \cdot 10^{-78}\ \text{J m}^6$$

To evaluate the London term we need the characteristic frequency ν. Using Equation (29), we obtain for this frequency

$$\nu = (1/2\pi)[(1.60 \cdot 10^{-19})^2/(10^{-39})(9.11 \cdot 10^{-31})]^{1/2}$$
$$= 8.44 \cdot 10^{14}$$

in terms of which the London energy is given by

$$[(3/4)h\nu\alpha^2/(4\pi\varepsilon_0)^2] = [3(6.63 \cdot 10^{-34})(8.44 \cdot 10^{14})(10^{-39})^2]/[4(1.11 \cdot 10^{-10})^2]$$
$$= 3.41 \cdot 10^{-77}\ \text{J m}^6$$

The sum of these three contributions gives the β_{11} value for this system $\beta_{11} = 3.76 \cdot 10^{-77}$ J m^6.

When the separation of the molecules is 0.3 nm, the energy is $3.76 \cdot 10^{-77}/(0.3 \cdot 10^{-9})^6$ $= 5.16 \cdot 10^{-20}$ J. When the distance of separation is infinity, the interaction energy is zero. Therefore scaling up $5.16 \cdot 10^{-20}$ J by Avogadro's number gives an estimate of the energy required to separate a mole of these molecules from a separation typical of a liquid to one that represents a gas: $(5.16 \cdot 10^{-20})(6.02 \cdot 10^{23}) = 31,000$ J. This final result is of the correct order of magnitude of heats of vaporization.   ∎

* * *

Dividing Equation (34) through by β_{11} gives the fractional contribution made to the total attraction by the Debye (D), Keesom (K), and London (L) components of potential energy:

$$f_D + f_K + f_L = 1 \tag{35}$$

Table 10.2 shows these fractions calculated for a variety of molecules. In virtually all cases except the highly polar water molecule, the London or dispersion component is the largest of the contributions to attraction. In the case of water, hydrogen bonding is also possible and contributes an additional strong interaction, so the role of dispersion is even less than shown in Table 10.2.

Before continuing with our development of van der Waals attraction, we examine how

TABLE 10.2 Percentage of the Debye, Keesom, and London Contributions to the van der Waals Attraction Between Various Molecules

Compound	μ (debye)	$\dfrac{\alpha}{4\pi\varepsilon_0} \times 10^{30}$ (m³)	$\beta \times 10^{77}$ (J m⁶)	Keesom (permanent-permanent)	Debye (permanent-induced)	London (induced-induced)
				Percentage contribution of		
CCl₄	0.00	10.70	4.41	0.0	0.0	100.0
Ethanol	1.73	5.49	3.40	42.6	9.7	47.6
Thiophene	0.51	9.76	3.90	0.3	1.3	98.5
t-Butanol	1.67	9.46	5.46	23.1	9.7	67.2
Ethyl ether	1.30	9.57	4.51	10.2	7.1	82.7
Benzene	0.00	10.50	4.29	0.0	0.0	100.0
Chlorobenzene	1.58	13.00	7.57	13.3	8.6	78.1
Fluorobenzene	1.35	10.30	5.09	10.6	7.5	81.9
Phenol	1.55	11.60	6.48	14.5	8.6	76.9
Aniline	1.56	12.40	7.06	13.6	8.5	77.9
Toluene	0.43	11.80	5.16	0.1	0.9	99.0
Anisole	1.25	13.70	7.22	5.5	6.0	88.5
Diphenylamine	1.08	22.60	14.25	1.5	3.7	94.7
Water	1.82	1.44	2.10	84.8	4.5	10.5

Source: Dipole moments and polarizibilities from A. L. McClellan, *Tables of Experimental Dipole Moments*, W. H. Freeman, San Francisco, CA, 1963.

van der Waals forces at the atomic or molecular level influence the macroscopic properties of materials.

10.4b Macroscopic Implications

We have already seen from Example 10.1 that van der Waals forces play a major role in the heat of vaporization of liquids, and it is not surprising, in view of our discussion in Section 10.2 about colloid stability, that they also play a significant part in (or at least influence) a number of macroscopic phenomena such as adhesion, cohesion, self-assembly of surfactants, conformation of biological macromolecules, and formation of biological cells. We see below in this chapter (Section 10.7) some additional examples of the relation between van der Waals forces and macroscopic properties of materials and investigate how, as a consequence, measurements of macroscopic properties could be used to determine the Hamaker constant, a material property that represents the strength of van der Waals attraction (or repulsion; see Section 10.8b) between macroscopic bodies. In this section, we present one illustration of the macroscopic implications of van der Waals forces in thermodynamics, namely, the relation between the interaction forces discussed in the previous section and the *van der Waals equation of state*. In particular, our objective is to relate the molecular van der Waals parameter (e.g., β_{11} in Equation (33)) to the parameter a that appears in the van der Waals equation of state:

$$[p + (a/v^2)](v - b) = RT \tag{36}$$

where p is the pressure and v is the molar volume of the gas. Incidentally, van der Waals's proposal of the above equation, presented as part of his doctoral dissertation, pre-dates the development of what are now known as van der Waals forces. Interested readers may wish to consult his dissertation, which is now available as a book (van der Waals 1988).

Let us begin with a collection of N atoms of diameter σ in a volume V. We shall assume that the atoms are spherical and mutually impenetrable, that the interparticle energy of attraction between any two particles is given by $(-\beta_{11} r^{-6})$ for $r > \sigma$, with r the center-to-center distance of separation between the two particles, and that the density (N/V) is sufficiently low that we can consider the total interatomic interaction energy as a sum of pair interactions (i.e., sum of interactions taken two atoms at a time). Given this information, our goal is to relate the parameters σ, β_{11}, N, and V to the van der Waals parameters a and b.

We first look into the "excluded-volume" effect, represented by the parameter b. Since the atoms are impenetrable, no atom can get closer than a distance σ from another. (Note that the location of each atom is represented by the location of the center of that atom.) This implies that each atom has a volume of $(4/3)\pi\sigma^3$ around its center that is not accessible to the center of any other atom; this volume is the excluded volume associated with each atom. (Figure 13.9 is an illustration of this situation for two colloidal particles of *different* radii.) The actual excluded volume per atom, b' (b, the excluded volume per mole, is equal to $N_A b'$, with N_A the Avogadro's number) is, however, smaller than $(4/3)\pi\sigma^3$ since the excluded volume of an atom as calculated above may overlap with that of other atoms. Therefore, to obtain an expression for b, we need to multiply the above value by N (since there are N atoms in the volume), take half of it since otherwise we will be "double counting" the excluded volumes, and divide by N to get excluded volume per atom, that is,

$$b' = (4/3)\pi\sigma^3 \cdot (N/2) \cdot (1/N) = (2/3)\pi\sigma^3 \tag{37}$$

Therefore, it follows that

$$b = (2/3)\pi\sigma^3 N_A \tag{38}$$

Now, we examine the effect of attraction. Each atom is surrounded by $(N - 1)$ atoms in the volume V. The number of atoms in a spherical shell of thickness dr at a distance r from any atom is $4\pi r^2(N/V)dr$. Therefore, the interaction energy u experienced by any atom is given by

$$u = \int_\sigma^S 4\pi \frac{N}{V}\left(-\frac{\beta_{11}}{r^6}\right)r^2\, dr = -\frac{4\pi N\beta_{11}}{V}\int_\sigma^S \frac{dr}{r^4} \tag{39}$$

where S is the radius of the container. The above equation can be integrated easily, and since S is usually very much larger than σ, the integration leads to

$$u = \frac{4\pi}{3}\frac{N}{V}\beta_{11}\left(\frac{1}{S^3} - \frac{1}{\sigma^3}\right) = -\frac{4\pi N\beta_{11}}{3\,V\sigma^3} \tag{40}$$

The total energy U for all the N atoms is then

$$U = -\frac{2\pi N^2\beta_{11}}{3\,V\sigma^3} \tag{41}$$

where we have again multiplied u by $(N/2)$ for the same reason as in the case of the expression for b.

Since pressure is equal to $-(\partial U/\partial V)_T$ and the deviation in pressure from ideality (i.e., from $pv = RT$) arising from the attractive forces in the van der Waals equation is equal to $(-a/v^2)$, we have

$$\left(\frac{\partial U}{\partial V}\right)_T = \frac{a}{v^2} = \frac{N^2 a}{N_A^2 V^2} \tag{42}$$

which we can combine with Equation (41) for U to obtain

$$a = \frac{2\pi N_A^2\beta_{11}}{3\sigma^3} \tag{43}$$

The above derivation shows that one can determine the van der Waals parameters a and b if σ (a measure of the size of the atom or molecules) and the van der Waals interaction energy parameter β_{11} are known. Alternatively, one can estimate β_{11} from known values of a and b. This is illustrated in Example 10.2.

* * *

EXAMPLE 10.2 *The Dispersion Force and Nonideality of Gases.*   The nonideality of gases arises from the repulsive and attractive forces between atoms. As a consequence, the deviation of the properties of a gas from ideal gas behavior can be traced to the interatomic or intermolecular forces. Assume that methane follows the van der Waals equation of state at sufficiently low densities. It is known from experiments that (see Israelachvili 1991)

$$a = 2.28 \cdot 10^{-1} \, m^3 \, J \, mol^{-2} \quad \text{and} \quad b = 4.28 \cdot 10^{-5} \, m^3 \, mol^{-1}$$

Assume that the intermolecular attraction in this case is dominated by the London interaction, that is, $\Phi_L = -Cr^{-6}$, for which C is the London parameter (Equation (25)). Estimate the London constant from the equation-of-state data and compare it with the coefficient from Equation (25). The polarizability $[\alpha_0/(4\pi\epsilon_0)]$ for CH_4 is $2.6 \cdot 10^{-30} \, m^3$. The ionization energy I is $2.0185 \cdot 10^{-18}$ J (Israelachvili 1991, Chapter 6 and Table 6.1).

Solution: One can eliminate the molecular diameter σ from the equation for the parameter a using the equation for b to get C in terms of a and b:

$$C = 9ab/(4\pi^2 N_A^3)$$

Using $N_A = 6.022 \cdot 10^{23} \, mol^{-1}$ and the given values of a and b, we get for C

$$C = 1.018 \cdot 10^{-77} \, J \, m^6 \quad \text{(from the experimental PV data)}$$

From Equation (27) for identical molecules, one has

$$\Phi_L = -(3/4)[\alpha_0/(4\pi\epsilon_0)]^2 I r^{-6} = -Cr^{-6}$$

Therefore, the London constant C is given by

$$C = (3/4)[\alpha_0/(4\pi\epsilon_0)]^2 I$$

From the given values of polarizability and ionization potential for CH_4, we have

$$C = 1.02 \cdot 10^{-77} \, J \, m^6 \quad \text{(from the polarizability data)}$$

The two values agree remarkably well, showing the relation between the London force and the nonideality of the gas.

This example illustrates how equation-of-state data may be used to estimate the strength of van der Waals forces or vice versa.   ∎

* * *

In this section we have examined the three major contributions to what is generally called the van der Waals attraction between molecules. All three originate in dipole-dipole interactions of one sort or another. There are two consequences of this: (a) all show the same functional dependence on the intermolecular separation, and (b) all depend on the same family of molecular parameters, especially dipole moment and polarizability, which are fairly readily available for many simple substances. Many of the materials we encounter in colloid science are not simple, however. Hence we must be on the lookout for other measurable quantities that depend on van der Waals interactions. Example 10.2 introduces one such possibility. We see in Section 10.7 that some other difficulties arise with condensed systems that do not apply to gases.

In the next section we take a preliminary look at the way van der Waals attractions scale up for macroscopic (i.e., colloidal) bodies. This will leave us in a better position to look for other measurements from which to estimate the van der Waals parameters.

10.5 VAN DER WAALS FORCES BETWEEN LARGE PARTICLES AND OVER LARGE DISTANCES

The interaction between individual molecules obviously plays an important role in determining, for example, the nonideality of gas, as illustrated in Example 10.2. It is less clear how to apply this insight to dispersed particles in the colloidal size range. If atomic interactions are assumed to be additive, however, then the extension to macroscopic particles is not particularly difficult. Moreover, when dealing with objects larger than atomic dimensions, we also have to consider interactions over appropriately large distances. In the case of the London attraction, forces over large distances show a more rapid decay than indicated by the inverse sixth-power equations derived in Section 10.4. This is known as (*electromagnetic*) *retardation*. We discuss these two important issues in this section before developing the equations for interactions between macroscopic bodies in Section 10.6.

10.5a Scaling van der Waals Interactions for Large Bodies

We begin by considering two spherical particles of the same composition and the same radius R_a. As Figure 10.5 shows, we consider two situations. In case (a) we consider spheres of radius R_a and in case (b) spheres of radius R_b. The two radii are related as follows:

$$R_b = fR_a \tag{44}$$

where f is a factor greater than unity.

Let us first focus on Figure 10.5a. Assume that every atom in sphere 1 attracts every atom in sphere 2 with an energy given by Equation (33). If $(\rho N_A/M)$ is the number of atoms per cubic centimeter in the material, then there are $(\rho N_A/M)dV_{1a}$ atoms in a volume element of sphere 1 and $(\rho N_A/M)dV_{2a}$ atoms in a volume element of sphere 2. The number of pairwise interactions between the two volume elements is $(1/2)\,(\rho N_A/M)^2\,dV_{1a}dV_{2a}$. The factor $(1/2)$ enters since each pair is counted twice. This number times the interaction per pair gives the increment in potential energy for the two interacting volume elements in case (a):

$$d\Phi_a = -\frac{1}{2}\left(\frac{\rho N_A}{M}\right)^2 \beta\, r_a^{-6}\, dV_{1a}\, dV_{2a} \tag{45}$$

By including additional geometrical considerations, the volume elements and their separation may be expressed in a common set of variables, and Equation (45) can be integrated over the volume of both spheres. Rather than complicate the issue by specific mathematical substitutions at this time, we indicate this procedure as follows:

$$\Phi_a = -\frac{1}{2}\left(\frac{\rho N_A}{M}\right)^2 \beta \int \int r_a^{-6}\, dV_{1a}\, dV_{2a} \tag{46}$$

Now let us turn our attention to Figure 10.5b. In this case the spheres are larger by a factor f as just noted. In addition to this, the distance between their centers is also assumed to

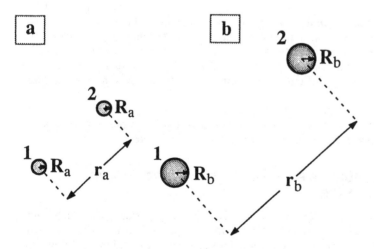

FIG. 10.5 Two spheres of equal radii separated by a distance r. All linear dimensions in (b) (i.e., R_b and r_b) are larger than those in (a) by a constant factor f.

be larger than that in case (a) by the factor f. If we let the separation of centers be represented by r, then

$$r_b = fr_a \tag{47}$$

Following the same procedure as used for case (a), we can evaluate the total interaction potential in case (b). It is given by

$$\Phi_b = -\frac{1}{2}\left(\frac{\rho N_A}{M}\right)^2 \beta \iint r_b^{-6} \, dV_{1b} \, dV_{2b} \tag{48}$$

Since the linear scale in case (b) is larger by the factor f than that in case (a), Equation (48) can also be written

$$\Phi_b = -\frac{1}{2}\left(\frac{\rho N_A}{M}\right)^2 \beta \iint (fr_a)^{-6} (f^3 \, dV_{1a})(f^3 \, dV_{2a}) \tag{49}$$

Note that the factor f cancels out of this expression entirely, so that

$$\Phi_a = \Phi_b \tag{50}$$

That is, the potential energy of attraction is identical in the two cases. This is an important result as far as the extension of molecular interactions to macroscopic spherical bodies is concerned. What it says is that two molecules, say, 0.3 nm in diameter and 1.0 nm apart, interact with exactly the same energy as two spheres of the same material that are 30.0 nm in diameter and 100 nm apart. Furthermore, an inspection of Equation (49) reveals that this is a direct consequence of the inverse sixth-power dependence of the energy on the separation. Therefore the conclusion applies equally to all three contributions to the van der Waals attraction. Precisely the same forces that are responsible for the association of individual gas molecules to form a condensed phase operate—over a suitably enlarged range—between colloidal particles and are responsible for their coagulation.

The analogy between the condensation of a gas and the coagulation of a dispersion is important. It points out that under certain circumstances the aggregation of the separated units is inevitable. However, there are also conditions under which the dispersed state is stable. We know from kinetic molecular theory that thermal energy—measured by $k_B T$ per molecule or RT per mole—supplies a reference level against which all energies are judged to be large or small.

* * *

EXAMPLE 10.3 *The Strength of van der Waals Forces and the Structure of Materials.* It is interesting to ask why some substances such as very small, nonpolar argon and methane are gaseous at room temperature while substances made up of larger molecules (e.g., high molecular weight hydrocarbons) are liquids or solids. Consider argon as an example and show that the typical attractive energy of two argon atoms separated by a distance of about 0.38 nm is of the order of thermal energy (i.e., about $k_B T$) at room temperature. Discuss the implication of this order-of-magnitude estimation to the structure of a substance at room temperature. The quantity $[\alpha_0/(4\pi\varepsilon_0)]$ for argon is about $1.6 \cdot 10^{-30}$ m^3. The ionization energy I for Ar is roughly $2.5 \cdot 10^{-18}$ J.

Solution: Using Equation (27), with $I_1 = I_2 = I$, one observes that for two argon atoms separated by the given (typical) distance of separation the dispersion energy is approximately equal to

$$\begin{aligned}
\Phi_L &= -(3/4)[\alpha_0/(4\pi\varepsilon_0)]^2 \, I \, x^{-6} \\
&= -0.75 \cdot 2.56 \cdot 10^{-60} \cdot 2.5 \cdot 10^{-18} \cdot (0.38 \cdot 10^{-9})^{-6} \text{J} \\
&= -1.6 \cdot 10^{-21} \text{ J}
\end{aligned}$$

Since $k_B = 1.381 \cdot 10^{-23}$ J K^{-1}, at room temperature (say, $T = 290$ K), one has

$$\Phi_L = -1.6 \cdot 10^{-21} \approx -0.4 \, k_B T$$

The above result implies that at room temperatures the attraction between two argon atoms is sufficiently weak relative to thermal energy ($k_B T$) that argon will remain as a gas rather than in a condensed state (liquid or solid). In contrast, in the case of larger molecules, the attractive energies can be sufficiently high to make the material condense into a liquid or solid. Although the above estimate is crude, it explains the role of van der Waals forces on condensation.

This example illustrates how the London forces influence whether a material exists as a gas, a liquid, or a solid at a given temperature. ∎

* * *

Thus, as illustrated in Example 10.3, a gas will condense if the energy of attraction between molecules is large compared to $k_B T$. Conversely, it remains dispersed if $k_B T$ is larger. The same is true for colloidal particles also, at least if they are not too large. For larger particles externally applied energy — such as mechanical stirring — joins thermal energy in establishing a reference level of energy. In general, however, the height of the barriers or the depth of the minima in the potential energy curves we discussed in Section 10.2 are considered large or small relative to $k_B T$. The argument leading to Equation (50) shows that van der Waals attraction is as important for colloidal particles as it is for individual molecules.

In reaching this conclusion we have assumed that no time lag affects the field that establishes the attraction between the particles. We have also considered particles under vacuum so no intervening medium enters the picture. Each of these simplifying approximations has the effect of overestimating the van der Waals attraction between particles at large separations from one another and embedded in a medium. We consider presently the effect of a time lapse between the interaction of a field with two different particles; the effect of the medium is discussed in Section 10.8.

10.5b Electromagnetic Retardation of the London Force

The electric field responsible for the London attraction between molecules propagates itself with the speed of light between the particles. Thus, if a pair of molecules is widely separated, a time lag or a phase difference develops between vibrations at the two locations. The situation is analogous in many ways to the scattering of light by particles with dimensions that are large compared to the wavelength of the light (see Section 5.5). In the present situation we find that the importance of this time lag or retardation increases as the separation becomes comparable to the wavelength of the propagating field.

Equation (29) provides us with an expression for the frequency of the interaction. Therefore its wavelength is given by

$$n\lambda = \frac{c}{\nu} = 2\pi c \sqrt{\frac{\alpha_0 m_e}{e^2}} \tag{51}$$

where n is the refractive index. For typical values of n and α_0, λ is about 200 nm. As was the case with light scattering, separations less than about 1/20 of this distance may be considered "small compared to the wavelength." This means that at separations of about 10 nm or so the effects of retardation begin to enter the picture. Our reason for interest in this is the fact that the comparison of attraction and repulsion between colloidal particles is made at this distance in some cases.

We shall not present the detailed analysis of this complication. In essence, it involves the time-dependent Schrödinger equation rather than the time-independent equation that resulted in Equation (22). Casimir and Polder have investigated this situation. They found that for values of $r \gg \lambda$, the potential energy of attraction according to the modified London treatment is given by

$$\Phi_A = -\frac{23}{8\pi^2 (4\pi\varepsilon_0)^2} \frac{hc\alpha_0^2}{r^7} \tag{52}$$

This is the one entry in Table 10.1 that has not yet been discussed. Direct measurement of the force of attraction between macroscopic bodies reveals a crossover from an inverse sixth-power to an inverse seventh-power law at separations in the range 10 to 100 nm.

Equation (49) is particularly convenient to show the effects of retardation on the attraction between spherical particles at large separations. On correcting the potential function for retardation according to Equation (52), Equation (49) becomes

$$\Phi_b \propto \iint_{1 \text{ and } 2} (f r_a)^{-7} (f^3 dV_{1a})(f^3 dV_{2a}) = \frac{1}{f}\Phi_a \tag{53}$$

at large separations. This result shows that the scale factor f does *not* drop out of the expression in the case of the retarded van der Waals forces. Therefore the potential energy of the attraction decreases as the scale increases (i.e., as f increases).

In subsequent discussions we shall not consider the effect of retardation any further. Additional details are given by Israelachvili (1991), Israelachvili and Tabor (1973), and Sonntag and Strenge (1964).

The development that led to Equation (50) was conducted without any actual integrations being performed. In order to generate exact expressions for the van der Waals attraction between bodies of specific geometry, it is necessary to carry out these operations. This is discussed in the next section.

10.6 CALCULATING VAN DER WAALS FORCES BETWEEN MACROSCOPIC BODIES

The strategy for scaling up the van der Waals attraction to macroscopic bodies requires that all pairwise combinations of intermolecular attraction between the two bodies be summed. This has been done for several different geometries by Hamaker. We consider only one example of the calculations involved, namely, the case of blocks of material with planar surfaces. This example serves to illustrate the method and also provides a foundation for connecting van der Waals forces with surface tension, the subject of the next section.

10.6a Attraction Between Two Semi-Infinite Blocks

Figure 10.6a represents a molecule at O located a normal distance z from the surface of a bulk sample of the same material. The bulk portion is assumed to have a planar face but otherwise is of infinite extension. The molecule is located a distance x from all the molecules in the ring-shaped volume element shown in the figure. The volume of this ring is given by $dV = 2\pi y\, dy\, d\zeta$. Therefore the increment of interaction between the molecule and the block due to the molecules a distance x from the point O is given by

$$d\Phi = -(\rho N_A/M)\beta(2\pi y/x^6)dy\, d\zeta \tag{54}$$

We assume the ring is located a distance ζ inside the surface of the block; then

$$x^2 = (z + \zeta)^2 + y^2 \tag{55}$$

Combining Equations (54) and (55) gives

$$d\Phi = -(\rho N_A/M)\,\beta\, 2\pi[(z + \zeta)^2 + y^2]^{-3}y\, dy\, d\zeta \tag{56}$$

Equation (56) is now integrated over the volume of the block, that is, for $0 < y < \infty$ and $0 < \zeta < \infty$.

Integration over y yields

$$\int_0^\infty \frac{y\, dy}{[(z + \zeta)^2 + y^2]^3} = \frac{1}{2}\int_0^\infty \frac{du}{[(z + \zeta)^2 + u]^3} = -\frac{1}{2}\left\{\frac{1}{2[(z + \zeta)^2 + u]^2}\right\}_0^\infty \tag{57}$$

$$= \frac{1}{4}\frac{1}{(z + \zeta)^4}$$

and integration over ζ yields

$$\frac{1}{4}\int_0^\infty \frac{d\zeta}{(z + \zeta)^4} = -\frac{1}{4}\left[\frac{1}{3(z + \zeta)^3}\right]_0^\infty = \frac{1}{12z^3} \tag{58}$$

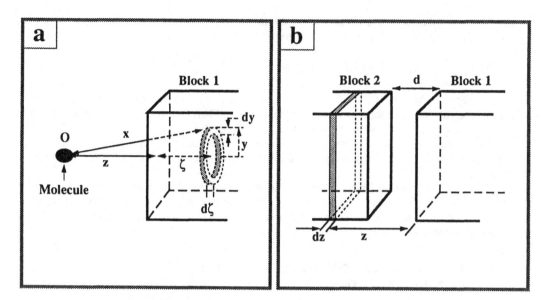

FIG. 10.6 Schematic illustration of interactions: (a) interaction between a molecule and a block of material; and (b) interaction between two blocks of material. (Redrawn with permission from P. C. Hiemenz, *J. Chem. Educ.*, **49**, 164 (1972).

Therefore the integration of Equation (56) over the block gives

$$\Phi = -(\rho N_A/M)\beta\pi/(6z^3) \tag{59}$$

Now suppose point O is located inside a second block of material as shown in Figure 10.6b. We recognize that all atoms in a slice of the second block a distance z from the first block will be attracted toward the second with an energy given by Equation (59). If we position a volume element of thickness dz at this location in the second block, we realize that it contains $(\rho N_A/M)$ molecules per unit area. This is $(\rho N_A/M)dz$ molecules; therefore the increment of attraction per unit area due to this slice is

$$d\Phi = -(\rho N_A/M)^2(\beta\pi/6)z^{-3}dz \tag{60}$$

Equation (60) may now be integrated over values of z between the distance of closest approach d and infinity. The result of this integration gives the potential energy of attraction per unit area between two blocks of infinite extension:

$$\Phi_A = -(\rho N_A/M)^2(\beta\pi/12)d^{-2} \tag{61}$$

where the subscript A has been appended to Φ to emphasize the fact that this energy is always attractive.

It is traditional to designate the cluster of constants $(\rho N_A/M)^2\pi^2\beta$ as the *Hamaker constant A*:

$$A = (\rho N_A\pi/M)^2\beta \tag{62}$$

With this change of notation. Equation (61) becomes

$$\Phi_A = -(A/12\pi)d^{-2} \tag{63}$$

The Hamaker constant has energy units since β has the units energy length^6 and the term in parentheses in Equation (62) has the units (volume^{-1})^2. The potential energy of attraction between blocks as calculated by Equation (63) is expressed per unit area of the facing surfaces. Note also that this attraction grows weaker as the distance increases. This is different from the conclusion drawn for spheres in Equation (50). The scaling up of van der Waals forces, therefore, depends on the geometry of the bodies involved.

10.6b Magnitudes of Hamaker Constants

Although we have considered numerical values of β before, we have not encountered the Hamaker constant previously and hence have no feel for its magnitude. A particularly convenient form for an order of magnitude estimation of A is obtained by making the following stipulations:

1. Assume that the dispersion component is the dominant contributor to the attraction; therefore $\beta = (3/4)h\nu(\alpha/4\pi\varepsilon_0)^2$.

2. Recognize that $(\rho N_A/M)$ is the reciprocal of the volume per molecule, and that $(\alpha/4\pi\varepsilon_0)$ is typically about 10% the magnitude of atomic volume; therefore $A \approx (3/4)\pi^2 h\nu(0.1)^2$.

3. The quantity $h\nu$ is the same order of magnitude as the ionization potential, typically about 10^{-18} J.

Combining these results leads us to estimate the Hamaker constant to lie in the range 10^{-20} to 10^{-19} J. If we take the midpoint value of $5 \cdot 10^{-20}$ J for A, Equation (63) predicts that Φ_A equals 1.33 and $1.33 \cdot 10^{-2}$ mJ m^{-2} when d equals 1.0 and 10 nm, respectively. Table 10.3 presents Hamaker constants for some materials of practical interest. Values for many other materials and reviews of methods of calculating Hamaker constants are available in Gregory (1969) and Visser (1972).

10.6c van der Waals Interaction Energies for Other Geometries

The interaction of two infinite blocks separated at the surface by a distance d is one of the easiest possible situations to consider and therefore was chosen as the example to develop in detail. Bodies of different geometries have also been analyzed in much the same way that we have the blocks. The results of several such derivations are shown in Table 10.4. The expressions for Φ_A become more complicated for more complex geometrical situations, but considerable simplification results for spheres in which the separation is much less than the radius. The feature described by Equation (50) for interacting spheres is also evident in the results given in Table 10.4. It is most readily seen by examining the case of equal spheres separated by small distances. Note that both R_s and d can be multiplied by any common factor without changing

TABLE 10.3 Hamaker Constants for Selected Materials

Material	$A \times 10^{20}$ Joules	Source
Acetone	4.2	Croucher and Hair
Alumina	15.4	Bargeman and van Voorst Vader
Gold	45.3	Bargeman and van Voorst Vader
Magnesia	10.5	Bargeman and van Voorst Vader
Metals	16–45	Visser
Natural rubber	8.58	Croucher and Hair
Polystyrene	7.8–9.8	Croucher and Hair ; Croucher
Silver	39.8	Bargeman and van Voorst Vader
Toluene	5.4	Croucher and Hair
Water	4.35	Bargeman and van Voorst Vader

Sources: D. Bargeman and F. van Voorst Vader, *J. Electroanal. Chem. Interfacial Electrochem.*, **37**, 45 (1972); Croucher and Hair 1977; Visser 1972; Croucher (1981).

TABLE 10.4 Potential Energy of Attraction Between Two Particles with the Indicated Geometries

Particles	Φ_A	Definitions/ Limitations
Two spheres	$-\dfrac{A}{6}\left[\dfrac{2R_1R_2}{f_1(R_1,R_2,d)} + \dfrac{2R_1R_2}{f_2(R_1,R_2,d)} + \ln\!\left(\dfrac{f_1(R_1,R_2,d)}{f_2(R_1,R_2,d)}\right)\right]$ $f_1(R_1,R_2,d) = d^2 + 2R_1d + 2R_2d$ $f_2(R_1,R_2,d) = d^2 + 2R_1d + 2R_2d + 4R_1R_2$	R_1, R_2 = radii; d = separation of surfaces along line of centers
Two spheres of equal radius	$-\dfrac{A}{6}\left[\dfrac{2R_s^2}{f_1(R_s,d)} + \dfrac{2R_s^2}{f_2(R_s,d)} + \ln\!\left(\dfrac{f_1(R_s,d)}{f_2(R_s,d)}\right)\right]$ $f_1(R_s,d) = d^2 + 4R_sd;\; f_2(R_s,d) = d^2 + 4R_sd + 4R_s^2$	$R_1 = R_2 = R_s$
Two spheres with equal radius	$-\dfrac{AR_s}{12d}$	$R_s \gg d$
Two spheres of unequal radius	$-\dfrac{A\,R_1R_2}{6d(R_1 + R_2)}$	R_1 and $R_2 \gg d$
Two plates of equal thickness	$-\dfrac{A}{12\pi}\left(\dfrac{1}{d^2} + \dfrac{1}{(d + 2\delta)^2} - \dfrac{2}{(d + \delta)^2}\right)$	δ = thickness of the plates; d = surface-to-surface distance
Identical blocks	$-\dfrac{A}{12\pi d^2}$	$\delta \to \infty$

the magnitude of the interaction energy. The expression for interacting plates of thickness δ in Table 10.4 is also seen to reduce to Equation (63) as $\delta \to \infty$.

10.7 THEORIES OF VAN DER WAALS FORCES BASED ON BULK PROPERTIES

Adding together molecular interactions to account for macroscopic attractions is undoubtedly an oversimplification. There are several things that this procedure overlooks that limit its validity. For example, even for two bodies under vacuum, those molecules nearer the surface screen the interactions of molecules buried more deeply in the material. Because of the inverse power dependence of the attraction, those molecules nearest the faces of the blocks make the predominant contribution to the interaction. If these molecules have permanent dipoles, they may experience orientation effects under the influence of the surface that are not described by the Debye and Keesom models. From a practical point of view, dipole moments and polarizabilities may not always be available for all substances of interest, as noted in Section 10.4. The possibility of surface heterogeneity confuses the choice of polarization parameters even further. Although we continue (for now) to assume a vacuum separates the bodies under consideration, the presence of an intervening medium can only aggravate all of the preceding points.

10.7a Dzyaloshinskii-Lifshitz-Pitaevskii Theory

For the above reasons a theory based entirely on measurable bulk properties rather than molecular parameters is a more powerful way of dealing with the interaction of macroscopic bodies. Dzyaloshinskii, Lifshitz, and Pitaevskii (DLP) have developed such a macroscopic theory in a very general form from quantum field theory. Bulk dielectric properties of matter are the basis for this theory, so screening effects are built into the solution. Unfortunately, in its general form, the DLP theory is complicated and difficult to apply. An indication of this is found in the following expression for the Hamaker constant A_{213}—as it would be used in Equation (63)—for the interaction of blocks of material 2 and material 3 when the two are separated by material 1:

$$A_{213} = \frac{3}{8\pi^2} h \int_0^\infty \left[\frac{\varepsilon_2(i\xi) - \varepsilon_1(i\xi)}{\varepsilon_2(i\xi) + \varepsilon_1(i\xi)} \right] \left[\frac{\varepsilon_3(i\xi) - \varepsilon_1(i\xi)}{\varepsilon_3(i\xi) + \varepsilon_1(i\xi)} \right] d\xi \tag{64}$$

In this expression $\varepsilon(i\xi)$ is the dielectric "constant"—it is a function of frequency—along the imaginary frequency axis $i\xi$; it is measurable as the dissipative part of the spectrum of dielectric constant for any material. The latter is an experimentally determined function of frequency for each of the three components, and the complicated expression in Equation (64) is integrated over all frequencies.

Although very general. the difficulty in applying the DLP theory has limited its use and encouraged the development of various approximations and special cases. The only use we make of the DLP theory is to point out that the differences between the $\varepsilon(i\xi)$ values for the blocks and the medium appear in the numerator of Equation (64). If these dissipative components of dielectric constant matched at all frequencies for substances 1, 2, and 3, then the Hamaker constant would equal zero according to Equation (64). Although an exact match seems unlikely, this is the first indication we have had that an intervening medium might have a compensating effect on van der Waals attraction. This idea will be developed more fully in Section 10.8.

10.7b Semiclassical Approaches Based on Bulk Properties Such as Surface Tension

Because of the complexity of the DLP theory, several semiclassical treatments have been developed to simplify and extend its findings. One such extension is applicable to blocks of nonpolar materials at small separations and is of interest to us because of the possibility it offers for relating surface tension to intermolecular forces.

For now we consider the interaction of two identical blocks of material separated by their own equilibrium vapor. From the point of view of modifying the force of attraction, the vapor is assumed to have a negligible effect, behaving essentially like a vacuum. The facing planes become equilibrium surfaces characterized by an interfacial free energy γ.

Our starting point in this discussion is the resemblance between Figure 6.8a and Figure 10.6b. The former illustrates the work of cohesion and the latter the interaction between two blocks of material. Suppose we identify by d_0 the equilibrium spacing between molecules of a bulk sample of the material under consideration. Then the cohesion process represented by Figure 6.8a can be viewed as one in which two blocks of material are separated from $d = d_0$ to $d = \infty$. In terms of Equation (63),

$$\Delta\Phi = \Phi_\infty - \Phi_0 = (A/12\pi) d_0^{-2} \tag{65}$$

Equating this with the work of cohesion given by Equation (6.56), we obtain

$$2\gamma = (A/12\pi) d_0^{-2} \tag{66}$$

or

$$A = 24\pi\gamma d_0^2 \tag{67}$$

A system for which this model is apt to work best is a nonpolar material such as an alkane for which $\gamma \approx 25$ mJ m^{-2} as seen from Table 6.1. Using 0.2 nm for d_0, we calculate $A = 24\pi(0.025)(0.2 \cdot 10^{-9})^2 = 7.5 \cdot 10^{-20}$ J, which is very close to the value predicted for this quantity from molecular parameters in the last section.

The same logic that we used to obtain the Girifalco-Good-Fowkes equation in Section 6.10 suggests that the dispersion component of the surface tension γ^d may be better to use than γ itself when additional interactions besides London forces operate between the molecules. Also, it has been suggested that intermolecular spacing should be explicitly considered within the bulk phases, especially when the interaction at $d = d_0$ is evaluated. The Hamaker approach, after all, treats matter as continuous, and at small separations the graininess of matter can make a difference in the attraction. The latter has been incorporated into one model, which results in the expression

$$A = (4\pi/1.2)\gamma^d d_0^2 \tag{68}$$

where d_0 is the intermolecular spacing in the bulk material. The contention is that this is more suitable than Equation (67) for relating the Hamaker constant to surface tension for materials with interactions other than London forces. In addition to only a fraction of the full value of γ being used, the numerical coefficient in Equation (68) is only about one-seventh that in Equation (67). Both of these considerations reinforce the idea that Equation (67) gives an upper limit for A.

Equations (67) and (68) provide alternatives to Equations (34) and (62) for the evaluation of the Hamaker constant. Although the last approach uses macroscopic properties and hence avoids some of the objections cited at the beginning of the section, the practical problem of computation is not solved by substituting one set of inaccessible parameters (γ^d and d_0) for another (α and μ).

The primary objective of the present discussion is to show the intrinsic connection between surface tension and the van der Waals energy of attraction between macroscopic bodies. The connection not only provides computational options but also — and more importantly — unites two apparently separate phenomena and strengthens our confidence in the correctness of our understanding.

Table 10.5 shows a few numerical examples of how well this attempt at unification succeeds. Equations (28), (32), and (62) have been used to calculate values of the Hamaker constant from refractive index data at visible wavelengths. These values have then been used along with γ^d values from Chapter 6 to calculate d_0 values according to Equation (68). The resulting values of d_0 are seen to be physically reasonable. That such plausible values for d_0 are obtained is especially noteworthy in view of the approximations made in the calculations.

The applicability of this procedure receives a far more stringent test in the case of water and SiO_2 than for the hydrocarbons. London forces are assumed to be the only contributors to γ (i.e., $\gamma^d = \gamma$) for hydrocarbons. This is definitely not the case for quartz or water, so the d_0 values obtained for these substances are quite satisfactory.

TABLE 10.5 Calculations Intended to Show that Equations (67) and (68) Predict Values of the Hamaker Constant Compatible with Those Evaluated from Equations (28), (32), and (62)[a]

Compound	M (g mole^{-1})	ρ (g cm^{-3})	n	A(J)	γ^d (mJ m^{-2})	d_0 (nm)
Heptane	100.2	0.684	1.39	1.05×10^{-20}	20.3	0.22
Dodecane	170.3	0.749	1.42	9.49×10^{-21}	25.4	0.18
Eicosane	282.5	0.789	1.44	2.07×10^{-20}	29.0	0.26
SiO_2 (quartz)	60	2.650	1.54	4.14×10^{-20}	78.0	0.22
Polystyrene	(104)[b]	1.050	1.59	2.2×10^{-20}	41.0	0.23
Water	18	1.000	1.33	2.43×10^{-20}	21.3	0.33

[a]Equations (28), (32), and (62) are used to evaluate A, then A and γ^d values from Chapter 6 are used to evaluate d_0.
[b]Monomer.

10.7c Method Based on the Thermodynamics of Liquids

The Hamaker constants of nonpolar fluids and polymeric liquids can be obtained using an expression similar to Equation (67) in combination with the corresponding state theory of thermodynamics and an expression for interfacial energy based on statistical thermodynamics (Croucher 1981). This leads to a simple, but reasonably accurate and useful, relation for Hamaker constants for nonpolar fluids and polymeric liquids. We present in this section the basic details and an illustration of the use of the equation derived by Croucher.

An expression for the Hamaker constant analogous to Equation (67) had been proposed by Fowkes (1964) for the case when only dispersion forces determine the surface tension. The Fowkes equation

$$A = 6\pi d_0^2 \gamma \tag{69}$$

is somewhat difficult to use since the choice of d_0, the intermolecular distance, is not always clear because of the asymmetric shape of most molecules. The above equation can be rewritten as

$$A = 6 a_m \gamma \tag{70}$$

where a_m is the surface area of a molecule. For spherical molecules, the area is well defined and the above equation reduces to Equation (67).

One can now substitute for the interfacial energy γ the following equation derived by Davis and Scriven (1976)

$$\gamma = (1/8)\rho^{-1/3} (\partial U/\partial V)_T \tag{71}$$

where ρ is the number density of the material and $(\partial U/\partial V)_T$ is the rate of change of the energy of the material with volume at constant temperature (i.e., the negative of internal pressure). This expression for γ can be used in combination with Equation (70) to obtain the Hamaker constant if the area a_m is known and if expressions or values for $(\partial U/\partial V)_T$ are available for any material for which the dispersion interaction determines γ. However, a simplified equation based on the above combination of equations and the corresponding state theory is possible (Croucher 1981; Croucher and Hair 1977), and one obtains

$$A = (3/4) k_B T [1 - (\xi V_r^{-1/3})]^{-1} \tag{72}$$

with $\xi = (m/n)^{1/(n-m)}$ \hfill (73)

where V_r is the reduced volume of the material and n and m are exponents of a power law potential similar to the one given in Equation (1). When $m = 6$ and $n = 12$, the intermolecular potential reduces to the Lennard-Jones potential mentioned in Section 10.3. Equation (72) is applicable to liquids (made up of simple molecules or polymers) in which dispersion forces dominate the attraction. It is not applicable to polar substances such as water or to glassy polymers.

The reduced volume V_r, which is needed for using Equation (72), can be determined from the relation

$$(\alpha T)^{-1} = -(m/3) + (n - m)[3(V_r^{(n-m)/3} - 1)]^{-1} + \xi[3(V_r^{1/3} - \xi)]^{-1} \tag{74}$$

where α is the coefficient of thermal expansion and *not* the polarizability. For the 6-∞ potential ($m = 6, n = \infty$), for which ξ is 1, Equation (74) becomes

$$V_r = \left(\frac{3 + 7\alpha T}{3 + 6\alpha T}\right)^3 \tag{75}$$

Equation (74) with $m = 6$ and $n = \infty$ has been tested for a range of α's by Croucher (1981) and leads to very good estimates for the Hamaker constant for a number of liquids. Moreover, the temperature dependence of the Hamaker constant is also accounted for explicitly. Example 10.4 illustrates the use of this procedure.

* * *

EXAMPLE 10.4 *Hamaker Constant of Liquid Polystyrene.* Estimate the Hamaker constant of liquid polystyrene at 298 K. The thermal expansion coefficient α for polystyrene at 298 K is approximately $5.7 \cdot 10^{-4}\,K^{-1}$. Compare your result with the experimentally obtained value of $A = 7.8 \cdot 10^{-20}$ J reported by Croucher (1981, Fig. 1).

Solution: From Equation (75), for $\alpha T = 5.7 \cdot 10^{-4} \cdot 298 \approx 0.17$, the reduced volume is given by

$$V_r = 1.132$$

One can obtain a good estimate for A using the above value of V_r and $\xi = 1$ from Equation (72). This leads to

$$A = (3/4) \cdot 1.38 \cdot 10^{-23} \cdot 298 \cdot [1 - (1/1.132)^{1/3}]^{-1}\,J$$
$$= 7.62 \cdot 10^{-20}\,\text{Joules} \quad \text{(from Croucher's equation)}$$
$$A = 7.8 \cdot 10^{-20}\,\text{Joules} \quad \text{(given experimental data)}$$

The estimate based on Croucher's equation is in excellent agreement with the reported result. ∎

* * *

10.7d Direct Measurement of van der Waals Forces

It is extremely difficult to measure the Hamaker constant directly, although this has been the object of considerable research efforts. Direct evaluation, however, is complicated either by experimental difficulties or by uncertainties in the values of other variables that affect the observations. The direct measurement of van der Waals forces has been undertaken by literally measuring the force between macroscopic bodies as a function of their separation. The distances, of course, must be very small, so optical interference methods may be used to evaluate the separation. The force has been measured from the displacement of a sensitive spring (or from capacitance-type measurements).

A well-known example of this type of measurement is shown in Figure 10.7 for attractive van der Waals force between two mica surfaces. The result shown is obtained using the surface force apparatus developed originally in the early 1970s for interaction between two thin cylindrical surfaces (of radius R_c) at 90° to each other in vacuum. (This geometry mimics the interaction between two flat surfaces; see Israelachvili 1991. The surface force apparatus (see Chapter 1, Section 1.6c) has since gone through many refinements and can now be used for measuring interactions in electrolyte solutions and for tangential force measurements.)

It is evident from Figure 10.7 that the measurements are consistent with both unretarded and retarded attractive forces at appropriate separation distances. It has also been possible to verify directly the functional dependence on radii for the attraction between dissimilar spheres (see Table 10.4), to determine the retardation of van der Waals forces (see Table 10.1), and to evaluate the Hamaker constant for several solids, including quartz. Values in the range of $6 \cdot 10^{-20}$ to $7 \cdot 10^{-20}$ J have been found for quartz by this method. This is remarkably close to the value listed in Table 10.5 for SiO_2.

A more common situation than two bodies interacting across a vacuum is the case in which some medium intervenes between the interacting bodies. Before proceeding any further, then, let us examine the effect of this medium on particle interactions.

10.8 EFFECT OF THE MEDIUM ON THE VAN DER WAALS ATTRACTION

Until now we have considered the interaction between isolated molecules or macroscopic bodies when the particles are separated by a vacuum. Interactions in a vacuum is reasonable for molecules in the gas phase. However, for dispersions of one phase in another, the effect of the medium must be taken into account. Accounting for the effects of the medium leads to some useful combining relations for the Hamaker constant A_{ijk}, which is the Hamaker constant for interaction between i and k in medium j. In addition, situations may arise in which A_{ijk} is negative, that is, the interaction is repulsive. We review these in this section.

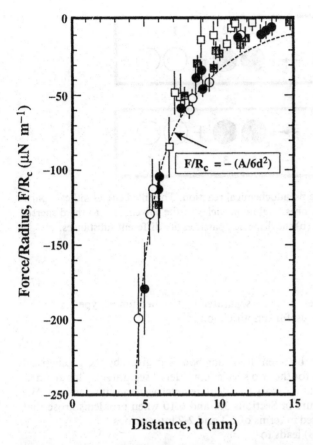

FIG. 10.7 Direct measurements of van der Waals dispersion forces. The measurements correspond to the force between two flat (mica) surfaces separated by a distance d. The line shown is the theoretical expression for unretarded van der Waals force. The figure shows that the unretarded expression describes the measurements sufficiently accurately for d about 6.5 nm or less. (Redrawn with permission of J. N. Israelachvili and G. E. Adams, *J. Chem. Soc., Faraday Trans. 1*, **78**, 975 (1978).)

10.8a Combining Relations for the Hamaker Constant

The easiest way to account for the effect of a medium is to consider the pseudochemical reaction illustrated in Figure 10.8a. The particles numbered 2 represent the dispersed phase, and those numbered 1 are the solvent. Note that both of the dispersed particles are of the same material in this reaction. In the initial condition, each dispersed particle and its satellite solvent particle comprise an independent kinetic unit. Figure 10.8a represents the process in which the two dispersed particles come together to form a doublet and the two solvent particles form a kinetically independent doublet.

The change in the potential energy that accompanies this process is given by

$$\Delta \Phi = \Phi_{11} + \Phi_{22} - 2\Phi_{12} \tag{76}$$

where the subscripts apply to the two types of particles. Each of the terms for Φ on the right-hand side of Equation (76) depends in the same way on the size and distance parameters and differs only in molecular parameters that are fully contained in the Hamaker constant. Therefore $\Delta \Phi$ follows the appropriate function for the interaction from Table 10.4 with the following value of the Hamaker constant:

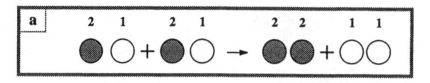

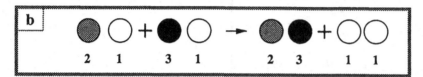

FIG. 10.8 The coagulation process as a pseudochemical reaction. The filled circles indicate particles of the dispersed phase and the open ones satellite particles of the solvent. In (a) the dispersed particles are chemically identical, and in (b) the dispersed particles are different substances.

$$A_{212} = A_{11} + A_{22} - 2A_{12} \tag{77}$$

The subscript 212 indicates two particles of type 2 separated by the medium of type 1.

An approximation that results in a useful simplification is

$$A_{12} \approx (A_{11}A_{22})^{1/2} \tag{78}$$

Equation (78) says that the interaction between dissimilar bodies is given by the geometrical mean of the homogeneous interactions for the two species considered separately. This geometrical *mixing rule* is widely used in solution theory to calculate heterogeneous interactions. We invoked this type of averaging procedure in Sections 3.4 and 6.10 when problems arose that required a 1-2 interaction to be expressed in terms of 1-1 and 2-2 interactions.

Combining Equations (77) and (78) leads to

$$A_{212} = (A_{11}^{1/2} - A_{22}^{1/2})^2 \tag{79}$$

This is the effective value of the Hamaker constant to be used in evaluating the attraction between (like) particles embedded in a medium. Equation (79) leads to three important generalizations about the value of A_{212}:

1. The effective Hamaker constant A_{212} is always positive, regardless of the relative magnitudes of A_{11} and A_{22}. Thus identical particles exert a net attraction on one another due to van der Waals forces in a medium as well as under vacuum.

2. Embedding particles in a medium generally diminishes the van der Waals attraction between them. Table 10.5 shows that the Hamaker constants for homogeneous interactions A_{ii} are generally of the same order of magnitude for different substances. Therefore the effective Hamaker constant—which depends on their difference according to Equation (79)—will be smaller than A_{ii} for either of the homogeneous interactions. For example, if A_{11} and A_{22} equal $8.1 \cdot 10^{-20}$ and $6.4 \cdot 10^{-20}$ J, then A_{212} is 10^{-21} J, almost two orders of magnitude less than the individual A_{ii} values.

3. For $A_{11} = A_{22}$, $A_{212} = 0$, and $\Phi_A = 0$, from the viewpoint of van der Waals forces, this condition corresponds to no net interaction between particles. By using experimental criteria to identify this state of affairs, it is possible to vary the medium in a disperse system until this condition is met and then use the surface tension of the medium (via Equation (67)) to evaluate A_{11} and, therefore A_{22}. Going further, Equation (67) can be applied again to estimate γ_{22} for the dispersed particles. This strategy implies that suitable values for d_0 are available.

Since van der Waals forces are responsible for the coagulation of lyophobic colloids, the mitigating influence of the continuous phase on the attraction between dispersed particles imparts a measure of stability to the system. This result was anticipated in our remarks about the DLP theory in Section 10.7. An even more dramatic modification of van der Waals attraction results when three different substances are involved: dispersed particles of 2 and 3 separated by medium 1. Figure 10.8b shows the coagulation of the dispersed particles for this situation. By analogy with the coagulation of identical particles, we write the following:

1. For the change in potential energy,

$$\Delta\Phi = \Phi_{11} + \Phi_{23} - \Phi_{12} - \Phi_{13} \tag{80}$$

2. For the contribution of molecular properties,

$$A_{312} = A_{11} + A_{23} - A_{12} - A_{13} \tag{81}$$

3. With the A_{ij} values replaced by $(A_{ii}A_{jj})^{1/2}$

$$A_{312} = (A_{11}^{1/2}A_{11}^{1/2}) + (A_{22}^{1/2}A_{33}^{1/2}) - (A_{11}^{1/2}A_{22}^{1/2}) - (A_{11}^{1/2}A_{33}^{1/2}) \tag{82}$$

which factors to

$$A_{312} = (A_{33}^{1/2} - A_{11}^{1/2})(A_{22}^{1/2} - A_{11}^{1/2}) \tag{83}$$

10.8b Negative Hamaker Constants

What makes the result expressed in Equation (83) particularly interesting is that one of the factors in the equation can be positive and one negative, in which case the effective Hamaker constant itself becomes negative. A negative proportionality factor in an expression based on attraction makes the potential energy change positive. The pairwise attraction between bodies results in a net repulsion between dissimilar particles. Since the factors in Equation (83) must have different signs, all that is required for this condition of a negative effective Hamaker constant is that the A value for the continuous phase be intermediate between the Hamaker constants of the two types of dispersed particles: $A_{22} > A_{11} > A_{33}$ or $A_{33} > A_{11} > A_{22}$.

For each of the A terms in Equation (83) we may substitute the corresponding version of Equation (67). Strictly speaking, each material is characterized by its own intermolecular spacing, and this as well as its γ value should be used in Equation (67). In a number of systems that have been investigated, however, the observed range of d_0 values is quite narrow, suggesting that d_0 can be regarded as a constant, at least as a first approximation. For a variety of polymers a value of about 0.2 nm appears to be a reasonable estimate for d_0. With this (assumed constant) value factored out, Equation (83) becomes

$$A_{312} = 24\pi d_0^2(\gamma_3^{1/2} - \gamma_1^{1/2})(\gamma_2^{1/2} - \gamma_1^{1/2}) \tag{84}$$

With the assumed uniformity of d_0 values, Equation (84) shows that the coagulation of dissimilar particles becomes energetically unfavorable when the surface tension of the medium is intermediate between the surface tensions of the two kinds of dispersed units.

An interesting variation on this idea is the study of particle engulfment by an advancing solidification front. In such an experiment, solid particles are dispersed in an appropriate medium that is then allowed to solidify in a channel along which a suitable temperature gradient is maintained. The fate of the dispersed units is monitored microscopically as the solidification front advances. What is observed is that the dispersed particles are either engulfed by the solid or pushed along in the liquid by the advancing front. These observations can be interpreted in terms of the formalism we have developed by considering the solid and the dispersed units as interacting through a medium composed of the melt. The fact that the solid and the melt are chemically identical is immaterial since they are in different physical states.

If we designate the melt as component 1, the dispersed particles as 2, and the solid front as 3, then engulfment is equivalent to coagulation and occurs spontaneously when A_{312} is positive. In this case $\Delta\Phi$ as given by Equation (80) provides ΔG for the engulfment process and, since the Φ's themselves are negative, A_{312} must be positive for spontaneous engulfment. Conversely, rejection by the front requires a negative value for A_{312}. Again under the assumption of the constancy of d_0, rejection is predicted so long as the surface tension of the melt is intermediate between the values of the solid and the dispersed particles. Example 10.5 considers an illustration of this kind of data.

* * *

EXAMPLE 10.5 *Particle Engulfment by An Advancing Solidification Front.* Experiments were conducted at 80°C in which the solidification front of naphthalene was observed to either engulf or reject dispersed particles of several solids. Table 10.6 lists the observed engulfment (E) or rejection (R) behavior for various systems as well as the surface tensions of the various substances. The surface tensions of solid and liquid naphthalene at 80°C are 26.4 and 32.8 mJ m^{-2}, respectively. Is the surface tension criterion cited above consistent with these observations? How might any inconsistencies be explained? Evaluate the product $(\gamma_3^{1/2} - \gamma_1^{1/2})(\gamma_2^{1/2} - \gamma_1^{1/2})$ for these systems.

Solution: The surface tension of liquid naphthalene (1) is greater than that of solid naphthalene (3). Therefore A_{312} is expected to be negative for all systems having γ values greater than γ_1. This is the case for the first six compounds listed in Table 10.6. Therefore these substances are expected to display rejection by the solidification front. This is indeed observed for five of the six cases. The case of nylon-6,12, which deviates from the predicted behavior, is best understood by examining the product $(\gamma_3^{1/2} - \gamma_1^{1/2})(\gamma_2^{1/2} - \gamma_1^{1/2})$. Values of this product for the various systems considered are listed in Table 10.6. The factor arising from the solid-liquid (3-1) naphthalene has the constant value -0.0186 for all cases, but differs when various solids are used as component 2. For nylon-6,12, the second factor becomes -0.0022, and the product of the two, $0.41 \cdot 10^{-4}$ mJ m^{-2}, is the smallest of all such products listed in the table. As the surface tension difference decreases, the sensitivity of the behavior to variations in d_0 increases. ∎

* * *

It is apparent that a single dispersed phase could be investigated with an assortment of matrix materials having known properties. The engulfment-rejection behavior of these systems may then be used to establish bracketing values of γ and A for the dispersed phase.

Repulsive van der Waals forces occur in a number of practically important cases, such as for different types of polymers in organic solvents (van Oss et al. 1980) and for certain hydrocarbon films on water (as mentioned in Vignette X). For example, the use of repulsive van der Waals forces has been suggested as a way to dissociate antigen-antibody complexes by van Oss et al. (1979). A much more detailed and quantitative introduction to repulsive van der Waals forces may be found in Israelachvili (1991).

Although we have not stopped dealing with van der Waals attraction, we have come a long

TABLE 10.6 Results of Engulfment Experiments Involving Various Solids Dispersed in Naphthalene[a]

Dispersed particles	Observed behavior[b]	γ (mJ m^{-2})	$(\gamma_3^{1/2} - \gamma_1^{1/2})(\gamma_2^{1/2} - \gamma_1^{1/2})$ $\times 10^4$ (mJ m^{-2})
Acetal	R	41.9	-4.39
Nylon-6	R	41.7	-4.30
Nylon-6,6	R	40.8	-3.89
Nylon-12	R	38.4	-2.77
Nylon-6,10	R	36.0	-1.60
Nylon-6,12	R	32.0	$+0.41$
Polystyrene	E	27.6	$+2.79$
Teflon	E	15.5	$+10.53$
Siliconed glass	E	11.5	$+13.75$

Source: Data of A. W. Neumann, S. N. Omenyi, and C. J. van Oss, *Colloid Polym. Sci.*, **257**, 413 (1979).
[a]Data discussed in Example 10.5.
[b]E, engulfment; R, rejection.

way from our initial point of view. Without any help from a repulsive mode of interaction, we are able to account for situations in which the dispersed state of a system is energetically favored over the coagulated state. The dispersion medium clearly plays an important role in this, since the process involves breaking "bonds" between the dispersed particles and the medium and replacing them with new "bonds." Just as in a metathesis reaction between chemical compounds, it is the difference in the "bond strength" between the final and initial states that determines the net interaction. If the attraction between molecules can thus override itself, this may be helped along by other mechanisms based on actual repulsion between approaching surfaces. In Chapter 11 we consider the overlap of ion atmospheres as the basis for such a repulsion.

REVIEW QUESTIONS

1. What are the three *van der Waals forces*, and what is the molecular origin of each of them?
2. Which of these is usually the most dominant?
3. Why is the *London force* also referred to as the *dispersion force*?
4. It is claimed that the dispersion component of the van der Waals forces is always present regardless of the nature of the molecules. Why?
5. List at least three macroscopic material properties that result from, or are strongly influenced by, van der Waals forces, and explain the reasons physically.
6. When would you expect the dispersion force to be *repulsive*? Why?
7. What is the *Hamaker "constant"*? How is it related to molecular-level van der Waals forces?
8. List some reasons why it is desirable to relate Hamaker constants to measurable *macroscopic properties* instead of relying entirely on molecular parameters.
9. List a few methods that could be used to determine Hamaker constants of materials from relevant macroscopic properties.
10. Explain how Hamaker constants for interaction between identical materials in vacuum can be used to determine Hamaker constants for interaction between dissimilar materials immersed in an arbitrary medium.
11. Describe the conditions under which the Hamaker constant between two interacting colloidal particles is always positive. When can it be negative?
12. List a few macroscopic phenomena that can be traced to van der Waals forces.
13. What is the connection between surface energy of a material and the Hamaker constant of that material?

REFERENCES

General References (with Annotations)

Atkins, P. W., *Physical Chemistry*, 5th ed., W. H. Freeman, New York, 1994. (Undergraduate level. A good textbook on basic physical chemistry and a convenient reference on concepts such as dipole interactions, polarizability of materials and van der Waals equation of state.)

Gregory, J., *Adv. Colloid Interface Sci.*, **2**, 396 (1969). (A review article on van der Waals forces with tables of Hamaker constants for a number of materials.)

Israelachvili, J. N., *Intermolecular and Surface Forces*, 2d ed., Academic Press, New York, 1991. (Graduate and undergraduate levels. An excellent source for the relation between molecular-level van der Waals interactions and macroscopic properties and phenomena such as surface tension, cohesive energies of materials, adhesion, and wetting. Also discusses direct measurement of van der Waals forces using the surface force apparatus.)

Lyklema, J., *Fundamentals of Interface and Colloid Science: Volume I. Fundamentals*, Academic Press, London, 1991. (A graduate-level treatment of the topics discussed in this chapter.)

Mahanty, J., and Ninham, B. W., *Dispersion Forces*, Academic Press, New York, 1976. (An advanced monograph on dispersion forces. Discusses topics such as London and Lifshitz theories.)

Ross, S., and Morrison, I. D., *Colloidal Systems and Interfaces*, Wiley, New York, 1988. (Undergraduate level. A textbook on colloids with Hamaker constant data for more materials than given in this chapter.)

Verwey, E. J. W., and Overbeek, J. Th. G., *Theory of the Stability of Lyophobic Colloids*,

Elsevier, Amsterdam, Netherlands, 1948. (Graduate and undergraduate levels. The classic reference on colloid stability.)

Visser, J., *Adv. Colloid Interface Sci.*, **3**, 331 (1972). (A review article on van der Waals forces with tables of Hamaker constants for a number of materials.)

Other References

Croucher, M. D., *Colloid and Polymer Sci.*, **259**, 462 (1981).

Croucher, M. D., and Hair, M. L., *J. Phys. Chem.*, **81**, 1631 (1977).

Davis, H. T., and Scriven, L. E., *J. Phys. Chem.*, **80**, 2805 (1976).

Fowkes, F. M., *Ind. Eng. Chem.*, **56**, 40 (1964).

Israelachvili, J. N., and Tabor, D., Van der Waals Forces: Theory and Experiment. In *Progress in Surface and Membrane Science*, Vol. 7 (J. F. Danielli, M. D. Rosenberg, and D. A. Cadenhead, Eds.), Academic Press, New York, 1973.

Moelwyn-Hughes, E. A., *Physical Chemistry*, 2d ed., Macmillan, New York, 1964.

Sonntag, H., and Strenge, K., *Coagulation and Stability of Disperse Systems*, Halsted, New York, 1964.

Van der Waals, J. D., *On the Continuity of the Gaseous and Liquid States*, North-Holland, Amsterdam, Netherlands, 1988.

Van Oss, C. J., Absolom, D. R., Grossberg, A. L., and Neumann, A. W., *Immunological Communications*, **8**, 11 (1979).

Van Oss, C. J., Absolom, D. R., and Neumann, A. W., *Colloids and Surfaces*, **1**, 45 (1980).

PROBLEMS

1. The parameter β_{12} for heterogeneous (12) interactions plays a similar role as β_{11} (Equation (34)) does for homogeneous (11) interactions. Use entries from Table 10.1 to write an expression for β_{12}. If Debye interaction makes a negligible contribution to β_{12} and $\nu_1\nu_2/(\nu_1 + \nu_2) \simeq \frac{1}{2}(\nu_1\nu_2)^{1/2}$, show that

$$\beta_{12} \simeq (f_{11L}\beta_{11})^{1/2}(f_{22L}\beta_{22})^{1/2} + (f_{11K}\beta_{11})^{1/2}(f_{22K}\beta_{22})^{1/2}$$

where the f terms are defined by Equation (35). If $f_{11L} = f_{22L}$ and $f_{11K} = f_{22K}$, show that this last result becomes

$$\beta_{12} \simeq (\beta_{11}\beta_{22})^{1/2}$$

Comment on the relevancy of this result to Equation (78). Criticize or defend the following proposition: The geometrical mixing rule does not require the absence of permanent dipoles, only that 11 and 22 interactions both consist of the same fraction of London and permanent dipole contributions. Specific interactions, such as hydrogen bonding, must also be absent in the 11, 22, and 12 systems.

2. Pressure is a manifestation of the kinetic energy of gas molecules. According to the van der Waals equation of state (see Eq. (36)), the pressure of 1 mole of gas must be increased by an amount a/v^2 due to intermolecular attractions that decrease the pressure from what it would be if ideal. Use the term a/v^2 as a general expression for the attraction between a pair of molecules and, based on this, reconstruct the argument leading to Equation (52).

3. A gas adsorption isotherm may be derived by comparing the adsorbed layer around a solid particle to a planetary atmosphere, with an equilibrium pressure p_0 at the surface. The change in free energy for a molecule going from the bulk pressure p to the surface is $k_B T \ln (p_0/p)$. Equating this with Equation (59), the potential energy of attraction responsible for the adsorption, gives

$$k_B T \ln \left(\frac{p_0}{p}\right) = \frac{\rho N_A}{M} \frac{\beta\pi}{6} z^{-3}$$

Since $z \propto V$, this may be written $\ln (p_0/p) = \text{const.} \ V^{-3}$, where V is the volume of gas adsorbed. A more general form of this isotherm, called the *Frenkel–Halsey–Hill (FHH) isotherm*, treats the power dependence of V as an unknown n and writes

$$\left(\frac{V}{V_m}\right)^n = \frac{K}{\ln (p_0/p)}$$

where V_m is the volume of gas at monolayer average.

Prepare a plot of the FHH isotherm using $n = 3$ and $K = 0.1$ and comment on the resemblance of this isotherm to actual gas adsorption isotherms as shown in Chapter 9.

4. The derivation of Equation (68) follows the same argument that leads to Equation (61), except that the blocks are assumed to be composed of stacks of matter of density ρ' in slices having a thickness δ and separated by a distance d. A molecule at O interacts with the ith slice in such a stack with an energy that is the analog of Equation (56):

$$-d\Phi_i = \frac{\rho' N_A \beta \delta}{M} \frac{2\pi y \, dy}{\{[z + i(d + \delta)]^2 + y^2\}^3}$$

Evaluate the attraction between the molecule and the ith layer by integrating this expression over all values of y between 0 and ∞. The assumption that $\rho'\delta = \rho(d + \delta)$ ensures that the blocks have the correct macroscopic density. If the separation between layers is the same as the separation of the surfaces of the blocks (i.e., $z = d$), then the equivalent of Equation (59) results from summing the Φ_i values for all i's between 0 and ∞ and taking the limit of $\delta \to 0$. Derive the analog of Equation (59) for matter with this hypothetical structure. Note that $\Sigma_i(1 + i)^{-4} = 1.082$. Equation (68) is obtained by assuming a similar structure for the second block and continuing along these lines.

5. Consider a body of water and a piece of styrofoam. Both are attracted by the earth because of gravity. However, when immersed in water, the styrofoam is effectively repelled by the earth due to the Archimedes principle. Use this analogy to explain the possibility of repulsive van der Waals forces between materials that, taken pairwise, experience only attractive forces.

6. Neumann et al.* have tabulated average values of experimentally determined Hamaker constants and then used surface tension data for the same systems to calculate d_0 values for the following materials:

Substance	$A_{22} \times 10^{20}$ (J)	γ (mJ m^{-2})
Polystyrene	8.47	33.0
Teflon	5.23	20.0
Nylon-6,6	12.05	46.0
Poly(methyl methacrylate)	8.83	39.0
n-Decane	5.13	23.9
Polyethylene	8.43	31.0
Poly(vinyl alcohol)	8.84	41.0
Poly(hexafluoropropylene)	5.20	17.0

Verify that the d_0 values thus calculated show a relatively narrow distribution around a mean value close to 0.2 nm. Criticize or defend the following proposition: As a mean center-to-center intermolecular spacing, this value is on the low side; as a back-calculated parameter, however, it probably compensates for deviations from the assumed geometry, breakdown of Equation (33) at short distances, or other shortcomings of the molecular additivity principle.

7. An extreme in sediment volume was used† as a criterion for the effective cancellation of interparticle attraction by the continuous phase. Nylon-6,6 dispersions consisting of 1.0 g of solid in 10 ml of n-propanol-thiodiethanol mixtures of various compositions were allowed to settle to sedimentation equilibrium. Listed here are the equilibrium sediment volumes, the volume/volume compositions, and the surface tensions of the media:

	n-Propanol-thiodiethanol (vol/vol)									
	20/20	60/40	45/55	35/65	25/75	20/80	15/85	10/90	7.5/92.5	5/95
V_{sed} (cm^3)	2.15	2.40	2.40	2.50	2.60	2.60	2.70	2.48	2.40	2.35
γ (mJ m^{-2})	26.20	28.70	30.90	33.60	36.20	38.20	39.60	40.70	44.00	47.10

*Neumann, A. W., Omenyi, S. N., and Van Oss, C. J., *Colloid Polym. Sci.*, **257**, 413 (1979).
†Neumann, A. W., Visser, J., Smith, R. P., Omenyi, S. N., Francis, D. W., Spelt, J. K., Vargha-Butler, E. B., Zingg, W., Van Oss, C. J., and Absolom, D. R., *Powder Technol.*, **37**, 229 (1984).

Plot the sediment volume versus the surface tension of the continuous phase for these dispersions. What is the apparent surface tension of the nylon-6,6? Briefly describe some precautions that must be observed in interpreting results such as these in terms of γ values for the dispersed solid.

8. African green monkey kidney cells (component 2) were cultured on suspensions of collagen-coated dextran particles (component 3). Harvesting such cells is traditionally accomplished by scraping, ultrasonication, or the use of chelating agents to disrupt cation bridging between surfaces 2 and 3. Van Oss et al.* reasoned that the cells might be eluted by lowering the surface tension of the suspending medium (component 1). What is the basis for this expectation? Dimethyl sulfoxide (DMSO) was incrementally added to cell-carrier particle dispersions in buffered aqueous solution. The surface tension of the eluting liquids and the percent yield of the harvested cells are tabulated:

Vol % DMSO	0.0	7.5	10.0	12.5	15.0	17.0	18.0	19.0
γ (mJ m^{-2})	73.0	67.0	65.0	63.6	62.1	61.2	60.7	60.3
Percent yield	8.4	15.1	16.5	35.6	42.3	59.3	66.7	80.1

For the solids γ_2 and γ_3 have been estimated to be 68.9 and 32 mJ m^{-2}, respectively. Is the observed behavior qualitatively consistent with expectations? Suggest some factors that might be responsible for any quantitative discrepancy.

---

*Van Oss, C. J., Charney, C. K., Absolom, D. R., and Flanagan, T. J., *BioTechniques,* 194 (Nov.–Dec. 1983).

11

The Electrical Double Layer and Double-Layer Interactions

Suppose that I had the power of passing through . . . things, so that I could penetrate my subjects, one after another, even to the number of a billion, verifying the size and distance of each by the sense of feeling.

From Abbott's *Flatland*

11.1 INTRODUCTION

11.1a What Is an Electrical Double Layer?

When ions are present in a system that contains an interface, there will be a variation in the ion density near that interface that is described by a profile like that shown in Figure 7.13. The boundary we identify as *the* surface defines the surface excess charge, as explained in Chapter 7. Suppose that it was possible to separate the two bulk phases at this boundary in the manner shown in Figure 6.8. Then, each of the separated phases would carry an equal and opposite charge. These two charged portions of the interfacial region are called the *electrical double layer*.

The purpose of this chapter is to introduce the basic ideas concerning electrical double layers and to develop equations for the distribution of charges and potentials in the double layers. We also develop expressions for the potential energies and forces that result from the overlap of double layers of different surfaces and the implication of these to colloid stability.

11.1b Why Are Electrostatic Effects Important?

Electrostatic and electrical double-layer forces play a very important role in a number of contexts in science and engineering. As we see in Chapter 13, the stability of a wide variety of colloids, ranging from food colloids, pharmaceutical dispersions, and paints, to colloidal contaminants in wastewater, is affected by surface charges on the particles. The filtration efficiency of submicron particles can be diminished considerably by electrical double-layer forces. As we point out in Chapter 13, coagulants are added to neutralize the electrostatic effects, to promote aggregation, and to enhance the ease of separation.

The electrostatic forces also play an important role in the conformation and structure of macromolecules such as polymers, polyelectrolytes, and proteins. The self-assembly of proteins from disks to virus is triggered by electrostatic interactions between neighboring subunits. In the case of polyelectrolytes (polymer molecules with charges) and charged colloids, transport behavior such as rheology is also affected significantly by charge effects, as we have already seen in Chapter 4.

Electrostatic and electrical double-layer interactions also create new opportunities in science and technology. We have already seen an example of this in a vignette in Chapter 1 on electrophoretic imaging devices, and another, on electrophotography, is described in the next chapter. The very basis of electrokinetic phenomena and their implications and uses arise from electrical double-layer interactions.

As an additional example, Vignette XI draws attention to the role of electrostatic effects in molecular recognition and specificity in the biosciences.

**VIGNETTE XI   LIFE SCIENCES, BIOTECHNOLOGY, AND
BIOMEDICAL ENGINEERING: Molecular Recognition
and Specificity**

Molecular recognition and specificity are the stuff of life. How are biological macromolecules able to recognize each other or recognize a membrane or a substrate and form specific associations? What do we need to know about the interplay between charged surface groups on interacting macromolecules such as proteins to design better drugs or to understand and modify enzyme catalysis? What determines the encounter between an influenza virus and its host site (a sialic acid residue) on a cell before the virus binds and enters the cell (by what is known as *endocytosis*)? These are the types of questions with answers that are central to the functioning of biological systems, design of drugs, development of artificial biomaterials, design of specific chromatographic techniques, and the like.

Although the answers to questions such as these depend on a complex array of factors ranging from the structure of the relevant molecules to their environment and the chemical activity of the medium containing the molecules, intermolecular (guest-host) interactions play a central role in determining the rate and the efficiency of the ultimate result. A major component of the many possible intermolecular forces is the electrostatic interaction, particularly because of the long-range nature of the Coulombic forces and the inevitable influence of the ionic atmospheres that surround the macromolecules and substrates.

The electrostatic interaction works in tandem with the diffusional (translational as well as rotational) motion of the macromolecules as the macromolecules find their way to their destination for that crucial first encounter. Moreover, the eventual mutual recognition of the guest and the host and the stability of the resulting binding depend on the changes in the free energy due to the encounter, contact, and binding. Such free energy changes can be sensitive to the details of the electrical double-layer contribution, and it is necessary to understand the electrostatic interactions between a target (i.e., a host such as a cell surface or an ion-exchange resin) and a guest (say, a protein) as a function of the ionic strength, separation distance, orientation, and details of the structure of the materials involved (e.g., both the guest and the host may have a mosaic of charged patches; see Figure 11.1). In many cases, a rather involved analysis of diffusional encounters (Brownian motion) in an ionic environment described by the nonlinear Poisson-Boltzmann equation may be necessary.

The purpose of the present chapter is to introduce some of the basic concepts essential for understanding electrostatic and electrical double-layer pheneomena that are important in problems such as the protein/ion-exchange surface pictured above. The scope of the chapter is of course considerably limited, and we restrict it to concepts such as the nature of surface charges in simple systems, the structure of the resulting electrical double layer, the derivation of the Poisson-Boltzmann equation for electrostatic potential distribution in the double layer and some of its approximate solutions, and the electrostatic interaction forces for simple geometric situations. Nonetheless, these concepts lay the foundatiorı on which the edifice needed for more complicated problems is built.

11.1c   Focus of This Chapter

This chapter focuses on some of the basic theories of electrical double layers near charged surfaces and develops the expressions for interaction energies when two electrical double layers overlap ("interact") with each other.

1.   The first topic we discuss is the origin of the charge at certain surfaces through ion

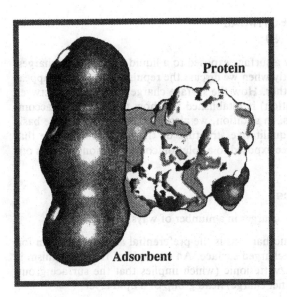

FIG. 11.1 Rat cytochrome b5 (a protein) in the apparent preferred chromatographic contact orientation with a simulated anion-exchange surface. Electrostatic equipotential surfaces are displayed for the protein and the ion-exchange surface according to the following scheme. The molecular surfaces of the protein and the resin are in an off-white color. The equipotential surfaces (on the protein as well as on the adsorbent) at a value of $+0.3\ k_BT/e$ $(+7.7\ \text{mV})$ are represented by the dark regions. The equipotential surface on the protein at $-1.2\ k_BT/e$ $(-30.8\ \text{mV})$ is in gray. (Reproduced with permission from D. J. Roush, D. S. Gill, and R. C. Willson, *Biophys. J.*, **66**, 1 (1994).)

adsorption (Section 11.2). A straightforward application of thermodynamics enables us to quantify this effect in terms of an electrical potential associated with the adsorbed ions.

2. After these phenomenological considerations, we turn our attention to various models for the distribution of charge near the surface, starting from the "capacitor model" in Section 11.3 and progressively refining it to more realistic approximations such as the Debye-Hückel model and Gouy-Chapman model (Sections 11.4–11.6). Although we examine several different models under several limiting conditions, most of the theoretical developments of this chapter will involve the following assumptions: (a) planar surfaces, (b) isolated surfaces, and (c) constant potential surfaces, examined specifically for (d) the variation of the potential with distance from the surface and (e) the effect of added electrolyte on the potential. However, expressions for spherical and cylindrical surfaces and for constant surface charge densities (rather than for constant surface potentials) are also summarized for some important cases.

3. To apply these ideas to coagulation phenomena, we must consider what happens to these distributions of potential when two similar surfaces approach one another (Section 11.7). To study coagulation phenomena, we need to compare the electrostatic effects of particle approach with the van der Waals effects discussed in the last chapter. This is done in terms of potential energy curves as discussed in Section 10.2. As we move through the chapter, our interest shifts from potential (volts) to potential energy (joules). It is important to keep track of the difference between the two as the development progresses.

4. Finally, a brief overview of the structure of the inner edge of the double layer (the so-called *Stern layer*), which accounts for the preferential adsorption of ions and the finite size of the ions, is presented in Section 11.8, along with a discussion of how the developments in previous sections can be modified to accommodate the variation in the surface potential because of the Stern layer.

11.2 SURFACE CHARGES AND ELECTRICAL DOUBLE LAYER: BACKGROUND

At the outset, we need to understand how a surface exposed to a liquid may acquire charges. This turns out to be important subsequently when we discuss the repulsive forces that appear as two charged surfaces approach each other. How the surface charge equilibrium (between the ions on the surface and those in solution) is established and how rapidly this is accomplished affect the magnitude of the forces. In addition, we also need to establish some basic concepts concerning the origin and the qualitative structure of the ionic atmosphere that develops in the vicinity of a charged surface exposed to a solution containing ions. This is our focus in this section.

11.2a Origin of Charges at a Surface

A surface immersed in a liquid can acquire charges in a number of ways. For example:

1. One of the common "charging" mechanisms is the preferential adsorption of an ion from a solution on an initially uncharged surface. An example of this mechanism is the binding of a Ca^{2+} ion on a zwitterionic (which implies that the surface group consists of surface dipoles but no net charge) head group of a lipid layer.
2. Another possible mechanism is the ionization or dissociation of a surface group (e.g., dissociation of a proton from a carboxylic group, namely, $-COOH \rightarrow -COO^- + H^+$, which leaves the surface with a negative charge).

The first of these is often the most common. Adsorption of Ag^+ and I^- on silver iodide particles is an example of this mechanism and is discussed in detail in the following section. Adsorption of H^+ and OH^- on insoluble oxides and adsorption of surfactants on mineral particles and air-water interfaces also fall under this category. Dissociation of surface groups is a common charging mechanism in the case of latex particles, which are frequently used in biomedical applications, as calibration standards in electron microscopy and as model particles in experiments on colloidal phenomena. In the case of clay minerals, one often encounters isomorphic substitution of ions (e.g., replacement of Si^{4+} ions in the crystalline mineral with other cations such as Al^{3+}, or replacement of Al^{3+} with Mg^{2+}). Accumulation of electrons is the main charge-inducing mechanism in the case of metal-solution interfaces. Table 11.1 presents a summary of these general charging mechanisms.

TABLE 11.1 Examples of Charging Mechanisms that Lead to Charges at an Interface[a]

	Nature of the interface	
	Air-Water Mercury-Water Oil-Water	Solid-Water
Preferential adsorption of ions	+	+
Dissociation of surface groups	−	+
Isomorphic substitution	−	+
Adsorption of polyelectrolytes	+	+
Accumulation of electrons	+	+

Source: J. Lyklema, Fundamentals of Electrical Double Layers in Colloidal Systems. In *Colloidal Dispersions* (J. Goodwin, Ed.), Royal Society of Chemistry, London, 1982, pp. 47–70.
[a]The signs in the table indicate the sign of the acquired charges.

In the following section we use the well-known and rather extensively studied silver iodide surface as a vehicle for introducing the basic concepts and terminology of charged surfaces and the nature of the ionic atmosphere that develops in the vicinity of such surfaces.

11.2b Reversible Electrodes: The Silver Iodide Electrode

To arrive at an understanding of the distribution of charge and potential near an interface, it is helpful to consider an electrode. A reversible electrode is one in which each of the phases contains a common ion that is free to cross the interface. The system Ag-AgI-aqueous solution is an example of a reversible electrode. A polarizable electrode, on the other hand, is imperme-able to charge carriers, although charge may be brought to the surface by the application of an external potential. The system metallic Hg-aqueous solution is an example of a polarizable electrode; we discussed the relationship among the applied potential, the interfacial tension, and the adsorption of ions in Chapter 7, Section 7.11.

It is also convenient to divide ions into two categories: *potential determining* and *indiffer-ent* ions. The terminology here is self-explanatory. For example, we can say that Ag^+ is potential determining for the Ag-Ag^+ electrode and that $NaNO_3$ is an indifferent electrolyte as far as this potential is concerned. This obviously neglects any effects of $NaNO_3$ on the activity of the Ag^+. Such an approximation increases in accuracy as the concentration of electrolyte decreases. We consistently neglect activity corrections in this chapter.

The solubility product constant for AgI is about $7.5 \cdot 10^{-17}$ at 25°C. This means that the equilibrium concentration of Ag^+ and I^- in a saturated solution of AgI in pure water equals about $8.7 \cdot 10^{-9}$ mole liter^{-1}. Electrokinetic experiments (Chapter 12) on AgI particles under these conditions reveal that the particles carry a negative charge in this case. Common ion sources such as $AgNO_3$ or KI may be added to the solution to vary the proportions of the Ag^+ and I^- ions in solution, subject to the condition that the ion product equals K_{sp}. When this is done, it is found that the AgI particles reverse charge at an Ag^+ concentration of about $3.0 \cdot 10^{-6}$ mole liter^{-1}. When the concentration of Ag^+ is greater than this, the particles are positively charged. For Ag^+ concentrations less than $3.0 \cdot 10^{-6}$ M, they are negatively charged.

One way of understanding these results is to consider the Ag^+ and I^- ions competing for adsorption sites on the surface. The tendency of both kinds of ions to adsorb at the AgI interface is not hard to understand. After all, the solid crystals would continue to grow if more ions were present. At the point of zero charge the two kinds of ions are adsorbed equally (in stoichiometric proportion). Negatively charged particles imply the adsorption of excess I^- ions, whereas positively charged particles imply excess Ag^+ adsorption. Since the zero point of charge and the saturation concentration in pure water do not coincide, we infer that the I^- ions have a greater affinity with the surface.

The Nernst equation provides us with a relationship that permits an electrical potential difference to be associated with a concentration difference. We adopt the convention that the potential at the AgI-solution interface is zero at the *zero point* (zp) of charge, a point at which the ion molarities will be symbolized c_{zp}. Our interest is to express the potential at the interface ψ_0 in terms of the concentration of ions in solution for conditions other than the zero point of charge. The Nernst equation gives

$$\psi_0 = (k_B T/e) \ln (c/c_{zp}) = (2.303 \, RT/\mathfrak{F}) \log (c/c_{zp}) \tag{1}$$

where $\mathfrak{F}$ is the Faraday constant. We are accustomed to using the second form of Equation (1) in physical and analytical chemistry. The quantity $(2.303 \, RT/\mathfrak{F})$ has the familiar numerical value 0.05917 V at 25°C. Multiplying this by 10^3 and dividing by 2.303 gives 25.7 mV as the value of $(k_B T/e)$. We verify this result shortly, but this is a convenient way to relate a familiar numerical constant to the units most often used in surface and colloid chemistry.

Suppose we apply Equation (1) to AgI in "pure water" (i.e., no common ion source present). We use $c_{Ag} = 8.7 \cdot 10^{-9}$ and $c_{Ag,zp} = 3.0 \cdot 10^{-6}$ to calculate

$$\psi_0 = 25.7 \ln (8.7 \cdot 10^{-9}/3.0 \cdot 10^{-6}) = -150 \, mV \tag{2}$$

Identical results would be obtained if the calculation had been based on I^- concentrations rather than Ag^+. As noted, the surface is negatively charged at this concentration since I^- is preferentially adsorbed.

In principle, part of the potential of any cell may be attributed to each interface; that is, if ψ_i is the potential drop associated with the ith interface, we can write for the total potential difference ψ_T

$$\psi_T = \psi_1 + \psi_2 + \ldots + \psi_i \tag{3}$$

Any electrochemical cell containing an Ag-AgI electrode automatically includes the AgI-solution interface and the potential associated with it. Generally speaking, we are not able to assign absolute numerical values to the various contributions in Equation (3). We can, however, design cells such that only one of the interfaces is sensitive to a particular ion. Clearly, Ag^+ and I^- are the potential-determining ions at the AgI-solution interface. If none of the other interfaces in the cell are appreciably affected by changes in the concentrations of these ions, then variations in the experimental cell potential ψ_T measure changes in ψ_0. This may be expressed

$$d\psi_T = d\psi_0 = (k_B T/e)(dc/c) \tag{4}$$

where c is the concentration of the potential-determining ion. This result is important because it shows how *changes* in ψ_0 can be measured even if ψ_0 itself is unknown. That is, to integrate Equation (4) back to an absolute value of ψ_0, an integration constant must be evaluated. According to our convention, this involves knowing the zero point of charge ($\psi_0 = 0$ at $c = c_{zp}$).

Although we have approached the potential ψ_0 from the perspective of electrodes, the discussion makes it clear that the potential given by Equation (1) applies to any AgI-aqueous solution interface, and not just to electrode surfaces. We may not always know the concentrations required to use Equation (1) numerically, but so long as the bulk concentration of potential-determining ions differs from c_{zp}, a potential difference exists at the surface. The ions of water itself are potential determining for many surfaces. These as well as added solutes or ions in equilibrium with the solid mean that surface potentials at (especially, but not exclusively) aqueous interfaces are the norm rather than something exceptional.

The potential at an interface can be related through the abundant relationships of thermodynamics to the concentration and adsorbability of ions, but thermodynamics provides no information as to how the potential varies as we move through a small distance perpendicular to the surface. This observation reminds us of Figure 7.13, in which the profile of the variation in some general property in the immediate vicinity of a surface is shown. Figure 11.2a is essentially the same drawing with the property under discussion specified as the potential.

The question we consider in the next few sections is this: How does the potential vary with distance across an interface? This question cannot be answered by thermodynamics alone, but it can be examined in terms of various models. We consider a succession of models for a planar surface between two phases. The models will become progressively more complex—and therefore realistic—as we proceed. As far as this presentation is concerned, models are proposed and modified in an intuitive way, rather than by critique of each in terms of experimental results.

11.3   THE CAPACITOR MODEL OF THE DOUBLE LAYER

Figure 11.2a shows schematically the situation in which the potential equals ψ_α at a position x_1, a small distance into the α phase, and equals ψ_β at x_2, a small distance into the β phase. One of our major goals in this chapter is to study the details of the potential variation between x_1 and x_2.

A vastly oversimplified model of how this potential variation might occur on a molecular scale (remember that the distance between x_1 and x_2 is of the order of molecular dimensions) is shown in Figure 11.2b. In this representation two smeared-out planes of charge are situated at

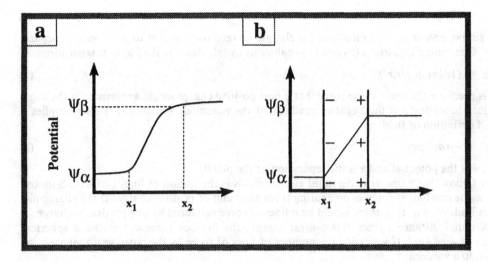

FIG. 11.2 The variation of electrochemical potential in the vicinity of the interface between two phases, α and β: (a) according to a schematic profile; and (b) according to the parallel plate capacitor model.

x_1 and x_2. Note that the model shown in Figure 11.2b resembles a parallel plate capacitor in which two charged conducting surfaces separated by a dielectric occur with a potential difference $\Delta\psi$ between them. Although this is certainly not a realistic picture of the actual distribution of charge at a solution interface, Figure 11.2b allows us to visualize a double layer of charge at an interface. Elsewhere in this book we have referred to the counterion atmosphere that adjoins a charged surface. Figure 11.2b represents this situation, and one of the layers may be regarded as a crude model of this ion atmosphere. This is the origin of the term *electrical double layer*, which is generally used to describe this physical situation, and we use this terminology from now on in preference to ion atmosphere.

Our interest in this chapter and in Chapter 12 is centered primarily on the part of the double layer that extends into the aqueous solution, which is the continuous phase in many important systems. There may be some interfaces between water and a second phase in which the charge on the nonaqueous phase is essentially concentrated on the surface plane. The rigid alignment of a second layer of counterions in the aqueous solution is implausible, however, because of thermal agitation, which tends to diffuse the ions throughout the solution. This leads to what is known as the *diffuse layer* (sometimes referred to as the *Gouy layer* or the Gouy-Chapman layer, in honor of G. Gouy and D. L. Chapman; see also Sections 11.4, 11.6, and 11.8). For now, the parallel plate capacitor model will get us started with the help of some basic relationships and units from elementary physics. The diffuse model of the double layer is discussed in other sections.

Coulomb's law is the basic point of departure. It may be written for the Coulombic force F_C as

$$F_C = (1/4\pi\varepsilon_0)(qq'/\varepsilon_r r^2) \tag{5}$$

to describe the force operating between two charges q and q' separated by a distance r. The factor ε_r is the dielectric constant of the medium, and the proportionality factor $(1/4\pi\varepsilon_0)$ implies that SI units are being used. Remember that ε_0 has the value $8.85 \cdot 10^{-12} \text{ C}^2 \text{ J}^{-1} \text{ m}^{-1}$, $4\pi\varepsilon_0 = 1.11 \cdot 10^{-10} \text{ C}^2 \text{ J}^{-1} \text{ m}^{-1}$, and $(1/4\pi\varepsilon_0) = 8.99 \cdot 10^9 \text{ J m C}^{-2}$. This factor does not appear when cgs units are used. Appendix B contains some additional remarks about these two systems of units, which can be especially troublesome in electrical calculations.

Next let us consider the definition of the strength of an electric field E. The field E describes the force per unit charge in an electrically influenced environment,

$$E = F_c/q \tag{6}$$

Now suppose we bring two identical $+q$ charges toward one another to a distance of separation r. Combining Equations (5) and (6) enables us to calculate the field at that separation:

$$E = (1/4\pi\varepsilon_0)(q/\varepsilon_r r^2) \tag{7}$$

This is precisely the same as the force that a unit positive charge would experience at the same location. Since force is the negative gradient of the potential, Equation (7) also supplies a second definition of field:

$$E = -(d\psi/dx) \tag{8}$$

where ψ is the potential and x is the separation of the plates.

A fiction that helps us understand electric fields is the notion of lines of force. Suppose we imagine one line of force as emanating from each unit of positive charge. If the charge has a magnitude of $+q$, then there would be q lines of force produced by this particular charge.

A radial distance r from this central charge, the lines of force cut across a spherical surface of area $4\pi r^2$. If we divide the number of lines of force by the cross-sectional area, we obtain, in a vacuum,

$$\text{(Number of lines/area)} = q/4\pi r^2 = \varepsilon_0 E \tag{9}$$

Equation (9) shows that the field E and the number of lines per area are directly proportional, with ε_0 the factor of proportionality. In the presence of a dielectric, ε_r is inserted into Equation (9) to bring it into conformity with Equation (7). Now suppose we apply this idea to a parallel plate capacitor.

In the case of a capacitor—taken here as a prototype of a double layer—the charges are not isolated. Instead, the lines of force emanating from one charged surface terminate at an opposite charge on the other plate of the capacitor. Figure 11.3a represents such a situation when the plates are separated by a vacuum. Suppose a plate of area A carries q charges; then we define the charge density σ^* as

$$\sigma^* = q/A \tag{10}$$

Since a line of force is associated with each unit of charge, there are (q/A) lines of force crossing the evacuated gap between the two plates of the capacitor. As we have already seen,

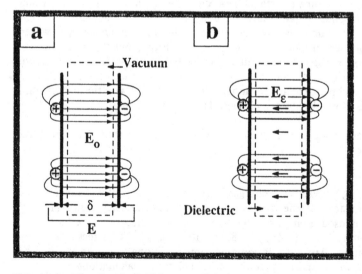

FIG. 11.3 The electric field in a parallel plate capacitor: (a) the dielectric is a vacuum; and (b) a material of dielectric constant ε_r is present.

the number of lines of force measures the field; therefore we write for a capacitor that contains a vacuum

$$E_0 = (q/\varepsilon_0 A) = (\sigma^*/\varepsilon_0) \tag{11}$$

If a substance with a relative dielectric constant ε_r is placed between the plates, the field will be less by this factor. This is because of the partially compensating field that is induced within the dielectric by dipole orientation, as suggested by Figure 11.3b. Therefore, in the presence of the dielectric, the field is given by

$$E_\varepsilon = (q/\varepsilon_r\varepsilon_0 A) = (\sigma^*/\varepsilon_r\varepsilon_0) \tag{12}$$

Next we ignore the directional (sign) aspect of the field and equate Equations (8) and (12) to obtain

$$(d\psi/dx) = (\Delta\psi/\delta) = (\sigma^*/\varepsilon_r\varepsilon_0) \tag{13}$$

where $\Delta\psi$ is the potential drop between plates separated by a distance δ. This equation relates the charge density, voltage difference, and distance of separation of the capacitor. Since this is the model we are using for the double layer, it is of interest to check whether Equation (13) agrees — at least qualitatively — with what we know about the double layer.

We saw in Chapter 7 that charged monolayers are likely to obey the two-dimensional ideal gas law, and we also saw that areas per molecule of 10 nm^2 or so were also required for this ideal law to apply. Hence we may estimate σ^* for a monovalent ion to be

$$\sigma^* = (ion/10\ nm^2)(10^{18}\ nm^2/1\ m^2)(1.60\cdot10^{-19}\ C/ion)$$
$$= 1.6\cdot10^{-2}\ C\,m^{-2} \tag{14}$$

Taking the dielectric constant of water to be about 80, its bulk value, Equation (12) permits the field strength to be estimated:

$$E = 1.6\cdot10^{-2}\ C\,m^{-2}/(80\cdot8.85\cdot10^{-12}\ C^2\,J^{-1}\,m^{-1})$$
$$= 2.26\cdot10^7\ V\,m^{-1} \tag{15}$$

Even allowing for an order of magnitude error in this estimate, we see that there is an exceptionally strong field in the vicinity of a charged interface. In Section 12.8 we examine this in greater detail and consider whether we are justified in using bulk values for such parameters as the permittivity ε and the viscosity η within the double layer.

If we estimate the potential drop between the two phases, we may determine the distance over which the potential drop occurs from the value of E given by Equation (15). If we take the potential difference to be 0.10 V — an arbitrary but reasonable value — Equation (13) shows that the plate separation of an equivalent capacitor is

$$\delta = \Delta\psi/E = (0.10/2.26\cdot10^7) = 4.4\cdot10^{-9}\ m = 4.4\ nm \tag{16}$$

Considering the simplicity of the model and the arbitrariness of the numerical estimates made in this calculation, 4.4 nm seems like a reasonable estimate of the distance over which surface charge neutralization is accomplished.

Throughout this section we have examined the distribution of charge at an interface as if the charge were constrained to two planes. When one of the phases is an aqueous electrolyte solution, the inadequacy of this model is apparent. An immediate improvement of the model is anticipated if we allow for a diffuse double layer, that is, a situation in which the charge density varies with distance from the interface, as shown in Figure 11.4a. Alternatively, we might combine features from both the parallel plate distribution and the diffuse distribution to give a still more elaborate picture of the double layer, as shown in Figure 11.4b. We consider this last situation in Section 11.8. According to this picture, each part of the double layer is analyzed independently, and the effects combined according to the rule for adding capacitors in series:

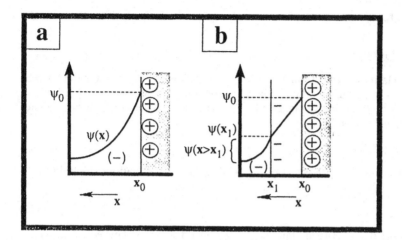

FIG. 11.4 Two models for the double layer: (a) a diffuse double layer; and (b) charge neutralization due partly to a parallel plate charge distribution and partly to a diffuse layer.

$$(1/C_T) = (1/C_1) + (1/C_2) + \ldots + (1/C_i) \tag{17}$$

where C_i is the capacitance of the ith element in series and C_T is the total capacitance.

11.4 THE DIFFUSE DOUBLE LAYER: THE DEBYE-HÜCKEL APPROXIMATION

In the preceding section we discussed the problem of the variation of potential with distance from an interface from the highly artificial perspective of a parallel plate capacitor. The variation of potential with distance from a charged surface of arbitrary shape is a classical electrostatic problem. The general problem is described by the *Poisson equation*,

$$(\partial^2\psi/\partial x^2) + (\partial^2\psi/\partial y^2) + (\partial^2\psi/\partial z^2) = -(\rho^*/\varepsilon) \tag{18}$$

or in terms of the Laplacian operator ∇^2,

$$\nabla^2\psi = -(\rho^*/\varepsilon) \tag{19}$$

where we define ε as the product $\varepsilon_r\varepsilon_0$ and thereby account for the effect of the medium. In these expressions, ρ^* is the charge density (i.e., C m^{-3}) in the system, a quantity that itself is a function of x, y, and z. The solution to this differential equation, therefore, is an expression for the potential that satisfies Equation (18) and also the boundary conditions of the specific problem. We adopt the convention of measuring all distances outward from the interface where the potential has the value ψ_0. As the distance from an isolated surface increases to infinity, the value of ψ approaches zero. The stipulation of an "isolated" surface means that we are concerned with only one interface at this time. In Section 11.7 we consider the case in which electrical double layers overlap.

11.4a Meaning of the Poisson Equation

The Poisson equation is a fundamental relationship of classical electrostatics and really need not be proved here. However, since we are using it as a starting point, it seems desirable to explore the meaning of this important equation to some extent.

Equation (7) describes the field a distance r from a charge $+q$. A basic law of electrostatics is that this field describes any distribution of charge that results in q units of positive charge being enclosed by a sphere of radius r. It is not critical that a single $+q$ charge be situated at the center for this expression to apply. Suppose, therefore, we consider a portion of solution in which the charge is distributed with a uniform density ρ^*. In this case

$$q = (4/3)\pi r^3 \rho^* \tag{20}$$

and, from Equation (7),

$$E = r\rho^*/3\varepsilon \tag{21}$$

Next we multiply both sides of Equation (21) by r^2 and then differentiate with respect to r:

$$d(r^2 E)/dr = r^2 \rho^*/\varepsilon \tag{22}$$

In the present notation, Equation (8) becomes

$$E = -(d\psi/dr) \tag{23}$$

where the minus sign is included since ψ decreases as r increases.

Substituting this result into Equation (22) gives

$$\frac{1}{r^2}\frac{\partial}{\partial r}\left(r^2\frac{\partial\psi}{\partial r}\right) = -\frac{\rho^*}{\varepsilon} \tag{24}$$

Remember that the operator $\nabla^2\psi$ in Equation (19) transforms into the following form in spherical coordinates:

$$\nabla^2\psi = \frac{1}{r^2}\frac{\partial}{\partial r}\left(r^2\frac{\partial\psi}{\partial r}\right) + \frac{1}{r^2\sin\theta}\frac{\partial}{\partial\theta}\left(\sin\theta\frac{\partial\psi}{\partial\theta}\right) + \frac{1}{r^2\sin^2\theta}\left(\frac{\partial^2\psi}{\partial\phi^2}\right) \tag{25}$$

Thus the left-hand side of Equation (24) is seen to be identical to $\nabla^2\psi$ for the case (spherical symmetry) in which ψ is independent of θ and ϕ. Although this presentation does not constitute the most general proof of the Poisson equation, it does give it some plausibility.

There are many situations in which the spherically symmetrical case is specifically invoked, as in the Debye-Hückel theory of electrolyte nonideality, for example. We consider situations for which this is the case in Chapter 12. For now, however, we consider the potential distribution adjacent to a planar wall that carries a positive charge.

11.4b Potential Distribution Near Planar Surfaces

We define the direction perpendicular to the wall as the x direction and consider the wall as extending to infinity in the positive and negative y and z directions. In this case the operation $\nabla^2\psi$ becomes $(d\psi^2/dx^2)$, and Equation (18) is written

$$(d\psi^2/dx^2) = -\rho^*/\varepsilon \tag{26}$$

The next problem is to express the charge density as a function of the potential so the differential equation (26) can be solved for ψ. The procedure is to describe the ion concentrations in terms of the potential by means of a Boltzmann factor in which the work required to bring an ion from infinity to a position at which the potential ψ is given by $z_ie\psi$. The probability of finding an ion at this position is given by the Boltzmann factor, with this work appearing as the exponential of energy:

$$n_i/n_{i\infty} = \exp(-z_ie\psi/k_BT) \tag{27}$$

In this expression n_i is the number of ions of type i per unit volume near the surface, and $n_{i\infty}$ is the concentration far from the surface, that is, the bulk concentration. The valence number z_i is either a positive or negative integer.

The charge density is related to the ion concentrations as follows:

$$\rho^* = \sum_i z_ie\,n_i = \sum_i z_ie\,n_{i\infty}\exp\left(-z_ie\psi/k_BT\right) \tag{28}$$

Combining Equations (26) and (28) gives a result known as the *Poisson-Boltzmann equation*:

$$(d^2\psi/dx^2) = -(e/\varepsilon) \sum_i z_i n_{i\infty} \exp(-z_i e\psi/k_B T) \tag{29}$$

This same relationship is the starting point of the Debye-Hückel theory of electrolyte nonideality, except that the Debye-Hückel theory uses the value of $\nabla^2\psi$ required for spherical symmetry. It is interesting to note that Gouy (in 1910) and Chapman (in 1913) applied this relationship to the diffuse double layer a decade before the Debye-Hückel theory appeared.

The derivation of the Poisson equation implies that the potentials associated with various charges combine in an additive manner. The Boltzmann equation, on the other hand, involves an exponential relationship between the charges and the potential. In this way a fundamental inconsistency is introduced when Equations (26) and (28) are combined. Equation (29) does not have an explicit general solution anyhow and must be solved for certain limiting cases. These involve approximations that — at the same time — overcome the objection just stated.

We introduce the first of the Debye-Hückel approximations by considering only those situations for which $(z_i e\psi < k_B T)$. In this case the exponentials in Equation (28) may be expanded (see Appendix A) as a power series. If only first-order terms in $(z_i e\psi/k_B T)$ are retained, Equation (28) becomes

$$\rho^* = \sum_i z_i e n_{i\infty} [1 - (z_i e\psi/k_B T)] \tag{30}$$

Because of electroneutrality, two of the terms in Equation (30) cancel:

$$\sum_i z_i e n_{i\infty} = 0 \tag{31}$$

so that Equation (30) becomes

$$\rho^* = -\sum_i (z_i^2 n_{i\infty} e^2 \psi/k_B T) \tag{32}$$

In this approximation, the ion potentials are additive, so Equation (32) may be consistently substituted into Equation (26) to give

$$(d^2\psi/dx^2) = [(e^2/\varepsilon k_B T) \sum_i z_i^2 n_{i\infty}]\psi \tag{33}$$

The above equation is known as the *linearized* Poisson-Boltzmann equation since the assumption of low potentials made in reaching this result from Equation (29) has allowed us make the right-hand side of the equation linear in ψ. This assumption is also made in the Debye-Hückel theory and prompts us to call this model the Debye-Hückel approximation. Equation (33) has an explicit solution. Since potential is the quantity of special interest in Equation (33), let us evaluate the potential at 25°C for a monovalent ion that satisfies the condition $e\psi = k_B T$:

$$\psi = (k_B T/e) = (1.38 \cdot 10^{-23})(298)/(1.60 \cdot 10^{-19})$$
$$= 0.0257 \, \text{V} = 25.7 \, \text{mV} \tag{34}$$

Thus at 25°C potentials may be regarded as low or high, depending on whether they are less or more than about 25 mV. The factor $(e\psi/k_B T)$ appears often in double-layer calculations, so this conversion factor is worth remembering. The relationship of $(k_B T/e)$ to $(RT/\mathfrak{F})$ was noted in Section 11.2.

It is convenient to identify the cluster of constants in Equation (33) by the symbol κ^2, which is defined as follows:

$$\kappa^2 = [(e^2/\varepsilon k_B T) \sum_i z_i^2 n_{i\infty}] \tag{35}$$

With this change in notation, Equation (33) becomes simply

$$(d^2\psi/dx^2) = \kappa^2\psi \tag{36}$$

The above equation will have to be solved under the conditions that $\psi \to \psi_0$ as $x \to 0$ and $\psi \to 0$ as $x \to \infty$. The solution that satisfies these conditions is given by

$$\psi = \psi_0 \exp\left(-\kappa x\right) \tag{37}$$

In Section 11.5 we examine the implications of Equation (37) in detail.

11.4c Potential Distribution Around Spherical Surfaces

For studying the stability of colloidal particles in suspension (Chapter 13) or for determining the potential at the surface of particles (Chapter 12), one often needs expressions for potential distributions around small particles that have curved surfaces. Solving the Poisson-Boltzmann equation for curved geometries is not a simple matter, and one often needs elaborate numerical methods. The linearized Poisson-Boltzmann equation (i.e., the Poisson-Boltzmann equation in the Debye-Hückel approximation) can, however, be solved for spherical electrical double layers relatively easily (see Section 12.3a), and one obtains, in place of Equation (37),

$$\psi = \psi_0(R_s/r) \exp\left[-\kappa(r - R_s)\right] \tag{38}$$

where R_s is the radius of the spherical particle and r is the distance of any point in the double layer from the *center of the particle*. Notice that the above equation satisfies the required boundary conditions $\psi \to \psi_0$ as $r \to R_s$ and $\psi \to 0$ as $r \to \infty$. We return to this topic in Chapter 12 when we discuss motion of charged particles in electric fields.

11.4d Potential Distribution Around Cylindrical Surfaces

The solution of the linearized Poisson-Boltzmann equation around cylinders also requires numerical methods, although when cylindrical symmetry *and* the Debye-Hückel approximation are assumed the equation can be solved. The solution, however, requires advanced mathematical techniques and we will not discuss it here. It is nevertheless useful to note the form of the solution. The potential ψ for symmetrical electrolytes has been given by Dube (1943) and is written in terms of the charge density σ^* as

$$\psi = \sigma^* K_0(\kappa r)/[\epsilon \kappa K_1(\kappa R_c)] \tag{39}$$

where r is the radial distance from the axis of the cylinder, R_c is the radius of the cylinder, and K_0 and K_1 are known as zero- and first-order modified Bessel functions of the second kind, respectively. A few values of these Bessel functions are tabulated in Table 11.2 and are useful for calculating electrophoretic mobilities (discussed in Chapter 12). A more detailed analysis of cylindrical double layers may be found in Hunter (1981).

TABLE 11.2 Bessel Functions $K_0(\kappa x)$ and $K_1(\kappa x)$ for Computing Potentials of Long Cylinders[a]

κx	$K_0(\kappa x)/[\kappa x K_1(\kappa x)]$	κx	$K_0(\kappa x)/[\kappa x K_1(\kappa x)]$
0.06	2.950	0.60	0.9942
0.08	2.675	1.00	0.7176
0.10	2.463	1.40	0.5426
0.14	2.151	2.00	0.4071
0.20	1.835	3.00	0.2872
0.40	1.275	4.00	0.2235

Source: H. A. Abramson, L. S. Moyer, and M. H. Gorin, *Electrophoresis of Proteins*, Reinhold, New York, 1942, p. 129.
[a]One may substitute either the radius of the cylinder R_c or the radial position r for the dummy variable x.

11.5   THE DEBYE-HÜCKEL APPROXIMATION: RESULTS

The Debye-Hückel approximation is strictly applicable only in the case of low potentials. Nevertheless, there are several reasons why the significance of Equation (37) should be fully appreciated:

1. It is simpler to understand than any of the modifications we consider subsequently.
2. It is a limiting result to which all equations that are more general must reduce in the limit of low potentials.
3. The effects of electrolyte concentration and valence in this approximation are qualitatively consistent with the results of more elaborate calculations.

11.5a   Physical Significance of the Debye-Hückel Parameter κ

One of the most important quantities to emerge from the Debye-Hückel approximation is the parameter κ. This quantity appears throughout double-layer discussions and not merely at this level of approximation. Since the exponent κx in Equation (37) is dimensionless, κ must have units of reciprocal length. This means that κ^{-1} has units of length. This last quantity is often (imprecisely) called the "thickness" of the double layer. All distances within the double layer are judged large or small relative to this length. Note that the exponent κx may be written x/κ^{-1}, a form that emphasizes the notion that distances are measured relative to κ^{-1} in the double layer.

Since κ and κ^{-1} are such important quantities, we examine them in greater detail, first verifying their dimensions and then considering their numerical magnitude. Especially important is the dependence of κ and κ^{-1} on the concentration and valence of the electrolyte in solution.

It is an easy matter to verify that κ^2 as defined by Equation (35) does indeed have units of length^{-2}, or m^{-2} in SI. This is seen by writing the SI units for the various factors appearing in Equation (35) as follows:

$$\kappa^2 = (C)^2 (m^{-3})/(C^2 J^{-1} m^{-1})(J K^{-1})(K) = m^{-2}$$

The parameter κ depends on concentration; accordingly, we must express it in practical concentration units. If n_i is expressed as the number of ions per cubic meter, then n_i is related to the molar concentration M_i of the ions and the Avogadro's number N_A by

$$n_i = 1000 M_i N_A \tag{40}$$

since $1000 \; dm^3 = 1 \; m^3$. Therefore Equation (35) yields

$$\kappa = [(1000 \, e^2 N_A / \varepsilon k_B T) \sum_i z_i^2 M_i]^{1/2} \tag{41}$$

The summation in this expression is twice the ionic strength of the solution. We examine the numerical substitutions into Equation (41) in Example 11.1.

* * *

EXAMPLE 11.1 *Dependence of the Debye-Hückel Parameter κ on Temperature and Type of Electrolytes.*   Evaluate the numerical factor in Equation (41) for aqueous solutions at 25°C. At this temperature $\varepsilon_r = 78.54$ for water. Calculate κ and κ^{-1} for 0.01 M solutions of 1 : 1, 2 : 1, and 3 : 1 electrolytes. Suggest how these values can be adapted to other temperatures (or media) without complete recalculation.

Solution: Recalling that $\varepsilon = \varepsilon_r \varepsilon_0$, we write

$$\kappa^2 = (1000)(1.60 \cdot 10^{-19})^2 (6.02 \cdot 10^{23})(2I)/(78.54)(8.85 \cdot 10^{-12})(1.38 \cdot 10^{-23})(298)$$
$$= 2.32 \cdot 10^9 (2I)^{1/2} m^{-1}$$

and

$$\kappa^{-1} = 4.31 \cdot 10^{-10} (2I)^{-1/2} \, \text{m}$$

where I is the ionic strength. Note that, for a $1:1$ electrolyte, the summation in Equation (41) gives $(2I)$. In general, for any symmetrical $(z:z)$ electrolyte, the ionic strength equals $z^2 M$ (or $I^{1/2} = |z| M^{1/2}$). Therefore in a 0.01 M solution of $1:1$ electrolyte

$$\kappa = 3.29 \cdot 10^8 \, \text{m}^{-1} \quad \text{and} \quad \kappa^{-1} = 3.04 \cdot 10^{-9} \, \text{m} = 3.04 \, \text{nm}$$

For *asymmetrical* electrolytes the ionic strength is the same for, say, $1:2$ as for $2:1$ solutes; namely, 3 M as verified by substitution into the summation (and remembering the stoichiometry of the dissociation!). Therefore in a 0.01 M solution of $2:1$ electrolyte

$$\kappa = 5.68 \cdot 10^8 \, \text{m}^{-1} \quad \text{and} \quad \kappa^{-1} = 1.76 \cdot 10^{-9} \, \text{m} = 1.76 \, \text{nm}$$

and in a 0.01 M solution of $3:1$ electrolyte ($I = 6$ M)

$$\kappa = 8.04 \cdot 10^8 \, \text{m}^{-1} \quad \text{and} \quad \kappa^{-1} = 1.24 \cdot 10^{-9} \, \text{m} = 1.24 \, \text{nm}$$

It is easy enough to evaluate κ and κ^{-1} for different concentrations and valence types; it is more of a nuisance to recalculate these quantities at different temperatures and/or in different media. An easy way to do this using the expressions given is to factor out $\varepsilon_{r,298}$ and $T = 298\text{K}$ and replace them with quantities pertinent to the problem at hand. For example, at 90°C, $\varepsilon_r = 57.98$ for water; therefore the value of κ at this temperature is given by

$$\kappa = 2.32 \cdot 10^9 [(78.54)(298)/(57.98)(363)]^{1/2} (2I)^{1/2}$$
$$= 2.32 \cdot 10^9 (1.05)(2I)^{1/2}$$
$$= 2.45 \cdot 10^9 (2I)^{1/2} \text{m}^{-1}$$

∎

* * *

Both κ and κ^{-1} are used extensively in this chapter and Chapters 12 and 13 (as well as in the sections on the rheology of charged dispersions in Chapter 4). Table 11.3 lists numerical values for these quantities and the pertinent equations for their calculation for aqueous solutions at 25°C. This table may be consulted as a source for κ and κ^{-1} values when these are required for exercises in these chapters.

Several things should be noted about Table 11.3:

1. The tabulated values of κ^{-1} multiplied by 10^9 give the double layer "thicknesses" in nanometers. For example, in a 0.01 M solution of a $1:1$ electrolyte, κ^{-1} equals 3.04 nm.

TABLE 11.3 Values of κ and κ^{-1} for Several Values of Electrolyte Concentrations and Valences for Aqueous Solutions at 25°C

		Symmetrical electrolyte		Asymmetrical electrolyte		
		$\kappa \, (\text{m}^{-1}) = 3.29 \times 10^9 |z| M^{1/2}$	$\kappa^{-1} \, (\text{m}) = 3.04 \times 10^{-10} |z|^{-1} M^{-1/2}$	$\kappa \, (\text{m}^{-1}) = 2.32 \times 10^9 (\Sigma_i z_i^2 M_i)^{1/2}$	$\kappa^{-1} \, (\text{m}) = 4.30 \times 10^{-10} (\Sigma_i z_i^2 M_i)^{-1/2}$	
Molarity	$z_+:z_-$			$z_+:z_-$		
0.001	1:1	1.04×10^8	9.61×10^{-9}	1:2, 2:1	1.80×10^8	5.56×10^{-9}
	2:2	2.08×10^8	4.81×10^{-9}	3:1, 1:3	2.54×10^8	3.93×10^{-9}
	3:3	3.12×10^8	3.20×10^{-9}	2:3, 3:2	4.02×10^8	2.49×10^{-9}
0.01	1:1	3.29×10^8	3.04×10^{-9}	1:2, 2:1	5.68×10^8	1.76×10^{-9}
	2:2	6.58×10^8	1.52×10^{-9}	1:3, 3:1	8.04×10^8	1.24×10^{-9}
	3:3	9.87×10^8	1.01×10^{-9}	2:3, 3:2	1.27×10^9	7.87×10^{-10}
0.1	1:1	1.04×10^9	9.61×10^{-10}	1:2, 2:1	1.80×10^9	5.56×10^{-10}
	2:2	2.08×10^9	4.81×10^{-10}	1:3, 3:1	2.54×10^9	3.93×10^{-10}
	3:3	3.12×10^9	3.20×10^{-10}	2:3, 3:2	4.02×10^9	2.49×10^{-10}

2. This "thickness" is about the same magnitude as the prediction based on the capacitor model (Equation (16)). The diffuse model is clearly superior, however, since it shows how the double layer "thickness" depends on the concentration and valence of the ions in the solution.

3. The "thickness" of the double layer varies inversely with z and inversely with $M^{1/2}$ for a symmetrical $z:z$ electrolyte. Therefore κ^{-1} equals 1.0 nm for a 0.01 M solution of a $3:3$ electrolyte and is about 10 nm for a 0.001 M solution of a $1:1$ electrolyte.

Figure 11.5 shows how the potential drops with distance from the surface according to Equation (37). The curves in this figure are drawn for two different variations in κ: in Figure 11.5a, a $1:1$ electrolyte at 0.1, 0.01, and 0.001 M concentrations; and in Figure 11.5b, a 0.001 M solution of $1:1$, $2:2$, and $3:3$ electrolytes. Again, it is important to recognize that the curves drop off more rapidly for either higher concentrations or higher valences of the electrolyte.

The curves in Figure 11.5 are marked at the x value that corresponds to κ^{-1}. Note that the potential has dropped to the value (ψ_0/e) at this point. Calling κ^{-1} the double layer "thickness" is clearly a misnomer. We see presently, however, that there is some logic underlying this terminology.

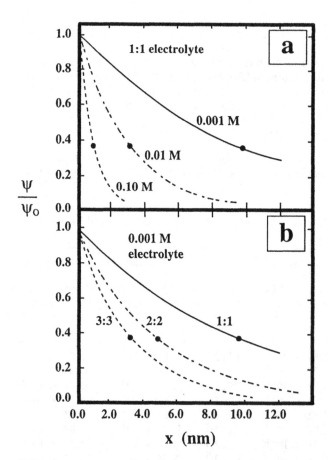

FIG. 11.5 Fraction of double-layer potential versus distance from a surface according to the Debye-Hückel approximation, Equation (37): (a) curves drawn for $1:1$ electrolyte at three concentrations; and (b) curves drawn for 0.001 M symmetrical electrolytes of three different valence types.

11.5b Comparison Between the Capacitor and Diffuse Models

Although the potential is fundamentally a more important quantity than charge density, examining the charge density will enable us to compare the capacitor and diffuse models for the double layer.

The condition of electroneutrality at a charged interface requires that the density of charge at the two faces be equal. Note that this does not require the charges to be physically situated at the interface. When one of the phases contains a diffuse layer, the total charge contained in a volume element of solution of unit cross section and extending from the wall to infinity must contain the same amount of charge—although of opposite sign—as a unit area of wall contains. Stated in formula, this becomes

$$\sigma^* = -\int_0^\infty \rho^* dx \tag{42}$$

We now examine the implications of Equation (42) for the situation in which one of the adjoining phases contains the diffuse half of a double layer. Combining Equations (26) and (42) gives

$$\sigma^* = \varepsilon \int_0^\infty (d^2\psi/dx^2)dx \tag{43}$$

a result that is easily integrated to yield

$$\sigma^* = \varepsilon(d\psi/dx)|_0^\infty \tag{44}$$

The quantity $d\psi/dx$ is zero at infinity, and we define its value at the wall by $(d\psi/dx)_0$; therefore Equation (44) becomes

$$\sigma^* = -\varepsilon(d\psi/dx)_0 \tag{45}$$

Equation (45) provides a relationship between the *surface charge density* and the slope of the potential at the surface. Next, we turn to Equation (37)—the Debye-Hückel approximation for ψ—to evaluate $(d\psi/dx)_0$. Differentiation leads to the value

$$(d\psi/dx)_0 = \lim_{x\to 0} [-\kappa\psi_0 \exp(-\kappa x)] = -\kappa\psi_0 \tag{46}$$

Substituting Equation (46) into Equation (45) gives

$$\sigma^* = \varepsilon\kappa\psi_0 \tag{47}$$

Rewriting Equation (47) in terms of κ^{-1}, the double-layer "thickness," yields

$$\sigma^* = \varepsilon(\psi_0/\kappa^{-1}) \tag{48}$$

Equation (48) is identical in form to Equation (13) for a parallel plate capacitor, with ψ_0 replacing $\Delta\psi$ and κ^{-1} replacing δ. This result shows that a diffuse double layer at low potentials behaves like a parallel plate capacitor in which the separation between the plates is given by κ^{-1}. This explains why κ^{-1} is called the double-layer thickness. It is important to remember, however, that the *actual* distribution of counterions in the vicinity of a charged wall is diffuse and approaches the unperturbed bulk value only at large distances from the surface.

Even allowing for the fact that the Debye-Hückel approximation applies only for low potentials, the above analysis reveals some features of the electrical double layer that are general and of great importance as far as stability with respect to coagulation of dispersions and electrokinetic phenomena are concerned. In summary, three specific items might be noted:

1. The distance (away from the wall) over which an electrostatic potential persists may be comparable to the dimensions of colloidal particles themselves.

2. The distance over which significant potentials exist decreases with increasing electrolyte concentration.

3. The range of electrostatic potentials decreases as the valence of the ions in solution increases.

The Debye-Hückel approximation to the diffuse double-layer problem produces a number of relatively simple equations that introduce a variety of double-layer topics as well as a number of qualitative generalizations. In order to extend the range of the quantitative relationships, however, it is necessary to return to the Poisson-Boltzmann equation and the unrestricted Gouy-Chapman theory, which we do in Section 11.6.

11.5c   Relation Between the Surface Charge and Surface Potential for Spherical Double Layers

Before we proceed to the Gouy-Chapman theory of electrical double layers, it is worthwhile to note that relations similar to Equations (45) and (47) can also be derived for double layers surrounding spherical particles. The equation for surface charge density takes the form

$$\sigma^* = -\varepsilon(d\psi/dr)_{R_s} \tag{49}$$

where the subscript R_s implies that the slope $(d\psi/dr)$ is evaluated at the particle surface, that is, $r = R_s$. When the expression for ψ given in Equation (38) is substituted in this expression, one derives for the following equation

$$q = 4\pi R_s^2 \sigma^* = 4\pi\varepsilon R_s(1 + \kappa R_s)\psi_0 \tag{50}$$

for the relation between the total charge q on the particle and the surface potential ψ_0.

11.5d   Relation Between the Surface Charge and Surface Potential for Cylindrical Double Layers

In the case of cylindrical particles (of radius R_c), if one again assumes that the charge is smeared uniformly on the surface of the cylinders (including the flat circular ends), we can use Equation (39) to obtain

$$q = [2\pi\varepsilon\kappa R_c(2R_c + L)K_1(\kappa R_c)/K_0(\kappa R_c)]\psi_0 \tag{51}$$

where L is the length of the cylinder, and K_0 and K_1 are the Bessel functions listed in Table 11.2. Additional details are available in the references cited in Section 11.4d.

11.6   THE ELECTRICAL DOUBLE LAYER: GOUY-CHAPMAN THEORY

The theoretical inconsistencies inherent in the Poisson-Boltzmann equation were shown in Section 11.4 to vanish in the limit of very small potentials. It may also be shown that errors arising from this inconsistency will not be too serious under the conditions that prevail in many colloidal dispersions, even though the potential itself may no longer be small. Accordingly, we return to the Poisson-Boltzmann equation as it applies to a planar interface, Equation (29), to develop the Gouy-Chapman result without the limitations of the Debye-Hückel approximation.

If both sides of Equation (29) are multiplied by $2\, d\psi/dx$, we obtain

$$2(d\psi/dx)(d^2\psi/dx^2) = 2(d\psi/dx)\{-(e/\varepsilon)\sum_i z_i n_{i\infty} \exp(-z_i e\psi/k_B T)\} \tag{52}$$

The left-hand side of this equation is the derivative of $(d\psi/dx)^2$; therefore

$$(d\psi/dx)^2 = (2k_B T/\varepsilon)\sum_i n_{i\infty} \exp(-z_i e\psi/k_B T) + \text{const.} \tag{53}$$

The integration constant in this expression is easily evaluated if we define the potential in the solution at $x = \infty$ to be zero. At the same limit, $d\psi/dx$ also equals zero. In view of these conventions, Equation (53) becomes

$$(d\psi/dx)^2 = (2k_B T/\varepsilon)\sum_i n_{i\infty} [\exp(-z_i e\psi/k_B T) - 1] \tag{54}$$

This result may be integrated further if we restrict the electrolyte in solution to the symmetrical $z : z$ type. In that case, Equation (54) can be written as

$$(d\psi/dx)^2 = (2k_BTn_\infty/\varepsilon)[\exp(-ze\psi/k_BT) + \exp(ze\psi/k_BT) - 2] \tag{55}$$

in which z is the absolute value of the valence number, the sign having been incorporated into the algebraic form. Note that in Equation (55) we have written $n_{i\infty}$ as simply n_∞ since we are considering symmetric $(z : z)$ electrolytes. The bracketed term is readily seen to equal $[\exp(-ze\psi/2k_BT) - \exp(ze\psi/2k_BT)]^2$; therefore Equation (55) may be written as

$$(d\psi/dx)^2 = (2k_BTn_\infty/\varepsilon)[\exp(-ze\psi/2k_BT) - \exp(ze\psi/2k_BT)]^2 \tag{56}$$

Identifying $(ze\psi/k_BT)$ as y permits the simplification of notation to

$$(dy/dx) = (2e^2z^2n_\infty/\varepsilon k_BT)^{1/2}[\exp(-y/2) - \exp(y/2)] \tag{57}$$
$$= \kappa[\exp(-y/2) - \exp(y/2)]$$

This last result may be written in an integrable form by defining u as $e^{y/2}$, in which case $dy = 2e^{-y/2}du$, and the following relationships hold:

$$\frac{dy}{e^{-y/2} - e^{y/2}} = \frac{2\,du}{e^{y/2}(e^{-y/2} - e^{y/2})} = \frac{2\,du}{1 - e^y} = \frac{2\,du}{1 - u^2} = \frac{du}{1 + u} + \frac{du}{1 - u} \tag{58}$$

Combining Equations (57) and (58) gives

$$\frac{du}{1 + u} + \frac{du}{1 - u} = \kappa\,dx \tag{59}$$

which is easily integrated to yield

$$\ln\left(\frac{1 + u}{1 - u}\right) = \kappa x + \text{const.} \tag{60}$$

The integration constant is evaluated from the fact that $\psi = \psi_0$, $y = y_0$, and $u = u_0$ at $x = 0$; therefore

$$\ln\left(\frac{(1 + u)(1 - u_0)}{(1 - u)(1 + u_0)}\right) = \kappa x \tag{61}$$

In terms of the physical variables, Equation (61) may be written

$$\ln\{[\exp(ze\psi/2k_BT) + 1][\exp(ze\psi_0/2k_BT) - 1]/\{[\exp(ze\psi/2k_BT) - 1]$$
$$[\exp(ze\psi_0/2k_BT) + 1]\}\} = \kappa x \tag{62}$$

Equation (62) describes the variation in potential with distance from the surface for a diffuse double layer without the simplifying assumption of low potentials. It is obviously far less easy to gain a "feeling" for this relationship than for the low-potential case. Anticipation of this fact is why so much attention was devoted to the Debye-Hückel approximation in the first place. Note that Equation (62) may be written

$$[\exp(ze\psi/2k_BT) - 1]/[\exp(ze\psi/2k_BT) + 1]$$
$$= \{[\exp(ze\psi_0/2k_BT) - 1]/[\exp(ze\psi_0/2k_BT) + 1]\}\exp(-\kappa x) \tag{63}$$

Equation (63) is the Gouy-Chapman expression for the variation of potential within the double layer. For simplicity, Equation (63) may be written

$$\Upsilon = \Upsilon_0 \exp(-\kappa x) \tag{64}$$

where Υ is defined by the relationship

$$\Upsilon = [\exp(ze\psi/2k_BT) - 1]/[\exp(ze\psi/2k_BT) + 1] \tag{65}$$

and Υ_0 is equal to Υ evaluated with $\psi = \psi_0$. Equation (64) shows that it is the ratio Υ that varies exponentially with x in the Gouy-Chapman theory rather than ψ, as is the case in the Debye-Hückel approximation. Some values of Υ_0 calculated for a variety of ψ_0 values are listed in Table 11.4.

As a check on the consistency of our mathematics, it is profitable to verify that Equation (63) reduces to Equation (37) in the limit of low potentials. Expanding the exponentials in Υ and truncating the series so that only one term survives in both the numerator and denominator results in the Debye-Hückel expression, Equation (37).

Another situation of interest in which Equation (63) simplifies considerably is the case of large values of x at which ψ has fallen to a small value regardless of its initial value. Under these conditions the exponentials of the left-hand side are expanded to give

$$ze\psi/4k_BT = \Upsilon_0 \exp(-\kappa x) \tag{66}$$

or

$$\psi = (4k_BT/ze)\, \Upsilon_0 \exp(-\kappa x) \tag{67}$$

For very large values of ψ_0, $\Upsilon_0 \to 1$. In this case, Equation (67) becomes

$$\psi = (4k_BT/ze) \exp(-\kappa x) \tag{68}$$

which shows that the potential in the outer (i.e., well removed from the wall) portion of the diffuse double layer is independent of the potential at the wall for larger potentials. In colloidal dispersions $(ze\psi_0/k_BT)$ is generally greater than unity, but not too much greater. This means that approximations represented by Equations (37) and (67) will generally bracket the true potential-versus-distance curve given by Equation (63). The situation is shown graphically in Figure 11.6. In drawing these curves, we chose a value of ψ_0 equal to 77.1 mV and arbitrarily selected the electrolyte to be a 0.01 M solution of a 1 : 1 electrolyte for which κ^{-1} is 3.04 nm. Equation (68) is a poor approximation in this case because ψ_0 is not large enough. Values of the abscissa are readily converted into dimensionless variables that apply to any solution by dividing the x coordinate in nanometers by 3.04 nm.

We conclude this section by considering the expression for charge density, Equation (42), as it applies in the Gouy-Chapman model. As we saw in the preceding section, the charge density expression integrates to Equation (45) with no assumptions as to the nature of the potential function. Accordingly, we may combine Equations (45) and (56) to obtain

$$\sigma^* = \varepsilon(2k_BTn_\infty/\varepsilon)^{1/2}[\exp(ze\psi_0/2k_BT) - \exp(-ze\psi_0/2k_BT)] \tag{69}$$

TABLE 11.4 Variation of the Parameter Υ_0 with ψ_0 at 25°C

ψ_0 (mV)	Υ_0
260	0.9874
240	0.9814
220	0.9727
200	0.9600
180	0.9415
160	0.9149
140	0.8765
120	0.8230
100	0.7500
80	0.6528
60	0.5249
40	0.3711
20	0.1968

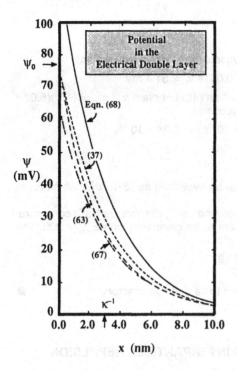

$\Psi_0 \rightarrow$

Potential
in the
Electrical Double Layer

Eqn. (68)

(37)

(63)

(67)

κ^{-1}

Ψ
(mV)

x (nm)

FIG. 11.6 Variation of the double-layer potential versus distance from the surface according to four expressions from this chapter. The figures are for $\psi_0 = 77.1$ mV and $\kappa = 3.29 \cdot 10^8$ m^{-1} (or 0.01 M solution of 1 : 1 electrolyte). Curves are drawn according to Equations (37), (63), (67), and (68).

Equation (69), known as the *Grahame equation*, describes the variation of charge density with potential at the surface with no limitations as to the value of the potential. Example 11.2 considers an application of this relationship.

* * * *

EXAMPLE 11.2 *Relation Between Surface Charge Density and Surface Potential.* Show that for a 1 : 1 electrolyte in water at 25°C, Equation (69) can be rearranged to give

$$\psi_0 = 51.4 \sinh^{-1}[137/(\sigma^0 \sqrt{c})]$$

in which ψ_0 is expressed in millivolts, c is in moles per liter, and σ^0 is the area (in Å^2) per charge at the surface. Davies (1951) measured the potential across an air-aqueous NaCl interface that carried a monolayer of $C_{18}H_{37}N(CH_3)_3{}^+$. When the quaternary octadecyl-amine was at a pressure corresponding to $\sigma^0 = 85$ Å^2, the following potentials were measured at different concentrations of NaCl (data from Davies 1951):

E_{obs} (mV)	240	280	325	340	380
c_{NaCl} (M)	2.0	0.5	0.1	0.033	0.01

About 200 mV of these potential differences arises from dipole effects at the interface and should be subtracted from each value to give the double-layer contribution to the measured potentials. Compare these corrected values with the values of ψ_0 calculated by the equation given.

Solution: First recognize that $2 \sinh x = e^x - e^{-x}$; therefore Equation (69) can be written

$$\sigma^* = (8\varepsilon k_B T n_\infty)^{1/2} \sinh(z e \psi_0 / 2 k_B T)$$

or

$$\sinh^{-1}[\sigma^*(8\varepsilon k_B T n_\infty)^{-1/2}] = z e \psi_0 / 2 k_B T$$

For a 1 : 1 electrolyte, this yields

$$\psi_0 = (2k_B T/e) \sinh^{-1}[\sigma^*(8\varepsilon k_B T n_\infty)^{-1/2}]$$

To assure proper units, we assemble the following substitutions for this expression:

$$(2k_B T/e) = 2(1.38 \cdot 10^{-23})(298)/1.60 \cdot 10^{-19} = 0.0514 \, V = 51.4 \, mV$$

which gives the desired coefficient. Next we use Equation (40) to obtain n_∞; $n_\infty = (10^3)(6.02 \cdot 10^{23})c = 6.02 \cdot 10^{26}c$. With this the factor, $8\varepsilon k_B T n_\infty$ becomes

$$8(78.54)(8.85 \cdot 10^{-12})(1.38 \cdot 10^{-23})(298)(6.02 \cdot 10^{26}c) = 1.38 \cdot 10^{-2}c$$

Finally, σ^* is related to σ^0 as follows:

$$\sigma^* = 1.60 \cdot 10^{-19}/\sigma^0(10^{-10})^2 = 16/\sigma^0$$

From these components the argument of the sinh can be evaluated as $137/\sigma^0 c^{1/2}$, which is the desired result.

By substituting the value of σ^0 into this expression and using the various NaCl concentrations given, ψ_0 values are readily calculated; these are to be compared with $E_{obs} - 200$. The following values are obtained:

ψ_0 (mV)	50.4	80.2	119.8	148.0	178.9
($E_{obs} - 200$) (mV)	40	80	125	140	180

The agreement between theory and experiment is seen to be quite satisfactory. ∎

* * *

11.7   OVERLAPPING DOUBLE LAYERS AND INTERPARTICLE REPULSION

From the viewpoint of the stability of lyophobic colloids discussed in Chapter 13, this section (and the developments based on this in Chapter 13) is of central importance. In this section we examine the force per unit area—that is, the pressure—that operates on two charged surfaces as a result of the overlapping of their double layers. As we saw in Chapter 10, it is more convenient to compare attraction and repulsion between particles in terms of potential energy rather than force. Therefore we see how to express double-layer repulsion in terms of potential energy; this will be developed further in Chapter 13. As was the case with a single double layer, it is easier to treat overlapping double layers if the potential is low. Accordingly, the detailed derivation we consider will assume this condition. We generalize to the case of higher potentials below, but without presenting all the mathematical details of that situation. Following this, we show how the results obtained for interaction forces between flat surfaces can be used to obtain forces between curved double layers. Since the basic approach is more easily demonstrated for spherical surfaces, the results will be given for interaction between spherical particles only.

11.7a   Repulsion Between Planar Double Layers

Figure 11.7 is a schematic representation of the situation with which we are concerned. It shows two plates of unspecified thickness; the planar faces of these plates are separated by a distance h. The plates are immersed in an infinite reservoir of electrolyte, the bulk concentration of which is n_∞. The potential at the surface of the plates is defined as ψ_0. It will be convenient to distinguish between the inner and outer regions of the solution. By the inner region, we refer to the region between the plates, and by the outer region we mean the region influenced by only one of the faces.

When the distance between the plates is large, the potential on both the inner and outer faces will drop with distance from the surface according to Equation (63) or one of its approximations. The profile of the potential drop in this case is shown by the solid curve in the outer region in Figure 11.7 and by the dashed line in the inner region. As the distance between the plates decreases, the potential from each of the inner faces begins to overlap in the inner region. Therefore the net potential in the inner region varies as the solid line in the figure for this region. The potential in the outer region is unaffected by the separation of the

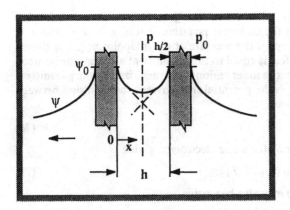

FIG. 11.7 Schematic representation of the overlap of two double layers when a pair of plates is brought to a surface separation h.

plates. Since we already know how to handle the potential in the outer region, our interest now focuses on the potential in the inner region in which double layer overlap occurs.

11.7a.1 Repulsive Force Between Interacting Surfaces

At equilibrium all forces on any volume element of a solution must balance. Suppose we apply this equilibrium criterion to a volume element of solution that lies in the plane parallel to the face of the plates in Figure 11.7 and lies a distance x from the face of one of these plates. Two kinds of force must operate on this volume element: osmotic pressure and electrostatic forces. According to Equations (4.25) and (4.27), the x component of the pressure force acting on the volume element (i.e., per unit volume) is given by

$$F_x = -dp/dx \tag{70}$$

The electrical force per unit volume is given by the product of the charge density times the field strength according to Equation (6):

$$F_{el} = -\rho^*(d\psi/dx) \tag{71}$$

Combining these results with the criterion for equilibrium leads to the expression

$$(dp/dx) + \rho^*(d\psi/dx) = 0 \tag{72}$$

When Equation (26) is substituted for ρ^*, this becomes

$$(dp/dx) - \varepsilon(d^2\psi/dx^2)(d\psi/dx) = 0 \tag{73}$$

Since

$$(d^2\psi/dx^2)(d\psi/dx) = (1/2)(d/dx)(d\psi/dx)^2$$

Equation (73) can be written

$$(d/dx)[p - (\varepsilon/2)(d\psi/dx)^2] = 0 \tag{74}$$

This result shows that the condition for equilibrium is equivalent to requiring that

$$[p - (\varepsilon/2)(d\psi/dx)^2] = \text{const.} \tag{75}$$

at all locations in the solution. Equation (75) shows that two influences—the pressure and the electric field—operate within the solution. The electric field contribution from $(d\psi/dx)^2$ is known as the *Maxwell pressure* (or, *Maxwell stress*). Equation (75) also shows that the difference between the two contributions is a constant that still remains to be evaluated. We are specifically interested in evaluating this constant in the inner region. We proceed in two steps: (i) first, we obtain p in terms of ψ, and (ii) then, we replace ψ as a function of the distance between the plates.

At this point, the symmetry of the situation shown in Figure 11.7 becomes helpful. We realize that ψ goes through a minimum at the midpoint position; that is, $d\psi/dx = 0$ at $x = h/2$. Thus the constant in Equation (75) equals the pressure at the midpoint $p_{h/2}$. The difference between the pressure and the field effect is equal to this quantity at all locations between the plates. Because of this constancy, the entire inner region is characterized by the parameters that apply at the midpoint. Therefore $\psi_{h/2}$ is the potential that governs the repulsion between the surfaces. Next, we write Equation (72) as

$$dp = -\rho^* d\psi \tag{76}$$

Now Equation (28) may be substituted for ρ^* for a $z:z$ electrolyte, giving

$$dp = -zen_\infty[\exp(-ze\psi/k_BT) - \exp(ze\psi/k_BT)]d\psi \tag{77}$$

Since $e^x - e^{-x} = 2\sinh x$, Equation (77) may also be written

$$dp = 2zen_\infty \sinh(ze\psi/k_BT)\,d\psi \tag{78}$$

Equation (78) is easily integrated between the following limits: $p = p_0$ (the outer reference pressure) at $\psi = 0$ and $p = p_{h/2}$ at $\psi = \psi_{h/2}$. Integration of Equation (78) gives

$$p_{h/2} - p_0 = 2k_BTn_\infty[\cosh(ze\psi_{h/2}/k_BT) - 1] = F_R \tag{79}$$

Equation (79) gives the excess pressure at $x = h/2$, and therefore the force per unit area with which the plates are pushed apart, F_R.

This analysis is one of several treatments of double-layer repulsion presented in Verwey and Overbeek's classic book cited at the end of this chapter. The reader will find the topics of this chapter developed in great detail in that source.

Although Equation (79) is correct, it is not particularly helpful. The problem is that F_R is expressed in terms of the potential at the midpoint, which is itself an unknown quantity. For the special case in which $h/2$ is relatively large, the approximation given by Equation (67) may be applied to the potential from each of the two approaching surfaces. The potential at the midpoint then becomes

$$\psi_{h/2} \approx \psi_1 + \psi_2 \approx 2(4k_BT\Upsilon_0/ze)\exp(-\kappa h/2) \tag{80}$$

according to this approximation. This approximation if often called the *superposition approximation* and is clearly admissible only when the surfaces are sufficiently far apart so that the overlap of the double layers is moderate. Since this result applies well away from the surface, the potential is low when Equation (80) holds. Therefore $\cosh(ze\psi_{h/2}/k_BT)$ in Equation (79) may be expanded as a power series (see Appendix A), with only the leading terms retained. This leads to the result

$$F_R \approx k_BTn_\infty(ze\psi_{h/2}/k_BT)^2 = k_BTn_\infty[8\Upsilon_0\exp(-\kappa h/2)]^2 \tag{81}$$

or

$$F_R \approx 64k_BTn_\infty\Upsilon_0^2\exp(-h/\kappa^{-1}) \tag{82}$$

An issue of practical importance is how the force of repulsion described by Equation (79) and its approximation Equation (82) varies with the electrolyte content of a solution. Since κ varies with $n_\infty^{1/2}$, Equation (82) is of the form

$$F_R = C_1 n_\infty \exp(-C_2 n_\infty^{1/2}) \tag{83}$$

where C_1 is a constant and C_2 remains fixed for a given distance of separation h. The exponential factor is clearly the more sensitive involvement of n_∞ in Equation (83). Therefore this expression shows that the force of repulsion decreases with increasing electrolyte concentration between two surfaces compared at the same separation, at least at relatively large separation. The addition of indifferent electrolyte to an aqueous dispersion of a lyophobic colloid may induce the coagulation of that colloid. Equation (83) is therefore an important step toward understanding this behavior. One interesting system to which these ideas have been applied is

ELECTRICAL DOUBLE LAYER AND ITS INTERACTIONS 523

the case of liquid films. An aqueous soap bubble, for example, reaches an equilibrium thickness at which the various forces acting on it balance out. Such forces are readily enumerated:

1. Gas pressure tends to squeeze the two faces of the film together.
2. Van der Waals attraction between the two gas masses is transmitted across the liquid surface.
3. Amphipathic surfactant ions adsorb at the liquid surface — tails out — so counterions form diffuse double layers that can overlap in a manner resembling Figure 11.7.
4. The equilibrium thickness of most bubbles is so much less than the radius of curvature of the bubble that the air masses can be regarded as blocks with planar faces like those in our models. In research studies on such systems, the bubbles are allowed to equilibrate on frames that make their compliance with the model even better.

In Example 11.3 we discuss the results of such a study with soap bubbles.

* * *

EXAMPLE 11.3 *Influence of Electrostatic Repulsion on the Thickness of Soap Bubbles.* Lyklema and Mysels (1965) measured the equilibrium thickness of a soap bubble as 73.1 nm when the bubble was stabilized by $8.7 \cdot 10^{-4}$ M sodium dodecyl sulfate and the hydrostatic pressure on the surface of the film was 66 N m^{-2}. In this experiment the thickness satisfies the condition of equilibrium between the hydrostatic pressure and the force of repulsion between double layers on the adjacent faces of the film, as given by Equation (82). Assuming that the adsorption of dodecyl sulfate ions at the air-solution interface gives a very high value of ψ_0, calculate the equilibrium bubble thickness predicted by this model. Criticize or defend the following propositions: If an appreciable concentration of indifferent electrolyte is added to the soap solution, the calculation just given is no longer feasible because (a) Equation (82) ceases to be valid for F_R and (b) van der Waals attraction between the two air masses also promotes film thinning.

Solution: The strategy here is to neglect van der Waals forces and to solve Equation (82) for h when $F_R = 66$ N m^{-2}. If ψ_0 is large enough, $\Upsilon_0 = 1$; furthermore, by Equation (40), $n_\infty = 1000$ $M N_A = 5.24 \cdot 10^{23}$ liter^{-1}. Assuming the bulk concentration is undiminished by adsorption, we obtain $\kappa = 3.29 \cdot 10^9 \cdot M^{1/2} = 9.70 \cdot 10^7$ m^{-1}. Therefore $66 = 64(5.24 \cdot 10^{23})(1.38 \cdot 10^{-23})(298) \exp[-(9.70 \cdot 10^7)h]$, from which we find $h = 7.88 \cdot 10^{-8}$ m $= 78.8$ nm. Considering that this calculation is insensitive to the actual potential at the surface, the agreement between this quantity and the experimental film thickness is acceptable.

The quantity F_R continues to be valid in the presence of indifferent electrolyte as long as the electrolyte is a z:z type of compound. As a matter of fact, added electrolyte helps meet the requirement that $\psi_{h/2}$ be low, which actually improves the applicability of Equation (82).

Adding electrolyte will cause the film to thin by shortening the range of the repulsive force. At smaller distances of separation the van der Waals attraction is definitely increased and should be added (as force area^{-2}) to the pressure before attempting this kind of calculation. Even without added electrolyte, the air masses attract each other; neglecting this is another possible source of discrepancy between theory and experiment in this example. ∎

* * *

11.7a.2 Energy of Repulsion Between Planar Double Layers

We discuss the topic of colloid stability and the role of electrostatic repulsion in imparting stability to a dispersion in Chapter 13. In order to evaluate fully the effect of electric charge on the stability of a colloidal dispersion, it is necessary to compare the magnitude of the repulsion between particles with the magnitude of the attraction between them. The attraction is most readily described in terms of potential energy; therefore the repulsion should be expressed in this form as well. For the approximation we have just discussed, this is not particularly difficult to evaluate. Since potential energy is given by the force times the distance through which it operates, we may write

$$d\Phi_R = -F_R dh \tag{84}$$

In this expression, $d\Phi_R$ is the increment in potential energy arising from a change in the separation. The minus sign arises from the fact that the potential energy decreases with increasing separation.

Substituting the approximation given by Equation (82) for F_R gives

$$d\Phi_R = -64k_B T n_\infty \Upsilon_0^2 \exp\left(-h/\kappa^{-1}\right)dh \tag{85}$$

a result that is easily integrated by recalling that $\Phi_R = 0$, when $h = \infty$. Integration yields

$$\Phi_R = 64k_B T n_\infty \kappa^{-1} \Upsilon_0^2 \exp\left(-h/\kappa^{-1}\right) \tag{86}$$

This particular form is limited in applicability to situations in which the separation of the surfaces is large compared to κ^{-1} and ψ_0 is large so that $\Upsilon_0 \approx 1$. As we did with the force of repulsion, we may write Φ_R as

$$\Phi_R = C_3 n_\infty^{1/2} \exp\left(-C_2 n_\infty^{1/2}\right) \tag{87}$$

since κ varies with $n_\infty^{1/2}$. In Equation (87), C_2 is the same as the one in Equation (83), but C_3 is a different constant that is proportional to C_1. Again we see that at large separations the potential energy of repulsion decreases with increasing electrolyte concentration. The continued emphasis on large separations is formally imposed by the use of approximation represented by Equation (67) in the development of this result. There are practical reasons for interest in this limit also. As colloidal particles approach one another, it is the outermost portions of their double layers that first interact. The outcome of such an encounter, then, is influenced by the interaction between the particles at large separations.

The derivation of Equation (86) is possible only because relatively simple approximations are available that permit $\psi_{h/2}$ to be solved explicitly and generate an integrable expression for Φ_R. This is not generally the case, however, so the approach used to reach Equation (86) is not applicable to most situations. Verwey and Overbeek have found another method for evaluating Φ_R, but the mathematics are tedious by this approach, involving numerical integrations. The method consists of calculating the free energy difference between particles separated by a distance h and infinitely separated. The interested reader will find details of this method discussed by Verwey and Overbeek in their 1948 book. As far as we are concerned, it is sufficient to note the following conclusions from the general theory:

1. A potential energy of repulsion may extend appreciable distances from surfaces, but its range is compressed (i.e., reduced) by increasing the electrolyte content of the system.

2. The conditions under which approaching particles first influence one another are at large distances of separation, for which the approximate relationship given by Equation (86) holds.

3. The sensitivity of aqueous lyophobic colloids to electrolyte content is due to the dependence of interparticle repulsion on this concentration.

What makes these generalizations significant is their relation to the discussion of the potential energy of attraction as developed in the last chapter. Item 1 means that, at least under some conditions, double-layer repulsion competes with van der Waals attraction in both magnitude and range to govern particle interactions. Item 2, we shall see, is similar to the attitude we take in discussing steric stabilization in Section 13.7, namely, that the interactions the particles experience at their first encounter are the easiest to handle and may determine their subsequent behavior. Item 3 indicates that the location and shape of the repulsion curves in Figure 10.1 are governed by the electrolyte concentration in the system. There are two aspects to this last point. The concentration of the potential-determining ions determines the potential at the wall via Equation (1), and the indifferent electrolyte content determines κ^{-1}, which in turn measures the range over which Φ_R decays. As we discussed in Section 10.2, the net potential energy curve—the result of attractive and repulsive components—is a convenient way to analyze coagulation phenomena. In Chapter 13 we combine the various results from this chapter and Chapter 10 to arrive at a quantitative theory for electrostatic stabilization against coagulation. The resulting theory is generally called the *DLVO theory* after B. Derjaguin, L. D. Landau, E. J. W. Verwey, and J. Th. G. Overbeek, who initially brought the diverse elements of this theory together.

11.7b Interaction Between Spherical Double Layers: The Derjaguin Approximation

As we mentioned in our discussion of the planar double layers, solving the Poisson-Boltzmann equation for the potential variation in curved double layers is complex and requires numerical solutions in the general case. One can see then that the same holds in the case of interacting double layers as well. In addition, the force of interaction between curved surfaces cannot be expressed in terms of the osmotic pressure at the midpoint of separation, as we did in Section 11.7a. This is because electrical stresses (i.e., the Maxwell stresses) arising from the derivative of ψ do not become zero as they did in the case of planar double-layer interaction (see Equation (75)). However, the solution we obtained for interacting planar double layers can be used in combination with what is known as the *Derjaguin approximation* to obtain the potential energy of interaction between two spherical particles. The basic geometric idea behind the Derjaguin approximation is that the surface of the spheres (or curved surfaces, in general) may be divided into a collection of small stepped surfaces of plane geometry. These infinitesimal planes will become ring-shaped flat plates in the case of spherical particles. The equations for plane geometry we developed above can then be used for determining the interactions between corresponding rings on the two spheres, and these can be integrated over all the rings. This is illustrated in Example 11.4.

* * *

EXAMPLE 11.4 *Interaction Between Spherical Particles: The Use of the Derjaguin Approxima-tion.* Spherical particles can be approximated by a stack of circular rings with planar faces as shown in Figure 11.8. Use Equation (86) to describe the repulsion between rings separated by a distance z and derive an expression for the repulsion between the two spheres of equal radius R_s. Assume that the strongest interaction occurs along the line of centers and make any approximations consistent with this to obtain the final result.

Solution: Indexing the various tiers of rings by i, we note that the increment of the interaction due to the ith ring is $d\Phi_R = \Phi_i dA_i$, with the area of the ring $2\pi h_i dh$. We eliminate dh as follows: The separation of the ith ring z_i is related to the distance of closest approach by the formula $(1/2)(z_i - s) = [R_s - (R_s^2 - h^2)^{1/2}]$, in which the factor $(1/2)$ arises because part of the effect occurs at each surface. Since R_s is a constant, this last result may be differentiated and rearranged to give $[R_s(1 - h^2/R_s^2)^{1/2}]dz = 2hdh$. This may be combined with the expression for

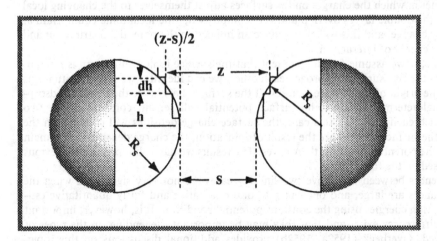

FIG. 11.8 Schematic illustration that shows how the repulsion between spheres may be calculated from the interaction between flat plates.

$d\Phi_R$ to yield $d\Phi_R = \pi R_s (1 - h^2/R_s^2)^{1/2} \Phi(z) dz$, where the functional notation has been added to Φ as a reminder that it is different for each ring. Substituting Equation (86) for $\Phi(z)$, we obtain

$$d\Phi_R = \pi R_s (1 - h^2/R_s^2)^{1/2} (64 k_B T n_\infty \kappa^{-1} \Upsilon_0^2) \exp(-\kappa z) dz$$

If we further assume that $h^2/R_s^2 \ll 1$, this simplifies to

$$d\Phi_R = (64 \pi R_s k_B T n_\infty \kappa^{-1} \Upsilon_0^2) \exp(-\kappa z) dz$$

The total potential energy of repulsion is given by integrating over all values of z. This is most readily done by assuming that z varies between s and infinity. This upper limit is justified because the potential function drops off exponentially with distance. Therefore large separations make a negligible contribution to the total. Integrating between these limits, we obtain

$$\Phi_R = (64 \pi R_s k_B T n_\infty \kappa^{-2} \Upsilon_0^2) \exp(-\kappa s)$$

Under various circumstances, it may be justified to replace either the exponential or Υ_0 by their series approximations, thereby simplifying this result still further.   ∎

* * *

The Derjaguin approximation illustrated in the above example is suitable when $\kappa R^* > 10$, that is, when the radius of curvature of the surface, denoted by the radius R^*, is much larger than the thickness of the double layer, denoted by κ^{-1}. (Note that for a spherical particle $R^* = R_s$, the radius of the particle.) Other approaches are required for thick double layers, and Verwey and Overbeek (1948) have tabulated results for this case. The results can be approximated by the following expression when the Debye-Hückel approximation holds:

$$\Phi_R = (4 \pi R_s^2 \varepsilon \psi_0^2) \exp(-\kappa s)/(s + 2R_s) \tag{88}$$

This expression, although seldom useful in the case of aqueous dispersions, is very useful in nonaqueous dispersions, for which the double layer is usually very thick.

11.7c   Interaction Between Double Layers: Other Considerations

The above discussions illustrate that the interactions between overlapping electrical double layers depend on a number of considerations, such as the magnitude of the surface potential, the thickness of the double layer, and the type of electrolyte, among others. Moreover, the expressions that have been obtained here (and others that are available in the literature) depend on additional conditions that are determined by the approximations made in deriving the expressions.

In addition to all these, it is also important to keep in mind that the results depend also on what types of surface equilibrium conditions exist as the double layers interact. For example, when two charged surfaces approach each other, the overlap of the double layers will also affect the manner in which the charges on the surfaces adjust themselves to the changing local conditions. As the double layers overlap and get "compressed," the local ionic equilibrium at the surface may change, and this will clearly have an impact on the potential distribution and on the potential energy of interaction.

For example, if we assume for the moment that the surface ionic equilibrium is reestablished rapidly as the two surfaces approach each other, we can identify two limiting situations depending on the origin of the surface charges. If the surface charges are the result of adsorption of potential-determining ions, the surface potential will remain constant as the two surfaces approach each other. In this case, the surface charge densities will change. On the other hand, if the surface charges are the result of ionization, the charge densities will remain constant while the potentials readjust themselves. The results we have obtained in the previous sections are based on the former.

The differences between the above two limiting cases are not very significant when the separation distances are large, and one can get good qualitative and fairly quantitative estimates of interaction energies using the constant-potential condition. It is, however, important to know that differences exist, and that they may be important, depending on the types of surfaces involved. Overbeek (1952a, 1952b) provides additional discussions on this topic. Clearly, intermediate conditions (referred to as *charge regulation*) are also possible (and are more likely), although very little is known currently about dealing with such cases.

In addition to the above considerations, practical situations often demand that we consider *dissimilar* surfaces (i.e., the potentials or surface charges of the interacting surfaces may differ from each other). In the case of particles, one may also have different sizes. The method of deriving expressions for interaction energies in these cases does not involve any new fundamental concepts. In fact, the expressions we have derived for interactions between identical planar surfaces can be extended using simple graphical and geometrical arguments to interactions between dissimilar surfaces.

A good discussion of this may be found in Hunter (1987, Chapter 6). It is, however, helpful to list some expressions for interactions between dissimilar spherical particles since they are useful in studying heterocoagulation problems. Hogg et al. (1966) and Wiese and Healy (1970) have derived the following equations for spherical particles for low surface potentials and thin double layers:

$$\Phi_R^\psi = N_1\{N_2 f(s) + \ln\left[1 - \exp\left(2\kappa s\right)\right]\} \tag{89}$$

and

$$\Phi_R^\sigma = N_1\{N_2 f(s) - \ln\left[1 - \exp\left(2\kappa s\right)\right]\} \tag{90}$$

with

$$N_1 = \pi\varepsilon R_{s,1} R_{s,2}(\psi_{0,1}^2 + \psi_{0,2}^2)/(R_{s,1} + R_{s,2}) \tag{91}$$

$$N_2 = 2\psi_{0,1}\psi_{0,2}/(\psi_{0,1}^2 + \psi_{0,2}^2) \tag{92}$$

and

$$f(s) = \ln\left(\frac{1 + e^{-\kappa s}}{1 - e^{-\kappa s}}\right) \tag{93}$$

In the above equations, s is the shortest surface-to-surface separation distance, $R_{s,i}$ is the radius of the ith particle, $\psi_{0,i}$ is the corresponding surface potential, and the superscripts of Φ_R indicate whether the particles interact at constant potential (ψ) or constant surface charge (σ). These equations can be reduced to appropriate simpler forms when $R_{s,1} = R_{s,2}$ or $\psi_{0,1} = \psi_{0,2}$. In addition, interactions between a flat surface and a particle (needed, for example, when considering filtration or deposition problems) can also be obtained by taking one of the radii to infinity (plane).

11.8 "NOT-QUITE-INDIFFERENT" ELECTROLYTES: STERN ADSORPTION

At the beginning of this chapter we divided electrolytes into categories of potential determining and indifferent. We saw in Section 11.2 that the potential-determining ions are adsorbed at surfaces and determine the value of ψ_0 according to Equation (1). Throughout our discussion of the diffuse double layer, we have treated indifferent electrolytes as point charges with no chemical uniqueness except for their valence number. While this is a useful simplifying approximation, it underestimates the complexity of the "real world." The assumption that ions have no volume is acceptable for the bulk region of dilute solutions, but real ions cannot be drawn toward charged surfaces without crowding becoming a problem. At the inner edge of the diffuse part of the double layer some sort of saturation limit must be approached.

One way of handling this—according to O. Stern—is to divide the aqueous part of the double layer by a hypothetical boundary known as the *Stern surface*. The Stern surface is situated a distance δ from the actual surface. Figure 11.9 schematically illustrates the way this surface intersects the double layer potential and how it divides the charge density of the double layer.

The Stern surface is drawn through the ions that are assumed to be adsorbed on the charged wall. (This surface is also known as the *inner Helmholtz plane* [IHP], and the surface running parallel to the IHP, through the surface of shear (see Chapter 12) shown in Figure 11.9, is called the *outer Helmholtz plane* [OHP]. Notice that the diffuse part of the ionic cloud beyond the OHP is the diffuse double layer, which is also known as the *Gouy-Chapman*

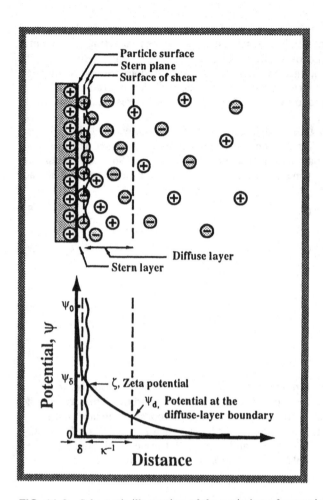

FIG. 11.9 Schematic illustration of the variation of potential with distance from a charged wall in the presence of a Stern layer. See Chapter 12 for discussion of surfaces of shear and zeta potential.

layer as mentioned in Section 11.3.) There are several consequences of assuming an adsorbed layer of ions at the surface:

1. An adsorption isotherm may be written for these ions that allows for surface saturation and thus introduces the idea of finite ionic size. The Langmuir isotherm (Equation (7.67)) is one expression that can be used for this purpose:

$$\theta = Kn_\infty/(1 + Kn_\infty) \tag{94}$$

    In this expression θ is the fraction of surface adsorption sites occupied, n_∞ is the concentration of the adsorbed ions in the solution, and K is a constant.

2. The constant in Equation (94) is easily shown to be proportional to a Boltzmann factor in which the exponential energy consists of two contributions: $ze\psi_\delta$, the electrical energy associated with the ion in the Stern layer, and ϕ, the specific chemical energy associated with the adsorption:

$$K \approx \exp\left[(ze\psi_\delta + \phi)/k_B T\right] \tag{95}$$

3. The Stern layer resembles the parallel plate capacitor model for the double layer. Therefore Equation (13) may be applied to this region:

$$(\psi_0 - \psi_\delta)/\delta = \sigma_\delta^*/\varepsilon_\delta \tag{96}$$

and ε_δ is the product of ε_0 and the relative dielectric constant that applies within the Stern layer.

4. The fraction of surface sites occupied equals the ratio $\sigma_\delta^*/\sigma_{\delta 0}^*$, where $\sigma_{\delta 0}^*$ is the charge density at surface saturation. Therefore, Equations (94) and (96) may be combined to give

$$(\psi_0 - \psi_\delta)/\delta = (1/\varepsilon_\delta)[\sigma_{\delta 0}^* Kn_\infty/(1 + Kn_\infty)] \tag{97}$$

This equation shows that the potential drop in the Stern layer increases with the concentration of the adsorbed ion and ultimately approaches a constant value when the surface is saturated.

Outside the Stern surface the double layer continues to be described by Equation (63) or one of its approximations. The only modifications of the analysis of the diffuse double layer required by the introduction of the Stern surface are that x be measured from δ rather than from the wall and that ψ_δ be used instead of ψ_0 as the potential at the inner boundary of the diffuse layer.

The Stern theory is difficult to apply quantitatively because several of the parameters it introduces into the picture of the double layer cannot be evaluated experimentally. For example, the dielectric constant of the water is probably considerably less in the Stern layer than it would be in bulk because the electric field is exceptionally high in this region. This effect is called *dielectric saturation* and has been measured for macroscopic systems, but it is difficult to know what value of ε_δ applies in the Stern layer. The constant K is also difficult to estimate quantitatively, principally because of the specific chemical interaction energy ϕ. Some calculations have been carried out, however, in which the various parameters in Equation (97) were systematically varied to examine the effect of these variations on the double layer. The following generalizations are based on these calculations:

1. As ϕ increases, K increases and the equilibrium amount adsorbed for any n_∞ value short of saturation also increases.
2. As the electrolyte concentration increases, increasing amounts of the potential drop occur in the Stern layer. This is true even if $\phi = 0$, which shows that specific chemical effects are not necessary for this result.
3. Values of ψ_δ that are much less than ψ_0 are possible in dilute solutions only if ϕ is relatively large.
4. The quantity ψ_δ varies only slightly with ϕ — although it is highly sensitive to n_∞ — until ϕ is relatively large.

Values of the parameter ϕ may be experimentally evaluated for the mercury-water surface from electrocapillary studies. The displacement of the coordinates of the electrocapillary maxima in Figure 7.23 reflects differences in the intrinsic adsorbability of various ions. Electrocapillary studies reveal that the strength of specific adsorption at the mercury-water interface for some monovalent anions follows the order

$$I^- > SCN^- > Br^- > Cl^- > OH^- > F^-$$

whereas for some monovalent cations the order is

$$N(C_2H_5)_4^+ > N(CH_3)_4^+ > Tl^+ > Cs^+ > Na^+$$

In general, the specific adsorption of an ion is enhanced by larger size — and therefore larger polarizability — and lower hydration, which itself is a function of ion size. For example, among the ions just listed, the large I^- ion is the most strongly adsorbed, and the small but highly hydrated Na^+ ion is adsorbed least.

By allowing for surface saturation, the Stern theory overcomes the objection to the Gouy-Chapman theory of excessive surface concentrations. In so doing, however, it trades off one set of difficulties for another. In the Gouy-Chapman theory the functional dependence of ψ on x involves only the parameters κ and ψ_0. The former is known and the latter may be evaluated — at least for some surfaces — by Equation (1) when the point of zero charge is known. The Stern modification of the double-layer picture introduces parameters that are not only difficult to estimate — such as δ and K (or ϕ) — but also specific characteristics of different ions. The generality of the Gouy-Chapman model is thus lost when the specific adsorption effects of the Stern theory are considered.

It is the outer portion of the double layer that interests us most as far as colloidal stability is concerned. The existence of a Stern layer does not invalidate the expressions for the diffuse part of the double layer. As a matter of fact, by lowering the potential at the inner boundary of the diffuse double layer, we enhance the validity of low-potential approximations. The only problem is that specific adsorption effects make it difficult to decide what value to use for ψ_δ.

In subsequent chapters it will be the potential in the diffuse double layer that concerns us. It can be described relative to its value at the inner limit of the diffuse double layer, which may be either the actual surface or the Stern surface. We continue to use the symbol ψ_0 for the potential at this inner limit. It should be remembered, however, that specific adsorption may make this quantity lower than the concentration of potential-determining ions in the solution would indicate. We see in Chapter 12 how the potential at some (unknown) location close to this inner limit can be measured. It is called the *zeta potential*.

REVIEW QUESTIONS

1. List some of the mechanisms by which a surface exposed to a liquid may acquire charges.
2. What is the *Helmholtz double layer*?
3. Explain the *Stern layer* and discuss its relation to the *inner Helmholtz plane*.
4. What is the *outer Helmholtz plane*?
5. What is the *diffuse layer*, and what is its relation to the *Gouy-Chapman theory* of electrical double layers?
6. Why is an electrical double layer called a "double layer"?
7. What is the physical significance of the *Poisson equation*?
8. What are the assumptions that are needed to obtain the *linearized Poisson-Boltzmann* (LPB) equation from the *Poisson-Boltzmann* equation, and under what conditions would you expect the LPB equation to be sufficiently accurate? What is the relation between the *Debye-Hückel approximation* and the LPB equation?
9. In what way does the Gouy-Chapman theory extend the Debye-Hückel approach?
10. Describe the physical significance of the Debye-Hückel parameter κ. How does it vary with the bulk concentration n_∞ for a 1 : 1 electrolyte? How does it vary with ionic charge for a constant bulk concentration?
11. What is the relation between the surface charge density and the potential in a double layer?
12. Why is there a repulsion between two surfaces with like charges in a liquid?
13. What is the so-called Derjaguin approximation?
14. Explain the difference between (electrical) potential and potential energy.
15. Why is it that the force of double-layer interactions for curved surfaces cannot be derived using osmotic pressure arguments as is done in the case of planar double layers?

REFERENCES

General References (with Annotations)

Crow, D. R., *Principles and Applications of Electrochemistry*, 4th ed., Blackie A and P, London, 1994. (Undergraduate level. An introduction to electrochemistry; requires only basic physical chemistry as prerequisite.)

Hunter, R. J., *Zeta Potentials in Colloid Science: Principles and Applications*, Academic Press, London, 1981. (Advanced level. The focus of this book is on the role of electrical double layers and zeta potential on electrophoresis and electroviscous effects. This volume presents some details on electrical double layers around nonspherical particles not discussed in the present book.)

Hunter, R. J., *Foundations of Colloid Science*, Vol. 1, Oxford University Press, Oxford, England, 1987. (Undergraduate level. Chapter 6 of this volume treats electrical double layers and interactions. Some additional details on interactions between dissimilar electrical double layers are presented.)

Israelachvili, J. N., *Intermolecular and Surface Forces*, 2d ed., Academic Press, New York, 1991. (Graduate and undergraduate levels. The objective of this excellent reference is to relate atomic-and molecular-level interactions to surface forces. Chapter 12 discusses electrical double-layer forces and how they can be measured using the surface force apparatus (which we have described in Chapter 1).)

Kruyt, H. R. (Ed.), *Colloid Science, Vol. 1. Irreversible Systems*, Elsevier, Amsterdam, Netherlands, 1952. (A classic reference on colloids. Chapters 4 and 6 by Overbeek, cited below, discuss the electrochemistry of the double layer and double-layer interactions.)

Overbeek, J. Th. G., Electrochemistry of Double Layer. In *Colloid Science*, Vol. 1 (H. R. Kruyt, Ed.), Elsevier, Amsterdam, Netherlands, 1952a, pp. 115–193. (See the annotation under Kruyt 1952.)

Overbeek, J. Th. G., The Interaction Between Colloidal Particles. In *Colloid Science*, Vol. 1 (H. R. Kruyt, Ed.), Elsevier, Amsterdam, Netherlands, 1952b, pp. 245–277. (See the annotation under Kruyt 1952).

Verwey, E. J. W., and Overbeek, J. Th. G., *Theory of the Stability of Lyophobic Colloids*, Elsevier, Amsterdam, Netherlands, 1948. (Another classic reference, by two of the originators of the DLVO theory of colloidal interactions.)

Other References

Davies, J. T., *Proc. R. Soc.*, **208A**, 224 (1951).
Dube, G. P., *Indian J. Phys.*, **17**, 189 (1943).
Goodwin, J. W. (Ed.), *Colloidal Dispersions*, Royal Society of Chemistry, London, 1982.
Hogg, R., Healy, T. W., and Fuerstenau, D. W., *Trans. Faraday Soc.*, **62**, 1638 (1966).
Loeb, A. L., Overbeek, J. Th. G., and Wiersema, P. H., *The Electrical Double Layer Around a Spherical Colloid Particle*, Massachusetts Institute of Technology Press, Cambridge, MA, 1960.
Lyklema, J., and Mysels, K. J., *J. Am. Chem. Soc.*, **87**, 2539 (1965).
Wiese, G. R., and Healy, T. W., *Trans. Faraday Soc.*, **66**, 490 (1970).

PROBLEMS

1. Show that the one-dimensional Poisson equation for planar electrical double layers discussed in the text has the following analog in the case of spherically symmetric double layers:

$$\frac{1}{r^2}\frac{\partial}{\partial r}\left(r^2\frac{\partial \psi}{\partial r}\right) = -\frac{\rho^*}{\varepsilon}$$

Obtain the corresponding Poisson-Boltzmann equation and the linearized version based on the Debye-Hückel approximation.

2. Solve the Poisson-Boltzmann equation for a spherically symmetric double layer surrounding a particle of radius R_s to obtain Equation (38) for the potential distribution in the double layer. Note that the required boundary conditions in this case are: $\psi = \psi_0$ at $r = R_s$, and $\psi \to 0$ as $r \to \infty$. (Hint: Transform $\psi(r)$ to a new function $y(r) = r\psi(r)$ before solving the LPB equation.)

3. Develop the equivalent of Equation (47) for the surface charge density of a spherical particle (i.e., the relation between σ^* and $(d\psi/dr)$ evaluated at the surface).

4. The viscosity of negatively charged colloidal agar (0.14% at 50°C) was studied with a variety of different electrolytes added.* The ratio of the specific viscosity ($\eta_{sp} = \eta/\eta_0 - 1$) in the presence of salt to η_{sp} without salt is given below for some of these salts at several concentrations:

---

*Kruyt, H. R., and deJong, H. G., *Kolloid Z.*, **100**, 250 (1922).

$$(\eta_{sp})_{salt}/(\eta_{sp})_{H_2O}$$

c (m Eq liter^{-1})	KCl	K$_2$SO$_4$	K$_4$Fe(CN)$_6$	BaCl$_2$	SrCl$_2$	MgSO$_4$	La(NO$_3$)$_3$	Pt(en)$_3$(NO$_3$)$_4$[a]
0.25	—	—	—	—	—	—	—	79.2
0.50	90.8	90.7	90.5	—	—	—	76.6	69.7
1.00	—	—	—	78.2	77.9	78.6	—	64.0
2.00	81.2	81.4	81.1	—	—	—	68.1	60.6
4.00	77.3	77.5	76.7	70.6	70.9	71.6	67.2	60.0

[a]en = ethylenediamine.

Discuss this electroviscous effect in terms of the concepts of this chapter and Chapter 4.

5. The deficiency of positive ions ($\Gamma_+ < 0$) adjacent to a positively charged planar surface may be evaluated as follows ($y = ze\psi/k_BT$):

$$\overset{(1)}{\Gamma_+ = \int_0^\infty (c_0 - c_x)dx} = \overset{(2)}{\int_0^\infty (1 - e^{-y})dx} = \overset{(3)}{c_0\int_0^\infty (1 - e^{-y})\frac{dx}{dy}dy} = \overset{(4)}{c_0\int_{\psi_0}^0 \frac{(1 - e^{-y})dy}{dy/dx}}$$

$$\overset{(5)}{= \frac{c_0}{\kappa}\int_{\psi_0}^0 \frac{1 - e^{-y}}{e^{-y/2} - e^{y/2}}dy} = \overset{(6)}{-\frac{c_0}{\kappa}\int_{\psi_0}^0 e^{-y/2}dy} = \overset{(7)}{\frac{2c_0}{\kappa}\left[1 - \exp\left(\frac{ze\psi_0}{2k_BT}\right)\right]}$$

Present the physical and/or mathematical justification for equalities (1)–(7) in this sequence. Show that the final result is equivalent, at high surface potentials, to emptying a region of thickness $2\kappa^{-1}$ of ions possessing the same charge as the wall.

6. If a soap film is sufficiently thin, its equilibrium thickness is the result of the double-layer repulsion, given by Equation (82), and van der Waals attraction, given by

$$F_A = \frac{\partial \Phi_A}{\partial d} = \frac{A}{6\pi}d^{-3}$$

according to Equation (10.63). These conditions are satisfied by certain films studied by Lyklema and Mysels,* who obtained the following results with 1 : 1 electrolyte:

Concentration (mole liter^{-1})	0.103	0.066	0.0197
Thickness of aqueous layer (Å)	91	94	153
Thickness of entire film (Å)	123	126	185

Use the thickness of the aqueous layer in Eq. (82) to calculate F_R per unit area (assume $\Upsilon_0 = 1$). By equating this quantity to F_A per unit area (just given) and using the total film thickness, estimate A for each of these data. Explain why two different values are used for d in this calculation.

7. Using the average value of A you determined in Problem 6, criticize or defend the proposition that van der Waals forces are negligible compared to hydrostatic forces when the hydrostatic forces equal 660 dyne cm^{-2} in a film for which the total thickness is 763 Å. Note that this is the assumption made in Example 11.3. Qualitatively reexamine the last question in light of the results of this problem.

8. Criticize or defend the following propositions: The DLVO theory should apply to particles dispersed in nonaqueous media once ε and A for the solvent have been included in the relevant expressions. Since ion concentrations are low in media with low dielectric constant, κ^{-1} will be very large for such systems. For a concentrated colloid the mean interparticle spacing may be less than κ^{-1}. In such a case it is more plausible to picture a particle approaching a "target" as traveling along a potential energy plateau rather than facing a potential energy barrier.

   Comment on the relevancy of these propositions to the observation† that a 15% (by volume) water-in-benzene emulsion stabilized by the calcium salt of didodecylsalicylic acid has a ψ_0 of ~130 mV, yet breaks immediately after preparation.

---

*Lyklema, J., and Mysels, K. J., *J. Am. Chem. Soc.*, **87**, 2539 (1965).
†Albers, W., and Overbeek, J. Th. G., *J. Colloid Sci.*, **14**, 501 (1959).

9. The interfacial tension at the electrocapillary maximum for several electrolytes in dimethyl-formamide (DMF) solutions has been measured as a function of the electrolyte concentration:*

log c	γ_{max} (dyne cm^{-1})		
	KI	LiCl	KSCN
0	354	366	370
−0.3	356	367	371
−1.0	361	369	373
−1.3	364	370	374

Use these results to estimate the relative adsorbabilities of the I$^-$, Cl$^-$, and SCN$^-$ ions from DMF. How does the sequence of anion adsorbabilities compare with that from aqueous solution as given in Section 11.8? More comprehensive electrocapillary data suggest that SCN$^-$ is more solvated in DMF than in water. Is this consistent with the adsorbability series just compared?

10. A negatively charged AgI dispersion was caused to coagulate by the addition of various electrolytes. The concentrations of several divalent metal nitrates needed to produce coagulation are as follows:†

Salt	$Mg(NO_3)_2$	$Ca(NO_3)_2$	$Sr(NO_3)_2$
$c \times 10^3$ (mole liter^{-1})	2.60	2.40	2.38
Salt	$Ba(NO_3)_2$	$Zn(NO_3)_2$	$Pb(NO_3)_2$
$c \times 10^3$ (mole liter^{-1})	2.26	2.50	2.43

That these different compounds produce the same effect at so nearly the same concentration argues that the principal cause of the effect is electrostatic. Use the average of these concentrations to calculate (a) the value of κ at which this system coagulates, (b) the force of repulsion (Equation (82)), and (c) the potential energy of repulsion (Equation (86)) when two planar surfaces are separated by a distance of 10 nm. For the purpose of calculation in parts (b) and (c), ψ_0 may be taken as 100 mV. Comment on the applicability of these equations to the physical system under consideration.

11. The slight differences in the concentrations of divalent cations required to flocculate the AgI sol of the preceding problem may be attributed to differences in the adsorbability of these cations at the AgI-solution interface. Use the data of Problem 10 to rank the cations with respect to their tendency to adsorb. Is there a correlation between adsorbability and ion size and/or hydration? List any references consulted for data concerning the last two quantities.

12. Once the significance of the midpoint between two parallel plates for the force between those plates is established, there are several ways of arriving at Equation (79). One argument is that the plates shown in Figure 11.7 function as a semipermeable membrane, sustaining a concentration difference between $x = h$ and the outer region of the solution. Use Equations (3.25) and (27) to show that the osmotic pressure across this "semipermeable membrane" is given by Equation (79).

*Bezuglyi, V. D., and Korshikov, L. A., *Electrokhimiya,* **3**, 390 (1967).
†Kruyt, H. R., and Klompe, M. A., *Kolloid Beihefte,* **54**, 484 (1942).

12

Electrophoresis and Other Electrokinetic Phenomena

There is a constant attraction to the South . . . yet the hampering effect of the southward attraction is quite sufficient to serve as a compass in most parts of our earth.

From Abbott's *Flatland*

12.1 INTRODUCTION

The word *electrokinetic* implies the combined effects of motion and electrical phenomena. Specifically, our interest in this chapter centers on those processes in which a relative velocity exists between two parts of the electrical double layer. This may arise from the migration of a particle relative to the continuous phase that surrounds it. Alternatively, it could be the solution phase that moves relative to stationary walls.

12.1a What Are Electrokinetic Phenomena?

There are four phenomena that are normally grouped under the term *electrokinetic phenomena*.

1. *Electrophoresis*: This refers to movement of a particle (and any material attached to the surface of the particle) relative to a stationary liquid under the influence of an applied electric field.
2. *Electroosmosis*: Here, the liquid (an electrolyte solution) moves past a charged surface (e.g., the surface of a capillary tube or through a porous plug) under the influence of an electric field. Thus, electroosmosis is the complement of electrophoresis. The pressure needed to balance the electroosmotic flow is known as *electroosmotic pressure*.
3. *Streaming potential*: This is a consequence of the electric field created when a liquid (an electrolyte) is forced to flow past a charged surface. The situation here is the opposite of electroosmosis.
4. *Sedimentation potential*: This is due to the electric field created by charged particles sedimenting in a liquid. This situation is the opposite of electrophoresis.

The first three electrokinetic processes are our concern in this chapter, with the emphasis on electrophoresis.

12.1b Why Study Electrokinetic Phenomena?

In each case the electrokinetic measurements can be interpreted to yield a quantity known as the *zeta* (ζ) *potential*. It is important to note that this is an experimentally determined potential measured in the double layer near the charged surface. Therefore it is the empirical equivalent to the surface potentials discussed in the Chapter 11. We saw in Chapter 10 how the stability of a hydrophobic colloid depends on the relative magnitude of the potential energies of attraction and repulsion between a pair of particles approaching a collision with each other. Therefore the electrokinetic or ζ potential has a direct bearing on the material of the previous two chapters and the following one as far as the theory and practice of colloid stability are concerned.

In addition to the above, there are a number of phenomena or processes of practical utility that are based on principles of electrokinetic processes. We already saw in Chapter 1 the possibility of developing image displays based on electrophoretic motion (i.e., electrophoretic image displays, Vignette I.4). There are so-called liquid immersion development devices in which the image that needs to be copied is transferred onto a plate or paper as a pattern of charges so that the transferred image could be developed by exposing the plate or paper to a dispersion of appropriately charged dye particles (Lyne and Aspler 1982). The method is actually an *electrophoretic* development process and is highlighted further in Vignette XII. Electroosmosis, for instance, has also been used in a number of practical applications such as dewatering of soils for construction purposes and dewatering of waste sludges and may also be useful for controlling toxic leakage in waste sites through the injection of detoxifying agents (Probstein 1994). We list some more applications in Section 12.11.

VIGNETTE XII ELECTROKINETIC PHENOMENA IN MODERN TECHNOLOGY: Electrophotography

When one thinks of electrokinetic phenomena in the context of a first-level course on colloid and surface chemistry, the first thought that probably comes to mind is the use of such phenomena to measure zeta potentials and charges of colloidal species. But, as we have already seen in Chapter 1 and as we see later in this chapter, electrokinetic phenomena play a significant role in many other applications. We take a look at one such application here and see why the topics we consider in this chapter and in others are important in that context.

Using photoconductive insulating surfaces to produce latent electrostatic images was suggested by Carlson over 50 years ago (Carlson 1938, 1940). The image thus produced is developed using toner particles of colloidal size and is subsequently transferred to paper. This forms the basis of *xerography*, a "dry" copying technique. (A modification of this process in which paper with a photoconductive coating containing a ZnO binder is used directly is known as *electrofax*. It is seldom used today.)

There are a number of advantages for using *liquid-developing* processes instead of "dry" processes: (a) the possibility of high resolution using fine-grain suspensions, (b) minimizing "edge effect" (namely, the partially higher image density caused by stronger electric field in the edge regions of the latent image) through the control of the conductivity of the developer, and (c) the possibility of compact equipment design. The liquid-development process was first proposed by Metcalfe (1955, 1956) and seeks to develop the latent image by *electrophoretic* particle deposition from a colloidal dispersion. Theories of the liquid-development process rest on the electrophoretic mobility equations (e.g., Henry's equation) we derive in this chapter and are presented by Kurita (1961) and Ohyama et al. (1961), among others (see Kitahara and Watanabe 1984).

Let us focus here on *some* of the *general issues* that a colloid scientist should be concerned with in this context: (a) the toner must have stable polarity and stable charges on the particles and must be stable against settling; (b) the liquid must have low ion concentration and a low dielectric constant to avoid "electrical leakage" of the electrostatic image; and (c) color, concentration, and distribution of the pigment particles inside the toner particles must be such that the quality of the developed image is high. In order to guarantee these, information on physicochemical properties of the liquid developer (e.g., viscosity, dielectric constant, etc.), the mechanisms responsible for the charges on the toner particles, and the interaction forces responsible for stabilization (usually steric hindrance and polymer-mediated forces arising from polymer additives) must be examined. These factors also determine the electrophoretic mobility of the particles and are, therefore, important since the electrophoretic mobility, in turn, determines the image density and the developing time (Kitahara and Watanabe 1984, Chapter 17).

12.1c   Focus of This Chapter

The focus of this chapter is to present the basics of electrokinetic phenomena and to provide a working knowledge of the subject. Some classical and some novel applications of these phenomena are also summarized at the end of the chapter.

1.   We begin with the definition of electrophoretic mobility, namely, the velocity of a charged particle per unit field strength (Section 12.2).

2.   Our next task is to relate this mobility to the zeta potential. This requires a number of assumptions, and we focus on the most important of these. We derive the equations for thick electrical double layers (Section 12.3) and for thin double layers (Section 12.4) first and then examine how intermediate cases can be studied (Section 12.5).

3.   Following this we derive the equation for electroosmotic flow and relate it to the zeta potential of the charged surface (Section 12.6). Section 12.7 focuses on the streaming potential and compares the zeta potentials obtained by the different methods.

4.   Although the ζ potential is undoubtedly an important quantity in colloid chemistry, it is not totally free of ambiguity. The problem is this: It is not clear at what location within the double layer the potential is measured. The derivations of this chapter show that the ζ potential is the double-layer potential close to the surface, but the precise quantitative meaning of close cannot be defined. We examine this briefly in Section 12.8.

5.   Following the above, we address some of the experimental aspects of electrophoresis. In this context, a few other forms of electrophoresis (e.g., moving boundary electrophoresis and zone electrophoresis) are described briefly. The last of these is used when separation of charged species, rather than the measurement of mobilities, is the item of interest (Section 12.9).

6.   One of the applications of zeta potential measurements is in the determination of surface charges of colloids such as proteins. We look into this briefly in Section 12.10 and derive the equation connecting the surface charge to the zeta potential under some simplifying assumptions.

7.   Finally, a number of examples of applications of electrokinetic phenomena are described briefly in Section 12.11.

Most textbooks on colloids usually devote no more than a chapter or two to electrokinetic phenomena since so many other topics also need to be covered for a balanced introduction to colloid and surface chemistry. A recent textbook (Masliyah 1994) focuses almost exclusively on electrokinetic phenomena and may be consulted for details not presented here.

12.2   MOBILITIES OF SMALL IONS AND MACROIONS IN ELECTRIC FIELDS: A COMPARISON

The fact that positive ions migrate toward the cathode and negative ions migrate toward the anode is so well known as to be virtually self-evident. It seems equally evident, therefore, that positively and negatively charged colloidal particles should display similar migrations. Indeed, this is the case. Because we are relatively familiar with the conductivity of simple electrolytes, we start our discussion of electrokinetic phenomena with a comparison of the mobilities of the particles in the small ion and macroion size domains.

12.2a Mobility of An Ion in An Electric Field

An isolated ion in an electric field experiences a force directed toward the oppositely charged electrode. This force is given by the product of the charge of the ion q times the electric field E:

$$F_{el} = qE \tag{1}$$

In SI units, E is expressed in volt meter^{-1} and q in coulombs, so F_{el} is correctly given in newtons, since C V = J = N m. The cgs unit system, which is widely encountered in older references, requires dividing the right-hand side of Equation (1) by a factor of about 300 since 299.8 V = 1.0 statvolt. Use of Equation (1) is limited to situations in which the electric field at the ion is due to the applied potential gradient only, undisturbed by the effects of other ions in the solution (i.e., infinite dilution).

   An ion in an electric field thus experiences an acceleration toward the oppositely charged electrode. However, its velocity does not increase without limit. An opposing force due to the viscous resistance of the medium increases as the particle velocity increases:

$$F_{vis} = fv \tag{2}$$

where f is the *friction factor* (see Equation (2.2)). A stationary-state velocity is established quite rapidly in which these two forces are equal:

$$v = qE/f \tag{3}$$

The situation is thus very much like the sedimentation velocity discussed in Chapter 2 in which the gravitational forces on a particle are opposed by viscous resistance.

   As a further development, we may tentatively substitute the value for f given by Stokes's law (Equation (2.8)) to obtain

$$v = qE/(6\pi\eta R_s) \tag{4}$$

where R_s is the radius of the particle, assumed to be a sphere by this substitution. The charge of a simple ion can be written as the product of its valence z times the electron charge e:

$$q = ze \tag{5}$$

Substitution of this result into Equation (4) yields

$$v = zeE/(6\pi\eta R_s) \tag{6}$$

The *velocity per unit field* is defined as the *mobility u* of the ion:

$$u = v/E \tag{7}$$

For simple ions mobilities are typically on the order of 10^{-8} m s^{-1}/V m^{-1} (m^2 V^{-1} s^{-1}). It is shown in physical chemistry (see, for example, Atkins 1994) that the mobility of an ion is directly proportional to its equivalent conductance λ_{i0}:

$$u_i = \lambda_{i0}/\mathfrak{F} \tag{8}$$

where $\mathfrak{F}$ is the *Faraday constant*. We have stipulated the conductance at infinite dilution (subscript 0) as a reminder that these relationships all refer to isolated ions. When ion mobilities are analyzed by Equation (6), quite reasonable values for the radii of the hydrated ions are obtained.

12.2b Electrophoretic Mobility of Macroions

The success and relative simplicity of conductivity as a method of study for small ions prompt us to extend these ideas to particles in the colloidal size range. For the purpose of our discussion here, we can treat a charged colloidal particle as an ion of large charge, hence the name *macroion*. However, we identify shortly some of the differences between such macroions and small ions with respect to their response to an applied electric field. For certain colloids the experimental aspects of studying mobilities are simpler than for small ions because of the

possibility of measuring the velocity of high-contrast particles by direct microscopic observation. If the velocity and the field responsible for the migration are known, the mobility of the
colloid may be evaluated directly from Equation (7). When the term is applied to colloidal
particles, the mobility is known specifically as the *electrophoretic mobility*. In this case the
overall phenomenon is known as *electrophoresis*, and the specific experimental technique of
direct microscopic observation of the electrophoretic mobility is called *microelectrophoresis*.
This and other electrophoretic techniques are described in more detail in Section 12.9.

Although the electrophoretic mobility is—at least in some cases—a readily measured
quantity, its interpretation is considerably more difficult for colloidal particles than for small
ions. First, we realize that the charge carried by a colloidal particle is not a constant known
quantity as is the case for simple ions. This prevents us from using Equation (6) to evaluate
R_s, but suggests instead a method by which the charge might be determined. Suppose, for
example, we substitute Equation (2.32) for f rather than use Stokes's law for this quantity.
Then the electrophoretic mobility is given by

$$u = ze/(k_BT/D) = zeD/(k_BT) \tag{9}$$

It appears that the combination of electrophoresis and diffusion experiments would allow
for the evaluation of the charge carried by the macroion. Again, the situation is reminiscent of
the procedures described in Chapter 2 in which sedimentation and diffusion experiments were
combined. However, this is only the beginning of the difficulty. The validity of Equation (9)
is limited to the situation in which a charged particle is considered in isolation from other ions.
A charged colloid will be surrounded by an electrical double layer, as we saw in Chapter 11.
Thus the field at the particle is modified by the potential of the double layer; that is, the
migrating unit is the charged colloidal particle along with its electrical double layer just as the
same composite is the kinetic unit in coagulation.

Therefore this strategy for determining the charge of a colloid from electrophoresis measurements is invalid except for the rather special case of determining the conditions of zero
charge for the colloid. We return to a discussion of this point in Section 12.10.

In Chapter 11 we discussed the structure of the double layer in terms of the potential at
the surface. This background plus the realization that the ion atmosphere also contributes to
the electrophoretic mobility of a colloid suggests that potential rather than charge is the more
useful parameter to pursue. This is the topic of the following section. In discussing the
migration of charged colloidal particles through a solution containing small ions, it is convenient to begin by distinguishing between two extremes of particle size. We saw in Chapter 11
that the parameter κ^{-1} (see Table 11.3) is a convenient way to characterize the "thickness" of
the ion atmosphere near a surface. Distances are regarded as large or small relative to this
quantity, as discussed in Section 4.8. For simplicity we restrict our consideration to spherical,
nonconducting particles (of radius R_s) and begin by examining the two extremes of very small
and very large particles. These designations acquire specific meaning when compared to κ^{-1},
taken as a standard length. Thus the two cases we consider first are those in which R_s/κ^{-1} (or
simply κR_s) is either small or large.

Figure 12.1 shows schematically the shape of the flow streamlines around the particle in
the two cases. The dashed line in the figure is displaced from the surface of the spherical
particles by an amount κ^{-1}. In Figure 12.1a, R_s is small (compared to κ^{-1}) and the streamlines
undergo negligible displacement. In Figure 12.1b, on the other hand, the streamlines follow
the contours of the particle nearly tangentially. Since matter is conserved, a flow streamline
that carries matter into a volume element must also carry matter out of that volume element.
Charge is also conserved, so the same number of lines of force must enter and leave a
volume element. Accordingly, the streamlines shown schematically in Figure 12.1 may also be
regarded as describing the electric field in the neighborhood of small and large particles.

12.3   ZETA POTENTIAL: THICK ELECTRICAL DOUBLE LAYERS

We know from Chapter 11 that the potential drops gradually with distance from a charged
surface, its range decreasing with increasing electrolyte content. Most of the expressions

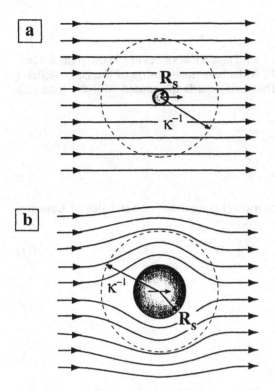

FIG. 12.1 Streamlines (which also represent the electric field) around spherical particles of radius R_s. The dashed lines are displaced from the surface of the spheres by the double-layer thickness κ^{-1}. In (a) κR_s is small; in (b) κR_s is large.

developed in Chapter 11, however, describe the situation adjacent to a planar wall. In the present context we need to know how the potential varies with distance from the surface of a sphere, as discussed briefly in Section 11.4c. We examine this a little closely before proceeding to discuss how the results are used for understanding motion of spherical colloidal particles in electric fields.

12.3a Potential Distribution Around a Spherical Particle

The Poisson equation (see Equation (11.18)) gives the fundamental differential equation for potential as a function of charge density. The Debye-Hückel approximation may be used to express the charge density as a function of potential as in Equation (11.28) if the potential is low. Combining Equations (11.24) and (11.32) gives

$$\frac{1}{r^2}\frac{d}{dr}\left(r^2\frac{d\psi}{dr}\right) = \frac{e^2}{\varepsilon k_B T}\left(\sum_i z_i^2 n_i\right)\psi = \kappa^2\,\psi \tag{10}$$

Remember that ε is equal to the product of ε_0 times ε_r with $\varepsilon_0 = 8.85 \cdot 10^{-12}\,\mathrm{C^2\,J^{-1}\,m^{-1}}$; for water at 25°C, $\varepsilon_r = 78.54$. Because of its importance to the present material, we repeat the defining expression for κ as given originally by Equation (11.41):

$$\kappa = \left(\frac{1000\,e^2 N_A}{\varepsilon k_B T}\sum_i z_i^2 M_i\right)^{1/2} \tag{11}$$

or, in terms of ionic strength $I = (1/2)\Sigma_i z_i^2 M_i$,

$$\kappa = \left(\frac{2000 \, e^2 N_A I}{\varepsilon k_B T} \right)^{1/2} \tag{12}$$

Remember that Table 11.3 contains some useful numerical values of κ at different concentrations of various electrolytes. Equation (10) is the basic relationship of the Debye-Hückel theory and may be integrated as follows. The variable x is introduced with the following definition:

$$x = r\psi \tag{13}$$

Thus Equation (10) may be written

$$\frac{d}{dr} \left(r^2 \frac{d\psi}{dr} \right) = \kappa^2 \, r \, x \tag{14}$$

Now let us consider the incorporation of Equation (13) into the left-hand side of Equation (14):

$$\frac{d\psi}{dr} = \frac{d(x/r)}{dr} = \frac{1}{r} \frac{dx}{dr} - \frac{x}{r^2} \tag{15}$$

and

$$\frac{d}{dr} \left(r^2 \frac{d\psi}{dr} \right) = \frac{d}{dr} \left(r \frac{dx}{dr} - x \right) = r \frac{d^2 x}{dr^2} \tag{16}$$

Combining Equations (14) and (16) gives

$$d^2 x/dr^2 = \kappa^2 x \tag{17}$$

for which a general solution is

$$x = A \exp(-\kappa r) + B \exp(\kappa r) \tag{18}$$

as may be readily verified by differentiation. Replacing x in this equation by its definition in Equation (13) gives

$$\psi = \frac{A \exp(-\kappa r)}{r} + \frac{B \exp(\kappa r)}{r} \tag{19}$$

Since $\psi \to 0$ as $r \to \infty$, it is apparent that $B = 0$.

To evaluate A we proceed as follows. In the limit of infinite dilution — that is, as $\kappa \to 0$ — the potential around the charged particle is given by the expression for the potential of an isolated charge. Elementary physics gives this as

$$\psi = \frac{1}{4\pi\varepsilon} \frac{q}{r} \tag{20}$$

a distance r from a charge q. As $\kappa \to 0$, Equations (19) and (20) must converge; therefore A must equal $(q/4\pi\varepsilon)$. The general expression for potential around a spherical particle at low potential may be written as

$$\psi = \frac{q}{4\pi\varepsilon r} \exp(-\kappa r) \tag{21}$$

Notice that if we use the condition that $\psi = \psi_0$ at $r = R_s$ we will recover the result developed in Section 11.4c for a spherical particle of finite size (given by radius R_s).

As a reminder that the level of approximation in Equation (21) is the same as that of the Debye-Hückel limiting law, the following example continues from this last result to the Debye-Hückel expression for the mean ionic activity coefficient of an electrolyte solution.

* * *

EXAMPLE 12.1 *Debye-Hückel Expression for Ionic Activity Coefficients.* The Debye-Hückel limiting law attributes all of the nonideality of an electrolyte solution to electrostatic effects associated with the diffuse double layer. As a way to isolate this effect, consider the hypothetical process of discharging an ion in a solution of concentration c_1 moving it to a solution of concentration c_2, and then recharging it in the new solution. The individual steps and general expressions for the associated free energy changes are listed below, with the subscripts 1 and 2 indicating the two concentration conditions:

$$q = ze \text{ at } c_1 \rightarrow q = 0 \text{ at } c_1 \qquad \Delta G = N_A \int_{ze}^{0} \psi_1 \, dq$$

$$q = 0 \text{ at } c_1 \rightarrow q = 0 \text{ at } c_2 \qquad \Delta G = RT \ln \, (c_2/c_1)$$

$$q = 0 \text{ at } c_2 \rightarrow q = ze \text{ at } c_2 \qquad \Delta G = N_A \int_{0}^{ze} \psi_2 \, dq$$

Net: $q = ze$ at $c_1 \rightarrow q = ze$ at $c_2 \qquad \Delta G_{net} = RT \ln \, (a_2/a_1)$

where a_i is the activity of component i and can be written as $a_i = \gamma_i c_i$ in terms of the *activity coefficient* γ_i. Derive the Debye-Hückel expression for the activity coefficient γ, assuming Equation (21) describes ψ and that solution 2 is dilute and solution 1 very dilute.

Solution: The activities in the expression for ΔG_{net} can be replaced by the product of concentrations and activity coefficients:

$$\Delta G_{net} = RT \ln \, (\gamma_2 c_2/\gamma_1 c_1) = RT \ln \, (\gamma_2/\gamma_1) + RT \ln \, (c_2/c_1)$$

$$= N_A \int_{ze}^{0} \psi_1 \, dq + RT \ln \, (c_2/c_1) + N_A \int_{0}^{ze} \psi_2 \, dq$$

Since solution 1 is very dilute, γ_1 can be set equal to unity, in which case

$$RT \ln \gamma_2 = N_A \int_{ze}^{0} \psi_1 \, dq + N_A \int_{0}^{ze} \psi_2 \, dq$$

The activity coefficient for (dilute) solution 2 is therefore obtained by evaluating the integrals based on Equation (21). Using a series approximation (see Appendix A) for the exponential in Equation (21) and retaining only the leading term for solution 1, where κ_1 (i.e., c_1) is very small, and the first two terms for solution 2, where κ_2 is small but larger than for 1, we obtain $\psi_1 = q/4\pi\varepsilon r$ and $\psi_2 = (q/4\pi\varepsilon r)(1 - \kappa_2 r)$. With these substitutions the integrals can be evaluated as follows:

$$RT \ln \gamma_2 = (N_A/4\pi\varepsilon r) \left(\int_{ze}^{0} q \, dq + \int_{0}^{ze} q \, dq - \kappa_2 r \int_{0}^{ze} q \, dq \right) = N_A \kappa_2 \, (ze)^2/8\pi\varepsilon$$

Substituting Equation (12) for κ yields

$$\ln \gamma_2 = - (N_A z^2 e^2/8\pi\varepsilon RT) \, (2000 e^2 \, N_A/\varepsilon k_B T)^{1/2} I^{1/2}$$

For aqueous solutions at 25°C this becomes $\log_{10} \gamma_2 = 0.0509 \, z^2 I^{1/2}$, or, with a bit more argumentation, $\log_{10} \gamma_\pm = -0.0509 \, z_+ z_- I^{1/2}$, which is the result sought. ∎

* * *

12.3b   Zeta Potential

Next, let us consider the application of Equation (21) to a particle migrating in an electric field. We recall from Chapter 4 that the layer of liquid immediately adjacent to a particle moves with the same velocity as the surface; that is, whatever the relative velocity between the particle and the fluid may be some distance from the surface, it is zero at the surface. What is not clear is the actual distance from the surface at which the relative motion sets in between the immobilized layer and the mobile fluid. This boundary is known as the *surface of shear*. Although the precise location of the surface of shear is not known, it is presumably within a couple of molecular diameters of the actual particle surface for smooth particles. Ideas about adsorption from solution (e.g., Section 7.7) in general and about the Stern layer (Section 11.8) in particular give a molecular interpretation to the stationary layer and lend plausibility to the statement about its thickness. What is most important here is the realization that the surface of shear occurs well within the double layer, probably at a location roughly equivalent to the Stern surface. Rather than identify the Stern surface as the surface of shear, we define the potential at the surface of shear to be the zeta potential ζ. It is probably fairly close to the

Stern potential ψ_δ in magnitude, and definitely less than the potential at the surface ψ_0. The relative values of these different potentials are shown in Figure 12.2.

Distances within the double layer are considered large or small, depending on their magnitude relative to κ^{-1}. Thus in dilute electrolyte solutions, in which κ^{-1} is large, the surface of shear — which is close to the particle surface even in absolute units — may be safely regarded as coinciding with the surface in units relative to the double-layer thickness. Therefore, in the case for which κ^{-1} is large (or κ small), Equation (21) becomes

$$\zeta = \frac{q}{4\pi\varepsilon R_s} \exp\,(-\kappa R_s) \tag{22}$$

where R_s is the actual radius of the particle.

Since this result applies only when κ is small, the exponential may be expanded (see Appendix A) to give

$$\zeta \approx \frac{q}{4\pi\varepsilon R_s} \frac{1}{\exp\,(\kappa R_s)} \approx \frac{q}{4\pi\varepsilon R_s} \frac{1}{(1\,+\,\kappa R_s)} \tag{23}$$

This result may also be written

$$\zeta = \frac{q}{4\pi\varepsilon R_s} \frac{\kappa^{-1}}{R_s\,+\,\kappa^{-1}} \tag{24}$$

which is the same as

$$\zeta = \frac{q}{4\pi\varepsilon R_s} - \frac{q}{4\pi\varepsilon\,(R_s\,+\,\kappa^{-1})} \tag{25}$$

This last result is interesting because it may be interpreted as the sum of two superimposed potentials: one arising from a charge q on a surface of radius R_s and a second arising from a charge $-q$ on a sphere of radius $(R_s\,+\,\kappa^{-1})$. This is the net potential between two concentric spheres carrying equal but opposite charges and differing in radius by an amount κ^{-1}. Such a

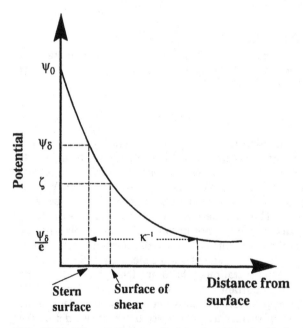

FIG. 12.2 The relative magnitudes of various double-layer potentials of interest.

situation corresponds to a concentric sphere capacitor. As in Chapter 11, we again see the double layer behaving as if it was a capacitor with a characteristic spacing κ^{-1}.

12.3c   Electrophoretic Mobility and the Hückel Equation

Having explored the capacitor analogy, we no longer need to retain the second term in the series expansion of the exponential in Equation (23). For our present purposes it is sufficient to note that for small values of κR_s, Equation (22) becomes

$$\zeta = \frac{q}{4\pi\varepsilon R_s} \tag{26}$$

Solving this result for q and combining with Equations (5)–(7), we obtain

$$u = \frac{2}{3}\frac{\varepsilon\,\zeta}{\eta} \tag{27}$$

The possible usefulness of this relationship—which is known as the *Hückel equation*—should not be overlooked. Throughout Chapter 11 we were concerned with the potential surrounding a charged particle. Equation (11.1) provides a way of evaluating the potential at the surface, ψ_0, in terms of the concentration of potential-determining ions. Owing to ion adsorption in the Stern layer, this may not be the appropriate value to use for the potential at the inner limit of the diffuse double layer. Although ζ is not necessarily identical to ψ_0, it is nevertheless a quantity of considerable interest.

More elaborate theory shows that Equation (27) is valid for spheres when κR_s is less than about 0.1. This imposes a rather severe restriction on the applicability of this result in aqueous systems since for $R_s = 10^{-8}$ m the corresponding concentration is about 10^{-5} M for a 1:1 electrolyte. In nonaqueous media, however, ion concentrations may be very low, and this result assumes increasing importance.

* * *

EXAMPLE 12.2   *Relation Between Electrophoretic Mobility and Zeta Potential.*   In many references the Hückel equation is written $u = \varepsilon_r\zeta/6\pi\eta$. How do you account for the difference between this expression and Equation (27)? What is the ζ potential of a particle that displays a mobility of 10^{-4} cm^2 V^{-1} s^{-1} in water at 20°C for which $\eta = 0.010$ P and $\varepsilon_r = 80.4$, assuming the Hückel potential applies?

Solution: The discrepancy between the equation given here and Equation (27) arises from the fact that the equation above is written for cgs units whereas Equation (27) applies to SI. Remember that $\varepsilon = \varepsilon_r\varepsilon_0$ in Equation (27) and that the vacuum permittivity ε_0 usually appears with the factor 4π. When we combine Equations (6) and (26), the ratio $4\pi/6\pi$ reduces to 2/3. Many electrokinetic formulas differ by a factor 4π, depending on the system of units being used, and the reader is cautioned to be aware of this difference. The presence or absence of the vacuum permittivity in the equation is the key to the system being used, although this factor is sometimes hidden, as in the case of Equation (27).

Either system of units can be used to calculate ζ from the mobility given; in either case some unit conversion must be done on the mobility since the units given are a hybrid of SI and cgs units.

In SI the mobility is given by

$$10^{-4}\ \text{cm}^2\,\text{V}^{-1}\,\text{s}^{-1}\cdot(1\ \text{m}/100\ \text{cm})^2 = 10^{-8}\ \text{m}^2\,\text{V}^{-1}\,\text{s}^{-1}$$

Therefore

$$\zeta = [(10^{-8})\,(3)\,(0.010\ \text{P}\cdot 1\ \text{kg}\ \text{m}^{-1}\,\text{s}^{-1}/10\ \text{P})]/[2\,(80.4)\,(8.85\cdot 10^{-12})]$$
$$= 2.11\cdot 10^{-2}\ \text{V} = 21.1\ \text{mV}$$

In cgs, the mobility is given by

$$10^{-4}\ \text{cm}\ \text{s}^{-1}/\text{V}\ \text{cm}^{-1}\cdot 300\ \text{V/statV} = 3\cdot 10^{-2}\ \text{cm}^2\ \text{statV}^{-1}\,\text{s}^{-1}$$

Therefore

$$\zeta = [6\pi(3\cdot 10^{-2})(0.010)]/80.4 = 7.03\cdot 10^{-5}\ \text{statV}$$

or

$$\zeta = 7.03 \cdot 10^{-5} \text{ statV} \cdot 300 \text{ V/statV} = 2.11 \cdot 10^{-2} \text{ V}$$

Note that the factor 300 V/statvolt enters the calculation twice if cgs units are used.   ■

* * *

The next question to be considered is the relationship between u and ζ when κR_s is not small.

12.4 ZETA POTENTIAL: THIN ELECTRICAL DOUBLE LAYERS

In this section we consider the situation in which the thickness of the double layer is negligible compared to the radius of curvature of the surface. The derivation is not limited to any particle geometry as long as the radius of curvature R^* is large compared to κ^{-1}. This situation may be brought about by making κ^{-1} small (i.e., κ large), which is equivalent to dealing with relatively high concentrations of electrolyte or with flat or slightly curved surfaces. For our purposes it is convenient to consider a planar surface, but the results will apply equally to any case for which the product κR^* is large.

12.4a Helmholtz-Smoluchowski Equation for Electrophoretic Mobility

Suppose we consider a volume element of area A and thickness dx situated a distance x from a planar surface as shown in Figure 12.3. The viscous force on the face nearest the surface is given by

$$F_x = \eta A \left(\frac{dv}{dx} \right)_x \tag{28}$$

and the force exerted on the face farther from the surface is given by

$$F_{x+dx} = \eta A \left(\frac{dv}{dx} \right)_{x+dx} \tag{29}$$

In these equations v is the relative velocity between the particle and the surrounding medium. The difference between Equations (28) and (29) therefore equals the net viscous force on the volume element:

$$F_{vis} = \eta A \left[\left(\frac{dv}{dx} \right)_{x+dx} - \left(\frac{dv}{dx} \right)_x \right] \tag{30}$$

Equation (4.12) can be used to write Equation (30) as

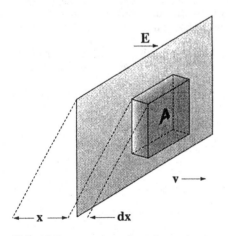

FIG. 12.3 Location of a volume element of solution adjacent to a planar wall.

$$F_{vis} = \eta \, A \, \frac{d^2v}{dx^2} \, dx \tag{31}$$

Under stationary-state conditions an equal and opposite force is exerted on the volume element by the electric field acting on the ions contained in the volume element. The force on the ions is given by the product of the field strength times the total charge. The latter equals the charge density ρ^* times the volume of the element; therefore

$$F_{el} = E\rho^* A \, dx \tag{32}$$

The Poisson equation for a planar surface (Equation (11.26)) may now be used as a substitution for ρ^* to yield

$$\rho^* = -\varepsilon \, \nabla^2 \psi = -\varepsilon \, \frac{d^2\psi}{dx^2} \tag{33}$$

where the second result applies specifically to the region adjacent to the planar surface and where ψ is the potential at a distance x from the surface.

Setting Equations (31) and (32) equal to each other, substituting Equation (33), and simplifying leads to the equation

$$\eta \, \frac{d^2v}{dx^2} = -\varepsilon \, E \, \frac{d^2\psi}{dx^2} \tag{34}$$

With certain assumptions this result may be integrated twice to give the relation between v and ψ.

The integration of Equation (34) is carried out by assuming that both η and ε are constants in the vicinity of the surface. We return to a discussion of this assumption in Section 12.8. Making this assumption, we can write Equation (34) as

$$\frac{d}{dx}\left(\eta \, \frac{dv}{dx}\right) = -E \, \frac{d}{dx}\left(\varepsilon \, \frac{d\psi}{dx}\right) \tag{35}$$

In this form the first integration is readily found to give

$$\eta \, \frac{dv}{dx} = -E \, \varepsilon \, \frac{d\psi}{dx} + C_1 \tag{36}$$

The constant of integration C_1 is evaluated by noting that both dv/dx and $d\psi/dx$ must equal zero at large distances from the surface; therefore $C_1 = 0$.

The resulting expression is easily integrated again with the following limits: (a) at the surface of shear $\psi = \zeta$ and $v = 0$; (b) at the outside edge of the double layer $\psi = 0$ and v equals the observed velocity of particle migration. Therefore

$$\eta \int_v^0 dv = -E \, \varepsilon \int_0^\zeta d\psi \tag{37}$$

or

$$\eta v = \varepsilon E \zeta \tag{38}$$

In terms of electrophoretic mobility, Equation (38) can be written as

$$u = v/E = (\varepsilon \zeta)/\eta \tag{39}$$

Equation (39) is known as the *Helmholtz-Smoluchowski equation*. No assumptions are made in its derivation as to the actual structure of the double layer, only that the Poisson equation applies and that bulk values of η and ε apply within the double layer. It has been shown that this result is valid for values of κR^* larger than about 100 (i.e., $\kappa R_s > 100$ for spherical particles).

12.4b   Hückel and Helmholtz-Smoluchowski Equations as Limiting Cases

We have now reached the position of having two expressions — Equations (27) and (39) — to describe the relationship between the mobility of a particle (an experimental quantity) and the zeta potential (a quantity of considerable theoretical interest). The situation may be summarized by noting that both the Hückel and the Helmholtz-Smoluchowski equations may be written as

$$u = C \frac{\varepsilon \zeta}{\eta} \tag{40}$$

where C is a constant with a numerical value that depends on the magnitude of κR_s. In the limit of both large and small values of κR_s, the value of C becomes independent of κR_s:

   1. For $\kappa R_s < 0.1$

$$C = (2/3) \tag{41}$$

   2. For $\kappa R_s > 100$

$$C = 1 \tag{42}$$

In view of the widely different pictures of the electric field surrounding the particles in the two extremes — as shown schematically in Figure 12.1 — it is not surprising that different results are obtained in the two limits.

A major remaining problem is that many systems of interest in colloid chemistry do not correspond to either of these two limiting cases. The situation is summarized in Figure 12.4, which maps the particle radii R_s and 1 : 1 electrolyte concentrations that correspond to various κR_s values. Clearly, there is a significant domain of particle size and/or electrolyte concentration for which neither the Hückel nor the Helmholtz-Smoluchowski equations can be used to evaluate ζ from experimental mobility values. The relationship between ζ and u for intermediate values of κR_s is the topic of the following section.

12.5   ZETA POTENTIAL: GENERAL THEORY FOR SPHERICAL PARTICLES

It is apparent from the above sections that the understanding of electrophoretic mobility involves both the phenomena of fluid flow as discussed in Chapter 4 and the double-layer potential as discussed in Chapter 11. In both places we see that theoretical results are dependent on the geometry chosen to describe the boundary conditions of the system under consideration. This continues to be true in discussing electrophoresis, for which these two topics are combined. As was the case in Chapters 4 and 11, solutions to the various differential equations that arise are possible only for rather simple geometries, of which the sphere is preeminent.

12.5a   General Formulation of Electrophoresis for Spherical Particles: Henry's Equation

The generalized electrophoresis problem has been solved for spherical and rod-shaped particles and, more approximately, for random coils. In this section we restrict our attention to spheres, although in the limit of large values of κR_s, the Helmholtz-Smoluchowski equation is obtained, a result that is independent of particle shape. In the general theory the conductivity of the particle is one of the parameters that must be considered. We discuss only the case of nonconducting spheres. It has been shown experimentally that mercury droplets for which κR_s is large follow Equation (39), even though — as conductors — the full theory predicts they should show zero mobility. The explanation of this anomaly is that the surface of the metallic drops becomes sufficiently polarized to block the passage of current through the particle. Thus even a metallic particle may behave as an insulator, thereby justifying our choice of the nonconducting particle as the model for consideration.

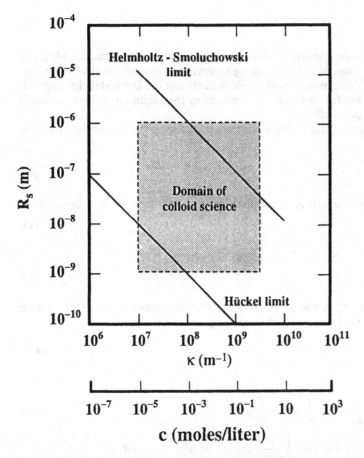

FIG. 12.4 The domain within which most investigations of aqueous colloidal systems lie in terms of particle radii and 1 : 1 electrolyte concentration. The diagonal lines indicate the limits of the Hückel and the Helmholtz-Smoluchowski equations. (Redrawn with permission from J. Th. G., Overbeek, Quantitative Interpretation of the Electrophoretic Velocity of Colloids. In *Advances in Colloid Science*, Vol. 3 (H. Mark and E. J. W. Verwey, Eds.), Wiley, New York, 1950.)

In addition, we consider only the case in which the colloid is present in small concentration so that colloid-colloid interactions can be ignored. We assume that the diffuse part of the double layer is adequately described by the Gouy-Chapman theory. Since the surface of shear more or less coincides with the Stern surface, it is the diffuse part of the double layer and not the Stern layer (where specific adsorption occurs) in which we are interested. Specific adsorption in the Stern layer may have a large effect on the zeta potential itself, but should be unimportant when it comes to establishing the connection between u and ζ. The Gouy-Chapman theory ignores the actual discreteness of electrical charges and is also subject to the objections against the Poisson-Boltzmann equation (see Section 11.4). An extensive body of research has been devoted either to circumventing these limitations or to estimating the approximation introduced by their use. Overbeek and Wiersma (1967) have rightly noted that it is rather futile to introduce one or two corrections to the theory while neglecting other approximations that are probably of the same magnitude. A safer procedure, they note, is to use the simpler theory, keeping in mind the semiquantitative nature of the result.

By assuming that the external field — deformed by the presence of the colloidal particle — and the field of the double layer are additive, D. C. Henry derived the following expression for mobility:

$$u = \frac{\varepsilon}{\eta} \left(\zeta + 5 R_s^5 \int_\infty^{R_s} \frac{\psi}{r^6} \, dr - 2 R_s^3 \int_\infty^{R_s} \frac{\psi}{r^4} \, dr \right) \tag{43}$$

where r is the radial distance from the center of the particle. To go beyond Equation (43), it is necessary to know ψ as a function of r. The resulting expressions are mathematically intractable unless a relatively simple expression is used for ψ. We may use the Debye-Hückel approximation given by Equation (19) for this, but the constant in that equation is best evaluated somewhat differently before proceeding.

We return to the solution of the Poisson-Boltzmann equation for a spherical particle, Equation (19), with $B = 0$:

$$\psi = \frac{A \exp{(-\kappa r)}}{r} \tag{44}$$

In the present development we evaluate A by recalling that $\psi = \zeta$ when $r = R_s$. Therefore

$$A = R_s \zeta \exp{(\kappa R_s)} \tag{45}$$

and Equation (44) becomes

$$\psi = \frac{R_s \, \zeta}{r} \exp{[-\kappa(r - R_s)]} \tag{46}$$

This equation, with ψ_0 instead of ζ, was given without proof as Equation (11.38) in Section 11.4c. Combining Equations (43) and (46) and integrating leads to the result

$$u = \frac{2 \, \varepsilon \, \zeta}{3 \, \eta} f(\alpha) \tag{47a}$$

where

$$f(\alpha) = \left(1 + \frac{1}{16} \alpha^2 - \frac{5}{48} \alpha^3 - \frac{1}{96} \alpha^4 - \frac{1}{96} \alpha^5 \right. \tag{47b}$$
$$\left. - \left[\frac{1}{8} \alpha^4 - \frac{1}{96} \alpha^6 \right] \exp{(\alpha)} \int_\infty^\alpha \frac{e^{-t}}{t} \, dt \right) \qquad \text{with} \qquad \alpha = \kappa R_s < 1$$

For $\kappa R_s > 1$, the function $f(\alpha)$ becomes (see Hunter 1981)

$$f(\alpha) = \left(\frac{3}{2} - \frac{9}{2} \alpha^{-1} + \frac{75}{2} \alpha^{-2} - 330 \, \alpha^{-3} \right) \tag{47c}$$

The above result (Equations (47a)–(47c)) is known as *Henry's equation*. Two specific assumptions underlying its derivation should be pointed out:

1. The ion atmosphere is undistorted by the external field.
2. The potential is low enough to justify writing $e\psi/k_B T < 1$, which is equivalent to requiring that $\psi < 25$ mV (Section 11.2b).

It should also be noted that in the limit of $\kappa R_s \rightarrow 0$, Equation (47a) reduces to the Hückel equation, and in the limit of $\kappa R_s \rightarrow \infty$, it reduces to the Helmholtz-Smoluchowski equation. Thus the general theory confirms the idea introduced in connection with the discussion of Figure 12.1, that the amount of distortion of the field surrounding the particles will be totally different in the case of large and small particles. The two values of C in Equation (40) are a direct consequence of this difference. Figure 12.5a shows how the constant C varies with κR_s (shown on a logarithmic scale) according to Henry's equation.

12.5b Effect of Double-Layer Relaxation

We noted above that many systems of interest in colloid chemistry involve intermediate values of κR_s, so Henry's equation fills an important gap. At the same time it explicitly introduces additional restrictions: (a) low potentials and (b) undistorted double layers. A topic of consid-

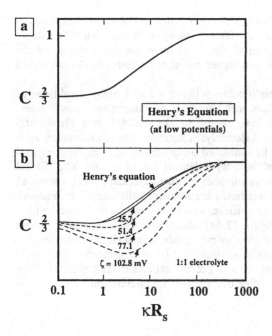

FIG. 12.5 Variation of the constant C (Equation (40)) with κR_s (log scale): (a) at low potentials according to Henry's equation; and (b) for various potentials. (Data from P. H. Wiersma, A. L. Loeb, and J. Th. G. Overbeek, *J. Colloid Interface Sci.*, **22**, 78 (1966); redrawn with permission from Shaw 1969.)

erable importance is the actual distortion of the double layer that accompanies particle migration. The consequences of this distortion — known as the *relaxation effect* — are known to be important in the conductivity of simple electrolytes. A remaining development, therefore, is to consider the relaxation effect in colloidal systems.

Because the charged particle and its ion atmosphere move in opposite directions, the center of positive charge and the center of negative charge do not coincide. If the external field is removed, this asymmetry disappears over a period of time known as the *relaxation time*. Therefore, in addition to the fact that the colloid and its atmosphere move countercurrent with respect to one another (which is called the *retardation effect*), there is a second inhibiting effect on the migration that arises from the tug exerted on the particle by its distorted atmosphere. Retardation and relaxation both originate with the double layer, then, but describe two different consequences of the ion atmosphere. The theories we have discussed until now have all correctly incorporated retardation, but relaxation effects have not been included in any of the models considered so far.

A number of workers have tackled the problem of relaxation. The use of computers has greatly assisted this area of research because of the complexity of the mathematics involved. Loeb et al. (1960) report the results of some numerical solutions to the mobility problem with relaxation specifically considered. Figure 12.5b summarizes some results from these studies for the case of a 1 : 1 electrolyte. The various curves correspond to values of zeta equaling 25.7, 51.4, 77.1, and 102.8 mV at 25 °C. It will be noted that the restriction to low potentials no longer applies in the theory from which these curves were evaluated. It is evident from the figure that the relaxation effect is negligible when $\zeta < 25$ mV, regardless of the value of κR_s, and in the limit of both large and small values of κR_s, regardless of the value of ζ. That is, intermediate values of κR_s and large potentials correspond to the condition of maximum resistance to flow arising from relaxation.

A family of curves qualitatively similar in appearance to those shown in Figure 12.5b results when C is plotted versus κR_s at constant ζ with the valence of the electrolyte taken as the variable parameter. In that case the relaxation effect is found to increase with the valence of the counterions. As the valence of those small ions that have the same charge as the macroion increases, the relaxation effect leads to a higher mobility (at constant ζ) than would be predicted from Henry's equation.

In this section we have considered the relationship between u and ζ under conditions of intermediate κR_s values, a wide range of ζ values, and a number of ionic valence possibilities. The relationship is seen to be quite complex, except in the Hückel and Helmholtz-Smoluchowski limits. When the particle size and electrolyte concentration conditions are such that one of these limits clearly applies, ζ can be evaluated unambiguously from experimental mobilities. The Helmholtz-Smoluchowski limit is independent of particle shape. The Hückel equation is equally free from ambiguity, although it does require spherical particles and—as already noted—the circumstances under which it holds are not especially useful for aqueous colloids. If a particle is of intermediate size with definite, known values of κ and R_s and with ζ known to be small, Henry's equation (or Figure 12.5a) could be used to evaluate ζ from mobility measurements. As the complexity (i.e., higher potentials, mixed electrolyte valences) of the system increases, however, the feasibility of evaluating ζ from experimental mobilities becomes increasingly tenuous. In these circumstances precise experimental results are best reported as mobilities, with the corresponding value of ζ only an approximation.

12.6 ELECTROOSMOSIS

In all the sections of this chapter until now we have focused attention on electrophoresis. We have seen that the potential at the surface of shear can be measured from electrophoretic mobility measurements, provided the system complies with the assumptions of a manageable model. One feature that has been conspicuously lacking from our discussions is any comparison between electrophoretically determined values of ζ and potential values determined by another method. The reasons for this are twofold:

1. Other techniques for measuring ζ are contingent on the same set of assumptions associated with electrophoresis and therefore do not constitute an independent determination.
2. Uncertainty as to the location within the double layer at which the shear surface is located makes it difficult to relate ζ to other double-layer potentials, such as ψ_0 as determined from knowledge of the concentration of potential-determining ions (see Equation (11.1)).

In this section we describe *electroosmosis* and in the following section the *streaming potential*. These two electrokinetic techniques also permit the evaluation of ζ, but are subject to objection 1. In Section 12.8 we examine in greater detail the location of the surface of shear, which is the essence of objection 2 above.

12.6a Difference Between Electroosmosis and Streaming Potential

Above we defined electrokinetic phenomena as arising from the relative motion of a charged surface and its associated double layer. In electrophoresis it is the dispersed phase that moves, with the continuous phase remaining (more or less) stationary. It is apparent that the required relative motion between a surface and its double layer could also be brought about by causing the electrolyte solution to flow past a stationary charged wall. The complements of electrophoresis are electroosmosis and streaming potential. These two measurements differ from each other as follows: (a) in *electroosmosis* it is an applied potential that induces the flow of solution, as illustrated in Figure 12.6; (b) in *streaming potential* the solution is made to flow by applying a pressure, and a potential is induced as a result. Cause and effect are thus interchanged in electroosmosis and streaming potential.

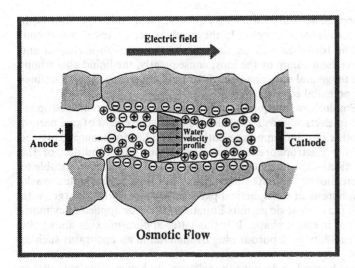

FIG. 12.6 Electroosmotic flow through a pore. If the fluid flow occurs as a result of applied pressure difference along the length of the pore, the resulting potential difference is known as the streaming potential. (Adapted with permission from Probstein 1994.)

12.6b Relating Electroosmotic Flow to Zeta Potential

The electroosmosis apparatus shown in Figure 12.7a consists of two capillaries in parallel attached at either end to reservoirs of electrolyte solution. One of the capillaries—the working capillary—is arranged with reversible electrodes at either end, while the measuring capillary contains an air bubble to indicate fluid displacement. It is the glass-solution interface in the working capillary at which the electroosmosis phenomenon originates. Substances other than glass may also be investigated by this method, a particularly useful variation being the replacement of the capillary by a plug of powdered material that cannot be fabricated into a cylindrical tube. For the purpose of discussion, we continue to refer to the capillary. The conditions under which the same analysis applies to a plug will be clear from the following discussion.

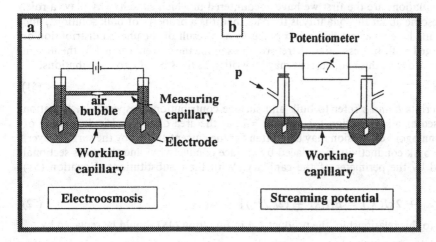

FIG. 12.7 Schematic illustrations of the apparatus used to measure (a) electroosmosis and (b) streaming potential.

When an electric field is applied across the working capillary, the double-layer ions begin to migrate and soon reach the steady-state velocity. In the steady state, electrical and viscous forces balance one another. The forces exerted on the ions by the medium are equal and opposite to the forces exerted on the medium by the ions; consequently, the liquid also attains a stationary-state velocity. The tangential displacement of the fluid relative to the wall defines a surface of shear at which the potential equals ζ.

Although the Helmholtz-Smoluchowski equation was derived in reference to electrophoretic mobility, it clearly applies to electroosmosis as well since the displacement of one part of the double layer relative to another part is common to both. Figure 12.3, for example, may be taken as an illustration of either electrophoresis or electroosmosis. The condition of the Helmholtz-Smoluchowski equation — that R_s is large compared to κ^{-1} — is clearly applicable to capillaries of macroscopic dimensions. We noted above that the Helmholtz-Smoluchowski equation applies to the electrophoresis of nonspherical particles as long as κR^* is large (with R^* the radius of the curvature); the same logic permits Equation (39) to be applied to cylindrical capillaries, as well as pores of irregular shape. It is this latter application that allows the replacement of a well-defined capillary by a porous plug of material in an apparatus such as that shown in Figure 12.7a.

Equation (38) may therefore be used to describe the relationship between the potential at the capillary wall and at the velocity of electroosmotic flow. The volume of liquid V displaced per unit time is given by multiplying both sides of Equation (38) by the cross-sectional area of the capillary:

$$V = vA = \varepsilon \zeta EA/\eta \tag{48}$$

Now suppose we apply Ohm's law to the capillary. The electric field is related to the current I and the conductivity k of the electrolyte solution as follows:

$$E = I/(Ak) \tag{49}$$

This result may be substituted into Equation (48) to yield

$$V = \varepsilon \zeta I/(\eta k) \tag{50}$$

This equation permits ζ to be evaluated from measurements of the rate of volume flow through the capillary; the measurements are made by observing the rate of displacement of the air bubble in the measuring capillary of Figure 12.7a.

12.6c Effect of Surface Conductivity on Electroosmosis

The last three equations are the first we have encountered in which conductivity plays a role. What is troublesome about this quantity is the fact that it is a property of bulk solutions, and we are considering here an effect that arises precisely as a result of the uneven distribution of ions near a charged wall. It is essential, therefore, to examine the current carried by the ions in the double layer. Toward this end, current may be written as the sum of two contributions:

$$I = I_b + I_s \tag{51}$$

where the subscripts b and s refer to bulk and surface contributions, respectively. Equation (49) may be used as a substitution for I_b, with πR_c^2 as the area of a cylindrical capillary of radius R_c. An analogous expression may be written for the current carried by the surface layer. In this case the bulk conductivity is replaced by surface conductivity, and the cross-sectional area is replaced by the perimeter of the capillary. With these substitutions, Equation (51) becomes

$$I = E(\pi R_c^2 k_b + 2\pi R_c k_s) = EA[k_b + (2k_s/R_c)] \tag{52}$$

According to this relationship, the product EA in Equation (48) should be replaced by

$$EA = I/[k_b + (2k_s/R_c)] \tag{53}$$

to give

$$V = \varepsilon\zeta I/\{\eta[k_b + (2k_s/R_c)]\} \tag{54}$$

It will be noted that the importance of the correction for surface conductivity increases as R_c decreases and vanishes as $R_c \to \infty$. Equation (54) also suggests that the numerical evaluation of k_s may be accomplished by studying electroosmosis in a set of capillaries identical in all respects except for variability in R_c. Finally, the expansion of Equation (50) to Equation (54) in correcting for surface conductivity explicitly assumes a cylindrical capillary. Experiments made with porous plugs cannot be corrected for surface conductivity by Equation (54), but the qualitative conclusion that the effect of surface conductivity increases as the pore radius decreases is valid in this case also.

It has already been noted that there is a close similarity between electroosmosis and streaming potential. Therefore we consider this additional electrokinetic phenomenon next.

12.7 STREAMING POTENTIAL

The definition of streaming potential was presented in the previous section. Here, we derive the relation between the streaming potential and the zeta potential and discuss some of the issues that must be considered in comparing zeta potentials obtained by different electrokinetic measurements.

12.7a Relation Between Streaming Potential and Zeta Potential

Figure 12.7b is a sketch of an apparatus that may be used to measure streaming potential. As was the case with electroosmosis, the capillary can be replaced by a plug of powdered material between perforated electrodes. An applied pressure difference p across the capillary causes the solution to flow through the capillary, thereby tangentially displacing the part of the double layer in the mobile phase from the stationary part.

The relationships developed in Chapter 4 for fluid flow through a capillary can be applied to this situation as follows:

1. The velocity of the fluid at radius r in a capillary of radius R_c and length ℓ is given by Equation (4.18):

$$v = [p/(4\eta\ell)](R_c^2 - r^2) \tag{55}$$

2. The volume flow rate through an elemental area $2\pi r dr$ is given by (see Equation (4.19)):

$$dV_F/dt = [p/(4\eta\ell)](R_c^2 - r^2)2\pi r\, dr \tag{56}$$

where V_F denotes the volume flow.

3. The current associated with this flow rate is

$$dI = \rho^*(dV_F/dt) = [\rho^*p/(4\eta\ell)](R_c^2 - r^2)2\pi r\, dr \tag{57}$$

where ρ^* is the charge density.

4. Next a change of variable is helpful. We replace r by a distance measured from the surface of shear x, where

$$x = R_c - r \tag{58}$$

In terms of this substitution, Equation (57) becomes

$$dI = -[\rho^*p/(4\eta\ell)](2R_c x - x^2)2\pi(R_c - x)\, dx \tag{59}$$

Our specific interest is in the region near the walls of the capillary where $x \ll R_c$. In this region Equation (59) may be approximated as

$$dI \approx -\pi[\rho^*p/(\eta\ell)]R_c^2 x\, dx \tag{60}$$

5. Substituting Equation (11.26) for ρ^* yields

$$dI = \frac{\pi \, \varepsilon \, p \, R_c^2}{\eta \ell} \frac{d^2\psi}{dx^2} \, x \, dx \tag{61}$$

6. The total current carried by the capillary is obtained by integrating the above equation. Integration by parts yields

$$I = \frac{\pi \, \varepsilon \, p \, R_c^2}{\eta \ell} \left[\left(x \frac{d\psi}{dx} \right)_0^{R_c} - \int_0^{R_c} \frac{d\psi}{dx} \, dx \right] = \frac{\pi \, \varepsilon \, p \, R_c^2 \, \zeta}{\eta \ell} \tag{62}$$

since $\psi = \zeta$ at $x = 0$ and $\psi = d\psi/dx = 0$ at $x = R_c$.

The quantity calculated by Equation (62) is known as the *streaming current*. It is specifically due to the net displacement of the mobile part of the double layer relative to the stationary part of the double layer. The field associated with this current is given by combining Equations (52) and (62):

$$E = \frac{\varepsilon \, \zeta}{\eta} \left(\frac{1}{k_b + 2k_s/R_c} \right) \frac{p}{\ell} \tag{63}$$

If both sides of Equation (63) are multiplied by the length of the capillary ℓ, the potential difference between the measuring electrodes — the *streaming potential E_{str}* — is obtained:

$$E_{str} = \frac{\varepsilon \, \zeta \, p}{\eta (k_b + 2k_s/R_c)} \tag{64}$$

The conditions under which Equation (64) for streaming potential and Equation (54) for electroosmosis were derived are comparable inasmuch as each applies to the case of large κR_c. Comparison of Equations (54) and (64) in the limit of large R_c shows that

$$(E_{str}/p) = (V/I) = \varepsilon \zeta /(\eta k) \tag{65}$$

The coupling of two different electrokinetic ratios (E_{str}/p and V/I) through Equation (65) is an illustration of a very general law of reciprocity due to L. Onsager (Nobel Prize, 1968). The general theory of the Onsager relations, of which Equation (65) is an example, is an important topic in nonequilibrium thermodynamics.

If the relationships shown in Equation (65) are to be used in computations, it is essential that proper units be used. Example 12.3 considers some numerical substitutions into Equation (65).

* * *

EXAMPLE 12.3 *Units of Electrokinetic Parameters*. Show that E_{str}/p, V/I, and $\varepsilon_0 \zeta / \eta k$ all have units $m^3 \, C^{-1}$. For water at 25°C, $\eta = 8.937 \cdot 10^{-4}$ kg m^{-1} s^{-1} and $\varepsilon_r = 78.54$. Evaluate the proportionality factor between V/I and ζ/k if V is expressed in cm^3 s^{-1} and I is expressed in milliamperes with all other quantities in SI units. Evaluate the proportionality factor between E_{str}/p and ζ/k if p is expressed in torr with all other quantities in SI units.

Solution: Examine the SI units of each term in Equation (65). The units of (V/I) are (m^{3}s^{-1})/ (Cs^{-1}) = m^3 C^{-1}. The units of E_{str}/p are V/(N m^{-2}); when multiplied by the conversion factor (J C^{-1})/V, this becomes m^3 C^{-1}. The units of $\varepsilon_0 \zeta /\pi k$ are (C^2 J^{-1} m^{-1})(V)/(kg m^{-1} s^{-1})(ohm^{-1} m^{-1}); when multiplied by the conversion factor V (C s)$^{-1}$/ohm, this becomes J C^{-1} s/kg m^{-1} = m^3 C^{-1}.

Substituting the SI values for all quantities, we obtain the following for $\varepsilon \zeta /\pi k$

$$(78.54)(8.85 \cdot 10^{-12}) \zeta /(8.937 \cdot 10^{-4})k = 7.777 \cdot 10^{-7} \zeta/k \, m^3 \, C^{-1}$$

from which

$$V/I = 7.777 \cdot 10^{-7} \, \zeta/k \, m^3 \, C^{-1} \cdot (10^2 \, cm/m)^3 \cdot 1 \, C \, s^{-1}/10^3 \, mA$$

$$= 7.777 \cdot 10^{-4} \, \zeta/k \, cm^3 \, s^{-1}/mA$$

and

$$E_{str}/p = [(7.777 \cdot 10^{-7} \, \zeta/k) \, m^3 \, C^{-1}] \cdot (10^3 \, liter/m^3) \cdot (1 \, C \, V/1 \, J)$$
$$\cdot (8.314 \, J/0.08205 \, liter \, atm) \cdot (1 \, atm/760 \, torr)$$
$$= (1.037 \cdot 10^{-4} \, \zeta/k) \, V \, torr^{-1}$$

∎

* * *

For 10^{-3} M NaCl, $k = 1.26 \cdot 10^{-2}$ ohm^{-1} m^{-1} and a surface with a ζ potential of 50 mV will displace about 11 cm^3 h^{-1} if a current of 1.0 mA flows through an electroosmosis apparatus. With the same electrolyte and the same value of ζ, an applied pressure of 760 mm Hg would produce a streaming potential of 313 mV.

In hydrocarbons the specific conductivity may be lower than that of aqueous solutions by many orders of magnitude, so the streaming potentials generated by the high-pressure pumping of these materials may be quite spectacular. The danger of sparking at such voltages plus the flammability of these substances makes the petroleum industry an area in which streaming potential finds important applications. For example, gasoline (for which the specific conductivity would be as low as 10^{-12} ohm^{-1} m^{-1} if untreated) pumping equipment must be grounded. In addition, a variety of organic-soluble electrolytes have been developed as antistatic additives for petroleum. Examples of two such compounds are tetraisoamyl ammonium picrate and calcium diisopropyl salicylate. Crude petroleum is less troublesome in this regard than refined products since the crude contains oxidation products, asphaltenes, and so on, which impart a natural conductivity to this material.

12.7b   Comparison of Zeta Potentials from Different Methods

The objective of comparing values of ζ determined from electrophoresis with those determined by other electrokinetic methods was stated at the beginning of Section 12.6. Enough experiments have been conducted in which at least two of the electrokinetic methods we have discussed are compared to leave no doubt as to the self-consistency of ζ as determined by these different methods. There is no guarantee, however, that self-consistent ζ potentials are correct. Consistency means only that ζ has been extracted from experimental quantities by a self-consistent set of approximations. It should be emphasized, however, that the existence of a potential at the surface of shear—which is the common component in all the electrokinetic analyses we have discussed—is more than amply confirmed by these observations.

Two conditions must be met to justify comparisons between ζ values determined by different electrokinetic measurements: (a) the effects of relaxation and surface conductivity must be either negligible or taken into account and (b) the surface of shear must divide comparable double layers in all cases being compared. This second limitation is really no problem when electroosmosis and streaming potential are compared since, in principle, the same capillary can be used for both experiments. However, obtaining a capillary and a migrating particle with identical surfaces may not be as readily accomplished. One means by which particles and capillaries may be compared is to coat both with a layer of adsorbed protein. It is an experimental fact that this procedure levels off differences between substrates: The surface characteristics of each are totally determined by the adsorbed protein. This technique also permits the use of microelectrophoresis for proteins since adsorbed and dissolved proteins have been shown to have nearly identical mobilities.

12.8   THE SURFACE OF SHEAR AND VISCOELECTRIC EFFECT

The surface of shear is the location within the electrical double layer at which the various electrokinetic phenomena measure the potential. We saw in Chapter 11 how the double layer extends outward from a charged wall. The potential at any particular distance from the wall can, in principle, be expressed in terms of the potential at the wall and the electrolyte content of the solution. In terms of electrokinetic phenomena, the question is: How far from the interface is the surface of shear situated and what implications does this have on the relation between measured zeta potential and the surface potential?

First, the very existence of a surface of shear implies some interesting behavior within the fluid phase of the system under consideration. In our discussion of all electrokinetic phenomena until now, we have assumed that the viscosity of the medium has its bulk value right up to the surface of shear. In addition, it has been implicitly assumed that the viscosity abruptly becomes infinite at the surface of shear.

At this point, it is convenient to recall Figure 7.13 and the discussion of it. In that context we observed that there is generally a variation of properties in the vicinity of an interface from the values that characterize one of the adjoining phases to those that characterize the other. This variation occurs over a distance τ measured perpendicular to the interface. In the present discussion viscosity is the property of interest and the surface of shear—rather than the interface per se—is the boundary of interest. The model we have considered until now has implied an infinite jump in viscosity, occurring so sharply that τ is essentially zero. From a molecular point of view such an abrupt transition is highly unrealistic. A gradual variation in η over a distance comparable to molecular dimensions is a far more realistic model.

With these ideas in mind, it is evident that we would do better to think of a *zone* of shear rather than a surface of shear. Although we continue to speak of the shear "surface," the term is not used in the mathematical sense of possessing zero thickness, but rather in the broader sense of Chapter 7. Moreover, under certain circumstances, the viscosity has a strong dependence on the electric field. This is known as *viscoelectric effect* and must be taken into account since the electric field changes sharply within the electrical double layer. We examine how the viscoelectric effect affects the Helmholtz-Smoluchowski results obtained above for the relation between the electrophoretic mobility and the zeta potential. Following this, we look at what conditions make the zeta potential obtained from the Helholtz-Smoluchowski result a good measure of the surface potential.

12.8a Viscoelectric Effect and Its Influence on Electrophoretic Mobility and Zeta Potential

How must the expressions derived in the sections above be modified to take into account the variation in η and the finite distance over which it increases? The answer is that η—the viscosity within the double layer—must be written as a function of location. Our objective in discussing this variation is not to examine in detail the efforts that have been directed along these lines. Instead, it is to arrive at a better understanding of the relationship between ζ and the potential at the inner limit of the diffuse double layer and a better appreciation of the physical significance of the surface of shear.

Measurements of the viscosity of organic liquids in the presence of an electric field reveal that there is an increase in viscosity in high electric fields that is described by the expression

$$(\eta_E - \eta_0)/\eta_0 = fE^2 \tag{66}$$

where the subscripts indicate the presence (E) or absence (0) of a field. The factor f is called the *viscoelectric constant* and has a value of about $2 \cdot 10^{-16}$ V^{-2} m^2 for several organic liquids. Thus a 10% increase in viscosity may be anticipated for a field strength of about $2 \cdot 10^7$ V m^{-1}.

An expression such as Equation (11.56) may be used to estimate E ($= d\psi/dx$) in the double layer. Table 12.1 shows values of E evaluated by means of this equation for a variety of ψ_0 values and 1 : 1 electrolyte concentrations. It will be noted that for high values of ψ_0 and high ionic strengths, the field in the double layer may be large enough to produce a very significant viscoelectric effect.

Now suppose we reexamine the derivation of the Helmholtz-Smoluchowski equation as given in Section 12.4. Returning to Equation (37), we note that the relationship between u and ζ is given by

$$u = \varepsilon \int_0^\zeta \frac{d\psi}{\eta} \tag{67}$$

where η has been left inside the integral this time since its value is assumed to vary with ψ. We continue to assume that ε is a constant since the effect of the field is known to be less for this quantity than for η.

TABLE 12.1 Values of the Electric Field ($V m^{-1}$)
Calculated in the Double Layer by Equation (11.56) for
Various ψ_0 Values and Concentrations of 1 : 1 Electrolyte

ψ_0, mV	c (mole liter^{-1})		
	10^{-3}	10^{-2}	10^{-1}
50	$6.36 \cdot 10^6$	$2.01 \cdot 10^7$	$6.36 \cdot 10^7$
100	$1.98 \cdot 10^7$	$6.24 \cdot 10^7$	$1.98 \cdot 10^8$
150	$5.49 \cdot 10^7$	$1.74 \cdot 10^8$	$5.49 \cdot 10^8$
200	$1.51 \cdot 10^8$	$4.77 \cdot 10^8$	$1.51 \cdot 10^9$

Now we substitute η_E from Equation (66) for the viscosity in the double layer in Equation (67) to obtain

$$u = \frac{\varepsilon}{\eta_0} \int_0^\zeta \frac{d\psi}{1 + f\,(d\psi/dx)^2} \tag{68}$$

Finally, Equations (11.40) and (11.56) may be used to evaluate $d\psi/dx$ in the double layer:

$$u = \frac{\varepsilon}{\eta_0} \int_0^\zeta \frac{d\psi}{1 + [f(8000\,c\,R\,T/\varepsilon)]\sinh^2\,(ze\psi/2k_BT)} \tag{69}$$

where c is in mole liter^{-1} and $2 \sinh x$ has been substituted for $e^x - e^{-x}$. If we define $A = (8000\,fcRT/\varepsilon)$ and $B = ze/2k_BT$, this result is more concisely written as

$$u = \frac{\varepsilon}{\eta_0} \int_0^\zeta \frac{d\psi}{1 + A\sinh^2\,(B\psi)} \tag{70}$$

If c (and therefore A) and ψ are small, Equation (70) becomes approximately

$$u \approx \frac{\varepsilon}{\eta_0} \int_0^\zeta [1 + A\sinh^2\,(B\psi)]\,d\psi \tag{71}$$

Under the same conditions the sinh function may be expanded (see Appendix A), with only the leading term retained, to obtain

$$u \approx \frac{\varepsilon}{\eta_0} \int_0^\zeta [1 - A\,(B\psi)^2]\,d\psi \tag{72}$$

This equation is readily integrated to yield

$$u \approx \frac{\varepsilon}{\eta_0}\left(\zeta - \frac{AB^2}{3}\zeta^3\right) = \frac{\varepsilon\,\zeta}{\eta_0}\left(1 - \frac{AB^2}{3}\zeta^2\right) \tag{73}$$

Under conditions in which the second term is negligibly small, Equation (73) becomes identical to Equation (39), the Helmholtz-Smoluchowski result. On the other hand, when the concentration and ζ increase, the value of ζ that would be associated with an observed mobility is larger than the Helmholtz-Smoluchowski equation would indicate.

12.8b Deviation from Helmholtz-Smoluchowski Predictions

Equation (69) may also be integrated analytically. Although we do not consider the actual solutions, which are rather complex, Figure 12.8 shows graphically the results of these integrations for water at 25°C, assuming $f = 10^{-15}$ V^{-2} m^2. The abscissa shows values of ψ_0, the potential at the inner limit of the diffuse double layer, with $\eta u/\varepsilon$ plotted on the ordinate. It must be remembered that this last quantity equals ζ according to Equation (39) — which we

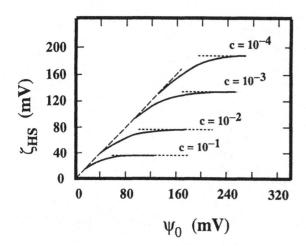

FIG. 12.8 Plot of $\eta u/\varepsilon$ versus ψ_0, that is, the zeta potential according to the Helmholtz-Smoluchowski equation, Equation (39), versus the potential at the inner limit of the diffuse part of the double layer. Curves are drawn for various concentrations of $1:1$ electrolyte with $f = 10^{-15}$ $V^{-2} m^2$. (Redrawn with permission from J. Lyklema and J. Th. G. Overbeek, *J. Colloid Sci.*, **16**, 501 (1961).)

designate ζ_{HS}—when the viscosity is assumed to be the bulk value throughout the double layer. The figure shows that $\zeta_{HS} = \psi_0$ at low values of the potential. As the potential increases, however, ζ_{HS} begins lagging behind ψ_0, the effect indicated by Equation (73) in a limiting approximation. At still higher potentials, ζ_{HS} eventually reaches a constant value that is independent of the actual value of ψ_0.

Note, further, that this leveling off occurs at progressively lower potentials as the concentration of electrolyte increases. Increasing both the potential and the electrolyte concentration tends to increase the field in the double layer (see Table 12.1), which in turn increases the viscosity of solvent in the double layer. As the effective viscosity of the medium increases, the surface of shear occurs progressively further from the surface. This accounts for the fact that ζ_{HS} falls behind ψ_0 as ψ_0 increases. These conclusions are consistent with the experimental observation that ζ_{HS} for AgI becomes independent of the concentration of the potential-determining Ag^+ and I^- ions once the concentrations of these ions are well removed from the conditions at which the particles are uncharged.

The results shown in Figure 12.8 illustrate quite clearly the relationship between ζ and ψ_0 and in this way reveal the dependence of the location of the surface of shear on the structure of the double layer. It might appear that one could consult curves such as those shown in Figure 12.8 to read from the appropriate plot the value of ψ_0 that corresponds to a particular ζ, at least for values of ζ less than the limiting value. Although semiquantitative interpretations based on this figure may be trusted, some caution must be expressed about the numerous assumptions and approximations inherent in Figure 12.8. In summary, the following may be cited as examples of such constraints:

1. The possible immobilization of solvent near the surface due to either chemical or mechanical (as opposed to viscoelectric) interaction with the solid phase has not been considered.

2. Use of the Gouy-Chapman theory (Equation (69)) overlooks any specific effects arising from differences between ions, especially with regard to hydration.

3. The validity of Equation (66) in electrolyte solutions, especially the dependence of f on concentration, has not been investigated as fully as might be desired.

12.9   EXPERIMENTAL ASPECTS OF ELECTROPHORESIS

Of the electrokinetic phenomena we have considered, electrophoresis is by far the most important. Until now our discussion of experimental techniques of electrophoresis has been limited to a brief description of microelectrophoresis, which is easily visualized and has provided sufficient background for our considerations to this point. Microelectrophoresis itself is subject to some complications that can be discussed now that we have some background in the general area of electrical transport phenomena. In addition, the methods of *moving-boundary electrophoresis* and *zone electrophoresis* are sufficiently important to warrant at least brief summaries.

12.9a   Microelectrophoresis

Microelectrophoresis depends on the visibility of the migrating particles under the microscope. As such, it is inapplicable to molecular colloids such as proteins. By adsorbing the protein molecules on suitable carrier particles, however, the range of utility for microelectrophoresis can be extended. Dark-field illumination (see Section 1.6a.1c) can sometimes be used to advantage to extend microelectrophoresis observations to small, high-contrast particles.

The migrating particles are observed in a cell that may be either cylindrical or rectangular. The walls must be optically uniform for observation and fewer optical corrections, and thermostating difficulties are encountered if the walls are thin. The working part of the apparatus is thus fragile, and auxiliary connecting rods are generally incorporated into the design to increase the mechanical strength of the cell. Figure 12.9 is a sketch of an electrophoresis apparatus with a rectangular working compartment.

The electric field in the cell is best established by means of reversible electrodes such as Ag-AgCl or Cu-CuSO$_4$. Care must be taken to prevent the electrolyte of the electrode from contaminating the dispersion. Platinized electrodes behave reversibly with low currents, but gas evolution causes troubles at higher currents.

The field strength is best obtained by including an accurate ammeter in the circuit to

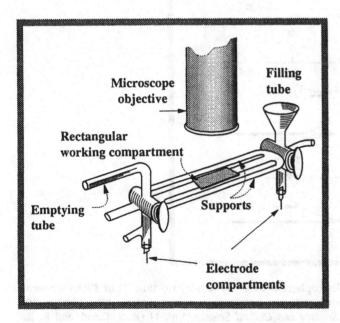

FIG. 12.9   Schematic illustration of a microelectrophoresis cell with a rectangular working compartment.

determine the current. Independent conductivity measurements in the cell with standard solutions permit the determination of the field through Equation (49).

The rate of particle migration is determined by measuring with a stopwatch the time required for a particle to travel between the marks of a calibrated graticule in the microscope eyepiece. If the objective of the microscope is immersed during the electrophoresis measurement, the calibration of the graticule should be made with the same immersion liquid.

Electrophoretic migrations are always superimposed on other displacements, which must either be eliminated or corrected to give accurate values for mobility. Examples of these other kinds of movement are Brownian motion, sedimentation, convection, and electroosmotic flow. Brownian motion, being random, is eliminated by averaging a series of individual observations. Sedimentation and convection, on the other hand, are systematic effects. Corrections for the former may be made by observing a particle with and without the electric field, and the latter may be minimized by effective thermostating and working at low current densities.

Even in the absence of a colloid, an electrolyte solution will display electroosmotic flow through a chamber of small dimensions. Therefore the observed particle velocity is the sum of two superimposed effects, namely, the true electrophoretic velocity relative to the stationary liquid and the velocity of the liquid relative to the stationary chamber. Figure 12.10a shows the results of this superpositioning for particles tracked at different depths in the cell. The particles used in this study are cells of the bacterium *Klebsiella aerogenes* in phosphate buffer. Rather than calculated velocities or mobilities, Figure 12.10a shows the reciprocal of the time

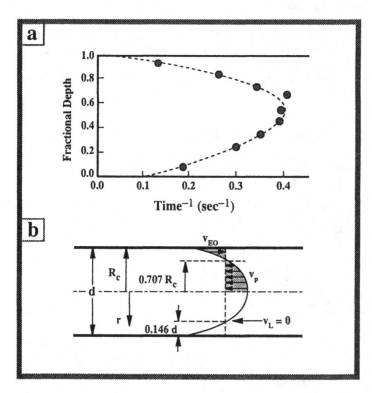

FIG. 12.10   Velocity profiles in electrophoresis cells: (a) velocity (as time^{-1}) of *Klebsiella aerogenes* particles as a function of their location in a rectangular electrophoresis cell (redrawn with permission from A. M. James, in *Surface and Colloid Science*, Vol. 11 (R. J. Good, and R. R. Stromberg, Eds.), Plenum, New York, 1979); (b) location of the surface of zero liquid velocity in a cylindrical capillary.

required for the particles to travel a fixed distance. Since the electrophoretic velocity is a single-valued property of the (uniform) particles under consideration, Figure 12.10a raises the question of which, if any, of these "velocities" represents the true mobility of these particles. Example 12.4 considers a possible interpretation of these results.

* * *

EXAMPLE 12.4 *Electrophoretic Mobility of Bacteria.* It is proposed to evaluate the electrophoretic mobility of the bacteria cells shown in Figure 12.10a by multiplying the appropriate value of time $^{-1}$ by the distance of particle displacement and then dividing by E. Criticize or defend the following proposition: It is appropriate to use the maximum apparent velocity since this is measured at the center of the cell and is therefore subject to the least interference by wall effects.

Solution: The observed effect is the sum of two contributions, one of which is the electroosmotic flow of the medium through the cell. The latter has its maximum value at the center since the layer of fluid adjacent to the walls is stationary. The particles tracked at the center of the cell therefore possess the maximum increment in velocity due to electroosmotic flow. Since the cell is a closed compartment, the liquid displaced by electroosmosis along the walls must circulate by a backflow down the center of the tube. Since the total liquid flow in a closed cell must be zero, the appropriate value from Figure 12.10a to use for the velocity is the average of observations made at all depths.   ∎

* * *

Since the liquid circulates, there must be certain locations in the cell at which the forward and backward flows of the liquid are equal. An alternative to the averaging procedure suggested in the example is to do the particle tracking at a location at which the medium experiences no net flow.

The analysis of this effect in a closed cylindrical cell is obtained by subtracting from the electroosmotic velocity v_{EO} the velocity of flow v_P through a capillary given by Poiseuille's equation (Equation (4.18)):

$$v_L = v_{EO} - v_P = v_{EO} - C(r^2 - R_c^2) \qquad (74)$$

where v_L is the velocity of the liquid and C is a constant. The requirement of no net displacement of liquid is incorporated by integrating Equation (74) over the cross section of the cylinder and setting the result equal to zero:

$$\int_0^{R_c} v_L (2\pi r) \, dr = 0 \qquad (75)$$

In this expression R_c is the radius of the capillary and r is the radial distance from the capillary axis as shown in Figure 12.10b. Substitution of Equation (74) into Equation (75) and integration gives

$$C = -2v_{EO}/R_c^2 \qquad (76)$$

This result may be substituted back into Equation (74) to evaluate that location in the cylinder at which the net liquid displacement is zero:

$$v_L = 0 = v_{EO}\left(1 + \frac{2}{R_c^2}(r^2 - R_c^2)\right) = v_{EO}\left(\frac{2\,r^2}{R_c^2} - 1\right) \qquad (77)$$

This result shows that electroosmotic flow and backflow in the capillary cancel when the factor $(2r^2/R_c^2 - 1)$ equals zero. This condition corresponds to $r/R_c = 0.707$. Thus at 70.7% of the radial distance from the center of the capillary lies a circular surface of zero liquid flow. Any particle tracked at this position in the capillary will display its mobility uncomplicated by the effects of electroosmosis. This location may also be described as lying 14.6% of the cell diameter inside the surface of the capillary. Experimentally, then, one establishes the inside diameter of the capillary and focuses the microscope 14.6% of this distance inside the walls of the capillary. Corrections for the effect of the refractive index must also be included. Additional details of this correction can be found in the book by Shaw (1969).

The location of the surface of zero liquid flow in cells of rectangular cross section has also been worked out. For a cell in which the direction of migration is very long compared to the width of the cell, the surface where $v = 0$ lies 21.1% of the cell depth above the bottom and below the top of the working compartment.

12.9b   Moving-Boundary Electrophoresis

In addition to microelectrophoresis, another important method for the determination of mobility is the *moving-boundary method* (Longsworth 1959). In essence, this is no different from the moving-boundary method as applied to simple ions. The apparatus most commonly used is that of Arne Tiselius (Nobel Prize, 1948), which is illustrated schematically in Figure 12.11. The Tiselius cell consists of a U tube of rectangular cross section that is segmented in such a way that the sections between the lines AA′ and BB′ in the figure can be laterally displaced with respect to the rest of the apparatus. The offset segments of the U tube are filled with the colloidal dispersion and, after thermal equilibration with the buffer solution contained in other parts of the apparatus, the various sections are aligned so that sharp boundaries are obtained. The location of the boundaries is usually observed by schlieren optics that identify refractive index gradients (see Section 2.4). As the macroions migrate in the electric field, the schlieren peak becomes displaced and the mobility of a colloidal component may be determined by measuring the rate of boundary movement per unit electric field. Relatively longer times are required for accurate mobility experiments than for microelectrophoresis since the particles must migrate over macroscopic distances rather than microscopic ones. To avoid contamination of the electrolyte in the U tube with electrode products, the electrodes are generally located near the bottom of large reservoirs as shown in Figure 12.11. Relatively concentrated salt solution is used to cover the electrodes, with the buffer solution layered on top.

Under optimum conditions the dimensions of the cross section of the cell are such that the effects of electroosmosis are minimal. The rectangular profile of the cross section allows for both good thermal equilibration (because one dimension is short) and good optical precision (because the other dimension is longer).

Moving-boundary electrophoresis is most widely applied to protein mixtures. In such a case each molecular species travels with a characteristic velocity. After sufficient time the various components in a mixture become effectively separated, and the percentage of each may be determined by measuring the areas under the schlieren peaks. Figure 12.12a shows a

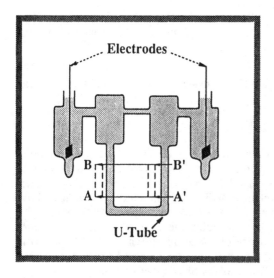

FIG. 12.11   Schematic illustration of a Tiselius-type moving boundary electrophoresis apparatus.

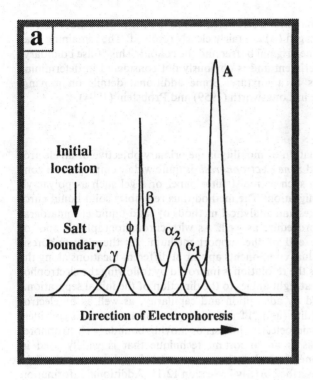

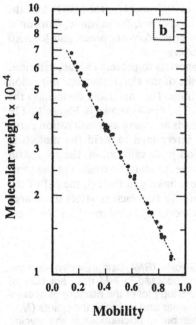

FIG. 12.12 Electrophoresis patterns for human serum: (a) schematic of schlieren profiles; and (b) semilog plot of protein molecular weight versus electrophoretic mobility for particles electrophoresed on cross-linked polyacrylamide. (Reprinted with permission from K. Weber and M. Osborn, *J. Biol. Chem.*, **244**, 4404 (1969).)

typical electrophoresis pattern for human blood serum. In this figure the protein albumin (A), α_1-, α_2-, β-, and γ-globulin, and fibrinogen (ϕ) are fairly clearly resolved. The remaining peak in the figure is the boundary between the original buffer and the colloid. This "false boundary" moves little in an electrophoresis experiment and is obviously not considered in determining the percentages of different proteins in a mixture. Some additional details on moving-boundary electrophoresis are available in Longsworth (1959) and Probstein (1994).

12.9c Zone Electrophoresis

When separation rather than determination of mobility is the primary objective of an electrophoresis experiment, a technique called *zone electrophoresis* is quite widely employed. In zone electrophoresis, a supporting medium such as moist filter paper or a gel such as polyacrylamide is the location of the particle migration. The method thus resembles solid-liquid chromatography, and many of the substrates and analytical methods of solid-liquid chromatography are used in this electrophoretic procedure as well. As with chromatography, a spot or band of a mixture is applied to one end of the support medium. As the electrophoresis proceeds, spots or bands of the individual components appear at different locations along the axis of the voltage gradient. Sometimes the resolution is improved by following the electrophoresis by a chromatographic separation at right angles to the direction of the initial separation.

Zone electrophoresis is influenced by adsorption and capillarity, as well as by electroosmosis. Therefore evaluation of mobility (and ζ) from this type of measurement is considerably more complex than from either microelectrophoresis or moving-boundary electrophoresis. Nevertheless, zone electrophoresis is an important technique that is widely used in biochemistry and clinical chemistry. One particularly important area of application is the field of immunoelectrophoresis, which is described briefly in Section 12.11. Additional information on zone electrophoresis may be obtained from Probstein (1994) and Hunter (1981) and the references given there. Variants of zone electrophoresis also exist; see, for example, Gordon et al. (1988) for information on a variant known as *capillary zone electrophoresis* and Righetti (1983) for information on what is known as *isoelectric focusing*.

Figure 12.12b illustrates the application of gel electrophoresis to protein characterization. In this illustration a cross-linked polyacrylamide gel is the site of the electrophoretic migration of proteins that have been treated with sodium dodecyl sulfate. The surfactant dissociates the protein molecules into their constituent polypeptide chains. The results shown in Figure 12.12b were determined with well-characterized polypeptide standards and serve as a calibration curve in terms of which the mobility of an unknown may be interpreted to yield the molecular weight of the protein. As with any experiment that relies on prior calibration, the successful application of this method requires that the unknown and the standard be treated in the same way. This includes such considerations as the degree of cross-linking in the gel, the pH of the medium, and the sodium dodecyl sulfate concentration. The last two factors affect the charge of the protein molecules by dissociation and adsorption, respectively. Example 12.5 considers a similar application of electrophoresis.

* * *

EXAMPLE 12.5   *Estimation of Number of Nucleotides in Glycine tRNA Using Electrophoresis.* Synthetic DNA standards and RNA molecules were electrophoresed in 7 M urea solution on cross-linked polyacrylamide gels (Maniatis et al. 1975). A semilog plot of the number of nucleotides versus the mobility relative to xylene cyanol FF dye is linear and includes the points ($N = 100$, $u_{rel} = 0.33$) and ($N = 50$, $u_{rel} = 0.55$). Estimate the number of nucleotides in the glycine tRNA molecule of *Staphylococcus epidermidis* if it shows a relative mobility of 0.16.

Solution: The linear semilog plot means that these data follow the equation $\ln N = b + mu_{rel}$. Since we know two points from this plot, we can evaluate the constants m and b by simultaneous equations. This procedure yields $b = 5.63$ and $m = -3.09$. Combining these constants with the observed mobility allows the number of nucleotides in the unknown to be calculated by the formula $\ln N = 5.63 + (-3.09)(0.16) = 170$.   ∎

* * *

12.10 DETERMINING THE SURFACE CHARGE FROM ELECTROKINETIC MEASUREMENTS

In the quantitative sections of this chapter the primary emphasis has been on establishing the relationship between the electrophoretic properties of the system and the zeta potential. We saw in Chapter 11 that potential is a particularly useful quantity for the characterization of lyophobic colloids. In this context, then, the ζ potential is a valuable property to measure for a lyophobic colloid. For lyophilic colloids such as proteins, on the other hand, the charge of the particle is a more useful way to describe the molecule. In this section we consider briefly what information may be obtained about the charge of a particle from electrophoresis measurements.

12.10a Relation Between Surface Charge and Zeta Potential

We have lamented the fact that electrokinetic potentials cannot be evaluated independently to check the correctness of various theories. However, the charge of a protein can be evaluated from its titration curve. Therefore, if we can find a way to evaluate particle charge from electrokinetic data, the long-sought independent verification will be established. The net charge q of a particle is equal and opposite to the total charge in the double layer. The increment of charge in a spherical shell of radius r and thickness dr in the double layer is given by the area of the shell times its thickness times the charge density:

$$dq = 4\pi r^2 \rho^* dr \tag{78}$$

Integrating this expression over the entire double layer gives

$$q = -\int_{R_s}^{\infty} 4\pi r^2 \rho^* \, dr = \int_{R_s}^{\infty} \varepsilon \nabla^2 \psi \, 4\pi r^2 \, dr = 4\pi\varepsilon \int_{R_s}^{\infty} \frac{d}{dr}\left(r^2 \frac{d\psi}{dr}\right) dr \tag{79}$$

where the Poisson equation (11.24) has been substituted for ρ^*. Integration yields

$$q = 4\pi\varepsilon \left(r^2 \frac{d\psi}{dr}\right)_{R_s}^{\infty} = -4\pi\varepsilon R_s^2 \left(\frac{d\psi}{dr}\right)_{R_s} \tag{80}$$

where the derivative is evaluated at $r = R_s$.

Now Equation (46) is used to evaluate the derivative in Equation (80):

$$\left(\frac{d\psi}{dr}\right)_{R_s} = -\frac{\zeta}{R_s}(1 + \kappa R_s) \tag{81}$$

Substituting this result into Equation (80) gives

$$q = 4\pi\varepsilon\zeta R_s(1 + \kappa R_s) \tag{82}$$

for the charge enclosed by the surface of shear. (Contrast this with Equation (11.50).)

This discussion shows that the evaluation of charge from electrokinetic measurements involves all the complications inherent in the evaluation of ζ plus the additional restrictions of low potentials and spherical particles. Additional relationships have been developed that permit these restrictions to be relaxed, but we do not discuss these here.

12.10b Charge of Protein Molecules

We conclude this section by comparing briefly the charge on protein molecules as determined by electrophoresis measurements through Equation (82) and as determined by titration. Protein molecules carry acid and base functions in side groups along the macromolecule. In a strongly acidic solution amine groups will be protonated, and the protein will carry a positive charge. Addition of a known number of equivalents of strong base to a measured volume of protein solution results in a change of pH and a change in the state of charge of the protein. From the volume of the solution, the change of pH, and a knowledge of activity coefficients, the number of added equivalents of base that react with the protein may be determined. It should be noted that the added base may remove protons from either neutral groups or cationic groups.

Thus in acid solution a protein may have a charge corresponding to the binding of z H^+ ions: $+z$. After z OH^- ions have reacted with it, the molecule will have a *net* charge of *zero*. If the reactions consist exclusively of the removal of bound H^+ ions, the net charge (zero) would correspond to the actual charge of the particle. If all the reacting OH^- ions remove H^+ ions from neutral groups on the other hand, the molecule would be twice as highly charged as the initial species, with an equal number of positive and negative charges. In reality, both processes occur together, so it cannot be inferred that the point of equivalency—called the *isoionic point*—corresponds to an uncharged state. All that can be said is that the net charge is zero at the isoionic point. Addition of more base beyond this point will increase (still by both processes) the negative charge of the molecule even further. The isoionic point corresponds to a point at which the polyelectrolyte changes sign. This discussion shows that the net charge relative to the initial condition of the colloid is readily determined from titration curves.

The electrophoretic mobility of a protein solution may also be measured as a function of pH. By this technique it may also be observed that the colloid passes through a point of zero net charge at which its mobility is zero. The point at which charge reversal is observed electrophoretically is called the *isoelectric point*.

Figure 12.13 shows the relationship between the charge of egg albumin as determined by titration and by electrophoresis. The points were determined electrophoretically, and the solid line was determined by titration. The titration curve has been shifted so that the isoionic point and the isoelectric point match. It will be observed that the two independent charge determinations led to slightly different values. The charges determined electrophoretically are 60% of those determined analytically. If the titration results are multiplied by 0.60, the dashed line in Figure 12.13 is obtained. This shows clearly that the two determinations are identical in pH dependence, but raises the question as to the origin of the constant percentage difference.

There are several minor corrections that tend to reduce the discrepancy between the two curves, for example, corrections for relaxation and finite ion size. It should also be remembered that electrophoresis measures the net charge inside the surface of shear. To the extent that this diverges from the "surface" of the molecule, the two techniques may very properly

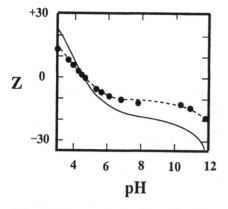

FIG. 12.13 Net charge of egg albumin versus pH. The points were determined by electrophoresis, and the solid line by titration; the broken line represents 60% of charge from titration. (Data from L. G. Longsworth, *Ann. NY Acad. Sci.*, **41**, 267 (1941). (Redrawn with permission from J. Th. G. Overbeek, Quantitative Interpretation of the Electrophoretic Velocity of Colloids. In *Advances in Colloid Science*, Vol. 3 (H. Mark and E. J. W. Verwey, Eds.), Wiley, New York 1950.)

"see" different charges for the colloid. Additional studies in this area, therefore, might help to clarify the relationship between the actual surface and the surface of shear.

We noted above that proteins display essentially the same mobility both as free molecules and when adsorbed on carrier particles. Adsorption clearly increases the radius of the kinetic unit appreciably, so this effect on mobility is unexpected. One way to rationalize this result is to assume that the protein adsorbs on the surface with very little alteration of the shape it has in free solution. Next assume that it is the radius of these molecular protuberances rather than the overall radius of curvature of the carrier that governs the mobility.

12.11 APPLICATIONS OF ELECTROKINETIC PHENOMENA

Throughout most of this chapter the emphasis has been on the evaluation of zeta potentials from electrokinetic measurements. This emphasis is entirely fitting in view of the important role played by the potential in the Derjaguin-Landau-Verwey-Overbeek (DLVO) theory of colloidal stability. From a theoretical point of view, a fairly complete picture of the stability of dilute dispersions can be built up from a knowledge of potential, electrolyte content, Hamaker constants, and particle geometry, as we discuss in Chapter 13. From this perspective the fundamental importance of the ζ potential is evident. Below we present a brief list of some of the applications of electrokinetic measurements.

12.11a Colloid Stability

As we have emphasized in some of the previous chapters, there are many practical situations in which coagulation is a process of considerable importance. Often all that is desired in these cases is either to maximize or minimize coagulation in some experimental system. Systems of practical interest are frequently so complex that theoretical models apply to them only qualitatively at best. In this context the concept of zeta potential emerges as a valuable practical parameter. If two systems of different ζ are compared — all other factors being equal — the one that has the higher ζ potential is expected to be more stable with respect to coagulation and the one with the lower potential less stable. The second case is particularly important. At the isoelectric point electrophoretic mobility is zero, ζ is zero, and the potential energy of repulsion between particles is minimal. Thus electrophoresis measurements can be used as an indicator for optimum conditions for coagulation. In this type of application the technique is used as a null detector; hence it is independent of any model or equation for interpretation.

12.11b Sewage Treatment

One important — if not so appealing — example of the above application is in sewage treatment. Industrial wastewater and domestic sewage contain an enormous assortment of hydrophilic and hydrophobic debris of technological and biological origin. The concentration of surface-active materials in sewage from household detergents alone is about 10 ppm. In addition, sewage abounds in amphipathic materials of natural and biological origin. These substances tend to adsorb on and impart a charge to the suspended solid and liquid particles in the polluted water. Negative zeta potentials in the range of 10 to 40 mV are fairly typical for the suspended particles in sewage.

A typical purification scheme consists of adding $NaHCO_3$ and $Al_2(SO_4)_3$ (alum) to water with agitation. The aluminum ion undergoes hydrolysis and precipitates as a gelatinous, polymeric hydrated oxide. Suspended material is enmeshed in this amorphous precipitate, which produces *flocs* by bridging the particles together. The polymeric nature of the "$Al(OH)_3$" precipitate permits us to compare it with protein in its ability to coat particles and impart to the carrier particles its own characteristic potential. Like proteins, $Al(OH)_3$ is also capable of reacting with both H^+ and OH^- so that these ions determine the charge of the suspended units, whether these are flocs formed by the $Al(OH)_3$ network or individual particles with an adsorbed layer of $Al(OH)_3$. In either case the charge is pH sensitive, the isoelectric point occurring near pH 6. It is under these pH conditions, then, that the flocculating effectiveness of the precipitate is optimum. In fact, the pH is often adjusted so that the hydrous aluminum oxide surface has a slightly positive value of ζ (about 5 mV). This promotes further interaction with slightly anionic polymeric materials that are also added to further build up and strengthen flocs. Once adequate flocculation has been accomplished, the dispersed particles are removed by sedimentation, centrifugation, or filtration.

Numerous other applications could be listed in which electrokinetic characterization provides a convenient experimental way of judging the relative stability of a system to coagulation. Paints, printing inks, drilling muds, and soils are examples of additional systems with properties that are extensively studied and controlled by means of the ζ potential.

12.11c Environmental Remediation

Contamination in low-permeability soil is a problem of major importance in environmental remediation. Traditional treatments of contaminated soils include bioremediation methods, vapor extraction, and what are known as "pump-and-treat" methods. However, poor accessibility to the contaminants and difficulties in delivering reagents used for treatment make these current *in situ* methods very ineffective. Electroosmosis (combined possibly with one or more of the traditional techniques) can potentially serve as an alternative *in situ* treatment process, as shown in Figure 12.14.

Electroosmosis has been used for dewatering fine sands, clays, and silts since the 1930s, and its application in environmental remediation is somewhat similar to its earlier uses. For example, as illustrated in Figure 12.14, an electrical potential difference is set up between two embedded granular electrodes on either side of the contaminated zone. Water injected into the soil at the anode is made to flow through the contaminated zone under the action of electroosmosis, thus bringing the contaminants (e.g., metals and organics) to the surface at the cathode region for further treatment and disposal. Advantages of such a technique include a relatively uniform flow through heterogeneous regions, a high degree of control of the flow direction, and very low power consumption.

Figure 12.14 illustrates a vertical flow arrangement, but other configurations are also possible. The process as illustrated in Figure 12.14 is being developed by Monsanto Company and is known as the "lasagna process" because of the layered structure of the treatment zones (Ho et al. 1993).

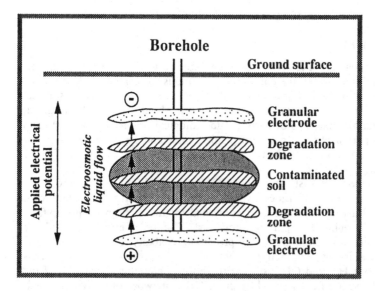

FIG. 12.14   The use of electroosmosis in environmental remediation (known as the lasagna process). (Redrawn with permission of Ho et al. 1993.)

The use of electroosmosis in groundwater remediation is not a well-established technology, and a number of technical issues are currently being investigated. However, it illustrates the types of large-scale practical applications of charge-induced phenomena in technology. Another interesting application of electroosmosis, for extracting bitumen from oil sands, is described by Masliyah (1994).

12.11d Immunoelectrophoresis

In addition to these applications in which ζ is used to monitor for optimum coagulation conditions, there are applications of electrophoresis that explicitly depend on mobility or differences in mobility for their usefulness. We have already noted that zone electrophoresis is similar in many ways to chromatography. One important application of the ability of electrophoresis to segregate materials by mobility is *immunoelectrophoresis*. This technique uses known immunochemical reactions between antigen and antibody for the identification of proteins separated electrophoretically. Experimentally, an antigen mixture is subjected to electrophoresis on a suitable medium (usually agar gel). Next the antibody mixture is introduced into a slit cut in the gel parallel to the axis of the separation. The antigen and antibody components then diffuse toward one another, producing an arc-shaped precipitate where the two fronts meet.

Tests of this sort are particularly useful for comparing either two antigen preparations (against a single antibody) or two antibody preparations (against a single antigen). In such a comparison one of the samples serves as a control, and differences between the two are revealed by an unpaired arc of precipitate at a particular location along the path of separation.

Alternatively, electrophoretic separation in one direction may be followed by a second electrophoresis in a perpendicular direction, the latter into a gel containing antibodies. This technique is called *crossed immunoelectrophoresis* and combines high resolution with the possibility of quantification by measuring the area of the precipitate formed. Figure 12.15 is a

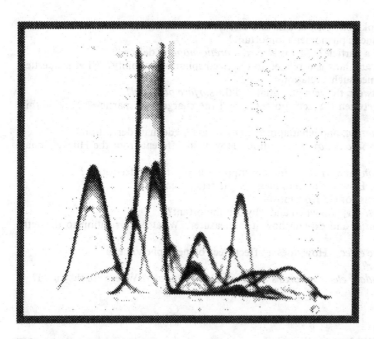

FIG. 12.15 Crossed immunoelectrophoresis of human serum with rabbit antihuman serum. (Redrawn with permission from B. Weeke, in *A Manual of Quantitative Immunoelectrophoresis: Methods and Applications* (N. H. Axelsen, J. Krøll, and B. Weeke, Eds.), Universitetsforlaget, Oslo, Norway, 1973.)

photograph of the peaks of antigen-antibody precipitates formed by crossed immunoelectro-phoresis of human serum interacting with rabbit antihuman serum.

12.11e Electrodeposition

Electrodeposition is another direct application of electrophoretic mobility. In this process, as in electroplating with metals, the substance to be coated is made into an electrode of opposite charge from the particles to be deposited. At one time, natural rubber latex was extensively fabricated in this way. Paint coatings that are quite dense and coherent with little tendency to sag or run can be prepared by electrodeposition. If the deposited layer has insulating proper-ties, this technique is also self-regulating, producing a uniform thin covering of very good quality.

12.11f Applications of Other Electrokinetic Phenomena

Although electrophoresis is the most important of the electrokinetic methods, it is not the only one with practical applications. We already noted in Section 12.7 that streaming potentials could be quite hazardous in low-conductivity, highly flammable substances such as purified hydrocarbons. In this case our knowledge of the effect enables us to minimize it. Dewatering fine suspensions that are not amenable to filtration is an application of electroosmosis, as mentioned in Section 12.1b. Peat, clay, and other minerals have been dewatered this way, and water may be removed from moist soil prior to excavation by electroosmosis. Electrodes are driven into the ground, with the cathode in the form of a perforated pipe. The surface of the soil particles carries a negative charge; therefore the diffuse part of the double layer is positive, and the solution moves toward the cathode. The water that collects in the cathode is subse-quently removed by pumping. Similar ideas have been suggested in studies of salt rejection characteristics of reverse osmosis membranes (Jacazio et al. 1972; Sonin 1976).

REVIEW QUESTIONS

1. List the different electrokinetic phenomena and discuss their similarities and differences.
2. Why are these electrokinetic phenomena important?
3. What is the *mobility* of a particle? How is *electrophoretic mobility* defined?
4. What is *zeta potential*, and how is it related to the electrophoretic mobility? What properties of the dispersion influence such a relation?
5. What is the relation between the zeta potential and the *surface potential*?
6. What is the relation between the zeta potential and the charge on a particle? How is this relation determined?
7. What is the *Hückel equation*, and what approximations are made in its derivation?
8. What is the *Helmholtz-Smoluchowski* equation? How is it different from the Hückel equa-tion?
9. What is *Henry's equation*, and what are the assumptions implicit in its derivation?
10. What is *electroosmosis*? How is it used to measure the zeta potential?
11. Give an example of the use of electroosmosis.
12. What is *surface conductivity*, and when and why is it important?
13. Define *streaming potential* and explain how it is measured. What is its relation to the zeta potential?
14. What is the *viscoelectric effect*? How does it affect electrokinetic phenomena?
15. What is *surface of shear*?
16. What are *moving-boundary electrophoresis* and *zone electrophoresis*? Where are they used?

REFERENCES

General References (with Annotations)

Crow, D. R., *Principles and Applications of Electrochemistry*, 4th ed., Blackie A and P, London, 1994. (Undergraduate level. An introduction to electrochemistry; requires only basic physical chemistry as prerequisite.)

Hunter, R. J., *Zeta Potentials in Colloid Science: Principles and Applications*, Academic Press, London, 1981. (Advanced. This is a research-level monograph on electrokinetic phenomena and electroviscous and viscoelectric effects and contains extensive details on the subject.)

Masliyah, J. H., *Electrokinetic Transport Phenomena*, Alberta Oil Sands Technology and Research Authority, Edmonton, Alberta, Canada, 1994. (Graduate and undergraduate levels. An excellent introduction to transport processes relevant to applications of electrokinetic phenomena. Most of the material is accessible to undergraduate students. Chapter 12 presents some very practical applications of electrokinetic phenomena of interest to engineers.)

Overbeek, J. Th. G., Electrokinetic phenomena. In *Colloid Science*, Vol. 1 (H. R. Kruyt, Ed.), Elsevier, Amsterdam, Netherlands, 1952.

Overbeek, J. Th. G., and Lyklema, J., Electric Potentials in Colloidal Systems. In *Electrophoresis*, Vol. 1 (M. Bier, Ed.), Academic Press, New York, 1959.

Overbeek, J. Th. G., and Wiersma, P. H., The Interpretation of Electrophoretic Mobilities. In *Electrophoresis*, Vol. 2 (M. Bier, Ed.), Academic Press, New York 1967. (Graduate and undergraduate levels. The above three articles, by three of the pioneers in colloid science, are some of the classical references on electrophoresis.)

Probstein, R. F., *Physicochemical Hydrodynamics*, 2d ed., Wiley-Interscience, New York, 1994. (A primarily first-year-graduate-level textbook on colloidal and hydrodynamic phenomena of interest in practice. Has good discussions on electrokinetic phenomena.)

Other References

Abramson, H. A., Moyer, L. S., and Gorin, M. H., *Electrophoresis of Proteins*, Hafner, New York, 1964.

Atkins, P. W., *Physical Chemistry*, 5th ed., W. H. Freeman, New York, 1994.

Carlson, C. F., U.S. Patent No. 2,221,776 (1938; 1940).

Gordon, M. J., Huang, X., Pentoney, S. L., Jr., and Zare, R. N., *Science*, **242**, 224 (1988).

Ho, S. V., Sheridan, P. W., Athmer, C. J., Brodsky, P. H., Heitkamp, M. A., and Brackin, J. M., Integrated *In Situ* Remediation Technology—The Lasagna Process, paper presented at the American Chemical Society meeting, Atlanta, GA, September 1993 (Paper 279, Session 43).

Jacazio, G., Probstein, R. F., Sonin, A. A., and Yung, D., *J. Phys. Chem.*, **76**, 4015 (1972).

Kitahara, A., and Watanabe, A. (Eds.), *Electrical Phenomena at Interfaces: Fundamentals, Measurements, and Applications*, Marcel Dekker, New York, 1984.

Kurita, T., *Electrophotography*, **3**, 16 (1961).

Loeb, A. L., Overbeek, J. Th. G., and Wiersema, P. H., *The Electrical Double Layer Around a Spherical Colloid Particle*, Massachusetts Institute of Technology Press, Cambridge, MA, 1960.

Longsworth, L. G., Moving Boundary Electrophoresis—Practice. In *Electrophoresis*, Vol. 1 (M. Bier, Ed.), Academic Press, New York, 1959.

Lyne, M. B., and Aspler, J. S., Ink-Paper Interactions in Printing—A Review. In *Colloids and Surfaces in Reprographic Technology* (M. Hair and M. D. Croucher, Eds.), American Chemical Society, Washington, DC, 1982.

Maniatis, T., Jeffrey, A., and van de Sande, H., *Biochemistry*, **14**, 3787 (1975).

Metcalfe, K. A., *J. Sci. Instrum.*, **32**, 74 (1955).

Metcalfe, K. A., *J. Sci. Instrum.*, **33**, 194 (1956).

Ohyama, Y., Kurita, T., and Takahashi, Y., *Electrophotography*, **3**, 26 (1961).

Righetti, P. G., *Isoelectric Focusing: Theory, Methodology and Applications*, Elsevier Biomedical, Amsterdam, Netherlands, 1983.

Shaw, D. J., *Electrophoresis*, Academic Press, New York, 1969.

Sonin, A. A., Osmosis and Ion Transport in Charged Porous Membranes: A Macroscopic Mechanistic Model. In *Charged Gels and Membranes I* (E. Sélégny, Ed.) D. Reidel, Dordrecht, Holland, 1976.

PROBLEMS

1. Particles of Fe_2O_3 with an average diameter of 1 μm were dispersed in xylene containing 5×10^{-3} mole liter^{-1} of copper(I) oleate. These showed an electrophoretic mobility of 0.110 μm s^{-1} V^{-1} cm. The conductivity of the solution was 4.7×10^{-10} ohm $^{-1}$ cm^{-1}, indicating an ion

concentration about 10^{-11} M.* Calculate κ^{-1} for this concentration. Which limiting form of Equation (40) is most applicable in this system? Would the same conclusion be true for a 5×10^{-3} M aqueous solution of a $1:1$ electrolyte? What is ζ for these particles? For xylene, $\varepsilon_r = 2.3$ and $\eta = 0.0065$ P.

2. Criticize or defend the following proposition: Zeta potentials for three different polystyrene latex preparations were calculated by the Helmholtz-Smoluchowski equation from electrophoresis measurements made in different concentrations of KCl.†

Latex designation	$R_s \times 10^8$ (cm)	ζ (mV)		
		10^{-1} M KCl	10^{-2} M KCl	10^{-3} M KCl
L	475	21	29	40
M	610	29	39	53
N	665	34	47	64

These zeta potentials are inaccurate because the range of κR_s values exceeds the range of validity for the Helmholtz-Smoluchowski equation. The nature of the error is such as to make the estimated values of ζ too low.

3. The electrophoretic mobility of sodium dodecyl sulfate micelles was determined by the moving-boundary method after the micelles were made visible by solubilizing dye in them. This quantity was measured at the critical micelle concentration (CMC) in the presence of various concentrations of NaCl. The radius of the micelles was determined by light scattering‡

Moles of NaCl liter^{-1}	$u \times 10^4$ (cm^2 s^{-1} V^{-1})	κR_s
0.00	4.55	0.61
0.05	3.63	1.69

Estimate from Figure 12.5a the appropriate value of C to be used in Equation (40) according to Henry's equation. Calculate ζ using these estimated C values. Figure 12.5b shows that Henry's equation overestimates C (and therefore underestimates ζ). Estimate C from Figure 12.5b using the curve for the ζ value that is nearest — on the high side — to the values obtained by using Henry's equation. Reevaluate ζ on the basis of these "constants." For this system, $\varepsilon_r = 78.5$ and $\eta = 0.0089$ P.

4. The accompanying mobility data for colloidal SiO_2 at a constant ionic strength of 10^{-3} M reveal the superpositioning of specific chemical effects on general electrostatic phenomena. Adjustment of pH was made by addition of HNO_3 or KOH, maintaining the ionic strength. The following results were obtained:§

pH of solution:	2.0	3.0	4.0	5.0	6.0	7.0	8.0	9.0	10.0
				$u \times 10^4$ (cm^2 s^{-1} V^{-1})					
SiO_2	0	-1.4	-1.7	-2.0	-2.3	-2.5	-2.6	-2.8	-3.0
$SiO_2 + 10^{-4}$ M La(NO$_3$)$_3$	0	-1.1	-1.2	-1.2	-1.1	-0.1	$+2.2$	$+0.5$	-1.2

Criticize or defend the following propositions: H^+ and OH^- are potential determining for SiO_2 — in the absence of a hydrolyzable cation — with an isoelectric point of 2.0. For solid La(OH)$_3$ the zero point of charge is known (by independent studies) to be 10.4. The solid

*Koelmans, H., and Overbeek, J. Th. G., *Discuss. Faraday Soc.*, **18**, 52 (1954).
†Kitahara, A., and Ushiyama, H., *J. Colloid Interface Sci.*, **43**, 73 (1973).
‡Stigter, D., and Mysels, K. J., *J. Phys. Chem.*, **59**, 45 (1955).
§James, R. O., and Healy, R. W., *J. Colloid Interface Sci.*, **40**, 42 (1972).

surface apparently becomes coated by $La(OH)_3$ at higher pH levels and goes through a transition from one character to another at intermediate pH values.

5. In their study of the effects of hydrolyzable cations on electrokinetic phenomena (see Problem 4), James and Healy compared the electrophoretic behavior of colloidal silica with the streaming potential through a silica capillary. In both sets of experiments the solution was 10^{-3} M KNO_3 and 10^{-4} M $Co(NO_3)_2$. The following results were obtained:

pH	6.0	7.0	7.5	8.0	9.0	10.0
ζ (mV) from streaming potential	-65	-55	-30	$+10$	$+25$	$+20$
$u \times 10^4$ (cm^2 s^{-1} V^{-1})	-2.5	-2.5	-2.2	-1.8	$+0.5$	$+0.3$

The silica surface area-to-solution volume ratio was 2×10^{-3} m^2 liter^{-1} for the streaming potential experiment and 1.0 m^2 liter^{-1} for the electrophoresis experiment. Calculate ζ_{HS} at each pH from the electrophoresis data ($\eta = 0.00894$ P, $\varepsilon_r = 78.5$). Propose an explanation for the charge reversal behavior of the silica. Discuss the origin of the difference between ζ_{HS} and $\zeta_{St\,Pot}$ in terms of this model.

6. Somasundaran and Kulkarni[*] measured the streaming potential of 10^{-3} N KNO_3 against quartz at 25°C, obtaining the following results:

E_{str} (mV)	-9.0	-18.0	-26.0	-35.0
p (mm Hg)	50	100	150	200

Use these data to evaluate ζ/k. What would be the value of the ratio V/I for this system? What would be the rate of volume displacement if a current of 1.0 mA flowed through the apparatus? Evaluate ζ for the quartz-solution interface, assuming $\Lambda \approx 145$ cm^2 eq^{-1} ohm^{-1} for 10^{-3} N KNO_3.

7. It has been estimated[†] that a specific conductivity of 10^3 picomho m^{-1} would provide an ample margin of safety against electrokinetic explosions for the handling of refined petroleum products. These authors also measured the concentrations of various additives needed to reach this level of conductivity:

Solvent	Additive	Concentration (kmol m^{-3})
Benzene	Tetraisoamyl ammonium picrate	1×10^{-4}
Benzene	Calcium diisopropylsalicylate	5×10^{-3}
Gasoline	Ca salt of di-(2-ethylhexyl)sulfosuccinic acid	1×10^{-3}
Gasoline	Cr salt of mono- and dialkyl (C_{14}–C_{18}) salicylic acid	2.5×10^{-6}

Calculate the apparent value of the equivalent conductance Λ for each of these electrolytes in the conventional units cm^2 eq^{-1} ohm^{-1}. How do the Λ values of these compounds compare with Λ_0 for simple electrolytes in aqueous solutions?

8. The pH variation of the electrophoretic mobility of solid $Th(OH)_4$ in 10^{-2} M HNO_3–KOH electrolyte is as follows:[‡]

pH	7.6	8.0	9.0	9.6	10.0	10.3	10.6	11.3
$u \times 10^4$ (cm^2 s^{-1} V^{-1})	$+2.4$	$+2.2$	$+1.3$	$+1.0$	-0.1	-1.1	-1.5	-1.8

Use these data to evaluate the isoelectric point for $Th(OH)_4$. Since H^+ and OH^- appear to be potential determining, ψ may be estimated at various pH levels according to Equation (11.1) if we identify the isoelectric point with the true point of zero charge. Compare these values

---

[*]Somasundaran, P., and Kulkarni, R. D., *J. Colloid Interface Sci.*, **45**, 591 (1973).

[†]Klinkenberg, A., and Poulston, B. V., *J. Inst. Pet.*, **44**, 379 (1958).

[‡]James, R. O., and Healy, T. W., *J. Colloid Interface Sci.*, **40**, 42 (1972).

with values of ζ calculated by means of the Helmholtz–Smoluchowski equation ($\eta = 0.0089$ P, $\varepsilon_r = 78.5$). Are the results qualitatively (quantitatively?) consistent with Figure 12.8?

9. The aggregation number n and radius of sodium dodecyl sulfate micelles (by light scattering) and the zeta potential (from electrophoresis, by an accurate formula) were determined in the presence of various concentrations of NaCl.*

Moles of NaCl liter^{-1}	ζ (mV)	$R_s \times 10^8$ (cm)	κR_s	n
0.01	92.3	22.1	0.86	89
0.03	80.9	23.0	1.32	100
0.10	68.3	24.0	2.40	112

Use Equation (82) to estimate the charge of the micelles. What approximation(s) in the derivation of Equation (82) prevents this expression from applying exactly to this system? On the basis of the charges evaluated by Equation (82), calculate the ratio of charge to aggregation number, the effective degree of dissociation, of these micelles. How do these results compare with the numbers given in Table 8.1?

10. An electrophoretic technique that is especially interesting for the study of proteins is called "isoelectric focusing." In this method electrophoresis is carried out across a medium that supports a pH gradient. The pH gradient and cell polarity are such that the cathode end of the column is relatively basic. Thus a positively charged protein gradually loses its charge as it migrates, finally coming to rest at a pH corresponding to its isoelectric point. Carlstrom and Vesterberg† used this method to study the heterogeneity of peroxidase from cow's milk. After focusing was achieved, the column was drained and the pH and absorbance (at 280 nm) of successive fractions of eluent were measured:

Fraction number	Absorbance	pH	Fraction number	Absorbance	pH
12	0.9	9.830	28	1.4	9.49
14	3.0	9.800	30	0.9	9.45
16	2.1	9.750	32	0.6	9.38
18	1.2	9.700	34	0.8	9.31
20	2.7	9.690	36	0.5	9.30
21	2.2	9.685	38	0.3	9.28
22	2.8	9.680	40	0.4	9.23
24	1.6	9.600	41	0.5	9.16
26	1.2	9.550	42	0.4	9.10

How many components does this sample apparently contain? What are the values of the isoelectric points for each?

*Stigter, D., and Mysels, K. J., *J. Phys. Chem.*, **59**, 45 (1955).
†Carlstrom, A., and Vesterberg, D., *Acta Chem. Scand.*, **21**, 271 (1967).

13

Electrostatic and Polymer-Induced
Colloid Stability

You, who are blessed with shade as well as light, you who are gifted with two eyes, endowed with a knowledge of perspective, and charmed with the enjoyment of various colours, you, who can actually see an angle, and contemplate the complete circumference of a Circle in the happy region of Three Dimensions—how shall I make clear to you the extreme difficulty which we in Flatland experience in recognizing one another's configuration?

From Abbott's *Flatland*

13.1 INTRODUCTION

13.1a What Is Colloid Stability?

The term *colloid stability* stands for the ability of a dispersion to resist coagulation. The stability of dispersions may be either *kinetic* or *thermodynamic* and has been traditionally the primary focus of colloid science at least dating back to the ancient Egyptians. Kinetic stability is a consequence of a force barrier against collisions between the particles and possible coagulation subsequently. As we saw in Section 10.2, in such cases, coagulation is preferred because of the resulting reduction in thermodynamic "free" energy, but the interaction energy barrier in the interparticle energy is larger than the thermal energy and any other applied energy of the particles. Dispersions of charged latex particles in low-electrolyte environments provide examples of kinetically stabilized colloids. In the case of thermodynamic stability, coagulated states correspond to an *increase* in free energy and, therefore, are thermodynamically unfavorable.

We saw in Chapter 10 that the van der Waals force between particles in a dispersion is usually attractive and is strong at short interparticle separations. Therefore, if there are no repulsive interactions between particles, the dispersion will be unstable and coagulate. The protection against van der Waals attraction is usually provided in one of two ways:

1.   As we saw in Chapter 11, surfaces of colloidal particles typically acquire charges for a number of reasons. The electrostatic force that results when the electrical double layers of two particles overlap, if repulsive, serves to counteract the attraction due to van der Waals force. The stability in this case is known as *electrostatic stability*, and our task is to understand how it depends on the relevant parameters.

2.   In many practical instances (see Vignette I.5), electrostatic repulsion is not a convenient option. In such cases, a suitable polymer that adsorbs on the particle surfaces may be added to the dispersion. The resulting polymer layer masks the attraction and may also provide a repulsive force, partly due to pure steric effect, when the polymer layers on two interacting particles attempt to overlap with each other. This is what is known as *polymer-induced stability*. Polymer-induced stability is often referred to as *steric stability* for the above

reason, but the role of polymers in a colloidal dispersion is actually much more complex than this, and we try to get a feel for this below.

Our primary focus in this chapter is on kinetic stability of dispersions arising due to either electrostatic forces or polymer-mediated forces.

13.1b Why Is Colloid Stability Important?

Colloid stability enters our daily life in many different ways. A visit to the kitchen provides numerous examples of food colloids with microstructure and stability that are, in no small measure, an important aspect of their appeal to the palate! For example, mayonnaise—a mixture of vegetable oil, egg yolk, and vinegar or lemon juice—is an emulsion of oil in water and is stable because the lecithin molecules in the egg yolk provide the needed stability. Milk is another example. We have seen others in the vignettes in Chapters 1 and 4.

Electrostatic stability plays a dominant role in many separation processes, such as filtration of industrial wastewaters. Coagulation aids (known as *coagulants*) are routinely used to improve the effectiveness of separation processes in such cases. Polymer-induced stability is often the method of choice, particularly in the case of concentrated dispersions; for example, many pharmaceutical preparations, paints, inks, and liquid toners depend on surfactants or polymer additives for ensuring stable preparations. We see in Section 13.2 that in the case of concentrated dispersions both thermodynamic and kinetic issues often become very important.

Colloid stability plays a role in many processing operations as well. Vignette XIII provides an example of current interest.

VIGNETTE XIII   DISPERSION-BASED PROCESSING:
Steric Stabilization and Environmentally
Friendly Polymerization Processing

Perhaps the term *steric stabilization* normally brings to mind dispersions such as paints and food colloids, which are usually stabilized against coagulation by a layer (sometimes a "brush") of polymer chains that mask the van der Waals attraction at short interparticle separations. However, there are other applications in which steric stabilization (and the steric effect, in general) provides interesting and exciting possibilities. We have already come across the role of the steric effect in the case of the so-called *stealth liposomes* as drug delivery vehicles in Vignette I.3 in Chapter 1. Here, let us look at another, the use of specially designed polymeric surfactants (*designer* surfactants?) for manufacturing polymer particles in an *environmentally friendly* fashion!

First, let us look at the classical manufacturing route to the large-scale production of many polymers of commercial importance (e.g., polystyrene, poly(vinyl chloride), poly(acrylic acid), etc.). These polymers are synthesized typically with water or an organic solvent as the dispersing medium (depending on whether the polymer is water insoluble or water soluble, respectively). This is a heterogeneous polymerization process that has two or more phases with the monomer or the polymer (or both) in a finely divided form (see Section 8.10b). The final product, in the form of a powder or a dispersion, is then used in subsequent fabrication steps to produce the end products in desired shapes through compaction, pressing, and so on. The particle sizes in the polymerization step are usually controlled by the addition of suitable surfactants that stabilize the particles against coagulation to prevent a large size distribution. To begin, this is already an example of steric stabilization in action! But what is wrong with it? The problem is that this processing route produces a large amount of hazardous waste, namely, either contaminated water or organic solvents (like chlorinated hydrocarbons and toluene), or both.

How can we minimize the damage to the environment? Common sense says that prevention is better than cure. If one can use an environmentally friendly solvent in the processing step, the discharge of hazardous chemicals and the need for environmental post-treatment can be avoided. This is an approach that is currently being explored in polymer processing as well as in other manufacturing technologies (DeSimone et al. 1994). In the case of the polymerization operations, one possibility is to use supercritical CO_2, which resembles a liquid but has low viscosity, as the solvent. However, most polymers, with the exception of fluoropolymers, do not dissolve in CO_2. Here is the point at which specially designed surfactants enter the picture. What one needs is a molecularly engineered polymer surfactant with a backbone that can adsorb on the growing polymer particle, but with side chains or "tentacles" that are CO_2 loving so that the growing polymer particles can stay dispersed in the solvent (see Figure 13.1). In addition, adsorption of the surfactant backbone anchors the polymer on the surface of the particles, and the loops and the tails of the polymer provide a steric barrier against coagulation as desired. Recent research shows that such an approach is feasible.

We are not concerned with molecular engineering of surfactants in this chapter; our objective is to introduce the basic concepts concerning colloid stability, including the role of polymer (adsorbed or grafted on the particles or simply dispersed in the solvent along with the particles) in imparting stability to colloidal dispersions. In addition to highlighting the role of polymeric stabilization in a novel (and *politically correct!*) context, the application described in this vignette brings to the forefront a number of topics of interest in colloid and surface chemistry we have discussed in previous chapters.

13.1b Focus of This Chapter

The stability and the structure of dispersions (*structure* here means the spatial organization of the colloidal particles) are topics of considerable research activity currently; there is a lot that we do not know despite the long-standing focus on these topics in colloid science. The first step in approaching problems in this area is to study the origin and the nature of the interparticle forces and how they affect coagulation in dilute dispersions. This is what we focus on in this chapter.

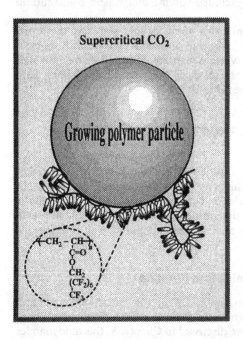

FIG. 13.1 A polymer particle in supercritical CO_2 is protected against coagulation by the steric action of specially designed polymer chains with fluorine-containing tails on the polymer backbone. (Redrawn with permission from K. C. Fox, *Science*, **265**, 321 (1994), and DeSimone et al. 1994.)

1.   In Section 13.2, we begin with a closer look at one of the applications we alluded to in Vignette I.5 in Chapter 1 (colloidal processing of ceramics) in order to gain some perspective on how interparticle forces influence the structure of a dispersion.

2.   The influence of interparticle interaction energies and how the energies vary with solution chemistry is the first key in understanding colloid stability. Since we have already developed a fairly detailed understanding of van der Waals and electrostatic forces in Chapters 10 and 11, we are already in a position to look at the role of interparticle energies on electrostatic stability. In Section 13.3, we investigate this and discuss the threshold value of electrolyte concentration, known as the *critical coagulation concentration* (CCC), beyond which a colloid coagulates rapidly. We also explore the role of ionic valences and a rule known as the Schulze-Hardy rule in this section.

3.   A more quantitative measure of stability, known as the *stability ratio*, can be obtained by setting up and solving the equation for diffusive collisions between the particles. Quantitative formulations of stability, known as the *Smoluchowski and Fuchs theories of colloid stability*, are the centerpieces of classical colloid science. These and related issues are covered in Section 13.4.

4.   Theoretical studies of the role of polymer additives lag behind their analogs in electrostatic stability since polymer molecules have considerably more configurational freedom and since the interaction of the polymer molecules with the solvent is an inseparable part of phenomena in polymer-colloid mixtures. We begin with some of the general issues and a thermodynamic analysis of the role of polymer on stability in Section 13.5.

5.   In contrast to the situation in the case of van der Waals and electrostatic forces, very little is known about polymer-induced forces. The development of the surface force apparatus and scanning tunneling and atomic force microscopies have begun to shed light on this very difficult topic. In Section 13.6, we take a brief look at some of the polymer-induced forces of interest in colloid stability and structure.

6.   In the final section, we build on the thermodynamic theories of polymer solutions developed in Chapter 3, Section 3.4, to provide an illustration of how a thermodynamic picture of steric stabilization can be built when excluded-volume and elastic contributions determine the interaction between polymer layers.

Despite the fact that there is much that is unknown about colloid stability, the topics covered in the chapter are sufficient to solve many routine problems of industrial interest, particularly in the case of electrostatic stability. More advanced information on polymer-induced forces is available in specialized monographs (Napper 1983; Israelachvili 1991; Sato and Ruch 1980) and in other texts on colloid science (Hunter 1987).

13.2   INTERPARTICLE FORCES AND THE STRUCTURE AND STABILITY OF DISPERSIONS

Our objective in this chapter is to establish the quantitative connections between interparticle forces and colloid stability. Before we consider this it is instructive to look at the role of interaction forces in a larger context, that is, the relation between interparticle forces and the microstructure of dispersions and the factors that determine such a relation. These aid us in appreciating the underlying theme of this chapter, namely, the manipulation of interparticle forces to control the properties of dispersions.

13.2a   Competition Between Thermodynamics and Kinetics

Before we proceed, however, it is important to review briefly the roles thermodynamic and kinetic considerations play in determining the structure. In some cases, the distinction is easy to establish. In the case of the association colloids we discussed in Chapter 8, thermodynamics determined the formation and the structure of the colloidal particles and their subsequent transformations to more complex structures at higher concentrations of the particles. In contrast, lyophobic colloids, such as the silica dispersion we saw in Example 5.4, are usually thermodynamically unstable, but are stable kinetically if the surface charges or potentials are sufficiently large. This is what we saw in Section 10.2 in our discussion of the relation between interparticle forces and colloid stability. In fact, arguments presented there form the backbone of the classical theory of electrostatic stability of colloids known as the *Derjaguin-Landau-Verwey-Overbeek* (DLVO) theory, based on kinetic arguments. We discuss this further in Sections 13.3 and 13.4.

Frequently, however, the stability and, more generally, the microstructure and the macroscopic states of dispersions are determined by kinetic *and* thermodynamic considerations. Thermodynamics dictates what the equilibrium state will be, but it is often the kinetics that determines if that equilibrium state will be reached and how fast. This becomes a consideration of special importance in practice since most processing operations involve dynamic variables such as flow, sedimentation, buoyancy, and the like. Although a detailed discussion of this is beyond our scope here, it is important that we consider at least one example so that we can place some of the topics we discuss in this chapter in proper context.

For this, let us consider the subject of Vignette I.5 on ultrastructural processing of ceramics, highlighted in Chapter 1.

13.2b   Interplay Between Interparticle Forces and Structure

We have already seen in previous chapters and in a number of vignettes (for example, Vignette I.4 on *electrophoretic imaging devices* in Chapter 1 and Vignette IV on the rheology of chocolate in Chapter 4) that the microstructure and stability of dispersions determine the quality, processability, and properties of many products and devices. The colloidal processing of ceramics is a particularly interesting example since our goal in such a processing route is to optimize the colloidal forces so that we can minimize the competition between thermodynamics and kinetics.

First, let us begin with a consideration of how interparticle forces and the concentration of the particles work together to determine the structure of dispersions. Figure 13.2 is a schematic representation of the microstructure (i.e., the local arrangement of the particles) in a monodisperse colloid in terms of the particle concentration as the relative magnitude of attractive and repulsive forces varies between the extremes. The figure is like a "phase" diagram and shows roughly three regions that differ from each other in terms of the expected microstructure of the dispersion:

1.   *Overall interparticle forces dominated by strong repulsion*: When the particles have large enough surface charges, the overall interaction force becomes strongly repulsive and could extend over large distances (of the order of particle dimensions or larger; i.e., the Debye-Hückel thickness, κ^{-1}, is very large). As we saw in Section 10.2, for all practical purposes this corresponds to a *thermodynamically* stable dispersion; in fact, because of the large repulsion, the particles can organize themselves in crystalline structures even at volume fractions of the order of 0.001 or lower, depending on the magnitude and the range of repulsion.

The type of crystalline structure that is formed depends on the concentration of the particles as well as the magnitude of the Debye-Hückel thickness. For large Debye-Hückel thicknesses a body-centered cubic crystal is formed, whereas for smaller values a face-centered cubic crystal is preferred. An example of the latter observed experimentally in a dispersion of latex spheres is shown in Figure 13.3. Note that this crystallization phenomenon is analogous to crystallization of simple atomic fluids, as is evident from Figure 13.3a, which shows the coexistence of a crystal with a "liquidlike" structure.

Because of the low volume fractions at which such transitions can occur, the typical interparticle spacing in these cases can be quite large. Therefore charged colloids can be used as model systems to study visually the influence of charge effects on phase transitions in colloidal as well as atomic systems (see Murray and Grier 1995 for another example).

The structure of the dispersion can be quite sensitive to the parameters that influence the

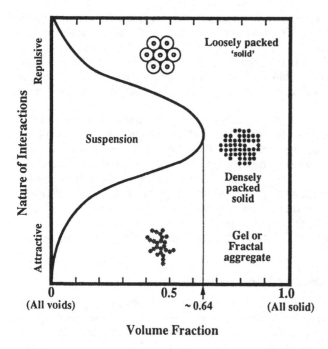

FIG. 13.2 Schematic illustration of the relation between the interparticle forces and the corresponding microstructure observed in dense, monodisperse colloids. (Adapted with permission from D. R. Ulrich, *Chem. and Eng. News*, 28–35, January 1, 1990.)

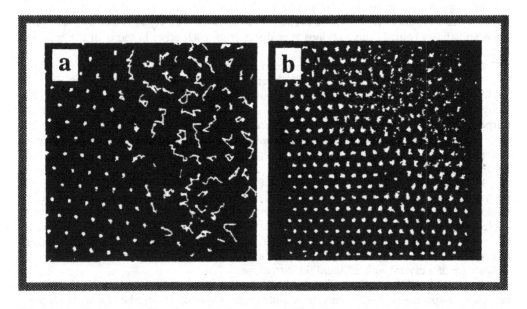

FIG. 13.3 Micrographs of local structure in a charged dispersion: (a) an ordered region coexisting with a liquidlike region; (b) an ordered crystalline structure, with a two-dimensional slice of the crystal shown. (Photographs courtesy of Dr. Norio Ise, Fukui Laboratory, Rengo Co., Fukui, Japan.)

interparticle forces near the liquid-crystal transition. This is illustrated in Figure 13.4, which presents stereograms based on computer simulations of a dispersion. The liquidlike isotropic structure seen in Figure 13.4a transforms to a crystalline (face-centered cubic) structure with a small change in the surface potential (Figure 13.4b).

2. *Overall interaction forces dominated by strong attraction*: At the other extreme, in Figure 13.2, attraction dominates the overall forces. Then the dispersion becomes thermodynamically and kinetically unstable and forms aggregates and an interconnected network known as colloidal "gels" as the particle concentration is increased. The aggregates in this case generally have a loose, fractal structure, similar to the one we saw in Example 5.4. Increases in the particle concentration do not eliminate the microvoids in the structure of the aggregates since strong van der Waals attraction leads to strong interparticle "bonds" that are difficult to break.

3. *Intermediate situations*: The situation in the intermediate region is much more complicated, but is very important in practice. When repulsion and attraction are comparable, the resulting microstructure depends on the details of the interaction forces more sensitively and can be manipulated by adjusting the relative magnitudes of the two forces. We already saw in Section 10.2 that – depending on the details of the interaction energy profile – aggregation in secondary or primary minima can occur in this case. More importantly, in contrast to cases 1 and 2 above, which do not permit closely packed structures, the intermediate situation can allow the formation of dense packing (see Fig. 13.2), and the solution chemistry and surface chemistry serve as important tools to achieve this goal.

In general, "soft" repulsion at short interparticle separation distances and mild attraction at larger distances promote better packing, and this can be used profitably in colloidal processing of ceramics (see Vignette I.5 in Chapter 1). In fact, this provides one of the strong incentives for studying the effects of polymer additives (discussed in Sections 13.5–13.7) since polymers adsorbed on (or end-grafted onto) particles can be tailored to provide the needed "soft" repulsive cushion and to mask out the strong van der Waals attraction, which, as we saw above, promotes the formation of fractal aggregates. Intuitively, one expects the weak "bonds" resulting from weak attraction to permit a particle in a cluster to break loose and, perhaps, reattach itself repeatedly until it finds a site with a larger number of neighbors (hence a stronger bond).

The picture gets even more complicated when we include other variables. For example, if minimizing microstructural voids (or, "defects") is our goal, the logical choice is to use a mixture of particle sizes since the smaller particles (in principle) can fill the voids created by the larger ones. However, the particle size also has a strong influence on the structure and stability of dispersions. As an illustration, Figure 13.5 shows the interaction energy profiles (calculated from the expressions we have developed in Chapters 10 and 11) for spherical particles of two different diameters d_1 and d_2. This figure illustrates two very important points of interest. First, it shows that, for otherwise identical conditions, the repulsive barrier is greater for larger particles than for smaller ones because of the different dependence of repulsion and attraction on particle size. Therefore, even if the small particles are unstable because of low barriers, once the resulting aggregates reach a large size the dispersion may become stable against further coagulation. The structure that develops with increasing particle concentration in this case is a complicated function of the kinetic stability of the dispersion. Second, in dispersions with a distribution of particle sizes, the smaller particles may aggregate with the larger ones, although a monodisperse system of small particles may be stable by itself.

13.2c Modeling Equilibrium Structures and Time Evolution of Structures

The discussions above highlight the following: (a) the structure of a dispersion is a complicated function of interaction forces; (b) equilibrium thermodynamics dictates what is possible and what is not, but (c) for lyophobic colloids, ultimately it is kinetics that determines whether the structures predicted by thermodynamics can be realized in practice. One of the major objectives of thermodynamic and kinetic studies of colloidal systems is to be able to predict and

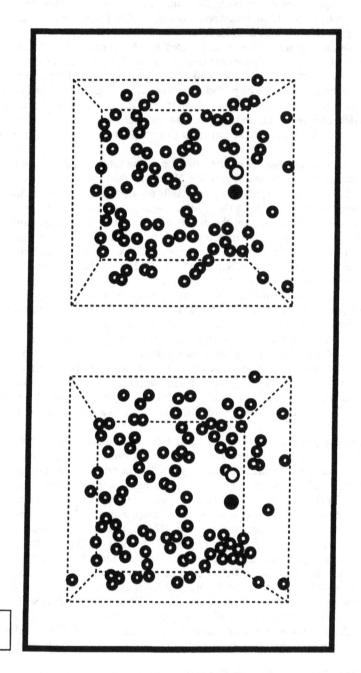

a

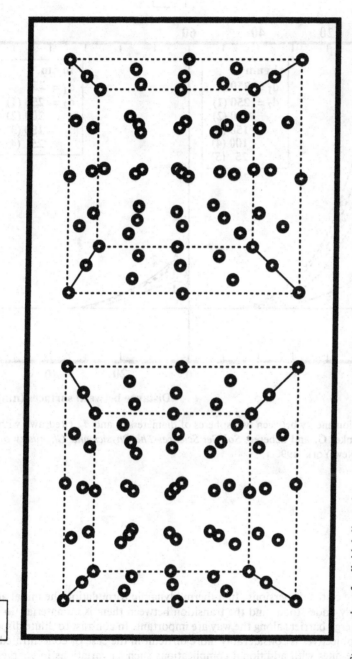

FIG. 13.4 Stereo pairs of colloidal dispersions generated using computer simulations. (a) Polystyrene latex particles at a volume fraction of 0.13 with a surface potential of 50 mV. The 1 : 1 electrolyte concentration is 10^{-7} mol/cm³. The structure shown is near crystallization. (The solid-black and solid-gray particles are in the back and in the front, respectively, in the three-dimensional view.) (b) A small increase in the surface potential changes the structure to face-centered cubic crystals. (Redrawn with permission from Hunter 1989.)

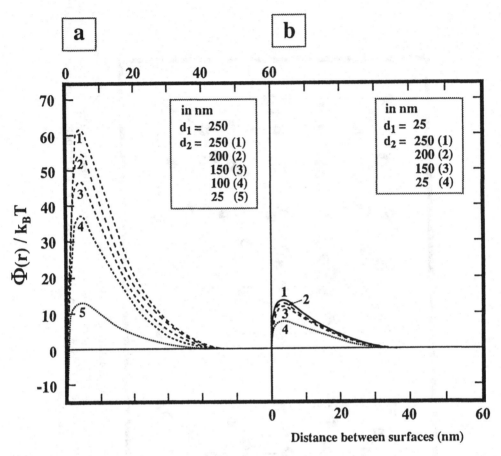

FIG. 13.5 Interaction energy between two spheres of diameters d_1 and d_2. (Redrawn with permission from C. J. Brinker G. and Scherer, *Sol-Gel Science: The Physics and Chemistry of Sol-Gel Processing*, Wiley, New York, 1990.)

control the structure of the dispersions. In a thermodynamic formulation the initial and final states must be clearly understood, and the transition between them is immaterial. In kinetics the path and any energy barriers along the way are important. In contrast to dilute dispersions, meeting the above objectives is considerably more difficult in the case of concentrated dispersions (especially the ones with additional complications such as variations in charges, sizes, and additives). This is an area of intense research activity currently, and radiation scattering techniques (Chapter 5), rheological behavior (Chapter 4), computer simulations (Figure 13.4 above), and statistical mechanical methods find considerable use in this context. The task is relatively more manageable in the case of dilute systems, and this forms the focus of this chapter. In the following sections we examine the classical approach to colloid stability and then consider the influence of polymers on stability.

13.3   THE DERJAGUIN-LANDAU-VERWEY-OVERBEEK THEORY OF COLLOID STABILITY

The interaction energy curves of the type shown in Figure 13.5 (and in Figure 10.1) are useful constructs for developing quantitative measures of kinetic stability. Such a study of stability is known as the *Derjaguin-Landau-Verwey-Overbeek* (DLVO) *theory*, in honor of the Russian physicists B. Derjaguin and L. Landau and the Dutch pioneers in colloid chemistry, E. Verwey and J. Th. G. Overbeek, who independently formulated theories of interaction forces between colloidal particles in the 1940s. In this section, we use the DLVO theory to examine, more quantitatively than we have done so far, the dependence of colloid stability on the various parameters that determine the shapes and the magnitudes of interaction energies between particles. A more elaborate formulation follows in Section 13.4.

13.3a   Interaction Energy Curves and Their Dependence on the Properties of the Dispersion

The exact shape of potential energy curves depends on the physical factors responsible for the interaction and also on the assumed geometry of the particles. We are mainly concerned with the case of two interacting planes since the expressions describing the various interactions are simpler in this case than for the more realistic case of interacting spheres. The more complicated spherical geometry contains no major fundamental insights beyond those already obtained from considering interactions between two planar surfaces.

A quantitative expression for the net interaction of two blocks of material separated by a distance d between their surfaces is obtained by combining Equations (10.63) and (11.86) to give

$$\Phi_{net} = 64k_B T n_\infty \kappa^{-1} \Upsilon_0^2 \exp(-\kappa d) - (A/12\pi)d^{-2} \tag{1}$$

where A is the Hamaker constant. In the next few paragraphs the effects of the Hamaker constant, the surface potential, and the electrolyte content — considered separately and in this order — on the net potential energy curves are examined.

13.3a.1   Effect of the Hamaker Constant A

It is understood that A in Equation (1) is the effective Hamaker constant A_{212} for the system. Of the variable parameters in this equation, it is the one over which we have least control; its value is determined by the chemical nature of the dispersed and continuous phases. The presence of small amounts of solute in the continuous phase leads to a negligible alteration of the value of A for the solvent.

The effect of variations in the value of A_{212} on the net potential energy is shown in Figure 13.6. Each of the curves in the figure is drawn for a different value of A, but at identical values of κ (10^9 m^{-1} or 0.093 M for 1 : 1 electrolyte) and ψ_0 (103 mV). As might be expected, the height of the potential energy barrier decreases and the depth of the secondary minimum increases with increasing values of A. If the cross-sectional area of interaction is 4.0 nm^2, each unit on the ordinate scale corresponds to $k_B T$ at 25°C. This is the unit of thermal energy against which all interactions are judged to be large or small. Thus for the curves shown in Figure 13.6, the depth of the secondary minimum is slight, and only for the smallest A value is the barrier height significant for particles with this interaction cross section.

13.3a.2   Effect of the Surface Potential ψ_0

The potential at the inner limit of the diffuse part of the double layer enters Equation (1) through Υ_0, defined by Equation (11.65) with ψ_0 in place of ψ. For large values of ψ_0, $\Upsilon_0 \approx 1$, so sensitivity to the value of ψ_0 decreases as ψ_0 increases. Figure 13.7 shows the effect of variations in the value of ψ_0 on the total interaction potential energy with κ (10^9 m^{-1} or 0.093 M for a 1 : 1 electrolyte) and A ($2 \cdot 10^{-19}$ J) constant. The height of the potential energy barrier is seen to increase with increasing values of ψ_0, as would be expected in view of the

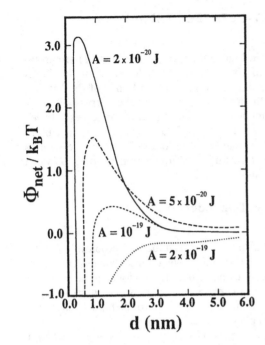

FIG. 13.6 Plot of Φ_{net} versus d according to Equation (1) for flat blocks. Curves are drawn for different values of A_{212} with constant values of κ (10^9 m^{-1}) and ψ_0 (103 mV). Units of ordinate are multiples of $k_B T$ at 25°C for an interaction area of 4.0 nm^2.

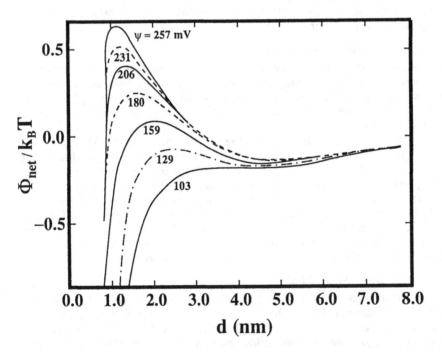

FIG. 13.7 Plot of Φ_{net} versus d according to Equation (1) for flat blocks. Curves are drawn for different values of ψ_0 with constant values of κ (10^9 m^{-1}) and A ($2 \cdot 10^{-19}$ J). Units of ordinate are multiples of $k_B T$ at 25°C for an interaction area of 4.0 nm^2.

increase of repulsion with this quantity. For some systems, ψ_0 is adjustable by varying the concentration of potential-determining ions, as described in Section 11.2 for the AgI surface. We have already seen in Section 11.8 that this quantity can be complicated by other adsorption phenomena; this is why we described it above as the potential at the "inner limit of the diffuse part of the double layer" rather than simply "at the wall." As we saw in Chapter 12, electrokinetic experiments measure a potential within the double layer—the ζ potential—but it is not entirely clear at which location within the double layer this potential applies. However, the experimental ζ potential does establish a lower limit for ψ_0.

13.3a.3   Effect of Electrolyte Concentration

Of the various quantities that affect the shape of the net interaction potential curve, none is as accessible to empirical adjustment as κ. This quantity depends on both the concentration and valence of the indifferent electrolyte, as shown by Equation (11.41). For the present we examine only the consequences of concentration changes on the total potential energy curve. We consider the valence of electrolytes in the following section. To consider the effect of electrolyte concentration on the potential energy of interaction, it is best to use the more elaborate expressions for interacting spheres. Figure 13.8 is a plot of Φ_{net} for this situation as a function of separation of surfaces with κ as the parameter that varies from one curve to another.

Figure 13.8 has been drawn for spheres with radius R_s = 100 nm, A = 10^{-19} J, and ψ_0 = 25.7 mV. The ordinate in this figure has been labeled both in joules and in multiples of $k_B T$ at 25°C. For the system described by these curves a significant energy barrier is present at all

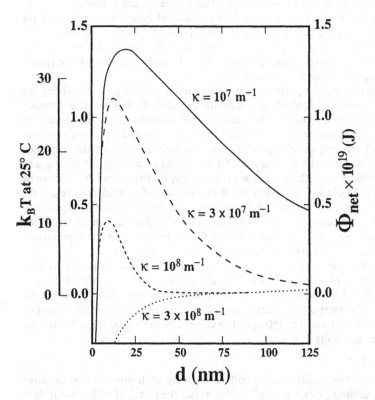

FIG. 13.8   Plot of Φ_{net} versus d, the minimum separation between surfaces, for two spheres of equal radius (100 nm). Curves are drawn for different values of κ with constant values of A (2 · 10^{-19} J) and ψ_0 (25.7 mV). (Redrawn with permission from E. J. W. Verwey and J. Th. G. Overbeek, *Theory of the Stability of Lyophobic Colloids*, Elsevier, Amsterdam, Netherlands, 1948.)

concentrations of a 1 : 1 electrolyte less than about 10^{-3} M. For concentrations between 10^{-3} and 10^{-2} M, however, the barrier vanishes. This particular colloid is thus expected to undergo a transition from a stable dispersion to a coagulated one with additions of an indifferent 1 : 1 electrolyte to a concentration in this range.

13.3b Implications for Colloid Stability

13.3b.1 Critical Coagulation Concentration

It has long been known that the addition of an indifferent electrolyte can cause a lyophobic colloid to undergo coagulation. The DLVO theory provides a quantitative explanation for this fact. Furthermore, it is known that, for a particular salt, a fairly sharply defined concentration is needed to induce coagulation. This concentration may be called the *critical coagulation concentration*. The DLVO theory in general and Figures 13.6–13.8 in particular can be summarized by the following statements:

1. The higher the potential at the surface of a particle—and therefore throughout the double layer—the larger the repulsion between the particles will be.

2. The lower the concentration of indifferent electrolyte, the longer is the distance from the surface before the repulsion drops significantly.

3. The larger the Hamaker constant, the larger is the attraction between macroscopic bodies.

Remember, the point of these figures is to see the effect on the potential energy curves of systematically varying one parameter at a time. It is the trends of behavior rather than the parameters themselves that are of greatest interest. In the following section we discuss the critical coagulation concentration as a simple quantitative test of the theory. In Section 13.4 we see how studies of the *rate* of coagulation provide still more stringent tests of the theory and the means for evaluating parameters of interest.

One of the easiest tests that can be performed on an aqueous colloid is to determine the critical concentration of electrolyte required to coagulate the colloid. We use the notation CCC (for critical coagulation concentration) to indicate this quantity. This experiment is conducted by introducing the dispersion into a series of test tubes and adding to each various proportions of water and electrolyte solution. In this way the total dilution of the dispersed particles is held constant while different amounts of salt are added to each. After mixing and waiting an arbitrary but consistent length of time, we visually inspect the tubes for evidence of the effect of the added salt. There will generally be clear evidence of coagulation (e.g., the settling out of the dispersed phase) in some of the tubes, whereas other tubes appear unchanged. Thus the highest concentration of salt that leaves the colloid unchanged and the lowest concentration that causes coagulation bracket the CCC. A second series of experiments may be conducted within this range to narrow the range of the CCC still further.

The actual concentration of electrolyte at the CCC depends on the following: (a) the time allowed to elapse before the evaluation is made, (b) the uniformity or, more likely, the polydispersity of the sample, (c) the potential at the surface, (d) the value of A, and (e) the valence of the ions. In a series of tests on any particular system, items (a)–(d) remain constant, so the CCC is a quantitative measure of the effect of the valence of the added ions. Table 13.1 summarizes some experimental results of this sort.

13.3b.2 Schulze-Hardy Rule

The results in Table 13.1 have been collected for colloids bearing both positive and negative surface charges. One of the earliest (1900) generalizations about the effect of added electrolyte is a result known as the *Schulze-Hardy rule*. This rule states that it is the valence of the ion of opposite charge to the colloid that has the principal effect on the stability of the colloid. The CCC value for a particular electrolyte is essentially determined by the valence of the counterion regardless of the nature of the ion with the same charge as the surface. The numbers listed in parentheses in Table 13.1 are the CCC values in moles per liter for counterions of the

TABLE 13.1 Critical Coagulation Concentration Values (in Moles Liter^{-1}) for Mono-, Di-, Tri-, and Tetravalent Ions Acting on Both Positive and Negative Colloids (Numbers in Parentheses) and CCC Values Relative to the Value for Monovalent Electrolytes in the Same System (Numbers Outside Parentheses)[a]

Valence of counterion	Negatively charged colloids			Positively charged colloids		
	As_2S_3	Au	AgI	Fe_2O_3	Al_2O_3	Theory
1	(5.5×10^{-2}) 1	(2.4×10^{-2}) 1	(1.42×10^{-1}) 1	(1.18×10^{-2}) 1	(5.2×10^{-2}) 1	1
2	(6.9×10^{-4}) 1.3×10^{-2}	(3.8×10^{-4}) 1.6×10^{-2}	(2.43×10^{-3}) 1.7×10^{-2}	(2.1×10^{-4}) 1.8×10^{-2}	(6.3×10^{-4}) 1.2×10^{-2}	1.56×10^{-2}
3	(9.1×10^{-5}) 1.7×10^{-3}	(6.0×10^{-6}) 0.3×10^{-3}	(6.8×10^{-5}) 0.5×10^{-3}	—	(8×10^{-5}) 1.5×10^{-3}	1.37×10^{-3}
4	(9.0×10^{-5}) 17×10^{-4}	(9.0×10^{-7}) 0.4×10^{-4}	(1.3×10^{-5}) 1×10^{-4}	—	(5.3×10^{-5}) 10×10^{-4}	2.44×10^{-4}
Potential-determining ion	S^{2-}	Cl^-	I^-	H^+	H^+	

Source: J. Th. G. Overbeek, in *Colloid Science*, Vol. I (H. R. Kruyt, Ed.), Elsevier, Amsterdam, Netherlands, 1952.
[a]Theoretical values are given by Equation (10).

indicated valence. That is, about $7 \cdot 10^{-4}$ M of divalent cation is needed to coagulate the negative As_2S_3 sols, whereas about $6 \cdot 10^{-4}$ M of divalent anion is required to coagulate positive Al_2O_3 sols.

The actual values of these concentrations depend on a whole array of unknown parameters, but their relative values depend only on the valence of the counterions. The entries outside parentheses in Table 13.1 are the values of the CCC relative to the value for the monovalent electrolyte in the same set of experiments. These are seen to be remarkably consistent for the divalent ions and acceptably close together for trivalent and tetravalent counterions.

Now let us see how this result is to be understood in terms of the DLVO theory. At first glance, it seems remarkable that any consistency at all can be found in tests as arbitrary as the CCC determination. It is not difficult, however, to show that these results are quite close to the values predicted in terms of the DLVO model for interacting blocks with flat faces. From an inspection of Figure 13.8, we concluded that the system at $\kappa = 10^8$ m^{-1} would be stable with respect to coagulation, whereas the one at $\kappa = 3 \cdot 10^8$ m^{-1} would coagulate. Furthermore, we examined the energy barrier to draw these conclusions. Next we must ask how the qualitative criteria we used in discussing the curves can be translated into an analytical expression.

One way of doing this is to assume that the demarcation between stable and unstable colloids occurs at the value of κ for which the height of the "barrier" is zero. Physically, this is a somewhat arbitrary choice: Thermal energy is sufficient to allow particles to overcome a barrier of low but nonzero height. Mathematically, however, the assumption that the maximum in the potential energy curve occurs at zero permits us to write

$$\Phi_{net} = 0 \qquad \text{at } d = d_m \tag{2}$$

and

$$d\Phi_{net}/dd = 0 \qquad \text{at } d = d_m \tag{3}$$

as the conditions for stability, where $d = d_m$ is the location of the maximum in the potential.

Applying Equations (2) and (3) to Equation (1) gives

$$64 \, k_B T n_\infty \kappa^{-1} \Upsilon_0^2 \exp \left(-\kappa d_m \right) = (A/12\pi) d_m^{-2} \tag{4}$$

and

$$64 k_B T n_\infty \Upsilon_0^2 \exp \left(-\kappa d_m \right) = 2(A/12\pi) d_m^{-3} \tag{5}$$

where the subscript m reminds us that this describes the maximum. From these equations it is readily apparent that

$$\kappa d_m = 2 \tag{6}$$

is the criterion for stability according to this model. This may also be written as

$$d_m = 2\kappa^{-1} \tag{7}$$

in terms of the "thickness" of the double layer. Again we see an important distance measured in terms of κ^{-1}. If we had used the interaction energy expression corresponding to Equation (1) for *spherical* particles of the same radii, the above result would have been $d_m = \kappa^{-1}$. The difference is thus only a numerical factor; the functional relationship does not change.

It is the dependence of the CCC values on the valence of the electrolyte that we seek to obtain rather than the absolute value of the CCC. Therefore it is sufficient to proceed from this point by merely retaining those factors that involve either the concentration ($n_\infty \propto c$, where c is the molar concentration) or the valence z. A more detailed expression can be obtained easily (see Problem 1 at the end of the chapter), but we shall not consider it here.

Substituting Equation (6) back into Equation (5) yields

$$n_\infty \propto \kappa^3 \tag{8}$$

The dependence of Υ_0 on z has been neglected in writing this result, a procedure that is entirely justified for the level of approximation involved here (recall that $\Upsilon_0 \approx 1$). From the definition of κ (Equation (11.41)), we obtain

$$n_\infty \propto z^3 n_\infty^{3/2} \qquad (9)$$

or

$$c \propto z^{-6} \qquad (10)$$

which is the desired result. According to Equation (10), the CCC value varies inversely with the sixth power of the valence of the ions in solution. The column of numbers in Table 13.1 labeled "theory" follows the progression z^{-6}: 1, 2^{-6}, 3^{-6}, 4^{-6}. The actual CCC values are seen to be in quite reasonable accord with these predictions.

Note that if ψ_0 is not assumed to be large, Υ_0 depends on both z and ψ_0. In this case the CCC is found to show a less sensitive dependence on the counterion valence than predicted by Equation (10). Example 13.1 examines this point.

* * *

EXAMPLE 13.1   *Critical Coagulation Concentration and ψ_0*   Use the accompanying data given below to criticize or defend the following proposition: Since the CCC for positively charged AgBr is less (regardless of counterion valence) than that for poly(vinyl chloride) (PVC) latex, ψ_0 must be less for AgBr.

Colloid	CCC (mole liter^{-1}) of ions opposite in charge to ψ_0	
	$z = 1$	$z = 2$
PVC latex	$2.3 \cdot 10^{-1}$	$1.2 \cdot 10^{-2}$
AgBr	$1.6 \cdot 10^{-2}$	$2.3 \cdot 10^{-4}$

Solution: Because of the arbitrary time of observation in CCC experiments, the absolute values of the CCCs have no significance. In this regard the proposition is wrong. It may be possible to rank the two colloids with respect to surface potential, however, by examining the order of the dependence of CCC on ion charge for each of the colloids:

For AgBr: $1.6 \cdot 10^{-2}/2.3 \cdot 10^{-4} = 69.6 = (1/2)^{-n}$ so $n = 6.12$

For PVC: $2.3 \cdot 10^{-1}/1.2 \cdot 10^{-2} = 19.2 = (1/2)^{-n}$ so $n = 4.26$

The fact that AgBr agrees with the prediction of Equation (10), which applies at high potentials, suggests that ψ_0 is greater for this colloid than for PVC.   ∎

* * *

Although this test of the DLVO theory itself introduces some additional approximations, it is a workable unification of experimental and theoretical points of view. The threshold of stability in terms of the concentration and valence of indifferent electrolyte is easily measured. Theoretical models describe the interaction between a pair of particles in terms of potential energy diagrams. The reconciliation of these two approaches constitutes an important step toward obtaining still more quantitative information from the study of coagulation. This is discussed in the following sections, which are concerned with the kinetics of coagulation. First, however, a few remaining comments about the critical coagulation concentrations must be made.

In Chapter 11 (Sections 11.4 and 11.6) we implicitly anticipated that the ion opposite in charge from the wall plays the predominant role in the double layer, the central observation of the Schulze-Hardy rule. This enters the mathematical formalism of the Gouy-Chapman theory in Equation (11.52), in which a Boltzmann factor is used to describe the relative concentration of the ions in the double layer compared to the bulk solution. For those ions that have the same charge as the surface (positive), the exponent in the Boltzmann factor is negative. This reflects the Coulombic repulsion of these ions from the wall. Ions with the same charge as the surface are thus present at lower concentration in the double layer than in the bulk solution.

The signs are reversed for oppositely charged ions; therefore the concentration of the latter is increased in the double layer. It may be shown — at least for high ψ_0 values — that the result of these considerations is essentially equivalent to emptying a region $2\kappa^{-1}$ thick of ions having the same charge as the wall (see Problem 11.5). Thus, in terms of the model for coagulation just presented, it is essentially only the counterions that contribute to the diffuse double layer at the critical separation for coagulation.

The CCC values reported in Table 13.1 in many cases are average values for several compounds of similar valence. The use of averages to compare CCC values is justified since the valence primarily determines the CCC for an electrolyte.

However, a closer inspection of the data reveals that there are second-order differences between different ions. For example, 0.058 and 0.051 M are the Li^+ and Na^+ concentrations required to coagulate As_2S_3 sols, and 0.165 and 0.140 M are the concentrations required to coagulate AgI sols. Although both sets of values are acceptably close to the mean (which includes a number of other compounds), it is also clear that Li^+ is consistently slightly less effective than Na^+ in inducing coagulation. A more complete sequence of these variations in effectiveness is as follows:

1. For monovalent cations,

$$Cs^+ > Rb^+ > NH_4^+ > K^+ > Na^+ > Li^+$$

2. For monovalent anions,

$$F^- > Cl^- > Br^- > NO_3^- > I^- > SCN^-$$

It is apparent from the correlation between the rankings of monatomic ions and their placement in the periodic table that some systematic effect is responsible for this ordering. This was the focus of Section 11.8 on Stern adsorption.

13.4 THEORY OF COAGULATION IN DILUTE DISPERSIONS

In this section we formulate the theory of kinetic stability of colloids. Our ultimate goal is to derive expressions for what is known as the stability ratio, defined in the following section, in terms of the interaction potential. The classical approach to this problem divides the process of coagulation into two steps. The first is the transport of particles toward each other, and the second is the eventual attachment on contact, for which we will assume that the particles stick to each other with a probability of unity on contact. The transport step can be driven by a number of mechanisms depending on the situation considered (e.g., externally imposed flow of fluids that carries the particles toward each other, differential sedimentation, and transport due to other forces such as magnetic or externally imposed electric fields), but for simplicity we consider only diffusion and attraction or repulsion due to interparticle forces. When the primary transport mechanism is diffusion, the coagulation is known as *perikinetic* coagulation. When velocity gradient is the dominant transport mechanism, the process is called *orthokinetic coagulation*.

13.4a Stability Ratio W

The stability of a dispersion against coagulation is expressed quantitatively by what is known as the *stability ratio*, usually denoted by W. The stability ratio is defined as

$$W = \frac{\text{Rate of diffusion-controlled interparticle collision}}{\text{Rate of interaction-force-controlled interparticle collision}} \tag{11}$$

If one assumes that a collision between two particles leads to permanent contact between the colliding particles, the diffusion-controlled collision rate in the numerator of the definition of W corresponds to "rapid" coagulation since the ever-present diffusion (i.e., the random Brownian motion) of the particles is unhindered by any energy barrier against contact. The denominator, the rate of collisions controlled by interaction forces (presumed to act against coagulation), then corresponds to "slow" coagulation. Thus large values of W imply that the dispersion is slow to coagulate (and is "stable"), whereas W of the order of unity implies that the dispersion coagulates rapidly.

Clearly, W is a function of any property of the dispersion that affects the strength of the interparticle forces and the energy barrier that slows down (or prevents) coagulation. A classical goal of colloid science has been to develop the equations necessary to predict the extent of stability of dispersions so that the results could be used in combination with the theories of interaction forces developed in previous chapters to promote or prevent the stability of dispersions.

We focus on the above goal in the following sections. First, we develop the equations for describing the mutual transport and collision of particles in the case of "rapid" coagulation. Then we modify the transport equation to reflect appropriately the influence of interparticle forces; the solution of this equation leads to the collision rate for "slow" coagulation. From these we then develop the equation for W in terms of the expressions for interparticle energies. Finally, the resulting expression for W is used, with a set of simplified electrostatic and van der Waals interaction energies, to show how colloidal stability can be predicted for known physicochemical conditions.

13.4b Theory of Rapid Coagulation

Let us now consider coagulation of particles in the absence of any repulsive barrier. In addition, we assume that, although there are no interparticle forces that contribute to the transport of particles toward each other, there is sufficient attraction between the particles on contact for them to form a permanent bond. As early as 1917, Smoluchowski formulated the equations for the collision rate for particles transported by diffusion alone (Smoluchowski 1917), and we develop the same idea here.

We begin by considering an array of spherical particles with motion that is totally governed by Brownian movement. Let us assume that there are particles of two different radii, $R_{s,1}$ and $R_{s,2}$. We assume the spheres interact on contact, in which case they adhere, forming a doublet. Although this is a highly oversimplified picture, it provides a model from which more realistic models can be developed in subsequent stages of the presentation.

Figure 13.9 shows a schematic illustration of the formation of a doublet. We have fixed the origin of the coordinate system (i.e., $r = 0$, with r the center-to-center distance between two particles) at the center of a particle of type 2, that is, a particle with radius $R_{s,2}$. Since the particles adhere on contact, the rate at which these particles disappear equals the rate at which they diffuse across the dashed surface in the figure. This surface corresponds to a spherical

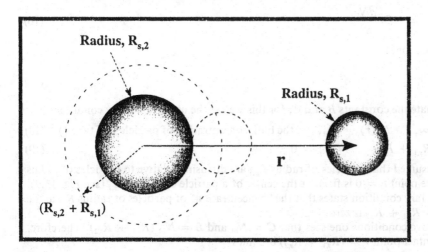

FIG. 13.9 Coagulation of two spherical particles of radii $R_{s,1}$ and $R_{s,2}$ to form a doublet.

surface of radius $(R_{s,1} + R_{s,2})$ inscribed around one of the spheres, which, for the present, is assumed to be stationary. After coagulation the number of independent kinetic units is locally decreased in the neighborhood of this coagulation site. Therefore we may imagine a concentration gradient around the fixed particle as responsible for the diffusion toward it.

As mentioned above, in 1917 M. Smoluchowski applied the theory of diffusion to this situation to evaluate the rate of doublet formation. According to Fick's first law (Equation (2.22)) J, the number of particles crossing a unit area toward the reference particle per unit of time is given by

$$J = -D_1(dN_1/dr) \tag{12}$$

where D_1 is the diffusion coefficient of the spheres of radius $R_{s,1}$, and N_1 is their total number of particles of type 1 per unit volume. This flux can be combined with a number or mass balance equation on a thin spherical shell to obtain the unsteady-state diffusion for the particles:

$$\frac{\partial N_1}{\partial t} = D_1 \frac{1}{r^2} \frac{\partial}{\partial r} \left(r^2 \frac{\partial N_1}{\partial r} \right) \tag{13}$$

where N_1 is the number concentration at position r and at time t. Note that this equation is the analog, in spherical coordinates, of Fick's second law of diffusion we derived in Chapter 2 (Equation (2.26)). The diffusion coefficient D_1 is given by the *Stokes-Einstein relation* (Equation (2.32)).

$$D = k_B T/6\pi\eta R_{s,1} \tag{14}$$

where T is the temperature of the dispersion (in K), η is the viscosity of the fluid (mass/(length · time))), and k_B is the Boltzmann constant. The diffusion equation for N_1 given above assumes that the spatial variation N_1 is spherically symmetric and depends on the radial distance only.

We restrict our attention to steady-state diffusion, for which the left-hand side of Equation (13) is zero. The resulting equation can be integrated easily, and one gets

$$N_1(r) = -B \frac{1}{r} + C \tag{15}$$

where B and C are integration constants that we determine shortly.

The flux at any distance r is obtained simply from the above solution using the Fick's law for flux:

$$J(r) = \text{Flux} = \text{Number of particles arriving at r per unit area per unit time}$$

$$= -D_1 \frac{dN_1}{dr} \tag{16}$$

which, for our solution, is

$$J(r) = -BD_1 \frac{1}{r^2} \tag{17}$$

Now, let us evaluate the constants B and C; for this we use the usual boundary conditions

(i) As $r \to \infty$, $N_1(r) \to N_{b1}$,   the bulk concentration of particles (of type 1)   (18)

(ii) At $r = R_{s,1} + R_{s,2}$, $N_1(r) = 0$ (19)

where we have assumed that particles of radius $R_{s,1}$ are diffusing relative to particles of radius $R_{s,2}$. The reference point $r = 0$ is fixed at the center of a particle of radius $R_{s,2}$ (see Fig. 13.9). The second boundary condition states that the "concentration" of particles of radius $R_{s,1}$ at the contact point $r = R_{s,1} + R_{s,2}$ is zero.

From the above conditions one sees that $C = N_{b1}$ and $B = N_{b1}(R_{s,1} + R_{s,2})$. Therefore, the flux at the contact point $r = (R_{s,1} + R_{s,2})$ becomes

$$J(\text{at } r = R_{s,1} + R_{s,2}) = -D_1 N_{b1}(R_{s,1} + R_{s,2})^{-1} \tag{20}$$

The negative sign indicates that the particles of radius $R_{s,1}$ are transported toward the particle of radius $R_{s,2}$ (which has been assumed to be fixed in its position at $r = 0$). The magnitude of the collision rate Z_1, that is, the number of collisions of type 1 particles with a stationary particle of type 2 per unit time, is then

$$Z_1 = |J|4\pi(R_{s,1} + R_{s,2})^2 = 4\pi(R_{s,1} + R_{s,2})D_1 N_{b1} \tag{21}$$

In general, particles of radius $R_{s,2}$ will also be executing random Brownian motion (i.e., diffusion). In such a case, D_1 should be replaced by $D_{12} = (D_1 + D_2)$. The collision rate Z_{12} (where the second subscript now reminds us that particle 2 is also executing diffusive motion) is then

$$Z_{12} = 4\pi(R_{s,1} + R_{s,2})D_{12}N_{b1} \tag{22}$$

The above result implies that the collision rate is of the form

$$Z_{12} = \alpha_r N_{b1} \tag{23}$$

where

$$\alpha_r = 4\pi(R_{s,1} + R_{s,2})D_{12} \tag{24}$$

with the subscript r indicating that the result is for rapid coagulation. One can now use this result to determine the reduction in N_{b1} as a result of the formation of 1-2 pairs (doublets). In particular, the rate of reduction of the bulk concentration N_{b1} with time t can be written as

$$dN_{b1}/dt = -\alpha_r N_{b1} N_{b2} \tag{25}$$

where we have multiplied Equation (23) by (a) N_{b2} since there are N_{b2} number of type 2 particles (used as reference particles in solving the diffusive collisions by particles of type 1) per unit volume of the dispersion and (b) -1 to indicate the reduction in the particles of type 1 due to coagulation. Note that the same equation with subscript 1 in place of 2 and vice versa describes the reduction in concentration N_{b2}.

For particles of identical radius R_s, one has

$$Z = 16\pi D R_s N_b = \alpha_r N_b \tag{26}$$

with

$$\alpha_r = 16\pi D R_s \tag{27}$$

Note that we no longer need the subscripts 1 and 2 on Z and N_b.

The coagulation rate for identical particles is then given by

$$\frac{dN_b}{dt} = -\frac{\alpha_r}{2} N_b^2 = -k_r N_b^2 \tag{28}$$

with $(\alpha_r/2)$ written as k_r, the *rate constant for rapid coagulation*. Notice that the factor $(1/2)$ appears in this case to avoid counting the same particle twice in the total collision rate; that is, collision of particle i with particle j accounts also for collision of j with i. Equations (25) and (28) correspond to the rate expressions for bimolecular "reactions" and, in this sense, the above description of coagulation is analogous to two reactant particles forming a doublet as the product of the reaction.

Equation (28) can be solved easily to obtain the concentration N_b as a function of time. If N_{b0} is the overall (bulk) concentration at $t = 0$, one gets from the above equation

$$N_b(t)/N_{b0} = (1 + t/t_{1/2})^{-1} \tag{29}$$

where

$$t_{1/2} = \frac{2}{\alpha_r N_{b0}} = \frac{1}{k_r N_{b0}} \tag{30}$$

is the so-called *half-life* for coagulation, that is, the time it takes for the overall concentration to reduce to half the initial concentration N_{b0}. Equation (29) can also be written as

$$\frac{1}{N(t)} - \frac{1}{N_{b0}} = k_r t \tag{31}$$

One should keep in mind the restrictions or limitations implicit in this development. It has been assumed, in effect, that only binary collisions occur. The result obtained is therefore strictly applicable to dilute dispersions, for which the probability of the formation of triplets, quadruplets, and so on, is negligible. We touch on the generalization of Equation (25) in Section 13.4f.

The most reliable way to evaluate a rate constant for coagulation, therefore, is to measure N_b as a function of time. Although this is an easy statement to make, it is not an easy thing to do experimentally. One technique for doing this is literally to count the particles microscopically. In addition to particle size limitations, this is an extraordinarily tedious procedure. Light scattering (Chapter 5) is particularly well suited to kinetic studies since, in principle, experimental turbidities can be interpreted in terms of the number and size of the scattering centers. A variety of additional techniques for following the rate of particle disappearance has been developed for specific systems. We do not pursue these, but merely note that experimental rate constants for coagulation can be determined.

Now, substituting the Stokes-Einstein equation (Equation (14)) for the diffusion coefficient in the expression for k_r leads to

$$k_r = 4k_B T / 3\eta \tag{32}$$

Note that the size of the particles drops out of the final expression for k_r; therefore the expression is equally valid for small molecules or colloidal particles so long as the various assumptions of the model apply. This constant describes the rate of diffusion-controlled reactions between molecules of the same size. In Example 13.2 we examine the numerical magnitude of the rate for the process we have been discussing.

* * *

EXAMPLE 13.2 *Variation of Particle Concentration Due to Rapid Coagulation.* An aqueous dispersion initially contains 10^9 particles cm^{-3}. Assuming rapid coagulation, calculate the time required for the concentration of the dispersed units to drop to 90% of the initial value. The viscosity of water is 0.010 P at 20°C, which may be used for the temperature of the experiment.

Solution: First we evaluate k_r, using Equation (32). It is convenient to use cgs units for this calculation; therefore we write $k_r = 4 \cdot (1.38 \cdot 10^{-16}) \cdot (293)/(3)(0.010) = 0.54 \cdot 10^{-11}$ cm^3 s^{-1}. Recall that the coefficient of viscosity has units (mass length^{-1} time^{-1}), so the cgs unit, the poise, is the same as (g cm^{-1} s^{-1}). As a second-order rate constant, k_r has units (concentration^{-1} time^{-1}), so we recognize that the value calculated for k_r gives this quantity per particle, or $k_r = 0.54 \cdot 10^{-11}$ cm^3 particle^{-1} s^{-1}. Note that multiplication by Avogadro's number of particles per mole and dividing by 10^3 cm^3 per liter gives $k_r = 3.25 \cdot 10^9$ liter mole^{-1} s^{-1} for the more familiar diffusion-controlled rate constant.

The 90% time is analogous to the half-life of the reaction. By considering a smaller extent of reaction, the assumptions of the model are more apt to remain valid. Substituting $N_b = 0.90$ N_{b0} into Equation (31), we obtain $(0.90\,N_{b0})^{-1} - (N_{b0})^{-1} = k_r t_{0.90}$ or $t_{0.90} = (1.00 - 0.90)/0.90$ $k_r N_{b0} = 0.10/(0.90)(10^9)(0.54 \cdot 10^{-11}) = 20.6$ s. Note the cancellation of concentration units in this last step. ∎

* * *

Examples are readily found in which the observed rate constant for coagulation is several orders of magnitude smaller than the rate constant we have been discussing. These are cases of slow coagulation and imply a component of net repulsion between the dispersed particles. We continue with the analysis of the kinetics of slow coagulation at this point. This is the topic of the next section.

13.4c Theory of Slow Coagulation

Attractive interparticle forces can enhance collision rates (although only moderately) and can cause more rapid coagulation; however, in most practical cases of interest we are concerned with the *reduction* in collision rates (and the consequent increase in stability) caused by energy barriers introduced through net interparticle repulsive forces (hence the name *slow coagulation*). The diffusion equation used in the last section can be modified to account for the presence of interparticle interaction forces, and a closed form expression can be developed for the collision rate. The corresponding analysis in this case is known as the *Fuchs theory of slow coagulation*, after Fuchs (1934), who addressed this problem first.

We saw above, from Fick's first law of diffusion, that the flux at a distance r from the central particle is given by

$$J(r) = -D_1 \frac{dN_1}{dr} \tag{33}$$

at steady state (when only diffusion is the transport mechanism). We must now add the flux due to the interparticle interaction energy $\Phi_{12}(r)$ between particles of type 1 and type 2 to the above expression. The interaction energy exerts a force given by $(-d\Phi_{12}/dr)$ on the diffusing particle. This force imposes on the particle an effective *drift velocity* v_{eff}, given by

$$v_{eff} = -\frac{1}{f} \frac{d\Phi_{12}}{dr} \tag{34}$$

where f is the *friction factor* (and $1/f$ is the *mobility* of the particle; see Atkins 1994). Note that for a spherical particle of radius $R_{s,1}$ in Stokes flow, $f = 6\pi\eta R_{s,1}$ as we introduced in Chapter 2 and used in the last section. The velocity v_{eff} can therefore be written as

$$v_{eff} = -D_1 \frac{d(\Phi_{12}/k_B T)}{dr} \tag{35}$$

by using the Stokes-Einstein relation. Equation (33) for the diffusive flux can now be modified to include the flux $v_{eff} N_1$ caused by the drift velocity of the particles in the direction of the force due to interparticle energy:

$$J(r) = -D_1 \left[\frac{dN_1}{dr} + N_1 \frac{d(\Phi_{12}/k_B T)}{dr} \right] \tag{36}$$

The magnitude of the number of particles transported through the spherical cross section of area $4\pi r^2$ is then equal to the collision rate Z_1, which at steady state becomes

$$Z_1 = -4\pi D_1 r^2 \left[\frac{dN_1}{dr} + N_1 \frac{d(\Phi_{12}/k_B T)}{dr} \right] = \text{Constant} \tag{37}$$

This equation can be simplified by defining

$$y(r) = \exp\left[\Phi_{12}(r)/k_B T\right] N_1(r) \tag{38}$$

to

$$\frac{Z_1}{4\pi D_1} \exp\left[-\Phi_{12}(r)/k_B T\right] \frac{dr}{r^2} = dy \tag{39}$$

On integration one has

$$N_1(r) = e^{-\Phi_{12}(r)/k_B T} \left[N_{b1} - \frac{Z_1}{4\pi D_1} \int_r^\infty e^{\Phi_{12}(r)/k_B T} \frac{dr}{r^2} \right] \tag{40}$$

where the conditions that $N = N_{b1}$ and $\Phi_{12} \to 0$ as $r \to \infty$ have been used.

The second boundary condition (i.e., $N_1 = 0$ at $r = R_{s,1} + R_{s,2}$, which also implies that $\{\exp\left[\Phi_{12}(r)/k_B T\right] N_1(r)\}$ is also equal to zero at $r = R_{s,1} + R_{s,2}$) can now be used to get the collision rate Z_1 (with a stationary reference particle of radius $R_{s,2}$):

$$Z_1 = 4\pi D_1 N_{b1} \div \int_{R_{s,1}+R_{s,2}}^\infty e^{\Phi_{12}(r)/k_B T} \frac{dr}{r^2} \tag{41}$$

As discussed in the previous section, D_1 will have to be replaced with D_{12} for two mutually diffusing particles of radii $R_{s,1}$ and $R_{s,2}$:

$$Z_{12} = 4\pi D_{12} N_{b1} \div \int_{R_{s,1}+R_{s,2}}^{\infty} e^{\Phi_{12}(r)/k_B T} \frac{dr}{r^2} \tag{42}$$

The total rate of collision (and, hence, the coagulation rate) is now given by

$$\frac{dN_{b1}}{dt} = -Z_{12} N_{b2} = -\left\{ 4\pi D_{12} N_{b1} \div \int_{R_{s,1}+R_{s,2}}^{\infty} e^{\Phi_{12}(r)/k_B T} \frac{dr}{r^2} \right\} N_{b2} \tag{43}$$

in analogy with Equation (25).

Equation (43) can be written as

$$dN_{b1}/dt = -\alpha_s N_{b1} N_{b2} \tag{44}$$

analogous to the case of rapid coagulation. The subscript s on the rate coefficient α_s draws attention to the fact that our focus here is slow coagulation:

$$\alpha_s = \left\{ 4\pi D_{12} \div \int_{R_{s,1}+R_{s,2}}^{\infty} e^{\Phi_{12}(r)/k_B T} \frac{dr}{r^2} \right\} \tag{45}$$

For identical particles of radius R_s, with $D_{12} = D_1 + D_2 = 2D$, one gets from Equation (42)

$$Z = 8\pi D R_s N_b \div \int_0^{\infty} e^{\Phi(s)/k_B T} \frac{ds}{(s+2)^2} \tag{46}$$

where s is the dimensionless surface-to-surface distance defined by $s = [(r/R_s) - 2]$. The corresponding coefficients α_s and k_s become

$$\alpha_s = 8\pi D R_s \div \int_0^{\infty} e^{\Phi(s)/k_B T} \frac{ds}{(s+2)^2} = 2 k_s \tag{47}$$

where the fact that $k_s = (\alpha_s/2)$ has been used.

When there is no interaction force between the particles (i.e., $\Phi(r) \equiv 0$), the above result reduces to α_r, corresponding to the rapid coagulation rate given by Equation (27) in the previous section. When there is a strong repulsive barrier, the integral in Equation (47) leads to a large value, thereby reducing the rate of coagulation.

13.4d Stability Ratio *W* and Its Dependence on $\Phi(r)$

Equations (27) and (47) show that

$$k_s = k_r/W \tag{48}$$

where W is the stability ratio defined in Equation (11). Thus

$$W = 2 \int_0^{\infty} e^{\Phi(s)/k_B T} \frac{ds}{(s+2)^2} \tag{49}$$

As we noted above, the evaluation of W for given values of dispersion properties such as surface potential, Hamaker constant, pH, electrolyte concentration, and so on, forms the goal of classical colloid stability analysis. Because of the complicated form of the expressions for electrostatic and van der Waals (and other relevant) energies of interactions, the above task is not a simple one and requires numerical evaluations of Equation (49). Under certain conditions, however, one can obtain a somewhat easier to use expression for W. This expression can be used to understand the qualitative (and, to some extent, quantitative) behavior of W with respect to the barrier against coagulation and the properties of the dispersion. We examine this in some detail below.

When the repulsion barrier is large (i.e., Φ_{max} is about 10 k_BT or larger), one can evaluate the integral in the expression for W using what are known as asymptotic techniques and obtain the following expression (Derjaguin 1989, p. 162):

$$W \approx 2\left(\frac{2\pi k_BT}{-\Phi''(s_m)}\right)^{1/2} \frac{e^{\Phi(s_m)/k_BT}}{(s_m + 2)^2} \tag{50}$$

where s_m is the value of s corresponding to the maximum in Φ [i.e., $\Phi(s_m) = \Phi_{max}$] and $\Phi''(s_m) = d^2\Phi/ds^2$ evaluated at $s = s_m$. Note that because $\Phi(s_m)$ is the maximum in the potential, the second derivative of Φ at $s = s_m$ is negative and $[-\Phi''(s_m)]$ is positive. Typically, the location of the maximum in $\Phi(s)$ occurs at distances of the order of a few nanometers. Therefore, for particle radii of the order of 100 nm or larger, $s_m \ll 1$, and the above equation simplifies to

$$W \approx \frac{1}{2}\left(\frac{2\pi k_BT}{-\Phi''(s_m)}\right)^{1/2} e^{\Phi_{max}/k_BT} \tag{51}$$

Equation (51) shows that W is a sensitive function of Φ_{max}, the maximum in the interaction potential, which in turn is a very sensitive function of properties such as ψ_0, electrolyte concentration, and so on. As a consequence, the stability ratio decreases rapidly with, for example, added electrolyte, and the dispersion coagulates beyond a threshold value of electrolyte concentration known as the critical coagulation concentration, as we saw in Section 13.3b.1.

Example 13.3 shows another way of deriving the above equation.

$$* \quad * \quad *$$

EXAMPLE 13.3 *Expression for Stability Ratio in Terms of Φ_{max}.* By replacing r by r_m, the center-to-center distance of separation at the maximum in the potential energy curve and using a truncated Taylor series about the maximum to estimate $\Phi(r)$, show that $W \propto \exp(\Phi_m/k_BT)$, with Φ_m the height of the potential energy barrier at the maximum. Comment briefly on these and any other assumptions or approximations involved.

Solution: The function $(r^{-2}) \exp(\Phi/k_BT)$ has its maximum at Φ_m and drops off rapidly for $\Phi < \Phi_m$. This justifies focusing attention on the maximum. Furthermore, the exponential term is more important than the r^{-2} factor; hence the latter is replaced by r_m^{-2}.

The Taylor series expansion of Φ can be written

$$\Phi \approx \Phi_m + (r - r_m)(\partial\Phi/\partial r)_m + [(r - r_m)^2/2](\partial^2\Phi/\partial r^2)_m + \cdots$$

according to Appendix A. The subscript m reminds us that the derivatives are evaluated at the maximum, and for this reason the term $(\partial\Phi/\partial r)_m$ equals zero. With these substitutions Equation (49) may be written as (in terms of the center-to-center distance r instead of the dimensionless surface-to-surface distance s)

$$W = \frac{2R_s}{r_m^2} e^{\Phi_m/k_BT} \int_{2R_s}^{\infty} \exp\left(\frac{(\partial^2\Phi/\partial r^2)(r - r_m)^2}{2k_BT}\right) dr$$

$$= \frac{2R_s}{r_m^2} e^{\Phi_m/k_BT} \int_{2R_s-r_m}^{\infty} \exp\left(-p^2(\Delta r)^2\right) d(\Delta r)$$

with $p^2 = -(\partial^2\Phi/\partial r^2)_m/2k_BT$. In the second part of the above equation, the variable has been changed to $\Delta r \, (= r - r_m)$ with the limits adjusted accordingly. We can rewrite the above result in terms of two integrals, one going from $\Delta r = (2R_s - r_m)$ to $\Delta r = 0$ (i.e., r going from $2R_s$ to r_m) and the second from $\Delta r = 0$ to $\Delta r = \infty$:

$$W = \frac{2R_s}{r_m^2} e^{\Phi_m/k_BT}\left\{\int_{2R_s-r_m}^{0} \exp\left(-p^2(\Delta r)^2\right)d(\Delta r) + \int_{0}^{\infty} \exp\left(-p^2(\Delta r)^2\right)d(\Delta r)\right\}$$

The second integral is one of those shown in Table 2.2 and is equal to $(1/2)(\pi/p^2)^{1/2}$. As noted above, $(\partial^2\Phi/\partial r^2)_m$ is negative since the point of evaluation is a maximum; therefore (p^2) is positive, and there is no sign problem with the square root. The fact that the function in question drops off sharply with distance from the maximum implies that the first integral can be written as

$$\int_0^{r_m - 2R_s} \exp\left(-p^2(\delta r)^2\right)d(\delta r) \approx \int_0^\infty \exp\left(-p^2(\delta r)^2\right)d(\delta r)$$

where δr is now $(r_m - r)$. Therefore, we again have the same result as above, that is, the first integral is also equal to $(1/2)(\pi/p^2)^{1/2}$. The final expression for W is thus

$$W = \frac{2\pi^{1/2}R_s}{r_m^2 p} e^{\Phi_m/k_B T}$$

It is a simple exercise to show that the above result is identical to Equation (50). If p can be regarded as a constant, then W plays a role in Equation (51) that converts the latter to the form of the Arrhenius expression for ordinary chemical rate constants.

Let us consider some variations of the above approximation.

Case 1.   The location of the maximum r_m is approximately $2R_s$ and the potential drops off relatively slowly.

In this case, the lower limit of the integral in

$$W = \frac{2R_s}{r_m^2} e^{\Phi_m/k_B T} \int_{2R_s - r_m}^\infty \exp\left(-p^2(\Delta r)^2\right)d(\Delta r)$$

can be replaced with zero and, using the arguments presented above, we obtain

$$W = \frac{\pi^{1/2}R_s}{r_m^2 p} e^{\Phi_m/k_B T} \approx \frac{\pi^{1/2}}{4R_s p} e^{\Phi_m/k_B T}$$

This expression can be used for arriving at the stability ratio for charged particles in nonaqueous media in which the repulsion can be modeled using a simple Coulombic expression (see Problem 3 at the end of the chapter).

Case 2.   Step function barrier.

A useful approximation for the interaction energy is a step function of the form

$$\Phi(r) = -\infty \qquad \text{for } r < 2R_s$$
$$= \Phi_m \qquad \text{for } 2R_s < r < 2R_s + \kappa^{-1}$$
$$= 0 \qquad \text{for } r > 2R_s + \kappa^{-1}$$

where κ^{-1} represents the range of the potential. In this case, the integral for the stability ratio can be evaluated easily to obtain

$$W = \frac{1}{2\kappa R_s} e^{\Phi_m/k_B T}$$

Although highly simplified, the above expression provides a simple way to get an estimate of W.

∎

* * *

The development sketched in Example 13.3 is not valid under all circumstances, but when it applies, it allows Equations (48) and (49) to be approximated by

$$k_s \propto k_r e^{-\Phi_m/k_B T} \tag{52}$$

This is the Arrhenius form to which the example refers. In it, the height of the maximum in a net potential energy curve plays the role of the activation energy. In the next chapter we see how this method has been used to evaluate W for systems in which the overlapping ion atmospheres of approaching colloidal particles provides the repulsion needed to give slow coagulation.

13.4e   Stability Ratio and Critical Coagulation Concentration

Using the approach developed in Example 13.3 and interaction energy expressions for spherical particles, it has been possible to predict how the stability ratio W varies with electrolyte concentration according to the DLVO theory. Since W can be measured by experimental studies of the rate of coagulation, this approach allows an even more stringent test of the DLVO theory than CCC values permit. We shall not bother with algebraic details, but instead go directly to the final result:

$$\log W = K_1 \log c + K_2 \tag{53}$$

where K_1 and K_2 are constants and c is the concentration of the ions in moles per liter. For water at 25°C the value of K_1 has been calculated as $[-2.15 \cdot 10^9 \, \Upsilon_0^2 R_s/z^2]$ where Υ_0 is given by Equation (11.65) with ψ_0 in place of ψ, z is the valence of the counterions, and R_s (in m) is the radius of the particles. Notice that because of the way Equation (53) has been written, the value of K_2 depends on the units used for c, although taken together the right-hand side is dimensionless since log W is dimensionless. We shall not go into this further since our primary objective here is to use Equation (53) to examine the dependence of W on c and its agreement with experimental results.

Figure 13.10 is a plot of log W versus log c for AgI sols of several different particle sizes. The experimental W values in this figure were determined from absorbance measurements. According to the preceding section, data of this sort not only test the DLVO theory but also permit the evaluation of several important colloidal parameters. From the data in Figure 13.10 the following conclusions can be drawn:

1.   A plot of log W versus log c is linear as required by Equation (53).
2.   The concentrations at which $W = 1$ (where the breaks in the curves appear) measure the CCC values for the electrolyte involved. The CCC values for mono-, di-, and trivalent ions are about 0.199, $2.82 \cdot 10^{-3}$, and $1.3 \cdot 10^{-4}$ mole liter^{-1}, respectively. These are in the ratio $1 : 1.42 \cdot 10^{-2} : 0.7 \cdot 10^{-3}$. These figures compare very favorably with the other experimental data for AgI and the theoretical values presented in Table 13.1.
3.   Slow coagulation is observed for log $W < 4$ or $W \approx 10^4$. For a typical potential energy curve, this corresponds to a value of Φ_m of about 15 $k_B T$. From this we may conclude that the height of an energy barrier must be at least 15 $k_B T$ if the colloid is to have any appreciable stability. Likewise, we may assume that unless the secondary minimum is approximately this deep, particles will be able to "escape" from it. In view of the general shape of the potential energy curves, the retardation effect, and this assessment of what constitutes a "high" barrier or a "deep" well, it seems likely that *rigid* aggregates are not formed in the secondary minimum.
4.   Equation (53) can be used to analyze the slopes of the curves in Figure 13.10 since the mean size of the AgI particles is known. In this way, Reerink and Overbeek (1954) found ψ_0 values in the range 12 to 53 mV and A values in the range $0.2 \cdot 10^{-20}$ to 10^{-19} J. Both of these

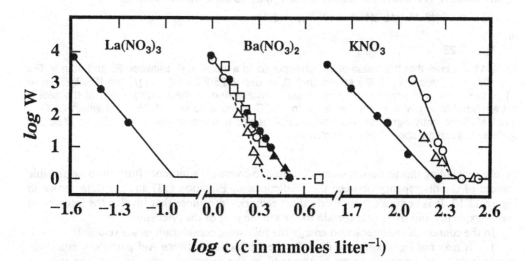

FIG. 13.10 Plot of log W versus log c for AgI sols of five different particle sizes coagulated with the electrolytes shown. The mean particle radii in the different sols are filled circles, 52.0 nm; open circles, 22.5 nm; squares, 53.5 nm; filled triangles, 65.0 nm; and open triangles, 158.0 nm. (Redrawn with permission from H. Reerink and J. Th. G. Overbeek, *Discuss. Faraday Soc.*, **18**, 74 (1954).)

are of the proper order of magnitude—no minor accomplishment in itself in light of the diverse assumptions required to get to this point.

5.   The values of A and ψ_0 obtained from this analysis are slightly less satisfying in detail: The values of A show a lot of scatter and ψ_0 appears to be too low. Recall that the variation of the CCC with z^{-6} implies large values of ψ_0 (see Section 13.3b.2); for lower ψ_0 values a different dependence on z is expected.

6.   Least satisfactory of all is the correlation with particle size. The results shown in Figure 13.10 were determined for AgI sols covering a 10-fold range of particle sizes.

It is evident from Figure 13.10 that the slopes do not vary over a similar range, as required by Equation (53). As a matter of fact, there are examples for which the steepest slope is associated with the coarsest particles (as required by theory) and others for which it occurs with the smallest particles. The quantitative predictions fail on this particular point, but, as we see below, there are some discrepancies between the theoretical model and the actual experimental system that may account for this apparent insensitivity to particle size. Example 13.4 considers another application of Equation (53) to an experimental system.

*   *   *

EXAMPLE 13.4   *Change of Stability Ratio with Ionic Concentration.*   Colloidal gold stabilized by citrate ions and having a mean particle radius of 103 Å was coagulated by the addition of $NaClO_4$. The kinetics of coagulation were studied colorimetrically and the stability ratio W for different $NaClO_4$ concentrations was determined (Enüstün and Turkevich 1963):

$c \cdot 10^3$ (mole liter^{-1})	2	3	5	8	10.5
W	48	31	17	8.9	0.84

When these data are plotted in the manner suggested by Equation (53), a straight line is obtained up to about 10^{-2} M, at which point a precipitous deviation from linearity sets in. The slope of the linear portion is about -1.20; estimate Υ_0 from this slope. Verify that this value of Υ_0 corresponds to a value of ψ_0 equal to about 25 mV. Estimate what the CCC value would be for this system if W continued to vary according to the same function of c both above and below 10^{-2} M. Suggest an explanation for the abrupt decrease in W near 10^{-2} M.

Solution: The slope of the linear portion equals K_1, the theoretical value of which is given above. Since R_s is known for these particles, Υ_0 may be calculated as follows:

$$\Upsilon_0^2 = 1.20/[(2.15 \cdot 10^9)(103 \cdot 10^{-10})] = 0.054$$

or

$$\Upsilon_0 = 0.23$$

Table 11.4 shows that this value of Υ_0 corresponds to a value of ψ_0 between 20 and 40 mV. For $\psi_0 = 25$ mV, Equation (11.65) shows that $\Upsilon_0 = \exp\{[25/2(25.7)] - 1\}/\{\exp[25/2(25.7)] + 1\} = 0.24$, which is very close to the experimental value. If the linear portion of the plot is extrapolated to $\log W = 0$, the value for $\log$ CCC is found to be -1.25, from which CCC $= 0.055$ M. Apparently significant Na^+ adsorption begins to occur at about 10^{-2} M, and W begins to decrease rapidly above this concentration.   ∎

*   *   *

Before concluding this section, it seems desirable to comment a bit more fully on some possible sources of the discrepancy between the predictions of Equation (53) and the data shown in Figure 13.10. It is convenient to divide these remarks into those that involve the interaction energy explicitly and those that pertain to the kinetic part of the discussion.

In the context of the interaction energy the following considerations are relevant:

1.   It may not be adequate to describe the interaction between AgI particles—especially at relatively close range—in terms of the radii of the dispersed units. In fact, the radii of surface protuberances rather than the dimensions of the particle as a whole may affect the short-range interaction.

2.   Throughout this discussion only nonspecific effects have been considered; that is, we have totally neglected to consider ion adsorption and the contribution of the Stern layer to the overall picture. This can be a serious source of complication, at least in some systems. This is

evident from the fact that some dispersions show a reversal of charge (from negative to positive) with the addition of La^{3+} and Th^{4+}, indicating the adsorption of these ions.

Several aspects of the kinetic part of the discussion above also warrant additional comments. These and a few other items are discussed in the following section.

13.4f Other Factors Affecting Coagulation Kinetics

There are a number of issues related to kinetics of coagulation that are not discussed in the previous sections. For example,

1. For highly asymmetrical particles, the probability of collision is greater than that predicted for identical particles. This may be understood by noting that the diffusion coefficient is most influenced by the smaller dimensions of the particles (therefore increased), and the "target radius" is most influenced by the longer dimension (also increased, relative to the case of symmetrical particles); see Equations (24) and (42).

2. The frequency of collisions is also expected to be greater in a polydisperse system than in a monodisperse system by the same logic as presented in item 1.

3. The presence of velocity gradients in the system may also increase the rate of coagulation above the value given by Equation (24) or (42).

The ratio of the probability of a collision induced by a fluid velocity gradient (dv/dx) (i.e., orthokinetic coagulation) to the collision probability under the influence of Brownian motion (perikinetic coagulation—what we have considered so far) has been shown to be (Probstein 1994)

$$k_{r, Flow}/k_{r, Diff.} = [\eta(2R_s)^3(dv/dx)]/(2k_BT) \tag{54}$$

(For particles of different sizes, $2R_s$ should be replaced with $R_{s,1} + R_{s,2}$.) Since this increases with the cube of particle size, it may be the dominant mechanism for the coagulation of larger particles. Note that all the kinetic complications—items 1 through 3—tend to cancel out of the evaluation of the stability ratio since the rates of both rapid and slow coagulation are determined on the same colloid. In spite of these complications, the value of the kinetic approach over, say, using Equation (11.1) or electrophoresis results (see Chapter 12) to determine ψ_0 is that the potential is evaluated in the actual coagulating system and hence includes any contribution from Stern adsorption. Once a suitable value for ψ_0 is known, potential energy curves for that value of ψ_0 and a measured concentration of electrolyte (i.e., κ) can be constructed for different values of A_{212}. That plot of Φ_{net} versus d that is most consistent with W (e.g., by the method of Example 13.3) may be used to identify A_{212} for the system under consideration.

For a polydisperse colloid (or for a monodisperse colloid that has evolved to a polydisperse suspension of aggregates of various sizes), we can develop an equation to calculate the concentrations of the aggregates as a function of time. For example, Equation (25) can be generalized to

$$\frac{dN_{bm}}{dt} = \frac{1}{2}\sum_{i=1}^{m-1} k_{i,(m-i)} N_{bi}N_{b(m-i)} - \sum_{i=1}^{\infty} k_{i,m} N_{bi}N_{bm} \qquad m > 1 \tag{55}$$

where N_{bj} is the concentration of particles of size j in the bulk and $k_{i,j}$ is the rate coefficient for collisions between particles of size i and size j. The first term on the right-hand side has been multiplied by $(1/2)$ to avoid double counting the same collisions (i.e., $i \leftrightarrow (m - i)$ collisions are the same as $(m - i) \leftrightarrow i$ collisions). The terms in the first summation on the right are known as the "birth" term since they represent the formation of particles of size m from the collisions of i and $(m - i)$. Similarly, the terms in the second summation are known as the "death" terms since they represent the disappearance of size m because of the formation of larger particles through $m \leftrightarrow i$ collisions. The above equation is therefore known as the *birth-death equation* or *population balance equation* (since it describes the populations of particles of various sizes). One can show that this equation reduces to the simpler ones we have used in the previous sections under appropriate conditions.

Although Equation (55) is formally correct, in practice one needs to make a number of assumptions to be able to use it for calculating the aggregate concentrations. For example, one usual assumption is that the radius of an aggregate of size m is simply the radius of a sphere with a volume that is equal to the volume of all the particles in the aggregate; that is, the likely nonspherical shape of the aggregate and the porosity are ignored. The use of the birth-death equation in the case of slow coagulation is complicated further by the fact that the rate coefficients are functions of the stability ratio W, which, in turn, is a complicated function of the interaction potential Φ_{ij} between particles of sizes i and j (see the discussion in Section 13.2b in the context of Fig. 13.5). The use of birth-death population balance equations is more common in aerosol science since some of the approximations needed (such as equating the volume of an aggregate to the volumes of all the particles in the aggregate and approximating the aggregates as spheres) are more easily met, particularly for liquid aerosol droplets. We shall not discuss these here further; more information on the above is available in Sonntag and Strenge (1987) and standard books on aerosols (e.g., Hidy and Brock 1970).

We shall now turn our attention to polymer-induced forces and their influence on colloid stability.

13.5 POLYMER-COLLOID MIXTURES: A PHENOMENOLOGICAL PERSPECTIVE

As mentioned in Section 13.1, polymers have been used since antiquity to stabilize dispersions of solids in liquids against coagulation. Paints and inks used by ancient civilizations were prepared by dispersing suitable pigments in water and "protecting" the resulting system by additives such as gum arabic, egg albumin, or casein. Gelatin has also been used extensively as a stabilizing agent. In molecular terms, these substances are charged polymers (polyelectrolytes) and their stabilizing influence is traceable to both electrostatic and polymeric effects. Each of these contributions is complicated in its own right, so we have divided the discussion into two parts. Charge effects have been discussed in Chapter 11 in terms of low molecular weight electrolytes. In this chapter, we consider the stabilizing effects of nonionic polymers. The advantages of polymer-induced stability over electrostatic stability imparted through low molecular weight electrolytes is summarized in Table 13.2.

13.5a General Considerations

The role of polymers on colloid stability is considerably more complicated than electrostatic stability due to low molecular weight electrolytes considered in Chapter 11. First, if the added polymer moieties are polyelectrolytes, then we clearly have a combination of electrostatic effects as well as effects that arise solely from the polymeric nature of the additive; this combined effect is referred to as *electrosteric* stabilization. Even in the case of nonionic

TABLE 13.2 Electrostatic and Steric Stabilization: A Comparison

Electrostatic stabilization	Steric stabilization
Addition of electrolytes causes coagulation.	Insensitive to electrolytes in the case of nonionic polymers.
Usually effective in aqueous systems.	Equally effective for both aqueous and nonaqueous dispersions.
More effective at low concentrations of the dispersion.	Effective at both low and high concentrations.
Coagulation is not always possible.	Reversible coagulation is more common.
Freezing of the dispersion induces irreversible coagulation.	Good freeze-thaw stability.

Source: Hunter, 1987.

polymers, addition of the polymer to a dispersion can promote stability or destabilize the dispersion, depending on the nature of interactions between the polymer and the solvent and between the polymer and the dispersed particles. As a result, both polymer solution thermodynamics and the thermodynamics of polymer-colloid interactions play important roles.

The importance of polymer solution thermodynamics in the present context will become evident when we discuss polymer-induced forces in Section 13.6. For the present, in order to illustrate the above point, we may consider a very highly simplified picture of *some* of the possible effects of polymer chains on a dispersion, as depicted in Figure 13.11.

1. In the case of very low polymer concentrations, *bridging flocculation* may occur as a polymer chain forms bridges by adsorbing on more than one particle (see also Fig. 13.12f).

2. At higher concentrations of the polymer, "brushlike" layers can form on the particles. These brushes can extend over sufficiently large distances to mask out the influence of van der Waals attraction between the particles, thereby imparting stability to the dispersion. This mechanism, already mentioned in Section 13.2, is known as *steric stabilization*. For steric stabilization, the polymer molecules must be adsorbed or anchored on the particle surfaces.

3. At moderate to high polymer concentrations, the "free" polymer chains in the solution may begin to exercise an influence. One such effect is the so-called *depletion flocculation* caused by the exclusion of polymer chains in the region between two particles when the latter are very close to each other (i.e., at surface-to-surface distances less than or equal to approximately the radius of gyration of the polymer chains). The depletion effect is an osmotic effect and is discussed further in Section 13.6.

4. At high polymer concentrations, one may also have what is known as *depletion stabilization*. The polymer-depleted regions between the particles can only be created by de-mixing the polymer chains and solvent. In good solvents the demixing process is thermody-namically unfavorable, and under such conditions one can have depletion stabilization.

We are concerned primarily with steric stabilization in this chapter.

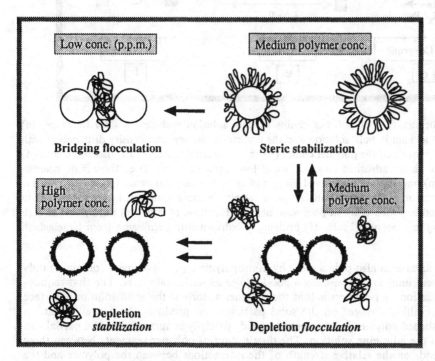

FIG. 13.11 A simplified representation of the effects of polymer additives on the stability of dispersions. See the text for explanation. (Redrawn with permission from Hunter 1987.)

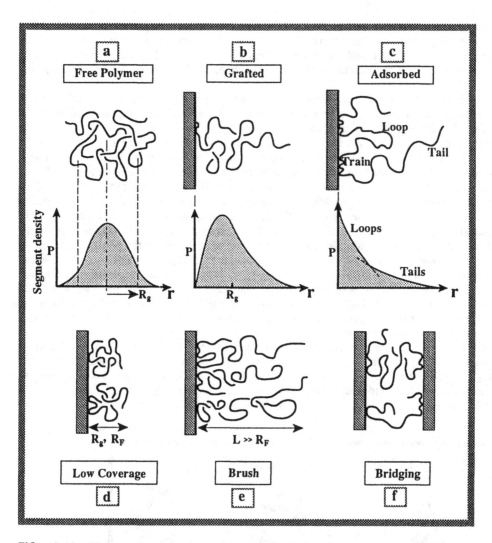

FIG. 13.12 The structure of polymer chains in bulk solution and near a solid interface: (a) configuration of a chain in bulk solution and the corresponding segment-density distribution with R_g the radius of gyration of the polymer chain; (b) an end-grafted chain; (c) an adsorbed chain; (d) configurations of chains adsorbed on a surface at low surface coverage (i.e., there is no nearest-neighbor overlap) and R_F is the so-called Flory radius of the polymer chain ($R_F = \alpha R_g$, with α known as the intramolecular expansion factor; α is unity in ideal solutions, larger than unity in good solvents and less than unity in poor solvents [see Equation (4.90)]); (e) adsorption at high coverage, leading to a polymer brush; (f) bridging. (Redrawn with permission from Israelachvili 1991.)

The above discussion also points out why homopolymers (i.e., polymers containing only one kind of repeat unit) are not usually a good choice as steric stabilizers. The first requirement for stabilization by polymers is that the polymer adsorb at the solid-solution interface since it is the resulting "fringe" on the solid particles that produces the desired result. In general, the adsorbed polymer is considered to reside partially at surface sites and partially in loops or tails in the adjoining solution. The distribution of polymer segments between these two states depends on the relative strength of the interactions between the polymer and the solid compared to those between the polymer and the solvent. Figure 13.12 illustrates some of the possible configurations of the polymer chains (including polymer brushes and bridging mentioned above) and the corresponding segment densities (i.e., number of monomer segments per unit area parallel to the surface). Synthetic polymers designed specifically as stabilizers are often block copolymers that contain two different kinds of repeat units, clustered in long sequences of one kind. In this type of polymer, one sequence is designed to optimize the adsorption, while the other gives maximum extension from the surface. In polymers made of a single kind of repeat unit, these two considerations tend to work in opposition.

Attachment of a single segment of the polymer chain is sufficient to confine the molecule to the layer of solution adjacent to the adsorbing surface. The solid exerts very little influence on the polymer molecule as a whole in such a case. In fact, the overall spatial extension of the polymer chain is expected to be about the same as that of an isolated molecule in this situation. The polymer-solvent interaction plays a more important role than the polymer-surface interaction in determining the thickness of the adsorbed layer in this case. This is only one of the relative interaction combinations possible, but it is the one that we consider in the greatest detail. As the number of polymer segments actually attached to the surface increases, the influence of the surface causes the spatial extension of the adsorbed chains to decrease.

The picture that emerges from the above discussion visualizes the layer adjacent to the solid surface as a polymer solution characterized by some average volume fraction of polymer ϕ^* and having an average thickness δR_s. If the interaction with the surface is not too strong, δR_s may be on the order of $2R_g$, twice the radius of gyration of the polymer in the solution under consideration. Example 13.5 considers how the thickness of such a layer can be determined experimentally.

* * *

EXAMPLE 13.5 *Determination of the Thickness of Adsorbed Polymer Layer from the Intrinsic Viscosity of the Dispersion.* An adsorbed layer of thickness δR_s on the surface of spherical particles of radius R_s increases the volume fraction occupied by the spheres and therefore makes the intrinsic viscosity of the dispersion greater than predicted by the Einstein theory. Derive an expression that allows the thickness of the adsorbed layer to be calculated from experimental values of intrinsic viscosity.

Solution: Equation (4.41) gives the Einstein relationship between $[\eta]$ and ϕ, the volume fraction occupied by the dispersed spheres. The volume fraction that should be used in this relationship is the value that describes the particles as they actually exist in the dispersion. In this case this includes the volume of the adsorbed layer. For spherical particles of radius R_s covered by a layer of thickness δR_s, the total volume of the particles is $(4/3) + 4\pi R_s^2 \delta R_s$. Factoring out the volume of the "dry" particle gives $V_{dry}(1 + 3\delta R_s/R_s)$, which shows by the second term how the volume is increased above the core volume by the adsorbed layer. Since it is the "dry" volume fraction that is used to describe the concentration of the dispersion and hence to evaluate $[\eta]$, the Einstein coefficient is increased above 2.5 by the factor $(1 + 3\delta R_s/R_s)$ by the adsorbed layer. The thickness of adsorbed layers can be extracted from experimental $[\eta]$ values by this formula. ∎

* * *

There are several additional points to be noted about this adsorbed layer:

1. The polymer concentration ϕ^* in the adsorbed layer is different from that in the bulk solution. The two are related through the adsorption isotherm of the polymer.
2. As illustrated in Figure 13.12, the concentration ϕ^* is not expected to be uniform at all distances outward from the wall, although we use some average value of this quantity as if it was uniform.
3. For steric stabilization, the surface layer rather than the bulk solution is the region of interest (see Fig. 13.13).

13.5b Thermodynamic Considerations

As the distance of separation between the core particles decreases in the coagulation step, the adsorbed layers begin to overlap as shown in Figure 13.13. Ultimately, it is the crowding of the polymer chains within this overlap volume that produces any stabilizing effect observed. Consequently, this mechanism for protecting against coagulation is called *steric stabilization*.

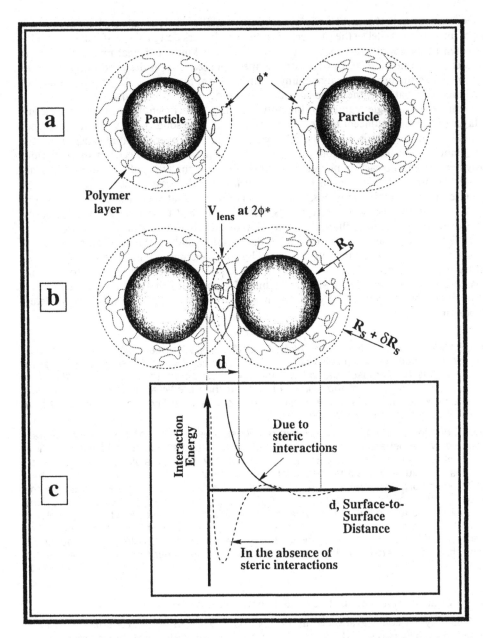

FIG. 13.13 Interaction between polymer-coated particles. Overlap of adsorbed polymer layers on close approach of dispersed solid particles (parts a and b). The figure also illustrates the repulsive interaction energy due to the overlap of the polymer layers (dark line in part c). Depending on the nature of the particles, a strong van der Waals attraction and perhaps electrostatic repulsion may exist between the particles in the absence of polymer layers (dashed line in part c), and the steric repulsion stabilizes the dispersion against coagulation in the primary minimum in the interaction potential.

In Section 13.7 we consider a specific model by which this can be accomplished. For now we take a more phenomenological point of view. It is an experimental fact that, at least under some circumstances, an adsorbed polymer layer stabilizes a dispersion against aggregation. This implies that the approaching particles in an aggregation step experience an increase in free energy ΔG_R (subscript R for repulsion) that prevents the completion of the step. Although the details of this repulsion are irrelevant for now, we associate ΔG_R with the overlap shown in Figure 13.13; that is, ΔG_R is zero when the overlap is zero and increases as the volume of the lens-shaped overlap region increases. As a free energy change, ΔG_R can be either positive or negative. In keeping with the usual sign conventions, a positive value for this quantity indicates repulsion (i.e., protection against aggregation), while a negative value contributes (along with van der Waals forces) to spontaneous aggregation.

As usual, ΔG_R can be broken into enthalpy and entropy contributions:

$$\Delta G_R = \Delta H_R - T\Delta S_R \tag{56}$$

where the individual terms describe changes in the respective property arising from the overlap. In principle, ΔH_R and ΔS_R can both be either positive or negative; therefore the possibility exists for ΔG_R to change sign with changes in temperature. In fact, it is observed experimentally that some sterically stabilized dispersions are caused to aggregate by increases in temperature and others by decreasing T. The threshold temperature for the onset of flocculation in these systems is called the *critical flocculation temperature* (CFT). Thermodynamics offers a formalism for interpreting these observations. With this in mind, let us consider some different sign combinations for ΔH_R and ΔS_R:

1. Suppose ΔH_R and ΔS_R are both positive. In this case the enthalpy change arising from the close approach of particles with adsorbed layers opposes aggregation while ΔS_R favors it. Since ΔS_R is weighted by T in ΔG_R, it follows that increasing T causes the entropy effect to become more important. The situation described here is one in which increasing temperature is expected to cause aggregation. Polyisobutylene adsorbed from 2-methyl butane and polyoxyethylene adsorbed from aqueous electrolyte are examples of systems that show a CFT with increasing temperature. Since it is the positive ΔH_R that is responsible for the stability in this case, such a system is said to display *enthalpic* stabilization.

2. Suppose ΔH_R and ΔS_R are both negative. In this case ΔS_R opposes aggregation while ΔH_R favors it. Since the resistance to aggregation decreases with decreasing temperature, aggregation is expected as T is lowered. Poly(12-hydroxystearic acid) adsorbed from n-heptane and polyoxyethylene adsorbed from methanol are examples of systems that display a CFT with decreasing temperature. Since ΔS_R is the source of the stabilization in these cases, this mechanism is called *entropic* stabilization.

3. If ΔH_R is positive and ΔS_R is negative, ΔG_R is positive at all temperatures, and the system is not subject to aggregation by changes in temperature.

Several things should be noted about the above discussion. First, both ΔH_R and ΔS_R are generally functions of temperature and may change signs themselves as T varies. Second, a given polymer may be governed by ΔH_R in one solvent and by ΔS_R in another. Finally, the addition of a second solvent to a dispersion can also induce aggregation in a polymer-stabilized system. We saw in Chapter 3 that changes of solvent goodness (as seen by the properties of polymer solutions) could be brought about by both temperature changes and addition of diluents.

Figure 13.14 illustrates the sort of data that can be used to determine the CFT for a dispersion. The absorbance of the system is measured as a function of temperature. A discontinuity in absorbance is observed to develop over a relatively narrow range of temperatures. The system shown in Figure 13.14 is an aqueous latex dispersion stabilized by polyoxyethylene. The threshold for aggregation is about 291 K. One characteristic of steric stabilization is that aggregated systems usually redisperse spontaneously if the goodness of the solvent is subsequently improved. In terms of the potential energy diagrams of Figure 10.1, this shows that aggregation in these systems does not occur by particles dropping into a deep primary minimum, but rather occurs in a shallow minimum at a larger distance from the surface.

Before we turn to a more detailed look at the origin of steric stabilization, it is informative

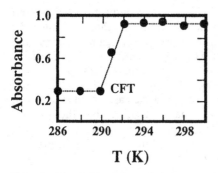

FIG. 13.14 Critical flocculation temperature of aqueous poly(vinyl acetate) dispersion stabilized by poly(oxyethylene) indicated by a sharp change in absorbance with temperature. (Redrawn with permission from D. H. Napper, in *Colloid and Interface Science* (M. Kerker, R. L. Rowell, and A. C. Zettlemoyer, Eds.), Academic Press, New York, 1977.)

to review the assortment of observations that are encountered if the polymer content of the continuous phase is varied from almost pure solvent to pure liquid polymer. As we have already seen, at very low polymer concentrations the solute can induce aggregation in a dispersion; at somewhat higher concentrations it can stabilize the dispersion against aggregation. At still higher concentrations there is a wide range of polymer concentrations in which the dispersion aggregates. Dispersions that are stable against aggregation have been prepared in polymer melts. This variety of behaviors with increasing polymer concentration suggests that the process under consideration is too complicated for explanation by a single model. In Section 13.7 we examine a mechanism for the stabilization that is observed at relatively low polymer concentrations. In such a case one can assume the adsorbed polymer molecule is not very different from an isolated molecule of the same polymer in solution. This assumption limits the applicability of our discussion but permits us to invoke some of the concepts of Chapter 3 for the problem at hand.

However, before we consider the above model, let us review briefly the types of interparticle forces that can result from the presence of polymer chains at particle-solvent interfaces.

13.6 POLYMER-INDUCED FORCES

In developing the expressions for van der Waals and electrical double-layer forces in Chapters 10 and 11 we have assumed that the surfaces involved are atomically smooth. Of course, in practice this is very seldom the case. One can divide "surface roughness" into two categories: *static roughness* and *dynamic roughness*. The former denotes surface roughness resulting from surface imperfections, as in the case of a solid surface. What is of interest to us, however, is the latter, which refers to thermally mobile surface molecules (usually adsorbed or grafted onto the surface) that may rearrange their positions and orientations in response to temperature, interactions with the solvent molecules, rearrangements of neighboring surface-bound molecules, proximity to another surface, and so on. The surface-bound molecules may be low molecular weight species or macromolecules such as neutral polymers or polyelectrolytes. The magnitude of the fluctuations depends on the size and the nature of the molecules. What is important is that, even if the scale of the fluctuations of these molecules is less than a nanometer (as one might expect in the case of simple molecules), the effects of such fluctuations on the roughness of the surface and the forces of interactions between two surfaces are not inconsequential. In fact, in many cases they play a very important role.

Our focus in this section is on the effects of polymer molecules adsorbed or grafted onto surfaces, but before we examine them in some depth it is good to get an idea of what phenomena cause the above molecular fluctuations to manifest themselves as forces (of attraction or repulsion). Even in the case of small molecules adsorbed strongly on a surface, the presence of another surface in proximity restricts the orientational freedom the molecules otherwise enjoy. This loss of entropy manifests itself as a repulsive force in the absence of any other interactions. The interaction force is then said to be *entropically driven*. The same principle applies in the case of polymer molecules adsorbed or grafted on a surface when another such surface is close by. When the second surface, with or without a polymer layer, is close enough to restrict the configurational freedom of the polymer chains dangling from the first surface, the net effect is to create an entropically driven repulsive or steric force between the surfaces.

What makes the understanding of such forces particularly difficult, especially in the case of polymer layers, is that the overall effect may be influenced by other factors such as the affinity of the polymer segments to each other and the affinity to solvent molecules. As a consequence the net force may be *enthalpically* (i.e., energetically) *driven* or entropically driven and may vary with other factors such as temperature (since, for example, temperature may change the nature of solvent-polymer-segment interactions; see Chapter 3) or depending on whether the polymer chains are physically *adsorbed* on the surfaces or *grafted* firmly to the surfaces (see Figure 13.12 for an illustration of the differences between the structures of adsorbed and grafted layers). The resulting net interaction can be attractive or repulsive, and the polymer additives may, therefore, serve as stabilizers or destabilizers.

Ideally, it would be desirable to be able to develop quantitative expressions for the interaction energies so that we can deal with coagulation or flocculation, at least in the case of fairly dilute dispersions, the way we did in Sections 13.3–13.4 for electrostatic stabilization. It is possible to develop approximate expressions for interaction energy due to various individual effects such as osmotic repulsion, attraction or repulsion due to the overlap of the tails of the adsorbed (or grafted) polymer layers, interaction of the loops in the layers, and so on (see Fig. 13.15). However, the complicated nature of polymer-induced interactions makes these tasks very difficult. In this section, we merely illustrate some of the issues that need to be considered in developing a fundamental quantitative understanding of polymer-induced forces. In Section

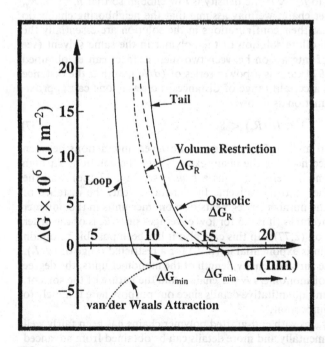

FIG. 13.15 Examples of the various contributions to the interaction energies between two polymer-coated particles. The curves shown illustrate qualitatively the change in Gibbs free energy due to the overlap of the tails, loops, and so on. (Redrawn with permission from Sato and Ruch 1980.)

13.7 we return to a thermodynamic analysis of stability along the lines of the discussion presented in Section 13.5 for a simple case to illustrate one of the approaches.

13.6a Repulsive Forces Due to Polymers

As mentioned above, there are a number of factors that affect the repulsive forces arising from the loss of configurational entropy of polymer chains.

1. The number of polymer chains per unit surface area (i.e., the density of surface coverage) determines whether or not interactions between the neighboring chains will have an influence on the extension of the chains from the surfaces. The affinity or the lack of affinity between the neighboring chains also affect the interaction forces.
2. Whether a polymer is grafted on the surface or simply adsorbed on the surface has a bearing on the force between two polymer-coated surfaces. Adsorption is a *reversible* process and can be affected by temperature, compression of the polymer layer, and the like.
3. The quality of solvent (i.e., solvent-polymer interactions) can also clearly affect the interaction forces. In a good solvent, polymer segments favor contacts with the solvent. Since the compression of the polymer layer by an approaching surface tends to squeeze out the solvent and force segment-segment interactions, the net result is a repulsion. In contrast, poor solvents produce an opposite effect, and a net attraction is possible for certain range of compression (see Section 13.6b).

A full theoretical treatment is therefore very difficult, and we restrict ourselves to two cases, primarily to illustrate some of the above points.

For example, consider a surface with a low coverage of end-grafted polymer chains under theta conditions (see Chapter 3). If the density of polymer ends per unit area of the surface is denoted by n_s (number per unit area), the typical magnitude of the distance between the grafted ends on the surface is equal to $n_s^{-1/2}$. If the density is low enough so that $n_s^{-1/2} > R_g$, the radius of gyration of the polymer chain, we may assume that the neighboring chains do not interfere with each other and that their configurations in the solution are essentially the same as what one would expect in a dilute solution of the polymer in the same solvent (see Fig. 13.12d). The repulsive energy of interaction between two such surfaces can be obtained theoretically (Dolan and Edwards 1974) and is a power series of (d/R_g), with d the distance between the interacting surfaces. In a certain range of distances, however, one can approximate the energy by an exponential function as below:

$$\Phi(r) \approx 36\, n_s k_B T \exp\left(-d/R_g\right) \qquad 2 < (d/R_g) < 8 \tag{57}$$

As one moves away from low to high surface coverages, however, interactions between neighboring chains begin to influence how far the chain extends into the solution. At high coverages, the surface layer is commonly referred to as a *brush* (see Fig. 13.12e), and the polymer chains tend to extend farther into the solvent. In a theta solvent, the distance of extension L increases as N^ν, with N the number of segments of monomer units in the polymer chain and ν roughly unity for high coverages (it is 0.5 for low coverages since R_g is of the order of $N^{0.5}$ for low coverages; see Equation (2.77)). In this case also it has been possible to obtain a quantitative expression, which again is exponential in (d/L) for a restricted range of (d/L). The length L itself is a function of the surface density, length of the monomer units, the degree of polymerization (i.e., the number of monomers N per chain), and the nature of the solvent. We shall not go into the theoretical and quantitative details since our purpose here is merely to point out the general issues that are important.

The repulsive force between surfaces with end-grafted polymer chains has been fairly well studied both theoretically and experimentally and more details can be obtained from advanced and specialized textbooks such as Israelachvili (1991) and Napper (1983). The situation is much more complicated in the case of chains adsorbed on surfaces. In this case, the segments adsorbed on the surfaces are held on the surfaces by relatively weaker forces (in contrast to end-grafted chains), and the adsorbed segments may continually detach from and reattach to the surfaces, especially as a second surface approaches the first. The calculation of the repulsive force is difficult because of this and because different segments of a polymer chain may adhere to both surfaces causing polymer "bridges" (see Fig. 13.12f). Moreover, the force between two surfaces at a given distance of separation may be time dependent since the time for equilibration between two surfaces can be rather large.

13.6b Attractive Forces Due to Polymers

As mentioned in the previous section, conditions corresponding to poor solvents give rise to attractive forces between polymer layer. At short enough distances, however, steric repulsion takes over, and the overall force has regions of repulsion as well as attraction. The attractive forces can come about for at least three reasons; we consider each briefly.

13.6b.1 Segment-Segment Attraction in Poor Solvents

In poor solvents, polymer segments prefer each other over the solvent. As polymer layers overlap with each other, segment-segment attraction occurs and draws the particles together. This occurs both in adsorbed layers and in grafted layers as the temperature is taken below the theta temperature (Θ) at low and high coverages and is illustrated in Figure 13.16. As illustrated in this figure, at close enough separations steric repulsion due to the overlap of the segments begins to dominate eventually, and one sees a net repulsion. As the temperature is taken above the theta temperature (i.e., as one goes from poor solvent to good solvent), the segment-segment attraction vanishes, and the force becomes repulsive at all separations (see Fig. 13.16).

13.6b.2 Bridging Attraction

Any polymer chain that has an affinity with a surface has the potential to form a bridge between the surfaces; this may give rise to an attraction known as *bridging attraction*. However, just as segment-segment forces, the segment-surface interaction can be attractive or repulsive depending on the affinity between the segments and the surfaces and the availability of free surface sites for adsorption. Strong bridging attraction is favored when a polymer is attracted to a surface that is not too highly or too sparsely covered with other polymer

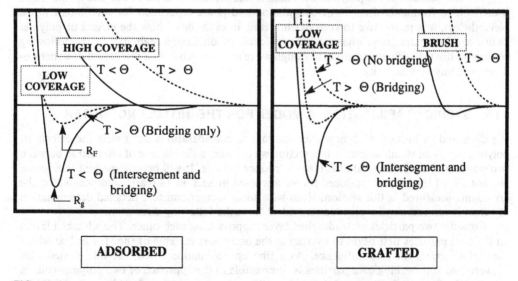

FIG. 13.16 Interaction forces between polymer-coated surfaces and their dependence on the type of layers and quality of the solvents. (Redrawn with permission from Israelachvili 1991.)

segments. In the case of densely covered surfaces, the availability of vacant surface sites is limited and bridging is unlikely. In the case of low coverages, the density of bridges will also be low. Although bridging attraction may exist in the last cases, the net attraction is relatively low because of the low number of bridges.

These are illustrated in Figure 13.16 for low and high surface coverages and for temperatures below and above the theta temperature. For instance, for adsorbed polymer layers at low coverage, the attraction is due to the formation of bridges for the temperature above the theta temperature. For $T < T_\theta$, both bridging and segment-segment attraction join together to increase the total attraction.

13.6b.3   Depletion Forces

Another source of attraction between two surfaces is possible when the surfaces are immersed in a solution of a nonadsorbing polymer (e.g., a polymer that does not adsorb on, or is repelled by, the surfaces). Although this force is generally weak, it can play a significant role in destabilization of colloidal particles under certain circumstances.

In order to understand the source of this force, consider two particles separated by a distance d as shown in Figure 13.17. The dispersed polymer molecules exert an osmotic pressure force on all sides of the particles when the particles are far apart, that is, when $d > R_g$. Then, there is no net force between the two particles. However, when $d < R_g$, there is a depletion of polymer molecules in the region between the particles since otherwise the polymer coils in that region lose configurational entropy. As a consequence, the osmotic pressure forces exerted by the molecules on the "external" sides of the particles exceed those on the interior (see Fig. 13.17), and there is a net force of attraction between the two particles. The range of this attraction is equal to R_g in our highly simplified model.

The depletion forces are relatively weak compared to the dispersion force and solvation forces but become significant with increases in the concentration and molecular weight of the polymer. In practice, however, the presence of strong depletion interaction requires a high polymer concentration (and therefore low molecular weights and R_g's). The existence of depletion forces has been predicted using computer simulations and has been verified experimentally using the surface force apparatus (SFA; see Section 1.6c.2) for surfaces immersed in a micellar solution, in which the spherical micelles play the same role as polymer molecules. However, theoretical treatments of depletion forces (and the bridging forces discussed above) are not as advanced as those for other polymer-induced forces.

The complexities of the polymer-mediated forces evident from the above discussions make it difficult to formulate theories of coagulation and phase separation for such interactions. Nevertheless, it is instructive to consider in detail an example of how the effects of polymer chains are incorporated in quantitative prediction of dispersion stability. In the following section we discuss such an example, although we restrict ourselves to a discussion of a thermodynamic analysis of stability.

13.7   STERIC STABILIZATION: A MODEL FOR THE INITIAL ENCOUNTER

We discussed in Section 13.5 how thermodynamic considerations determine the effects of polymer layers on stabilization. In this section we consider a simple case of interaction between two polymer layers to illustrate the ideas introduced above. It will become evident that some, but not all, of the polymer-induced forces discussed in Section 13.6 are encompassed by the arguments presented in this section. Even with some factors omitted, detailed determination of forces is considerably involved, as is evident from the following discussion.

Consider two particles with adsorbed layers approaching each other. The adsorbed layers on the core particles first begin to overlap at the outermost extreme of the "fringe," at which the surface exerts the least influence. As a first approximation, then, the initial encounter between two approaching core particles is comparable to the approach of two polymer coils in solution. In Chapter 3, Section 3.4a, we saw that the concept of excluded volume could be

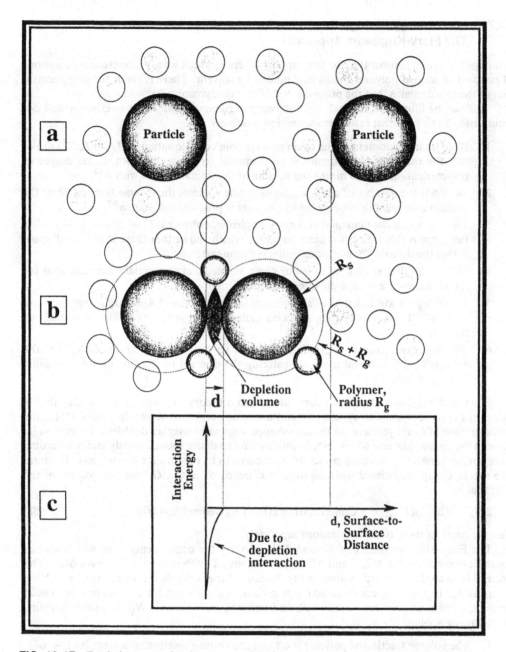

FIG. 13.17 Depletion attraction between two surfaces immersed in a polymer solution.

used to account for the initial nonideality in a polymer solution. Remember that the excluded volume is that region of space from which one molecule is denied access because of the presence of a second molecule. This also describes what happens as the adsorbed layers on two interacting particles approach one another. To quantify this idea, we are interested in an expression for ΔG_R as a function of the distance of separation between core particles as shown in Figure 13.13.

13.7a   Free-Energy Change Due to Excluded-Volume Interactions: The Flory-Krigbaum Approach

It is important to remember that the overlapping layers in which we are interested are solutions of pendant or looped polymer chains surrounded by solvent. There is room for interpenetration of the two domains, but the process is uphill thermodynamically.

This can be illustrated through the following arguments against the overlap of coil domains using some pertinent results from previous chapters:

1.   If we treat the domain of the chain as a random coil, Equation (2.73) or (2.77) shows that the radius of this domain is proportional to the square root of the degree of polymerization n. This means the volume of the domain varies with $n^{3/2}$.

2.   Since n is the number of chain segments in this volume, the volume fraction ϕ^* of the domain that is actually occupied by polymer is proportional to $n/n^{3/2} = n^{-1/2}$.

3.   The fraction of the domain that is *not* occupied by polymer is $(1 - \phi^*) = (1 - n^{-1/2})$. For large n this fraction is close to unity, which shows that there is plenty of space within the domain for additional polymer segments.

4.   Those sites that are not occupied by chain segments are potential placement sites for the segments of a second polymer molecule.

5.   Following the kind of statistical argument used in Section 3.4, we argue that there are $(1 - n^{-1/2})^n$ ways of placing a second molecule of n segments within the domain of the first.

6.   This last expression is approximately equal to $\exp(-n^{1/2})$, which shows that the probability of a second molecule entering the domain of the first decreases rapidly with increasing n.

Flory and Krigbaum developed these ideas into a theory of solution nonideality that is useful in the present context. As the two coil domains overlap as shown in Figure 13.13, the concentration of chain segments in the lens-shaped volume of overlap doubles compared to its value in the separate layers: $\phi^* \rightarrow 2\phi^*$. Solutions tend to dilute spontaneously and not become more concentrated; therefore we expect ΔG in the lens to be positive for this process. The total free energy change associated with the overlap depends on both ΔG_{ov} and the volume of the lens; that is,

$$\Delta G_R = (\Delta G_{ov})(V_{lens}) = (\text{Concentration effect}) \cdot (\text{Geometrical effect}) \tag{58}$$

Next we consider these two contributions separately.

The Flory-Huggins theory of Section 3.4b provides the components from which we can assemble expressions for $\Delta G_{m,\phi^*}$ and $\Delta G_{m,2\phi^*}$, the weighted difference of which gives ΔG_{ov}. The strategy is to apply these expressions to the domain of the individual polymer coil, which we define as V_d. If it was necessary to arrive at a numerical estimate for this volume, we could treat the domain as a sphere of radius R_g and write $V_d = (4/3)\pi R_g^3$. Within such a domain, the following applies:

1.   The volume fraction of polymer is ϕ^* and the volume fraction of solvent is $(1 - \phi^*)$.

2.   The number of polymer molecules N_2 is unity, and the number of solvent molecules is $[(1 - \phi^*)V_d/(\overline{V}_1/N_A)]$, with $\overline{V}_1$ the partial molar volume of the solvent. Therefore, $N_1 = (1 - \phi^*)V_d N_A/\overline{V}_1$.

3.   The Flory-Huggins expression for the enthalpy of mixing is given by Equation (3.71). In the present notation it becomes

$$\Delta H_m = (1/2)zN_1\phi_2\Delta w = \chi RTN_1\phi^*$$

4.   The Flory-Huggins expression for the entropy of mixing is given by Equation (3.65). In the present notation it becomes

$$-T\Delta S_m = RT(N_1 \ln \phi_1 + N_2 \ln \phi_2) = RT[N_1 \ln(1 - \phi^*) + N_2 \ln \phi^*]$$

Since $N_2 \ll N_1$, the second term can be neglected compared to the first to give

$$-T\Delta S_m \approx RTN_1 \ln(1 - \phi^*)$$

5. Within the volume of the coil domain we can write

$$\Delta G_m = \Delta H_m - T\Delta S_m = RTN_1[\chi\phi^* + \ln(1 - \phi^*)]$$

Substituting for N_1 from item 2, we then have

$$\Delta G_m = (RTV_dN_A/\overline{V}_1)(1 - \phi^*)[\chi\phi^* + \ln(1 - \phi^*)]$$

which may be approximated as

$$\Delta G_m = (RTV_dN_A/\overline{V}_1)[(\chi - 1)\phi^* + (1/2 - \chi)\phi^{*2}]$$

by expanding the logarithm (see Appendix A) and retaining no terms higher than second order in ϕ^*.

6. For the process $\phi^* \rightarrow 2\phi^*$, $\Delta G_{ov} = (\Delta G_{m,2\phi^*} - 2\Delta G_{m,\phi^*})$. Therefore

$$\Delta G_{ov} = (RTV_dN_A/\overline{V}_1)\{[(\chi - 1)(2\phi^*) + (1/2 - \chi)(2\phi^*)^2]$$
$$- 2[(\chi - 1)\phi^* + (1/2 - \chi)\phi^{*2}]\}$$
$$= (2RTV_dN_A/\overline{V}_1)(1/2 - \chi)\phi^{*2}$$

7. We can write $\phi^* = \overline{V}_2/N_AV_d$, with $\overline{V}_2$ the partial molar volume of the polymer. With this substitution

$$\Delta G_{ov} = 2k_BT(\overline{V}_2^2/V_d\overline{V}_1)(1/2 - \chi)$$

8. Finally, we give without proof the expression for the volume of the lens that experiences the free energy change given by item 7 (see Figure 13.13b):

$$V_{lens}^* = \frac{V_{lens}}{V_d} = \frac{2\pi}{3}\left(\delta R_s - \frac{d}{2}\right)^2\left(3R_s + 2\delta R_s + \frac{d}{2}\right)V_d^{-1}$$

Substituting the results in items 7 and 8 into Equation (58) gives

$$\Delta G_R = \frac{4\pi}{3}k_BT(\overline{V}_2^2/V_d^2\overline{V}_1)(1/2 - \chi)\left(\delta R_s - \frac{d}{2}\right)^2\left(3R_s + 2\delta R_s + \frac{d}{2}\right) \tag{59}$$

which is the result toward which we have been working.

Two of the terms in Equation (59) are easy to interpret in relation to dispersion stability. Although it is not the entire geometrical effect, the term containing $(\delta R_s - d/2)$ — which is half the maximum width of the lens — shows that the repulsion increases with the square of this quantity. This shows how the repulsion develops as the overlap increases.

The term $[(1/2) - \chi]$ is the only factor in Equation (59) that can have either positive or negative values, depending on whether χ is less than or greater than $1/2$. Again, we review some ideas from Section 3.4b pertaining to the Flory-Huggins interaction parameter χ:

1. In units of k_BT, χ measures the energy change per pair for the process $(1,1) + (2,2) \rightarrow 2(1,2)$ in which the indices refer to different pairwise interactions.
2. $\chi = 1/2$ corresponds to the critical point on a miscibility diagram for a polymer of infinite molecular weight.
3. $\chi < 1/2$ corresponds to a good solvent, and $\chi > 1/2$ corresponds to a poor solvent.
4. According to Equation (3.81),

$$\frac{1}{2} - \chi = \psi\left(1 - \frac{\Theta}{T}\right)$$

so $T = \Theta$ divides the temperature range into good $(T > \Theta)$ and poor $(T < \Theta)$ regions.

This review indicates that good solvent conditions (in terms of either χ or Θ) result in a positive value for ΔG_R. This is what would be expected from a model that assumes that the first encounter between particles with adsorbed layers is dominated by the polymers. Conversely, in a poor solvent ΔG_R is negative and amounts to a contribution to the attraction between the core particles as far as flocculation is concerned. Under these conditions the polymer itself is at the threshold of phase separation. Van der Waals attraction between the core particles further promotes aggregation, but it is possible that coagulation could be induced in a poor solvent even if the medium decreases the effective Hamaker constant to zero.

Replacing $[(1/2) - \chi]$ by $[1 - \Theta/T]$ offers an opportunity to test the ideas of the Flory-Krigbaum theory against experimental observations of the CFT. It is apparent from the material presented here that the CFT and the Θ temperature correspond to the same condition. Table 13.3 lists CFT values and independently determined Θ temperatures for several systems. The agreement between the two is quite satisfactory for these systems. Incidentally, electrolytes are added to the aqueous media in Table 13.3 to suppress the ion atmosphere mechanism for stabilization.

13.7b Inclusion of Elastic Contributions

There is another type of free energy change that can be considered within the overlap volume in addition to the concentration effect considered by the Flory-Krigbaum theory. This additional contribution to ΔG_{lens} is likely to be more important for $d < \delta R_s$ and should be considered when the outcome of the encounter is not determined by the initial approach of the colliding particles. This contribution arises from an elastic response by the adsorbed polymer, effectively pushing the approaching particles apart.

The elasticity of polymer coils is a well-known phenomenon and is involved in many important mechanical properties of bulk polymers. Stated briefly, it arises from a difference in conformational entropy between stretched and randomly jumbled chains. A statistical theory that counts the number of ways the two conformations can come about can be combined with the Boltzmann entropy equation (Equation (3.45)) to give an expression for the

TABLE 13.3 Critical Flocculation Temperatures for Various Polymer-Solvent Systems, with Θ Temperatures Included for Comparison

Stabilizer	Molecular weight	Dispersion medium	CFT (in K)	Θ (in K)
Poly(ethylene oxide)	10,000	0.39 M MgSO$_4$	318	315
	96,000		316	315
	1,000,000		317	315
Poly(acrylic acid)	9,800	0.2 M HCl	287	287
	51,900		283	287
	89,700		281	287
Poly(vinyl alcohol)	26,000	2 M NaCl	302	300
	57,000		301	300
	270,000		312	300
Polyacrylamide	18,000	2.1 M (NH4)$_2$SO$_4$	292	—
	60,000		295	—
	180,000		280	—
Polyisobutylene	23,000	2-Methylbutane	325	325
	150,000		325	325

Source: Reprinted with permission of D. H. Napper, in *Colloid and Interface Science*, Vol. 1 (M. Kerker, R. L. Rowell, and A. C. Zettlemoyer, Eds.), Academic Press, New York, 1977.

entropy change associated with stretching a chain or its "snapping" back to equilibrium. At small deformations, individual chains obey Hooke's law with the force constant

$$k_H = 3k_B T/(n\ell_0^2) \tag{60}$$

where n is the degree of polymerization and ℓ_0 is the step length in the random walk analysis of Section 2.7. By Equation (2.77), the denominator in this expression can be replaced with $6R_g^2$. In the context of steric stabilization, we can readily picture the close approach of two core particles as compressing a layer of polymeric "springs." According to Hooke's law, these springs will oppose compression with a restoring force that is proportional to the extent of compression with a force constant given by Equation (60). This is the origin of the elastic mechanism for interparticle repulsion from adsorbed polymers.

One of the first theoretical attempts to understand steric stabilization of dispersions was based on an entropic mechanism that resembles the elastic contribution to ΔG_R. We consider this mechanism in Example 13.3.

* * *

EXAMPLE 13.6 *An Entropic Model for Steric Stabilization Due to Adsorbed Polymer Layers.* Picture a flat surface to which rigid rods are attached by ball-and-socket-type joints. The free ends of the rods can lie anywhere on the surface of a hemisphere. The approach of a second surface blocks access to some of the sites on the cap of the hemisphere. Outline the qualitative argument that converts this physical picture to a theory for stabilization. What are some of the shortcomings of the model?

Solution: When the two solid surfaces are far apart ($d = \infty$), the free end of each adsorbed rod has access to Ω_∞ sites, with $\Omega_\infty = 2\pi L^2$ and L the length of the rod. When the separation is such that the second surface cuts off access to some of these sites, the number of accessible sites becomes Ω_d. The subscript here indicates a separation less than some critical distance that is the threshold for interaction. The exact form of Ω_d and the critical separation at which it begins to apply depends on whether one or both surfaces carry the adsorbed rods. The fraction of the area of the hemisphere that remains accessible to the free ends of the rods could be calculated from geometrical considerations. Using these Ω values as substitutions in Equation (3.46), we obtain $\Delta S_R = k_B \ln (\Omega_d/\Omega_\infty)$. Since $\Omega_d < \Omega_\infty$, ΔS_R is negative. This gives the effect per rod; if there are N such rods per unit area and if $\Delta H_R = 0$, then we obtain $\Delta G_R = -Nk_B T \ln (\Omega_d/\Omega_\infty)$, which gives a positive repulsion as required. This model has several deficiencies:

1. It is unrealistic for polymers, although it may be a reasonable approximation for linear amphipathic molecules.
2. It assumes very low surface coverage so that the individual rods do not interfere laterally.
3. It is a purely entropic mechanism and makes no provision for enthalpy effects. ■

* * * *

What we have covered in this chapter barely scratches the surface of a vast area of applications of colloidal phenomena in chemical and materials processing industries and in environmental and other operations. There are many fundamental, as well as practical, problems in the above topics (especially ones involving polymers, polyelectrolytes, and polymer-colloid and polymer-surfactant mixtures) that are currently areas of active research in engineering, chemistry, physics, and biology. Some of the references cited at the end of this chapter contain good reviews of topics that are extensions of what we have covered in this chapter (see, e.g., Elimelech et al. 1995, Hirtzel and Rajagopalan 1985, Israelachvili 1991, Gregory 1989, and O'Melia 1990).

REVIEW QUESTIONS

1. What is meant by the *microstructure* of a dispersion?
2. What is meant by the *stability* of a dispersion? Why is it important? Give a few examples for which stability is important and a few for which it is not desirable.
3. Discuss the differences between the *thermodynamic* and *kinetic* factors that determine the structure and stability of dispersions. Give examples of dispersions with stability that are

controlled by thermodynamic factors. Give a few examples for which kinetic considerations determine stability.

4. How does the interparticle interaction energy influence thermodynamic or kinetic stability? Discuss how variations in dispersion properties influence thermodynamic or kinetic stability.
5. What is the *critical coagulation concentration*? How does it vary with interaction energies?
6. What is the *Derjaguin-Landau-Verwey-Overbeek theory*?
7. Explain the *Schulze-Hardy rule*. Is it empirical or based on theory?
8. What are *perikinetic* and *orthokinetic* coagulation?
9. What is the *stability ratio W*? What values of *W* correspond to a stable dispersion? Why?
10. What is meant by *rapid coagulation*? What is the basic principle behind the *Smoluchowski theory of rapid coagulation*? What is the rate coefficient for rapid coagulation? How is it defined, and what properties of the dispersion determine its magnitude? What are the limitations of this theory as presented in the text?
11. What is meant by *slow coagulation*? What is the basic principle behind the *Fuchs theory* of slow coagulation? What is the rate coefficient for slow coagulation? How is it defined, and what properties of the dispersion determine its magnitude? What are the limitations of this theory as presented in the text?
12. How is the stability ratio related to the interparticle interaction energy? Under what conditions can it be written in *Arrhenius form*?
13. What is the relation between the stability ratio and the DLVO theory? How would you use the DLVO theory to predict the stability ratio of a dispersion?
14. Discuss how the stability ratio varies with dispersion properties such as electrolyte concentration, pH, surface potential, Hamaker constant, particle size, and so on.
15. Discuss the agreements and disagreements between the theoretical predictions for the stability ratio with what is observed experimentally.
16. What is a *birth-death* (or *population balance*) equation?
17. What are the factors that are relevant for extending the theories of coagulation presented here to (a) polydisperse colloids, (b) nonspherical colloids, and (c) conditions for which fluid flow is important?
18. Discuss how polymers influence colloid stability.
19. For polymer-induced stability, would you recommend the use of a homopolymer or a block copolymer? Why?
20. Discuss polymer-colloid interactions and steric stability from a thermodynamic perspective. What is *enthalpic* stabilization? What is *entropic* stabilization? What is the *critical flocculation temperature* (CFT)?
21. Does the adjective *steric* fully justify the role of polymers in imparting stability to a dispersion? Discuss.
22. List the types of polymer-induced forces one can expect when two particles with polymer coatings interact with each other. Discuss what factors influence the existence and the magnitude of such forces.
23. Why can polymer-induced forces be functions of solution temperature?
24. Why does polymer solution thermodynamics play an important role in polymer-colloid mixtures?
25. What is the *depletion force*? What is the *bridging force*?

REFERENCES

General References (with Annotations)

Elimelech, M., Gregory, J., Jia, X., and Williams, R., *Particle Deposition and Aggregation: Measurement, Modelling and Simulation*, Butterworth-Heinemann, Oxford, England, 1995. (Graduate and research levels. A state-of-the-art treatment of deposition of colloidal particles and their dependence on colloidal forces. Includes theoretical, computational, and experimental approaches.)

Hirtzel, C. S., and Rajagopalan, R., *Colloidal Phenomena: Advanced Topics*, Noyes, Park Ridge, NJ, 1985. (Research monograph. A broad, qualitative review of colloidal phenomena. Discusses colloid stability as well as deposition phenomena and structural evolution in concentrated dispersions. Contains an extensive collection of references prior to 1985.)

Hunter, R. J., *Foundations of Colloid Science*, Vol. 1, Clarendon Press, Oxford, England, 1987.

Hunter, R. J., *Foundations of Colloid Science*, Vol. 2, Clarendon Press, Oxford, England, 1989. (Undergraduate and graduate levels. Along with Volume 1, these two volumes cover almost all the topics covered in the present chapter at a more advanced level. Volume 1 discusses DLVO theory and thermodynamic approaches to polymer-induced stability or instability and is at the undergraduate level. Volume 2 presents advanced topics (e.g., statistical mechanics of concentrated dispersions, rheology of dispersions, etc.).)

Israelachvili, J. N., *Intermolecular and Surface Forces*, 2d ed., Academic Press, London, 1991. (Graduate and undergraduate levels. An excellent and intuitive introduction to polymer-induced forces is given in this volume, which also discusses direct measurement of such forces using the surface force apparatus.)

Kruyt, H. R. (Ed.), *Colloid Science. Vol. 1. Irreversible Systems*, Elsevier, Amsterdam, Netherlands, 1952. (Graduate and undergraduate levels. A classic reference on colloids. Chapters 6–8, by Professor J. Th. G. Overbeek, present the classical DLVO theory of colloidal forces and their application to kinetics of coagulation.)

Murray, C. A., and Grier, D. G., Colloidal Crystals, *American Scientist*, **83**, 238 (1995). (Graduate-level concepts, but some of the general ideas presented are accessible to undergraduates. This is a moderately technical, but qualitative, overview of the use of charged latex particles for studying phase transitions in atomic solids and alloys. The color illustrations presented provide a glimpse of the numerous possibilities made possible by "model colloids' for studying structure of materials.)

Napper, D. H., *Polymeric Stabilization of Colloidal Dispersions*, Academic Press, London, 1983. (Graduate-level monograph. An advanced and in-depth treatment of the role of polymers in colloid stability.)

Sato, T., and Ruch, R., *Stabilization of Colloidal Dispersions by Polymer Adsorption*, Marcel Dekker, New York, 1980. (Research monograph. An advanced treatment of polymer-induced forces.)

Sonntag, H., and Strenge, K., *Coagulation Kinetics and Structure Formation*, Plenum Press, New York, 1987. (Undergraduate level. A concise volume devoted to classical theories of coagulation kinetics. Contains more details than given in this book. Accessible to advanced undergraduates.)

Other References

Atkins, P. W., *Physical Chemistry*, 5th ed., W. H. Freeman, New York, 1994.

Derjaguin, B. V., *Theory of Stability of Colloids and Thin Films*, Consultants Bureau, New York, 1989.

DeSimone, J. M., Maury, E. E., Menceloglu, Y. Z., McClain, J. B., Romack, T. J., and Combes, J. R., *Science*, **265**, 356 (1994).

Dolan, A. K., and Edwards, S. F., *Proc. Roy. Soc. London*, **A337**, 509 (1974).

Enüstün, B. V., and Turkevich, J., *J. Am. Chem. Soc.*, **85**, 3317 (1963).

Fuchs, N. A., *Z. Phys.*, **89**, 736 (1934).

Gregory, J., *CRC Critical Reviews in Environmental Control*, **19**, 185 (1989).

Hidy, G. M., and Brock, J. R., *The Dynamics of Aerocolloidal Systems*, Pergamon Press, Oxford, England, 1970.

Matijevic, E., *J. Colloid Interface Sci.*, **43**, 217 (1973).

O'Melia, C. R., Kinetics of Colloidal Chemical Processes in Aquatic Systems. In *Aquatic Chemical Kinetics: Reaction Rates of Processes in Natural Water* (W. Stumm, Ed.), Wiley-Interscience, New York, 1990

Probstein, R. F., *Physicochemical Hydrodynamics: An Introduction*, 2d ed., Wiley-Interscience, New York, 1994.

Reerink, H., and Overbeek, J. Th. G., *Discuss. Faraday Soc.*, **18**, 74 (1954).

Smoluchowski, M., *Z. Phys. Chem.*, **92**, 129 (1917).

Tien, C., *Granular Filtration of Aerosols and Hydrosols*, Butterworth, Stoneham, MA, 1989.

PROBLEMS

1. Analogous to the planar case considered in Section 13.3, using the Derjaguin approximation, the interaction energy between two spherical particles of radius R_s can be written as

$$\Phi(d) = \pi R_s \{ -(A/12\pi)d^{-1} + 64 k_B T n_\infty \kappa^{-2} \Upsilon_0^2 \exp(-\kappa d) \}$$

Show that the location d_m of the maximum in the potential in this case is given by

$d_m = \kappa^{-1}$ if $\Phi(d_m)$ is also zero.

Substitute this in Equation (2) or (3) and solve for the critical coagulation concentration (CCC):

$$CCC \sim 10^5 [(k_B T)^5 (\varepsilon_r \varepsilon_0)^3 \Upsilon_0^4] / [(ze)^6 A^2]$$

2. Draw a sketch of the integral for the stability ratio W in Example 13.3 and verify that the assumptions made there to evaluate the integral are correct when the maximum in the interaction energy is large.

3. Morrison* uses the following simple Coulombic form of repulsion and the Hamaker expression for attraction for spherical particles of radius R_s for interaction between the particles in nonaqueous dispersions:

$$\Phi(r) = -\frac{A R_s}{12 d} + \frac{4\pi\varepsilon_r \varepsilon_0 R_s^2 \psi_0^2}{r}$$

where A is the Hamaker constant for interaction between two particles through the fluid, d is the shortest surface-to-surface distance between the particles, r is the center-to-center separation, and ψ_0 is the surface potential. The location of the maximum in the interaction energy may be assumed to be approximately $2R_s$.

a. Show that the stability ratio in this case may be approximated by

$$W \approx \left(\frac{A k_B^2 T^2}{3072\pi (\varepsilon_r\varepsilon_0 R_s\psi_0^2)^3}\right)^{1/4} \exp\left(\frac{2\pi \varepsilon_r\varepsilon_0 R_s\psi_0^2}{k_B T}\right)$$

b. For $A = 10^{-20}$ J and $T = 295$ K, show that the criterion for stability can be written as

$$\varepsilon_r R_s \psi_0^2 > 10^3$$

(with R_s in μm and ψ_0 in mV) if we assume that the dispersion is stable for $W > 10^5$.

4. Obtain the result for the stability ratio

$$W \approx \frac{1}{2 \kappa R_s} e^{\Phi_m/k_B T}$$

given in Example 13.3 for the step function energy barrier given by

$$\Phi(r) = -\infty \qquad \text{for } r < 2R_s$$
$$= \Phi_m \qquad \text{for } 2R_s < r < 2R_s + \kappa^{-1}$$
$$= 0 \qquad \text{for } r > 2R_s + \kappa^{-1}$$

where κ^{-1} represents the range of the potential.

5. It has been observed† that AgI sols at pH 3.5 are coagulated by 4.5×10^{-5} M Al $(NO_3)_3$ in the absence of a second salt and by 1.7×10^{-4} M Al$(NO_3)_3$ in the presence of 0.009 M K_2SO_4. Calculate the value of κ for each of these solutions. Is the threshold of instability consistent with your expectations in terms of the value of κ for these two systems? The authors of this research suggest that $K = 370$ for the equilibrium $Al^{3+} + SO_4^{2-} \rightleftharpoons AlSO_4^+$. Calculate the concentration of Al^{3+} in the system containing K_2SO_4. Does the behavior of the AgI appear to be correlated with the concentration of "free" Al^{3+}? Is the specific adsorption of Al^{3+} expected on AgI in the presence of 4×10^{-4} M excess KI? Discuss in terms of ψ_0 and Φ.

6. Verify that combining Eqs. (5) and (6) with the definition of κ [Eq. (11.35)] leads to the following expression (purely numerical factors may be omitted):

$$CCC \propto n_\infty \propto \frac{\varepsilon^3 (k_B T)^5 \Upsilon_0^4}{e^6 A^2 z^6}$$

By the series expansion of Υ_0 verify that $\Upsilon_0^4 \propto (ze\psi_0/k_B T)^4$ if ψ_0 is low. Use these two results to predict the dependence of the CCC on the ionic valence if ψ_0 is small. Compare this result with the same quantity in the limit of large ψ_0.

---

*Morrison, I. D., *Langmuir*, **7**, 1920 (1991); Morrison, I. D., The Influence of Electric Charges in Nonaqueous Dispersions, in *Dispersion and Aggregation: Fundamentals and Applications* (B. M. Moudgil and P. Somasundaran, Eds.) Engineering Foundation, New York, NY 1994.
†Stryker, L. J., and Matijević, E., *J. Phys. Chem.*, **73**, 1484 (1969).

7. Arachidic acid sols were studied with different concentrations of La^{3+} added. The stability ratio W and the direction of particle migration in an electric field (i.e., particle charge) were observed* and the following results obtained:

c (mole La^{3+} liter^{-1})	10^{-5}	3×10^{-5}	10^{-4}	3×10^{-3}	10^{-3}
W	7.9	4.5	~1	1.6	15.8
Particle charge	−	−	~0	+	+

Taking 10^{-4} M as the CCC value for La^{3+}, CCC values of 7.29×10^{-2} M and 1.1×10^{-3} M would be predicted for monovalent and divalent cations, respectively, according to Equation (10). In view of the observed behavior of La^{3+}, would you expect these calculated CCC values to be correct or too low or too high? Explain briefly.

8. Kitahara and Ushiyama† flocculated a polystyrene latex of radius 665 Å with KCl. The stability ratio W was found to vary with the KCl concentration as follows:

$\log c$ (c in mole liter^{-1})	0	−0.13	−0.33	−0.44	−0.60
$\log W$	0	0	0.30, 0.46	0.73	1.20

From a plot of $\log W$ versus $\log c$ determine the CCC value and Υ_0 [by means of Eq. (53)]. Use the approximation for Υ_0 given in Problem 6 to estimate ψ_0 for this colloid. Use the values of the CCC and Υ_0 determined in Eqs. (5) and (6) to estimate the effective Hamaker constant A_{212} for polystyrene dispersed in water. Describe how A might be estimated using a more realistic model than that used in the derivation of Eqs. (5) and (6).

9. Müller‡ studied by dark-field microscopy the flocculation of colloidal gold upon the addition of NaCl to the aqueous sol. For a sample in which the gold particles have a 36.9-Å radius, the following particle counts were observed at different times after the colloid was made about 0.2 M with NaCl:

t(s)	120	195	270	390	450	570
$N \times 10^{-8}$ (cm^{-3})	11.2	7.3	5.4	4.5	3.7	2.7

Determine the second-order rate constant k_{exp} which describes this flocculation process. How does k_{exp} compare with k_r as given by Eq. (32)?

10. Polystyrene latex particles were coagulated by the addition of $Ba(NO_3)_2$. The number of dispersed particles deposited onto a planar polystyrene surface was determined 15 min after the addition of salt by optical microscopy. The light microscope does not permit the aggregation of the deposited particles to be determined; subsequent examination by the electron microscope gives this information. Clint et al.§ obtained the following results:

Ba $(No_3)_2$ concentration $\times 10^3$ (mole liter^{-1})	Total deposition cm$^{-2} \times 10^{-5}$ (after 15 min)	Percent deposit		
		Single	Double	Triple
9.1	8.04	94.7	3.3	0.3
15.4	14.25	95.0	4.3	0.4
22.7	14.43	82.9	11.3	3.5
57.0	11.25	75.0	15.8	5.6

Discuss these data in terms of the following points:

(a) For all salt concentrations the order of particle abundance in the deposit is single > double > triple.

(b) The decrease in total deposition with increasing concentration is not offset by the higher aggregation state of the deposit, but arises from the slower diffusion of more highly aggregated kinetic units.

---

*Ottewill, R. H., and Wilkins, D. J., *Trans. Faraday Soc.*, **58**, 608 (1962).
†Kitahara, A., and Ushiyama, H., *J. Colloid Interface Sci.*, **43**, 73 (1973).
‡Müller, H., *Kolloid Z.*, **38**, 1 (1926).
§Clint, G. E., Clint, J. H., Corkil, J. M., and Walker, T., *J. Colloid Interface Sci.*, **44**, 121 (1973).

11. If ccoagulation involves two noninteracting spheres of different radii R_i and R_j, Equations (24) and (28) predict

$$k_r = \frac{2}{3} \frac{k_B T}{\eta} (R_i + R_j) \left(\frac{1}{R_i} + \frac{1}{R_j} \right)$$

Show that this expression is identical to

$$k_r = \frac{2}{3} \frac{k_B T}{\eta} \left[4 + \left(\sqrt{\frac{R_i}{R_j}} - \sqrt{\frac{R_j}{R_i}} \right)^2 \right]$$

Estimate the ratio R_i/R_j needed to account for a k_r value of 2.9×10^{-11} cm^3 s^{-1} as observed for arachidic acid sols.* Does this expression reduce to the proper limit when $R_i = R_j$?

12. Verify (a) that Equation (40) is a solution to Equation (37), (b) that Equation (41) reduces to Equation (21) if $\Phi = 0$, and (c) that Equation (46) leads to a rate constant larger than k_r if $\Phi = 0$ for $r > \Delta$ and $\Phi = -\infty$ for $r < \Delta$, where $\Delta > 2R_s$. Show that the experimental k_r value cited in Problem 11 is consistent with a value of Δ equaling $5.4 R_s$ for an aqueous colloid at 20°C ($\eta = 0.01$ P). Discuss the relevancy of this last result to the rapid flocculation of particles between which van der Waals attraction exists.

13. Ahmed et al.† measured η_{red}/ϕ for polystyrene latexes with adsorbed layers of commercial poly(vinyl alcohol) (PVA) samples of different molecular weights. The latex particles were 190 nm in diameter and the limiting values of η_{red}/ϕ as $\phi \to 0$ had the following values for PVA samples of the indicated molecular weight:

M_{PVA} (g mole^{-1}):	None	26,400	23,000	80,000	79,100
R_g (nm)	—	6.5	7.4	13.2	13.0
$[\eta]$	3.0	5.0	5.5	8.2	9.1

Use the experimental value of $[\eta]$ for the bare particles and the relationship given in Example 13.5 to estimates δR_s for these particles. How do the layer thicknesses compare with $2R_g$, for which the given radii of gyration were determined for the polymers in bulk solution?

14. The parameter χ is proportional to the energy of interaction per 1-2 pair. To allow for solvent molecules of various sizes, we can write $\chi \propto n_1^* \Delta w$, where n_1^* is the number of segments in a solvent molecule, or $\bar{V}_1/V_1^*$, where V_1^* is the molar volume of a segment. The ratio $(1/2 - \chi)/\bar{V}_1$ that appears in the expression for ΔG_R in Equation (59) can be written $(1/2\bar{V}_1) -$ (const. $\Delta w/V_1^*$). Criticize or defend the following proposition: For a specific polymer in a homologous series of solvent molecules of different sizes, the second term in this expression should remain constant while the first decreases with increasing $\bar{V}_1$. When the "solvent" is the melt of the polymer under consideration, the first term is negligible because $\bar{V}_1$ becomes the partial molar volume of the polymer and is very large. Therefore ΔG_R becomes negative for such a system, and flocculation is predicted. Since this contradicts experimental evidence, the Flory–Krigbaum model is seen to break down at this limit.

15. Mackor‡ used the model outlined in Example 13.6 to derive the expression $\Delta G_R = Nk_B T(1 - d/L)$ for the repulsion per unit area of particles carrying N rods of length L when the surfaces are separated by a distance d. Assuming this repulsion equals the van der Waals attraction when the particle separation is 1.5 nm, calculate the effective Hamaker constant in this system if $L = 2.5$ nm. Select a reasonable value for N in this calculation and justify your choice.

16. In view of the model used in Problem 15, criticize or defend the following proposition: If one surface carries adsorbed rods and the other is bare, the system could be stabilized against flocculation by dispersing the particles in a medium of intermediate γ. Such a system would remain dispersed indefinitely since both steric considerations and a negative Hamaker constant oppose flocculation.

---

*Ottewill, R. H., and Wilkins, D. J., *Trans. Faraday Soc.,* **58**, 608 (1962).
†Ahmed, M. S., El-Aasser, M. S., and Vanderhoff, J. W., in *Polymer Adsorption and Dispersion Stability* (E. D. Goddard and B. Vincent, Eds.), American Chemical Society, Washington, DC, 1964.
‡Mackor, E. L., *J. Colloid Sci.,* **6**, 492 (1951).

Appendix A

Examples of Expansions Encountered in This Book

1. Inverse of $(1 - x)$:

$$1/(1 - x) = 1 + x + x^2 + x^3 + \ldots$$

2. Logarithm:

$$\ln(1 + x) = x - (1/2)x^2 + (1/3)x^3 - \ldots \qquad (-1 < x < +1)$$

3. Binomial:

$$(1 \pm x)^n = 1 \pm nx + [n(n - 1)/2!]x^2 \pm [n(n - 1)(n - 2)/3!]x^3$$
$$+ \ldots \qquad (x^2 < 1)$$

4. Taylor's series for $f(x)$ around $x = x_0$:

$$f(x) = f(x_0) + (x - x_0)f'(x_0) + [(x - x_0)^2/2!]f''(x_0) + \ldots$$

where each prime on f indicates a differentiation with respect to x.

5. Sine function:

$$\sin x = x - x^3/3! + x^5/5! - \ldots$$

6. Hyperbolic sine function:

$$\sinh x = x + x^3/3! + x^5/5! + \ldots$$

7. Hyperbolic cosine function:

$$\cosh x = 1 + x^2/2! + x^4/4! + \ldots$$

8. Exponential function:

$$e^x = 1 + x + x^2/2! + x^3/3! + \ldots$$

REFERENCE

Gradshteyn, I. S., and Ryzhik, I. M., *Table of Integrals, Series, and Products*, 5th ed., Academic Press, New York, 1993. (There are numerous excellent reference books available on mathematical formulas. This book has a reasonably large collection, far more than what is needed for the present book. A CD-ROM version of this book is also available: CD-ROM Version 1, Alan Jeffrey, Ed.)

Appendix B

Units: CGS-SI Interconversions

From time to time, probably all science students find themselves entangled in a problem of units. For those who have advanced through physical chemistry to the level of this book, these problems have obviously not been insurmountable. It is likely, however, that — along with feelings of frustration — these students have been left with the wish that everyone used the same units, specifically those units with which they are most comfortable. In response to the recognized need for uniformity, IUPAC recommends the use of *Système international d'unités (International System of Units, SI)* units, which are essentially standardized mks units.

The SI system is based on mutually consistent units assigned to the nine physical quantities listed in Table B.1. In addition to the SI units for these nine quantities, the table also lists cgs or other commonly encountered units, as well as the conversion factors between the two. In this table the headings at the top of the table indicate how the conversion factors are to be used in going from SI to cgs/common units, whereas the bottom headings indicate the use of these factors for calculations in the reverse direction.

From these nine basic quantities, numerous other SI units may be derived. Table B.2 lists a number of these derived units, particularly those relevant to colloid and surface chemistry. The table is arranged alphabetically according to the name of the physical quantity involved. Note that instructions for the use of the conversion factors — depending on the direction of the conversion — are given in the top and bottom headings of the columns. Table B.2 is by no means an exhaustive list of the various derived SI units; Hopkins (1973) reports on many additional conversions, as do most handbooks and numerous other references.

One reason for the great diversity of units in existence is the fact that quantities of such diverse magnitudes are measured. A general rule is that the unit should be appropriate in magnitude to the quantity being measured. To obtain a dimension of convenient size in SI units, the SI unit is multiplied by a power of 10 and the prefixes listed in Table B.3 are affixed to the unit.

REFERENCES

Hopkins, R. A., *The International (SI) Metric System and How It Works*, Polymeric Services, Reseda, CA, 1973.
McGlashan, M. L., *Pure Appl. Chem.*, **21**, 577 (1970).
Page, C. H., and Vigoureux, P. (Eds.), *The International System of Units (SI)*, National Bureau of Standards, Special Publication 330, Washington, DC, 1974.
Paul, M. A., *J. Chem. Educ.*, **48**, 569 (1971).

TABLE B.1 Basic SI Units and Their Relation to cgs or Other Common Units[a]

Physical quantity	SI unit		×	Conversion factor	→	cgs/common unit	
	Name	Symbol				Name	Symbol
Length	Meter	m		10^2		Centimeter	cm
Mass	Kilogram	kg		10^3		Gram	g
Time	Second	s		1		Second	s
Electric current	Ampere	A		2.998×10^9		Statampere	statamp
Thermodynamic temperature	Kelvin	K		1		Kelvin	K
Luminous intensity	Candela	cd		π		Lambert	(cm^2)
Amount of substance	Mole	mole		1		Mole	mole
Plane angle	Radian	rad		$180/\pi$		Degree (angle)	°
Solid angle	Steradian	sr		1		Steradian	sr
	SI unit	←		Conversion factor	÷	cgs/common unit	

[a]Note that different column headings are given at the top and bottom of the table to facilitate conversions from SI to cgs and from cgs to SI, respectively.

TABLE B.2　Derived SI Units and Their Relation to cgs or Other Common Units[a]

Physical quantity	SI unit	×	Conversion factor	→	cgs/common unit
Acceleration	$m\,s^{-2}$		10^2		$cm\,s^{-2}$
Acceleration, angular	$rad\,s^{-1}$		1		$rad\,s^{-1}$
Area	m^2		10^4		cm^2
			10^{20}		Å^2
Capacitance (farad)	$F = m^{-2}\,kg^{-1}\,s^4\,A^2 = C\,V^{-1}$		8.99×10^{11}		statfarad
Charge (Coulomb)	$C = A\,s = J\,V^{-1}$		3.00×10^9		statcoulomb (esu)
Charge density, surface	$C\,m^{-2}$		3.00×10^5		$statcoul\,cm^{-2}$
Charge density, volume	$C\,m^{-3}$		3.00×10^3		$statcoul\,cm^{-3}$
Conductance (siemens)	$S = m^{-2}\,kg^{-1}\,s^3\,A^2 = ohm^{-1}$		8.99×10^{11}		$statmho\,(statohm^{-1})$
Conductivity	$ohm^{-1}\,m^{-1}$		8.99×10^9		$statmho\,cm^{-1}$
			10^{-2}		$mho\,cm^{-1}\,(ohm^{-1}\,cm^{-1})$
Density	$kg\,m^{-3}$		10^{-3}		$g\,cm^{-3}$
Diffusion coefficient	$m^2\,s^{-1}$		10^4		$cm^2\,s^{-1}$
Dipole moment	$C\,m$		3.00×10^{11}		$statcoul\,cm$
Electric field	$V\,m^{-1}$		3.34×10^{-5}		$statvolt\,cm^{-1}$
			10^{-2}		$V\,cm^{-1}$
Electric potential (volt)	$V = m^2\,kg\,s^{-3}\,A^{-1} = J\,C^{-1}$		3.34×10^{-3}		statvolt
Energy (joule)	$J = m^2\,kg\,s^{-2} = N\,m$		10^7		erg
			0.2390		calorie
Entropy	$J\,K^{-1}$		0.2390		$cal\,K^{-1}$
Force (newton)	$N = m\,kg\,s^{-2}$		10^5		dyne
Frequency (hertz)	$Hz = s^{-1}$		1		s^{-1}
Friction factor	$kg\,s^{-1}$		10^3		$g\,s^{-1}$
Heat capacity	$J\,K^{-1}$		0.2390		$cal\,K^{-1}$
Molarity	$mole\,dm^{-3}$		1		$mole\,liter^{-1}$
Moment, dipole	$C\,m$		3.00×10^{11}		$statcoul\,cm$

	SI unit	←	Conversion factor	÷	cgs/common unit
Moment, force	N m		10^7		dyne cm
Moment, inertia	$kg\ m^2$		10^7		$g\ cm^2$
Momentum	N s		10^5		dyne s
Momentum, angular	J s		10^7		erg s
Period	s		1		s
Permittivity	$F\ m^{-1}$		8.99×10^9		statfarad cm^{-1}
Polarization, electric	$C\ m^{-2}$		3.00×10^5		statcoul cm^{-2}
Potential (volt)	V		3.34×10^{-3}		statvolt
Power (watt)	$W = m^2\ kg\ s^{-3} = J\ s^{-1}$		10^7		erg s^{-1}
Pressure (pascal)	$Pa = m^{-1}\ kg\ s^{-2} = N\ m^{-2}$		10		dyne cm^{-2}
			9.87×10^{-6}		atm
Radius of gyration	m		10^2		cm
Resistance (ohm)	$ohm = m^2\ kg\ s^{-3}\ A^{-2} = VA^{-1}$		1.11×10^{-12}		statohm
Specific heat capacity	$J\ kg^{-1}\ K^{-1}$		2.39×10^{-4}		cal $g^{-1}\ K^{-1}$
Stress	$N\ m^{-2}$		10		dyne cm^{-2}
Surface energy	$J\ m^{-2}$		10^3		erg cm^{-2}
Surface tension	$N\ m^{-1}$		10^3		dyne cm^{-1}
Torque	N m		10^7		dyne cm
Velocity	$m\ s^{-1}$		10^2		cm s^{-1}
Velocity, angular	rad s^{-1}		1		rad s^{-1}
Viscosity	$N\ s\ m^{-2}$		10		dyne s cm^{-2} (P)
Volume	m^3		10^6		cm^3
			10^3		dm^3 (liter)
Wave number	m^{-1}		10^{-2}		cm^{-1}
Weight	N		10^5		dyne

[a]Note that different column headings are given at the top and bottom of the table to facilitate conversions from SI to cgs and from cgs to SI, respectively. The cgs units must be divided *by* the conversion factor to get the SI units.

TABLE B.3   Multiples of Units, Their Names, and Symbols

Multiple	Prefix	Symbol
10^{12}	tera	T
10^9	giga	G
10^6	mega	M
10^3	kilo	k
10^2	hecto	h
10	deca	da
10^{-1}	deci	d
10^{-2}	centi	c
10^{-3}	milli	m
10^{-6}	micro	μ
10^{-9}	nano	n
10^{-12}	pico	p
10^{-15}	femto	f
10^{-18}	atto	a

Appendix C

Statistics of Discrete and Continuous Distributions of Data

C.1   INTRODUCTION

Students of the physical sciences generally encounter statistics in two different places. One of these deals with the treatment of experimental data. From this viewpoint, all measured quantities contain some error, which raises questions concerning the best way to report the results of multiple measurements. This is obviously related to our problem of describing, for example, the "characteristic dimension" of the particles discussed in Chapter 1 (Figure 1.10). Another place in which the science student encounters statistics is in the theoretical description of large populations or populations that change through a large number of states, as, for example, in the kinetic molecular theory of gases. We are concerned with this aspect of statistics also. For example, the randomly coiled piece of string we considered in Chapter 1 (Section 1.5a.2) changes size and shape continually while it is being shaken.

These two applications of statistics, from our point of view, will differ primarily in the kind of information we have available about our system. Sometimes, as when measuring micrographs, we have individual information on a large number of particles. Our question under these circumstances is how to condense these data into a few key parameters. In other circumstances, the experimental quantity itself will be an "average" quantity. Our question, then, is what kind of distribution is consistent with this average. In both cases, the underlying fact is the existence of a distribution of values for the quantity in question. We consider some aspects of these statistical topics here; references in statistics should be consulted if additional information is needed.

C.2   DISCRETE DISTRIBUTIONS

Suppose we have just measured the diameters of a field of polydisperse spheres in an electron micrograph. The data in such cases consist of sets of numbers such as the ones we saw in Table 1.5. Our objective is to devise reasonable ways of presenting a description of the system in terms of the measured data. A fairly large number of observations is required for any statistical approach to be valid; therefore merely to tabulate the measurements is inadequate. Some condensation of the data is clearly required. Generally, the first step along these lines is a device known as *classification* of the data. Classification consists of sorting the observed quantities into 10–20 categories called *classes*. Having fewer than 10 categories results in a loss of detail in the description of the distribution; having more than 20 categories does not improve the representation in proportion to the extra effort it requires. The frequency distribution of such a sample is a tabulation of the number of particles in each class. Table 1.5 in Chapter 1, for example, represents the frequency distribution for a hypothetical array of spheres; all the numerical examples of this section are based on this sample of 400 particles. Each class is represented by the midpoint of the interval, a quantity called the *class mark*, symbolized by d_i for class i. Similarly, we define the number of particles in each class as n_i.

We discussed in Chapter 1 how discrete data — such as the ones reported in Table 1.5 — can be represented graphically as a histogram, that is, a bar graph in which the class marks are plotted as the abscissa and the height of the bar is proportional to the number of particles in the class (Fig. 1.18a). The corresponding cumulative curve, that is, the total number (or fraction) of particles $n_{T,i}$ having diameters less (sometimes more) than and including a particular d_i are plotted versus d_i. This was illustrated in Figure 1.18b.

In this appendix, we summarize how the discrete data sets can be represented in terms of averages, standard deviation, and moments in a concise way. In principle, under certain conditions, these quantities represent the same information represented through histograms.

C.2a   Average and Standard Deviation

In Section 1.5c.1 we presented the following general definition of an average:

$$\text{Average} = \sum_i (\text{Weighting factor})_i \, (\text{Quantity being averaged})_i \tag{1}$$

where the weighting factor f_i and the quantity being averaged ϕ_i can both mean different things in different contexts. The most common definition of a weighting factor is the number fraction of the particles in each class i. We indicate this by the additional subscript n of f:

$$f_{n,i} = n_i / \sum_i n_i \tag{2}$$

Then the *mean* or the average of ϕ_i denoted by $\bar{\phi}$ is given by

$$\bar{\phi} = \sum_i f_{n,i} \phi_i = \frac{\sum_i n_i \phi_i}{\sum_i n_i} \tag{3}$$

and the *standard deviation* σ

$$\sigma = \left(\sum_i f_{n,i} (\phi_i - \bar{\phi})^2 \right)^{1/2} = \left(\frac{\sum_i n_i (\phi_i - \bar{\phi})^2}{\sum_i n_i} \right)^{1/2} \tag{4}$$

of the distribution. The significance of the standard deviation should be noted. The quantity $[\phi_i - \bar{\phi}]$ is the deviation of a particular value from the mean, and the deviations can be positive or negative. Comparing Equation (1) with Equation (4), we see that σ^2 is actually the *number average of the square of the deviations* $[\phi_i - \bar{\phi}]$. The square root of this average, then, is a measure of the spread of the data in a particular sample. For this reason, σ is often called the *root mean square* (rms) deviation. From a computational point of view, the following formula, which is a different way of writing Equation (4), provides an easier means for evaluating σ:

$$\sigma = \left[\overline{\phi^2} - (\bar{\phi})^2 \right]^{1/2} \tag{5}$$

This result was used in Section 1.5c.3b to relate $(\overline{M}_w / \overline{M}_n)$ to σ.

In general, the *number average* (i.e., based on number fraction $f_{n,i}$ as the weight function) of any property, say, ψ, of the particles is defined as

$$\overline{\psi} = \sum_i f_{n,i} \psi_i \tag{6}$$

where ψ_i is the value of the property in class i. If ψ is taken to be the area A of the particles, one then has for the average area, that is, area per particle $\overline{A}$:

$$\overline{A} = \frac{\pi \sum_i n_i d_i^2}{\sum_i n_i} = \pi \sum_i f_{n,i} d_i^2 \tag{7}$$

The diameter of a sphere having this "average" area is the surface area average diameter $\overline{d}_s$ and equals

$$\overline{d}_s = \left(\sum_i f_{n,i} d_i^2 \right)^{1/2} \tag{8}$$

The surface average diameter is always larger than the direct average of the diameters for a polydisperse system since the larger diameters contribute relatively more to the sum of the squares than they would if totaled directly. Example C.1 illustrates such a calculation.

* * *

EXAMPLE C.1 *Average Surface Area Versus Surface Area Based on Average-Size Particles.* Using the distribution of spheres in Table 1.5, calculate the area of each size sphere in the distribution and average these areas, using the number of spheres in each class as weighting factors. How is this result related to $\bar{d}_s$ for this population?

Solution: To calculate the area of each size sphere, we apply the formula $A = \pi d^2$ to each class in Table 1.5:

$d_i\,(\mu m)$	$A_i\,(\mu m^2)$	$n_i A_i\,(\mu m^2)$
0.05	$7.85 \cdot 10^{-3}$	$5.50 \cdot 10^{-2}$
0.15	$7.07 \cdot 10^{-2}$	1.06
0.25	$1.96 \cdot 10^{-1}$	3.53
0.35	$3.85 \cdot 10^{-1}$	$1.08 \cdot 10^{1}$
0.45	$6.36 \cdot 10^{-1}$	$2.04 \cdot 10^{1}$
0.55	$9.50 \cdot 10^{-1}$	$6.65 \cdot 10^{1}$
0.65	1.33	$8.63 \cdot 10^{1}$
0.75	1.77	$1.04 \cdot 10^{2}$
0.85	2.27	$1.02 \cdot 10^{2}$
0.95	2.84	$1.08 \cdot 10^{2}$
1.05	3.46	$6.58 \cdot 10^{1}$
1.15	4.15	$1.66 \cdot 10^{1}$

$$\text{Sum} = 5.85 \cdot 10^2$$

Therefore the "average" area is $5.85 \cdot 10^2/400 = 1.46 \;\mu m^2$.
   This is the value used in Section 1.5c.1. This average area divided by π gives $(\bar{d}_s)^2$. ∎

* * *

The general formulas for calculating averages are given by Equations (1) and (2). Equation (2) represents the weighting factor in normalized form. A weighting factor is said to be *normalized* if the sum of weighting factors for the whole population equals unity, a consideration that must be introduced at some point in the calculation. When fractions of the whole are used as weighting factors, normalization is introduced through the definition of the weighting factor.

C.2b   Higher-Order Moments of the Distribution

In general one can define the kth *moment* of the distribution about a point d_0 as

$$k\text{th moment} = \sum_i f_{n,i}\,(d_i - d_0)^k \tag{9}$$

Note that Equation (9) implies that the square of the standard deviation σ^2 is the *second moment* of d relative to the mean $\bar{d}$. Higher order moments can be used to represent additional information about the shape of a distribution. For example, the third moment is a measure of the "skewness" or lopsidedness of a distribution. It equals zero for symmetrical distributions and is positive or negative, depending on whether a distribution contains a higher proportion of particles larger or smaller, respectively, than the mean. The fourth moment (called *kurtosis*) purportedly measures peakedness, but this quantity is of questionable value.

   In discussing the random walk and diffusion in Chapter 2, we use continuous functions as the weighting factors for calculating averages (e.g., Equation (2.63)). In the following section we discuss in greater detail the use of continuous weighting functions in statistics.

C.3   THEORETICAL DISTRIBUTION FUNCTIONS

We noted in Section 1.5c that histograms of distributions of quantities such as particle size approach smooth distribution curves as the number of classes is increased to a very large number. Sometimes it is desirable to represent a distribution function by an analytical expression that is a continuous function of the measured variable. We consider only a few examples of such distribution functions here.

C.3a   Normal or Gaussian Distribution Function

The most familiar of such functions is the *normal*, or *Gaussian*, distribution function:

$$f(x)\,dx = \frac{1}{\sigma\sqrt{2\pi}} \exp\left[-\frac{1}{2}\left(\frac{x - \bar{x}}{\sigma}\right)^2\right] dx \tag{10}$$

In this equation, $f(x)dx$ expresses the fraction of particles having x values between x and $x + dx$; it replaces $f_{n,i}$, which plays a corresponding role in discrete distributions. Some characteristics of the normal distribution are the following:

1.   The function $f(x)$ has its maximum value at $x = \bar{x}$ and drops off exponentially with the square of the deviation of x from the mean, when such deviations are measured as fractions or multiples of the standard deviation.

2.   The preexponential factor accomplishes the normalization of the function; that is, the integral of the function over all possible values of x ($-\infty$ to ∞) equals unity. In a broad distribution σ is large, and the exponential does not drop off as rapidly as in a narrow distribution (recall that all deviations are measured relative to the standard deviation).

3.   Since the area under the curve is always unity, a narrow distribution will show larger values of $f(x)$ at the maximum, whereas a broader distribution will have a smaller value for the function at the maximum. This is why the standard deviation appears in the denominator of the preexponential normalization factor.

The normal distribution is the "curve" over which students and teachers alike agonize in connection with course grades. We discuss this distribution function in greater detail in Chapter 2. For the present we are concerned only with its descriptive capabilities. For this purpose it is sufficient to note that tables are available (e.g., Beyer 1987; also see the other references at the end of this Appendix) that supply the value of this function in terms of the standard unit,

$$f(t)\,dt = \frac{1}{\sqrt{2\pi}} \exp\left(-\frac{t^2}{2}\right) dt \tag{11}$$

The normal distribution is commonly encountered in the cumulative form, that is, as the fraction of particles larger (oversized) or smaller (undersized) than a particular t_i value. Since the total area under the normal curve equals unity, the area under one "tail" of the curve from t_i to ∞ gives the fraction of the population having t values greater than the integration limit t_i:

$$f_{t>t_i} = \frac{1}{\sqrt{2\pi}} \int_{t_i}^{\infty} \exp\left(-\frac{t^2}{2}\right) dt \tag{12}$$

Likewise, the cumulative fraction of particles smaller than t_i equals

$$f_{t<t_i} = 1 - \frac{1}{\sqrt{2\pi}} \int_{t_i}^{\infty} \exp\left(-\frac{t^2}{2}\right) dt \tag{13}$$

Tables are also available (e.g., see Beyer 1987) for the area under the normal curve between $\bar{t}$ ($\bar{t} = 0$) and one value of t_i (i.e., they apply to one tail only). This area is known as the *error function* and is often symbolized as erf (t_i):

$$\text{erf } (t_i) = \frac{1}{\sqrt{2\pi}} \int_0^{t_i} \exp\left(-\frac{t^2}{2}\right) dt \tag{14}$$

In terms of the error function, Equation (12) becomes

$$f_{t > t_i} = \frac{1}{2} \pm \text{erf } (t_i) \tag{15}$$

where the positive value is used if t_i is negative and the negative value is used if t_i is positive. These signs are reversed if the cumulative fraction of undersized particles is to be determined. Consulting the tables, we see that in a normally distributed sample 15.87% of the particles will have t values greater than $+1.0$. This is the percentage of particles for which the deviation from the mean is greater than one standard deviation unit.

Example C.2 considers a numerical application of these ideas.

* * *

EXAMPLE C.2 *Polydispersity in Size: Gaussian Distribution.* A sample of 75 particles follows the normal distribution with respect to particle diameters, with $\bar{d} = 1.140$ μm and $\sigma = 0.311$ μm. Consult a table of the normal distribution to evaluate the fraction of particles in the population having diameters less than 1.0 μm. What is the probability that a particle picked at random from this population lies in the class for which $0.95 < d < 1.05$ in the histogram of this distribution?

Solution: Evaluate t for $d = 1.00$: $t = (d - \bar{d})/\sigma = (1.00 - 1.14)/0.311 = -0.450$. Tables of the normal distribution (see Beyer 1987) indicate that the area under the curve is 0.1736 at $t = 0.45$. To interpret this, picture the familiar bell-shaped curve centered at $t = 0$ ($d = \bar{d} = 1.14$ μm); we are interested in an abscissa value of $t = -0.45$ ($d = 1.0$). The figure 0.1736 gives the fraction of the area under the whole curve that lies between $t = -0.45$ and $t = 0.0$. This means that $(0.5 - 0.1736) = 0.3264$ is the fraction of the area that lies beyond $t = -0.45$. This means that about 33% of the particles have diameters less than 1.00 μm.

For $d = 0.95$, $(d - \bar{d})/\sigma = -0.61$; the area between $t = -0.61$ and $t = 0.0$ is 0.229 according to the tables. For $d = 1.05$, $(d - \bar{d})/\sigma = 0.29$; the area between $t = -0.29$ and $t = 0.0$ is 0.114. Therefore the probability of a particle falling between 0.95 and 1.05 is $(0.229 - 0.114) = 11.5\%$. ∎

* * *

Suppose a polydisperse system is investigated experimentally by measuring the number of particles in a set of different classes of diameter or molecular weight. Suppose further that these data are believed to follow a normal distribution function. To test this hypothesis rigorously, the chi-squared test from statistics should be applied. A simple graphical examination of the hypothesis can be conducted by plotting the cumulative distribution data on probability paper as a rapid, preliminary way to evaluate whether the data conform to the requirements of the normal distribution.

Probability paper is a commercially available graph paper that has one coordinate subdivided in ordinary arithmetic units and the other coordinate subdivided into cumulative probability units. The latter are spaced in such a way that normally distributed data will produce a straight line graph when the cumulative percentage of undersize (or oversize) particles is plotted on the probability coordinate, and the size variable (diameters, weights, etc.) is plotted on the arithmetic scale. Figure C.1 shows schematically how (a) normally distributed data are transformed when replotted as (b) a cumulative distribution and, finally, when (c) graphed on probability paper. The x value corresponding to $y = 50\%$ on probability paper gives the mean value. The x value at $y = 15.87\%$ gives $(\bar{x} - \sigma)$, and the x value at $y = 84.13\%$ ($100 - 15.87$) gives $(\bar{x} - \sigma)$ when the percentage of undersize particles is plotted (the signs are reversed when the percentage of oversize particles is plotted). From these values σ may be determined. Thus a linear plot on probability paper suggests conformity to the normal distribution and also permits the graphical evaluation of $\bar{x}$ and σ.

As we see in Chapter 2, the normal distribution comes about when a large number of purely random factors is responsible for the distribution. It is mainly applicable to particles

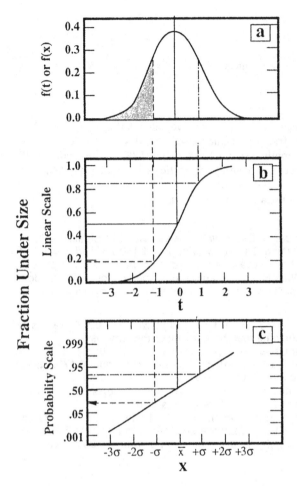

FIG. C.1 A normal, or Gaussian, distribution: (a) represented as a frequency function; (b) represented as a cumulative function; and (c) represented as a cumulative function linearized by plotting on probability paper.

that are produced by condensation, precipitation, or polymerization processes, which are purely random.

C.3b   Log-Normal Distribution Function

Dispersions that are produced by comminution — the mechanical subdivision of larger chunks — are more likely to produce a linear graph on probability paper if the logarithm of the variable rather than the variable itself is plotted against the probability. Graph paper graduated this way is called *log-probability paper*. The logarithmic scale implies a much wider range of values for the variable, and also an asymmetrical distribution function may be written by analogy with Equation (10):

$$f(\ln x) = \frac{1}{\ln \sigma_g \sqrt{2\pi}} \exp\left[-\frac{1}{2}\left(\frac{\ln x_i - \ln \bar{x}_g}{\ln \sigma_g} \right)^2 \right] \tag{16}$$

However, an important difference also emerges from this analogy. The quantities that are normally distributed are logarithms of variables, not the variables themselves. This means that the mean and standard deviation obtained from log-probability plots are geometric averages rather than arithmetic averages. This is the significance of the subscript g in Equation (16). This is most easily understood by writing the expression for the number average value for the quantity $\ln x_i$:

$$\overline{\ln x} = \sum_i f_i \ln x_i = \sum_i \ln x_i^{f_i} \tag{17}$$

Taking the antilog of this quantity converts the summation into a product over all terms, indicated by Π:

$$\text{antiln} \, (\overline{\ln x}) = \prod_i x_i^{f_i} = \bar{x}_g \tag{18}$$

When the averaging is carried out in this way, the result is known as the geometric mean x_g. The coordinate corresponding to the 15.87% undersize y value equals $(\ln x_g - \ln \sigma_g)$. Because of the properties of logarithms, this is the same as the logarithm of the ratio (x_g/σ_g). From this, σ_g may be evaluated. We shall not concern ourselves further with these geometrical averages except to note that

$$\bar{x}_g < \bar{x}_n \tag{19}$$

for any polydisperse system. (The subscript n on the $\bar{x}$ on the right-hand side of Equation (19) stands for *arithmetic* average of the type in Equation (3).) The validity of this relationship is easily demonstrated by calculating the two averages for a hypothetical distribution.

REFERENCES

Beyer, W. H. (Ed.), *CRC Handbook of Mathematical Sciences*, 6th ed., CRC Press, Boca Raton, FL, 1987. (One of the standard references for tables of mathematical concepts and formulas.)

Crow, E. L., Davis, F. A., and Maxfield, M. W., *Statistics Manual*, Dover, New York, 1960. (There are numerous excellent books on statistics available in print. This book is a handy, affordable manual that is sufficient for most purposes.)

Meyer, S. L., *Data Analysis for Scientists and Engineers*, Wiley, New York, 1975. (This is a more detailed manual on statistics and contains very useful collections of tables, distribution functions, data analysis procedures, and graphical techniques.)

Appendix D

List of Worked-Out Examples

Index

The notation *e*, *f*, *t* or *v* accompanying certain entries implies that the examples, figures, tables or vignettes, respectively, appearing on the indicated pages *also* contain relevant information.

Printed in the United States
by Baker & Taylor Publisher Services

Printed in the United States
by Baker & Taylor Publisher Services